Frequently Used Symbols and Formulas

A Key to the Icons Appearing in the Exercise Sets

Concept reinforcement exercises, indicated by purple exercise numbers, provide basic practice with the new concepts and vocabulary.

1 Following most examples, students are directed to **TRY EXERCISES**. These selected exercises are identified with a color block around the exercise numbers.

Aha! Exercises labeled **Aha!** can often be solved quickly with the proper insight.

Calculator exercises are designed to be worked using a scientific or graphing calculator.

Graphing calculator exercises are designed to be worked using a graphing calculator and often provide practice for concepts discussed in the Technology Connections.

Writing exercises are designed to be answered using one or more complete sentences.

Symbols

$=$	Is equal to
$\approx$	Is approximately equal to
$>$	Is greater than
$<$	Is less than
$\geq$	Is greater than or equal to
$\leq$	Is less than or equal to
$\lvert x \rvert$	The absolute value of x
$\{x \mid x \dots\}$	The set of all x such that $x \dots$
$-x$	The opposite of x
$\sqrt{x}$	The square root of x
$\sqrt[n]{x}$	The nth root of x
LCM	Least Common Multiple
LCD	Least Common Denominator
π	Pi
i	$\sqrt{-1}$
$f(x)$	f of x, or f at x

Formulas

$m = \dfrac{y_2 - y_1}{x_2 - x_1}$	Slope of a line
$y = mx + b$	Slope–intercept form of a linear equation
$y - y_1 = m(x - x_1)$	Point–slope form of a linear equation
$(A + B)(A - B) = A^2 - B^2$	Product of the sum and difference of the same two terms
$(A + B)^2 = A^2 + 2AB + B^2,$ $(A - B)^2 = A^2 - 2AB + B^2$	Square of a binomial
$d = rt$	Formula for distance traveled
$\dfrac{1}{a} \cdot t + \dfrac{1}{b} \cdot t = 1$	Work principle
$s = 16t^2$	Free-fall distance
$y = kx$	Direct variation
$y = \dfrac{k}{x}$	Inverse variation
$x = \dfrac{-b \pm \sqrt{b^2 - 4ac}}{2a}$	Quadratic formula

Resources Designed with You in Mind

This textbook was designed with features and applications to help make learning easier for you, but there's more than just the textbook if you want additional help. At www.mypearsonstore.com, you can check out these and other supplemental materials that can help you pass your math course.

Worksheets for Classroom or Lab Practice
(ISBN-10: 0-321-59931-4; ISBN-13: 978-0-321-59931-5)

Need more practice? Each worksheet provides key terms, fill-in-the-blank vocabulary practice, and exercises for each objective. There are two worksheets for every section of the text.

Videos on DVD
(ISBN-10: 0-321-59932-2; ISBN-13: 978-0-321-59932-2)

Miss a lecture? Need some extra help studying the night before an exam? Having a tough time with a certain topic? The Videos on DVD are here to help. Watch an experienced math instructor present important definitions, procedures, and concepts from each section of the book. The instructor will show you how to solve examples and exercises taken straight from the text.

Student's Solutions Manual
(ISBN-10: 0-321-56733-1; ISBN-13: 978-0-321-56733-8)

Looking for more than just the answer in the back of the book? The *Student's Solutions Manual* contains step-by-step solutions for all the odd-numbered exercises in the text as well as step-by-step solutions for Connecting the Concepts, Chapter Review Exercises, and Chapter Test exercises.

Elementary Algebra

CONCEPTS AND APPLICATIONS

EDITION

8

MARVIN L. BITTINGER
Indiana University Purdue University Indianapolis

DAVID J. ELLENBOGEN
Community College of Vermont

ADDISON-WESLEY

Boston San Francisco New York
London Toronto Sydney Tokyo Singapore Madrid
Mexico City Munich Paris Cape Town Hong Kong Montreal

Editorial Director	Christine Hoag
Editor in Chief	Maureen O'Connor
Executive Project Manager	Kari Heen
Associate Editor	Joanna Doxey
Editorial Assistant	Jonathan Wooding
Production Manager	Ron Hampton
Editorial and Production Services	Martha K. Morong/Quadrata, Inc.
Art Editor and Photo Researcher	The Davis Group, Inc.
Compositor	Pre-Press PMG
Senior Media Producer	Ceci Fleming
Associate Producer	Jennifer Thomas
Software Development	Eileen Moore and Marty Wright
Marketing Manager	Marlana Voerster
Marketing Coordinator	Nathaniel Koven
Prepress Supervisor	Caroline Fell
Manufacturing Manager	Evelyn Beaton
Senior Manufacturing Buyer	Carol Melville
Senior Media Buyer	Ginny Michaud
Text Designer	The Davis Group, Inc.
Cover Designer	Beth Paquin
Cover Photograph	Papua, New Guinea, false clown anemone fish and sea anemone. © Darryl Leniuk/Digital Vision/Getty Images

Photo Credits
Photo credits appear on page xviii.

Library of Congress Cataloging-in-Publication Data
Bittinger, Marvin L.
    Elementary algebra : concepts and applications.
    — Eighth ed. / Marvin L. Bittinger, David J. Ellenbogen.
        p. cm.
    Includes indexes.

1. Algebra—Textbooks.   I. Ellenbogen, David.   II. Title.
    QA152.3.B53 2010
    512. 9—dc22   2008024067

1 2 3 4 5 6 7 8 9 10—RRDJC—12 11 10 09 08

© 2010, 2006, 2002, 1998, 1994, 1990, 1986, 1982 Pearson Education, Inc.

Addison-Wesley
is an imprint of

www.pearsonhighered.com

ISBN-13: 978-0-321-55717-9
ISBN-10: 0-321-55717-4

Contents

3   Introduction to Graphing   145

4   Polynomials   221

7 Systems and More Graphing 429

8   Radical Expressions and Equations   483

9   Quadratic Equations   535

Appendixes

Tables

Preface

It is with great pleasure that we introduce you to the eighth edition of *Elementary Algebra: Concepts and Applications*. Our goal, as always, is to present content that is easy to understand and has the depth required for success in this and future courses. In this edition, faculty will recognize features, applications, and explanations that they have come to rely on and expect. Students and faculty will also find many changes resulting from our own ideas for improvement as well as insights from faculty and students throughout North America. Thus this new edition contains exciting new features and applications, along with updates and refinements to those from previous editions.

Appropriate for a one-term course in elementary algebra, this text is intended for those students who have a firm background in arithmetic. It is the first of three texts in an algebra series that also includes *Intermediate Algebra: Concepts and Applications*, Eighth Edition, by Bittinger/Ellenbogen, and *Elementary and Intermediate Algebra: Concepts and Applications*, Fifth Edition, by Bittinger/Ellenbogen/Johnson.

Approach

Our goal, quite simply, is to help today's students both learn and retain mathematical concepts. To achieve this goal, we feel that we must prepare developmental-mathematics students for the transition from "skills-oriented" elementary algebra courses to more "concept-oriented" college-level mathematics courses. This requires that we teach these same students critical thinking skills: to reason mathematically, to communicate mathematically, and to identify and solve mathematical problems. Following are three aspects of our approach that we use to help meet the challenges we all face when teaching developmental mathematics.

Problem Solving

One distinguishing feature of our approach is our treatment of and emphasis on problem solving. We use problem solving and applications to motivate the material wherever possible, and we include real-life applications and problem-solving techniques throughout the text. Problem solving not only encourages students to think about how mathematics can be used, it helps to prepare them for more advanced material in future courses.

In Chapter 2, we introduce our five-step process for solving problems: (1) Familiarize, (2) Translate, (3) Carry out, (4) Check, and (5) State the answer. These steps are then used consistently throughout the text when encountering a problem-solving situation. Repeated use of this problem-solving strategy helps provide students with a starting point for any type of problem they encounter, and frees them to focus on the unique aspects of the particular problem situation. We often use estimation and carefully checked guesses to help with the *Familiarize* and *Check* steps (see pp. 109 and 452–453). We also use dimensional analysis as a quick check of certain problems (see pp. 175 and 412).

Applications

Interesting applications of mathematics help motivate both students and instructors. Solving applied problems gives students the opportunity to see their conceptual understanding put to use in a real way. In the eighth edition of *Elementary Algebra: Concepts and Applications*, we have increased the number of applications, the number of real-data problems, and the number of reference lines that specify the sources of the real-world data. As in the past, art is integrated into the applications and exercises to aid the student in visualizing the mathematics. (See pp. 110, 188, 244, 348, and 417.)

Pedagogy

New!

TRY EXERCISES ▶

Try Exercises. This icon concludes nearly every example by pointing students to one or more parallel exercises from the corresponding exercise set so that they can immediately reinforce the concepts and skills presented in the examples. For easy identification in the exercise sets, the "Try" exercises have a shaded block on the exercise number. (See pp. 65, 240, and 441.)

New!

Translating for Success and **Visualizing for Success.** These matching exercises help students learn to associate word problems (through translation) and graphs (through visualization) with their appropriate mathematical equations. (See pp. 132, 350, and 415 (Translating); 208, 259 (Visualizing).)

Revised!

Connecting the Concepts. Revised and expanded to include new Mixed Review exercises, this midchapter review helps students understand the big picture and prepare for chapter tests and cumulative reviews by relating the concept at hand to previously learned and upcoming concepts. (See pp. 202, 263, and 451.)

Revised!

Study Summary. Found at the end of each chapter and now presented in a two-column format organized by section, this synopsis gives students a fast and effective review of key chapter terms and concepts paired with accompanying examples. (See pp. 137, 214, and 356.)

Revised!

Cumulative Review. This review now appears after every chapter to help students retain and apply their knowledge from previous chapters. (See pp. 143, 294, and 532.)

Algebraic–Graphical Connections. This feature provides students with a way to visualize concepts that might otherwise prove elusive. (See pp. 339, 404, and 433.)

Study Skills. This feature in the margin provides tips for successful study habits that even experienced students will appreciate. Ranging from time management to test preparation, these study skills can be applied in any college course. (See pp. 86, 298, and 541.)

Student Notes. These notes in the margin give students extra explanation of the mathematics appearing on that page. These comments are more casual in format than the typical exposition and range from suggestions for avoiding common mistakes to how to best read new notation. (See pp. 79, 310, and 524.)

Technology Connection. These optional boxes in each chapter help students use a graphing calculator to better visualize a concept that they have just learned. To connect this optional instruction to the exercise sets, certain exercises are marked with a graphing calculator icon ▨ to indicate the optional use of technology. (See pp. 162, 340, and 467.)

Revised! ↪

Concept Reinforcement Exercises. Now with all answers listed in the answer section at the back of the book, these section and review exercises build students' confidence and comprehension through true/false, matching, and fill-in-the-blank exercises at the start of most exercise sets. To help further student understanding, emphasis is given to new vocabulary and notation developed in the section. (See pp. 10, 199, and 260.)

Aha! **Aha! Exercises.** These exercises are not more difficult than their neighboring exercises and can be solved quickly, without going through a lengthy computation, if the student has the proper insight. Designed to reward students who "look before they leap," the icon indicates the first time a new insight applies, and then it is up to the student to determine when to use the Aha! method on subsequent exercises. (See pp. 172, 284, and 558.)

Revised! **Skill Review Exercises.** These exercises, included in Section 1.2 and every section thereafter, review skills and concepts from preceding sections of the text. In most cases, these exercises prepare students for the next section. An introduction to each set directs students to the appropriate sections to review if necessary. On occasion, Skill Review exercises focus on a single topic in greater depth and from multiple perspectives. (See pp. 155, 237, and 329.)

Synthesis Exercises. Synthesis exercises follow the Skill Review exercises at the end of each exercise set. Generally more challenging, these exercises synthesize skills and concepts from earlier sections with the present material, often providing students with deeper insight into the current topic. Aha! exercises are sometimes included as Synthesis exercises. (See pp. 98, 384, and 435.)

Writing Exercises. These appear just before the Skill Review exercises (two basic writing exercises) and also in the Synthesis exercises (at least two more challenging exercises). Writing exercises aid student comprehension by requiring students to use critical thinking to provide explanations of concepts in one or more complete sentences. Because some instructors may collect answers to writing exercises and because more than one answer can be correct, only answers to writing exercises in the review section are included at the back of the text. (See pp. 58, 136, and 370.)

Collaborative Corner. These optional activities for students to explore together usually appear two to three times per chapter at the end of an exercise set. Studies show that students who study in groups generally outperform those who do not, so these exercises are for students who want to solve mathematical problems together. Additional collaborative activities and suggestions for directing collaborative learning appear in the *Instructor and Adjunct Support Manual.* (See pp. 272, 313, and 436.)

What's New in the Eighth Edition?

We have rewritten many key topics in response to user and reviewer feedback and have made significant improvements in design, art, pedagogy, and an expanded supplements package. Detailed information about the content changes is available in the form of a conversion guide. Please ask your local Pearson sales consultant for more information. Following is a list of the major changes in this edition.

NEW DESIGN

While incorporating a new layout, a fresh palette of colors, and new features, we have a larger page dimension for an open look and a typeface that is easy to read. As always, it is our goal to make the text look mature without being intimidating. In addition, we continue to pay close attention to the pedagogical use of color to make sure that it is used to present concepts in the clearest possible manner.

CONTENT CHANGES

A variety of content changes have been made throughout the text. Some of the more significant changes are listed below.

- Examples and exercises that use real data are updated or replaced with current applications.
- Over 35% of the exercises are new or updated.
- Quick-glance reminders for multistep processes are included next to examples. These appear by one multistep example of each type. (See pp. 158, 325, and 388.)
- Chapter 2 now includes increased practice of solving for y in a formula.
- Chapter 3 now gives increased emphasis to units when finding a rate of change.
- Chapter 5 now makes greater use of prime factorizations as a tool for finding the largest common factor.
- Chapter 8 now includes the distance formula as one application of the Pythagorean theorem.

ANCILLARIES

The following ancillaries are available to help both instructors and students use this text more effectively.

STUDENT SUPPLEMENTS

New! Chapter Test Prep Video CD

- Watch instructors work through step-by-step solutions to all the chapter test exercises from the textbook. The Chapter Test Prep Video CD is included with each new student text.

New! Worksheets for Classroom or Lab Practice

by Carrie Green
These lab- and classroom-friendly workbooks offer the following resources for every section of the text:

- A list of learning objectives;
- Vocabulary practice problems;
- Extra practice exercises with ample work space.

ISBNs: 0-321-59931-4 and 978-0-321-59931-5

Student's Solutions Manual

by Christine S. Verity

- Contains completely worked-out solutions with step-by-step annotations for all the odd-numbered exercises in the text, with the exception of the writing exercises.
- **New!** Now contains all solutions to Chapter Review, Chapter Test, and Connecting the Concepts exercises.

ISBNs: 0-321-56733-1 and 978-0-321-56733-8

INSTRUCTOR SUPPLEMENTS

Annotated Instructor's Edition

- Provides answers to all text exercises in color next to the corresponding problems.
- Includes Teaching Tips.
- Icons identify writing and graphing calculator exercises.

ISBNs: 0-321-55945-2 and 978-0-321-55945-6

Instructor's Solutions Manual

by Christine S. Verity

- Contains fully worked-out solutions to the odd-numbered exercises and brief solutions to the even-numbered exercises in the exercise sets.
- Available for download at www.pearsonhighered.com

ISBNs: 0-321-56732-3 and 978-0-321-56732-1

Instructor and Adjunct Support Manual

- Includes resources designed to help both new and adjunct faculty with course preparation and classroom management.
- Offers helpful teaching tips correlated to the sections of the text.

ISBNs: 0-321-56740-4 and 978-0-321-56740-6

Videos on DVD

- A complete set of digitized videos on DVD for use at home or on campus.
- Includes a full lecture for each section of the text, many presented by author team members David J. Ellenbogen and Barbara Johnson.
- Optional subtitles in English are available.

ISBNs: 0-321-59932-2 and 978-0-321-59932-2

InterAct Math® Tutorial Website

www.interactmath.com

- Online practice and tutorial help.
- Retry an exercise with new values each time for unlimited practice and mastery.
- Every exercise is accompanied by an interactive guided solution that gives helpful feedback when an incorrect answer is entered.
- View the steps of a worked-out sample problem similar to those in the text.

Printable Test Bank

by Laurie Hurley

- Contains two multiple-choice tests per chapter, six free-response tests per chapter, and eight final exams.
- Available for download at www.pearsonhighered.com

PowerPoint® Lecture Slides

- Present key concepts and definitions from the text.
- Available for download at www.pearsonhighered.com

TestGen

www.pearsonhighered.com/testgen

- Enables instructors to build, edit, print, and administer tests using a computerized bank of questions developed to cover all text objectives.
- Algorithmically based, TestGen allows instructors to create multiple but equivalent versions of the same question or test with the click of a button.
- Instructors can also modify test bank questions or add new questions.
- Tests can be printed or administered online.

Pearson Math Adjunct Support Center

http://www.pearsontutorservices.com/math-adjunct.html

Staffed by qualified instructors with more than 50 years of combined experience at both the community college and university levels, this center provides assistance for faculty in the following areas:

- Suggested syllabus consultation;
- Tips on using materials packed with the text;
- Book-specific content assistance;
- Teaching suggestions, including advice on classroom strategies.

AVAILABLE FOR STUDENTS AND INSTRUCTORS

MyMathLab® Online Course (access code required)

MyMathLab is a series of text-specific, easily customizable online courses for Pearson Education's textbooks in mathematics and statistics. Powered by CourseCompass™ (our online teaching and learning environment) and MathXL® (our online homework, tutorial, and assessment system), MyMathLab gives you the tools you need to deliver all or a portion of your course online, whether your students are in a lab setting or working from home. MyMathLab provides a rich and flexible set of course materials, featuring free-response exercises that are algorithmically generated for unlimited practice and mastery. Students can also use online tools, such as video lectures, animations, and a multimedia textbook, to independently improve their understanding and performance. Instructors can use MyMathLab's homework and test managers to select and assign online exercises correlated directly to the textbook, and they can also create and assign their own online exercises and import TestGen tests for added flexibility. MyMathLab's online gradebook—designed specifically for mathematics and statistics—automatically tracks students' homework and test results and gives the instructor control over how to calculate final grades. Instructors can also add offline (paper-and-pencil) grades to the gradebook. MyMathLab also includes access to the **Pearson Tutor Center** (www.pearsontutorservices.com). The Tutor Center is staffed by qualified mathematics instructors who provide textbook-specific tutoring for students via toll-free phone, fax, e-mail, and interactive Web sessions. MyMathLab is available to qualified adopters. For more information, visit our website at www.mymathlab.com or contact your sales representative.

MathXL® Online Course (access code required)

MathXL® is a powerful online homework, tutorial, and assessment system that accompanies Pearson Education's textbooks in mathematics or statistics. With MathXL, instructors can create, edit, and assign online homework and tests using algorithmically generated exercises correlated at the objective level to the textbook. They can also create and assign their own online excrcises and import TestGen tests for added flexibility. All student work is tracked in MathXL's online gradebook. Students can take chapter tests in MathXL and receive personalized study plans based on their test results. The study plan diagnoses weaknesses and links students directly to tutorial exercises for the objectives they need to study and retest. Students can also access supplemental animations and video clips directly from selected exercises. MathXL is available to qualified adopters. For more information, visit our website at www.mathxl.com, or contact your Pearson sales representative.

MathXL® Tutorials on CD

This interactive tutorial CD-ROM provides algorithmically generated practice exercises that are correlated at the objective level to the exercises in the textbook. Every practice exercise is accompanied by an example and a guided solution designed to involve students in the solution process. Selected exercises may also include a video clip to help students visualize concepts. The software provides helpful feedback for incorrect answers and can generate printed summaries of students' progress.

Acknowledgments

No book can be produced without a team of professionals who take pride in their work and are willing to put in long hours. Laurie Hurley, in particular, deserves extra thanks for her work as developmental editor. Rebecca Hubiak, Laurie Hurley, Holly Martinez, and Christine Verity also deserve special thanks for their careful accuracy checks, well-thought-out suggestions, and uncanny eye for detail. Thanks to Carrie Green, Laurie Hurley, and Christine Verity for their outstanding work in preparing supplements.

We are also indebted to Chris Burditt and Jann MacInnes for their many fine ideas that appear in our Collaborative Corners and Vince McGarry and Janet Wyatt for their recommendations for Teaching Tips featured in the Annotated Instructor's Edition.

Martha Morong, of Quadrata, Inc., provided editorial and production services of the highest quality imaginable—she is amazing and a joy to work with. Geri Davis, of the Davis Group, Inc., performed superb work as designer, art editor, and photo researcher, and is always a pleasure to work with. Network Graphics generated the graphs, charts, and many of the illustrations. Not only are the people at Network reliable, but they clearly take pride in their work. The many illustrations appear thanks to Bill Melvin—an artist with insight and creativity.

Our team at Pearson deserves special thanks. Acquisitions Editor Randy Welch provided many fine suggestions, remaining involved and accessible throughout the project. Executive Project Manager Kari Heen carefully coordinated tasks and schedules, keeping a widely spread team working together. Associate Editor Joanna Doxey coordinated reviews and assisted in a variety of tasks with patience and creativity. Editorial Assistant Jonathan Wooding responded quickly to all requests, always in a pleasant manner. Production Manager Ron Hampton's attention to detail, willingness to listen, and creative responses helped result in a book that is beautiful to look at. Marketing Manager Marlana Voerster and Marketing Assistant Nathaniel Koven skillfully kept us in touch with the needs of faculty. Our Editor in Chief, Maureen O'Connor, and Editorial Director, Chris Hoag, deserve credit for assembling this fine team.

We also thank the students at Indiana University Purdue University Indianapolis and the Community College of Vermont and the following professors for their thoughtful reviews and insightful comments.

Roberta Abarca, *Centralia College*
Darla J. Aguilar, *Pima Community College, Desert Vista Campus*
Bonnie Alcorn, *Waubonsee College*
Eugene Alderman, *South University*
Joseph Berland, *Chabot College*
Paul Blankenship, *Lexington Community College*
Susan Caldiero, *Cosumnes River College*
David Casey, *Citrus College*
Henri Feiner, *Coastline University*
Janet Hansen, *Dixie State College*
Elizabeth Hodes, *Santa Barbara City College*
Weilin Jang, *Austin Community College*
Paulette Kirkpatrick, *Wharton County Junior College*
Susan Knights, *Boise State University*
Jeff Koleno, *Lorain County Community College*
Julianne Labbiento, *Lehigh Carbon Community College*
Kathryn Lavelle, *Westchester Community College*
Amy Marolt, *Northeastern Mississippi Community College*
Rogers Martin, *Louisiana State University, Shreveport*
Ben Mayo, *Yakima Valley Community College*
Laurie McManus, *St. Louis Community College–Meramac*
Carol Metz, *Westchester Community College*
Anne Marie Mosher, *St. Louis Community College–Florissant Valley*
Pedro Mota, *Austin Community College, South Austin Campus*
Brenda M. Norman, *Tidewater Community College*
Kim Nunn, *Northeast State Technical College*
Michael Oppedisano, *Morrisville College, State University of New York*
Zaddock B. Reid, *San Bernardino Valley College*
Terri Seiver, *San Jacinto College–Central*
Timothy Thompson, *Oregon Institute of Technology*
Diane Trimble, *Tulsa Community College, West Campus*
Jennifer Vanden Eynden, *Grossmont College*
Beverly Vredevelt, *Spokane Falls Community College*
Michael Yarbrough, *Cosumnes River College*

Finally, a special thank-you to all those who so generously agreed to discuss their professional use of mathematics in our chapter openers. These dedicated people all share a desire to make math more meaningful to students. We cannot imagine a finer set of role models.

M.L.B.
D.J.E.

Photo Credits

Introduction to Algebraic Expressions

BRIAN BUSBY
CHIEF METEOROLOGIST
Kansas City, Missouri

All weather measurements are a series of numbers and values. Temperature, relative humidity, wind speed and direction, precipitation amount, and air pressure are all expressed in various numbers and percentages. Because weather systems move north and south, east and west, up and down, *and* over time, high-level math like calculus is the only way to represent that movement. But before you study calculus, you must begin with algebra.

AN APPLICATION

On December 10, Jenna notes that the temperature is −3°F at 6:00 A.M. She predicts that the temperature will rise at a rate of 2° per hour for 3 hr, and then rise at a rate of 3° per hour for 6 hr. She also predicts that the temperature will then fall at a rate of 2° per hour for 3 hr, and then fall at a rate of 5° per hour for 2 hr. What is Jenna's temperature forecast for 8:00 P.M.?

This problem appears as Exercise 135 in Section 1.7.

P roblem solving is the focus of this text. Chapter 1 presents important preliminaries that are needed for the problem-solving approach that is developed in Chapter 2 and used throughout the rest of the book. These preliminaries include a review of arithmetic, a discussion of real numbers and their properties, and an examination of how real numbers are added, subtracted, multiplied, divided, and raised to powers.

1.1   Introduction to Algebra

Algebraic Expressions   •   Translating to Algebraic Expressions   •   Translating to Equations

This section introduces some basic concepts and expressions used in algebra. Solving real-world problems is an important part of algebra, so we will focus on the wordings and mathematical expressions that often arise in applications.

Algebraic Expressions

Probably the greatest difference between arithmetic and algebra is the use of *variables* in algebra. When a letter can be any one of a set of numbers, that letter is a **variable**. For example, if n represents the number of tickets purchased for a Maroon 5 concert, then n will vary, depending on factors like price and day of the week. This makes n a variable. If each ticket costs \$40, then 3 tickets cost $40 \cdot 3$ dollars, 4 tickets cost $40 \cdot 4$ dollars, and n tickets cost $40 \cdot n$, or $40n$ dollars. Note that both $40 \cdot n$ and $40n$ mean 40 *times n*. The number 40 is an example of a **constant** because it does not change.

Price per Ticket (in dollars)	Number of Tickets Purchased	Total Paid (in dollars)
40	n	$40n$

The expression $40n$ is a **variable expression** because its value varies with the replacement for n. In this case, the total amount paid, $40n$, will change with the number of tickets purchased. In the following chart, we replace n with a variety of values and compute the total amount paid. In doing so, we are **evaluating the expression** $40n$.

Price per Ticket (in dollars), 40	Number of Tickets Purchased, n	Total Paid (in dollars), $40n$
40	400	$16,000
40	500	20,000
40	600	24,000

Variable expressions are examples of *algebraic expressions*. An **algebraic expression** consists of variables and/or numerals, often with operation signs and grouping symbols. Examples are

$$t + 97, \quad 5 \cdot x, \quad 3a - b, \quad 18 \div y, \quad \frac{9}{7}, \quad \text{and} \quad 4r(s + t).$$

Recall that a fraction bar is a division symbol: $\frac{9}{7}$, or 9/7, means $9 \div 7$. Similarly, multiplication can be written in several ways. For example, "5 times x" can be written as $5 \cdot x$, $5 \times x$, $5(x)$, or simply $5x$. On many calculators, this appears as $5 * x$.

To **evaluate** an algebraic expression, we substitute a number for each variable in the expression. We then calculate the result.

EXAMPLE   1

Evaluate each expression for the given values.

a) $x + y$ for $x = 37$ and $y = 28$

b) $5ab$ for $a = 2$ and $b = 3$

SOLUTION

a) We substitute 37 for x and 28 for y and carry out the addition:

$$x + y = 37 + 28 = 65.$$

The number 65 is called the **value** of the expression.

b) We substitute 2 for a and 3 for b and multiply:

$$5ab = 5 \cdot 2 \cdot 3 = 10 \cdot 3 = 30. \qquad \text{5ab means 5 times } a \text{ times } b.$$

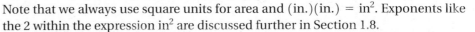 TRY EXERCISE 17

STUDENT NOTES

As we will see later, it is sometimes necessary to use parentheses when substituting a number for a variable. You may wish to use parentheses whenever you substitute. In Example 1, we could write

$$x + y = (37) + (28) = 65$$

and

$$5ab = 5(2)(3) = 30.$$

EXAMPLE   2

The area A of a rectangle of length l and width w is given by the formula $A = lw$. Find the area when l is 17 in. and w is 10 in.

SOLUTION   We evaluate, using 17 in. for l and 10 in. for w, and carry out the multiplication:

$$A = lw$$
$$A = (17 \text{ in.})(10 \text{ in.})$$
$$A = (17)(10)(\text{in.})(\text{in.})$$
$$A = 170 \text{ in}^2, \text{ or } 170 \text{ square inches.}$$

Note that we always use square units for area and $(\text{in.})(\text{in.}) = \text{in}^2$. Exponents like the 2 within the expression in^2 are discussed further in Section 1.8.

TRY EXERCISE 25

EXAMPLE **3**   The area of a triangle with a base of length b and a height of length h is given by the formula $A = \frac{1}{2}bh$. Find the area when b is 8 m (meters) and h is 6.4 m.

SOLUTION   We substitute 8 m for b and 6.4 m for h and then multiply:

$$A = \frac{1}{2}bh$$
$$A = \frac{1}{2}(8\,\text{m})(6.4\,\text{m})$$
$$A = \frac{1}{2}(8)(6.4)(\text{m})(\text{m})$$
$$A = 4(6.4)\,\text{m}^2$$
$$A = 25.6\,\text{m}^2, \text{ or } 25.6 \text{ square meters.}$$

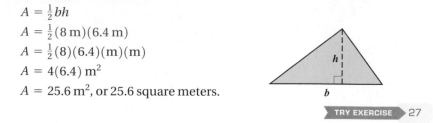

TRY EXERCISE   27

Translating to Algebraic Expressions

Before attempting to translate problems to equations, we need to be able to translate certain phrases to algebraic expressions.

Important Words	Sample Phrase or Sentence	Translation
Addition (+)		
added to	700 pounds was added to the car's weight.	$w + 700$
sum of	The sum of a number and 12	$n + 12$
plus	53 plus some number	$53 + x$
more than	800 more than Biloxi's population	$p + 800$
increased by	Ty's original estimate, increased by 4	$n + 4$
Subtraction (−)		
subtracted from	2 ounces was subtracted from the bag's weight.	$w - 2$
difference of	The difference of two scores	$m - n$
minus	A team of size s, minus 2 injured players	$s - 2$
less than	9 less than the number of volunteers last month	$v - 9$
decreased by	The car's speed, decreased by 8 mph	$s - 8$
Multiplication (·)		
multiplied by	The number of reservations, multiplied by 3	$r \cdot 3$
product of	The product of two numbers	$m \cdot n$
times	5 times the dog's weight	$5w$
twice	Twice the wholesale cost	$2c$
of	$\frac{1}{2}$ of Amelia's salary	$\frac{1}{2}s$
Division (÷)		
divided by	A 2-pound coffee cake, divided by 3	$2 \div 3$
quotient of	The quotient of 14 and 7	$14 \div 7$
divided into	4 divided into the delivery fee	$f \div 4$
ratio of	The ratio of $500 to the price of a new car	$500/p$
per	There were 18 computers per class of size s.	$18/s$

Any variable can be used to represent an unknown quantity; however, it is helpful to choose a descriptive letter. For example, w suggests weight and p suggests population or price. It is important to write down what the chosen variable represents.

EXAMPLE **4**   Translate each phrase to an algebraic expression.

a) Four less than Ava's height, in inches
b) Eighteen more than a number
c) A day's pay, in dollars, divided by eight

SOLUTION   To help think through a translation, we sometimes begin with a specific number in place of a variable.

a) If the height were 60, then 4 less than 60 would mean $60 - 4$. If the height were 65, the translation would be $65 - 4$. If we use h to represent "Ava's height, in inches," the translation of "Four less than Ava's height, in inches" is $h - 4$.

b) If we knew the number to be 10, the translation would be $10 + 18$, or $18 + 10$. If we use t to represent "a number," the translation of "Eighteen more than a number" is

$$t + 18, \quad \text{or} \quad 18 + t.$$

c) We let d represent "a day's pay, in dollars." If the pay were \$78, the translation would be $78 \div 8$, or $\frac{78}{8}$. Thus our translation of "A day's pay, in dollars, divided by eight" is

$$d \div 8, \quad \text{or} \quad \frac{d}{8}.$$

> TRY EXERCISE   31

CAUTION!   The order in which we subtract and divide affects the answer! Answering $4 - h$ or $8 \div d$ in Examples 4(a) and 4(c) is incorrect.

EXAMPLE **5**   Translate each phrase to an algebraic expression.

a) Half of some number
b) Seven more than twice the weight
c) Six less than the product of two numbers
d) Nine times the difference of a number and 10
e) Eighty-two percent of last year's enrollment

SOLUTION

Phrase	Variable(s)	Algebraic Expression
a) Half of some number	Let n represent the number.	$\frac{1}{2}n$, or $\frac{n}{2}$, or $n \div 2$
b) Seven more than twice the weight	Let w represent the weight.	$2w + 7$, or $7 + 2w$
c) Six less than the product of two numbers	Let m and n represent the numbers.	$mn - 6$
d) Nine times the difference of a number and 10	Let a represent the number.	$9(a - 10)$
e) Eighty-two percent of last year's enrollment	Let r represent last year's enrollment.	82% of r, or $0.82r$

> TRY EXERCISE   45

Translating to Equations

The symbol = ("equals") indicates that the expressions on either side of the equals sign represent the same number. An **equation** is a number sentence with the verb =. Equations may be true, false, or neither true nor false.

EXAMPLE **6**   Determine whether each equation is true, false, or neither.

a) $8 \cdot 4 = 32$ **b)** $7 - 2 = 4$ **c)** $x + 6 = 13$

SOLUTION

a) $8 \cdot 4 = 32$ The equation is *true*.

b) $7 - 2 = 4$ The equation is *false*.

c) $x + 6 = 13$ The equation is *neither* true nor false, because we do not know what number x represents.

> ### Solution
>
> A replacement or substitution that makes an equation true is called a *solution*. Some equations have more than one solution, and some have no solution. When all solutions have been found, we have *solved* the equation.

To see if a number is a solution, we evaluate all expressions in the equation. If the values on both sides of the equation are the same, the number is a solution.

EXAMPLE **7**   Determine whether 7 is a solution of $x + 6 = 13$.

SOLUTION   We evaluate $x + 6$ and compare both sides of the equation.

$$x + 6 = 13 \qquad \text{Writing the equation}$$
$$7 + 6 \mid 13 \qquad \text{Substituting 7 for } x$$
$$13 \overset{?}{=} 13 \qquad 13 = 13 \text{ is TRUE.}$$

Since the left-hand side and the right-hand side are the same, 7 is a solution.

▶ TRY EXERCISE ▶ 57

Although we do not study solving equations until Chapter 2, we can translate certain problem situations to equations now. The words "is the same as," "equal," "is," and "are" often translate to "=."

> **Words indicating equality, =** : "is the same as," "equal," "is," "are"

When translating a problem to an equation, we translate phrases to algebraic expressions, and the entire statement to an equation containing those expressions.

EXAMPLE 8   Translate the following problem to an equation.

What number plus 478 is 1019?

SOLUTION   We let y represent the unknown number. The translation then comes almost directly from the English sentence.

$$
\begin{array}{ccccc}
\text{What number} & \text{plus} & 478 & \text{is} & 1019? \\
\downarrow & \downarrow & \downarrow & \downarrow & \downarrow \\
y & + & 478 & = & 1019
\end{array}
$$

Note that "what number plus 478" translates to "$y + 478$" and "is" translates to "=."

TRY EXERCISE 63

Sometimes it helps to reword a problem before translating.

EXAMPLE 9   Translate the following problem to an equation.

The Taipei Financial Center, or Taipei 101, in Taiwan is the world's tallest building. At 1666 ft, it is 183 ft taller than the Petronas Twin Towers in Kuala Lumpur. How tall are the Petronas Twin Towers?

Source: *Guinness World Records* 2007

SOLUTION   We let h represent the height, in feet, of the Petronas Towers. A rewording and translation follow:

$$
\begin{array}{lccc}
\textit{Rewording:} & \text{The height of} & & \text{183 ft more than the height} \\
 & \text{Taipei 101} & \text{is} & \text{of the Petronas Towers} \\
 & \downarrow & \downarrow & \downarrow \\
\textit{Translating:} & 1666 & = & h + 183
\end{array}
$$

TRY EXERCISE 69

TECHNOLOGY CONNECTION

Technology Connections are activities that make use of features that are common to most graphing calculators. In some cases, students may find the user's manual for their particular calculator helpful for exact keystrokes.

Although all graphing calculators are not the same, most share the following characteristics.

Screen.   The large screen can show graphs and tables as well as the expressions entered. The screen has a different layout for different functions. Computations are performed in the **home screen**. On many calculators, the home screen is accessed by pressing **2ND** (QUIT). The **cursor** shows location on the screen, and the **contrast** (set by **2ND** ︿ or **2ND** ﹀) determines how dark the characters appear.

Keypad.   There are options written above the keys as well as on them. To access those above the keys, we press **2ND** or **ALPHA** and then the key. Expressions are usually entered as they would appear in print. For example, to evaluate $3xy + x$ for $x = 65$ and $y = 92$, we press 3 (×) 65 (×) 92 (+) 65 and then **ENTER**. The value of the expression, 18005, will appear at the right of the screen.

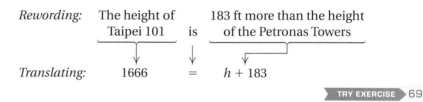

Evaluate each of the following.

1. $27a - 18b$, for $a = 136$ and $b = 13$
2. $19xy - 9x + 13y$, for $x = 87$ and $y = 29$

STUDY SKILLS

Get the Facts

Throughout this textbook, you will find a feature called Study Skills. These tips are intended to help improve your math study skills. On the first day of class, you should complete this chart.

Instructor: Name _____

Office hours and location _____

Phone number _____

Fax number _____

E-mail address _____

Find the names of two students whom you could contact for information or study questions:

1. Name _____

   Phone number _____

   E-mail address _____

2. Name _____

   Phone number _____

   E-mail address _____

Math lab on campus:

Location _____

Hours _____

Phone _____

Tutoring:

Campus location _____

Hours _____

Important supplements:

(See the preface for a complete list of available supplements.)

Supplements recommended by the instructor.

_____

_____

Translating for Success

1. Twice the difference of a number and 11

2. The product of a number and 11 is 2.

3. Twice the difference of two numbers is 11.

4. The quotient of twice a number and 11

5. The quotient of 11 and the product of two numbers

Translate to an expression or an equation and match that translation with one of the choices A–O below. Do not solve.

A. $x = 0.2(11)$

B. $\dfrac{2x}{11}$

C. $2x + 2 = 11$

D. $2(11x + 2)$

E. $11x = 2$

F. $0.2x = 11$

G. $11(2x - y)$

H. $2(x - 11)$

I. $11 + 2x = 2$

J. $2x + y = 11$

K. $2(x - y) = 11$

L. $11(x + 2x)$

M. $2(x + y) = 11$

N. $2 + \dfrac{x}{11}$

O. $\dfrac{11}{xy}$

Answers on page A-1

An additional, animated version of this activity appears in MyMathLab. To use MyMathLab, you need a course ID and a student access code. Contact your instructor for more information.

6. Eleven times the sum of a number and twice the number

7. Twice the sum of two numbers is 11.

8. Two more than twice a number is 11.

9. Twice the sum of 11 times a number and 2

10. Twenty percent of some number is 11.

1.1 EXERCISE SET

🦢 **Concept Reinforcement** *Classify each of the following as either an expression or an equation.*

1. $10n - 1$
2. $3x = 21$
3. $2x - 5 = 9$
4. $5(x + 2)$
5. $38 = 2t$
6. $45 = a - 1$
7. $4a - 5b$
8. $3s + 4t = 19$
9. $2x - 3y = 8$
10. $12 - 4xy$
11. $r(t + 7) + 5$
12. $9a + b$

To the student and the instructor: The **TRY EXERCISES** *for examples are indicated by a shaded block on the exercise number. Complete step-by-step solutions for these exercises appear online at www.pearsonhighered.com/ bittingerellenbogen.*

Evaluate.

13. $5a$, for $a = 9$
14. $11y$, for $y = 7$
15. $12 - r$, for $r = 4$
16. $t + 8$, for $t = 2$
17. $\dfrac{a}{b}$, for $a = 45$ and $b = 9$
18. $\dfrac{c + d}{3}$, for $c = 14$ and $d = 13$
19. $\dfrac{x + y}{4}$, for $x = 2$ and $y = 14$
20. $\dfrac{m}{n}$, for $m = 54$ and $n = 9$
21. $\dfrac{p - q}{7}$, for $p = 55$ and $q = 20$
22. $\dfrac{9m}{q}$, for $m = 6$ and $q = 18$
23. $\dfrac{5z}{y}$, for $z = 9$ and $y = 15$
24. $\dfrac{m - n}{2}$, for $m = 20$ and $n = 8$

Substitute to find the value of each expression.

25. *Hockey.* The area of a rectangle with base b and height h is bh. A regulation hockey goal is 6 ft wide and 4 ft high. Find the area of the opening.

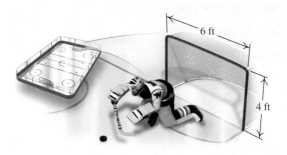

26. *Orbit time.* A communications satellite orbiting 300 mi above the earth travels about 27,000 mi in one orbit. The time, in hours, for an orbit is

$$\frac{27,000}{v},$$

where v is the velocity, in miles per hour. How long will an orbit take at a velocity of 1125 mph?

27. *Zoology.* A great white shark has triangular teeth. Each tooth measures about 5 cm across the base and has a height of 6 cm. Find the surface area of the front side of one such tooth. (See Example 3.)

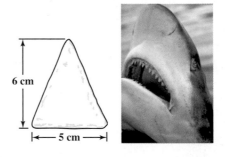

28. *Work time.* Javier takes three times as long to do a job as Luis does. Suppose t represents the time it takes Luis to do the job. Then $3t$ represents the time it takes Javier. How long does it take Javier if Luis takes **(a)** 30 sec? **(b)** 90 sec? **(c)** 2 min?

29. *Women's softball.* A softball player's batting average is h/a, where h is the number of hits and a is the number of "at bats." In the 2007 Women's College World Series, Caitlin Lowe of the Arizona Wildcats had 10 hits in 29 at bats. What was her batting average? Round to the nearest thousandth.

30. *Area of a parallelogram.*   The area of a parallelogram with base b and height h is bh. Find the area of the parallelogram when the height is 6 cm (centimeters) and the base is 7.5 cm.

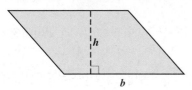

Translate to an algebraic expression.

31. 5 more than Ron's age

32. The product of 4 and a

33. 6 times b

34. 7 more than Lori's weight

35. 9 less than c

36. 4 less than d

37. 6 increased by q

38. 11 increased by z

39. 8 times Mai's speed

40. m subtracted from n

41. x less than y

42. 2 less than Than's age

43. x divided by w

44. The quotient of two numbers

45. The sum of the box's length and height

46. The sum of d and f

47. The product of 9 and twice m

48. Pemba's speed minus twice the wind speed

49. Thirteen less than one quarter of some number

50. Four less than ten times a number

51. Five times the difference of two numbers

52. One third of the sum of two numbers

53. 64% of the women attending

54. 38% of a number

Determine whether the given number is a solution of the given equation.

55. 25; $x + 17 = 42$

56. 75; $93 - y = 28$

57. 93; $a - 28 = 75$

58. 12; $8t = 96$

59. 63; $\dfrac{t}{7} = 9$

60. 52; $\dfrac{x}{8} = 6$

61. 3; $\dfrac{108}{x} = 36$

62. 7; $\dfrac{94}{y} = 12$

Translate each problem to an equation. Do not solve.

63. What number added to 73 is 201?

64. Seven times what number is 1596?

65. When 42 is multiplied by a number, the result is 2352. Find the number.

66. When 345 is added to a number, the result is 987. Find the number.

67. *Chess.*   A chess board has 64 squares. If pieces occupy 19 squares, how many squares are unoccupied?

68. *Hours worked.*   A carpenter charges \$35 an hour. How many hours did she work if she billed a total of \$3640?

69. *Recycling.*   Currently, Americans recycle or compost 32% of all municipal solid waste. This is the same as recycling or composting 79 million tons. What is the total amount of waste generated?
Source: U.S. EPA, Municipal Solid Waste Department

70. *Travel to work.*   In 2005, the average commuting time to work in New York was 31.2 min. The average commuting time in North Dakota was 14.9 min shorter. How long was the average commute in North Dakota?
Source: American Community Survey

In each of Exercises 71–78, match the phrase or sentence with the appropriate expression or equation from the column on the right.

71. ____ Twice the sum of two numbers

a) $\dfrac{x}{y} + 6$

72. ____ Five less than a number is twelve.

b) $2(x + y) = 48$

73. ____ Twelve more than a number is five.

c) $\dfrac{1}{2} \cdot a \cdot b$

74. ____ Half of the product of two numbers

d) $t + 12 = 5$

75. ____ Three times the sum of a number and five

e) $ab - 1 = 48$

76. ____ Twice the sum of two numbers is 48.

f) $2(m + n)$

77. ____ One less than the product of two numbers is 48.

g) $3(t + 5)$

78. ____ Six more than the quotient of two numbers

h) $x - 5 = 12$

To the student and the instructor: Writing exercises, denoted by 📜, should be answered using one or more English sentences. Because answers to many writing exercises will vary, solutions are not listed in the answers at the back of the book.

📜 **79.** What is the difference between a variable, a variable expression, and an equation?

📜 **80.** What does it mean to evaluate an algebraic expression?

Synthesis

To the student and the instructor: Synthesis exercises are designed to challenge students to extend the concepts or skills studied in each section. Many synthesis exercises will require the assimilation of skills and concepts from several sections.

📜 **81.** If the lengths of the sides of a square are doubled, is the area doubled? Why or why not?

📜 **82.** Write a problem that translates to $1998 + t = 2006$.

83. Signs of Distinction charges $120 per square foot for handpainted signs. The town of Belmar commissioned a triangular sign with a base of 3 ft and a height of 2.5 ft. How much will the sign cost?

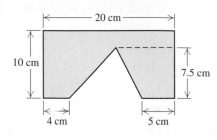

84. Find the area that is shaded.

85. Evaluate $\dfrac{x - y}{3}$ when x is twice y and $x = 12$.

86. Evaluate $\dfrac{x + y}{2}$ when y is twice x and $x = 6$.

87. Evaluate $\dfrac{a + b}{4}$ when a is twice b and $a = 16$.

88. Evaluate $\dfrac{a - b}{3}$ when a is three times b and $a = 18$.

Answer each question with an algebraic expression.

89. If $w + 3$ is a whole number, what is the next whole number after it?

90. If $d + 2$ is an odd number, what is the preceding odd number?

Translate to an algebraic expression.

91. The perimeter of a rectangle with length l and width w (perimeter means distance around)

92. The perimeter of a square with side s (perimeter means distance around)

93. Ellie's race time, assuming she took 5 sec longer than Joe and Joe took 3 sec longer than Molly. Assume that Molly's time was t seconds.

94. Ray's age 7 yr from now if he is 2 yr older than Monique and Monique is a years old

📜 **95.** If the length of the height of a triangle is doubled, is its area also doubled? Why or why not?

CORNER

Teamwork

Focus: Group problem solving; working collaboratively

Time: 15 minutes

Group size: 2

Working and studying as a team often enables students to solve problems that are difficult to solve alone.

ACTIVITY

1. The left hand column below contains the names of 12 colleges. A scrambled list of the names of their sports teams is on the right. As a group, match the names of the colleges to the teams.

   1. University of Texas
   2. Western State College of Colorado
   3. University of North Carolina
   4. University of Massachusetts
   5. Hawaii Pacific University
   6. University of Nebraska
   7. University of California, Santa Cruz
   8. University of Louisiana at Lafayette
   9. Grand Canyon University
   10. Palm Beach Atlantic University
   11. University of Alaska, Anchorage
   12. University of Florida

   a) Antelopes
   b) Banana Slugs
   c) Sea Warriors
   d) Gators
   e) Mountaineers
   f) Sailfish
   g) Longhorns
   h) Tar Heels
   i) Seawolves
   j) Ragin' Cajuns
   k) Cornhuskers
   l) Minutemen

2. After working for 5 min, confer with another group and reach mutual agreement.

3. Does the class agree on all 12 pairs?

4. Do you agree that group collaboration enhances our ability to solve problems?

1.2   The Commutative, Associative, and Distributive Laws

Equivalent Expressions • The Commutative Laws • The Associative Laws • The Distributive Law • The Distributive Law and Factoring

In order to solve equations, we must be able to manipulate algebraic expressions. The commutative, associative, and distributive laws discussed in this section enable us to write *equivalent expressions* that will simplify our work. Indeed, much of this text is devoted to finding equivalent expressions.

Equivalent Expressions

The expressions $4 + 4 + 4, 3 \cdot 4$, and $4 \cdot 3$ all represent the same number, 12. Expressions that represent the same number are said to be **equivalent**. The equivalent expressions $t + 18$ and $18 + t$ were used on p. 5 when we translated "eighteen more than a number." These expressions are equivalent because they represent the same number for any value of t. We can illustrate this by making some choices for t.

$$\text{When } t = 3, \quad t + 18 = 3 + 18 = 21$$
$$\text{and} \quad 18 + t = 18 + 3 = 21.$$
$$\text{When } t = 40, \quad t + 18 = 40 + 18 = 58$$
$$\text{and} \quad 18 + t = 18 + 40 = 58.$$

The Commutative Laws

Recall that changing the order in addition or multiplication does not change the result. Equations like $3 + 78 = 78 + 3$ and $5 \cdot 14 = 14 \cdot 5$ illustrate this idea and show that addition and multiplication are **commutative**.

> ### The Commutative Laws
>
> *For Addition.*   For any numbers a and b,
>
> $$a + b = b + a.$$
>
> (Changing the order of addition does not affect the answer.)
>
> *For Multiplication.*   For any numbers a and b,
>
> $$ab = ba.$$
>
> (Changing the order of multiplication does not affect the answer.)

EXAMPLE **1**   Use the commutative laws to write an expression equivalent to each of the following: **(a)** $y + 5$; **(b)** $9x$; **(c)** $7 + ab$.

SOLUTION

a) $y + 5$ is equivalent to $5 + y$ by the commutative law of addition.

b) $9x$ is equivalent to $x \cdot 9$ by the commutative law of multiplication.

c) $7 + ab$ is equivalent to $ab + 7$ by the commutative law of *addition*.

$7 + ab$ is also equivalent to $7 + ba$ by the commutative law of *multiplication*.

$7 + ab$ is also equivalent to $ba + 7$ by the two commutative laws, used together.

▶ **TRY EXERCISE** ▶ 11

The Associative Laws

Parentheses are used to indicate groupings. We generally simplify within the parentheses first. For example,

$$3 + (8 + 4) = 3 + 12 = 15$$

and

$$(3 + 8) + 4 = 11 + 4 = 15.$$

Similarly,

$$4 \cdot (2 \cdot 3) = 4 \cdot 6 = 24$$

and

$$(4 \cdot 2) \cdot 3 = 8 \cdot 3 = 24.$$

Note that, so long as only addition or only multiplication appears in an expression, changing the grouping does not change the result. Equations such as $3 + (7 + 5) = (3 + 7) + 5$ and $4(5 \cdot 3) = (4 \cdot 5)3$ illustrate that addition and multiplication are **associative**.

> ## The Associative Laws
>
> *For Addition.*   For any numbers a, b, and c,
>
> $$a + (b + c) = (a + b) + c.$$
>
> (Numbers can be grouped in any manner for addition.)
>
> *For Multiplication.*   For any numbers a, b, and c,
>
> $$a \cdot (b \cdot c) = (a \cdot b) \cdot c.$$
>
> (Numbers can be grouped in any manner for multiplication.)

EXAMPLE **2**   Use an associative law to write an expression equivalent to each of the following: **(a)** $y + (z + 3)$; **(b)** $(8x)y$.

SOLUTION

a) $y + (z + 3)$ is equivalent to $(y + z) + 3$ by the associative law of addition.

b) $(8x)y$ is equivalent to $8(xy)$ by the associative law of multiplication.

TRY EXERCISE ▶ 27

When only addition or only multiplication is involved, parentheses do not change the result. For that reason, we sometimes omit them altogether. Thus,

$$x + (y + 7) = x + y + 7 \quad \text{and} \quad l(wh) = lwh.$$

A sum such as $(5 + 1) + (3 + 5) + 9$ can be simplified by pairing numbers that add to 10. The associative and commutative laws allow us to do this:

$$(5 + 1) + (3 + 5) + 9 = 5 + 5 + 9 + 1 + 3$$
$$= 10 + 10 + 3 = 23.$$

EXAMPLE **3**   Use the commutative and/or associative laws of addition to write two expressions equivalent to $(7 + x) + 3$. Then simplify.

SOLUTION

$$(7 + x) + 3 = (x + 7) + 3 \qquad \text{Using the commutative law;}$$
$(x + 7) + 3$ is one equivalent expression.

$$= x + (7 + 3) \qquad \text{Using the associative law; } x + (7 + 3)$$
is another equivalent expression.

$$= x + 10 \qquad \text{Simplifying} \qquad \text{TRY EXERCISE ▶ 39}$$

EXAMPLE 4   Use the commutative and/or associative laws of multiplication to write two expressions equivalent to $2(x \cdot 3)$.

SOLUTION

$$2(x \cdot 3) = 2(3x) \qquad \text{Using the commutative law; } 2(3x) \text{ is one equivalent expression.}$$
$$= (2 \cdot 3)x \qquad \text{Using the associative law; } (2 \cdot 3)x \text{ is another equivalent expression.}$$
$$= 6x \qquad \text{Simplifying}$$

> **TRY EXERCISE** 41

The Distributive Law

The *distributive law* is probably the single most important law for manipulating algebraic expressions. Unlike the commutative and associative laws, the distributive law uses multiplication together with addition.

You have already used the distributive law although you may not have realized it at the time. To illustrate, try to multiply $3 \cdot 21$ mentally. Many people find the product, 63, by thinking of 21 as $20 + 1$ and then multiplying 20 by 3 and 1 by 3. The sum of the two products, $60 + 3$, is 63. Note that if the 3 does not multiply *both* 20 and 1, the result will not be correct.

EXAMPLE 5   Compute in two ways: $4(7 + 2)$.

SOLUTION

a) As in the discussion of $3(20 + 1)$ above, to compute $4(7 + 2)$, we can multiply both 7 and 2 by 4 and add the results:

$$4(7 + 2) = 4 \cdot 7 + 4 \cdot 2 \qquad \text{Multiplying both 7 and 2 by 4}$$
$$= 28 + 8 = 36. \qquad \text{Adding}$$

b) By first adding inside the parentheses, we get the same result in a different way:

$$4(7 + 2) = 4(9) \qquad \text{Adding; } 7 + 2 = 9$$
$$= 36. \qquad \text{Multiplying}$$

STUDENT NOTES

To remember the names *commutative*, *associative*, and *distributive*, first understand the concept. Next, use everyday life to link each word to the concept. For example, think of commuting to and from college as changing the order of appearance.

> **The Distributive Law**
>
> For any numbers a, b, and c,
>
> $$a(b + c) = ab + ac.$$
>
> (The product of a number and a sum can be written as the sum of two products.)

EXAMPLE 6   Multiply: $3(x + 2)$.

SOLUTION   Since $x + 2$ cannot be simplified unless a value for x is given, we use the distributive law:

$$3(x + 2) = 3 \cdot x + 3 \cdot 2 \qquad \text{Using the distributive law}$$
$$= 3x + 6. \qquad \text{Note that } 3 \cdot x \text{ is the same as } 3x.$$

> **TRY EXERCISE** 47

The expression $3x + 6$ has two *terms*, $3x$ and 6. In general, a **term** is a number, a variable, or a product or a quotient of numbers and/or variables. Thus, $t, 29, 5ab$, and $2x/y$ are terms in $t + 29 + 5ab + 2x/y$. Note that terms are separated by plus signs.

EXAMPLE 7   List the terms in the expression $7s + st + \dfrac{3}{t}$.

SOLUTION   Terms are separated by plus signs, so the terms in $7s + st + \dfrac{3}{t}$ are

$7s, st$, and $\dfrac{3}{t}$.

TRY EXERCISE 61

The distributive law can also be used when more than two terms are inside the parentheses.

EXAMPLE 8   Multiply: $6(s + 2 + 5w)$.

SOLUTION

$$6(s + 2 + 5w) = 6 \cdot s + 6 \cdot 2 + 6 \cdot 5w \qquad \text{Using the distributive law}$$
$$= 6s + 12 + (6 \cdot 5)w \qquad \text{Using the associative law for multiplication}$$
$$= 6s + 12 + 30w$$

TRY EXERCISE 55

Because of the commutative law of multiplication, the distributive law can be used on the "right": $(b + c)a = ba + ca$.

EXAMPLE 9   Multiply: $(c + 4)5$.

SOLUTION

$$(c + 4)5 = c \cdot 5 + 4 \cdot 5 \qquad \text{Using the distributive law on the right}$$
$$= 5c + 20 \qquad \text{Using the commutative law; } c \cdot 5 = 5c$$

TRY EXERCISE 57

CAUTION!   To use the distributive law for removing parentheses, be sure to multiply *each* term inside the parentheses by the multiplier outside. Thus,

$$\overline{a(b + c)} \neq \overline{ab + c} \quad \text{but} \quad a(b + c) = ab + ac.$$

The Distributive Law and Factoring

If we use the distributive law in reverse, we have the basis of a process called **factoring**: $ab + ac = a(b + c)$. To **factor** an expression means to write an equivalent expression that is a product. The parts of the product are called **factors**. Note that "factor" can be used as either a verb or a noun. Thus in the expression $5t$, the factors are 5 and t. In the expression $4(m + n)$, the factors are 4 and $(m + n)$. A **common factor** is a factor that appears in every term in an expression.

EXAMPLE 10   Use the distributive law to factor each of the following.

a) $3x + 3y$ **b)** $7x + 21y + 7$

SOLUTION

a) By the distributive law,

$$3x + 3y = 3(x + y). \quad \text{The common factor for } 3x \text{ and } 3y \text{ is } 3.$$

b) $7x + 21y + 7 = 7 \cdot x + 7 \cdot 3y + 7 \cdot 1$ The common factor is 7.

$$= 7(x + 3y + 1) \quad \text{Using the distributive law.}$$
Be sure to include both the 1 and the common factor, 7.

TRY EXERCISE 69

To check our factoring, we multiply to see if the original expression is obtained. For example, to check the **factorization** in Example 10(b), note that

$$7(x + 3y + 1) = 7 \cdot x + 7 \cdot 3y + 7 \cdot 1$$
$$= 7x + 21y + 7.$$

Since $7x + 21y + 7$ is what we started with in Example 10(b), we have a check.

CAUTION!   Do not confuse **terms** with **factors**. Terms are separated by plus signs, and factors are parts of products. The distributive law is used when there are two or more terms inside parentheses. For example, in the expression $a(b \cdot c)$, b and c are factors, not terms. We can use the commutative and associative laws to reorder and regroup the factors, but the distributive law does not apply here. Thus,

$$a(b \cdot c) \neq a \cdot b \cdot a \cdot c \quad \text{but} \quad a(b \cdot c) = (a \cdot b) \cdot c.$$

1.2   EXERCISE SET

For Extra Help   MyMathLab   Math XL PRACTICE   WATCH   DOWNLOAD

Concept Reinforcement   *Complete each sentence using one of these terms:* commutative, associative, *or* distributive.

1. $8 + t$ is equivalent to $t + 8$ by the _____ law for addition.

2. $3(xy)$ is equivalent to $(3x)y$ by the _____ law for multiplication.

3. $(5b)c$ is equivalent to $5(bc)$ by the _____ law for multiplication.

4. mn is equivalent to nm by the _____ law for multiplication.

5. $x(y + z)$ is equivalent to $xy + xz$ by the _____ law.

6. $(9 + a) + b$ is equivalent to $9 + (a + b)$ by the _____ law for addition.

7. $a + (6 + d)$ is equivalent to $(a + 6) + d$ by the _____ law for addition.

8. $3(t + 4)$ is equivalent to $3(4 + t)$ by the _____ law for addition.

9. $5(x + 2)$ is equivalent to $(x + 2)5$ by the _____ law for multiplication.

10. $2(a + b)$ is equivalent to $2 \cdot a + 2 \cdot b$ by the _____ law.

Use the commutative law of addition to write an equivalent expression.

11. $11 + t$

12. $a + 2$

13. $4 + 8x$

14. $ab + c$

15. $9x + 3y$

16. $3a + 7b$

17. $5(a + 1)$

18. $9(x + 5)$

Use the commutative law of multiplication to write an equivalent expression.

19. $7x$

20. xy

21. st

22. $13m$

23. $5 + ab$

24. $x + 3y$

25. $5(a + 1)$

26. $9(x + 5)$

Use the associative law of addition to write an equivalent expression.

27. $(x + 8) + y$

28. $(5 + m) + r$

29. $u + (v + 7)$

30. $x + (2 + y)$

31. $(ab + c) + d$

32. $(m + np) + r$

Use the associative law of multiplication to write an equivalent expression.

33. $(8x)y$

34. $(4u)v$

35. $2(ab)$

36. $9(7r)$

37. $3[2(a + b)]$

38. $5[x(2 + y)]$

Use the commutative and/or associative laws to write two equivalent expressions. Answers may vary.

39. $s + (t + 6)$

40. $7 + (v + w)$

41. $(17a)b$

42. $x(3y)$

Use the commutative and/or associative laws to show why the expression on the left is equivalent to the expression on the right. Write a series of steps with labels, as in Example 4.

43. $(1 + x) + 2$ is equivalent to $x + 3$

44. $(2a)4$ is equivalent to $8a$

45. $(m \cdot 3)7$ is equivalent to $21m$

46. $4 + (9 + x)$ is equivalent to $x + 13$

Multiply.

47. $2(x + 15)$

48. $3(x + 5)$

49. $4(1 + a)$

50. $6(v + 4)$

51. $8(3 + y)$

52. $7(s + 1)$

53. $10(9x + 6)$

54. $9(6m + 7)$

55. $5(r + 2 + 3t)$

56. $4(5x + 8 + 3p)$

57. $(a + b)2$

58. $(x + 2)7$

59. $(x + y + 2)5$

60. $(2 + a + b)6$

List the terms in each expression.

61. $x + xyz + 1$

62. $9 + 17a + abc$

63. $2a + \dfrac{a}{3b} + 5b$

64. $3xy + 20 + \dfrac{4a}{b}$

65. $4(x + y)$

66. $(7 + y)2$

67. $4x + 4y$

68. $14 + 2y$

Use the distributive law to factor each of the following. Check by multiplying.

69. $2a + 2b$

70. $5y + 5z$

71. $7 + 7y$

72. $13 + 13x$

73. $32x + 4$

74. $20a + 5$

75. $5x + 10 + 15y$

76. $3 + 27b + 6c$

77. $7a + 35b$

78. $8x + 24y$

79. $44x + 11y + 22z$

80. $14a + 56b + 7$

List the factors in each expression.

81. $5n$

82. uv

83. $3(x + y)$

84. $(a + b)12$

85. $7 \cdot a \cdot b$

86. $m \cdot n \cdot 2$

87. $(a - b)(x - y)$

88. $(3 - a)(b + c)$

89. Is subtraction commutative? Why or why not?

90. Is division associative? Why or why not?

Skill Review

To the student and the instructor: Exercises included for Skill Review include skills previously studied in the text. Often these exercises provide preparation for the next section of the text. The numbers in brackets immediately following the directions or exercise indicate the section in which the skill was introduced. The answers to all Skill Review exercises appear at the back of the book. If a Skill Review exercise gives you difficulty, review the material in the indicated section of the text.

Translate to an algebraic expression. [1.1]

91. Half of Kara's salary

92. Twice the sum of m and 3

Synthesis

93. Give an example illustrating the distributive law, and identify the terms and the factors in your example. Explain how you can determine terms and factors in an expression.

94. Explain how the distributive, commutative, and associative laws can be used to show that $2(3x + 4y)$ is equivalent to $6x + 8y$.

Tell whether the expressions in each pairing are equivalent. Then explain why or why not.

95. $8 + 4(a + b)$ and $4(2 + a + b)$

96. $5(a \cdot b)$ and $5 \cdot a \cdot 5 \cdot b$

97. $7 \div 3m$ and $m \cdot 3 \div 7$

98. $(rt + st)5$ and $5t(r + s)$

99. $30y + x \cdot 15$ and $5[2(x + 3y)]$

100. $[c(2 + 3b)]5$ and $10c + 15bc$

101. Evaluate the expressions $3(2 + x)$ and $6 + x$ for $x = 0$. Do your results indicate that $3(2 + x)$ and $6 + x$ are equivalent? Why or why not?

102. Factor $15x + 40$. Then evaluate both $15x + 40$ and the factorization for $x = 4$. Do your results *guarantee* that the factorization is correct? Why or why not? (*Hint:* See Exercise 101.)

COLLABORATIVE CORNER

Mental Addition

Focus: Application of commutative and associative laws

Time: 10 minutes

Group size: 2–3

Legend has it that while still in grade school, the mathematician Carl Friedrich Gauss (1777–1855) was able to add the numbers from 1 to 100 mentally. Gauss did not add them sequentially, but rather paired 1 with 99, 2 with 98, and so on.

ACTIVITY

1. Use a method similar to Gauss's to simplify the following:

    $$1 + 2 + 3 + 4 + 5 + 6 + 7 + 8 + 9 + 10.$$

    One group member should add from left to right as a check.

2. Use Gauss's method to find the sum of the first 25 counting numbers:

    $$1 + 2 + 3 + \cdots + 23 + 24 + 25.$$

    Again, one student should add from left to right as a check.

3. How were the associative and commutative laws applied in parts (1) and (2) above?

4. Now use a similar approach involving both addition and division to find the sum of the first 10 counting numbers:

    $$\begin{array}{l} 1 + 2 + 3 + \cdots + 10 \\ \underline{+ 10 + 9 + 8 + \cdots + 1} \end{array}$$

5. Use the approach in step (4) to find the sum of the first 100 counting numbers. Are the associative and commutative laws applied in this method, too? How is the distributive law used in this approach?

1.3   Fraction Notation

Factors and Prime Factorizations   •   Fraction Notation   •   Multiplication, Division, and Simplification   •
More Simplifying   •   Addition and Subtraction

This section covers multiplication, addition, subtraction, and division with fractions. Although much of this may be review, note that fraction expressions that contain variables are also included.

Factors and Prime Factorizations

In preparation for work with fraction notation, we first review how *natural numbers* are factored. **Natural numbers** can be thought of as the counting numbers:

$$1, 2, 3, 4, 5, \ldots .^*$$

(The dots indicate that the established pattern continues without ending.)

Since factors are parts of products, to factor a number, we express it as a product of two or more numbers.

Several factorizations of 12 are

$$1 \cdot 12, \qquad 2 \cdot 6, \qquad 3 \cdot 4, \qquad 2 \cdot 2 \cdot 3.$$

It is easy to miss a factor of a number if the factorizations are not written methodically.

EXAMPLE 1   List all factors of 18.

SOLUTION   Beginning at 1, we check all natural numbers to see if they are factors of 18. If they are, we write the factorization. We stop when we have already included the next natural number in a factorization.

1 is a factor of every number. $1 \cdot 18$

2 is a factor of 18. $2 \cdot 9$

3 is a factor of 18. $3 \cdot 6$

4 is *not* a factor of 18.

5 is *not* a factor of 18.

6 is the next natural number, but we have already listed 6 as a factor in the product $3 \cdot 6$.

We need check no additional numbers, because any natural number greater than 6 must be paired with a factor less than 6.

We now write the factors of 18 beginning with 1, going down the list of factorizations writing the first factor, then up the list of factorizations writing the second factor:

$$1, \quad 2, \quad 3, \quad 6, \quad 9, \quad 18.$$

> TRY EXERCISE 15

Some numbers have only two different factors, the number itself and 1. Such numbers are called **prime**.

---

*A similar collection of numbers, the **whole numbers,** includes 0: 0, 1, 2, 3, ...

> ### Prime Number
>
> A *prime number* is a natural number that has exactly two different factors: the number itself and 1. The first several primes are 2, 3, 5, 7, 11, 13, 17, 19, and 23.

If a natural number other than 1 is not prime, we call it **composite**.

EXAMPLE 2   Label each number as prime, composite, or neither: 29, 4, 1.

SOLUTION

29 is prime. It has exactly two different factors, 29 and 1.

4 is not prime. It has three different factors, 1, 2, and 4. It is composite.

1 is not prime. It does not have two *different* factors. The number 1 is not considered composite. It is neither prime nor composite.

TRY EXERCISE 5

Every composite number can be factored into a product of prime numbers. Such a factorization is called the **prime factorization** of that composite number.

EXAMPLE 3   Find the prime factorization of 36.

SOLUTION   We first factor 36 in any way that we can. One way is like this:

$$36 = 4 \cdot 9.$$

The factors 4 and 9 are not prime, so we factor them:

$$36 = 4 \cdot 9$$
$$= 2 \cdot 2 \cdot 3 \cdot 3.   \text{2 and 3 are both prime.}$$

The prime factorization of 36 is $2 \cdot 2 \cdot 3 \cdot 3$.

TRY EXERCISE 25

STUDENT NOTES

When writing a factorization, you are writing an equivalent expression for the original number. Some students do this with a tree diagram:

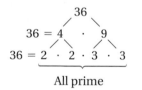

Fraction Notation

An example of **fraction notation** for a number is

$$\frac{2}{3}. \quad \begin{array}{l} \leftarrow \text{Numerator} \\ \leftarrow \text{Denominator} \end{array}$$

The top number is called the **numerator**, and the bottom number is called the **denominator**. When the numerator and the denominator are the same nonzero number, we have fraction notation for the number 1.

> ### Fraction Notation for 1
>
> For any number *a*, except 0,
>
> $$\frac{a}{a} = 1.$$
>
> (Any nonzero number divided by itself is 1.)

Note that in the definition for fraction notation for the number 1, we have excluded 0. In fact, 0 cannot be the denominator of *any* fraction. In this section, we limit our discussion to natural numbers, so this situation does not arise. Later in this chapter, we will discuss why denominators cannot be 0.

Multiplication, Division, and Simplification

Recall from arithmetic that fractions are multiplied as follows.

> ### Multiplication of Fractions
> For any two fractions a/b and c/d,
> $$\frac{a}{b} \cdot \frac{c}{d} = \frac{ac}{bd}.$$
> (The numerator of the product is the product of the two numerators. The denominator of the product is the product of the two denominators.)

EXAMPLE 4   Multiply: **(a)** $\dfrac{2}{3} \cdot \dfrac{5}{7}$; **(b)** $\dfrac{4}{x} \cdot \dfrac{8}{y}$.

SOLUTION   We multiply numerators as well as denominators.

a) $\dfrac{2}{3} \cdot \dfrac{5}{7} = \dfrac{2 \cdot 5}{3 \cdot 7} = \dfrac{10}{21}$

b) $\dfrac{4}{x} \cdot \dfrac{8}{y} = \dfrac{4 \cdot 8}{x \cdot y} = \dfrac{32}{xy}$

> TRY EXERCISE 53

Two numbers whose product is 1 are **reciprocals**, or **multiplicative inverses**, of each other. All numbers, except zero, have reciprocals. For example,

the reciprocal of $\dfrac{2}{3}$ is $\dfrac{3}{2}$ because $\dfrac{2}{3} \cdot \dfrac{3}{2} = \dfrac{6}{6} = 1$;

the reciprocal of 9 is $\dfrac{1}{9}$ because $9 \cdot \dfrac{1}{9} = \dfrac{9}{9} = 1$; and

the reciprocal of $\dfrac{1}{4}$ is 4 because $\dfrac{1}{4} \cdot 4 = 1$.

Reciprocals are used to rewrite division in an equivalent form that uses multiplication.

> ### Division of Fractions
> To divide two fractions, multiply by the reciprocal of the divisor:
> $$\frac{a}{b} \div \frac{c}{d} = \frac{a}{b} \cdot \frac{d}{c}.$$

EXAMPLE 5   Divide: $\dfrac{1}{2} \div \dfrac{3}{5}$.

SOLUTION

$$\frac{1}{2} \div \frac{3}{5} = \frac{1}{2} \cdot \frac{5}{3} \qquad \frac{5}{3} \text{ is the reciprocal of } \frac{3}{5}.$$

$$= \frac{5}{6}$$

> TRY EXERCISE 73

When one of the fractions being multiplied is 1, multiplying yields an equivalent expression because of the *identity property of* 1. A similar property could be stated for division, but there is no need to do so here.

The Identity Property of 1

For any number a,

$$a \cdot 1 = 1 \cdot a = a.$$

(Multiplying a number by 1 gives that same number.) The number 1 is called the *multiplicative identity*.

EXAMPLE 6   Multiply $\dfrac{4}{5} \cdot \dfrac{6}{6}$ to find an expression equivalent to $\dfrac{4}{5}$.

SOLUTION   Since $\frac{6}{6} = 1$, the expression $\frac{4}{5} \cdot \frac{6}{6}$ is equivalent to $\frac{4}{5} \cdot 1$, or simply $\frac{4}{5}$. We have

$$\frac{4}{5} \cdot \frac{6}{6} = \frac{4 \cdot 6}{5 \cdot 6} = \frac{24}{30}.$$

Thus, $\frac{24}{30}$ is equivalent to $\frac{4}{5}$.   ■

The steps of Example 6 are reversed by "removing a factor equal to 1"—in this case, $\frac{6}{6}$. By removing a factor that equals 1, we can *simplify* an expression like $\frac{24}{30}$ to an equivalent expression like $\frac{4}{5}$.

To simplify, we factor the numerator and the denominator, looking for the largest factor common to both. This is sometimes made easier by writing prime factorizations. After identifying common factors, we can express the fraction as a product of two fractions, one of which is in the form a/a.

EXAMPLE 7   Simplify: **(a)** $\dfrac{15}{40}$; **(b)** $\dfrac{36}{24}$.

SOLUTION

a) Note that 5 is a factor of both 15 and 40:

$$\frac{15}{40} = \frac{3 \cdot 5}{8 \cdot 5} \qquad \text{Factoring the numerator and the denominator, using the common factor, 5}$$

$$= \frac{3}{8} \cdot \frac{5}{5} \qquad \text{Rewriting as a product of two fractions; } \frac{5}{5} = 1$$

$$= \frac{3}{8} \cdot 1 = \frac{3}{8}. \qquad \text{Using the identity property of 1 (removing a factor equal to 1)}$$

STUDENT NOTES

The following rules can help you quickly determine whether 2, 3, or 5 is a factor of a number.

2 is a factor of a number if the number is even (the ones digit is 0, 2, 4, 6, or 8).

3 is a factor of a number if the sum of its digits is divisible by 3.

5 is a factor of a number if its ones digit is 0 or 5.

b) $\dfrac{36}{24} = \dfrac{2 \cdot 2 \cdot 3 \cdot 3}{2 \cdot 2 \cdot 2 \cdot 3}$ Writing the prime factorizations and identifying common factors; 12/12 could also be used.

$= \dfrac{3}{2} \cdot \dfrac{2 \cdot 2 \cdot 3}{2 \cdot 2 \cdot 3}$ Rewriting as a product of two fractions; $\dfrac{2 \cdot 2 \cdot 3}{2 \cdot 2 \cdot 3} = 1$

$= \dfrac{3}{2} \cdot 1 = \dfrac{3}{2}$ Using the identity property of 1   **TRY EXERCISE** ▶ 35

It is always wise to check your result to see if any common factors of the numerator and the denominator remain. (This will never happen if prime factorizations are used correctly.) If common factors remain, repeat the process by removing another factor equal to 1 to simplify your result.

More Simplifying

"Canceling" is a shortcut that you may have used for removing a factor equal to 1 when working with fraction notation. With *great* concern, we mention it as a possible way to speed up your work. Canceling can be used only when removing common factors in numerators and denominators. Canceling *cannot* be used in sums or differences. Our concern is that "canceling" be used with understanding. Example 7(b) might have been done faster as follows:

$$\frac{36}{24} = \frac{\cancel{2} \cdot \cancel{2} \cdot 3 \cdot \cancel{3}}{\cancel{2} \cdot \cancel{2} \cdot 2 \cdot \cancel{3}} = \frac{3}{2}, \quad \text{or} \quad \frac{36}{24} = \frac{3 \cdot \cancel{12}}{2 \cdot \cancel{12}} = \frac{3}{2}, \quad \text{or} \quad \frac{\overset{3}{\cancel{\underset{\cancel{36}}{18}}}}{\underset{\underset{2}{\cancel{12}}}{\cancel{24}}} = \frac{3}{2}.$$

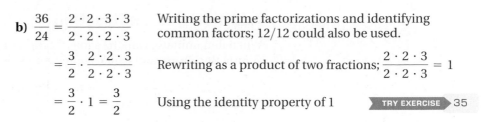

> ***CAUTION!***   Unfortunately, canceling is often performed incorrectly:
>
> $$\frac{\cancel{2} + 3}{\cancel{2}} = 3, \qquad \frac{\cancel{4} - 1}{\cancel{4} - 2} = \frac{1}{2}, \qquad \frac{1\cancel{5}}{\cancel{5}4} = \frac{1}{4}.$$
>
> The above cancellations are incorrect because the expressions canceled are *not* factors. For example, in $2 + 3$, the 2 and the 3 are not factors. Correct simplifications are as follows:
>
> $$\frac{2 + 3}{2} = \frac{5}{2}, \qquad \frac{4 - 1}{4 - 2} = \frac{3}{2}, \qquad \frac{15}{54} = \frac{5 \cdot \cancel{3}}{18 \cdot \cancel{3}} = \frac{5}{18}.$$
>
> ***Remember*: If you can't factor, you can't cancel! If in doubt, don't cancel!**

Sometimes it is helpful to use 1 as a factor in the numerator or the denominator when simplifying.

EXAMPLE 8   Simplify: $\dfrac{9}{72}$.

SOLUTION

$\dfrac{9}{72} = \dfrac{1 \cdot 9}{8 \cdot 9}$ Factoring and using the identity property of 1 to write 9 as $1 \cdot 9$

$= \dfrac{1 \cdot \cancel{9}}{8 \cdot \cancel{9}} = \dfrac{1}{8}$ Simplifying by removing a factor equal to 1: $\dfrac{9}{9} = 1$ **TRY EXERCISE** ▶ 39

Addition and Subtraction

When denominators are the same, fractions are added or subtracted by adding or subtracting numerators and keeping the same denominator.

> ### Addition and Subtraction of Fractions
> For any two fractions a/d and b/d,
> $$\frac{a}{d} + \frac{b}{d} = \frac{a+b}{d} \quad \text{and} \quad \frac{a}{d} - \frac{b}{d} = \frac{a-b}{d}.$$

EXAMPLE 9   Add and simplify: $\dfrac{4}{8} + \dfrac{5}{8}$.

SOLUTION   The common denominator is 8. We add the numerators and keep the common denominator:

$$\frac{4}{8} + \frac{5}{8} = \frac{4+5}{8} = \frac{9}{8}. \qquad \begin{array}{l}\text{You can think of this as} \\ 4 \cdot \dfrac{1}{8} + 5 \cdot \dfrac{1}{8} = 9 \cdot \dfrac{1}{8}, \text{or } \dfrac{9}{8}.\end{array}$$

TRY EXERCISE 63

In arithmetic, we often write $1\frac{1}{8}$ rather than the "improper" fraction $\frac{9}{8}$. In algebra, $\frac{9}{8}$ is generally more useful and is quite "proper" for our purposes.

When denominators are different, we use the identity property of 1 and multiply to find a common denominator. Then we add, as in Example 9.

EXAMPLE 10   Add or subtract as indicated: **(a)** $\dfrac{7}{8} + \dfrac{5}{12}$; **(b)** $\dfrac{9}{8} - \dfrac{4}{5}$.

SOLUTION

a) The number 24 is divisible by both 8 and 12. We multiply both $\frac{7}{8}$ and $\frac{5}{12}$ by suitable forms of 1 to obtain two fractions with denominators of 24:

$$\frac{7}{8} + \frac{5}{12} = \frac{7}{8} \cdot \frac{3}{3} + \frac{5}{12} \cdot \frac{2}{2} \qquad \begin{array}{l}\text{Multiplying by 1. Since } 8 \cdot 3 = 24, \text{ we} \\ \text{multiply } \frac{7}{8} \text{ by } \frac{3}{3}. \text{ Since } 12 \cdot 2 = 24, \text{ we} \\ \text{multiply } \frac{5}{12} \text{ by } \frac{2}{2}.\end{array}$$

$$= \frac{21}{24} + \frac{10}{24} \qquad \text{Performing the multiplication}$$

$$= \frac{31}{24}. \qquad \text{Adding fractions}$$

b) $\dfrac{9}{8} - \dfrac{4}{5} = \dfrac{9}{8} \cdot \dfrac{5}{5} - \dfrac{4}{5} \cdot \dfrac{8}{8}$ \qquad Using 40 as a common denominator

$$= \frac{45}{40} - \frac{32}{40} = \frac{13}{40} \qquad \text{Subtracting fractions}$$

TRY EXERCISE 69

After adding, subtracting, multiplying, or dividing, we may still need to simplify the answer.

EXAMPLE 11   Perform the indicated operation and, if possible, simplify.

a) $\dfrac{7}{10} - \dfrac{1}{5}$

b) $8 \cdot \dfrac{5}{12}$

c) $\dfrac{\frac{5}{6}}{\frac{25}{9}}$

SOLUTION

a) $\dfrac{7}{10} - \dfrac{1}{5} = \dfrac{7}{10} - \dfrac{1}{5} \cdot \dfrac{2}{2}$ Using 10 as the common denominator

$= \dfrac{7}{10} - \dfrac{2}{10}$

$= \dfrac{5}{10} = \dfrac{1 \cdot \cancel{5}}{2 \cdot \cancel{5}} = \dfrac{1}{2}$ Removing a factor equal to 1: $\dfrac{5}{5} = 1$

b) $8 \cdot \dfrac{5}{12} = \dfrac{8 \cdot 5}{12}$ Multiplying numerators and denominators. Think of 8 as $\frac{8}{1}$.

$= \dfrac{2 \cdot 2 \cdot 2 \cdot 5}{2 \cdot 2 \cdot 3}$ Factoring; $\dfrac{4 \cdot 2 \cdot 5}{4 \cdot 3}$ can also be used.

$= \dfrac{\cancel{2} \cdot \cancel{2} \cdot 2 \cdot 5}{\cancel{2} \cdot \cancel{2} \cdot 3}$ Removing a factor equal to 1: $\dfrac{2 \cdot 2}{2 \cdot 2} = 1$

$= \dfrac{10}{3}$ Simplifying

c) $\dfrac{\frac{5}{6}}{\frac{25}{9}} = \dfrac{5}{6} \div \dfrac{25}{9}$ Rewriting horizontally. Remember that a fraction bar indicates division.

$= \dfrac{5}{6} \cdot \dfrac{9}{25}$ Multiplying by the reciprocal of $\frac{25}{9}$

$= \dfrac{5 \cdot 3 \cdot 3}{2 \cdot 3 \cdot 5 \cdot 5}$ Writing as one fraction and factoring

$= \dfrac{\cancel{5} \cdot \cancel{3} \cdot 3}{2 \cdot \cancel{3} \cdot \cancel{5} \cdot 5}$ Removing a factor equal to 1: $\dfrac{5 \cdot 3}{3 \cdot 5} = 1$

$= \dfrac{3}{10}$ Simplifying

TRY EXERCISE 65

TECHNOLOGY CONNECTION

Some graphing calculators can perform operations using fraction notation. Others can convert answers given in decimal notation to fraction notation. Often this conversion is done using a command found in a **menu** of options that appears when a key is pressed. To select an item from a menu, we highlight its number and press **ENTER** or simply press the number of the item.

For example, to find fraction notation for $\frac{2}{15} + \frac{7}{12}$, we enter the expression as $2/15 + 7/12$. The answer is given in decimal notation. To convert this to fraction notation, we press **MATH** and select the Frac option. In this case, the notation Ans ▸ Frac shows that the graphing calculator will convert .7166666667 to fraction notation.

```
2/15+7/12
              .7166666667
Ans▶Frac
                   43/60
```

We see that $\frac{2}{15} + \frac{7}{12} = \frac{43}{60}$.

1.3 **EXERCISE SET**

For Extra Help
MyMathLab   Math XL PRACTICE   WATCH   DOWNLOAD

To the student and the instructor: Beginning in this section, selected exercises are marked with the symbol **Aha!**. *Students who pause to inspect an Aha! exercise should find the answer more readily than those who proceed mechanically. This is done to discourage rote memorization. Some later "Aha!" exercises in this exercise set are unmarked, to encourage students to always pause before working a problem.*

✎ **Concept Reinforcement**   *In each of Exercises 1–4, match the description with a number from the list on the right.*

1. ____ A factor of 35   **a)** 2

2. ____ A number that has 3 as a factor   **b)** 7

3. ____ An odd composite number   **c)** 60

4. ____ The only even prime number   **d)** 65

Label each of the following numbers as prime, composite, or neither.

5. 9   **6.** 15   **7.** 41   **8.** 49

9. 77   **10.** 37   **11.** 2   **12.** 1

13. 0   **14.** 16

Write all two-factor factorizations of each number. Then list all the factors of the number.

15. 50   **16.** 70   **17.** 42   **18.** 60

Find the prime factorization of each number. If the number is prime, state this.

19. 39   **20.** 34   **21.** 30

22. 55   **23.** 27   **24.** 98

25. 150   **26.** 54   **27.** 40

28. 56   **29.** 31   **30.** 180

31. 210   **32.** 79   **33.** 115

34. 143

Simplify.

35. $\dfrac{21}{35}$ **36.** $\dfrac{20}{26}$ **37.** $\dfrac{16}{56}$

38. $\dfrac{72}{27}$ **39.** $\dfrac{12}{48}$ **40.** $\dfrac{18}{84}$

41. $\dfrac{52}{13}$ **42.** $\dfrac{132}{11}$ **43.** $\dfrac{19}{76}$

44. $\dfrac{17}{51}$ **45.** $\dfrac{150}{25}$ **46.** $\dfrac{180}{36}$

47. $\dfrac{42}{50}$ **48.** $\dfrac{75}{80}$ **49.** $\dfrac{120}{82}$

50. $\dfrac{75}{45}$ **51.** $\dfrac{210}{98}$ **52.** $\dfrac{140}{350}$

Perform the indicated operation and, if possible, simplify.

53. $\dfrac{1}{2} \cdot \dfrac{3}{5}$ **54.** $\dfrac{11}{10} \cdot \dfrac{8}{5}$ **55.** $\dfrac{9}{2} \cdot \dfrac{4}{3}$

Aha! **56.** $\dfrac{11}{12} \cdot \dfrac{12}{11}$ **57.** $\dfrac{1}{8} + \dfrac{3}{8}$ **58.** $\dfrac{1}{2} + \dfrac{1}{8}$

59. $\dfrac{4}{9} + \dfrac{13}{18}$ **60.** $\dfrac{4}{5} + \dfrac{8}{15}$ **61.** $\dfrac{3}{a} \cdot \dfrac{b}{7}$

62. $\dfrac{x}{5} \cdot \dfrac{y}{z}$ **63.** $\dfrac{4}{n} + \dfrac{6}{n}$ **64.** $\dfrac{9}{x} - \dfrac{5}{x}$

65. $\dfrac{3}{10} + \dfrac{8}{15}$ **66.** $\dfrac{7}{8} + \dfrac{5}{12}$ **67.** $\dfrac{11}{7} - \dfrac{4}{7}$

68. $\dfrac{12}{5} - \dfrac{2}{5}$ **69.** $\dfrac{13}{18} - \dfrac{4}{9}$ **70.** $\dfrac{13}{15} - \dfrac{11}{45}$

Aha! **71.** $\dfrac{20}{30} - \dfrac{2}{3}$ **72.** $\dfrac{5}{7} - \dfrac{5}{21}$ **73.** $\dfrac{7}{6} \div \dfrac{3}{5}$

74. $\dfrac{7}{5} \div \dfrac{10}{3}$ **75.** $\dfrac{8}{9} \div \dfrac{4}{15}$ **76.** $\dfrac{9}{4} \div 9$

77. $12 \div \dfrac{4}{9}$ **78.** $\dfrac{1}{10} \div \dfrac{1}{5}$ Aha! **79.** $\dfrac{7}{13} \div \dfrac{7}{13}$

80. $\dfrac{17}{8} \div \dfrac{5}{6}$ **81.** $\dfrac{\frac{2}{7}}{\frac{5}{3}}$ **82.** $\dfrac{\frac{3}{8}}{\frac{1}{5}}$

83. $\dfrac{\frac{9}{1}}{2}$ **84.** $\dfrac{\frac{3}{7}}{6}$

85. Under what circumstances would the sum of two fractions be easier to compute than the product of the same two fractions?

86. Under what circumstances would the product of two fractions be easier to compute than the sum of the same two fractions?

Skill Review

Use a commutative law to write an equivalent expression. There can be more than one correct answer. [1.2]

87. $5(x + 3)$ **88.** $7 + (a + b)$

Synthesis

89. Bryce insists that $(2 + x)/8$ is equivalent to $(1 + x)/4$. What mistake do you think is being made and how could you demonstrate to Bryce that the two expressions are not equivalent?

90. Why are 0 and 1 considered neither prime nor composite?

91. In the following table, the top number can be factored in such a way that the sum of the factors is the bottom number. For example, in the first column, 56 is factored as $7 \cdot 8$, since $7 + 8 = 15$, the bottom number. Find the missing numbers in each column.

Product	56	63	36	72	140	96	168
Factor	7						
Factor	8						
Sum	15	16	20	38	24	20	29

92. *Packaging.* Tritan Candies uses two sizes of boxes, 6 in. long and 8 in. long. These are packed end to end in bigger cartons to be shipped. What is the shortest-length carton that will accommodate boxes of either size without any room left over? (Each carton must contain boxes of only one size; no mixing is allowed.)

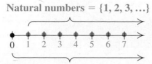

Simplify.

93. $\dfrac{16 \cdot 9 \cdot 4}{15 \cdot 8 \cdot 12}$

94. $\dfrac{9 \cdot 8xy}{2xy \cdot 36}$

95. $\dfrac{45pqrs}{9prst}$

96. $\dfrac{247}{323}$

97. $\dfrac{15 \cdot 4xy \cdot 9}{6 \cdot 25x \cdot 15y}$

98. $\dfrac{10x \cdot 12 \cdot 25y}{2z \cdot 30x \cdot 20y}$

99. $\dfrac{\frac{27ab}{15mn}}{\frac{18bc}{25np}}$

100. $\dfrac{\frac{45xyz}{24ab}}{\frac{30xz}{32ac}}$

101. $\dfrac{5\frac{3}{4}rs}{4\frac{1}{2}st}$

102. $\dfrac{3\frac{5}{7}mn}{2\frac{4}{5}np}$

Find the area of each figure.

103.

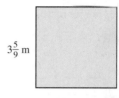

$\frac{7}{9}$ m $\frac{7}{9}$ m $\frac{4}{5}$ m

104.

$\frac{5}{4}$ m $\frac{10}{7}$ m

105. Find the perimeter of a square with sides of length $3\frac{5}{9}$ m.

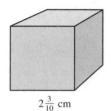

$3\frac{5}{9}$ m

106. Find the perimeter of the rectangle in Exercise 103.

107. Find the total length of the edges of a cube with sides of length $2\frac{3}{10}$ cm.

$2\frac{3}{10}$ cm

1.4   Positive and Negative Real Numbers

The Integers • The Rational Numbers • Real Numbers and Order • Absolute Value

A **set** is a collection of objects. The set containing 1, 3, and 7 is usually written $\{1, 3, 7\}$. In this section, we examine some important sets of numbers. More on sets can be found in Appendix C.

Natural numbers = $\{1, 2, 3, \ldots\}$

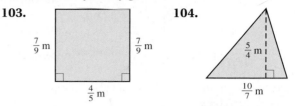

0 1 2 3 4 5 6 7

Whole numbers = $\{0, 1, 2, 3, \ldots\}$

The Integers

Two sets of numbers were mentioned in Section 1.3. We represent these sets using dots on a number line, as shown at left.

To create the set of *integers,* we include all whole numbers, along with their *opposites.* To find the opposite of a number, we locate the number that is the same distance from 0 but on the other side of the number line. For example,

the opposite of 1 is negative 1, written -1;

and

the opposite of 3 is negative 3, written -3.

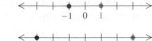

The **integers** consist of all whole numbers and their opposites.

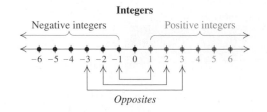

Integers

Negative integers　　Positive integers

-6 -5 -4 -3 -2 -1 0 1 2 3 4 5 6

Opposites

STUDENT NOTES

It is not uncommon in mathematics for a symbol to have more than one meaning in different contexts. The symbol "$-$" in $5 - 3$ indicates subtraction. The same symbol in -10 indicates the opposite of 10, or negative 10.

Opposites are discussed in more detail in Section 1.6. Note that, except for 0, opposites occur in pairs. Thus, 5 is the opposite of -5, just as -5 is the opposite of 5. Note that 0 acts as its own opposite.

Set of Integers

The set of integers $= \{ \ldots, -4, -3, -2, -1, 0, 1, 2, 3, 4, \ldots \}$.

Integers are associated with many real-world problems and situations.

EXAMPLE　**1**　State which integer(s) corresponds to each situation.

a) In 2006, there was \$13 trillion in outstanding mortgage debt in the United States.
   Source: Board of Governors of the Federal Reserve System

b) Part of Death Valley is 200 ft below sea level.

c) To lose one pound of fat, it is necessary for most people to create a 3500-calorie deficit.
   Source: World Health Organization

SOLUTION

a) The integer $-13{,}000{,}000{,}000{,}000$ corresponds to a debt of \$13 trillion.

b) The integer -200 corresponds to 200 ft below sea level.

c) The integer -3500 corresponds to a deficit of 3500 calories.

> **TRY EXERCISE** 9

The Rational Numbers

A number like $\frac{5}{9}$, although built out of integers, is not itself an integer. Another set of numbers, the **rational numbers**, contains integers, fractions, and decimals. Some examples of rational numbers are

$$\frac{5}{9}, \quad -\frac{4}{7}, \quad 95, \quad -16, \quad 0, \quad \frac{-35}{8}, \quad 2.4, \quad -0.31.$$

In Section 1.7, we show that $-\frac{4}{7}$ can be written as $\frac{-4}{7}$ or $\frac{4}{-7}$. Indeed, every number listed above can be written as an integer over an integer. For example, 95 can be written as $\frac{95}{1}$ and 2.4 can be written as $\frac{24}{10}$. In this manner, any *rational* number can be expressed as the *ratio* of two integers. Rather than attempt to list all rational numbers, we use this idea of ratio to describe the set as follows.

Set of Rational Numbers

The set of rational numbers $= \left\{ \dfrac{a}{b} \,\middle|\, a \text{ and } b \text{ are integers and } b \neq 0 \right\}$.

This is read "the set of all numbers a over b, where a and b are integers and b does not equal zero."

In Section 1.7, we explain why b cannot equal 0.
To *graph* a number is to mark its location on a number line.

EXAMPLE   **2**   Graph each of the following rational numbers: **(a)** $\frac{5}{2}$; **(b)** -3.2; **(c)** $\frac{11}{8}$.

SOLUTION

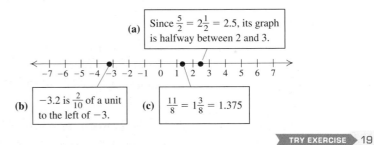

(a) Since $\frac{5}{2} = 2\frac{1}{2} = 2.5$, its graph is halfway between 2 and 3.

(b) -3.2 is $\frac{2}{10}$ of a unit to the left of -3.

(c) $\frac{11}{8} = 1\frac{3}{8} = 1.375$

> TRY EXERCISE   19

It is important to remember that every rational number can be written using fraction notation or decimal notation.

EXAMPLE   **3**   Convert to decimal notation: $-\frac{5}{8}$.

SOLUTION   We first find decimal notation for $\frac{5}{8}$. Since $\frac{5}{8}$ means $5 \div 8$, we divide.

$$
\begin{array}{r}
0.6\,2\,5 \\
8\overline{)5.0\,0\,0} \\
\underline{4\,8\,0\,0} \\
2\,0\,0 \\
\underline{1\,6\,0} \\
4\,0 \\
\underline{4\,0} \\
0 \quad \leftarrow \text{The remainder is 0.}
\end{array}
$$

Thus, $\frac{5}{8} = 0.625$, so $-\frac{5}{8} = -0.625$.

> TRY EXERCISE   25

Because the division in Example 3 ends with the remainder 0, we consider −0.625 a **terminating decimal**. If we are "bringing down" zeros and a remainder reappears, we have a **repeating decimal**, as shown in the next example.

 **4**   Convert to decimal notation: $\frac{7}{11}$.

SOLUTION   We divide:

$$
\begin{array}{r}
0.6\,3\,6\,3\ldots \\
11\overline{)7.0\,0\,0\,0} \\
6\,6 \\
\overline{4\,0} \\
3\,3 \\
\overline{7\,0} \\
6\,6 \\
\overline{4\,0}
\end{array}
$$

4 reappears as a remainder, so the pattern of 6's and 3's in the quotient will continue.

We abbreviate repeating decimals by writing a bar over the repeating part—in this case, $0.\overline{63}$. Thus, $\frac{7}{11} = 0.\overline{63}$.

TRY EXERCISE ▶ 29

Although we do not prove it here, every rational number can be expressed as either a terminating or repeating decimal, and every terminating or repeating decimal can be expressed as a ratio of two integers.

Real Numbers and Order

Some numbers, when written in decimal form, neither terminate nor repeat. Such numbers are called **irrational numbers**.

What sort of numbers are irrational? One example is π (the Greek letter *pi*, read "pie"), which is used to find the area and the circumference of a circle: $A = \pi r^2$ and $C = 2\pi r$.

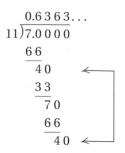

Another irrational number, $\sqrt{2}$ (read "the square root of 2"), is the length of the diagonal of a square with sides of length 1. It is also the number that, when multiplied by itself, gives 2. No rational number can be multiplied by itself to get 2, although some approximations come close:

1.4 is an *approximation* of $\sqrt{2}$ because $(1.4)(1.4) = 1.96$;

1.41 is a better approximation because $(1.41)(1.41) = 1.9881$;

1.4142 is an even better approximation because $(1.4142)(1.4142) = 1.99996164$.

To approximate $\sqrt{2}$ on some calculators, we simply press ②ꞏ and then ☑ꞏ. With other calculators, we press ☑ꞏ, ②ꞏ, and **ENTER**, or consult a manual.

EXAMPLE 5   Graph the real number $\sqrt{3}$ on the number line.

SOLUTION   We use a calculator and approximate: $\sqrt{3} \approx 1.732$ ("$\approx$" means "approximately equals"). Then we locate this number on the number line.

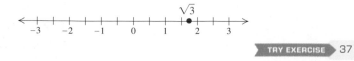

TRY EXERCISE ▶ 37

The rational numbers and the irrational numbers together correspond to all the points on the number line and make up what is called the **real-number system**.

To approximate $\sqrt{3}$ on most graphing calculators, we press $\boxed{\sqrt{}}$ and then enter 3 enclosed by parentheses. Some graphing calculators will supply the left parenthesis automatically when $\boxed{\sqrt{}}$ is pressed.

Approximate each of the following to nine decimal places.

1. $\sqrt{5}$ **2.** $\sqrt{7}$

3. $\sqrt{13}$ **4.** $\sqrt{27}$

5. $\sqrt{38}$ **6.** $\sqrt{50}$

Set of Real Numbers

The set of real numbers = The set of all numbers corresponding to
points on the number line.

The following figure shows the relationships among various kinds of numbers.

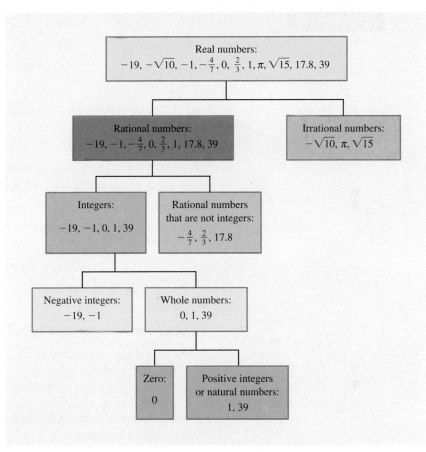

 **EXAMPLE 6**   Which numbers in the following list are **(a)** whole numbers? **(b)** integers? **(c)** rational numbers? **(d)** irrational numbers? **(e)** real numbers?

$$-38, \quad -\frac{8}{5}, \quad 0, \quad 0.\overline{3}, \quad 4.5, \quad \sqrt{30}, \quad 52$$

SOLUTION

a) 0 and 52 are whole numbers.

b) -38, 0, and 52 are integers.

c) -38, $-\frac{8}{5}$, 0, $0.\overline{3}$, 4.5, and 52 are rational numbers.

d) $\sqrt{30}$ is an irrational number.

e) -38, $-\frac{8}{5}$, 0, $0.\overline{3}$, 4.5, $\sqrt{30}$, and 52 are real numbers.   ▶ TRY EXERCISE ▶ 75

Real numbers are named in order on the number line, with larger numbers further to the right. For any two numbers, the one to the left is less than the one to the right. We use the symbol $<$ to mean "**is less than**." The sentence $-8 < 6$ means "-8 is less than 6." The symbol $>$ means "**is greater than**." The sentence $-3 > -7$ means "-3 is greater than -7."

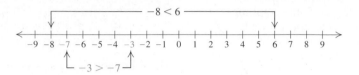

EXAMPLE **7**

Use either $<$ or $>$ for ■ to write a true sentence.

a) 2 ■ 9   **b)** -3.45 ■ 1.32 **c)** 6 ■ -12

d) -18 ■ -5 **e)** $\frac{7}{11}$ ■ $\frac{5}{8}$

SOLUTION

a) Since 2 is to the left of 9 on the number line, we know that 2 is less than 9, so $2 < 9$.

b) Since -3.45 is to the left of 1.32, we have $-3.45 < 1.32$.

c) Since 6 is to the right of -12, we have $6 > -12$.

d) Since -18 is to the left of -5, we have $-18 < -5$.

e) We convert to decimal notation: $\frac{7}{11} = 0.\overline{63}$ and $\frac{5}{8} = 0.625$. Thus, $\frac{7}{11} > \frac{5}{8}$.

We also could have used a common denominator: $\frac{7}{11} = \frac{56}{88} > \frac{55}{88} = \frac{5}{8}$.

TRY EXERCISES 41 and 45

Sentences like "$a < -5$" and "$-3 > -8$" are **inequalities**. It is useful to remember that every inequality can be written in two ways. For example,

$$-3 > -8 \quad \text{has the same meaning as} \quad -8 < -3.$$

It may be helpful to think of an inequality sign as an "arrow" with the smaller side pointing to the smaller number.

Note that $a > 0$ means that a represents a positive real number and $a < 0$ means that a represents a negative real number.

Statements like $a \leq b$ and $b \geq a$ are also inequalities. We read $a \leq b$ as "a **is less than or equal to** b" and $a \geq b$ as "a **is greater than or equal to** b."

EXAMPLE **8**

Classify each inequality as true or false.

a) $-3 \leq 5$ **b)** $-3 \leq -3$ **c)** $-5 \geq 4$

SOLUTION

STUDENT NOTES

It is important to remember that just because an equation or inequality is written or printed, it is not necessarily *true*. For instance, $6 = 7$ is an equation and $2 > 5$ is an inequality. Of course, both statements are *false*.

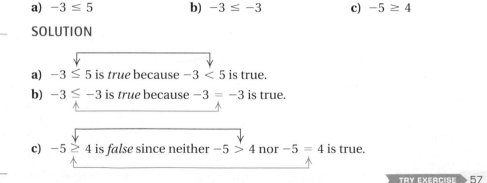

a) $-3 \leq 5$ is *true* because $-3 < 5$ is true.

b) $-3 \leq -3$ is *true* because $-3 = -3$ is true.

c) $-5 \geq 4$ is *false* since neither $-5 > 4$ nor $-5 = 4$ is true.

TRY EXERCISE 57

Absolute Value

There is a convenient terminology and notation for the distance a number is from 0 on the number line. It is called the **absolute value** of the number.

> **Absolute Value**
>
> We write $|a|$, read "the absolute value of a," to represent the number of units that a is from zero.

EXAMPLE 9   Find each absolute value: **(a)** $|-3|$; **(b)** $|7.2|$; **(c)** $|0|$.

TECHNOLOGY CONNECTION

Most graphing calculators use the notation abs (2) to indicate the absolute value of 2. This is often accessed using the NUM option of the MATH key. When using a graphing calculator for this, be sure to distinguish between the $\boxed{-}$ and $\boxed{(-)}$ keys (see p. 46 in Section 1.6).

SOLUTION

a) $|-3| = 3$ since -3 is 3 units from 0.

b) $|7.2| = 7.2$ since 7.2 is 7.2 units from 0.

c) $|0| = 0$ since 0 is 0 units from itself.

TRY EXERCISE 63

Distance is never negative, so numbers that are opposites have the same absolute value. If a number is nonnegative, its absolute value is the number itself. If a number is negative, its absolute value is its opposite.

1.4   EXERCISE SET

For Extra Help MyMathLab Math XL PRACTICE WATCH DOWNLOAD

↪ *Concept Reinforcement   In each of Exercises 1–8, fill in the blank using one of the following words:* natural number, whole number, integer, rational number, terminating, repeating, irrational number, absolute value.

1. Division can be used to show that $\frac{4}{7}$ can be written as a(n) _____ decimal.

2. Division can be used to show that $\frac{3}{20}$ can be written as a(n) _____ decimal.

3. If a number is a(n) _____, it is either a whole number or the opposite of a whole number.

4. 0 is the only _____ that is not a natural number.

5. Any number of the form a/b, where a and b are integers, with $b \neq 0$, is an example of a(n) _____.

6. A number like $\sqrt{5}$, which cannot be written precisely in fraction notation or decimal notation, is an example of a(n) _____.

7. If a number is a(n) _____, then it can be thought of as a counting number.

8. When two numbers are opposites, they have the same _____.

State which real number(s) correspond to each situation.

9. *Student loans and grants.* The maximum amount that a student may borrow each year with a Stafford Loan is $10,500. The maximum annual award for the Nurse Educator Scholarship Program is $27,482.
Sources: www.studentaid.ed.gov and www.collegezone.com

10. Using a NordicTrack exercise machine, LaToya burned 150 calories. She then drank an isotonic drink containing 65 calories.

11. The highest temperature ever recorded in a desert was 136 degrees Fahrenheit (°F) at Al-Aziziyah in the Sahara, Libya. The coldest temperature recorded in a desert was 4°F below zero in the McMurdo Dry Valleys, Antarctica.
Source: *Guinness World Records* 2007

12. The Dead Sea is 1312 ft below sea level, whereas Mt. Everest is 29,035 ft above sea level.
Source: *Guinness World Records* 2007

13. *Stock market.* The Dow Jones Industrial Average is an indicator of the stock market. On October 12, 1997, the Dow Jones fell a record 554 points. On March 16, 2002, the Dow Jones gained a record 499.19 points.
Source: www.finfacts.ie

14. Ignition occurs 10 sec before liftoff. A spent fuel tank is detached 235 sec after liftoff.

15. Kim deposited $650 in a savings account. Two weeks later, she withdrew $180.

16. *Birth and death rates.* Recently, the world birth rate was 20.09 per thousand. The death rate was 8.37 per thousand.
Source: Central Intelligence Agency, 2007

17. The halfback gained 8 yd on the first play. The quarterback was tackled for a 5-yd loss on the second play.

18. In the 2007 Masters Tournament, golfer Tiger Woods finished 3 over par. In the World Golf Championship, he finished 10 under par.
Source: PGA Tour Inc.

Graph each rational number on a number line.

19. $\frac{10}{3}$ **20.** $-\frac{17}{5}$

21. -4.3 **22.** 3.87

23. -2 **24.** 5

Write decimal notation for each number.

25. $\frac{7}{8}$ **26.** $-\frac{1}{8}$ **27.** $-\frac{3}{4}$

28. $\frac{11}{6}$ **29.** $-\frac{7}{6}$ **30.** $-\frac{5}{12}$

31. $\frac{2}{3}$ **32.** $\frac{1}{4}$ **33.** $-\frac{1}{2}$

34. $-\frac{1}{9}$ Aha! **35.** $\frac{13}{100}$ **36.** $-\frac{9}{20}$

Graph each irrational number on a number line.

37. $\sqrt{5}$ **38.** $\sqrt{92}$

39. $-\sqrt{22}$ **40.** $-\sqrt{54}$

Write a true sentence using either < or >.

41. $5 \ \square \ 0$ **42.** $8 \ \square \ -8$

43. $-9 \ \square \ 9$ **44.** $0 \ \square \ -7$

45. $-8 \ \square \ -5$ **46.** $-4 \ \square \ -3$

47. $-5 \ \square \ -11$ **48.** $-3 \ \square \ -4$

49. $-12.5 \ \square \ -10.2$ **50.** $-10.3 \ \square \ -14.5$

51. $\frac{5}{12} \ \square \ \frac{11}{25}$ **52.** $-\frac{14}{17} \ \square \ -\frac{27}{35}$

For each of the following, write a second inequality with the same meaning.

53. $-2 > x$ **54.** $a > 9$

55. $10 \le y$ **56.** $-12 \ge t$

Classify each inequality as either true or false.

57. $-3 \ge -11$ **58.** $5 \le -5$

59. $0 \ge 8$ **60.** $-5 \le 7$

61. $-8 \le -8$ **62.** $10 \ge 10$

Find each absolute value.

63. $|-58|$ **64.** $|-47|$

65. $|-12.2|$ **66.** $|4.3|$

67. $|\sqrt{2}|$ **68.** $|-456|$

69. $\left|-\frac{9}{7}\right|$ **70.** $|-\sqrt{3}|$

71. $|0|$ **72.** $\left|-\frac{3}{4}\right|$

73. $|x|$, for $x = -8$ **74.** $|a|$, for $a = -5$

For Exercises 75–80, consider the following list:
$$-83, \quad -4.7, \quad 0, \quad \tfrac{5}{9}, \quad 2.\overline{16}, \quad \pi, \quad \sqrt{17}, \quad 62.$$

75. List all rational numbers.

76. List all natural numbers.

77. List all integers.

78. List all irrational numbers.

79. List all real numbers.

80. List all nonnegative integers.

81. Is every integer a rational number? Why or why not?

82. Is every integer a natural number? Why or why not?

Skill Review

83. Evaluate $3xy$ for $x = 2$ and $y = 7$. [1.1]

84. Use a commutative law to write an expression equivalent to $ab + 5$. [1.2]

Synthesis

85. Is the absolute value of a number always positive? Why or why not?

86. How many rational numbers are there between 0 and 1? Justify your answer.

87. Does "nonnegative" mean the same thing as "positive"? Why or why not?

List in order from least to greatest.

88. $13, -12, 5, -17$

89. $-23, 4, 0, -17$

90. $-\frac{2}{3}, \frac{1}{2}, -\frac{3}{4}, -\frac{5}{6}, \frac{3}{8}, \frac{1}{6}$

91. $\frac{4}{5}, \frac{4}{3}, \frac{4}{8}, \frac{4}{6}, \frac{4}{9}, \frac{4}{2}, -\frac{4}{3}$

Write a true sentence using either $<$, $>$, or $=$.

92. $|-5| \ \blacksquare \ |-2|$

93. $|4| \ \blacksquare \ |-7|$

94. $|-8| \ \blacksquare \ |8|$

95. $|23| \ \blacksquare \ |-23|$

96. $|-11| \ \blacksquare \ |5|$

Solve. Consider only integer replacements.

Aha!   **97.** $|x| = 19$

98. $|x| < 3$

99. $2 < |x| < 5$

Given that $0.3\overline{3} = \frac{1}{3}$ and $0.6\overline{6} = \frac{2}{3}$, express each of the following as a ratio of two integers.

100. $0.1\overline{1}$

101. $0.9\overline{9}$

102. $5.5\overline{5}$

103. $7.7\overline{7}$

Translate to an inequality.

104. A number a is negative.

105. A number x is nonpositive.

106. The distance from x to 0 is no more than 10.

107. The distance from t to 0 is at least 20.

To the student and the instructor: The calculator icon, $\blacksquare$, is used to indicate those exercises designed to be solved with a calculator.

108. When Helga's calculator gives a decimal value for $\sqrt{2}$ and that value is promptly squared, the result is 2. Yet when that same decimal approximation is entered by hand and then squared, the result is not exactly 2. Why do you suppose this is?

109. Is the following statement true? Why or why not?
$$\sqrt{a^2} = |a| \quad \text{for any real number } a.$$

1.5   Addition of Real Numbers

Adding with the Number Line   ▪   Adding Without the Number Line   ▪   Problem Solving   ▪
Combining Like Terms

We now consider addition of real numbers. To gain understanding, we will use the number line first. After observing the principles involved, we will develop rules that allow us to work more quickly without the number line.

Adding with the Number Line

To add $a + b$ on the number line, we start at a and move according to b.

a) If b is positive, we move to the right (the positive direction).

b) If b is negative, we move to the left (the negative direction).

c) If b is 0, we stay at a.

EXAMPLE 1

Add: $-4 + 9$.

SOLUTION   To add on the number line, we locate the first number, -4, and then move 9 units to the right. Note that it requires 4 units to reach 0. The difference between 9 and 4 is where we finish.

$$-4 + 9 = 5$$

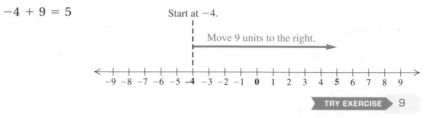

TRY EXERCISE ▶ 9

EXAMPLE 2

STUDENT NOTES ———

Parentheses are essential when a negative sign follows an operation. Just as we would never write $8 \div \times 2$, it is improper to write $3 + -5$.

Add: $3 + (-5)$.

SOLUTION   We locate the first number, 3, and then move 5 units to the left. Note that it requires 3 units to reach 0. The difference between 5 and 3 is 2, so we finish 2 units to the left of 0.

$$3 + (-5) = -2$$

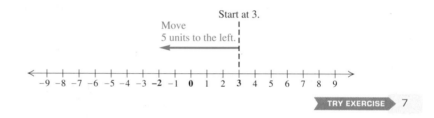

TRY EXERCISE ▶ 7

EXAMPLE 3

Add: $-4 + (-3)$.

SOLUTION   After locating -4, we move 3 units to the left. We finish a total of 7 units to the left of 0.

$$-4 + (-3) = -7$$

Start at -4.

Move
3 units to the left.

TRY EXERCISE ▶ 13

EXAMPLE 4

Add: $-5.2 + 0$.

SOLUTION   We locate -5.2 and move 0 units. Thus we finish where we started, at -5.2.

$$-5.2 + 0 = -5.2$$

TRY EXERCISE ▶ 11

From Examples 1–4, the following rules emerge.

Rules for Addition of Real Numbers

1. *Positive numbers*: Add as usual. The answer is positive.
2. *Negative numbers*: Add absolute values and make the answer negative (see Example 3).
3. *A positive number and a negative number*: Subtract the smaller absolute value from the greater absolute value. Then:

   a) If the positive number has the greater absolute value, the answer is positive (see Example 1).
   b) If the negative number has the greater absolute value, the answer is negative (see Example 2).
   c) If the numbers have the same absolute value, the answer is 0.

4. *One number is zero*: The sum is the other number (see Example 4).

Rule 4 is known as the **identity property of 0**.

Identity Property of 0

For any real number a,

$$a + 0 = 0 + a = a.$$

(Adding 0 to a number gives that same number.) The number 0 is called the *additive identity*.

Adding Without the Number Line

The rules listed above can be used without drawing the number line.

EXAMPLE 5   Add without using the number line.

a) $-12 + (-7)$

b) $-1.4 + 8.5$

c) $-36 + 21$

d) $1.5 + (-1.5)$

e) $-\frac{7}{8} + 0$

f) $\frac{2}{3} + \left(-\frac{5}{8}\right)$

SOLUTION

a) $-12 + (-7) = -19$ Two negatives. *Think:* Add the absolute values, 12 and 7, to get 19. Make the answer *negative*, -19.

b) $-1.4 + 8.5 = 7.1$ A negative and a positive. *Think:* The difference of absolute values is $8.5 - 1.4$, or 7.1. The positive number has the greater absolute value, so the answer is *positive*, 7.1.

c) $-36 + 21 = -15$ A negative and a positive. *Think:* The difference of absolute values is $36 - 21$, or 15. The negative number has the greater absolute value, so the answer is *negative*, -15.

d) $1.5 + (-1.5) = 0$ A negative and a positive. *Think:* Since the numbers are opposites, they have the same absolute value and the answer is 0.

e) $-\dfrac{7}{8} + 0 = -\dfrac{7}{8}$ One number is zero. The sum is the other number, $-\frac{7}{8}$.

f) $\dfrac{2}{3} + \left(-\dfrac{5}{8}\right) = \dfrac{16}{24} + \left(-\dfrac{15}{24}\right)$ This is similar to part (b) above. We find a common denominator and then add.

$\qquad = \dfrac{1}{24}$

> TRY EXERCISES ▸ 15 and 21

If we are adding several numbers, some positive and some negative, the commutative and associative laws allow us to add all the positives, then add all the negatives, and then add the results. Of course, we can also add from left to right, if we prefer.

EXAMPLE 6   Add: $15 + (-2) + 7 + 14 + (-5) + (-12)$.

SOLUTION

$$15 + (-2) + 7 + 14 + (-5) + (-12)$$

$$= 15 + 7 + 14 + (-2) + (-5) + (-12) \qquad \text{Using the commutative law of addition}$$

$$= (15 + 7 + 14) + [(-2) + (-5) + (-12)] \qquad \text{Using the associative law of addition}$$

$$= 36 + (-19) \qquad \text{Adding the positives; adding the negatives}$$

$$= 17 \qquad \text{Adding a positive and a negative}$$

> TRY EXERCISE ▸ 55

Problem Solving

EXAMPLE 7   *Interest rates.*   Between 1994 and 2007, the average interest rate for a 30-yr fixed-rate mortgage dropped 2.5 percent, rose 1.75 percent, dropped 3.25 percent, and rose 1 percent. By how much did the average interest rate change?

Source: Mortgage-X.com

SOLUTION   The problem translates to a sum:

Rewording: The 1st change plus the 2nd change plus the 3rd change plus the 4th change is the total change.

Translating: $-2.5 \;+\; 1.75 \;+\; (-3.25) \;+\; 1 \;=\;$ Total change

Adding from left to right, we have

$$-2.5 + 1.75 + (-3.25) + 1 = -0.75 + (-3.25) + 1 = -4 + 1 = -3.$$

The average interest rate dropped 3 percent between 1994 and 2007.

> TRY EXERCISE ▸ 59

Combining Like Terms

When two terms have variable factors that are exactly the same, like $5a$ and $-7a$, the terms are called **like**, or **similar**, **terms**.* The distributive law enables us to **combine**, or **collect**, **like terms**. The above rules for addition will again apply.

EXAMPLE **8**   Combine like terms.

a) $-7x + 9x$

b) $2a + (-3b) + (-5a) + 9b$

c) $6 + y + (-3.5y) + 2$

SOLUTION

a) $-7x + 9x = (-7 + 9)x$ Using the distributive law

$= 2x$ Adding -7 and 9

b) $2a + (-3b) + (-5a) + 9b$

$= 2a + (-5a) + (-3b) + 9b$ Using the commutative law of addition

$= (2 + (-5))a + (-3 + 9)b$ Using the distributive law

$= -3a + 6b$ Adding

c) $6 + y + (-3.5y) + 2 = y + (-3.5y) + 6 + 2$ Using the commutative law of addition

$= (1 + (-3.5))y + 6 + 2$ Using the distributive law

$= -2.5y + 8$ Adding

TRY EXERCISE 69

With practice we can omit some steps, combining like terms mentally. Note that numbers like 6 and 2 in the expression $6 + y + (-3.5y) + 2$ are constants and are also considered to be like terms.

1.5   EXERCISE SET

For Extra Help

MyMathLab   Math XL   PRACTICE   WATCH   DOWNLOAD

Concept Reinforcement   *In each of Exercises 1–6, match the term with a like term from the column on the right.*

1. _____ $8n$

2. _____ $7m$

3. _____ 43

4. _____ $28z$

5. _____ $-2x$

6. _____ $-9t$

a) $-3z$

b) $5x$

c) $2t$

d) $-4m$

e) 9

f) $-3n$

Add using the number line.

7. $5 + (-8)$

8. $2 + (-5)$

9. $-6 + 10$

10. $-3 + 8$

11. $-7 + 0$

12. $-6 + 0$

13. $-3 + (-5)$

14. $-4 + (-6)$

Add. Do not use a number line except as a check.

15. $-35 + 0$

16. $-68 + 0$

17. $0 + (-8)$

18. $0 + (-2)$

19. $12 + (-12)$

20. $17 + (-17)$

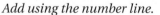

---

*Like terms are discussed in greater detail in Section 1.8.

21. $-24 + (-17)$

22. $-17 + (-25)$

23. $-13 + 13$

24. $-31 + 31$

25. $20 + (-11)$

26. $8 + (-5)$

27. $10 + (-12)$

28. $9 + (-13)$

29. $-3 + 14$

30. $25 + (-6)$

31. $-24 + (-19)$

32. $11 + (-9)$

33. $19 + (-19)$

34. $-20 + (-6)$

35. $23 + (-5)$

36. $-15 + (-7)$

37. $-31 + (-14)$

38. $40 + (-8)$

39. $40 + (-40)$

40. $-25 + 25$

41. $85 + (-69)$

42. $63 + (-13)$

43. $-3.6 + 2.8$

44. $-6.5 + 4.7$

45. $-5.4 + (-3.7)$

46. $-3.8 + (-9.4)$

47. $\frac{4}{5} + \left(\frac{-1}{5}\right)$

48. $\frac{-2}{7} + \frac{3}{7}$

49. $\frac{-4}{7} + \frac{-2}{7}$

50. $\frac{-5}{9} + \frac{-2}{9}$

51. $-\frac{2}{5} + \frac{1}{3}$

52. $-\frac{4}{13} + \frac{1}{2}$

53. $\frac{-4}{9} + \frac{2}{3}$

54. $\frac{1}{9} + \left(\frac{-1}{3}\right)$

55. $35 + (-14) + (-19) + (-5)$

56. $-28 + (-44) + 17 + 31 + (-94)$

Aha! **57.** $-4.9 + 8.5 + 4.9 + (-8.5)$

58. $24 + 3.1 + (-44) + (-8.2) + 63$

Solve. Write your answer as a complete sentence.

59. *Gasoline prices.* In a recent year, the price of a gallon of 87-octane gasoline was $2.89. The price rose 15¢, then dropped 3¢, and then rose 17¢. By how much did the price change during that period?

60. *Natural gas prices.* In a recent year, the price of a gallon of natural gas was $1.88. The price dropped 2¢, then rose 25¢, and then dropped 43¢. By how much did the price change during that period?

61. *Telephone bills.* Chloe's cell-phone bill for July was $82. She sent a check for $50 and then ran up $63 in charges for August. What was her new balance?

62. *Profits and losses.* The following table lists the profits and losses of Premium Sales over a 3-yr period. Find the profit or loss after this period of time.

Year	Profit or loss
2006	−$26,500
2007	−$10,200
2008	+$32,400

63. *Yardage gained.* In an intramural football game, the quarterback attempted passes with the following results.

First try	13-yd loss
Second try	12-yd gain
Third try	21-yd gain

Find the total gain (or loss).

64. *Account balance.* Aiden has $450 in a checking account. He writes a check for $530, makes a deposit of $75, and then writes a check for $90. What is the balance in the account?

65. *Lake level.* Between October 2003 and February 2005, the south end of the Great Salt Lake dropped $\frac{2}{5}$ ft, rose $1\frac{1}{5}$ ft, and dropped $\frac{1}{2}$ ft. By how much did the level change?
Source: U.S. Geological Survey

66. *Peak elevation.* The tallest mountain in the world, as measured from base to peak, is Mauna Kea in Hawaii. From a base 19,684 ft below sea level, it rises 33,480 ft. What is the elevation of its peak?
Source: *Guinness World Records* 2007

67. *Credit-card bills.* Logan's credit-card bill indicates that he owes $470. He sends a check to the credit-card company for $45, charges another $160 in merchandise, and then pays off another $500 of his bill. What is Logan's new balance?

68. *Class size.* During the first two weeks of the semester, 5 students withdrew from Hailey's algebra class, 8 students were added to the class, and 4 students were dropped as "no-shows." By how many students did the original class size change?

Combine like terms.

69. $7a + 10a$

70. $3x + 8x$

71. $-3x + 12x$

72. $-2m + (-7m)$

73. $4t + 21t$

74. $5a + 8a$

75. $7m + (-9m)$

76. $-4x + 4x$

77. $-8y + (-2y)$

78. $10n + (-17n)$

79. $-3 + 8x + 4 + (-10x)$

80. $8a + 5 + (-a) + (-3)$

Find the perimeter of each figure.

81.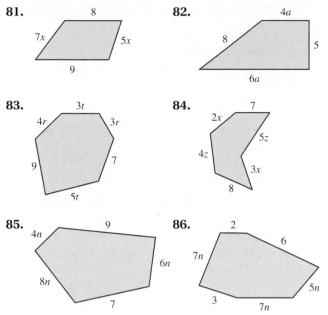

82.

83.

84.

85.

86.

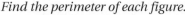

87. Explain in your own words why the sum of two negative numbers is negative.

88. Without performing the actual addition, explain why the sum of all integers from -10 to 10 is 0.

Skill Review

89. Multiply: $7(3z + 2y + 1)$. [1.2]

90. Divide and simplify: $\frac{7}{2} \div \frac{3}{8}$. [1.3]

Synthesis

91. Under what circumstances will the sum of one positive number and several negative numbers be positive?

92. Is it possible to add real numbers without knowing how to calculate $a - b$ with a and b both nonnegative and $a \geq b$? Why or why not?

93. *Banking.* Travis had $257.33 in his checking account. After depositing $152 in the account and writing a check, his account was overdrawn by $42.37. What was the amount of the check?

94. *Sports-card values.* The value of a sports card dropped $12 and then rose $17.50 before settling at $61. What was the original value of the card?

Find the missing term or terms.

95. $4x + \underline{\quad} + (-9x) + (-2y) = -5x - 7y$

96. $-3a + 9b + \underline{\quad} + 5a = 2a - 6b$

97. $3m + 2n + \underline{\quad} + (-2m) = 2n + (-6m)$

98. $\underline{\quad} + 9x + (-4y) + x = 10x - 7y$

Aha! **99.** $7t + 23 + \underline{\quad} + \underline{\quad} = 0$

100. *Geometry.* The perimeter of a rectangle is $7x + 10$. If the length of the rectangle is 5, express the width in terms of x.

101. *Golfing.* After five rounds of golf, a golf pro was 3 under par twice, 2 over par once, 2 under par once, and 1 over par once. On average, how far above or below par was the golfer?

1.6   Subtraction of Real Numbers

Opposites and Additive Inverses ▪ Subtraction ▪ Problem Solving

In arithmetic, when a number b is subtracted from another number a, the difference, $a - b$, is the number that when added to b gives a. For example, $45 - 17 = 28$ because $28 + 17 = 45$. We will use this approach to develop an efficient way of finding the value of $a - b$ for any real numbers a and b. Before doing so, however, we must develop some terminology.

Opposites and Additive Inverses

Numbers such as 6 and -6 are *opposites*, or *additive inverses*, of each other. Whenever opposites are added, the result is 0; and whenever two numbers add to 0, those numbers are opposites.

EXAMPLE **1** Find the opposite of each number: **(a)** 34; **(b)** −8.3; **(c)** 0.

SOLUTION

a) The opposite of 34 is −34: 34 + (−34) = 0.

b) The opposite of −8.3 is 8.3: −8.3 + 8.3 = 0.

c) The opposite of 0 is 0: 0 + 0 = 0.

TRY EXERCISE 19

To write the opposite, we use the symbol −, as follows.

> **Opposite**
>
> The *opposite*, or *additive inverse*, of a number a is written $-a$ (read "the opposite of a" or "the additive inverse of a").

Note that if we take a number, say 8, and find its opposite, −8, and then find the opposite of the result, we will have the original number, 8, again.

EXAMPLE **2** Find $-x$ and $-(-x)$ when $x = 16$.

SOLUTION

If $x = 16$, then $-x = -16$. The opposite of 16 is −16.

If $x = 16$, then $-(-x) = -(-16) = 16$. The opposite of the opposite of 16 is 16.

TRY EXERCISE 25

> **The Opposite of an Opposite**
>
> For any real number a,
>
> $$-(-a) = a.$$
>
> (The opposite of the opposite of a is a.)

EXAMPLE **3** Find $-x$ and $-(-x)$ when $x = -3$.

SOLUTION

If $x = -3$, then $-x = -(-3) = 3$. The opposite of −3 is 3.

Since $-(-x) = x$, it follows that $-(-(-3)) = -3$. Finding the opposite of an opposite

TRY EXERCISE 31

Note in Example 3 that an extra set of parentheses is used to show that we are substituting the negative number −3 for x. The notation $- -x$ is not used.

A symbol such as −8 is usually read "negative 8." It could be read "the additive inverse of 8," because the additive inverse of 8 is negative 8. It could also be read "the opposite of 8," because the opposite of 8 is −8.

A symbol like $-x$, which has a variable, should be read "the opposite of x" or "the additive inverse of x," and *not* "negative x," since to do so suggests that $-x$ represents a negative number.

The symbol "$-$" is read differently depending on where it appears. For example, $-5 - (-x)$ should be read "negative five minus the opposite of x."

EXAMPLE 4

Write each of the following in words.

a) $2 - 8$ **b)** $5 - (-4)$ **c)** $-6 - (-x)$

STUDENT NOTES ──────

As you read mathematics, it is important to verbalize correctly the words and symbols to yourself. Consistently reading the expression $-x$ as "the opposite of x" is a good step in this direction.

SOLUTION

a) $2 - 8$ is read "two minus eight."

b) $5 - (-4)$ is read "five minus negative four."

c) $-6 - (-x)$ is read "negative six minus the opposite of x."

As we saw in Example 3, $-x$ can represent a positive number. This notation can be used to restate a result from Section 1.5 as *the law of opposites*.

> ### The Law of Opposites
> For any two numbers a and $-a$,
> $$a + (-a) = 0.$$
> (When opposites are added, their sum is 0.)

A negative number is said to have a "negative *sign*." A positive number is said to have a "positive *sign*." If we change a number to its opposite, or additive inverse, we say that we have "changed or reversed its sign."

EXAMPLE 5

Change the sign (find the opposite) of each number: **(a)** -3; **(b)** -10; **(c)** 14.

SOLUTION

a) When we change the sign of -3, we obtain 3.

b) When we change the sign of -10, we obtain 10.

c) When we change the sign of 14, we obtain -14.   TRY EXERCISE 35

Subtraction

Opposites are helpful when subtraction involves negative numbers. To see why, look for a pattern in the following:

Subtracting		*Adding the Opposite*
$9 - 5 = 4$	since $4 + 5 = 9$	$9 + (-5) = 4$
$5 - 8 = -3$	since $-3 + 8 = 5$	$5 + (-8) = -3$
$-6 - 4 = -10$	since $-10 + 4 = -6$	$-6 + (-4) = -10$
$-7 - (-10) = 3$	since $3 + (-10) = -7$	$-7 + 10 = 3$
$-7 - (-2) = -5$	since $-5 + (-2) = -7$	$-7 + 2 = -5$

The matching results suggest that we can subtract by adding the opposite of the number being subtracted. This can always be done and often provides the easiest way to subtract real numbers.

> ## Subtraction of Real Numbers
>
> For any real numbers a and b,
>
> $$a - b = a + (-b).$$
>
> (To subtract, add the opposite, or additive inverse, of the number being subtracted.)

EXAMPLE **6**   Subtract each of the following and then check with addition.

a) $2 - 6$ **b)** $4 - (-9)$ **c)** $-4.2 - (-3.6)$

d) $-1.8 - (-7.5)$ **e)** $\frac{1}{5} - \left(-\frac{3}{5}\right)$

SOLUTION

> ### TECHNOLOGY CONNECTION
>
> On nearly all graphing calculators, it is essential to distinguish between the key for negation and the key for subtraction. To enter a negative number, we use (−) and to subtract, we use (−). This said, be careful not to rely on a calculator for computations that you will be expected to do by hand.

a) $2 - 6 = 2 + (-6) = -4$ The opposite of 6 is -6. We change the subtraction to addition and add the opposite. *Check:* $-4 + 6 = 2$.

b) $4 - (-9) = 4 + 9 = 13$ The opposite of -9 is 9. We change the subtraction to addition and add the opposite. *Check:* $13 + (-9) = 4$.

c) $-4.2 - (-3.6) = -4.2 + 3.6$ Adding the opposite of -3.6
$ = -0.6$ *Check:* $-0.6 + (-3.6) = -4.2$.

d) $-1.8 - (-7.5) = -1.8 + 7.5$ Adding the opposite
$ = 5.7$ *Check:* $5.7 + (-7.5) = -1.8$.

e) $\dfrac{1}{5} - \left(-\dfrac{3}{5}\right) = \dfrac{1}{5} + \dfrac{3}{5}$ Adding the opposite

$\phantom{\dfrac{1}{5} - \left(-\dfrac{3}{5}\right)} = \dfrac{1 + 3}{5}$ A common denominator exists so we add in the numerator.

$\phantom{\dfrac{1}{5} - \left(-\dfrac{3}{5}\right)} = \dfrac{4}{5}$

Check: $\dfrac{4}{5} + \left(-\dfrac{3}{5}\right) = \dfrac{4}{5} + \dfrac{-3}{5} = \dfrac{4 + (-3)}{5} = \dfrac{1}{5}$.

TRY EXERCISES 39 and 47

EXAMPLE **7**   Simplify: $8 - (-4) - 2 - (-5) + 3$.

SOLUTION

$8 - (-4) - 2 - (-5) + 3 = 8 + 4 + (-2) + 5 + 3$ To subtract, we add the opposite.

$ = 18$

TRY EXERCISE 109

Recall from Section 1.2 that the terms of an algebraic expression are separated by plus signs. This means that the terms of $5x - 7y - 9$ are $5x$, $-7y$, and -9, since $5x - 7y - 9 = 5x + (-7y) + (-9)$.

EXAMPLE 8

Identify the terms of $4 - 2ab + 7a - 9$.

SOLUTION   We have

$$4 - 2ab + 7a - 9 = 4 + (-2ab) + 7a + (-9), \qquad \text{Rewriting as addition}$$

so the terms are 4, $-2ab$, $7a$, and -9.

TRY EXERCISE 117

EXAMPLE 9

Combine like terms.

a) $1 + 3x - 7x$

b) $-5a - 7b - 4a + 10b$

c) $4 - 3m - 9 + 2m$

SOLUTION

a)
$$
\begin{aligned}
1 + 3x - 7x &= 1 + 3x + (-7x) && \text{Adding the opposite} \\
&= 1 + (3 + (-7))x && \text{Using the distributive law.} \\
&= 1 + (-4)x && \text{Try to do this mentally.} \\
&= 1 - 4x && \text{Rewriting as subtraction to be more concise}
\end{aligned}
$$

b)
$$
\begin{aligned}
-5a - 7b - 4a + 10b &= -5a + (-7b) + (-4a) + 10b && \text{Adding the opposite} \\
&= -5a + (-4a) + (-7b) + 10b && \text{Using the commutative law of addition} \\
&= -9a + 3b && \text{Combining like terms mentally}
\end{aligned}
$$

c)
$$
\begin{aligned}
4 - 3m - 9 + 2m &= 4 + (-3m) + (-9) + 2m && \text{Rewriting as addition} \\
&= 4 + (-9) + (-3m) + 2m && \text{Using the commutative law of addition} \\
&= -5 + (-1m) && \text{We can write } -1m \\
& && \text{as } -m. \\
&= -5 - m
\end{aligned}
$$

TRY EXERCISE 121

Problem Solving

We use subtraction to solve problems involving differences. These include problems that ask "How much more?" or "How much higher?"

EXAMPLE 10

Record elevations.   The current world records for the highest parachute jump and the lowest manned vessel ocean dive were both set in 1960. On August 16 of that year, Captain Joseph Kittinger jumped from a height of 102,800 ft above sea level. Earlier, on January 23, Jacques Piccard and Navy Lieutenant Donald Walsh descended in a bathyscaphe 35,797 ft below sea level. What was the difference in elevation between the highest parachute jump and the lowest ocean dive?

Sources: www.firstflight.org and www.seasky.org

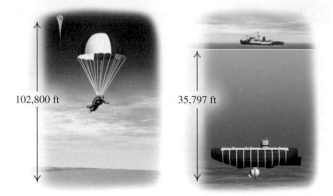

SOLUTION   To find the difference between two elevations, we always subtract the lower elevation from the higher elevation:

$$\underbrace{\text{Higher elevation}}_{\downarrow} \quad - \quad \underbrace{\text{Lower elevation}}_{\downarrow}$$

$$102{,}800 \qquad - \qquad (-35{,}797)$$
$$= 102{,}800 + 35{,}797$$
$$= 138{,}597.$$

The parachute jump began 138,597 ft higher than the ocean dive ended.

TRY EXERCISE ▶135

1.6	**EXERCISE SET**

For Extra Help
MyMathLab   *Math XL* PRACTICE   WATCH   DOWNLOAD

Concept Reinforcement   *In each of Exercises 1–8, match the expression with the appropriate wording from the column on the right.*

1. _____ $-x$

2. _____ $12 - x$

3. _____ $12 - (-x)$

4. _____ $x - 12$

5. _____ $x - (-12)$

6. _____ $-x - 12$

7. _____ $-x - x$

8. _____ $-x - (-12)$

a) x minus negative twelve

b) The opposite of x minus x

c) The opposite of x minus twelve

d) The opposite of x

e) The opposite of x minus negative twelve

f) Twelve minus the opposite of x

g) Twelve minus x

h) x minus twelve

Write each of the following in words.

9. $6 - 10$

10. $5 - 13$

11. $2 - (-12)$

12. $4 - (-1)$

13. $9 - (-t)$

14. $8 - (-m)$

15. $-x - y$

16. $-a - b$

17. $-3 - (-n)$

18. $-7 - (-m)$

Find the opposite, or additive inverse.

19. 51

20. -17

21. $-\frac{11}{3}$

22. $\frac{7}{2}$

23. -3.14

24. 48.2

Find $-x$ when x is each of the following.

25. -45

26. 26

27. $-\frac{14}{3}$

28. $\frac{1}{328}$

29. 0.101

30. 0

Find $-(-x)$ when x is each of the following.

31. 37   **32.** 29

33. $-\frac{2}{5}$ **34.** -9.1

Change the sign. (Find the opposite.)

35. -1 **36.** -7

37. 15   **38.** 10

Subtract.

39. $7 - 10$ **40.** $4 - 13$

41. $0 - 6$ **42.** $0 - 8$

43. $2 - 5$ **44.** $3 - 13$

45. $-4 - 3$ **46.** $-5 - 6$

47. $-9 - (-3)$ **48.** $-9 - (-5)$

Aha! **49.** $-8 - (-8)$ **50.** $-10 - (-10)$

51. $14 - 19$ **52.** $12 - 16$

53. $30 - 40$ **54.** $20 - 27$

55. $0 - 11$ **56.** $0 - 31$

57. $-9 - (-9)$ **58.** $-40 - (-40)$

59. $5 - 5$ **60.** $7 - 7$

61. $4 - (-4)$ **62.** $6 - (-6)$

63. $-7 - 4$ **64.** $-6 - 8$

65. $6 - (-10)$ **66.** $3 - (-12)$

67. $-4 - 15$ **68.** $-14 - 2$

69. $-6 - (-7)$ **70.** $-4 - (-7)$

71. $5 - (-12)$ **72.** $5 - (-6)$

73. $0 - (-3)$ **74.** $0 - (-5)$

75. $-5 - (-2)$ **76.** $-3 - (-1)$

77. $-7 - 14$ **78.** $-9 - 16$

79. $0 - (-10)$ **80.** $0 - (-1)$

81. $-8 - 0$ **82.** $-9 - 0$

83. $-52 - 8$ **84.** $-63 - 11$

85. $2 - 25$ **86.** $18 - 63$

87. $-4.2 - 3.1$ **88.** $-10.1 - 2.6$

89. $-1.3 - (-2.4)$ **90.** $-5.8 - (-7.3)$

91. $3.2 - 8.7$ **92.** $1.5 - 9.4$

93. $0.072 - 1$ **94.** $0.825 - 1$

95. $\frac{2}{11} - \frac{9}{11}$ **96.** $\frac{3}{7} - \frac{5}{7}$

97. $\frac{-1}{5} - \frac{3}{5}$ **98.** $\frac{-2}{9} - \frac{5}{9}$

99. $-\frac{4}{17} - \left(-\frac{9}{17}\right)$ **100.** $-\frac{2}{13} - \left(-\frac{5}{13}\right)$

In each of Exercises 101–104, translate the phrase to mathematical language and simplify. See the solution to Example 10.

101. The difference between 3.8 and -5.2

102. The difference between -2.1 and -5.9

103. The difference between 114 and -79

104. The difference between 23 and -17

105. Subtract 32 from -8.

106. Subtract 19 from -7.

107. Subtract -25 from 18.

108. Subtract -31 from -5.

Simplify.

109. $16 - (-12) - 1 - (-2) + 3$

110. $22 - (-18) + 7 + (-42) - 27$

111. $-31 + (-28) - (-14) - 17$

112. $-43 - (-19) - (-21) + 25$

113. $-34 - 28 + (-33) - 44$

114. $39 + (-88) - 29 - (-83)$

Aha! **115.** $-93 + (-84) - (-93) - (-84)$

116. $84 + (-99) + 44 - (-18) - 43$

Identify the terms in each expression.

117. $-3y - 8x$

118. $7a - 9b$

119. $9 - 5t - 3st$

120. $-4 - 3x + 2xy$

Combine like terms.

121. $10x - 13x$

122. $3a - 14a$

123. $7a - 12a + 4$

124. $-9x - 13x + 7$

125. $-8n - 9 + 7n$

126. $-7 + 9n - 8n$

127. $5 - 3x - 11$

128. $2 + 3a - 7$

129. $2 - 6t - 9 - 2t$

130. $-5 + 4b - 7 - 5b$

131. $5y + (-3x) - 9x + 1 - 2y + 8$

132. $14 - (-5x) + 2z - (-32) + 4z - 2x$

133. $13x - (-2x) + 45 - (-21) - 7x$

134. $8t - (-2t) - 14 - (-5t) + 53 - 9t$

Solve.

135. *Temperature extremes.*   The highest temperature ever recorded in the United States is 134°F in Greenland Ranch, California, on July 10, 1913. The lowest temperature ever recorded is −80°F in Prospect Creek, Alaska, on January 23, 1971. How much higher was the temperature in Greenland Ranch than that in Prospect Creek?
Source: Information Please Database 2007, Pearson Education, Inc.

136. *Temperature change.*   In just 12 hr on February 21, 1918, the temperature in Granville, North Dakota, rose from −33°F to 50°F. By how much did the temperature change?
Source: Information Please Database 2007, Pearson Education, Inc.

137. *Elevation extremes.*   The lowest elevation in Asia, the Dead Sea, is 1312 ft below sea level. The highest elevation in Asia, Mount Everest, is 29,035 ft. Find the difference in elevation.
Source: *Guinness World Records* 2007

138. *Elevation extremes.*   The elevation of Mount Whitney, the highest peak in California, is 14,776 ft more than the elevation of Death Valley, California.

If Death Valley is 282 ft below sea level, find the elevation of Mount Whitney.
Source: *The Columbia Electronic Encyclopedia*, 6th ed., 2007 (New York: Columbia University Press)

139. *Changes in elevation.*   The lowest point in Africa is Lake Assal, which is 156 m below sea level. The lowest point in South America is the Valdes Peninsula, which is 40 m below sea level. How much lower is Lake Assal than the Valdes Peninsula?
Source: Information Please Database 2007, Pearson Education, Inc.

140. *Underwater elevation.*   The deepest point in the Pacific Ocean is the Marianas Trench, with a depth of 10,911 m. The deepest point in the Atlantic Ocean is the Puerto Rico Trench, with a depth of 8648 m. What is the difference in elevation of the two trenches?
Source: *Guinness World Records* 2007

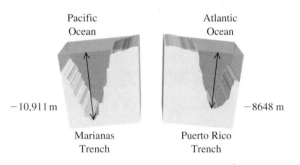

141. Jeremy insists that if you can *add* real numbers, then you can also *subtract* real numbers. Do you agree? Why or why not?

142. Are the expressions $-a + b$ and $a + (-b)$ opposites of each other? Why or why not?

Skill Review

143. Find the area of a rectangle when the length is 36 ft and the width is 12 ft. [1.1]

144. Find the prime factorization of 864. [1.3]

Synthesis

145. Explain the different uses of the symbol "−". Give examples of each and how they should be read.

146. If a and b are both negative, under what circumstances will $a - b$ be negative?

147. *Power outages.*   During the Northeast's electrical blackout of August 14, 2003, residents of Bloomfield, New Jersey, lost power at 4:00 P.M. One resident returned from vacation at 3:00 P.M. the following day to find the clocks in her apartment reading 8:00 A.M. At what time, and on what day, was power restored?

Tell whether each statement is true or false for all real numbers m and n. Use various replacements for m and n to support your answer.

148. If $m > n$, then $m - n > 0$.

149. If $m > n$, then $m + n > 0$.

150. If m and n are opposites, then $m - n = 0$.

151. If $m = -n$, then $m + n = 0$.

152. A gambler loses a wager and then loses "double or nothing" (meaning the gambler owes twice as much) twice more. After the three losses, the gambler's assets are $-\$20$. Explain how much the gambler originally bet and how the $20 debt occurred.

153. List the keystrokes needed to compute $-9 - (-7)$.

154. If n is positive and m is negative, what is the sign of $n + (-m)$? Why?

1.7   Multiplication and Division of Real Numbers

Multiplication • Division

We now develop rules for multiplication and division of real numbers. Because multiplication and division are closely related, the rules are quite similar.

Multiplication

We already know how to multiply two nonnegative numbers. To see how to multiply a positive number and a negative number, consider the following pattern in which multiplication is regarded as repeated addition:

This number → $4(-5) = (-5) + (-5) + (-5) + (-5) = -20$ ← This number
decreases by $\quad 3(-5) = \qquad\qquad (-5) + (-5) + (-5) = -15$ $\quad$ increases by
1 each time. $\quad 2(-5) = \qquad\qquad\qquad\qquad (-5) + (-5) = -10$ $\quad$ 5 each time.
$\qquad\qquad\quad 1(-5) = \qquad\qquad\qquad\qquad\qquad\qquad (-5) = -5$
$\qquad\qquad\quad 0(-5) = \qquad\qquad\qquad\qquad\qquad\qquad\qquad 0 = 0$

This pattern illustrates that the product of a negative number and a positive number is negative.

> **The Product of a Negative Number and a Positive Number**
>
> To multiply a positive number and a negative number, multiply their absolute values. The answer is negative.

EXAMPLE **1**   Multiply: **(a)** $8(-5)$; **(b)** $-\frac{1}{3} \cdot \frac{5}{7}$.

SOLUTION

a) $8(-5) = -40$ *Think:* $8 \cdot 5 = 40$; make the answer negative.

b) $-\frac{1}{3} \cdot \frac{5}{7} = -\frac{5}{21}$ *Think:* $\frac{1}{3} \cdot \frac{5}{7} = \frac{5}{21}$; make the answer negative.

▸ TRY EXERCISE 11

The pattern developed above includes not just products of positive numbers and negative numbers, but a product involving zero as well.

> ### The Multiplicative Property of Zero
> For any real number a,
>
> $$0 \cdot a = a \cdot 0 = 0.$$
>
> (The product of 0 and any real number is 0.)

EXAMPLE **2**   Multiply: $173(-452)0$.

SOLUTION   We have

$$173(-452)0 = 173[(-452)0]$$ Because of the associative law of multiplication, we can multiply the last two factors first.

$$= 173[0]$$ Using the multiplicative property of zero

$$= 0.$$ Using the multiplicative property of zero again

Note that whenever 0 appears as a factor, the product is 0.

▸ TRY EXERCISE 33

We can extend the above pattern still further to examine the product of two negative numbers.

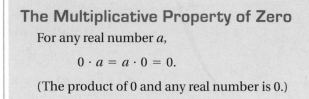

According to the pattern, the product of two negative numbers is positive.

> ### The Product of Two Negative Numbers
> To multiply two negative numbers, multiply their absolute values. The answer is positive.

EXAMPLE 3   Multiply: **(a)** $(-6)(-8)$; **(b)** $(-1.2)(-3)$.

SOLUTION

a) The absolute value of -6 is 6 and the absolute value of -8 is 8. Thus,

$$(-6)(-8) = 6 \cdot 8 \qquad \text{Multiplying absolute values. The answer is positive.}$$
$$= 48.$$

b) $(-1.2)(-3) = (1.2)(3) \qquad \text{Multiplying absolute values. The answer is positive.}$

$\qquad\qquad\quad = 3.6 \qquad \text{Try to go directly to this step.}$

> TRY EXERCISE 17

When three or more numbers are multiplied, we can order and group the numbers as we please, because of the commutative and associative laws.

EXAMPLE 4   Multiply: **(a)** $-3(-2)(-5)$; **(b)** $-4(-6)(-1)(-2)$.

SOLUTION

a) $-3(-2)(-5) = 6(-5) \qquad \text{Multiplying the first two numbers. The product of two negatives is positive.}$

$\qquad\qquad\qquad = -30 \qquad \text{The product of a positive and a negative is negative.}$

b) $-4(-6)(-1)(-2) = 24 \cdot 2 \qquad \text{Multiplying the first two numbers and the last two numbers}$

$\qquad\qquad\qquad\quad = 48$

> TRY EXERCISE 43

We can see the following pattern in the results of Example 4.

The product of an even number of negative numbers is positive.

The product of an odd number of negative numbers is negative.

Division

Recall that $a \div b$, or $\dfrac{a}{b}$, is the number, if one exists, that when multiplied by b gives a. For example, to show that $10 \div 2$ is 5, we need only note that $5 \cdot 2 = 10$. Thus division can always be checked with multiplication.

EXAMPLE 5   Divide, if possible, and check your answer.

a) $14 \div (-7)$ 　　　　　　　　　　　**b)** $\dfrac{-32}{-4}$

c) $\dfrac{-10}{9}$ 　　　　　　　　　　　　　**d)** $\dfrac{-17}{0}$

SOLUTION

a) $14 \div (-7) = -2 \qquad$ We look for a number that when multiplied by -7 gives 14. That number is -2. *Check:* $(-2)(-7) = 14$.

b) $\dfrac{-32}{-4} = 8 \qquad$ We look for a number that when multiplied by -4 gives -32. That number is 8. *Check:* $8(-4) = -32$.

c) $\dfrac{-10}{9} = -\dfrac{10}{9} \qquad$ We look for a number that when multiplied by 9 gives -10. That number is $-\frac{10}{9}$. *Check:* $-\frac{10}{9} \cdot 9 = -10$.

d) $\dfrac{-17}{0}$ is **undefined**. 　 We look for a number that when multiplied by 0 gives -17. There is no such number because if 0 is a factor, the product is 0, not -17.

> TRY EXERCISE 57

STUDENT NOTES ————

Try to regard "undefined" as a mathematical way of saying "we do not give any meaning to this expression."

The rules for signs for division are the same as those for multiplication: The quotient of a positive number and a negative number is negative; the quotient of two negative numbers is positive.

> ## Rules for Multiplication and Division
> To multiply or divide two nonzero real numbers:
> 1. Using the absolute values, multiply or divide, as indicated.
> 2. If the signs are the same, the answer is positive.
> 3. If the signs are different, the answer is negative.

Had Example 5(a) been written as $-14 \div 7$ or $-\frac{14}{7}$, rather than $14 \div (-7)$, the result would still have been -2. Thus from Examples 5(a)–5(c), we have the following:

$$\frac{-a}{b} = \frac{a}{-b} = -\frac{a}{b} \quad \text{and} \quad \frac{-a}{-b} = \frac{a}{b}.$$

EXAMPLE 6 Rewrite each of the following in two equivalent forms: **(a)** $\frac{5}{-2}$; **(b)** $-\frac{3}{10}$.

SOLUTION We use one of the properties just listed.

a) $\frac{5}{-2} = \frac{-5}{2}$ and $\frac{5}{-2} = -\frac{5}{2}$

b) $-\frac{3}{10} = \frac{-3}{10}$ and $-\frac{3}{10} = \frac{3}{-10}$

Since $\frac{-a}{b} = \frac{a}{-b} = -\frac{a}{b}$

TRY EXERCISE 81

When a fraction contains a negative sign, it can be helpful to rewrite (or simply visualize) the fraction in an equivalent form.

EXAMPLE 7 Perform the indicated operation: **(a)** $\left(-\frac{4}{5}\right)\left(\frac{-7}{3}\right)$; **(b)** $-\frac{2}{7} + \frac{9}{-7}$.

SOLUTION

a) $\left(-\frac{4}{5}\right)\left(\frac{-7}{3}\right) = \left(-\frac{4}{5}\right)\left(-\frac{7}{3}\right)$ Rewriting $\frac{-7}{3}$ as $-\frac{7}{3}$

$= \frac{28}{15}$ Try to go directly to this step.

b) Given a choice, we generally choose a positive denominator:

$-\frac{2}{7} + \frac{9}{-7} = \frac{-2}{7} + \frac{-9}{7}$ Rewriting both fractions with a common denominator of 7

$= \frac{-11}{7}$, or $-\frac{11}{7}$.

TRY EXERCISE 101

To divide with fraction notation, it is usually easiest to find a reciprocal and then multiply.

EXAMPLE **8**   Find the reciprocal of each number, if it exists.

a) -27 **b)** $\frac{-3}{4}$

c) $-\frac{1}{5}$ **d)** 0

SOLUTION   Recall from Section 1.3 that we can check that two numbers are reciprocals of each other by confirming that their product is 1.

a) The reciprocal of -27 is $\frac{1}{-27}$. More often, this number is written as $-\frac{1}{27}$.

Check: $(-27)\left(-\frac{1}{27}\right) = \frac{27}{27} = 1$.

b) The reciprocal of $\frac{-3}{4}$ is $\frac{4}{-3}$, or, equivalently, $-\frac{4}{3}$.   Check: $\frac{-3}{4} \cdot \frac{4}{-3} = \frac{-12}{-12} = 1$.

c) The reciprocal of $-\frac{1}{5}$ is -5.   Check: $-\frac{1}{5}(-5) = \frac{5}{5} = 1$.

d) The reciprocal of 0 does not exist. To see this, recall that there is no number r for which $0 \cdot r = 1$.   **TRY EXERCISE** ▶ 89

EXAMPLE **9**   Divide: **(a)** $-\frac{2}{3} \div \left(-\frac{5}{4}\right)$; **(b)** $-\frac{3}{4} \div \frac{3}{10}$.

SOLUTION   We divide by multiplying by the reciprocal of the divisor.

a) $-\dfrac{2}{3} \div \left(-\dfrac{5}{4}\right) = -\dfrac{2}{3} \cdot \left(-\dfrac{4}{5}\right) = \dfrac{8}{15}$ Multiplying by the reciprocal

Be careful not to change the sign when taking a reciprocal!

b) $-\dfrac{3}{4} \div \dfrac{3}{10} = -\dfrac{3}{4} \cdot \left(\dfrac{10}{3}\right) = -\dfrac{30}{12} = -\dfrac{5}{2} \cdot \dfrac{6}{6} = -\dfrac{5}{2}$ Removing a factor equal to 1: $\frac{6}{6} = 1$

TRY EXERCISE ▶ 109

To divide with decimal notation, it is usually easiest to carry out the division.

EXAMPLE **10**   Divide: $27.9 \div (-3)$.

SOLUTION

$$27.9 \div (-3) = \frac{27.9}{-3} = -9.3 \qquad \text{Dividing: } 3\overline{)27.9}\,. \\ \text{The answer is negative.}$$

TRY EXERCISE ▶ 67

In Example 5(d), we explained why we cannot divide -17 by 0. To see why *no* nonzero number b can be divided by 0, remember that $b \div 0$ would have to be the number that when multiplied by 0 gives b. But since the product of 0 and any number is 0, not b, we say that $b \div 0$ is **undefined** for $b \neq 0$. In the special case of $0 \div 0$, we look for a number r such that $0 \div 0 = r$ and $r \cdot 0 = 0$. But, $r \cdot 0 = 0$ for *any* number r. For this reason, we say that $b \div 0$ is undefined for any choice of b.*

Finally, note that $0 \div 7 = 0$ since $0 \cdot 7 = 0$. This can be written $0/7 = 0$. It is important not to confuse division *by* 0 with division *into* 0.

---

*Sometimes $0 \div 0$ is said to be *indeterminate*.

 **11**   Divide, if possible: **(a)** $\frac{0}{-2}$; **(b)** $\frac{5}{0}$.

SOLUTION

a) $\dfrac{0}{-2} = 0$ We can divide 0 by a nonzero number.
           Check: $0(-2) = 0$.

b) $\dfrac{5}{0}$ is undefined.   We cannot divide by 0.   ▶ **TRY EXERCISE** 73

Division Involving Zero

For any real number a,

$$\frac{a}{0} \text{ is undefined,}$$

and for $a \neq 0$,

$$\frac{0}{a} = 0.$$

It is important *not* to confuse *opposite* with *reciprocal*. Keep in mind that the opposite, or additive inverse, of a number is what we add to the number to get 0. The reciprocal, or multiplicative inverse, is what we multiply the number by to get 1.

Compare the following.

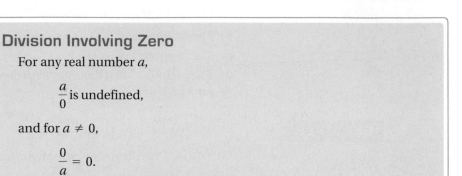

Number	Opposite (Change the sign.)	Reciprocal (Invert but do not change the sign.)
$-\dfrac{3}{8}$	$\dfrac{3}{8}$	$-\dfrac{8}{3}$
19	-19	$\dfrac{1}{19}$
$\dfrac{18}{7}$	$-\dfrac{18}{7}$	$\dfrac{7}{18}$
-7.9	7.9	$-\dfrac{1}{7.9}$, or $-\dfrac{10}{79}$
0	0	Undefined

$\left(-\dfrac{3}{8}\right)\left(-\dfrac{8}{3}\right) = 1$

$-\dfrac{3}{8} + \dfrac{3}{8} = 0$

1.7   EXERCISE SET

For Extra Help   **MyMathLab**   Math XL PRACTICE   WATCH   DOWNLOAD

↪ *Concept Reinforcement*   *In each of Exercises 1–10, replace the blank with either* 0 *or* 1 *to match the description given.*

1. The product of two reciprocals ____

2. The sum of a pair of opposites ____

3. The sum of a pair of additive inverses ____

4. The product of two multiplicative inverses ____

5. This number has no reciprocal. ____

6. This number is its own reciprocal. ____

7. This number is the multiplicative identity. ____

8. This number is the additive identity. ____

9. A nonzero number divided by itself ____

10. Division by this number is undefined. ____

Multiply.

11. $-4 \cdot 10$
12. $-5 \cdot 6$

13. $-8 \cdot 7$
14. $-9 \cdot 2$

15. $4 \cdot (-10)$
16. $9 \cdot (-5)$

17. $-9 \cdot (-8)$
18. $-10 \cdot (-11)$

19. $-6 \cdot 7$
20. $-2 \cdot 5$

21. $-5 \cdot (-9)$
22. $-9 \cdot (-2)$

23. $-19 \cdot (-10)$
24. $-12 \cdot (-10)$

25. $11 \cdot (-12)$
26. $-13 \cdot (-15)$

27. $-25 \cdot (-48)$
28. $15 \cdot (-43)$

29. $4.5 \cdot (-28)$
30. $-49 \cdot (-2.1)$

31. $-5 \cdot (-2.3)$
32. $-6 \cdot 4.8$

33. $(-25) \cdot 0$
34. $0 \cdot (-4.7)$

35. $\frac{2}{5} \cdot \left(-\frac{5}{7}\right)$
36. $\frac{5}{7} \cdot \left(-\frac{2}{3}\right)$

37. $-\frac{3}{8} \cdot \left(-\frac{2}{9}\right)$
38. $-\frac{5}{8} \cdot \left(-\frac{2}{5}\right)$

39. $(-5.3)(2.1)$
40. $(9.5)(-3.7)$

41. $-\frac{5}{9} \cdot \frac{3}{4}$
42. $-\frac{8}{3} \cdot \frac{9}{4}$

43. $3 \cdot (-7) \cdot (-2) \cdot 6$
44. $9 \cdot (-2) \cdot (-6) \cdot 7$

Aha! 45. $27 \cdot (-34) \cdot 0$
46. $-43 \cdot (-74) \cdot 0$

47. $-\frac{1}{3} \cdot \frac{1}{4} \cdot \left(-\frac{3}{7}\right)$
48. $-\frac{1}{2} \cdot \frac{3}{5} \cdot \left(-\frac{2}{7}\right)$

49. $-2 \cdot (-5) \cdot (-3) \cdot (-5)$

50. $-3 \cdot (-5) \cdot (-2) \cdot (-1)$

51. $(-31) \cdot (-27) \cdot 0 \cdot (-13)$

52. $7 \cdot (-6) \cdot 5 \cdot (-4) \cdot 3 \cdot (-2) \cdot 1 \cdot 0$

53. $(-8)(-9)(-10)$

54. $(-7)(-8)(-9)(-10)$

55. $(-6)(-7)(-8)(-9)(-10)$

56. $(-5)(-6)(-7)(-8)(-9)(-10)$

Divide, if possible, and check. If a quotient is undefined, state this.

57. $18 \div (-2)$
58. $\frac{24}{-3}$

59. $\frac{36}{-9}$
60. $26 \div (-13)$

61. $\frac{-56}{8}$
62. $\frac{-35}{-7}$

63. $\frac{-48}{-12}$
64. $-63 \div (-9)$

65. $-72 \div 8$
66. $\frac{-50}{25}$

67. $-10.2 \div (-2)$
68. $-2 \div 0.8$

69. $-100 \div (-11)$
70. $\frac{-64}{-7}$

71. $\frac{400}{-50}$
72. $-300 \div (-13)$

73. $\frac{48}{0}$
74. $\frac{0}{-5}$

75. $-4.8 \div 1.2$
76. $-3.9 \div 1.3$

77. $\frac{0}{-9}$
78. $0 \div 18$

Aha! 79. $\frac{9.7(-2.8)0}{4.3}$
80. $\frac{(-4.9)(7.2)}{0}$

Write each expression in two equivalent forms, as in Example 6.

81. $\frac{-8}{3}$
82. $\frac{18}{-7}$

83. $\frac{29}{-35}$
84. $\frac{-10}{3}$

85. $-\frac{7}{3}$
86. $-\frac{4}{15}$

87. $\frac{-x}{2}$
88. $\frac{9}{-a}$

Find the reciprocal of each number, if it exists.

89. $-\frac{4}{5}$
90. $-\frac{13}{11}$

91. $\frac{51}{-10}$
92. $\frac{43}{-24}$

93. -10
94. 34

95. 4.3
96. -1.7

97. $\frac{-9}{4}$
98. $\frac{-6}{11}$

99. 0
100. -1

Perform the indicated operation and, if possible, simplify. If a quotient is undefined, state this.

101. $\left(\frac{-7}{4}\right)\left(-\frac{3}{5}\right)$
102. $\left(-\frac{5}{6}\right)\left(\frac{-1}{3}\right)$

103. $\frac{-3}{8} + \frac{-5}{8}$
104. $\frac{-4}{5} + \frac{7}{5}$

Aha! **105.** $\left(\frac{-9}{5}\right)\left(\frac{5}{-9}\right)$ **106.** $\left(-\frac{2}{7}\right)\left(\frac{5}{-8}\right)$

107. $\left(-\frac{3}{11}\right) - \left(-\frac{6}{11}\right)$ **108.** $\left(-\frac{4}{7}\right) - \left(-\frac{2}{7}\right)$

109. $\frac{7}{8} \div \left(-\frac{1}{2}\right)$ **110.** $\frac{3}{4} \div \left(-\frac{2}{3}\right)$

Aha! **111.** $-\frac{5}{9} \div \left(-\frac{5}{9}\right)$ **112.** $\frac{-5}{12} \div \frac{15}{7}$

113. $\frac{-3}{10} + \frac{2}{5}$ **114.** $\frac{-5}{9} + \frac{2}{3}$

115. $\frac{7}{10} \div \left(\frac{-3}{5}\right)$ **116.** $\left(\frac{-3}{5}\right) \div \frac{6}{15}$

117. $\frac{14}{-9} \div \frac{0}{3}$ **118.** $\frac{0}{-10} \div \frac{-3}{8}$

119. $\frac{-4}{15} + \frac{2}{-3}$ **120.** $\frac{3}{-10} + \frac{-1}{5}$

121. Most calculators have a key, often appearing as **1/x**, for finding reciprocals. To use this key, we enter a number and then press **1/x** to find its reciprocal. What should happen if we enter a number and then press the reciprocal key twice? Why?

122. Multiplication can be regarded as repeated addition. Using this idea and a number line, explain why $3 \cdot (-5) = -15$.

Skill Review

123. Simplify: $\frac{264}{468}$. [1.3]

124. Combine like terms: $x + 12y + 11x - 13y - 9$. [1.5]

Synthesis

125. If two nonzero numbers are opposites of each other, are their reciprocals opposites of each other? Why or why not?

126. If two numbers are reciprocals of each other, are their opposites reciprocals of each other? Why or why not?

Translate to an algebraic expression or equation.

127. The reciprocal of a sum

128. The sum of two reciprocals

129. The opposite of a sum

130. The sum of two opposites

131. A real number is its own opposite.

132. A real number is its own reciprocal.

133. Show that the reciprocal of a sum is *not* the sum of the two reciprocals.

134. Which real numbers are their own reciprocals?

135. Jenna is a meteorologist. On December 10, she notes that the temperature is $-3°\text{F}$ at 6:00 A.M. She

predicts that the temperature will rise at a rate of 2° per hour for 3 hr, and then rise at a rate of 3° per hour for 6 hr. She also predicts that the temperature will then fall at a rate of 2° per hour for 3 hr, and then fall at a rate of 5° per hour for 2 hr. What is Jenna's temperature forecast for 8:00 P.M?

Tell whether each expression represents a positive number or a negative number when m and n are negative.

136. $\frac{m}{-n}$

137. $\frac{-n}{-m}$

138. $-m \cdot \left(\frac{-n}{m}\right)$

139. $-\left(\frac{n}{-m}\right)$

140. $(m + n) \cdot \frac{m}{n}$

141. $(-n - m)\frac{n}{m}$

142. What must be true of m and n if $-mn$ is to be **(a)** positive? **(b)** zero? **(c)** negative?

143. The following is a proof that a positive number times a negative number is negative. Provide a reason for each step. Assume that $a > 0$ and $b > 0$.

$$a(-b) + ab = a[-b + b]$$
$$= a(0)$$
$$= 0$$

Therefore, $a(-b)$ is the opposite of ab.

144. Is it true that for any numbers a and b, if a is larger than b, then the reciprocal of a is smaller than the reciprocal of b? Why or why not?

CONNECTING ↕ the CONCEPTS

The rules for multiplication and division of real numbers differ significantly from the rules for addition and subtraction. When simplifying an expression, look at the operation first to determine which set of rules to follow.

Addition

If the signs are the same, add absolute values.

- *Both numbers are positive:* Add as usual. The answer is positive.
- *Both numbers are negative:* Add absolute values and make the answer negative.

If the signs are different, subtract absolute values.

- *The positive number has the greater absolute value:* The answer is positive.
- *The negative number has the greater absolute value:* The answer is negative.
- *The numbers have the same absolute value:* The answer is 0.

If one number is zero, the sum is the other number.

Subtraction

Add the opposite of the number being subtracted.

Multiplication

If the signs are the same, multiply absolute values. The answer is positive.

If the signs are different, multiply absolute values. The answer is negative.

If one number is zero, the product is 0.

Division

Multiply by the reciprocal of the divisor.

MIXED REVIEW

Perform the indicated operation and, if possible, simplify.

1. $-8 + (-2)$ **2.** $-8 \cdot (-2)$

3. $-8 \div (-2)$ **4.** $-8 - (-2)$

5. $12 \cdot (-10)$ **6.** $13 - 20$

7. $-5 - 18$ **8.** $-12 \div 4$

9. $\dfrac{3}{5} - \dfrac{8}{5}$ **10.** $\dfrac{-12}{5} + \left(\dfrac{-3}{5}\right)$

11. $-5.6 + 4.8$ **12.** $1.3 \cdot (-2.9)$

13. $-44.1 \div 6.3$ **14.** $6.6 + (-10.7)$

15. $\dfrac{9}{5} \cdot \left(-\dfrac{20}{3}\right)$ **16.** $-\dfrac{5}{4} \div \left(-\dfrac{3}{4}\right)$

17. $38 - (-62)$ **18.** $-17 + 94$

19. $(-15) \cdot (-12)$ **20.** $-26 - 26$

1.8 Exponential Notation and Order of Operations

Exponential Notation ▪ Order of Operations ▪ Simplifying and the Distributive Law ▪
The Opposite of a Sum

Algebraic expressions often contain *exponential notation*. In this section, we learn how to use exponential notation as well as rules for the *order of operations* in performing certain algebraic manipulations.

Exponential Notation

A product like $3 \cdot 3 \cdot 3 \cdot 3$, in which the factors are the same, is called a **power**. Powers occur often enough that a simpler notation called **exponential notation** is used. For

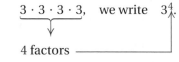

Because $3^4 = 81$, we can say that 81 "is a power of 3."

This is read "three to the fourth power," or simply, "three to the fourth." The number 4 is called an **exponent** and the number 3 a **base**.

Expressions like s^2 and s^3 are usually read "s squared" and "s cubed," respectively. This comes from the fact that a square with sides of length s has an area A given by $A = s^2$ and a cube with sides of length s has a volume V given by $V = s^3$.

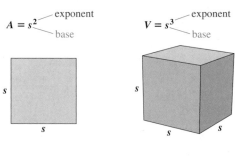

EXAMPLE 1 Write exponential notation for $10 \cdot 10 \cdot 10 \cdot 10 \cdot 10$.

SOLUTION

Exponential notation is 10^5. 5 is the exponent. 10 is the base. **TRY EXERCISE** 3

EXAMPLE 2 Simplify: **(a)** 5^2; **(b)** $(-5)^3$; **(c)** $(2n)^3$.

SOLUTION

a) $5^2 = 5 \cdot 5 = 25$ The exponent 2 indicates two factors of 5.

b) $(-5)^3 = (-5)(-5)(-5)$ The exponent 3 indicates three factors of -5.
$= 25(-5)$ Using the associative law of multiplication
$= -125$

c) $(2n)^3 = (2n)(2n)(2n)$ The exponent 3 indicates three factors of $2n$.
$= 2 \cdot 2 \cdot 2 \cdot n \cdot n \cdot n$ Using the associative and commutative laws of multiplication
$= 8n^3$ **TRY EXERCISE** 13

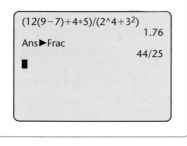
STUDENT NOTES

Although most scientific and graphing calculators follow the rules for order of operations when evaluating expressions, some calculators do not. Try calculating $4 + 2 \times 5$ on your calculator. If the result shown is 30, your calculator does not follow the rules for order of operations. In this case, you will have to multiply 2×5 first and then add 4.

To determine what the exponent 1 will mean, look for a pattern in the following:

$$7 \cdot 7 \cdot 7 \cdot 7 = 7^4$$
$$7 \cdot 7 \cdot 7 = 7^3$$
$$7 \cdot 7 = 7^2$$
$$? = 7^1$$

The exponent decreases by 1 each time.

The number of factors decreases by 1 each time. To extend the pattern, we say that

$$7 = 7^1.$$

> ### Exponential Notation
> For any natural number n,
>
> $$b^n \quad \text{means} \quad \overbrace{b \cdot b \cdot b \cdot b \cdots b}^{n \text{ factors}}.$$

Order of Operations

How should $4 + 2 \times 5$ be computed? If we multiply 2 by 5 and then add 4, the result is 14. If we add 2 and 4 first and then multiply by 5, the result is 30. Since these results differ, the order in which we perform operations matters. If grouping symbols such as parentheses ( ), brackets [ ], braces { }, absolute-value symbols | |, or fraction bars appear, they tell us what to do first. For example,

$$(4 + 2) \times 5 \quad \text{indicates} \quad 6 \times 5, \quad \text{resulting in } 30,$$

and

$$4 + (2 \times 5) \quad \text{indicates} \quad 4 + 10, \quad \text{resulting in } 14.$$

Besides grouping symbols, the following conventions exist for determining the order in which operations should be performed.

> ### Rules for Order of Operations
> 1. Calculate within the innermost grouping symbols, ( ), [ ], { }, | |, and above or below fraction bars.
> 2. Simplify all exponential expressions.
> 3. Perform all multiplications and divisions, working from left to right.
> 4. Perform all additions and subtractions, working from left to right.

Thus the correct way to compute $4 + 2 \times 5$ is to first multiply 2 by 5 and then add 4. The result is 14.

EXAMPLE   3   Simplify: $15 - 2 \cdot 5 + 3$.

SOLUTION   When no groupings or exponents appear, we *always* multiply or divide before adding or subtracting:

$$15 - 2 \cdot 5 + 3 = 15 - 10 + 3 \qquad \text{Multiplying}$$
$$= 5 + 3$$
$$= 8. \qquad \text{Subtracting and adding from left to right}$$

TRY EXERCISE   31

Always calculate within parentheses first. When there are exponents and no parentheses, we simplify powers before multiplying or dividing.

EXAMPLE 4

Simplify: **(a)** $(3 \cdot 4)^2$; **(b)** $3 \cdot 4^2$.

SOLUTION

a) $(3 \cdot 4)^2 = (12)^2$ Working within parentheses first

              $= 144$

b) $3 \cdot 4^2 = 3 \cdot 16$ Simplifying the power

        $= 48$ Multiplying

Note that $(3 \cdot 4)^2 \neq 3 \cdot 4^2$.

> **TRY EXERCISE** 37

> *CAUTION!* Example 4 illustrates that, in general, $(ab)^2 \neq ab^2$.

EXAMPLE 5

Evaluate when $x = 5$: **(a)** $(-x)^2$; **(b)** $-x^2$.

SOLUTION

a) $(-x)^2 = (-5)^2 = (-5)(-5) = 25$ We square the opposite of 5.

b) $-x^2 = -5^2 = -25$ We square 5 and then find the opposite.

> **TRY EXERCISE** 15

> *CAUTION!* Example 5 illustrates that, in general, $(-x)^2 \neq -x^2$.

To simplify $-x^2$, it may help to write

$$-x^2 = (-1)x^2.$$

EXAMPLE 6

Evaluate $-15 \div 3(6 - a)^3$ when $a = 4$.

SOLUTION

$$-15 \div 3(6 - a)^3 = -15 \div 3(6 - 4)^3 \qquad \text{Substituting 4 for } a$$
$$= -15 \div 3(2)^3 \qquad \text{Working within parentheses first}$$
$$= -15 \div 3 \cdot 8 \qquad \text{Simplifying the exponential expression}$$
$$= -5 \cdot 8 \,\Big\} \qquad \text{Dividing and multiplying from left to right}$$
$$= -40 \,\Big/$$

> **TRY EXERCISE** 67

STUDENT NOTES

When simplifying an expression, it is important to copy the entire expression on each line, not just the parts that have been simplified in a given step. As shown in Examples 6 and 7, each line should be equivalent to the line above it.

The symbols ( ), [ ], and { } are all used in the same way. Used inside or next to each other, they make it easier to locate the left and right sides of a grouping. When combinations of grouping symbols are used, we begin with the innermost grouping symbols and work to the outside.

EXAMPLE 7   Simplify: $8 \div 4 + 3[9 + 2(3 - 5)^3]$.

SOLUTION

$8 \div 4 + 3[9 + 2(3 - 5)^3] = 8 \div 4 + 3[9 + 2(-2)^3]$ Doing the calculations in the innermost grouping symbols first

$= 8 \div 4 + 3[9 + 2(-8)]$ $(-2)^3 = (-2)(-2)(-2)$
$= -8$

$= 8 \div 4 + 3[9 + (-16)]$

$= 8 \div 4 + 3[-7]$ Completing the calculations within the brackets

$= 2 + (-21)$ Multiplying and dividing from left to right

$= -19$ TRY EXERCISE 47

EXAMPLE 8   Calculate: $\dfrac{12(9 - 7) + 4 \cdot 5}{3^4 + 2^3}$.

SOLUTION   An equivalent expression with brackets is

$[12(9 - 7) + 4 \cdot 5] \div [3^4 + 2^3]$.   Here the grouping symbols are necessary.

In effect, we need to simplify the numerator, simplify the denominator, and then divide the results:

$\dfrac{12(9 - 7) + 4 \cdot 5}{3^4 + 2^3} = \dfrac{12(2) + 4 \cdot 5}{81 + 8}$

$= \dfrac{24 + 20}{89} = \dfrac{44}{89}$.   TRY EXERCISE 55

Simplifying and the Distributive Law

Sometimes we cannot simplify within grouping symbols. When a sum or a difference is being grouped, the distributive law provides a method for removing the grouping symbols.

EXAMPLE 9   Simplify: $5x - 9 + 2(4x + 5)$.

SOLUTION

$5x - 9 + 2(4x + 5) = 5x - 9 + 8x + 10$ Using the distributive law

$= 13x + 1$ Combining like terms

TRY EXERCISE 85

Now that exponents have been introduced, we can make our definition of *like* or *similar terms* more precise. **Like**, or **similar**, **terms** are either constant terms or terms containing the same variable(s) raised to the same power(s). Thus, 5 and -7, $19xy$ and $2yx$, and $4a^3b$ and a^3b are all pairs of like terms.

EXAMPLE 10

Simplify: $7x^2 + 3[x^2 + 2x] - 5x$.

SOLUTION

$$7x^2 + 3[x^2 + 2x] - 5x = 7x^2 + 3x^2 + 6x - 5x \qquad \text{Using the distributive law}$$

$$= 10x^2 + x \qquad \text{Combining like terms}$$

TRY EXERCISE 91

The Opposite of a Sum

When a number is multiplied by -1, the result is the opposite of that number. For example, $-1(7) = -7$ and $-1(-5) = 5$.

> ### The Property of -1
> For any real number a,
> $$-1 \cdot a = -a.$$
> (Negative one times a is the opposite of a.)

An expression such as $-(x + y)$ indicates the *opposite*, or *additive inverse*, of the sum of x and y. When a sum within grouping symbols is preceded by a "$-$" symbol, we can multiply the sum by -1 and use the distributive law. In this manner, we can find an equivalent expression for the opposite of a sum.

EXAMPLE 11

Write an expression equivalent to $-(3x + 2y + 4)$ without using parentheses.

SOLUTION

$$-(3x + 2y + 4) = -1(3x + 2y + 4) \qquad \text{Using the property of } -1$$

$$= -1(3x) + (-1)(2y) + (-1)4 \qquad \text{Using the distributive law}$$

$$= -3x - 2y - 4 \qquad \text{Using the associative law and the property of } -1$$

TRY EXERCISE 73

Example 11 illustrates an important property of real numbers.

> ### The Opposite of a Sum
> For any real numbers a and b,
> $$-(a + b) = -a + (-b) = -a - b.$$
> (The opposite of a sum is the sum of the opposites.)

To remove parentheses from an expression like $-(x - 7y + 5)$, we can first rewrite the subtraction as addition:

$$-(x - 7y + 5) = -(x + (-7y) + 5) \qquad \text{Rewriting as addition}$$

$$= -x + 7y - 5. \qquad \text{Taking the opposite of a sum}$$

This procedure is normally streamlined to one step in which we find the opposite by "removing parentheses and changing the sign of every term":

$$-(x - 7y + 5) = -x + 7y - 5.$$

EXAMPLE 12   Simplify: $3x - (4x + 2)$.

SOLUTION

$$
\begin{aligned}
3x - (4x + 2) &= 3x + [-(4x + 2)] && \text{Adding the opposite of } 4x + 2 \\
&= 3x + [-4x - 2] && \text{Taking the opposite of } 4x + 2 \\
&= 3x + (-4x) + (-2) && \\
&= 3x - 4x - 2 && \text{Try to go directly to this step.} \\
&= -x - 2 && \text{Combining like terms}
\end{aligned}
$$

TRY EXERCISE 81

In practice, the first three steps of Example 12 are generally skipped.

EXAMPLE 13   Simplify: $5t^2 - 2t - (-4t^2 + 9t)$.

SOLUTION

$$
\begin{aligned}
5t^2 - 2t - (-4t^2 + 9t) &= 5t^2 - 2t + 4t^2 - 9t && \text{Removing parentheses and changing the sign of each term inside} \\
&= 9t^2 - 11t && \text{Combining like terms}
\end{aligned}
$$

TRY EXERCISE 89

Expressions such as $7 - 3(x + 2)$ can be simplified as follows:

$$
\begin{aligned}
7 - 3(x + 2) &= 7 + [-3(x + 2)] && \text{Adding the opposite of } 3(x + 2) \\
&= 7 + [-3x - 6] && \text{Multiplying } x + 2 \text{ by } -3 \\
&= 7 - 3x - 6 && \text{Try to go directly to this step.} \\
&= 1 - 3x. && \text{Combining like terms}
\end{aligned}
$$

EXAMPLE 14   Simplify: **(a)** $3n - 2(4n - 5)$; **(b)** $7x^3 + 2 - [5(x^3 - 1) + 8]$.

SOLUTION

a)
$$
\begin{aligned}
3n - 2(4n - 5) &= 3n - 8n + 10 && \text{Multiplying each term inside the parentheses by } -2 \\
&= -5n + 10 && \text{Combining like terms}
\end{aligned}
$$

b)
$$
\begin{aligned}
7x^3 + 2 - [5(x^3 - 1) + 8] &= 7x^3 + 2 - [5x^3 - 5 + 8] && \text{Removing parentheses} \\
&= 7x^3 + 2 - [5x^3 + 3] && \\
&= 7x^3 + 2 - 5x^3 - 3 && \text{Removing brackets} \\
&= 2x^3 - 1 && \text{Combining like terms}
\end{aligned}
$$

TRY EXERCISE 93

As we progress through our study of algebra, it is important that we be able to distinguish between the two tasks of **simplifying an expression** and **solving an equation**. In Chapter 1, we have not solved equations, but we have simplified expressions. This enabled us to write *equivalent expressions* that were simpler than the given expression. In Chapter 2, we will continue to simplify expressions, but we will also begin to solve equations.

1.8   EXERCISE SET

For Extra Help
MyMathLab   Math XL PRACTICE   WATCH   DOWNLOAD

Concept Reinforcement *In each part of Exercises 1 and 2, name the operation that should be performed first. Do not perform the calculations.*

1. a) $4 + 8 \div 2 \cdot 2$

   b) $7 - 9 + 15$

   c) $5 - 2(3 + 4)$

   d) $6 + 7 \cdot 3$

   e) $18 - 2[4 + (3 - 2)]$

   f) $\dfrac{5 - 6 \cdot 7}{2}$

2. a) $9 - 3 \cdot 4 \div 2$

   b) $8 + 7(6 - 5)$

   c) $5 \cdot [2 - 3(4 + 1)]$

   d) $8 - 7 + 2$

   e) $4 + 6 \div 2 \cdot 3$

   f) $\dfrac{37}{8 - 2 \cdot 2}$

Write exponential notation.

3. $x \cdot x \cdot x \cdot x \cdot x \cdot x$

4. $y \cdot y \cdot y \cdot y \cdot y \cdot y$

5. $(-5)(-5)(-5)$

6. $(-7)(-7)(-7)(-7)$

7. $3t \cdot 3t \cdot 3t \cdot 3t \cdot 3t$

8. $5m \cdot 5m \cdot 5m \cdot 5m \cdot 5m$

9. $2 \cdot n \cdot n \cdot n \cdot n$

10. $8 \cdot a \cdot a \cdot a$

Simplify.

11. 4^2

12. 5^3

13. $(-3)^2$

14. $(-7)^2$

15. -3^2

16. -7^2

17. 4^3

18. 9^1

19. $(-5)^4$

20. 5^4

21. 7^1

22. $(-1)^7$

23. $(-2)^5$

24. -2^5

25. $(3t)^4$

26. $(5t)^2$

27. $(-7x)^3$

28. $(-5x)^4$

29. $5 + 3 \cdot 7$

30. $3 - 4 \cdot 2$

31. $10 \cdot 5 + 1 \cdot 1$

32. $19 - 5 \cdot 4 + 3$

33. $6 - 70 \div 7 - 2$

34. $12 \div 3 + 18 \div 2$

Aha! **35.** $14 \cdot 19 \div (19 \cdot 14)$

36. $18 - 6 \div 3 \cdot 2 + 7$

37. $3(-10)^2 - 8 \div 2^2$

38. $9 - 3^2 \div 9(-1)$

39. $8 - (2 \cdot 3 - 9)$

40. $(8 - 2 \cdot 3) - 9$

41. $(8 - 2)(3 - 9)$

42. $32 \div (-2)^2 \cdot 4$

43. $13(-10)^2 + 45 \div (-5)$

44. $2^4 + 2^3 - 10 \div (-1)^4$

45. $5 + 3(2 - 9)^2$

46. $9 - (3 - 5)^3 - 4$

47. $[2 \cdot (5 - 8)]^2$

48. $3(5 - 7)^3 \div 4$

49. $\dfrac{7 + 2}{5^2 - 4^2}$

50. $\dfrac{(5^2 - 3^2)^2}{2 \cdot 6 - 4}$

51. $8(-7) + |3(-4)|$

52. $|10(-5)| + 1(-1)$

53. $36 \div (-2)^2 + 4[5 - 3(8 - 9)^5]$

54. $-48 \div (7 - 9)^3 - 2[1 - 5(2 - 6) + 3^2]$

55. $\dfrac{7^2 - (-1)^7}{5 \cdot 7 - 4 \cdot 3^2 - 2^2}$

56. $\dfrac{(-2)^3 + 4^2}{2 \cdot 3 - 5^2 + 3 \cdot 7}$

57. $\dfrac{-3^3 - 2 \cdot 3^2}{8 \div 2^2 - (6 - |2 - 15|)}$

58. $\dfrac{(-5)^2 - 3 \cdot 5}{3^2 + 4 \cdot |6 - 7| \cdot (-1)^5}$

Evaluate.

59. $9 - 4x$, for $x = 7$

60. $1 + x^3$, for $x = -2$

61. $24 \div t^3$, for $t = -2$

62. $-100 \div a^2$, for $a = -5$

63. $45 \div a \cdot 5$, for $a = -3$

64. $50 \div 2 \cdot t$, for $t = 5$

65. $5x \div 15x^2$, for $x = 3$

66. $6a \div 12a^3$, for $a = 2$

67. $45 \div 3^2 x(x - 1)$, for $x = 3$

68. $-30 \div t(t + 4)^2$, for $t = -6$

69. $-x^2 - 5x$, for $x = -3$

70. $(-x)^2 - 5x$, for $x = -3$

71. $\dfrac{3a - 4a^2}{a^2 - 20}$, for $a = 5$

72. $\dfrac{a^3 - 4a}{a(a - 3)}$, for $a = -2$

Write an equivalent expression without using grouping symbols.

73. $-(9x + 1)$

74. $-(3x + 5)$

75. $-[-7n + 8]$

76. $-(6x - 7)$

77. $-(4a - 3b + 7c)$

78. $-[5n - m - 2p]$

79. $-(3x^2 + 5x - 1)$

80. $-(-9x^3 + 8x + 10)$

Simplify.

81. $8x - (6x + 7)$

82. $2a - (5a - 9)$

83. $2x - 7x - (4x - 6)$

84. $2a + 5a - (6a + 8)$

85. $9t - 7r + 2(3r + 6t)$

86. $4m - 9n + 3(2m - n)$

87. $15x - y - 5(3x - 2y + 5z)$

88. $4a - b - 4(5a - 7b + 8c)$

89. $3x^2 + 11 - (2x^2 + 5)$

90. $5x^4 + 3x - (5x^4 + 3x)$

91. $5t^3 + t + 3(t - 2t^3)$

92. $8n^2 - 3n + 2(n - 4n^2)$

93. $12a^2 - 3ab + 5b^2 - 5(-5a^2 + 4ab - 6b^2)$

94. $-8a^2 + 5ab - 12b^2 - 6(2a^2 - 4ab - 10b^2)$

95. $-7t^3 - t^2 - 3(5t^3 - 3t)$

96. $9t^4 + 7t - 5(9t^3 - 2t)$

97. $5(2x - 7) - [4(2x - 3) + 2]$

98. $3(6x - 5) - [3(1 - 8x) + 5]$

99. Some students use the mnemonic device PEMDAS to help remember the rules for the order of operations. Explain how this can be done and how the order of the letters in PEMDAS could lead a student to a wrong conclusion about the order of some operations.

100. Jake keys $18/2 \cdot 3$ into his calculator and expects the result to be 3. What mistake is he probably making?

Skill Review

Translate to an algebraic expression. [1.1]

101. Nine less than twice a number

102. Half of the sum of two numbers

Synthesis

103. Write the sentence $(-x)^2 \neq -x^2$ in words. Explain why $(-x)^2$ and $-x^2$ are not equivalent.

104. Write the sentence $-|x| \neq -x$ in words. Explain why $-|x|$ and $-x$ are not equivalent.

Simplify.

105. $5t - \{7t - [4r - 3(t - 7)] + 6r\} - 4r$

106. $z - \{2z - [3z - (4z - 5z) - 6z] - 7z\} - 8z$

107. $\{x - [f - (f - x)] + [x - f]\} - 3x$

108. Is it true that for all real numbers a and b,
$$ab = (-a)(-b)?$$
Why or why not?

109. Is it true that for all real numbers a, b, and c,
$$a|b - c| = ab - ac?$$
Why or why not?

If $n > 0$, $m > 0$, and $n \neq m$, classify each of the following as either true or false.

110. $-n + m = -(n + m)$

111. $m - n = -(n - m)$

112. $n(-n - m) = -n^2 + nm$

113. $-m(n - m) = -(mn + m^2)$

114. $-n(-n - m) = n(n + m)$

Evaluate.

Aha! **115.** $[x + 3(2 - 5x) \div 7 + x](x - 3)$, for $x = 3$

Aha! **116.** $[x + 2 \div 3x] \div [x + 2 \div 3x]$, for $x = -7$

117. $\dfrac{x^2 + 2^x}{x^2 - 2^x}$, for $x = 3$

118. $\dfrac{x^2 + 2^x}{x^2 - 2^x}$, for $x = 2$

119. In Mexico, between 500 B.C. and 600 A.D., the Mayans represented numbers using powers of 20 and certain symbols. For example, the symbols

represent $4 \cdot 20^3 + 17 \cdot 20^2 + 10 \cdot 20^1 + 0 \cdot 20^0$. Evaluate this number.
Source: National Council of Teachers of Mathematics, 1906 Association Drive, Reston, VA 22091

120. Examine the Mayan symbols and the numbers in Exercise 119. What numbers do

 and

each represent?

121. Calculate the volume of the tower shown below.

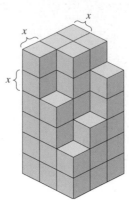

CORNER

Select the Symbols

Focus: Order of operations
Time: 15 minutes
Group size: 2

One way to master the rules for the order of operations is to insert symbols within a display of numbers in order to obtain a predetermined result. For example, the display

$$1 \quad 2 \quad 3 \quad 4 \quad 5$$

can be used to obtain the result 21 as follows:

$$(1 + 2) \div 3 + 4 \cdot 5.$$

Note that without an understanding of the rules for the order of operations, solving a problem of this sort is impossible.

ACTIVITY

1. Each group should prepare an exercise similar to the example shown above. (Exponents are not allowed.) To do so, first select five single-digit numbers for display. Then insert operations and grouping symbols and calculate the result.

2. Pair with another group. Each group should give the other its result along with its five-number display, and challenge the other group to insert symbols that will make the display equal the result given.

3. Share with the entire class the various mathematical statements developed by each group.

Study Summary

KEY TERMS AND CONCEPTS	EXAMPLES

SECTION 1.1: INTRODUCTION TO ALGEBRA

An **algebraic expression** is a collection of **variables** and **constants** on which operations are performed.

$5ab^3$ is an algebraic expression; 5 is a constant; and a and b are variables.

To **evaluate** an algebraic expression, substitute a number for each variable and carry out the operations. The result is a **value** of that expression.

Evaluate $\dfrac{x + y}{8}$ for $x = 15$ and $y = 9$.

$$\frac{x + y}{8} = \frac{15 + 9}{8} = \frac{24}{8} = 3$$

To find the area of a rectangle, a triangle, or a parallelogram, evaluate the appropriate formula for the given values.

Find the area of a triangle with base 3.1 m *and height* 6 m.

$$A = \tfrac{1}{2}bh = \tfrac{1}{2}(3.1\,\text{m})(6\,\text{m}) = \tfrac{1}{2}(3.1)(6)(\text{m} \cdot \text{m}) = 9.3\,\text{m}^2$$

Many problems can be solved by **translating** phrases to algebraic expressions and then forming an equation. The table on p. 4 shows translations of many words that occur in problems.

Translate to an equation. Do not solve.

When 34 is subtracted from a number, the result is 13. What is the number?

Let n represent the number.

Rewording: 34 subtracted from a number   is   13

Translating: $n - 34$ $=$ 13

An **equation** is a number sentence with the verb $=$. A substitution for the variable in an equation that makes the equation true is a **solution** of the equation.

Determine whether 9 *is a solution of* $47 - n = 38$.

$$47 - n = 38$$

$$\frac{47 - 9 \mid 38}{38 \stackrel{?}{=} 38 \quad \text{TRUE}}$$

Since $38 = 38$ is true, 9 is a solution.

SECTION 1.2: THE COMMUTATIVE, ASSOCIATIVE, AND DISTRIBUTIVE LAWS

Equivalent expressions represent the same value for any replacement of the variable.

$x + 10$ and $3 + 7 + x$ are equivalent expressions.

The Commutative Laws

$a + b = b + a;$

$ab = ba$

$3 + (-5) = -5 + 3;$

$8(10) = 10(8)$

The Associative Laws

$a + (b + c) = (a + b) + c;$

$a \cdot (b \cdot c) = (a \cdot b) \cdot c$

$-5 + (5 + 6) = (-5 + 5) + 6;$

$2 \cdot (5 \cdot 9) = (2 \cdot 5) \cdot 9$

The Distributive Law

$$a(b + c) = ab + ac$$

$$4(x + 2) = 4 \cdot x + 4 \cdot 2 = 4x + 8$$

The distributive law is used to multiply and to **factor** expressions.

Multiply: $3(2x + 5y)$.

$$3(2x + 5y) = 3 \cdot 2x + 3 \cdot 5y = 6x + 15y$$

Factor: $16x + 24y + 8$.

$$16x + 24y + 8 = 8(2x + 3y + 1)$$

SECTION 1.3: FRACTION NOTATION

Natural numbers: $\{1, 2, 3, \ldots\}$
Whole numbers: $\{0, 1, 2, 3, \ldots\}$

15, 39, and 1567 are some natural numbers.
0, 5, 16, and 2890 are some whole numbers.

A **prime** number has only two different factors, the number itself and 1. Natural numbers that have factors other than 1 and the number itself are **composite** numbers.

2, 3, 5, 7, 11, and 13 are the first six prime numbers.
4, 6, 8, 24, and 100 are examples of composite numbers.

The **prime factorization** of a composite number expresses that number as a product of prime numbers.

The prime factorization of 136 is $2 \cdot 2 \cdot 2 \cdot 17$.

For any nonzero number a,

$$\frac{a}{a} = 1.$$

$$\frac{15}{15} = 1 \quad \text{and} \quad \frac{2x}{2x} = 1$$

The Identity Property of 1

$$a \cdot 1 = 1 \cdot a = a$$

The number 1 is called the **multiplicative identity**.

$$\frac{2}{3} = \frac{2}{3} \cdot \frac{5}{5} \quad \text{since} \quad \frac{5}{5} = 1.$$

$$\frac{a}{d} + \frac{b}{d} = \frac{a + b}{d}$$

$$\frac{a}{d} - \frac{b}{d} = \frac{a - b}{d}$$

$$\frac{a}{b} \cdot \frac{c}{d} = \frac{a \cdot c}{b \cdot d}$$

$$\frac{a}{b} \div \frac{c}{d} = \frac{a}{b} \cdot \frac{d}{c}$$

$$\frac{1}{6} + \frac{3}{8} = \frac{4}{24} + \frac{9}{24} = \frac{13}{24}$$

$$\frac{5}{12} - \frac{1}{6} = \frac{5}{12} - \frac{2}{12} = \frac{3}{12} = \frac{1 \cdot 3}{4 \cdot 3} = \frac{1}{4} \cdot \frac{3}{3} = \frac{1}{4} \cdot 1 = \frac{1}{4}$$

$$\frac{2}{5} \cdot \frac{7}{8} = \frac{2 \cdot 7}{5 \cdot 2 \cdot 4} = \frac{7}{20} \qquad \text{Removing a factor equal to 1: } \frac{2}{2} = 1$$

$$\frac{10}{9} \div \frac{4}{15} = \frac{10}{9} \cdot \frac{15}{4} = \frac{2 \cdot 5 \cdot 3 \cdot 5}{3 \cdot 3 \cdot 2 \cdot 2} = \frac{25}{6} \qquad \text{Removing a factor equal to 1: } \frac{2 \cdot 3}{2 \cdot 3} = 1$$

SECTION 1.4: POSITIVE AND NEGATIVE REAL NUMBERS

Integers:
$\{\ldots, -3, -2, -1, 0, 1, 2, 3, \ldots\}$

$-25, -2, 0, 1$, and 2000 are some integers.

Rational numbers:

$\left\{ \dfrac{a}{b} \,\middle|\, a \text{ and } b \text{ are integers and } b \neq 0 \right\}$

$\dfrac{1}{6}, \dfrac{-3}{7}, 0, 17, 0.758$, and $9.\overline{608}$ are some rational numbers.

The rational numbers and the **irrational numbers** make up the set of **real numbers**.

$\sqrt{7}$ and π are some irrational numbers.

Every rational number can be written using fraction notation or decimal notation. When written in decimal notation, a rational number either **repeats** or **terminates**.

$-\dfrac{1}{16} = -0.0625$ This is a terminating decimal.

$\dfrac{5}{6} = 0.8333\ldots = 0.8\overline{3}$ This is a repeating decimal.

Every real number corresponds to a point on the number line. For any two numbers, the one to the left is less than the one to the right. The symbol $<$ means "**is less than**" and the symbol $>$ means "**is greater than.**"

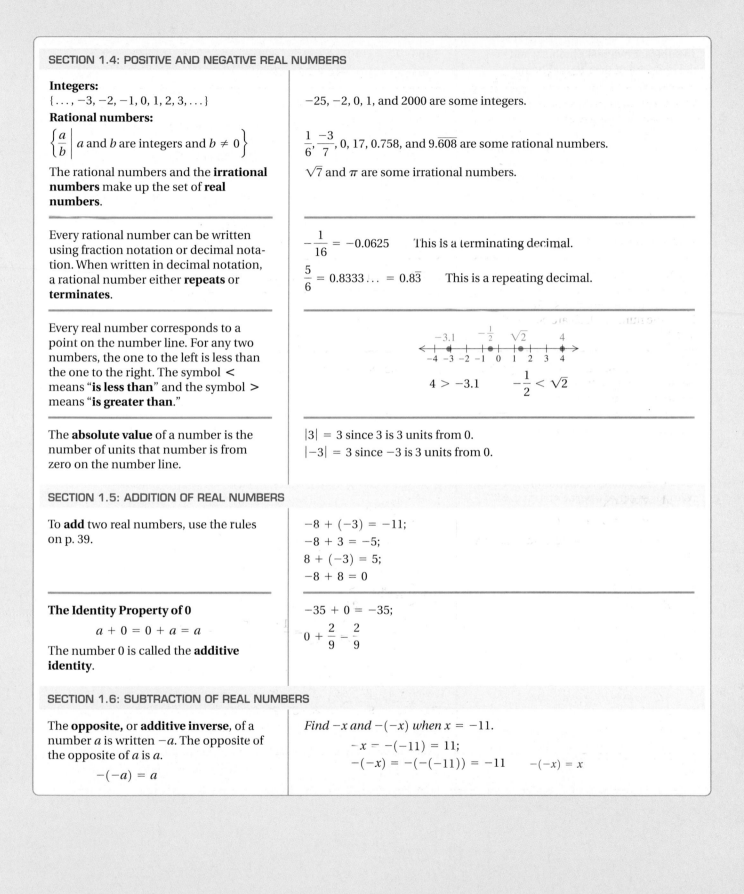

$4 > -3.1$ $-\dfrac{1}{2} < \sqrt{2}$

The **absolute value** of a number is the number of units that number is from zero on the number line.

$|3| = 3$ since 3 is 3 units from 0.
$|-3| = 3$ since -3 is 3 units from 0.

SECTION 1.5: ADDITION OF REAL NUMBERS

To **add** two real numbers, use the rules on p. 39.

$-8 + (-3) = -11;$
$-8 + 3 = -5;$
$8 + (-3) = 5;$
$-8 + 8 = 0$

The Identity Property of 0
$$a + 0 = 0 + a = a$$
The number 0 is called the **additive identity**.

$-35 + 0 = -35;$
$0 + \dfrac{2}{9} = \dfrac{2}{9}$

SECTION 1.6: SUBTRACTION OF REAL NUMBERS

The **opposite**, or **additive inverse**, of a number a is written $-a$. The opposite of the opposite of a is a.
$$-(-a) = a$$

Find $-x$ and $-(-x)$ when $x = -11$.
$$-x = -(-11) = 11;$$
$$-(-x) = -(-(-11)) = -11 \qquad -(-x) = x$$

To **subtract** two real numbers, add the opposite of the number being subtracted.

$-10 - 12 = -10 + (-12) = -22;$
$-10 - (-12) = -10 + 12 = 2$

The **terms** of an expression are separated by plus signs. **Like terms** either are constants or have the same variable factors raised to the same power. Like terms can be **combined** using the distributive law.

In the expression $-2x + 3y + 5x - 7y$:

The terms are $-2x, 3y, 5x,$ and $-7y$.
The like terms are $-2x$ and $5x$, and $3y$ and $-7y$.

Combining like terms gives

$$-2x + 3y + 5x - 7y = -2x + 5x + 3y - 7y$$
$$= (-2 + 5)x + (3 - 7)y = 3x - 4y.$$

SECTION 1.7: MULTIPLICATION AND DIVISION OF REAL NUMBERS

To **multiply** or **divide** two real numbers, use the rules on p. 54.

Division by 0 is undefined.

$(-5)(-2) = 10;$
$30 \div (-6) = -5;$
$0 \div (-3) = 0;$
$-3 \div 0$ is undefined.

SECTION 1.8: EXPONENTIAL NOTATION AND ORDER OF OPERATIONS

Exponential notation:

Exponent
n factors
$b^n = \overbrace{b \cdot b \cdot b \cdots b}$
Base

$6^2 = 6 \cdot 6 = 36;$
$(-6)^2 = (-6) \cdot (-6) = 36;$
$-6^2 = -6 \cdot 6 = -36;$
$(6x)^2 = (6x) \cdot (6x) = 36x^2$

To perform multiple operations, use the rules for **order of operations** on p. 61.

$$-3 + (3 - 5)^3 \div 4(-1) = -3 + (-2)^3 \div 4(-1)$$
$$= -3 + (-8) \div 4(-1)$$
$$= -3 + (-2)(-1)$$
$$= -3 + 2$$
$$= -1$$

The Property of -1

For any real number a,

$$-1 \cdot a = -a.$$

$-1 \cdot 5x = -5x$ and $-5x = -1(5x)$

The Opposite of a Sum

For any real numbers a and b,

$$-(a + b) = -a - b.$$

$-(2x - 3y) = -(2x) - (-3y) = -2x + 3y$

Expressions containing parentheses can be simplified by removing parentheses using the distributive law.

Simplify: $3x^2 - 5(x^2 - 4xy + 2y^2) - 7y^2.$

$$3x^2 - 5(x^2 - 4xy + 2y^2) - 7y^2 = 3x^2 - 5x^2 + 20xy - 10y^2 - 7y^2$$
$$= -2x^2 + 20xy - 17y^2$$

Review Exercises: Chapter 1

➥ *Concept Reinforcement* *In each of Exercises 1–10, classify the statement as either true or false.*

1. $4x - 5y$ and $12 - 7a$ are both algebraic expressions containing two terms. [1.2]

2. $3t + 1 = 7$ and $8 - 2 = 9$ are both equations. [1.1]

3. The fact that $2 + x$ is equivalent to $x + 2$ is an illustration of the associative law for addition. [1.2]

4. The statement $4(a + 3) = 4 \cdot a + 4 \cdot 3$ illustrates the distributive law. [1.2]

5. The number 2 is neither prime nor composite. [1.3]

6. Every irrational number can be written as a repeating decimal or a terminating decimal. [1.4]

7. Every natural number is a whole number and every whole number is an integer. [1.4]

8. The expressions $9r^2 s$ and $5rs^2$ are like terms. [1.8]

9. The opposite of x, written $-x$, never represents a positive number. [1.6]

10. The number 0 has no reciprocal. [1.7]

Evaluate.

11. $8t$, for $t = 3$ [1.1]

12. $\dfrac{x - y}{3}$, for $x = 17$ and $y = 5$ [1.1]

13. $9 - y^2$, for $y = -5$ [1.8]

14. $-10 + a^2 \div (b + 1)$, for $a = 5$ and $b = -6$ [1.8]

Translate to an algebraic expression. [1.1]

15. 7 less than y

16. 10 more than the product of x and z

17. 15 times the difference of Brandt's speed and the wind speed

18. Determine whether 35 is a solution of $x/5 = 8$. [1.1]

19. Translate to an equation. Do not solve. [1.1]

   According to Photo Marketing Association International, in 2006, 14.1 billion prints were made from film. This number is 3.2 billion more than the number of digital prints made. How many digital prints were made in 2006?

20. Use the commutative law of multiplication to write an expression equivalent to $3t + 5$. [1.2]

21. Use the associative law of addition to write an expression equivalent to $(2x + y) + z$. [1.2]

22. Use the commutative and associative laws to write three expressions equivalent to $4(xy)$. [1.2]

Multiply. [1.2]

23. $6(3x + 5y)$

24. $8(5x + 3y + 2)$

Factor. [1.2]

25. $21x + 15y$

26. $22a + 99b + 11$

27. Find the prime factorization of 56. [1.3]

Simplify. [1.3]

28. $\dfrac{20}{48}$

29. $\dfrac{18}{8}$

Perform the indicated operation and, if possible, simplify. [1.3]

30. $\dfrac{5}{12} + \dfrac{3}{8}$

31. $\dfrac{9}{16} \div 3$

32. $\dfrac{2}{3} - \dfrac{1}{15}$

33. $\dfrac{9}{10} \cdot \dfrac{6}{5}$

34. Tell which integers correspond to this situation. [1.4]

   Natalie borrowed \$3600 for an entertainment center. Sean has \$1350 in his savings account.

35. Graph on a number line: $\frac{-1}{3}$. [1.4]

36. Write an inequality with the same meaning as $-3 < x$. [1.4]

37. Classify as true or false: $2 \geq -8$. [1.4]

38. Classify as true or false: $0 \leq -1$. [1.4]

39. Find decimal notation: $-\dfrac{4}{9}$. [1.4]

40. Find the absolute value: $|-1|$. [1.4]

41. Find $-(-x)$ when x is -12. [1.6]

Simplify.

42. $-3 + (-7)$ [1.5]

43. $-\frac{2}{3} + \frac{1}{12}$ [1.5]

44. $10 + (-9) + (-8) + 7$ [1.5]

45. $-3.8 + 5.1 + (-12) + (-4.3) + 10$ [1.5]

46. $-2 - (-10)$ [1.6]

47. $-\frac{9}{10} - \frac{1}{2}$ [1.6]

48. $-3.8 - 4.1$ [1.6]

49. $-9 \cdot (-7)$ [1.7]

50. $-2.7(3.4)$ [1.7]

51. $\frac{2}{3} \cdot \left(-\frac{3}{7}\right)$ [1.7]

52. $2 \cdot (-7) \cdot (-2) \cdot (-5)$ [1.7]

53. $35 \div (-5)$ [1.7]

54. $-5.1 \div 1.7$ [1.7]

55. $-\frac{3}{5} \div \left(-\frac{4}{15}\right)$ [1.7]

56. $120 - 6^2 \div 4 \cdot 8$ [1.8]

57. $(120 - 6^2) \div 4 \cdot 8$ [1.8]

58. $(120 - 6^2) \div (4 \cdot 8)$ [1.8]

59. $16 \div (-2)^3 - 5[3 - 1 + 2(4 - 7)]$ [1.8]

60. $|-3 \cdot 5 - 4 \cdot 8| - 3(-2)$ [1.8]

61. $\dfrac{4(18 - 8) + 7 \cdot 9}{9^2 - 8^2}$ [1.8]

Combine like terms.

62. $11a + 2b + (-4a) + (-3b)$ [1.5]

63. $7x - 3y - 11x + 8y$ [1.6]

64. Find the opposite of -7. [1.6]

65. Find the reciprocal of -7. [1.7]

66. Write exponential notation for $2x \cdot 2x \cdot 2x \cdot 2x$. [1.8]

67. Simplify: $(-5x)^3$. [1.8]

Remove parentheses and simplify. [1.8]

68. $2a - (5a - 9)$

69. $5b + 3(2b - 9)$

70. $11x^4 + 2x + 8(x - x^4)$

71. $2n^2 - 5(-3n^2 + m^2 - 4mn) + 6m^2$

72. $8(x + 4) - 6 - [3(x - 2) + 4]$

Synthesis

73. Explain the difference between a constant and a variable. [1.1]

74. Explain the difference between a term and a factor. [1.2]

75. Describe at least three ways in which the distributive law was used in this chapter. [1.2]

76. Devise a rule for determining the sign of a negative number raised to a power. [1.8]

77. Evaluate $a^{50} - 20a^{25}b^4 + 100b^8$ for $a = 1$ and $b = 2$. [1.8]

78. If $0.090909\ldots = \frac{1}{11}$ and $0.181818\ldots = \frac{2}{11}$, what rational number is named by each of the following?

a) $0.272727\ldots$ [1.4]   **b)** $0.909090\ldots$ [1.4]

Simplify. [1.8]

79. $-\left|\frac{7}{8} - \left(-\frac{1}{2}\right) - \frac{3}{4}\right|$

80. $(|2.7 - 3| + 3^2 - |-3|) \div (-3)$

Match each phrase in the left column with the most appropriate choice from the right column.

81. ____ A number is nonnegative. [1.4]

82. ____ The product of a number and its reciprocal is 1. [1.7]

83. ____ A number squared [1.8]

84. ____ A sum of squares [1.8]

85. ____ The opposite of an opposite is the original number. [1.6]

86. ____ The order in which numbers are added does not change the result. [1.2]

87. ____ A number is positive. [1.4]

88. ____ The absolute value of a product [1.4]

89. ____ A sum of a number and its reciprocal [1.7]

90. ____ The square of a sum [1.8]

91. ____ The absolute value of one number is less than the absolute value of another number. [1.4]

a) a^2

b) $a + b = b + a$

c) $a > 0$

d) $a + \dfrac{1}{a}$

e) $|ab|$

f) $(a + b)^2$

g) $|a| < |b|$

h) $a^2 + b^2$

i) $a \geq 0$

j) $a \cdot \dfrac{1}{a} = 1$

k) $-(-a) = a$

Test: Chapter 1

1. Evaluate $\dfrac{2x}{y}$ for $x = 10$ and $y = 5$.

2. Write an algebraic expression: Nine less than the product of two numbers.

3. Find the area of a triangle when the height h is 30 ft and the base b is 16 ft.

4. Use the commutative law of addition to write an expression equivalent to $3p + q$.

5. Use the associative law of multiplication to write an expression equivalent to $x \cdot (4 \cdot y)$.

6. Determine whether 7 is a solution of $65 - x = 69$.

7. Translate to an equation. Do not solve.

   In the summer of 2005, member utilities of the Florida Reliability Coordinating Council had a demand of 45,950 megawatts. This is only 4250 megawatts less than its maximum production capability. What is the maximum capability of production?
   Source: Energy Information Administration

Multiply.

8. $7(5 + x)$

9. $-5(y - 2)$

Factor.

10. $11 + 44x$

11. $7x + 7 + 49y$

12. Find the prime factorization of 300.

13. Simplify: $\dfrac{24}{56}$.

Write a true sentence using either $<$ or $>$.

14. $-4 \ \blacksquare \ 0$

15. $-3 \ \blacksquare \ -8$

Find the absolute value.

16. $\left| \frac{9}{4} \right|$

17. $|-3.8|$

18. Find the opposite of $-\frac{2}{3}$.

19. Find the reciprocal of $-\frac{4}{7}$.

20. Find $-x$ when x is -10.

21. Write an inequality with the same meaning as $x \le -5$.

Perform the indicated operations and, if possible, simplify.

22. $3.1 - (-4.7)$

23. $-8 + 4 + (-7) + 3$

24. $3.2 - 5.7$

25. $-\frac{1}{8} - \frac{3}{4}$

26. $4 \cdot (-12)$

27. $-\frac{1}{2} \cdot \left(-\frac{4}{9} \right)$

28. $-66 \div 11$

29. $-\frac{3}{5} \div \left(-\frac{4}{5} \right)$

30. $4.864 \div (-0.5)$

31. $-2(16) - |2(-8) - 5^3|$

32. $9 + 7 - 4 - (-3)$

33. $256 \div (-16) \cdot 4$

34. $2^3 - 10[4 - (-2 + 18)3]$

35. Combine like terms: $18y + 30a - 9a + 4y$.

36. Simplify: $(-2x)^4$.

Remove parentheses and simplify.

37. $4x - (3x - 7)$

38. $4(2a - 3b) + a - 7$

39. $3[5(y - 3) + 9] - 2(8y - 1)$

Synthesis

40. Evaluate $\dfrac{5y - x}{2}$ when $x = 20$ and y is 4 less than half of x.

41. Insert one pair of parentheses to make the following a true statement:
$$9 - 3 - 4 + 5 = 15.$$

Simplify.

42. $|-27 - 3(4)| - |-36| + |-12|$

43. $a - \{3a - [4a - (2a - 4a)]\}$

44. Classify the following as either true or false:
$$a|b - c| = |ab| - |ac|.$$

Equations, Inequalities, and Problem Solving

DEBORAH ELIAS
EVENT COORDINATOR
Houston, Texas

As an event planner, I am constantly using math. Calculations range from figuring the tax and gratuity percentage to add to the final bill to finding dimensions of linens to fit a table properly. For every client, I also determine a budget with income and expenses.

AN APPLICATION

Event promoters use the formula

$$p = \frac{1.2x}{s}$$

to determine a ticket price p for an event with x dollars of expenses and s anticipated ticket sales. Grand Events expects expenses for an upcoming concert to be $80,000 and anticipates selling 4000 tickets. What should the ticket price be?

Source: *The Indianapolis Star*, 2/27/03

This problem appears as Example 1 in Section 2.3.

S olving equations and inequalities is a recurring theme in much of mathematics. In this chapter, we will study some of the principles used to solve equations and inequalities. We will then use equations and inequalities to solve applied problems.

2.1   Solving Equations

Equations and Solutions   ●   The Addition Principle   ●   The Multiplication Principle   ●   Selecting the Correct Approach

Solving equations is essential for problem solving in algebra. In this section, we study two of the most important principles used for this task.

Equations and Solutions

We have already seen that an equation is a number sentence stating that the expressions on either side of the equals sign represent the same number. Some equations, like $3 + 2 = 5$ or $2x + 6 = 2(x + 3)$, are *always* true and some, like $3 + 2 = 6$ or $x + 2 = x + 3$, are *never* true. In this text, we will concentrate on equations like $x + 6 = 13$ or $7x = 141$ that are *sometimes* true, depending on the replacement value for the variable.

> ### Solution of an Equation
>
> Any replacement for the variable that makes an equation true is called a *solution* of the equation. To *solve* an equation means to find all of its solutions.

To determine whether a number is a solution, we substitute that number for the variable throughout the equation. If the values on both sides of the equals sign are the same, then the number that was substituted is a solution.

EXAMPLE **1**   Determine whether 7 is a solution of $x + 6 = 13$.

SOLUTION   We have

$$
\begin{array}{ll}
x + 6 = 13 & \text{Writing the equation} \\
\overline{7 + 6 \mid 13} & \text{Substituting 7 for } x \\
13 \stackrel{?}{=} 13 \quad \text{TRUE} & 13 = 13 \text{ is a true statement.}
\end{array}
$$

Since the left-hand side and the right-hand side are the same, 7 is a solution.

CAUTION!   Note that in Example 1, the solution is 7, not 13.

EXAMPLE   2   Determine whether $\frac{2}{3}$ is a solution of $156x = 117$.

SOLUTION   We have

$$156x = 117 \qquad \text{Writing the equation}$$

$$156\left(\frac{2}{3}\right) \;\bigg|\; 117 \qquad \text{Substituting } \tfrac{2}{3} \text{ for } x$$

$$104 \overset{?}{=} 117 \quad \text{FALSE} \qquad \text{The statement } 104 = 117 \text{ is false.}$$

Since the left-hand side and the right-hand side differ, $\frac{2}{3}$ is not a solution. ∎

The Addition Principle

Consider the equation

$$x = 7.$$

We can easily see that the solution of this equation is 7. Replacing x with 7, we get

$$7 = 7, \quad \text{which is true.}$$

Now consider the equation

$$x + 6 = 13.$$

In Example 1, we found that the solution of $x + 6 = 13$ is also 7. Although the solution of $x = 7$ may seem more obvious, because $x + 6 = 13$ and $x = 7$ have identical solutions, the equations are said to be **equivalent**.

Equivalent Equations

Equations with the same solutions are called *equivalent equations*.

STUDENT NOTES

Be sure to remember the difference between an expression and an equation. For example, $5a - 10$ and $5(a - 2)$ are *equivalent expressions* because they represent the same value for all replacements for a. The *equations* $5a = 10$ and $a = 2$ are *equivalent* because they have the same solution, 2.

There are principles that enable us to begin with one equation and end up with an equivalent equation, like $x = 7$, for which the solution is obvious. One such principle concerns addition. The equation $a = b$ says that a and b stand for the same number. Suppose this is true, and some number c is added to a. We get the same result if we add c to b, because a and b are the same number.

The Addition Principle

For any real numbers a, b, and c,

$$a = b \quad \text{is equivalent to} \quad a + c = b + c.$$

To visualize the addition principle, consider a balance similar to one a jeweler might use. When the two sides of a balance hold equal weight, the balance is level. If weight is then added or removed, equally, on both sides, the balance will remain level.

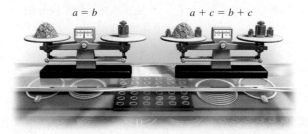

$a = b \qquad\qquad a + c = b + c$

When using the addition principle, we often say that we "add the same number to both sides of an equation." We can also "subtract the same number from both sides," since subtraction can be regarded as the addition of an opposite.

EXAMPLE **3**

Solve: $x + 5 = -7$.

SOLUTION   We can add any number we like to both sides. Since -5 is the opposite, or additive inverse, of 5, we add -5 to each side:

$$x + 5 = -7$$
$$x + 5 - 5 = -7 - 5 \qquad \text{Using the addition principle: adding } -5 \text{ to both sides or subtracting 5 from both sides}$$
$$x + 0 = -12 \qquad \text{Simplifying; } x + 5 - 5 = x + 5 + (-5) = x + 0$$
$$x = -12. \qquad \text{Using the identity property of 0}$$

The equation $x = -12$ is equivalent to the equation $x + 5 = -7$ by the addition principle, so the solution of $x = -12$ is the solution of $x + 5 = -7$.

It is obvious that the solution of $x = -12$ is the number -12. To check the answer in the original equation, we substitute.

Check:
$$\frac{x + 5 = -7}{-12 + 5 \mid -7}$$
$$-7 \overset{?}{=} -7 \quad \text{TRUE} \qquad -7 = -7 \text{ is true.}$$

The solution of the original equation is -12.   **TRY EXERCISE** ⟩ 11

In Example 3, note that because we added the *opposite*, or *additive inverse*, of 5, the left side of the equation simplified to x plus the *additive identity*, 0, or simply x. These steps effectively replaced the 5 on the left with a 0. To solve $x + a = b$ for x, we add $-a$ to (or subtract a from) both sides.

EXAMPLE **4**

Solve: $-6.5 = y - 8.4$.

SOLUTION   The variable is on the right side this time. We can isolate y by adding 8.4 to each side:

$$-6.5 = y - 8.4 \qquad y - 8.4 \text{ can be regarded as } y + (-8.4).$$
$$-6.5 + 8.4 = y - 8.4 + 8.4 \qquad \text{Using the addition principle: Adding 8.4 to both sides "eliminates" } -8.4 \text{ on the right side.}$$
$$1.9 = y. \qquad y - 8.4 + 8.4 = y + (-8.4) + 8.4 = y + 0 = y$$

STUDENT NOTES

We can also think of "undoing" operations in order to isolate a variable. In Example 4, we began with $y - 8.4$ on the right side. To undo the subtraction, we *add* 8.4.

Check:
$$\frac{-6.5 = y - 8.4}{-6.5 \mid 1.9 - 8.4}$$
$$-6.5 \overset{?}{=} -6.5 \qquad \text{TRUE} \qquad -6.5 = -6.5 \text{ is true.}$$

The solution is 1.9.   **TRY EXERCISE** ⟩ 15

Note that the equations $a = b$ and $b = a$ have the same meaning. Thus, $-6.5 = y - 8.4$ could have been rewritten as $y - 8.4 = -6.5$.

The Multiplication Principle

A second principle for solving equations concerns multiplying. Suppose a and b are equal. If a and b are multiplied by some number c, then ac and bc will also be equal.

> ### The Multiplication Principle
> For any real numbers a, b, and c, with $c \neq 0$,
> $$a = b \quad \text{is equivalent to} \quad a \cdot c = b \cdot c.$$

EXAMPLE 5

Solve: $\frac{5}{4}x = 10$.

SOLUTION   We can multiply both sides by any nonzero number we like. Since $\frac{4}{5}$ is the reciprocal of $\frac{5}{4}$, we decide to multiply both sides by $\frac{4}{5}$:

$$\frac{5}{4}x = 10$$

$$\frac{4}{5} \cdot \frac{5}{4}x = \frac{4}{5} \cdot 10 \qquad \text{Using the multiplication principle: Multiplying both sides by } \tfrac{4}{5} \text{ “eliminates” the } \tfrac{5}{4} \text{ on the left.}$$

$$1 \cdot x = 8 \qquad \text{Simplifying}$$

$$x = 8. \qquad \text{Using the identity property of 1}$$

Check:
$$\frac{5}{4}x = 10$$

$$\frac{\frac{5}{4} \cdot 8 \ \Big|\ 10}{}$$
$$\frac{40}{4} \qquad \text{Think of 8 as } \tfrac{8}{1}.$$
$$10 \overset{?}{=} 10 \quad \text{TRUE} \qquad 10 = 10 \text{ is true.}$$

The solution is 8.

TRY EXERCISE 49

In Example 5, to get x alone, we multiplied by the *reciprocal*, or *multiplicative inverse* of $\frac{5}{4}$. We then simplified the left-hand side to x times the *multiplicative identity*, 1, or simply x. These steps effectively replaced the $\frac{5}{4}$ on the left with 1.

Because division is the same as multiplying by a reciprocal, the multiplication principle also tells us that we can "divide both sides by the same nonzero number." That is,

$$\text{if } a = b, \text{ then } \quad \frac{1}{c} \cdot a = \frac{1}{c} \cdot b \quad \text{and} \quad \frac{a}{c} = \frac{b}{c} \qquad (\text{provided } c \neq 0).$$

In a product like $3x$, the multiplier 3 is called the **coefficient**. *When the coefficient of the variable is an integer or a decimal, it is usually easiest to solve an equation by dividing on both sides. When the coefficient is in fraction notation, it is usually easiest to multiply by the reciprocal.*

EXAMPLE 6

Solve: **(a)** $-4x = 9$; **(b)** $-x = 5$; **(c)** $\dfrac{2y}{9} = \dfrac{8}{3}$.

SOLUTION

a) In $-4x = 9$, the coefficient of x is an integer, so we *divide* on both sides:

$$\frac{-4x}{-4} = \frac{9}{-4} \qquad \text{Using the multiplication principle: Dividing both sides by } -4 \text{ is the same as multiplying by } -\tfrac{1}{4}.$$

$$1 \cdot x = -\frac{9}{4} \qquad \text{Simplifying}$$

$$x = -\frac{9}{4}. \qquad \text{Using the identity property of 1}$$

STUDENT NOTES

In Example 6(a), we can think of undoing the multiplication $-4 \cdot x$ by *dividing* by -4.

Check: $-4x = 9$

$$\frac{-4\left(-\frac{9}{4}\right) \mid 9}{}$$

$9 \overset{?}{=} 9$ TRUE $9 = 9$ is true.

The solution is $-\frac{9}{4}$.

b) To solve an equation like $-x = 5$, remember that when an expression is multiplied or divided by -1, its sign is changed. Here we divide both sides by -1 to change the sign of $-x$:

$-x = 5$ Note that $-x = -1 \cdot x$.

$\dfrac{-x}{-1} = \dfrac{5}{-1}$ Dividing both sides by -1. (Multiplying by -1 would also work. Note that the reciprocal of -1 is -1.)

$x = -5.$ Note that $\dfrac{-x}{-1}$ is the same as $\dfrac{x}{1}$.

Check: $-x = 5$

$$\frac{-(-5) \mid 5}{}$$

$5 \overset{?}{=} 5$ TRUE $5 = 5$ is true.

The solution is -5.

c) To solve an equation like $\dfrac{2y}{9} = \dfrac{8}{3}$, we rewrite the left-hand side as $\dfrac{2}{9} \cdot y$ and then use the multiplication principle, multiplying by the reciprocal of $\dfrac{2}{9}$:

$$\frac{2y}{9} = \frac{8}{3}$$

$$\frac{2}{9} \cdot y = \frac{8}{3} \qquad \text{Rewriting } \frac{2y}{9} \text{ as } \frac{2}{9} \cdot y$$

$$\frac{9}{2} \cdot \frac{2}{9} \cdot y = \frac{9}{2} \cdot \frac{8}{3} \qquad \text{Multiplying both sides by } \frac{9}{2}$$

$$1y = \frac{3 \cdot \cancel{3} \cdot \cancel{2} \cdot 4}{\cancel{2} \cdot \cancel{3}} \qquad \text{Removing a factor equal to 1: } \frac{3 \cdot 2}{2 \cdot 3} = 1$$

$$y = 12.$$

Check: $$\frac{2y}{9} = \frac{8}{3}$$

$$\frac{\dfrac{2 \cdot 12}{9} \mid \dfrac{8}{3}}{}$$

$$\frac{24}{9}$$

$$\frac{8}{3} \overset{?}{=} \frac{8}{3} \quad \text{TRUE} \qquad \frac{8}{3} = \frac{8}{3} \text{ is true.}$$

The solution is 12.   TRY EXERCISE ▸ 35

Selecting the Correct Approach

It is important that you be able to determine which principle should be used to solve a particular equation.

EXAMPLE 7

Solve: **(a)** $-\frac{2}{3} + x = \frac{5}{2}$; **(b)** $12.6 = 3t$.

SOLUTION

a) To undo addition of $-\frac{2}{3}$, we subtract $-\frac{2}{3}$ from both sides. Subtracting $-\frac{2}{3}$ is the same as adding $\frac{2}{3}$.

$$-\frac{2}{3} + x = \frac{5}{2}$$

$$-\frac{2}{3} + x + \frac{2}{3} = \frac{5}{2} + \frac{2}{3} \qquad \text{Using the addition principle}$$

$$x = \frac{5}{2} + \frac{2}{3}$$

$$x = \frac{5}{2} \cdot \frac{3}{3} + \frac{2}{3} \cdot \frac{2}{2} \qquad \text{Finding a common denominator}$$

$$x = \frac{15}{6} + \frac{4}{6}$$

$$x = \frac{19}{6}$$

Check:

$$\begin{array}{c|c} -\frac{2}{3} + x = \frac{5}{2} \\ \hline -\frac{2}{3} + \frac{19}{6} & \frac{5}{2} \\ -\frac{4}{6} + \frac{19}{6} & \\ \frac{15}{6} & \\ \frac{5 \cdot \cancel{3}}{2 \cdot \cancel{3}} & \\ \frac{5}{2} \overset{?}{=} \frac{5}{2} \quad \text{TRUE} \end{array}$$

$-\frac{2}{3} \cdot \frac{2}{2} = -\frac{4}{6}$

Removing a factor equal to 1: $\frac{3}{3} = 1$

$\frac{5}{2} = \frac{5}{2}$ is true.

The solution is $\frac{19}{6}$.

b) To undo multiplication by 3, we either divide both sides by 3 or multiply both sides by $\frac{1}{3}$:

$$12.6 = 3t$$

$$\frac{12.6}{3} = \frac{3t}{3} \qquad \text{Using the multiplication principle}$$

$$4.2 = t. \qquad \text{Simplifying}$$

Check:

$$\begin{array}{c|c} 12.6 = 3t \\ \hline 12.6 & 3(4.2) \\ 12.6 \overset{?}{=} 12.6 \quad \text{TRUE} \end{array}$$

$12.6 = 12.6$ is true.

The solution is 4.2.

TRY EXERCISES 59 and 67

2.1 EXERCISE SET

↪ **Concept Reinforcement** *For each of Exercises 1–6, match the statement with the most appropriate choice from the column on the right.*

1. ____ The equations $x + 3 = 7$ and $6x = 24$

2. ____ The expressions $3(x - 2)$ and $3x - 6$

3. ____ A replacement that makes an equation true

4. ____ The role of 9 in $9ab$

5. ____ The principle used to solve $\frac{2}{3} \cdot x = -4$

6. ____ The principle used to solve $\frac{2}{3} + x = -4$

a) Coefficient

b) Equivalent expressions

c) Equivalent equations

d) The multiplication principle

e) The addition principle

f) Solution

For each of Exercises 7–10, match the equation with the step, from the column on the right, that would be used to solve the equation.

7. $6x = 30$

8. $x + 6 = 30$

9. $\frac{1}{6}x = 30$

10. $x - 6 = 30$

a) Add 6 to both sides.

b) Subtract 6 from both sides.

c) Multiply both sides by 6.

d) Divide both sides by 6.

To the student and the instructor: *The* **TRY EXERCISES** *for examples are indicated by a shaded block* ☐ *on the exercise number. Complete step-by-step solutions for these exercises appear online at www.pearsonhighered.com/ bittingerellenbogen.*

Solve using the addition principle. Don't forget to check!

11. $x + 10 = 21$

12. $t + 9 = 47$

13. $y + 7 = -18$

14. $x + 12 = -7$

15. $-6 = y + 25$

16. $-5 = x + 8$

17. $x - 18 = 23$

18. $x - 19 = 16$

19. $12 = -7 + y$

20. $15 = -8 + z$

21. $-5 + t = -11$

22. $-6 + y = -21$

23. $r + \frac{1}{3} = \frac{8}{3}$

24. $t + \frac{3}{8} = \frac{5}{8}$

25. $x - \frac{3}{5} = -\frac{7}{10}$

26. $x - \frac{2}{3} = -\frac{5}{6}$

27. $x - \frac{5}{6} = \frac{7}{8}$

28. $y - \frac{3}{4} = \frac{5}{6}$

29. $-\frac{1}{5} + z = -\frac{1}{4}$

30. $-\frac{2}{3} + y = -\frac{3}{4}$

31. $m - 2.8 = 6.3$

32. $y - 5.3 = 8.7$

33. $-9.7 = -4.7 + y$

34. $-7.8 = 2.8 + x$

Solve using the multiplication principle. Don't forget to check!

35. $8a = 56$

36. $6x = 72$

37. $84 = 7x$

38. $45 = 9t$

39. $-x = 38$

40. $100 = -x$

Aha! **41.** $-t = -8$

42. $-68 = -r$

43. $-7x = 49$

44. $-4x = 36$

45. $-1.3a = -10.4$

46. $-3.4t = -20.4$

47. $\frac{y}{8} = 11$

48. $\frac{a}{4} = 13$

49. $\frac{4}{5}x = 16$

50. $\frac{3}{4}x = 27$

51. $\frac{-x}{6} = 9$

52. $\frac{-t}{4} = 8$

53. $\frac{1}{9} = \frac{z}{-5}$

54. $\frac{2}{7} = \frac{x}{-3}$

Aha! **55.** $-\frac{3}{5}r = -\frac{3}{5}$

56. $-\frac{2}{5}y = -\frac{4}{15}$

57. $\frac{-3r}{2} = -\frac{27}{4}$

58. $\frac{5x}{7} = -\frac{10}{14}$

Solve. The icon ▤ *indicates an exercise designed to give practice using a calculator.*

59. $4.5 + t = -3.1$

60. $\frac{3}{4}x = 18$

61. $-8.2x = 20.5$

62. $t - 7.4 = -12.9$

63. $x - 4 = -19$

64. $y - 6 = -14$

65. $t - 3 = -8$

66. $t - 9 = -8$

67. $-12x = 14$

68. $-15x = 20$

69. $48 = -\frac{3}{8}y$

70. $14 = t + 27$

71. $a - \frac{1}{6} = -\frac{2}{3}$

72. $-\frac{x}{6} = \frac{2}{9}$

73. $-24 = \frac{8x}{5}$

74. $\frac{1}{5} + y = -\frac{3}{10}$

75. $-\frac{4}{3}t = -12$

76. $\frac{17}{35} = -x$

▤ **77.** $-483.297 = -794.053 + t$

▤ **78.** $-0.2344x = 2028.732$

✍ **79.** When solving an equation, how do you determine what number to add, subtract, multiply, or divide by on both sides of that equation?

✍ **80.** What is the difference between equivalent expressions and equivalent equations?

Skill Review

To prepare for Section 2.2, review the rules for order of operations (Section 1.8).

Simplify. [1.8]

81. $3 \cdot 4 - 18$

82. $14 - 2(7 - 1)$

83. $16 \div (2 - 3 \cdot 2) + 5$

84. $12 - 5 \cdot 2^3 + 4 \cdot 3$

Synthesis

✍ **85.** To solve $-3.5 = 14t$, Anita adds 3.5 to both sides. Will this form an equivalent equation? Will it help solve the equation? Explain.

✍ **86.** Explain why it is not necessary to state a subtraction principle: For any real numbers a, b, and c, $a = b$ is equivalent to $a - c = b - c$.

Some equations, like $3 = 7$ or $x + 2 = x + 5$, have no solution and are called **contradictions**. Other equations, like $7 = 7$ or $2x = 2x$, are true for all numbers and are called **identities**. Solve each of the following and if an identity or contradiction is found, state this.

87. $2x = x + x$

88. $x + 5 + x = 2x$

89. $9x = 0$

90. $4x - x = 2x + x$

91. $x + 8 = 3 + x + 7$

92. $3x = 0$

Aha! **93.** $2|x| = -14$

94. $|3x| = 12$

Solve for x. Assume $a, c, m \neq 0$.

95. $mx = 11.6m$

96. $x - 4 + a = a$

97. $cx + 5c = 7c$

98. $c \cdot \dfrac{21}{a} = \dfrac{7cx}{2a}$

99. $7 + |x| = 30$

100. $ax - 3a = 5a$

101. If $t - 3590 = 1820$, find $t + 3590$.

102. If $n + 268 = 124$, find $n - 268$.

103. Lydia makes a calculation and gets an answer of 22.5. On the last step, she multiplies by 0.3 when she should have divided by 0.3. What should the correct answer be?

104. Are the equations $x = 5$ and $x^2 = 25$ equivalent? Why or why not?

2.2   Using the Principles Together

Applying Both Principles ▪ Combining Like Terms ▪ Clearing Fractions and Decimals

An important strategy for solving new problems is to find a way to make a new problem look like a problem that we already know how to solve. This is precisely the approach taken in this section. You will find that the last steps of the examples in this section are nearly identical to the steps used for solving the equations of Section 2.1. What is new in this section appears in the early steps of each example.

Applying Both Principles

The addition and multiplication principles, along with the laws discussed in Chapter 1, are our tools for solving equations. In this section, we will find that the sequence and manner in which these tools are used is especially important.

EXAMPLE 1   Solve: $5 + 3x = 17$.

SOLUTION   Were we to evaluate $5 + 3x$, the rules for the order of operations direct us to *first* multiply by 3 and *then* add 5. Because of this, we can isolate $3x$ and then x by reversing these operations: We first subtract 5 from both sides and then divide both sides by 3. Our goal is an equivalent equation of the form $x = a$.

$$5 + 3x = 17$$

$$5 + 3x - 5 = 17 - 5 \qquad \text{Using the addition principle: subtracting 5 from both sides (adding } -5\text{)}$$

$$5 + (-5) + 3x = 12 \qquad \text{Using a commutative law. Try to perform this step mentally.}$$

Isolate the x-term.
$$3x = 12 \qquad \text{Simplifying}$$

$$\frac{3x}{3} = \frac{12}{3} \qquad \text{Using the multiplication principle: dividing both sides by 3 (multiplying by } \tfrac{1}{3}\text{)}$$

Isolate x.
$$x = 4 \qquad \text{Simplifying}$$

Check:

$$5 + 3x = 17$$

$5 + 3 \cdot 4$	17
$5 + 12$	
$17 \overset{?}{=} 17$	TRUE

We use the rules for order of operations: Find the product, $3 \cdot 4$, and then add.

The solution is 4.

 TRY EXERCISE 7

EXAMPLE 2

Solve: $\frac{4}{3}x - 7 = 1$.

SOLUTION   In $\frac{4}{3}x - 7$, we multiply first and then subtract. To reverse these steps, we first add 7 and then either divide by $\frac{4}{3}$ or multiply by $\frac{3}{4}$.

$$\frac{4}{3}x - 7 = 1$$

$$\frac{4}{3}x - 7 + 7 = 1 + 7 \qquad \text{Adding 7 to both sides}$$

$$\frac{4}{3}x = 8$$

$$\frac{3}{4} \cdot \frac{4}{3}x = \frac{3}{4} \cdot 8 \qquad \text{Multiplying both sides by } \tfrac{3}{4}$$

$$1 \cdot x = \frac{3 \cdot 4 \cdot 2}{4} \left.\begin{array}{c} \\ \\ \end{array}\right\} \text{Simplifying}$$

$$x = 6$$

STUDY SKILLS

Use the Answer Section Carefully

When using the answers listed at the back of this book, try not to "work backward" from the answer. If you frequently require two or more attempts to answer an exercise correctly, you probably need to work more carefully and/or reread the section preceding the exercise set. Remember that on quizzes and tests you have only one attempt per problem and no answer section to consult.

Check:

$$\frac{4}{3}x - 7 = 1$$

$\frac{4}{3} \cdot 6 - 7$	1
$8 - 7$	
$1 \overset{?}{=} 1$	TRUE

The solution is 6.

TRY EXERCISE 27

EXAMPLE 3

Solve: $45 - t = 13$.

SOLUTION   We have

$$45 - t = 13$$

$$45 - t - 45 = 13 - 45 \qquad \text{Subtracting 45 from both sides}$$

$$\left.\begin{array}{c} 45 + (-t) + (-45) = 13 - 45 \\ 45 + (-45) + (-t) = 13 - 45 \end{array}\right\} \quad \text{Try to do these steps mentally.}$$

$$-t = -32 \qquad \text{Try to go directly to this step.}$$

$$(-1)(-t) = (-1)(-32) \qquad \begin{array}{l} \text{Multiplying both sides by } -1 \\ \text{(Dividing by } -1 \text{ would also} \\ \text{work.)} \end{array}$$

$$t = 32.$$

Check:

$$45 - t = 13$$

$45 - 32$	13
$13 \overset{?}{=} 13$	TRUE

The solution is 32.

TRY EXERCISE 19

As our skills improve, certain steps can be streamlined.

EXAMPLE 4

Solve: $16.3 - 7.2y = -8.18$.

SOLUTION   We have

$$16.3 - 7.2y = -8.18$$

$$16.3 - 7.2y - 16.3 = -8.18 - 16.3 \qquad \text{Subtracting 16.3 from both sides}$$

$$-7.2y = -24.48 \qquad \text{Simplifying}$$

$$\frac{-7.2y}{-7.2} = \frac{-24.48}{-7.2} \qquad \text{Dividing both sides by } -7.2$$

$$y = 3.4. \qquad \text{Simplifying}$$

Check:

$$\begin{array}{c|c} 16.3 - 7.2y = -8.18 \\ \hline 16.3 - 7.2(3.4) & -8.18 \\ 16.3 - 24.48 & \\ -8.18 \stackrel{?}{=} -8.18 & \text{TRUE} \end{array}$$

The solution is 3.4.

TRY EXERCISE 23

Combining Like Terms

If like terms appear on the same side of an equation, we combine them and then solve. Should like terms appear on both sides of an equation, we can use the addition principle to rewrite all like terms on one side.

EXAMPLE 5

Solve.

a) $3x + 4x = -14$ **b)** $-x + 5 = -8x + 6$

c) $6x + 5 - 7x = 10 - 4x + 7$ **d)** $2 - 5(x + 5) = 3(x - 2) - 1$

SOLUTION

a) $3x + 4x = -14$

$$7x = -14 \qquad \text{Combining like terms}$$

$$\frac{7x}{7} = \frac{-14}{7} \qquad \text{Dividing both sides by 7}$$

$$x = -2 \qquad \text{Simplifying}$$

The check is left to the student. The solution is -2.

b) To solve $-x + 5 = -8x + 6$, we must first write only variable terms on one side and only constant terms on the other. This can be done by subtracting 5 from both sides, to get all constant terms on the right, and adding $8x$ to both sides, to get all variable terms on the left.

$$-x + 5 = -8x + 6$$

Isolate variable terms on one side and constant terms on the other side.

$$-x + 8x + 5 = -8x + 8x + 6 \qquad \text{Adding } 8x \text{ to both sides}$$

$$7x + 5 = 6 \qquad \text{Simplifying}$$

$$7x + 5 - 5 = 6 - 5 \qquad \text{Subtracting 5 from both sides}$$

$$7x = 1 \qquad \text{Combining like terms}$$

$$\frac{7x}{7} = \frac{1}{7} \qquad \text{Dividing both sides by 7}$$

$$x = \frac{1}{7}$$

The check is left to the student. The solution is $\frac{1}{7}$.

Most graphing calculators have a TABLE feature that lists the value of a variable expression for different choices of x. For example, to evaluate $6x + 5 - 7x$ for $x = 0, 1, 2, \ldots$, we first use (Y=) to enter $6x + 5 - 7x$ as y_1. We then use (2ND) (TBLSET) to specify $\text{TblStart} = 0$, $\Delta\text{Tbl} = 1$, and select AUTO twice. By pressing (2ND) (TABLE), we can generate a table in which the value of $6x + 5 - 7x$ is listed for values of x starting at 0 and increasing by ones.

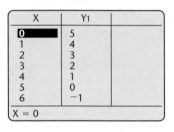

X	Y₁	
0	5	
1	4	
2	3	
3	2	
4	1	
5	0	
6	−1	
X = 0		

1. Create the above table on your graphing calculator. Scroll up and down to extend the table.
2. Enter $10 - 4x + 7$ as y_2. Your table should now have three columns.
3. For what x-value is y_1 the same as y_2? Compare this with the solution of Example 5(c). Is this a reliable way to solve equations? Why or why not?

c)
$$6x + 5 - 7x = 10 - 4x + 7$$

$$-x + 5 = 17 - 4x \qquad \text{Combining like terms within each side}$$

$$-x + 5 + 4x = 17 - 4x + 4x \qquad \text{Adding } 4x \text{ to both sides}$$

$$5 + 3x = 17 \qquad \text{Simplifying. This is identical to Example 1.}$$

$$3x = 12 \qquad \text{Subtracting 5 from both sides}$$

$$\frac{3x}{3} = \frac{12}{3} \qquad \text{Dividing both sides by 3}$$

$$x = 4$$

Check:

$$
\begin{array}{c|c}
6x + 5 - 7x &= 10 - 4x + 7 \\
\hline
6 \cdot 4 + 5 - 7 \cdot 4 & 10 - 4 \cdot 4 + 7 \\
24 + 5 - 28 & 10 - 16 + 7 \\
1 \overset{?}{=} 1 & \qquad \text{TRUE}
\end{array}
$$

The student can confirm that 4 checks and is the solution.

d)
$$2 - 5(x + 5) = 3(x - 2) - 1$$

$$2 - 5x - 25 = 3x - 6 - 1 \qquad \text{Using the distributive law. This is now similar to part (c) above.}$$

$$-5x - 23 = 3x - 7 \qquad \text{Combining like terms on each side}$$

$$
\left.
\begin{array}{l}
-5x - 23 + 7 = 3x \\
-23 + 7 = 3x + 5x
\end{array}
\right\}
\quad
\begin{array}{l}
\text{Adding 7 and } 5x \text{ to both sides. This isolates} \\
\text{the } x\text{-terms on one side and the constant} \\
\text{terms on the other.}
\end{array}
$$

$$-16 = 8x \qquad \text{Simplifying}$$

$$\frac{-16}{8} = \frac{8x}{8} \qquad \text{Dividing both sides by 8}$$

$$-2 = x \qquad \text{This is equivalent to } x = -2.$$

The student can confirm that -2 checks and is the solution.

> **TRY EXERCISE** 39

Clearing Fractions and Decimals

Equations are generally easier to solve when they do not contain fractions or decimals. The multiplication principle can be used to "clear" fractions or decimals, as shown here.

Clearing Fractions	Clearing Decimals
$\frac{1}{2}x + 5 = \frac{3}{4}$	$2.3x + 7 = 5.4$
$4\left(\frac{1}{2}x + 5\right) = 4 \cdot \frac{3}{4}$	$10(2.3x + 7) = 10 \cdot 5.4$
$2x + 20 = 3$	$23x + 70 = 54$

In each case, the resulting equation is equivalent to the original equation, but easier to solve.

The easiest way to clear an equation of fractions is to multiply *both sides* of the equation by the smallest, or *least*, common denominator of the fractions in the equation.

EXAMPLE 6 Solve: **(a)** $\frac{2}{3}x - \frac{1}{6} = 2x$; **(b)** $\frac{2}{5}(3x + 2) = 8$.

SOLUTION

a) We multiply both sides by 6, the least common denominator of $\frac{2}{3}$ and $\frac{1}{6}$.

$$6\left(\frac{2}{3}x - \frac{1}{6}\right) = 6 \cdot 2x \qquad \text{Multiplying both sides by 6}$$

$$6 \cdot \frac{2}{3}x - 6 \cdot \frac{1}{6} = 6 \cdot 2x \longleftarrow$$

> **CAUTION!** Be sure the distributive law is used to multiply *all* the terms by 6.

$$4x - 1 = 12x \qquad \text{Simplifying. Note that the fractions are cleared: } 6 \cdot \frac{2}{3} = 4, 6 \cdot \frac{1}{6} = 1, \text{ and } 6 \cdot 2 = 12.$$

$$-1 = 8x \qquad \text{Subtracting } 4x \text{ from both sides}$$

$$\frac{-1}{8} = \frac{8x}{8} \qquad \text{Dividing both sides by 8}$$

$$-\frac{1}{8} = x$$

The student can confirm that $-\frac{1}{8}$ checks and is the solution.

b) To solve $\frac{2}{5}(3x + 2) = 8$, we can multiply both sides by $\frac{5}{2}$ (or divide by $\frac{2}{5}$) to "undo" the multiplication by $\frac{2}{5}$ on the left side.

$$\frac{5}{2} \cdot \frac{2}{5}(3x + 2) = \frac{5}{2} \cdot 8 \qquad \text{Multiplying both sides by } \frac{5}{2}$$

$$3x + 2 = 20 \qquad \text{Simplifying; } \frac{5}{2} \cdot \frac{2}{5} = 1 \text{ and } \frac{5}{2} \cdot \frac{8}{1} = 20$$

$$3x = 18 \qquad \text{Subtracting 2 from both sides}$$

$$x = 6 \qquad \text{Dividing both sides by 3}$$

The student can confirm that 6 checks and is the solution.

▶ **TRY EXERCISE** 63

To clear an equation of decimals, we count the greatest number of decimal places in any one number. If the greatest number of decimal places is 1, we multiply both sides by 10; if it is 2, we multiply by 100; and so on. This procedure is the same as multiplying by the least common denominator after converting the decimals to fractions.

EXAMPLE 7 Solve: $16.3 - 7.2y = -8.18$.

SOLUTION The greatest number of decimal places in any one number is *two*. Multiplying by 100 will clear all decimals.

$$100(16.3 - 7.2y) = 100(-8.18) \qquad \text{Multiplying both sides by 100}$$

$$100(16.3) - 100(7.2y) = 100(-8.18) \qquad \text{Using the distributive law}$$

$$1630 - 720y = -818 \qquad \text{Simplifying}$$

$$-720y = -818 - 1630 \qquad \text{Subtracting 1630 from both sides}$$

$$-720y = -2448 \qquad \text{Combining like terms}$$

$$y = \frac{-2448}{-720} \qquad \text{Dividing both sides by } -720$$

$$y = 3.4$$

STUDENT NOTES ———

Compare the steps of Examples 4 and 7. Note that although the two approaches differ, they yield the same solution. Whenever you can use two approaches to solve a problem, try to do so, both as a check and as a valuable learning experience.

In Example 4, the same solution was found without clearing decimals. Finding the same answer in two ways is a good check. The solution is 3.4.

▶ **TRY EXERCISE** 69

An Equation-Solving Procedure

1. Use the multiplication principle to clear any fractions or decimals. (This is optional, but can ease computations. See Examples 6 and 7.)
2. If necessary, use the distributive law to remove parentheses. Then combine like terms on each side. (See Example 5.)
3. Use the addition principle, as needed, to isolate all variable terms on one side. Then combine like terms. (See Examples 1–7.)
4. Multiply or divide to solve for the variable, using the multiplication principle. (See Examples 1–7.)
5. Check all possible solutions in the original equation. (See Examples 1–4.)

2.2   EXERCISE SET

For Extra Help   MyMathLab   Math XL PRACTICE   WATCH   DOWNLOAD

🖐 **Concept Reinforcement**   *In each of Exercises 1–6, match the equation with an equivalent equation from the column on the right that could be the next step in finding a solution.*

1. ____ $3x - 1 = 7$
2. ____ $4x + 5x = 12$
3. ____ $6(x - 1) = 2$
4. ____ $7x = 9$
5. ____ $4x = 3 - 2x$
6. ____ $8x - 5 = 6 - 2x$

a) $6x - 6 = 2$
b) $4x + 2x = 3$
c) $3x = 7 + 1$
d) $8x + 2x = 6 + 5$
e) $9x = 12$
f) $x = \frac{9}{7}$

Solve and check.

7. $2x + 9 = 25$
8. $3x - 11 = 13$
9. $6z + 5 = 47$
10. $5z + 2 = 57$
11. $7t - 8 = 27$
12. $6x - 5 = 2$
13. $3x - 9 = 1$
14. $5x - 9 = 41$
15. $8z + 2 = -54$
16. $4x + 3 = -21$
17. $-37 = 9t + 8$
18. $-39 = 1 + 5t$
19. $12 - t = 16$
20. $9 - t = 21$
21. $-6z - 18 = -132$
22. $-7x - 24 = -129$
23. $5.3 + 1.2n = 1.94$
24. $6.4 - 2.5n = 2.2$
25. $32 - 7x = 11$
26. $27 - 6x = 99$
27. $\frac{3}{5}t - 1 = 8$
28. $\frac{2}{3}t - 1 = 5$

29. $6 + \frac{7}{2}x = -15$
30. $6 + \frac{5}{4}x = -4$
31. $-\dfrac{4a}{5} - 8 = 2$
32. $-\dfrac{8a}{7} - 2 = 4$
33. $-5z - 6z = -44$
34. $-3z + 8z = 45$
35. $4x - 6 = 6x$
36. $9y - 35 = 4y$
37. $2 - 5y = 26 - y$
38. $6x - 5 = 7 + 2x$
39. $7(2a - 1) = 21$
40. $5(3 - 3t) = 30$

Aha! 41. $11 = 11(x + 1)$
42. $9 = 3(5x - 2)$
43. $2(3 + 4m) - 6 = 48$
44. $3(5 + 3m) - 8 = 7$
45. $2r + 8 = 6r + 10$
46. $3b - 2 = 7b + 4$
47. $6x + 3 = 2x + 3$
48. $5y + 3 = 2y + 15$
49. $5 - 2x = 3x - 7x + 25$
50. $10 - 3x = x - 2x + 40$
51. $7 + 3x - 6 = 3x + 5 - x$
52. $5 + 4x - 7 = 4x - 2 - x$
53. $4y - 4 + y + 24 = 6y + 20 - 4y$

54. $5y - 10 + y = 7y + 18 - 5y$

55. $19 - 3(2x - 1) = 7$

56. $5(d + 4) = 7(d - 2)$

57. $7(5x - 2) = 6(6x - 1)$

58. $5(t + 1) + 8 = 3(t - 2) + 6$

59. $2(3t + 1) - 5 = t - (t + 2)$

60. $4x - (x + 6) = 5(3x - 1) + 8$

61. $19 - (2x + 3) = 2(x + 3) + x$

62. $13 - (2c + 2) = 2(c + 2) + 3c$

Clear fractions or decimals, solve, and check.

63. $\frac{2}{3} + \frac{1}{4}t = 2$

64. $-\frac{5}{6} + x = -\frac{1}{2} - \frac{2}{3}$

65. $\frac{2}{3} + 4t = 6t - \frac{2}{15}$

66. $\frac{1}{2} + 4m = 3m - \frac{5}{2}$

67. $\frac{1}{3}x + \frac{2}{5} = \frac{4}{5} + \frac{3}{5}x - \frac{2}{3}$

68. $1 - \frac{2}{3}y = \frac{9}{5} - \frac{1}{5}y + \frac{3}{5}$

69. $2.1x + 45.2 = 3.2 - 8.4x$

70. $0.91 - 0.2z = 1.23 - 0.6z$

71. $0.76 + 0.21t = 0.96t - 0.49$

72. $1.7t + 8 - 1.62t = 0.4t - 0.32 + 8$

73. $\frac{2}{5}x - \frac{3}{2}x = \frac{3}{4}x + 3$

74. $\frac{5}{16}y + \frac{3}{8}y = 2 + \frac{1}{4}y$

75. $\frac{1}{3}(2x - 1) = 7$

76. $\frac{1}{5}(4x - 1) = 7$

77. $\frac{3}{4}(3t - 4) = 15$

78. $\frac{3}{2}(2x + 5) = -\frac{15}{2}$

79. $\frac{1}{6}\left(\frac{3}{4}x - 2\right) = -\frac{1}{5}$

80. $\frac{2}{3}\left(\frac{7}{8} - 4x\right) - \frac{5}{8} = \frac{3}{8}$

81. $0.7(3x + 6) = 1.1 - (x - 3)$

82. $0.9(2x - 8) = 4 - (x + 5)$

83. $a + (a - 3) = (a + 2) - (a + 1)$

84. $0.8 - 4(b - 1) = 0.2 + 3(4 - b)$

85. Tyla solves $45 - t = 13$ (Example 3) by adding $t - 13$ to both sides. Is this approach preferable to the one used in Example 3? Why or why not?

86. Why must the rules for the order of operations be understood before solving the equations in this section?

Skill Review

To prepare for Section 2.3, review evaluating algebraic expressions (Section 1.8).

Evaluate. [1.8]

87. $3 - 5a$, for $a = 2$

88. $12 \div 4 \cdot t$, for $t = 5$

89. $7x - 2x$, for $x = -3$

90. $t(8 - 3t)$, for $t = -2$

Synthesis

91. What procedure would you use to solve an equation like $0.23x + \frac{17}{3} = -0.8 + \frac{3}{4}x$? Could your procedure be streamlined? If so, how?

92. Dave is determined to solve $3x + 4 = -11$ by first using the multiplication principle to "eliminate" the 3. How should he proceed and why?

Solve. If an equation is an identity or a contradiction (see p. 85), state this.

93. $8.43x - 2.5(3.2 - 0.7x) = -3.455x + 9.04$

94. $0.008 + 9.62x - 42.8 = 0.944x + 0.0083 - x$

95. $-2[3(x - 2) + 4] = 4(5 - x) - 2x$

96. $0 = t - (-6) - (-7t)$

97. $3(x + 5) = 3(5 + x)$

98. $5(x - 7) = 3(x - 2) + 2x$

99. $2x(x + 5) - 3(x^2 + 2x - 1) = 9 - 5x - x^2$

100. $x(x - 4) = 3x(x + 1) - 2(x^2 + x - 5)$

101. $9 - 3x = 2(5 - 2x) - (1 - 5x)$

102. $2(7 - x) - 20 = 7x - 3(2 + 3x)$

Aha! **103.** $[7 - 2(8 \div (-2))]x = 0$

104. $\dfrac{x}{14} - \dfrac{5x + 2}{49} = \dfrac{3x - 4}{7}$

105. $\dfrac{5x + 3}{4} + \dfrac{25}{12} = \dfrac{5 + 2x}{3}$

COLLABORATIVE CORNER

Step-by-Step Solutions

Focus: Solving linear equations
Time: 20 minutes
Group size: 3

In general, there is more than one correct sequence of steps for solving an equation. This makes it important that you write your steps clearly and logically so that others can follow your approach.

ACTIVITY

1. Each group member should select a different one of the following equations and, on a fresh sheet of paper, perform the first step of the solution.

$$4 - 3(x - 3) = 7x + 6(2 - x)$$
$$5 - 7[x - 2(x - 6)] = 3x + 4(2x - 7) + 9$$
$$4x - 7[2 + 3(x - 5) + x] = 4 - 9(-3x - 19)$$

2. Pass the papers around so that the second and third steps of each solution are performed by the other two group members. Before writing, make sure that the previous step is correct. If a mistake is discovered, return the problem to the person who made the mistake for repairs. Continue passing the problems around until all equations have been solved.

3. Each group should reach a consensus on what the three solutions are and then compare their answers to those of other groups.

2.3   Formulas

Evaluating Formulas • Solving for a Variable

Many applications of mathematics involve relationships among two or more quantities. An equation that represents such a relationship will use two or more letters and is known as a **formula**. Most of the letters in this book are variables, but some are constants. For example, c in $E = mc^2$ represents the speed of light.

Evaluating Formulas

EXAMPLE 1   *Event promotion.*   Event promoters use the formula

$$p = \frac{1.2x}{s}$$

to determine a ticket price p for an event with x dollars of expenses and s anticipated ticket sales. Grand Events expects expenses for an upcoming concert to be $80,000 and anticipates selling 4000 tickets. What should the ticket price be?

Source: *The Indianapolis Star,* 2/27/03

SOLUTION   We substitute 80,000 for x and 4000 for s in the formula and calculate p:

$$p = \frac{1.2x}{s} = \frac{1.2(80{,}000)}{4000} = 24.$$

The ticket price should be $24.

 TRY EXERCISE 1

Solving for a Variable

In the Northeast, the formula $B = 30a$ is used to determine the minimum furnace output B, in British thermal units (Btu's), for a well-insulated home with a square feet of flooring. Suppose that a contractor has an extra furnace and wants to determine the size of the largest (well-insulated) house in which it can be used. The contractor can substitute the amount of the furnace's output in Btu's—say, 63,000—for B, and then solve for a:

$$63{,}000 = 30a \qquad \text{Replacing } B \text{ with } 63{,}000$$
$$2100 = a. \qquad \text{Dividing both sides by 30}$$

The home should have no more than 2100 ft² of flooring.

Were these calculations to be performed for a variety of furnaces, the contractor would find it easier to first solve $B = 30a$ for a, and *then* substitute values for B. Solving for a variable can be done in much the same way that we solved equations in Sections 2.1 and 2.2.

EXAMPLE 2

Solve for a: $B = 30a$.

SOLUTION   We have

$$B = 30\overset{\downarrow}{a} \qquad \text{We want this letter alone.}$$
$$\frac{B}{30} = a. \qquad \text{Dividing both sides by 30}$$

The equation $a = B/30$ gives a quick, easy way to determine the floor area of the largest (well-insulated) house that a furnace supplying B Btu's could heat.

TRY EXERCISE 9

To see how solving a formula is just like solving an equation, compare the following. In (A), we solve as usual; in (B), we show steps but do not simplify; and in (C), we *cannot* simplify because a, b, and c are unknown.

A. $5x + 2 = 12$
$$5x = 12 - 2$$
$$5x = 10$$
$$x = \frac{10}{5} = 2$$

B. $5x + 2 = 12$
$$5x = 12 - 2$$
$$x = \frac{12 - 2}{5}$$

C. $ax + b = c$
$$ax = c - b$$
$$x = \frac{c - b}{a}$$

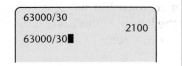
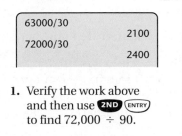

EXAMPLE **3**

Circumference of a circle. The formula $C = 2\pi r$ gives the *circumference C* of a circle with radius r. Solve for r.

SOLUTION The **circumference** is the distance around a circle.

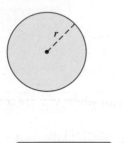

Given a radius r, we can use this equation to find a circle's circumference C.

Given a circle's circumference C, we can use this equation to find the radius r.

$$C = 2\pi r \qquad \text{We want this variable alone.}$$

$$\frac{C}{2\pi} = \frac{2\pi r}{2\pi} \qquad \text{Dividing both sides by } 2\pi$$

$$\frac{C}{2\pi} = r$$

TRY EXERCISE 13

EXAMPLE **4**

Solve for y: $3x - 4y = 10$.

SOLUTION There is one term that contains y, so we begin by isolating that term on one side of the equation.

$$3x - 4y = 10 \qquad \text{We want this variable alone.}$$

$$-4y = 10 - 3x \qquad \text{Subtracting } 3x \text{ from both sides}$$

$$-\tfrac{1}{4}(-4y) = -\tfrac{1}{4}(10 - 3x) \qquad \text{Multiplying both sides by } -\tfrac{1}{4}$$

$$y = -\tfrac{10}{4} + \tfrac{3}{4}x \qquad \text{Multiplying using the distributive law}$$

$$y = -\tfrac{5}{2} + \tfrac{3}{4}x \qquad \text{Simplifying the fraction}$$

TRY EXERCISE 33

EXAMPLE **5**

Nutrition. The number of calories K needed each day by a moderately active woman who weighs w pounds, is h inches tall, and is a years old, can be estimated using the formula

$$K = 917 + 6(w + h - a).^*$$

Solve for w.

SOLUTION We reverse the order in which the operations occur on the right side:

$$K = 917 + 6(w + h - a) \qquad \text{We want } w \text{ alone.}$$

$$K - 917 = 6(w + h - a) \qquad \text{Subtracting 917 from both sides}$$

$$\frac{K - 917}{6} = w + h - a \qquad \text{Dividing both sides by 6}$$

$$\frac{K - 917}{6} + a - h = w. \qquad \text{Adding } a \text{ and subtracting } h \text{ on both sides}$$

This formula can be used to estimate a woman's weight, if we know her age, height, and caloric needs.

TRY EXERCISE 43

STUDY SKILLS

Pace Yourself

Most instructors agree that it is better for a student to study for one hour four days in a week, than to study once a week for four hours. Of course, the total weekly study time will vary from student to student. It is common to expect an average of two hours of homework for each hour of class time.

---

*Based on information from M. Parker (ed.), *She Does Math!* (Washington D.C.: Mathematical Association of America, 1995), p. 96.

The above steps are similar to those used in Section 2.2 to solve equations. We use the addition and multiplication principles just as before. An important difference that we will see in the next example is that we will sometimes need to factor.

> **To Solve a Formula for a Given Variable**
> 1. If the variable for which you are solving appears in a fraction, use the multiplication principle to clear fractions.
> 2. Isolate the term(s), with the variable for which you are solving on one side of the equation.
> 3. If two or more terms contain the variable for which you are solving, factor the variable out.
> 4. Multiply or divide to solve for the variable in question.

We can also solve for a letter that represents a constant.

EXAMPLE   6

Surface area of a right circular cylinder.   The formula $A = 2\pi rh + 2\pi r^2$ gives the surface area A of a right circular cylinder of height h and radius r. Solve for π.

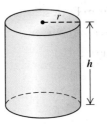

SOLUTION   We have

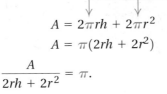

$$A = 2\pi rh + 2\pi r^2 \qquad \text{We want this letter alone.}$$
$$A = \pi(2rh + 2r^2) \qquad \text{Factoring}$$
$$\frac{A}{2rh + 2r^2} = \pi. \qquad \text{Dividing both sides by } 2rh + 2r^2,$$
$$\text{or multiplying both sides by } 1/(2rh + 2r^2)$$

We can also write this as

$$\pi = \frac{A}{2rh + 2r^2}.$$

> **TRY EXERCISE** 47

CAUTION!   Had we performed the following steps in Example 6, we would *not* have solved for π:

$$A = 2\pi rh + 2\pi r^2 \qquad \text{We want } \pi \text{ alone.}$$
$$A - 2\pi r^2 = 2\pi rh \qquad \text{Subtracting } 2\pi r^2 \text{ from both sides}$$

Two occurrences of π

$$\frac{A - 2\pi r^2}{2rh} = \pi. \qquad \text{Dividing both sides by } 2rh$$

The mathematics of each step is correct, but because π occurs on both sides of the formula, *we have not solved the formula for π.* Remember that the letter being solved for should be alone on one side of the equation, with no occurrence of that letter on the other side!

2.3   **EXERCISE SET**

For Extra Help   MyMathLab   Math XL PRACTICE   WATCH   DOWNLOAD

1. *Outdoor concerts.* The formula $d = 344t$ can be used to determine how far d, in meters, sound travels through room-temperature air in t seconds. At a large concert, fans near the back of the crowd experienced a 0.9-sec time lag between the time each word was pronounced on stage (as shown on large video monitors) and the time the sound reached their ears. How far were these fans from the stage?

2. *Furnace output.* Contractors in the Northeast use the formula $B = 30a$ to determine the minimum furnace output B, in British thermal units (Btu's), for a well-insulated house with a square feet of flooring. Determine the minimum furnace output for an 1800-ft^2 house that is well insulated.
   Source: U.S. Department of Energy

3. *College enrollment.* At many colleges, the number of "full-time-equivalent" students f is given by
   $$f = \frac{n}{15},$$
   where n is the total number of credits for which students have enrolled in a given semester. Determine the number of full-time-equivalent students on a campus in which students registered for a total of 21,345 credits.

4. *Distance from a storm.* The formula $M = \frac{1}{5}t$ can be used to determine how far M, in miles, you are from lightning when its thunder takes t seconds to reach your ears. If it takes 10 sec for the sound of thunder to reach you after you have seen the lightning, how far away is the storm?

5. *Federal funds rate.* The Federal Reserve Board sets a target f for the federal funds rate, that is, the interest rate that banks charge each other for overnight borrowing of Federal funds. This target rate can be estimated by
   $$f = 8.5 + 1.4(I - U),$$
   where I is the core inflation rate over the previous 12 months and U is the seasonally adjusted unemployment rate. If core inflation is 0.025 and unemployment is 0.044, what should the federal funds rate be?
   Source: Greg Mankiw, Harvard University, www.gregmankiw .blogspot.com/2006/06/what-would-alan-do.html

6. *Calorie density.* The calorie density D, in calories per ounce, of a food that contains c calories and weighs w ounces is given by
   $$D = \frac{c}{w}.^*$$
   Eight ounces of fat-free milk contains 84 calories. Find the calorie density of fat-free milk.

7. *Absorption of ibuprofen.* When 400 mg of the painkiller ibuprofen is swallowed, the number of milligrams n in the bloodstream t hours later (for $0 \le t \le 6$) is estimated by
   $$n = 0.5t^4 + 3.45t^3 - 96.65t^2 + 347.7t.$$
   How many milligrams of ibuprofen remain in the blood 1 hr after 400 mg has been swallowed?

8. *Size of a league schedule.* When all n teams in a league play every other team twice, a total of N games are played, where
   $$N = n^2 - n.$$
   If a soccer league has 7 teams and all teams play each other twice, how many games are played?

In Exercises 9–48, solve each formula for the indicated letter.

9. $A = bh$, for b
   (Area of parallelogram with base b and height h)

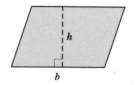

10. $A = bh$, for h

11. $d = rt$, for r
    (A distance formula, where d is distance, r is speed, and t is time)

Source: Nutrition Action Healthletter, March 2000, p. 9. Center for Science in the Public Interest, Suite 300; 1875 Connecticut Ave NW, Washington, D.C. 20008.

12. $d = rt$, for t

13. $I = Prt$, for P
(Simple-interest formula, where I is interest, P is principal, r is interest rate, and t is time)

14. $I = Prt$, for t

15. $H = 65 - m$, for m
(To determine the number of heating degree days H for a day with m degrees Fahrenheit as the average temperature)

16. $d = h - 64$, for h
(To determine how many inches d above average an h-inch-tall woman is)

17. $P = 2l + 2w$, for l
(Perimeter of a rectangle of length l and width w)

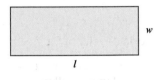

18. $P = 2l + 2w$, for w

19. $A = \pi r^2$, for π
(Area of a circle with radius r)

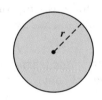

20. $A = \pi r^2$, for r^2

21. $A = \frac{1}{2}bh$, for h
(Area of a triangle with base b and height h)

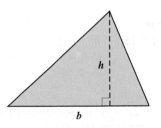

22. $A = \frac{1}{2}bh$, for b

23. $E = mc^2$, for c^2
(A relativity formula from physics)

24. $E = mc^2$, for m

25. $Q = \dfrac{c + d}{2}$, for d **26.** $Q = \dfrac{p - q}{2}$, for p

27. $A = \dfrac{a + b + c}{3}$, for b **28.** $A = \dfrac{a + b + c}{3}$, for c

29. $w = \dfrac{r}{f}$, for r
(To compute the wavelength w of a musical note with frequency f and speed of sound r)

30. $M = \dfrac{A}{s}$, for A
(To compute the Mach number M for speed A and speed of sound s)

31. $F = \dfrac{9}{5}C + 32$, for C
(To convert the Celsius temperature C to the Fahrenheit temperature F)

32. $M = \dfrac{5}{9}n + 18$, for n

33. $2x - y = 1$, for y

34. $3x - y = 7$, for y

35. $2x + 5y = 10$, for y

36. $3x + 2y = 12$, for y

37. $4x - 3y = 6$, for y

38. $5x - 4y = 8$, for y

39. $9x + 8y = 4$, for y

40. $x + 10y = 2$, for y

41. $3x - 5y = 8$, for y

42. $7x - 6y = 7$, for y

43. $z = 13 + 2(x + y)$, for x

44. $A = 115 + \dfrac{1}{2}(p + s)$, for s

45. $t = 27 - \dfrac{1}{4}(w - l)$, for l

46. $m = 19 - 5(x - n)$, for n

47. $A = at + bt$, for t

48. $S = rx + sx$, for x

49. *Area of a trapezoid.*   The formula
$$A = \tfrac{1}{2}ah + \tfrac{1}{2}bh$$
can be used to find the area A of a trapezoid with bases a and b and height h. Solve for h. (*Hint*: First clear fractions.)

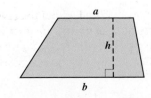

50. *Compounding interest.* The formula

$$A = P + Prt$$

is used to find the amount A in an account when simple interest is added to an investment of P dollars (see Exercise 13). Solve for P.

51. *Chess rating.* The formula

$$R = r + \frac{400(W - L)}{N}$$

is used to establish a chess player's rating R after that player has played N games, won W of them, and lost L of them. Here r is the average rating of the opponents. Solve for L.

Source: The U.S. Chess Federation

52. *Angle measure.* The angle measure S of a sector of a circle is given by

$$S = \frac{360A}{\pi r^2},$$

where r is the radius, A is the area of the sector, and S is in degrees. Solve for r^2.

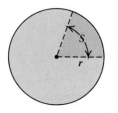

53. Naomi has a formula that allows her to convert Celsius temperatures to Fahrenheit temperatures. She needs a formula for converting Fahrenheit temperatures to Celsius temperatures. What advice can you give her?

54. Under what circumstances would it be useful to solve $d = rt$ for r? (See Exercise 11.)

Skill Review

Review simplifying expressions (Sections 1.6, 1.7, and 1.8).

Perform the indicated operations.

55. $-2 + 5 - (-4) - 17$ [1.6]

56. $-98 \div \frac{1}{2}$ [1.7]

Aha! **57.** $4.2(-11.75)(0)$ [1.7]

58. $(-2)^5$ [1.8]

Simplify. [1.8]

59. $20 \div (-4) \cdot 2 - 3$

60. $5|8 - (2 - 7)|$

Synthesis

61. The equations

$$P = 2l + 2w \quad \text{and} \quad w = \frac{P}{2} - l$$

are equivalent formulas involving the perimeter P, length l, and width w of a rectangle. Devise a problem for which the second of the two formulas would be more useful.

62. While solving $2A = ah + bh$ for h, Lea writes $\frac{2A - ah}{b} = h$. What is her mistake?

63. The Harris–Benedict formula gives the number of calories K needed each day by a moderately active man who weighs w kilograms, is h centimeters tall, and is a years old as

$$K = 21.235w + 7.75h - 10.54a + 102.3.$$

If Janos is moderately active, weighs 80 kg, is 190 cm tall, and needs to consume 2852 calories a day, how old is he?

64. *Altitude and temperature.* Air temperature drops about 1° Celsius (C) for each 100-m rise above ground level, up to 12 km. If the ground level temperature is t°C, find a formula for the temperature T at an elevation of h meters.

Source: *A Sourcebook of School Mathematics*, Mathematical Association of America, 1980

65. *Surface area of a cube.* The surface area A of a cube with side s is given by

$$A = 6s^2.$$

If a cube's surface area is 54 in^2, find the volume of the cube.

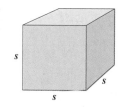

66. *Weight of a fish.* An ancient fisherman's formula for estimating the weight of a fish is

$$w = \frac{lg^2}{800},$$

where w is the weight, in pounds, l is the length, in inches, and g is the girth (distance around the midsection), in inches. Estimate the girth of a 700-lb yellow tuna that is 8 ft long.

67. *Dosage size.*   Clark's rule for determining the size of a particular child's medicine dosage c is

$$c = \frac{w}{a} \cdot d,$$

where w is the child's weight, in pounds, and d is the usual adult dosage for an adult weighing a pounds. Solve for a.

Source: Olsen, June Looby, et al., *Medical Dosage Calculations.* Redwood City, CA: Addison-Wesley, 1995

Solve each formula for the given letter.

68. $\dfrac{y}{z} \div \dfrac{z}{t} = 1$, for y

69. $ac = bc + d$, for c

70. $qt = r(s + t)$, for t

71. $3a = c - a(b + d)$, for a

72. *Furnace output.*   The formula

$$B = 50a$$

is used in New England to estimate the minimum furnace output B, in Btu's, for an old, poorly insulated house with a square feet of flooring. Find an equation for determining the number of Btu's saved by insulating an old house. (*Hint:* See Exercise 2.)

73. Revise the formula in Exercise 63 so that a man's weight in pounds (2.2046 lb = 1 kg) and his height in inches (0.3937 in. = 1 cm) are used.

74. Revise the formula in Example 5 so that a woman's weight in kilograms (2.2046 lb = 1 kg) and her height in centimeters (0.3937 in. = 1 cm) are used.

2.4   Applications with Percent

Converting Between Percent Notation and Decimal Notation   •   Solving Percent Problems

Percent problems arise so frequently in everyday life that most often we are not even aware of them. In this section, we will solve some real-world percent problems. Before doing so, however, we need to review a few basics.

Converting Between Percent Notation and Decimal Notation

Nutritionists recommend that no more than 30% of the calories in a person's diet come from fat. This means that of every 100 calories consumed, no more than 30 should come from fat. Thus, 30% is a ratio of 30 to 100.

Calories consumed

Calories from fat
30%

The percent symbol % means "per hundred." We can regard the percent symbol as part of a name for a number. For example,

$$30\% \quad \text{is defined to mean} \quad \frac{30}{100}, \quad \text{or} \quad 30 \times \frac{1}{100}, \quad \text{or} \quad 30 \times 0.01.$$

> **Percent Notation**
>
> $n\%$ means $\dfrac{n}{100}$,   or $n \times \dfrac{1}{100}$,   or $n \times 0.01$.

EXAMPLE   **1**   Convert to decimal notation: **(a)** 78%; **(b)** 1.3%.

SOLUTION

a) 78% = 78 × 0.01   Replacing % with × 0.01

= 0.78

b) 1.3% = 1.3 × 0.01   Replacing % with × 0.01

= 0.013

TRY EXERCISE   19

As shown above, multiplication by 0.01 simply moves the decimal point two places to the left.

To convert from percent notation to decimal notation, move the decimal point two places to the left and drop the percent symbol.

EXAMPLE   **2**   Convert the percent notation in the following sentence to decimal notation: Only 20% of teenagers get 8 hr of sleep a night.

Source: National Sleep Foundation

SOLUTION

20% = 20.0%   0.20.0   20% = 0.20, or simply 0.2

Move the decimal point two places to the left.

TRY EXERCISE   11

The procedure used in Examples 1 and 2 can be reversed:

0.38 = 38 × 0.01

= 38%.   Replacing × 0.01 with %

To convert from decimal notation to percent notation, move the decimal point two places to the right and write a percent symbol.

EXAMPLE   **3**   Convert to percent notation: **(a)** 1.27; **(b)** $\frac{1}{4}$; **(c)** 0.3.

SOLUTION

a) We first move the decimal point
two places to the right: 1.27.

and then write a % symbol: 127%   This is the same as multiplying 1.27 by 100 and writing %.

b) Note that $\frac{1}{4}$ = 0.25. We move the
decimal point two places to the
right: 0.25.

and then write a % symbol: 25%   Multiplying by 100 and writing %

c) We first move the decimal point
two places to the right (recall that
0.3 = 0.30):   0.30.

and then write a % symbol: 30%   Multiplying by 100 and writing %

TRY EXERCISE   33

Solving Percent Problems

In solving percent problems, we first *translate* the problem to an equation. Then we *solve* the equation using the techniques discussed in Sections 2.1–2.3. The key words in the translation are as follows.

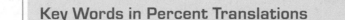

> **Key Words in Percent Translations**
>
> "**Of**" translates to " · " or " × ".   "**Is**" or "**Was**" translates to " = ".
> "**What**" translates to a variable.   "**%**" translates to "× $\frac{1}{100}$" or "× 0.01".

EXAMPLE 4

STUDENT NOTES ─────

A way of checking answers is by estimating as follows:

$$11\% \times 49 \approx 10\% \times 50$$
$$= 0.10 \times 50 = 5.$$

Since 5 is close to 5.39, our answer is reasonable.

What is 11% of 49?

SOLUTION

Translate:
What	is	11%	of	49?
↓	↓	↓	↓	↓
a	=	0.11	·	49

"of" means multiply; 11% = 0.11

$$a = 5.39$$

Thus, 5.39 is 11% of 49. The answer is 5.39.

TRY EXERCISE 51

EXAMPLE 5

3 is 16 percent of what?

SOLUTION

Translate:
3	is	16 percent	of	what?
↓	↓	↓	↓	↓
3	=	0.16	·	y

$$\frac{3}{0.16} = y \qquad \text{Dividing both sides by 0.16}$$

$$18.75 = y$$

Thus, 3 is 16 percent of 18.75. The answer is 18.75.

TRY EXERCISE 47

EXAMPLE 6

What percent of $50 is $34?

SOLUTION

Translate:
What percent	of	$50	is	$34?
↓		↓	↓	↓
n	·	50	=	34

$$n = \frac{34}{50} \qquad \text{Dividing both sides by 50}$$

$$n - 0.68 = 68\% \qquad \text{Converting to percent notation}$$

Thus, $34 is 68% of $50. The answer is 68%.

TRY EXERCISE 43

Examples 4–6 represent the three basic types of percent problems. Note that in all the problems, the following quantities are present:

- a percent, expressed in decimal notation in the translation,
- a base amount, indicated by "of" in the problem, and
- a percentage of the base, found by multiplying the base times the percent.

EXAMPLE 7

Discount stores. In 2006, there were 300 million people in the United States, and 62.2% of them lived within 5 mi of a Wal-Mart store. How many lived within 5 mi of a Wal-Mart store?

Source: *The Wall Street Journal, 9/25/06*

SOLUTION   We first reword and then translate. We let a = the number of people in the United States, in millions, who live within 5 mi of a Wal-Mart store.

Rewording: What   is   62.2%   of   300?
$\downarrow$ $\downarrow$ $\downarrow$ $\downarrow$ $\downarrow$
Translating: a =   0.622 $\times$ 300

The letter is by itself. To solve the equation, we need only multiply:

$$a = 0.622 \times 300 = 186.6.$$

Since 186.6 million is 62.2% of 300 million, we have found that in 2006 about 186.6 million people in the United States lived within 5 mi of a Wal-Mart store.

TRY EXERCISE ▶ 65

EXAMPLE 8

College enrollment. About 1.6 million students who graduated from high school in 2006 were attending college in the fall of 2006. This was 66% of all 2006 high school graduates. How many students graduated from high school in 2006?

Source: U.S. Bureau of Labor Statistics

SOLUTION   Before translating the problem to mathematics, we reword and let S represent the total number of students, in millions, who graduated from high school in 2006.

Rewording: 1.6   is   66%   of S.
$\downarrow$ $\downarrow$ $\downarrow$ $\downarrow$ $\downarrow$
Translating: 1.6   =   0.66   · S

$$\frac{1.6}{0.66} = S \qquad \text{Dividing both sides by 0.66}$$

$$2.4 \approx S \qquad \text{The symbol } \approx \text{ means } \textit{is approximately equal to.}$$

About 2.4 million students graduated from high school in 2006.

TRY EXERCISE ▶ 67

EXAMPLE 9

Automobile prices. Recently, Harken Motors reduced the price of a Flex Fuel 2007 Chevy Impala from the manufacturer's suggested retail price (MSRP) of $20,830 to $18,955.

a) What percent of the MSRP does the sale price represent?
b) What is the percent of discount?

SOLUTION

a) We reword and translate, using n for the unknown percent.

Rewording: What percent   of   20,830   is   18,955?
$\downarrow$ $\downarrow$ $\downarrow$ $\downarrow$ $\downarrow$
Translating: n ·   20,830   =   18,955

$$n = \frac{18,955}{20,830} \qquad \text{Dividing both sides by 20,830}$$

$$n \approx 0.91 = 91\% \qquad \text{Converting to percent notation}$$

The sale price is about 91% of the MSRP.

b) Since the original price of $20,830 represents 100% of the MSRP, the sale price represents a discount of $(100 - 91)\%$, or 9%.

Alternatively, we could find the amount of discount and then calculate the percent of discount:

Amount of discount: $\$20,830 - \$18,955 = \$1875$.

Rewording: $\underbrace{\text{What percent}}$ of 20,830 is 1875?

Translating: n $\cdot$ 20,830   =   1875

$$n = \frac{1875}{20,830}$$ Dividing both sides by 20,830

$$n \approx 0.09 = 9\%$$ Converting to percent notation

Again we find that the percent of discount is 9%.   ▷ **TRY EXERCISE** ▶ 69

2.4   EXERCISE SET

For Extra Help **MyMathLab** | **Math XL** PRACTICE   WATCH   DOWNLOAD

↪ *Concept Reinforcement*   *In each of Exercises 1–10, match the question with the most appropriate translation from the column on the right. Some choices are used more than once.*

1. ____ What percent of 57 is 23?

2. ____ What percent of 23 is 57?

3. ____ 23 is 57% of what number?

4. ____ 57 is 23% of what number?

5. ____ 57 is what percent of 23?

6. ____ 23 is what percent of 57?

7. ____ What is 23% of 57?

8. ____ What is 57% of 23?

9. ____ 23% of what number is 57?

10. ____ 57% of what number is 23?

a) $a = (0.57)23$

b) $57 = 0.23y$

c) $n \cdot 23 = 57$

d) $n \cdot 57 = 23$

e) $23 = 0.57y$

f) $a = (0.23)57$

Convert the percent notation in each sentence to decimal notation.

11. *Energy use.* Heating accounts for 49% of all household energy use.
Source: Chevron

12. *Energy use.* Water heating accounts for 15% of all household energy use.
Source: Chevron

13. *Drinking water.* Only 1% of the water on earth is suitable for drinking.
Source: www.drinktap.org

Drinking water, 1%

14. *Dehydration.* A 2% drop in water content of the body can affect one's ability to study mathematics.
Source: High Performance Nutrition

15. *College tuition.* Tuition and fees at two-year public colleges increased 4.1 percent in 2006.
Source: College Board 2006 tuition survey

16. *Plant species.* Trees make up about 3.5% of all plant species found in the United States.
Source: South Dakota Project Learning Tree

17. *Women in the workforce.* Women comprise 20% of all database administrators.
Source: U.S. Census Bureau

18. *Women in the workforce.* Women comprise 60% of all accountants and auditors.
Source: U.S. Census Bureau

Convert to decimal notation.

19. 6.25%   **20.** 8.375%

21. 0.2%   **22.** 0.8%

23. 175%   **24.** 250%

Convert the decimal notation in each sentence to percent notation.

25. *NASCAR fans.* Auto racing is the seventh most popular sport in the United States, with 0.38 of the adult population saying they are NASCAR fans.
Source: ESPN Sports poll

26. *Baseball fans.* Baseball is the second most popular sport in the United States, with 0.61 of the adult population saying they are baseball fans.
Source: ESPN Sports poll

27. *Food security.* The USDA defines food security as access to enough nutritious food for a healthy life. In 2005, 0.039 of U.S. households had very low food security.
Source: USDA

28. *Poverty rate.* In 2005, 0.199 of Americans age 65 and older were under the poverty level.
Source: www.census.gov

29. *Music downloads.* In 2006, 0.45 of Americans downloaded music.
Source: Solutions Research Group

30. *Music downloads.* In 2006, 0.23 of Americans paid to download a song.
Source: Solutions Research Group

31. *Composition of the sun.* The sun is 0.7 hydrogen.

32. *Jupiter's atmosphere.* The atmosphere of Jupiter is 0.1 helium.

Convert to percent notation.

33. 0.0009   **34.** 0.0056

35. 1.06   **36.** 1.08

37. 1.8   **38.** 2.4

39. $\dfrac{3}{5}$ **40.** $\dfrac{3}{4}$

41. $\dfrac{8}{25}$ **42.** $\dfrac{5}{8}$

Solve.

43. What percent of 76 is 19?

44. What percent of 125 is 30?

45. What percent of 150 is 39?

46. What percent of 360 is 270?

47. 14 is 30% of what number?

48. 54 is 24% of what number?

49. 0.3 is 12% of what number?

50. 7 is 175% of what number?

51. What number is 1% of one million?

52. What number is 35% of 240?

53. What percent of 60 is 75?

Aha! **54.** What percent of 70 is 70?

55. What is 2% of 40?

56. What is 40% of 2?

Aha! **57.** 25 is what percent of 50?

58. 0.8 is 2% of what number?

59. What percent of 69 is 23?

60. What percent of 40 is 9?

Riding bicycles.   There are 57 million Americans who ride a bicycle at least occasionally. The following circle graph shows the reasons people ride. In each of Exercises 61–64, determine the number of Americans who ride a bicycle for the given reason.

Reasons people rode bicycles, 2003

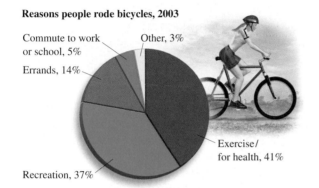

Commute to work or school, 5%

Other, 3%

Errands, 14%

Exercise/ for health, 41%

Recreation, 37%

Sources: U.S. Census Bureau; Bureau of Transportation Statistics

61. Commute to school or work

62. Run errands

63. Exercise for health

64. Recreation

65. *College graduation.*   To obtain his bachelor's degree in nursing, Cody must complete 125 credit hours of instruction. If he has completed 60% of his requirement, how many credits did Cody complete?

66. *College graduation.*   To obtain her bachelor's degree in journalism, Addy must complete 125 credit hours of instruction. If 20% of Addy's credit hours remain to be completed, how many credits does she still need to take?

67. *Batting average.*   In the 2007 season, Magglio Ordonez of the Detroit Tigers had 216 hits. His batting average was 0.363, the highest in major league baseball for that season. This means that of the total number of at bats, 36.3% were hits. How many at bats did he have?
Source: ESPN

68. *Pass completions.*   At one point in a recent season, Peyton Manning of the Indianapolis Colts had com-

pleted 357 passes. This was 62.5% of his attempts. How many attempts did he make?
Source: National Football League

69. *Tipping.*   Trent left a $4 tip for a meal that cost $25.

   a) What percent of the cost of the meal was the tip?

   b) What was the total cost of the meal including the tip?

70. *Tipping.*   Selena left a $12.76 tip for a meal that cost $58.

   a) What percent of the cost of the meal was the tip?

   b) What was the total cost of the meal including the tip?

71. *Crude oil imports.*   In April 2007, crude oil imports to the United States averaged 10.2 million barrels per day. Of this total, 3.4 million came from Canada and Mexico. What percent of crude oil imports came from Canada and Mexico? What percent came from the rest of the world?
Source: Energy Information Administration

72. *Alternative-fuel vehicles.*   Of the 550,000 alternative-fuel vehicles produced in the United States in 2004, 150,000 were E85 flexible-fuel vehicles. What percent of alternative-fuel vehicles used E85? What percent used other alternative fuels?
Source: Energy Information Administration

73. *Student loans.*   Glenn takes out a subsidized federal Stafford loan for $2400. After a year, Glenn decides to pay off the interest, which is 7% of $2400. How much will he pay?

74. *Student loans.*   To finance her community college education, LaTonya takes out a Stafford loan for $3500. After a year, LaTonya decides to pay off the interest, which is 8% of $3500. How much will she pay?

75. *Infant health.*   In a study of 300 pregnant women with "good-to-excellent" diets, 95% had babies in good or excellent health. How many women in this group had babies in good or excellent health?

76. *Infant health.*   In a study of 300 pregnant women with "poor" diets, 8% had babies in good or excellent health. How many women in this group had babies in good or excellent health?

77. *Cost of self-employment.*   Because of additional taxes and fewer benefits, it has been estimated that a self-employed person must earn 20% more than a non–self-employed person performing the same task(s). If Tia earns $16 an hour working for Village

Copy, how much would she need to earn on her own for a comparable income?

78. Refer to Exercise 77. Rik earns $18 an hour working for Round Edge stairbuilders. How much would Rik need to earn on his own for a comparable income?

79. *Budget overruns.* The Indianapolis Central Library expansion, begun in 2002, was expected to cost $103 million. By 2006, library officials estimated the cost would be $45 million over budget. By what percent did the actual cost exceed the initial estimate?
Source: *The Indianapolis Star*, 5/23/06

80. *Fastest swimmer.* In 1990, Tom Jager of the United States set a world record by swimming 50 m at a rate of 2.29 m/s. Previously, the fastest swimming rate on record was 2.26 m/s, set in 1975 by David Holmes Edgar, also of the United States. Calculate the percentage by which the rate increased.
Source: *Guinness Book of World Records* 1975 and 1998

81. A bill at Officeland totaled $47.70. How much did the merchandise cost if the sales tax is 6%?

82. Marta's checkbook shows that she wrote a check for $987 for building materials. What was the price of the materials if the sales tax is 5%?

83. *Deducting sales tax.* A tax-exempt school group received a bill of $157.41 for educational software. The bill incorrectly included sales tax of 6%. How much should the school group pay?

84. *Deducting sales tax.* A tax-exempt charity received a bill of $145.90 for a sump pump. The bill incorrectly included sales tax of 5%. How much does the charity owe?

85. *Body fat.* One author of this text exercises regularly at a local YMCA that recently offered a body-fat percentage test to its members. The device used measures the passage of a very low voltage of electricity through the body. The author's body-fat percentage was found to be 16.5% and he weighs 191 lb. What part, in pounds, of his body weight is fat?

86. *Areas of Alaska and Arizona.* The area of Arizona is 19% of the area of Alaska. The area of Alaska is 586,400 mi². What is the area of Arizona?

87. *Direct mail.* Only 2.15% of mailed ads lead to a sale or a response from customers. In 2006, businesses sent out 114 billion pieces of direct mail (catalogs, coupons, and so on). How many pieces of mail led to a response from customers?
Sources: Direct Marketing Association; U.S. Postal Service

88. *Kissing and colds.* In a medical study, it was determined that if 800 people kiss someone else who has a cold, only 56 will actually catch the cold. What percent is this?

89. *Calorie content.* Pepperidge Farm Light Style 7 Grain Bread® has 140 calories in a 3-slice serving. This is 15% less than the number of calories in a serving of regular bread. How many calories are in a serving of regular bread?

90. *Fat content.* Peek Freans Shortbread Reduced Fat Cookies® contain 35 calories of fat in each serving. This is 40% less than the fat content in the leading imported shortbread cookie. How many calories of fat are in a serving of the leading shortbread cookie?

91. Campus Bookbuyers pays $30 for a book and sells it for $60. Is this a 100% markup or a 50% markup? Explain.

92. If Julian leaves a $12 tip for a $90 dinner, is he being generous, stingy, or neither? Explain.

Skill Review

To prepare for Section 2.5, review translating to algebraic expressions and equations (Section 1.1).

Translate to an algebraic expression or equation. [1.1]

93. Twice the length plus twice the width

94. 5% of $180

95. 5 fewer than the number of points Tino scored

96. 15 plus the product of 1.5 and *x*

97. The product of 10 and half of *a*

98. 10 more than three times a number

99. The width is 2 in. less than the length.

100. A number is four times as large as a second number.

Synthesis

101. How is the use of statistics in the following misleading?

    a) A business explaining new restrictions on sick leave cited a recent survey indicating that 40% of all sick days were taken on Monday or Friday.

    b) An advertisement urging summer installation of a security system quoted FBI statistics stating that over 26% of home burglaries occur between Memorial Day and Labor Day.

102. Erin is returning a tent that she bought during a 25%-off storewide sale that has ended. She is offered store credit for 125% of what she paid (not to be used on sale items). Is this fair to Erin? Why or why not?

103. The community of Bardville has 1332 left-handed females. If 48% of the community is female and 15% of all females are left-handed, how many people are in the community?

104. It has been determined that at the age of 10, a girl has reached 84.4% of her final adult height. Dana is 4 ft 8 in. at the age of 10. What will her final adult height be?

105. It has been determined that at the age of 15, a boy has reached 96.1% of his final adult height. Jaraan is 6 ft 4 in. at the age of 15. What will his final adult height be?

106. *Dropout rate.* Between 2002 and 2004, the high school dropout rate in the United States decreased from 105 to 103 per thousand. Calculate the percent by which the dropout rate decreased and use that percentage to estimate dropout rates for the United States in 2005 and in 2006.
Source: www.childrendsdatabank.org

107. *Photography.* A 6-in. by 8-in. photo is framed using a mat meant for a 5-in. by 7-in. photo. What percentage of the photo will be hidden by the mat?

108. Would it be better to receive a 5% raise and then, a year later, an 8% raise or the other way around? Why?

109. Jorge is in the 30% tax bracket. This means that 30¢ of each dollar earned goes to taxes. Which would cost him the least: contributing $50 that is tax-deductible or contributing $40 that is not tax-deductible? Explain.

COLLABORATIVE CORNER

Sales and Discounts

Focus: Applications and models using percent

Time: 15 minutes

Group size: 3

Materials: Calculators are optional.

Often a store will reduce the price of an item by a fixed percentage. When the sale ends, the items are returned to their original prices. Suppose a department store reduces all sporting goods 20%, all clothing 25%, and all electronics 10%.

ACTIVITY

1. Each group member should select one of the following items: a $50 basketball, an $80 jacket, or a $200 MP3 player. Fill in the first three columns of the first three rows of the chart below.

2. Apply the appropriate discount and determine the sale price of your item. Fill in the fourth column of the chart.

3. Next, find a multiplier that can be used to convert the sale price back to the original price and fill in the remaining column of the chart. Does this multiplier depend on the price of the item?

4. Working as a group, compare the results of part (3) for all three items. Then develop a formula for a multiplier that will restore a sale price to its original price, p, after a discount r has been applied. Complete the fourth row of the table and check that your formula will duplicate the results of part (3).

5. Use the formula from part (4) to find the multiplier that a store would use to return an item to its original price after a "30% off" sale expires. Fill in the last line on the chart.

6. Inspect the last column of your chart. How can these multipliers be used to determine the percentage by which a sale price is increased when a sale ends?

Original Price, p	Discount, r	$1 - r$	Sale Price	Multiplier to convert back to p
p	r	$1 - r$		
	0.30			

2.5   Problem Solving

Five Steps for Problem Solving   •   Applying the Five Steps

Probably the most important use of algebra is as a tool for problem solving. In this section, we develop a problem-solving approach that is used throughout the remainder of the text.

Five Steps for Problem Solving

In Section 2.4, we solved several real world problems. To solve them, we first *familiarized* ourselves with percent notation. We then *translated* each problem into an equation, *solved* the equation, *checked* the solution, and *stated* the answer.

Five Steps for Problem Solving in Algebra

1. *Familiarize* yourself with the problem.
2. *Translate* to mathematical language. (This often means writing an equation.)
3. *Carry out* some mathematical manipulation. (This often means *solving* an equation.)
4. *Check* your possible answer in the original problem.
5. *State* the answer clearly, using a complete English sentence.

Of the five steps, the most important is probably the first one: becoming familiar with the problem. Here are some hints for familiarization.

To Become Familiar with a Problem

1. Read the problem carefully. Try to visualize the problem.
2. Reread the problem, perhaps aloud. Make sure you understand all important words and any symbols or abbreviations.
3. List the information given and the question(s) to be answered. Choose a variable (or variables) to represent the unknown and specify exactly what the variable represents. For example, let L = length in centimeters, d = distance in miles, and so on.
4. Look for similarities between the problem and other problems you have already solved. Ask yourself what type of problem this is.
5. Find more information. Look up a formula in a book, at a library, or online. Consult a reference librarian or an expert in the field.
6. Make a table that uses all the information you have available. Look for patterns that may help in the translation.
7. Make a drawing and label it with known and unknown information, using specific units if given.
8. Think of a possible answer and check the guess. Note the manner in which the guess is checked.

Applying the Five Steps

EXAMPLE   1

Bicycling.   After finishing college, Nico spent a week touring Tuscany, Italy, by bicycle. He biked 260 km from Pisa through Siena to Florence. At Siena, he had biked three times as far from Pisa as he would then bike to Florence. How far had he biked, and how far did he have left to go?

SOLUTION

1. **Familiarize.** It is often helpful to make a drawing. In this case, we can use a map of Nico's trip.

To gain familiarity, let's suppose that Nico has 50 km to go. Then he would have traveled three times 50 km, or 150 km, already. Since 50 km + 150 km = 200 km and 200 km < 260 km, we see that our guess is too small. Rather than guess again, we let

d = the distance, in kilometers, from Siena to Florence

and

$3d$ = the distance, in kilometers, from Siena to Pisa.

(We could also let x = the distance to Pisa; then the distance to Florence would be $\frac{1}{3}x$.)

2. **Translate.** The lengths of the two parts of the trip must add up to 260 km. This leads to our translation.

Rewording:	Distance to Florence	plus	distance to Pisa	is	260 km
	↓	↓	↓	↓	↓
Translating:	d	$+$	$3d$	$=$	260

3. **Carry out.** We solve the equation:

$$d + 3d = 260$$
$$4d = 260 \qquad \text{Combining like terms}$$
$$d = 65. \qquad \text{Dividing both sides by 4}$$

4. **Check.** As predicted in the *Familiarize* step, d is greater than 50 km. If d = 65 km, then $3d$ = 195 km. Since 65 km + 195 km = 260 km, we have a check.

5. **State.** At Siena, Nico had biked 195 km and had 65 km left to go to arrive in Florence.

TRY EXERCISE 9

Before we solve the next problem, we need to learn some additional terminology regarding integers.

The following are examples of **consecutive integers:** 16, 17, 18, 19, 20; and −31, −30, −29, −28. Note that consecutive integers can be represented in the form $x, x + 1, x + 2$, and so on.

The following are examples of **consecutive even integers:** 16, 18, 20, 22, 24; and −52, −50, −48, −46. Note that consecutive even integers can be represented in the form x, $x + 2$, $x + 4$, and so on.

The following are examples of **consecutive odd integers:** 21, 23, 25, 27, 29; and −71, −69, −67, −65. Note that consecutive odd integers can be also represented in the form x, $x + 2$, $x + 4$, and so on.

EXAMPLE 2

Interstate mile markers. U.S. interstate highways post numbered markers at every mile to indicate location in case of an emergency. The sum of two consecutive mile markers on I-70 in Kansas is 559. Find the numbers on the markers.

Source: Federal Highway Administration, Ed Rotalewski

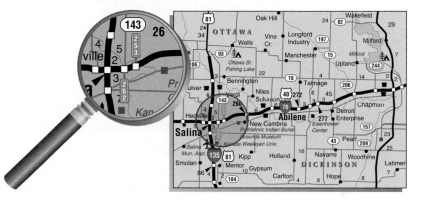

x	$x + 1$	Sum of x and $x + 1$
114	115	229
252	253	505
302	303	605

SOLUTION

1. **Familiarize.** The numbers on the mile markers are consecutive positive integers. Thus if we let $x =$ the smaller number, then $x + 1 =$ the larger number.

   To become familiar with the problem, we can make a table, as shown at left. First, we guess a value for x; then we find $x + 1$. Finally, we add the two numbers and check the sum.

   From the table, we see that the first marker will be between 252 and 302. We could continue guessing and solve the problem this way, but let's work on developing our algebra skills.

2. **Translate.** We reword the problem and translate as follows.

   Rewording: First integer   plus   second integer   is   559.

   Translating: x $+$ $(x + 1)$ $=$ 559

3. **Carry out.** We solve the equation:

$$x + (x + 1) = 559$$
$$2x + 1 = 559 \quad \text{Using an associative law and combining like terms}$$
$$2x = 558 \quad \text{Subtracting 1 from both sides}$$
$$x = 279. \quad \text{Dividing both sides by 2}$$

   If x is 279, then $x + 1$ is 280.

4. **Check.** Our possible answers are 279 and 280. These are consecutive positive integers and $279 + 280 = 559$, so the answers check.

5. **State.** The mile markers are 279 and 280.

TRY EXERCISE 13

EXAMPLE **3**

Color printers. Egads Computer Corporation rents a Xerox Phaser 8400 Color Laser Printer for $300 a month. A new art gallery is leasing a printer for a 2-month advertising campaign. The ink and paper for the brochures will cost an additional 21.5¢ per copy. If the gallery allots a budget of $3000, how many brochures can they print?

Source: egadscomputer.com

SOLUTION

1. **Familiarize.** Suppose that the art gallery prints 20,000 brochures. Then the cost is the monthly charges plus ink and paper cost, or

$$
\underbrace{2(\$300)}_{\$600} \quad \text{plus} \atop + \quad \underbrace{\text{cost per brochure}}_{\$0.215} \quad \text{times} \atop \cdot \quad \underbrace{\text{number of brochures}}_{20{,}000,}
$$

which is $4900. Our guess of 20,000 is too large, but we have familiarized ourselves with the way in which a calculation is made. Note that we convert 21.5¢ to $0.215 so that all information is in the same unit, dollars. We let $c =$ the number of brochures that can be printed for $3000.

2. **Translate.** We reword the problem and translate as follows.

$$
\begin{array}{ccccc}
\textit{Rewording}: & \underline{\text{Monthly cost}} & \text{plus} & \underline{\text{ink and paper cost}} & \text{is} & \$3000 \\
& \downarrow & \downarrow & \downarrow & \downarrow & \downarrow \\
\textit{Translating}: & 2(\$300) & + & (\$0.215)c & = & \$3000
\end{array}
$$

STUDENT NOTES

For most students, the most challenging step is step (2), "Translate." The table on p. 4 (Section 1.1) can be helpful in this regard.

3. **Carry out.** We solve the equation:

$$2(300) + 0.215c = 3000$$
$$600 + 0.215c = 3000$$
$$0.215c = 2400 \qquad \text{Subtracting 600 from both sides}$$
$$c = \frac{2400}{0.215} \qquad \text{Dividing both sides by 0.215}$$
$$c \approx 11{,}162. \qquad \text{We round } \textit{down} \text{ to avoid going over the budget.}$$

4. **Check.** We check in the original problem. The cost for 11,162 brochures is $11,162(\$0.215) = \$2399.83$. The rental for 2 months is $2(\$300) = \$600$. The total cost is then $\$2399.83 + \$600 = \$2999.83$, which is just under the amount that was allotted. Our answer is less than 20,000, as we expected from the *Familiarize* step.

5. **State.** The art gallery can make 11,162 brochures with the rental allotment of $3000.

TRY EXERCISE 37

EXAMPLE **4**

Perimeter of NBA court. The perimeter of an NBA basketball court is 288 ft. The length is 44 ft longer than the width. Find the dimensions of the court.

Source: National Basketball Association

SOLUTION

1. **Familiarize.** Recall that the perimeter of a rectangle is twice the length plus twice the width. Suppose the court were 30 ft wide. The length would then be $30 + 44$, or 74 ft, and the perimeter would be $2 \cdot 30\,\text{ft} + 2 \cdot 74\,\text{ft}$, or 208 ft. This shows that in order for the perimeter to be 288 ft, the width must exceed 30 ft. Instead of guessing again, we let $w =$ the width of the court, in feet.

Since the court is "44 ft longer than it is wide," we let $w + 44 =$ the length of the court, in feet.

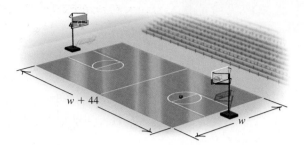

2. **Translate.** To translate, we use $w + 44$ as the length and 288 as the perimeter. To double the length, $w + 44$, parentheses are essential.

Rewording:	Twice the length	plus	twice the width	is	288 ft.
Translating:	$2(w + 44)$	$+$	$2w$	$=$	288

3. **Carry out.** We solve the equation:

$$2(w + 44) + 2w = 288$$
$$2w + 88 + 2w = 288 \qquad \text{Using the distributive law}$$
$$4w + 88 = 288 \qquad \text{Combining like terms}$$
$$4w = 200$$
$$w = 50.$$

The dimensions appear to be $w = 50\,\text{ft}$, and $l = w + 44 = 94\,\text{ft}$.

4. **Check.** If the width is 50 ft and the length is 94 ft, then the court is 44 ft longer than it is wide. The perimeter is $2(50\,\text{ft}) + 2(94\,\text{ft}) = 100\,\text{ft} + 188\,\text{ft}$, or 288 ft, as specified. We have a check.

5. **State.** An NBA court is 50 ft wide and 94 ft long.

> **TRY EXERCISE** ▶ 25

> **CAUTION!**   Always be sure to answer the original problem completely. For instance, in Example 1 we needed to find *two* numbers: the distances from *each* city to Siena. Similarly, in Example 4 we needed to find two dimensions, not just the width. Be sure to label each answer with the proper unit.

EXAMPLE   5

Selling at an auction.   Jared is selling his collection of Transformers at an auction. He wants to be left with $1150 after paying a seller's premium of 8% on the final bid (hammer price) for the collection. What must the hammer price be in order for him to clear $1150?

SOLUTION

1. **Familiarize.** Suppose the collection sells for $1200. The 8% seller's premium can be determined by finding 8% of $1200:

$$8\% \text{ of } \$1200 = 0.08(\$1200) = \$96.$$

Subtracting this premium from $1200 would leave Jared with

$$\$1200 - \$96 = \$1104.$$

This shows that in order for Jared to clear $1150, the collection must sell for more than $1200. We let $x =$ the hammer price, in dollars. Jared then must pay a seller's premium of $0.08x$.

2. Translate. We reword the problem and translate as follows.

Rewording: Hammer price   less   seller's premium   is   amount remaining.

Translating: x $-$ $0.08x$ $=$ $\$1150$

3. Carry out. We solve the equation:

$$x - 0.08x = 1150$$
$$1x - 0.08x = 1150$$
$$0.92x = 1150 \quad \text{Combining like terms. Had we noted that after the premium has been paid, 92\% remains, we could have begun with this equation.}$$

$$x = \frac{1150}{0.92} \quad \text{Dividing both sides by 0.92}$$

$$x = 1250.$$

4. Check. To check, we first find 8% of $1250:

$$8\% \text{ of } \$1250 = 0.08(\$1250) = \$100. \qquad \text{This is the premium.}$$

Next, we subtract the premium to find the remaining amount:

$$\$1250 - \$100 = \$1150.$$

Since, after Jared pays the seller's premium, he is left with $1150, our answer checks. Note that the $1250 hammer price is greater than $1200, as predicted in the *Familiarize* step.

5. State. Jared's collection must sell for $1250 in order for him to be left with $1150.

▶ **TRY EXERCISE** ▷ 7

EXAMPLE ▌ **6**   *Cross section of a roof.* In a triangular gable end of a roof, the angle of the peak is twice as large as the angle on the back side of the house. The measure of the angle on the front side is 20° greater than the angle on the back side. How large are the angles?

SOLUTION

1. Familiarize. We make a drawing. In this case, the measure of the back angle is x, the measure of the front angle is $x + 20$, and the measure of the peak angle is $2x$.

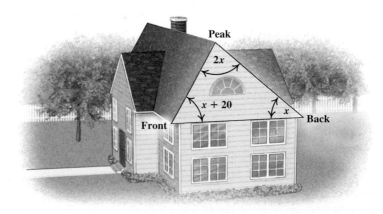

2. **Translate.** To translate, we need to recall that the sum of the measures of the angles in a triangle is 180°.

Rewording: Measure of back angle + measure of front angle + measure of peak angle is 180°

Translating: $x + (x + 20) + 2x = 180$

3. **Carry out.** We solve:

$$x + (x + 20) + 2x = 180$$
$$4x + 20 = 180$$
$$4x = 160$$
$$x = 40.$$

The measures for the angles appear to be:

Back angle: $x = 40°$,

Front angle: $x + 20 = 40 + 20 = 60°$,

Peak angle: $2x = 2(40) = 80°$.

4. **Check.** Consider 40°, 60°, and 80°, as listed above. The measure of the front angle is 20° greater than the measure of the back angle, the measure of the peak angle is twice the measure of the back angle, and the sum is 180°. These numbers check.

5. **State.** The measures of the angles are 40°, 60°, and 80°.   TRY EXERCISE 31

We close this section with some tips to aid you in problem solving.

Problem-Solving Tips

1. The more problems you solve, the more your skills will improve.
2. Look for patterns when solving problems. Each time you study an example or solve an exercise, you may observe a pattern for problems found later.
3. Clearly define variables before translating to an equation.
4. Consider the dimensions of the variables and constants in the equation. The variables that represent length should all be in the same unit, those that represent money should all be in dollars or all in cents, and so on.
5. Make sure that units appear in the answer whenever appropriate and that you completely answer the original problem.

2.5 EXERCISE SET

For Extra Help MyMathLab Math XL PRACTICE WATCH DOWNLOAD

Solve. Even though you might find the answer quickly in some other way, practice using the five-step problem-solving process in order to build the skill of problem solving.

1. Three less than twice a number is 19. What is the number?

2. Two fewer than ten times a number is 78. What is the number?

3. Five times the sum of 3 and twice some number is 70. What is the number?

4. Twice the sum of 4 and three times some number is 34. What is the number?

5. *Price of an iPod.* Kyle paid $120 for an iPod nano during a 20%-off sale. What was the regular price?

6. *Price of sneakers.* Amy paid $102 for a pair of New Balance 1122 running shoes during a 15%-off sale. What was the regular price?

7. *Price of a calculator.* Kayla paid $137.80, including 6% tax, for her graphing calculator. How much did the calculator itself cost?

8. *Price of a printer.* Laura paid $219.45, including 5% tax, for an all-in-one color printer. How much did the printer itself cost?

9. *Unicycling.* In 2005, Ken Looi of New Zealand set a record by covering 235.3 mi in 24 hr on his unicycle. After 8 hr, he was approximately twice as far from the finish line as he was from the start. How far had he traveled?
Source: *Guinness World Records* 2007

10. *Sled-dog racing.* The Iditarod sled-dog race extends for 1049 mi from Anchorage to Nome. If a musher is twice as far from Anchorage as from Nome, how many miles has the musher traveled?

11. *Indy Car racing.* In April 2008, Danica Patrick won the Indy Japan 300 with a time of 01:51:02.6739 for the 300-mi race. At one point, Patrick was 20 mi closer to the finish than to the start. How far had Patrick traveled at that point?

12. *NASCAR racing.* In June 2007, Carl Edwards won the Michigan 400 with a time of 2:42:5 for the 400-mi race. At one point, Edwards was 80 mi closer to the finish than to the start. How far had Edwards traveled at that point?

13. *Apartment numbers.* The apartments in Erica's apartment house are consecutively numbered on each floor. The sum of her number and her next-door neighbor's number is 2409. What are the two numbers?

14. *Apartment numbers.* The apartments in Brian's apartment house are numbered consecutively on each floor. The sum of his number and his next-door neighbor's number is 1419. What are the two numbers?

15. *Street addresses.* The houses on the west side of Lincoln Avenue are consecutive odd numbers. Sam and Colleen are next-door neighbors and the sum of their house numbers is 572. Find their house numbers.

16. *Street addresses.* The houses on the south side of Elm Street are consecutive even numbers. Wanda and Larry are next-door neighbors and the sum of their house numbers is 794. Find their house numbers.

17. The sum of three consecutive page numbers is 99. Find the numbers.

18. The sum of three consecutive page numbers is 60. Find the numbers.

19. *Longest marriage.* As half of the world's longest-married couple, the woman was 2 yr younger than her husband. Together, their ages totaled 204 yr. How old were the man and the woman?
Source: *Guinness World Records* 2007

20. *Oldest bride.* The world's oldest bride was 19 yr older than her groom. Together, their ages totaled 185 yr. How old were the bride and the groom?
Source: *Guinness World Records* 2007

21. *e-mail.* In 2006, approximately 125 billion e-mail messages were sent each day. The number of spam messages was about four times the number of non-spam messages. How many of each type of message were sent each day in 2006?
Source: Ferris Research

22. *Home remodeling.* In 2005, Americans spent a total of $26 billion to remodel bathrooms and kitchens. They spent $5 billion more on kitchens than on bathrooms. How much was spent on each?
Source: Joint Center for Housing Studies, Harvard University

23. *Page numbers.* The sum of the page numbers on the facing pages of a book is 281. What are the page numbers?

24. *Perimeter of a triangle.* The perimeter of a triangle is 195 mm. If the lengths of the sides are consecutive odd integers, find the length of each side.

25. *Hancock Building dimensions.* The top of the John Hancock Building in Chicago is a rectangle whose length is 60 ft more than the width. The perimeter is 520 ft. Find the width and the length of the rectangle. Find the area of the rectangle.

26. *Dimensions of a state.* The perimeter of the state of Wyoming is 1280 mi. The width is 90 mi less than the length. Find the width and the length.

27. A rectangular community garden is to be enclosed with 92 m of fencing. In order to allow for compost storage, the garden must be 4 m longer than it is wide. Determine the dimensions of the garden.

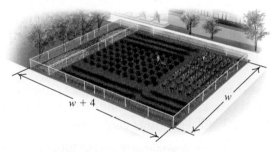

28. *Perimeter of a high school basketball court.* The perimeter of a standard high school basketball court is 268 ft. The length is 34 ft longer than the width. Find the dimensions of the court.

Source: Indiana High School Athletic Association

29. *Two-by-four.* The perimeter of a cross section of a "two-by-four" piece of lumber is $10\frac{1}{2}$ in. The length is twice the width. Find the actual dimensions of the cross section of a two-by-four.

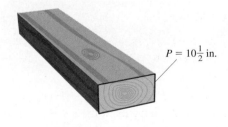

$P = 10\frac{1}{2}$ in.

30. *Standard billboard sign.* A standard rectangular highway billboard sign has a perimeter of 124 ft. The length is 6 ft more than three times the width. Find the dimensions of the sign.

31. *Angles of a triangle.* The second angle of an architect's triangle is three times as large as the first. The third angle is 30° more than the first. Find the measure of each angle.

32. *Angles of a triangle.* The second angle of a triangular garden is four times as large as the first. The third angle is 45° less than the sum of the other two angles. Find the measure of each angle.

33. *Angles of a triangle.* The second angle of a triangular kite is four times as large as the first. The third angle is 5° more than the sum of the other two angles. Find the measure of the second angle.

34. *Angles of a triangle.* The second angle of a triangular building lot is three times as large as the first. The third angle is 10° more than the sum of the other two angles. Find the measure of the third angle.

35. *Rocket sections.* A rocket is divided into three sections: the payload and navigation section in the top, the fuel section in the middle, and the rocket engine section in the bottom. The top section is one-sixth the length of the bottom section. The middle section is one-half the length of the bottom section. The total length is 240 ft. Find the length of each section.

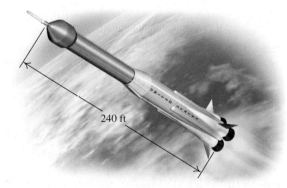

36. *Gourmet sandwiches.*   Jenny, Demi, and Drew buy an 18-in. long gourmet sandwich and take it back to their apartment. Since they have different appetites, Jenny cuts the sandwich so that Demi gets half of what Jenny gets and Drew gets three-fourths of what Jenny gets. Find the length of each person's sandwich.

37. *Taxi rates.*   In Chicago, a taxi ride costs $2.25 plus $1.80 for each mile traveled. Debbie has budgeted $18 for a taxi ride (excluding tip). How far can she travel on her $18 budget?
Source: City of Chicago

38. *Taxi fares.*   In New York City, taxis charge $2.50 plus $2.00 per mile for off-peak fares. How far can Ralph travel for $17.50 (assuming an off-peak fare)?
Source: New York City Taxi and Limousine Commission

39. *Truck rentals.*   Truck-Rite Rentals rents trucks at a daily rate of $49.95 plus 39¢ per mile. Concert Productions has budgeted $100 for renting a truck to haul equipment to an upcoming concert. How far can they travel in one day and stay within their budget?

40. *Truck rentals.*   Fine Line Trucks rents an 18-ft truck for $42 plus 35¢ per mile. Judy needs a truck for one day to deliver a shipment of plants. How far can she drive and stay within a budget of $70?

41. *Complementary angles.*   The sum of the measures of two *complementary* angles is 90°. If one angle measures 15° more than twice the measure of its complement, find the measure of each angle.

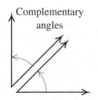

Complementary angles

42. *Complementary angles.*   Two angles are complementary. (See Exercise 41.) The measure of one angle is $1\frac{1}{2}$ times the measure of the other. Find the measure of each angle.

43. *Supplementary angles.*   The sum of the measures of two *supplementary* angles is 180°. If the measure of one angle is $3\frac{1}{2}$ times the measure of the other, find the measure of each angle.

Supplementary angles

44. *Supplementary angles.*   Two angles are supplementary. (See Exercise 43.) If one angle measures 45° less than twice the measure of its supplement, find the measure of each angle.

45. *Copier paper.*   The perimeter of standard-size copier paper is 99 cm. The width is 6.3 cm less than the length. Find the length and the width.

46. *Stock prices.*   Sarah's investment in Jet Blue stock grew 28% to $448. How much did she originally invest?

47. *Savings interest.*   Janeka invested money in a savings account at a rate of 6% simple interest. After 1 yr, she has $6996 in the account. How much did Janeka originally invest?

48. *Credit cards.*   The balance in Will's Mastercard® account grew 2%, to $870, in one month. What was his balance at the beginning of the month?

49. *Scrabble®.*   In a single game on October 12, 2006, Michael Cresta and Wayne Yorra set three North American Scrabble records: the most points in one game by one player, the most total points in the game, and the most points on a single turn. Cresta scored 340 points more than Yorra, and together they scored 1320 points. What was the winning score?
Source: www.slate.com

50. *Color printers.*   The art gallery in Example 3 decides to raise its budget to $5000 for the 2-month period. How many brochures can they print for $5000?

51. *Selling a home.*   The Brannons are planning to sell their home. If they want to be left with $117,500 after paying 6% of the selling price to a realtor as a commission, for how much must they sell the house?

52. *Budget overruns.*   The massive roadworks project in Boston known as The Big Dig cost approximately $14.6 billion. This cost was 484% more than the original estimate. What was the original estimate of the cost of The Big Dig?
Sources: Taxpayers for Common Sense; www.msnbc.cmsn.com

53. *Cricket chirps and temperature.*   The equation $T = \frac{1}{4}N + 40$ can be used to determine the temperature T, in degrees Fahrenheit, given the number of times N a cricket chirps per minute. Determine the number of chirps per minute for a temperature of 80°F.

54. *Race time.*   The equation $R = -0.028t + 20.8$ can be used to predict the world record in the 200-m dash, where R is the record in seconds and t is the number of years since 1920. In what year will the record be 18.0 sec?

55. Sean claims he can solve most of the problems in this section by guessing. Is there anything wrong with this approach? Why or why not?

56. When solving Exercise 20, Beth used a to represent the bride's age and Ben used a to represent the groom's age. Is one of these approaches preferable to the other? Why or why not?

Skill Review

To prepare for Section 2.6, review inequalities (Section 1.4).

Write a true sentence using either $<$ or $>$. [1.4]

57. $-8 \blacksquare 1$

58. $-2 \blacksquare -5$

59. $\frac{1}{2} \blacksquare 0$

60. $-3 \blacksquare -1$

Write a second inequality with the same meaning. [1.4]

61. $x \geq -4$

62. $x < 5$

63. $5 > y$

64. $-10 \leq t$

Synthesis

65. Write a problem for a classmate to solve. Devise it so that the problem can be translated to the equation $x + (x + 2) + (x + 4) = 375$.

66. Write a problem for a classmate to solve. Devise it so that the solution is "Audrey can drive the rental truck for 50 mi without exceeding her budget."

67. *Discounted dinners.*   Kate's "Dining Card" entitles her to $10 off the price of a meal after a 15% tip has been added to the cost of the meal. If, after the discount, the bill is $32.55, how much did the meal originally cost?

68. *Test scores.*   Pam scored 78 on a test that had 4 fill-in questions worth 7 points each and 24 multiple-choice questions worth 3 points each. She had one fill-in question wrong. How many multiple-choice questions did Pam get right?

69. *Gettysburg Address.*   Abraham Lincoln's 1863 Gettysburg Address refers to the year 1776 as "four *score* and seven years ago." Determine what a score is.

70. One number is 25% of another. The larger number is 12 more than the smaller. What are the numbers?

71. A storekeeper goes to the bank to get $10 worth of change. She requests twice as many quarters as half dollars, twice as many dimes as quarters, three times as many nickels as dimes, and no pennies or dollars. How many of each coin did the storekeeper get?

72. *Perimeter of a rectangle.*   The width of a rectangle is three fourths of the length. The perimeter of the rectangle becomes 50 cm when the length and the width are each increased by 2 cm. Find the length and the width.

73. *Discounts.*   In exchange for opening a new credit account, Macy's Department Stores® subtracts 10% from all purchases made the day the account is established. Julio is opening an account and has a coupon for which he receives 10% off the first day's reduced price of a camera. If Julio's final price is $77.75, what was the price of the camera before the two discounts?

74. *Sharing fruit.*   Apples are collected in a basket for six people. One third, one fourth, one eighth, and one fifth of the apples are given to four people, respectively. The fifth person gets ten apples, and one apple remains for the sixth person. Find the original number of apples in the basket.

75. *eBay purchases.*   An eBay seller charges $9.99 for the first DVD purchased and $6.99 for all others. For shipping and handling, he charges the full shipping fee of $3 for the first DVD, one half of the shipping charge for the second item, and one third of the shipping charge per item for all remaining items. The total cost of a shipment (excluding tax) was $45.45. How many DVDs were in the shipment?

76. *Winning percentage.*   In a basketball league, the Falcons won 15 of their first 20 games. In order to win 60% of the total number of games, how many more games will they have to play, assuming they win only half of the remaining games?

77. *Taxi fares.*   In New York City, a taxi ride costs $2.50 plus 40¢ per $\frac{1}{5}$ mile and 40¢ per minute stopped in traffic. Due to traffic, Glenda's taxi took 20 min to complete what is usually a 10-min drive. If she is charged $18.50 for the ride, how far did Glenda travel?
Source: New York City Taxi and Limousine Commission

78. *Test scores.*   Ella has an average score of 82 on three tests. Her average score on the first two tests is 85. What was the score on the third test?

79. A school purchases a piano and must choose between paying $2000 at the time of purchase or $2150 at the end of one year. Which option should the school select and why?

80. Annette claims the following problem has no solution: "The sum of the page numbers on facing pages is 191. Find the page numbers." Is she correct? Why or why not?
Aha!

81. The perimeter of a rectangle is 101.74 cm. If the length is 4.25 cm longer than the width, find the dimensions of the rectangle.

82. The second side of a triangle is 3.25 cm longer than the first side. The third side is 4.35 cm longer than the second side. If the perimeter of the triangle is 26.87 cm, find the length of each side.

2.6   Solving Inequalities

Solutions of Inequalities   •   Graphs of Inequalities   •   Solving Inequalities Using the Addition Principle
•   Solving Inequalities Using the Multiplication Principle   •   Using the Principles Together

Many real-world situations translate to *inequalities*. For example, a student might need to register for *at least* 12 credits; an elevator might be designed to hold *at most* 2000 pounds; a tax credit might be allowable for families with incomes of *less than* $25,000; and so on. Before solving applications of this type, we must adapt our equation-solving principles to the solving of inequalities.

Solutions of Inequalities

Recall from Section 1.4 that an inequality is a number sentence containing > (is greater than), < (is less than), ≥ (is greater than or equal to), or ≤ (is less than or equal to). Inequalities like

$$-7 > x, \qquad t < 5, \qquad 5x - 2 \geq 9, \quad \text{and} \quad -3y + 8 \leq -7$$

are true for some replacements of the variable and false for others.

EXAMPLE   **1**   Determine whether the given number is a solution of $x < 2$: **(a)** -3; **(b)** 2.

SOLUTION

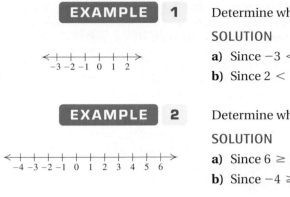

a) Since $-3 < 2$ is true, -3 is a solution.

b) Since $2 < 2$ is false, 2 is not a solution.

> TRY EXERCISE   9

EXAMPLE   **2**   Determine whether the given number is a solution of $y \geq 6$: **(a)** 6; **(b)** -4.

SOLUTION

a) Since $6 \geq 6$ is true, 6 is a solution.

b) Since $-4 \geq 6$ is false, -4 is not a solution.

> TRY EXERCISE   11

Graphs of Inequalities

Because the solutions of inequalities like $x < 2$ are too numerous to list, it is helpful to make a drawing that represents all the solutions. The **graph** of an inequality is such a drawing. Graphs of inequalities in one variable can be drawn on the number line by shading all points that are solutions. Open dots are used to indicate endpoints that are *not* solutions and closed dots to indicate endpoints that *are* solutions.*

EXAMPLE   **3**   Graph each inequality: **(a)** $x < 2$; **(b)** $y \geq -3$; **(c)** $-2 < x \leq 3$.

SOLUTION

a) The solutions of $x < 2$ are those numbers less than 2. They are shown on the graph by shading all points to the left of 2. The open dot at 2 and the shading to its left indicate that 2 is *not* part of the graph, but numbers like 1.2 and 1.99 are.

b) The solutions of $y \geq -3$ are shown on the number line by shading the point for -3 and all points to the right of -3. The closed dot at -3 indicates that -3 *is* part of the graph.

STUDENT NOTES

Note that $-2 < x < 3$ means $-2 < x$ *and* $x < 3$. Because of this, statements like $2 < x < 1$ make no sense—no number is both greater than 2 and less than 1.

c) The inequality $-2 < x \leq 3$ is read "-2 is less than x *and* x is less than or equal to 3," or "x is greater than -2 *and* less than or equal to 3." To be a solution of $-2 < x \leq 3$, a number must be a solution of both $-2 < x$ *and* $x \leq 3$. The number 1 is a solution, as are -0.5, 1.9, and 3. The open dot indicates that -2 is *not* a solution, whereas the closed dot indicates that 3 *is* a solution. The other solutions are shaded.

> TRY EXERCISE   17

---

*An alternative notation uses parentheses to indicate endpoints that are not solutions and brackets to indicate endpoints that are solutions. Using this notation, the solutions of $x < 2$ are graphed as 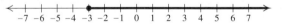 and the solutions of $y \geq -3$ are graphed as
 .

Solving Inequalities Using the Addition Principle

Consider a balance similar to one that appears in Section 2.1. When one side of the balance holds more weight than the other, the balance tips in that direction. If equal amounts of weight are then added to or subtracted from both sides of the balance, the balance remains tipped in the same direction.

The balance illustrates the idea that when a number, such as 2, is added to (or subtracted from) both sides of a true inequality, such as $3 < 7$, we get another true inequality:

$$3 + 2 < 7 + 2, \quad \text{or} \quad 5 < 9.$$

Similarly, if we add -4 to both sides of $x + 4 < 10$, we get an *equivalent* inequality:

$$x + 4 + (-4) < 10 + (-4), \quad \text{or} \quad x < 6.$$

We say that $x + 4 < 10$ and $x < 6$ are **equivalent**, which means that both inequalities have the same solution set.

The Addition Principle for Inequalities

For any real numbers a, b, and c:

$$a < b \quad \text{is equivalent to} \quad a + c < b + c;$$
$$a \le b \quad \text{is equivalent to} \quad a + c \le b + c;$$
$$a > b \quad \text{is equivalent to} \quad a + c > b + c;$$
$$a \ge b \quad \text{is equivalent to} \quad a + c \ge b + c.$$

As with equations, our goal is to isolate the variable on one side.

EXAMPLE 4   Solve $x + 2 > 8$ and then graph the solution.

SOLUTION   We use the addition principle, subtracting 2 from both sides:

$$x + 2 - 2 > 8 - 2 \qquad \text{Subtracting 2 from, or adding } -2 \text{ to, both sides}$$
$$x > 6.$$

From the inequality $x > 6$, we can determine the solutions easily. Any number greater than 6 makes $x > 6$ true and is a solution of that inequality as well as the inequality $x + 2 > 8$. The graph is as follows:

$$\xleftarrow{\hspace{2cm}} \overset{\qquad\qquad\qquad\qquad\qquad\qquad\;\circ}{\underset{-5\;\;-4\;\;-3\;\;-2\;\;-1\;\;\;0\;\;\;1\;\;\;2\;\;\;3\;\;\;4\;\;\;5\;\;\;6\;\;\;7\;\;\;8\;\;\;9}{\rule{10cm}{0.4pt}}} \xrightarrow{\hspace{1cm}}$$

Because most inequalities have an infinite number of solutions, we cannot possibly check them all. A partial check can be made using one of the possible solutions. For this example, we can substitute any number greater than 6—say, 6.1—into the original inequality:

$$\frac{x + 2 > 8}{6.1 + 2 \mid 8}$$
$$8.1 \overset{?}{>} 8 \quad \text{TRUE} \qquad 8.1 > 8 \text{ is a true statement.}$$

Since $8.1 > 8$ is true, 6.1 is a solution. Any number greater than 6 is a solution. ∎

The solutions of an inequality are numbers. In Example 4, $x > 6$ is an *inequality*, not a *solution*. To describe the set of all solutions, we will use **set-builder notation** to write the *solution set* of Example 4 as

$$\{x \mid x > 6\}.$$

This notation is read

"The set of all x such that x is greater than 6."

Thus a number is in $\{x \mid x > 6\}$ if that number is greater than 6. From now on, solutions of inequalities will be written using set-builder notation.

EXAMPLE 5

Solve $3x - 1 \leq 2x - 5$ and then graph the solution.

SOLUTION We have

$$3x - 1 \leq 2x - 5$$
$$3x - 1 + 1 \leq 2x - 5 + 1 \qquad \text{Adding 1 to both sides}$$
$$3x \leq 2x - 4 \qquad \text{Simplifying}$$
$$3x - 2x \leq 2x - 4 - 2x \qquad \text{Subtracting } 2x \text{ from both sides}$$
$$x \leq -4. \qquad \text{Simplifying}$$

The graph is as follows:

The student should check that any number less than or equal to -4 is a solution. The solution set is $\{x \mid x \leq -4\}$.

TRY EXERCISE 41

Solving Inequalities Using the Multiplication Principle

There is a multiplication principle for inequalities similar to that for equations, but it must be modified when multiplying both sides by a negative number. Consider the true inequality

$$3 < 7.$$

If we multiply both sides by a *positive* number—say, 2—we get another true inequality:

$$3 \cdot 2 < 7 \cdot 2, \quad \text{or} \quad 6 < 14. \qquad \text{TRUE}$$

If we multiply both sides by a negative number—say, -2—we get a *false* inequality:

$$3 \cdot (-2) < 7 \cdot (-2), \quad \text{or} \quad -6 < -14. \qquad \text{FALSE}$$

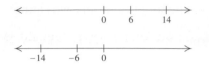

The fact that $6 < 14$ is true, but $-6 < -14$ is false, stems from the fact that the negative numbers, in a sense, *mirror* the positive numbers. Whereas 14 is to the *right* of 6, the number -14 is to the *left* of -6. Thus if we reverse the inequality symbol in $-6 < -14$, we get a true inequality:

$$-6 > -14. \qquad \text{TRUE}$$

The Multiplication Principle for Inequalities

For any real numbers a and b, and for any *positive* number c:

$$a < b \quad \text{is equivalent to} \quad ac < bc, \quad \text{and}$$
$$a > b \quad \text{is equivalent to} \quad ac > bc.$$

For any real numbers a and b, and for any *negative* number c:

$$a < b \quad \text{is equivalent to} \quad ac > bc, \quad \text{and}$$
$$a > b \quad \text{is equivalent to} \quad ac < bc.$$

Similar statements hold for $\leq$ and $\geq$.

CAUTION! When multiplying or dividing both sides of an inequality by a negative number, don't forget to reverse the inequality symbol!

EXAMPLE **6**　Solve and graph each inequality: **(a)** $\frac{1}{4}x < 7$; **(b)** $-2y \leq 18$.

SOLUTION

a) 　$\frac{1}{4}x < 7$

　$4 \cdot \frac{1}{4}x < 4 \cdot 7$　　Multiplying both sides by 4, the reciprocal of $\frac{1}{4}$

　　　⤒———— The symbol stays the same, since 4 is positive.

　$x < 28$　　Simplifying

The solution set is $\{x \mid x < 28\}$. The graph is shown at left.

b) 　$-2y \leq 18$

　$\dfrac{-2y}{-2} \geq \dfrac{18}{-2}$　　Multiplying both sides by $-\frac{1}{2}$, or dividing both sides by -2

　　　└———— *At this step*, we reverse the inequality, because $-\frac{1}{2}$ is negative.

　$y \geq -9$　　Simplifying

As a partial check, we substitute a number greater than -9, say -8, into the original inequality:

$$\frac{-2y \leq 18}{-2(-8) \mid 18}$$
$$16 \overset{?}{\leq} 18 \quad \text{TRUE} \qquad 16 \leq 18 \text{ is a true statement.}$$

The solution set is $\{y \mid y \geq -9\}$. The graph is shown at left.

▶ **TRY EXERCISE** 53

Using the Principles Together

We use the addition and multiplication principles together to solve inequalities much as we did when solving equations.

EXAMPLE 7 Solve: **(a)** $6 - 5y > 7$; **(b)** $2x - 9 < 7x + 1$.

SOLUTION

a)
$$6 - 5y > 7$$
$$-6 + 6 - 5y > -6 + 7 \qquad \text{Adding } -6 \text{ to both sides}$$
$$-5y > 1 \qquad \text{Simplifying}$$
$$-\tfrac{1}{5} \cdot (-5y) < -\tfrac{1}{5} \cdot 1 \qquad \text{Multiplying both sides by } -\tfrac{1}{5}, \text{ or dividing both sides by } -5$$

Remember to reverse the inequality symbol!

$$y < -\tfrac{1}{5} \qquad \text{Simplifying}$$

As a partial check, we substitute a number smaller than $-\tfrac{1}{5}$, say -1, into the original inequality:

$$\frac{6 - 5y > 7}{\begin{array}{c|c} 6 - 5(-1) & 7 \\ 6 - (-5) & \end{array}}$$
$$11 \overset{?}{>} 7 \quad \text{TRUE} \qquad 11 > 7 \text{ is a true statement.}$$

The solution set is $\left\{y \mid y < -\tfrac{1}{5}\right\}$. We show the graph in the margin for reference.

b)
$$2x - 9 < 7x + 1$$
$$2x - 9 - 1 < 7x + 1 - 1 \qquad \text{Subtracting 1 from both sides}$$
$$2x - 10 < 7x \qquad \text{Simplifying}$$
$$2x - 10 - 2x < 7x - 2x \qquad \text{Subtracting } 2x \text{ from both sides}$$
$$-10 < 5x \qquad \text{Simplifying}$$
$$\frac{-10}{5} < \frac{5x}{5} \qquad \text{Dividing both sides by 5}$$
$$-2 < x \qquad \text{Simplifying}$$

The solution set is $\{x \mid -2 < x\}$, or $\{x \mid x > -2\}$.   **TRY EXERCISE** 65

All of the equation-solving techniques used in Sections 2.1 and 2.2 can be used with inequalities provided we remember to reverse the inequality symbol when multiplying or dividing both sides by a negative number.

EXAMPLE 8 Solve: **(a)** $16.3 - 7.2p \le -8.18$; **(b)** $3(x - 9) - 1 \le 2 - 5(x + 6)$.

SOLUTION

a) The greatest number of decimal places in any one number is *two*. Multiplying both sides by 100 will clear decimals. Then we proceed as before.

$$16.3 - 7.2p \le -8.18$$
$$100(16.3 - 7.2p) \le 100(-8.18) \qquad \text{Multiplying both sides by 100}$$
$$100(16.3) - 100(7.2p) \le 100(-8.18) \qquad \text{Using the distributive law}$$
$$1630 - 720p \le -818 \qquad \text{Simplifying}$$
$$-720p \le -818 - 1630 \qquad \text{Subtracting 1630 from both sides}$$

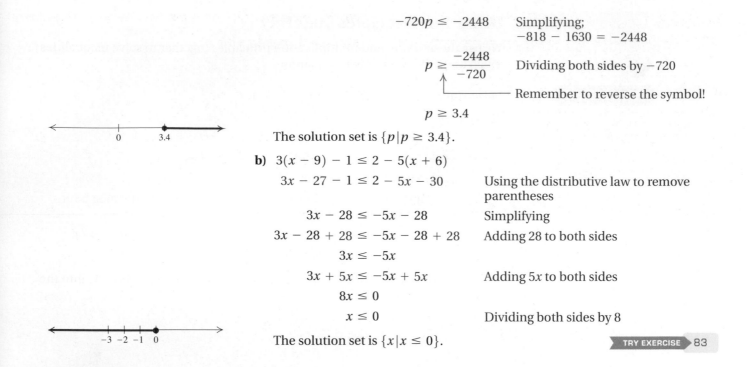

$$-720p \le -2448 \qquad \text{Simplifying;} \\ -818 - 1630 = -2448$$

$$p \ge \frac{-2448}{-720} \qquad \text{Dividing both sides by } -720$$

⎯⎯⎯⎯ Remember to reverse the symbol!

$$p \ge 3.4$$

The solution set is $\{p \mid p \ge 3.4\}$.

b) $3(x - 9) - 1 \le 2 - 5(x + 6)$

$$3x - 27 - 1 \le 2 - 5x - 30 \qquad \text{Using the distributive law to remove parentheses}$$

$$3x - 28 \le -5x - 28 \qquad \text{Simplifying}$$

$$3x - 28 + 28 \le -5x - 28 + 28 \qquad \text{Adding 28 to both sides}$$

$$3x \le -5x$$

$$3x + 5x \le -5x + 5x \qquad \text{Adding } 5x \text{ to both sides}$$

$$8x \le 0$$

$$x \le 0 \qquad \text{Dividing both sides by 8}$$

The solution set is $\{x \mid x \le 0\}$.

TRY EXERCISE ▶ 83

2.6   EXERCISE SET

For Extra Help
MyMathLab   Math XL
PRACTICE   WATCH   DOWNLOAD

↪ **Concept Reinforcement**   *Insert the symbol*
$<, >, \le$, or $\ge$ to make each pair of inequalities
equivalent.

1. $-5x \le 30$; $x \,\blacksquare\, -6$

2. $-7t \ge 56$; $t \,\blacksquare\, -8$

3. $-2t > -14$; $t \,\blacksquare\, 7$

4. $-3x < -15$; $x \,\blacksquare\, 5$

*Classify each pair of inequalities as "equivalent" or "not
equivalent."*

5. $x < -2$; $-2 > x$

6. $t > -1$; $-1 < t$

7. $-4x - 1 \le 15$;
$-4x \le 16$

8. $-2t + 3 \ge 11$;
$-2t \ge 14$

*Determine whether each number is a solution of the
given inequality.*

9. $x > -4$

   a) 4   **b)** -6 **c)** -4

10. $t < 3$

   a) -3 **b)** 3   **c)** $2\frac{19}{20}$

11. $y \le 19$

   a) 18.99   **b)** 19.01   **c)** 19

12. $n \ge -4$

   a) 0   **b)** -4.1 **c)** -3.9

13. $c \ge -7$

   a) 0   **b)** -5.4 **c)** 7.1

14. $a > 6$

   a) 6   **b)** -6.7 **c)** 0

15. $z < -3$

   a) 0   **b)** $-3\frac{1}{3}$ **c)** 1

16. $m \le -2$

   a) $-1\frac{9}{10}$ **b)** 0   **c)** $-2\frac{1}{3}$

Graph on a number line.

17. $y < 2$

18. $x \le 7$

19. $x \ge -1$

20. $t > -2$

21. $0 \le t$

22. $1 \le m$

23. $-5 \le x < 2$

24. $-3 < x \le 5$

25. $-4 < x < 0$

26. $0 \le x \le 5$

Describe each graph using set-builder notation.

27.

28.

29.

30.

31.

32.

33.

34.

Solve using the addition principle. Graph and write set-builder notation for each answer.

35. $y + 6 > 9$

36. $y + 2 > 9$

37. $x + 9 \le -12$

38. $x + 8 \le -10$

39. $n - 6 < 11$

40. $n - 4 > -3$

41. $2x \le x - 9$

42. $3x \le 2x + 7$

43. $y + \frac{1}{3} \le \frac{5}{6}$

44. $x + \frac{1}{4} \le \frac{1}{2}$

45. $t - \frac{1}{8} > \frac{1}{2}$

46. $y - \frac{1}{3} > \frac{1}{4}$

47. $-9x + 17 > 17 - 8x$

48. $-8n + 12 > 12 - 7n$

Aha! **49.** $-23 < -t$

50. $19 < -x$

51. $10 - y \le -12$

52. $3 - y \ge -6$

Solve using the multiplication principle. Graph and write set-builder notation for each answer.

53. $4x < 28$

54. $3x \ge 24$

55. $-7x < 13$

56. $8y < 17$

57. $-24 > 8t$

58. $-16x < -64$

59. $1.8 \ge -1.2n$

60. $9 \le -2.5a$

61. $-2y \le \frac{1}{5}$

62. $-2x \ge \frac{1}{5}$

63. $-\frac{8}{5} > 2x$

64. $-\frac{5}{8} < -10y$

Solve using the addition and multiplication principles.

65. $2 + 3x < 20$

66. $7 + 4y < 31$

67. $4t - 5 \le 23$

68. $15x - 7 \le -7$

69. $16 \le 6 - 10y$

70. $22 < 6 - 8x$

71. $39 > 3 - 9x$

72. $5 > 5 - 7y$

73. $5 - 6y > 25$

74. $8 - 2y > 9$

75. $-3 < 8x + 7 - 7x$

76. $-5 < 9x + 8 - 8x$

77. $6 - 4y > 6 - 3y$

78. $7 - 8y > 5 - 7y$

79. $7 - 9y \le 4 - 7y$

80. $6 - 13y \le 4 - 12y$

81. $33 - 12x < 4x + 97$

82. $27 - 11x > 14x - 18$

83. $2.1x + 43.2 > 1.2 - 8.4x$

84. $0.96y - 0.79 \le 0.21y + 0.46$

85. $1.7t + 8 - 1.62t < 0.4t - 0.32 + 8$

86. $0.7n - 15 + n \ge 2n - 8 - 0.4n$

87. $\frac{x}{3} + 4 \le 1$

88. $\frac{2}{3} - \frac{x}{5} < \frac{4}{15}$

89. $3 < 5 - \frac{t}{7}$

90. $2 > 9 - \frac{x}{5}$

91. $4(2y - 3) \le -44$

92. $3(2y - 3) > 21$

93. $8(2t + 1) > 4(7t + 7)$

94. $3(t - 2) \ge 9(t + 2)$

95. $3(r - 6) + 2 < 4(r + 2) - 21$

96. $5(t + 3) + 9 \ge 3(t - 2) - 10$

97. $\frac{4}{5}(3x + 4) \le 20$

98. $\frac{2}{3}(2x - 1) \ge 10$

99. $\frac{2}{3}\left(\frac{7}{8} - 4x\right) - \frac{5}{8} < \frac{3}{8}$

100. $\frac{3}{4}\left(3x - \frac{1}{2}\right) - \frac{2}{3} < \frac{1}{3}$

101. Are the inequalities $x > -3$ and $x \ge -2$ equivalent? Why or why not?

102. Are the inequalities $t < -7$ and $t \le -8$ equivalent? Why or why not?

Skill Review

Review simplifying expressions (Section 1.8).

Simplify. [1.8]

103. $5x - 2(3 - 6x)$

104. $8m - n - 3(2m + 5n)$

105. $x - 2[4y + 3(8 - x) - 1]$

106. $5 - 3t - 4[6 + 5(2t - 1) + t]$

107. $3[5(2a - b) + 1] - 5[4 - (a - b)]$

108. $9x - 2\{4 - 5[6 - 2(x + 1) - x]\}$

Synthesis

109. Explain how it is possible for the graph of an inequality to consist of just one number. (*Hint*: See Example 3c.)

110. The statements of the addition and multiplication principles begin with *conditions* set for the variables. Explain the conditions given for each principle.

Solve.

Aha! 111. $x < x + 1$

112. $6[4 - 2(6 + 3t)] > 5[3(7 - t) - 4(8 + 2t)] - 20$

113. $27 - 4[2(4x - 3) + 7] \geq 2[4 - 2(3 - x)] - 3$

Solve for x.

114. $\frac{1}{2}(2x + 2b) > \frac{1}{3}(21 + 3b)$

115. $-(x + 5) \geq 4a - 5$

116. $y < ax + b$ (Assume $a < 0$.)

117. $y < ax + b$ (Assume $a > 0$.)

118. Graph the solutions of $|x| < 3$ on a number line.

Aha! 119. Determine the solution set of $|x| > -3$.

120. Determine the solution set of $|x| < 0$.

CONNECTING the CONCEPTS

The procedure for solving inequalities is very similar to that used to solve equations. There are, however, two important differences.

- The multiplication principle for inequalities differs from the multiplication principle for equations: When we multiply or divide on both sides of an inequality by a *negative* number, we must *reverse* the direction of the inequality.

- The solution set of an equation like those we solved in this chapter typically consists of one number. The solution set of an inequality typically consists of a set of numbers and is written using set-builder notation.

Compare the following solutions.

Solve: $2 - 3x = x + 10$.

SOLUTION

$2 - 3x = x + 10$

$-3x = x + 8$ Subtracting 2 from both sides

$-4x = 8$ Subtracting x from both sides

$x = -2$ Dividing both sides by -4

The solution is -2.

Solve: $2 - 3x > x + 10$.

SOLUTION

$2 - 3x > x + 10$

$-3x > x + 8$ Subtracting 2 from both sides

$-4x > 8$ Subtracting x from both sides

$x < -2$ Dividing both sides by -4 and reversing the direction of the inequality symbol

The solution is $\{x | x < -2\}$.

MIXED REVIEW

Solve.

1. $x - 6 = 15$

2. $x - 6 \leq 15$

3. $3x = -18$

4. $3x > -18$

5. $-3x > -18$

6. $5x + 2 = 17$

7. $7 - 3x = 8$

8. $4y - 7 < 5$

9. $3 - t \geq 19$

10. $2 + 3n = 5n - 9$

11. $3 - 5a > a + 9$

12. $1.2x - 3.4 < 0.4x + 5.2$

13. $\frac{2}{3}(x + 5) \geq -4$

14. $\frac{n}{5} - 6 = 15$

15. $0.5x - 2.7 = 3x + 7.9$

16. $5(6 - t) = -45$

17. $8 - \frac{y}{3} \leq 7$

18. $\frac{1}{3}x - \frac{5}{6} = \frac{3}{2} - \frac{1}{6}x$

19. $-15 > 7 - 5x$

20. $10 \geq -2(a - 5)$

2.7   Solving Applications with Inequalities

Translating to Inequalities • Solving Problems

The five steps for problem solving can be used for problems involving inequalities.

Translating to Inequalities

Before solving problems that involve inequalities, we list some important phrases to look for. Sample translations are listed as well.

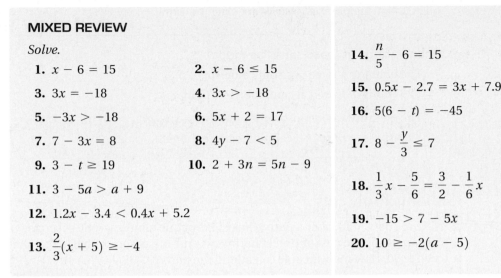

Important Words	Sample Sentence	Translation
is at least	Ming walks at least 2 mi a day.	$m \geq 2$
is at most	At most 5 students dropped the course.	$n \leq 5$
cannot exceed	The width cannot exceed 40 ft.	$w \leq 40$
must exceed	The speed must exceed 15 mph.	$s > 15$
is less than	Kamal's weight is less than 120 lb.	$w < 120$
is more than	Boston is more than 200 mi away.	$d > 200$
is between	The film was between 90 and 100 min long.	$90 < t < 100$
minimum	Ned drank a minimum of 5 glasses of water a day.	$w \geq 5$
maximum	The maximum penalty is $100.	$p \leq 100$
no more than	Alan weighs no more than 90 lb.	$w \leq 90$
no less than	Mallory scored no less than 8.3.	$s \geq 8.3$

The following phrases deserve special attention.

Translating "at least" and "at most"

The quantity x is at least some amount q: $x \geq q$.

(If x is *at least q*, it cannot be less than q.)

The quantity x is at most some amount q: $x \leq q$.

(If x is *at most q*, it cannot be more than q.)

Solving Problems

EXAMPLE **1**

Catering costs. To cater a party, Curtis' Barbeque charges a $50 setup fee plus $15 per person. The cost of Hotel Pharmacy's end-of-season softball party cannot exceed $450. How many people can attend the party?

SOLUTION

1. **Familiarize.** Suppose that 20 people were to attend the party. The cost would then be $50 + $15 · 20, or $350. This shows that more than 20 people could attend without exceeding $450. Instead of making another guess, we let n = the number of people in attendance.

2. **Translate.** The cost of the party will be $50 for the setup fee plus $15 times the number of people attending. We can reword as follows:

Rewording:	The setup fee	plus	the cost of the meals	cannot exceed	$450.
	↓	↓	↓	↓	↓
Translating:	50	+	15 · n	≤	450

3. **Carry out.** We solve for n:

$$50 + 15n \leq 450$$
$$15n \leq 400 \qquad \text{Subtracting 50 from both sides}$$
$$n \leq \frac{400}{15} \qquad \text{Dividing both sides by 15}$$
$$n \leq 26\frac{2}{3}. \qquad \text{Simplifying}$$

4. **Check.** Although the solution set of the inequality is all numbers less than or equal to $26\frac{2}{3}$, since n represents the number of people in attendance, we round *down* to 26. If 26 people attend, the cost will be $50 + $15 · 26, or $440, and if 27 attend, the cost will exceed $450.

5. **State.** At most 26 people can attend the party.

▶ **TRY EXERCISE** ▶ 23

CAUTION! Solutions of problems should always be checked using the original wording of the problem. In some cases, answers might need to be whole numbers or integers or rounded off in a particular direction.

Some applications with inequalities involve *averages*, or *means*. You are already familiar with the concept of averages from grades in courses that you have taken.

Average, or Mean

To find the **average**, or **mean**, of a set of numbers, add the numbers and then divide by the number of addends.

EXAMPLE 2

Financial aid.   Full-time students in a health-care education program can receive financial aid and employee benefits from Covenant Health System by working at Covenant while attending school and also agreeing to work there after graduation. Students who work an average of at least 16 hr per week receive extra pay and part-time employee benefits. For the first three weeks of September, Dina worked 20 hr, 12 hr, and 14 hr. How many hours must she work during the fourth week in order to average at least 16 hr per week for the month?

Source: Covenant Health Systems

SOLUTION

1. **Familiarize.** Suppose Dina works 10 hr during the fourth week. Her average for the month would be

$$\frac{20\ \text{hr} + 12\ \text{hr} + 14\ \text{hr} + 10\ \text{hr}}{4} = 14\ \text{hr}.$$ There are 4 addends, so we divide by 4.

This shows that Dina must work more than 10 hr during the fourth week, if she is to average at least 16 hr of work per week. We let x represent the number of hours Dina works during the fourth week.

2. **Translate.** We reword the problem and translate as follows:

Rewording: The average number of hours worked   should be at least   16 hr.

Translating: $\dfrac{20 + 12 + 14 + x}{4}$ $\geq$ 16

3. **Carry out.** Because of the fraction, it is convenient to use the multiplication principle first:

$$\frac{20 + 12 + 14 + x}{4} \geq 16$$

$$4\left(\frac{20 + 12 + 14 + x}{4}\right) \geq 4 \cdot 16$$ Multiplying both sides by 4

$$20 + 12 + 14 + x \geq 64$$

$$46 + x \geq 64$$ Simplifying

$$x \geq 18.$$ Subtracting 46 from both sides

4. **Check.** As a partial check, we show that if Dina works 18 hr, she will average at least 16 hr per week:

$$\frac{20 + 12 + 14 + 18}{4} = \frac{64}{4} = 16.$$ Note that 16 is at least 16.

5. **State.** Dina will average at least 16 hr of work per week for September if she works at least 18 hr during the fourth week.

TRY EXERCISE 27

Translating for Success

1. **Consecutive integers.** The sum of two consecutive even integers is 102. Find the integers.

2. **Salary increase.** After Susanna earned a 5% raise, her new salary was $25,750. What was her former salary?

3. **Dimensions of a rectangle.** The length of a rectangle is 6 in. more than the width. The perimeter of the rectangle is 102 in. Find the length and the width.

4. **Population.** The population of Kelling Point is decreasing at a rate of 5% per year. The current population is 25,750. What was the population the previous year?

5. **Reading assignment.** Quinn has 6 days to complete a 150-page reading assignment. How many pages must he read the first day so that he has no more than 102 pages left to read on the 5 remaining days?

Translate each word problem to an equation or an inequality and select a correct translation from A–O.

A. $0.05(25{,}750) = x$

B. $x + 2x = 102$

C. $2x + 2(x + 6) = 102$

D. $150 - x \le 102$

E. $x - 0.05x = 25{,}750$

F. $x + (x + 2) = 102$

G. $x + (x + 6) > 102$

H. $x + 5x = 150$

I. $x + 0.05x = 25{,}750$

J. $x + (2x + 6) = 102$

K. $x + (x + 1) = 102$

L. $102 + x > 150$

M. $0.05x = 25{,}750$

N. $102 + 5x > 150$

O. $x + (x + 6) = 102$

Answers on page A-5

An additional, animated version of this activity appears in MyMathLab. To use MyMathLab, you need a course ID and a student access code. Contact your instructor for more information.

6. **Numerical relationship.** One number is 6 more than twice another. The sum of the numbers is 102. Find the numbers.

7. **DVD collections.** Together Mindy and Ken have 102 DVDs. If Ken has 6 more DVDs than Mindy, how many does each have?

8. **Sales commissions.** Kirk earns a commission of 5% on his sales. One year he earned commissions totaling $25,750. What were his total sales for the year?

9. **Fencing.** Jess has 102 ft of fencing that he plans to use to enclose two dog runs. The perimeter of one run is to be twice the perimeter of the other. Into what lengths should the fencing be cut?

10. **Quiz scores.** Lupe has a total of 102 points on the first 6 quizzes in her sociology class. How many total points must she earn on the 5 remaining quizzes in order to have more than 150 points for the semester?

2.7	EXERCISE SET

🖐 **Concept Reinforcement** *In each of Exercises 1–8, match the sentence with one of the following:*

$$a < b; \quad a \le b; \quad b < a; \quad b \le a.$$

1. a is at least b

2. a exceeds b.

3. a is at most b.

4. a is exceeded by b.

5. b is no more than a.

6. b is no less than a.

7. b is less than a.

8. b is more than a.

Translate to an inequality.

9. A number is less than 10.

10. A number is greater than or equal to 4.

11. The temperature is at most $-3°C$.

12. The average credit-card debt is more than $8000.

13. To rent a car, a driver must have a minimum of 5 yr driving experience.

14. The Barringdean Shopping Center is no more than 20 mi away.

15. The age of the Mayan altar exceeds 1200 yr.

16. The maximum safe exposure limit of formaldehyde is 2 parts per million.

17. Tania earns between $12 and $15 an hour.

18. Leslie's test score was at least 85.

19. Wind speeds were greater than 50 mph.

20. The costs of production of that software cannot exceed $12,500.

21. A room at Pine Tree Bed and Breakfast costs no more than $120 a night.

22. The cost of gasoline was at most $4 per gallon.

Use an inequality and the five-step process to solve each problem.

23. *Furnace repairs.* RJ's Plumbing and Heating charges $55 plus $40 per hour for emergency service. Gary remembers being billed over $150 for an emergency call. How long was RJ's there?

24. *College tuition.* Karen's financial aid stipulates that her tuition not exceed $1000. If her local community college charges a $35 registration fee plus $375 per course, what is the greatest number of courses for which Karen can register?

25. *Graduate school.* An unconditional acceptance into the Master of Business Administration (MBA) program at Arkansas State University will be given to students whose GMAT score plus 200 times the undergraduate grade point average is at least 950. Robbin's GMAT score was 500. What must her grade point average be in order to be unconditionally accepted into the program?
Source: graduateschool.astate.edu

26. *Car payments.* As a rule of thumb, debt payments (other than mortgages) should be less than 8% of a consumer's monthly gross income. Oliver makes $54,000 a year and has a $100 student-loan payment every month. What size car payment can he afford?
Source: money.cnn.com

27. *Quiz average.* Rod's quiz grades are 73, 75, 89, and 91. What scores on a fifth quiz will make his average quiz grade at least 85?

28. *Nutrition.* Following the guidelines of the U.S. Department of Agriculture, Dale tries to eat at least 5 half-cup servings of vegetables each day. For the first six days of one week, she had 4, 6, 7, 4, 6, and 4 servings. How many servings of vegetables should Dale eat on Saturday, in order to average at least 5 servings per day for the week?

29. *College course load.* To remain on financial aid, Millie needs to complete an average of at least 7 credits per quarter each year. In the first three quarters of 2008, Millie completed 5, 7, and 8 credits. How many credits of course work must Millie complete in the fourth quarter if she is to remain on financial aid?

30. *Music lessons.* Band members at Colchester Middle School are expected to average at least 20 min of practice time per day. One week Monroe practiced 15 min, 28 min, 30 min, 0 min, 15 min, and 25 min. How long must he practice on the seventh day if he is to meet expectations?

31. *Baseball.* In order to qualify for a batting title, a major league baseball player must average at least 3.1 plate appearances per game. For the first nine games of the season, a player had 5, 1, 4, 2, 3, 4, 4, 3, and 2 plate appearances. How many plate appearances must the player have in the tenth game in order to average at least 3.1 per game?
Source: Major League Baseball

32. *Education.* The Mecklenberg County Public Schools stipulate that a standard school day will average at least $5\frac{1}{2}$ hr, excluding meal breaks. For the first four days of one school week, bad weather resulted in school days of 4 hr, $6\frac{1}{2}$ hr, $3\frac{1}{2}$ hr, and $6\frac{1}{2}$ hr. How long must the Friday school day be in order to average at least $5\frac{1}{2}$ hr for the week?
Source: www.meck.k12.va.us

33. *Perimeter of a triangle.* One side of a triangle is 2 cm shorter than the base. The other side is 3 cm longer than the base. What lengths of the base will allow the perimeter to be greater than 19 cm?

34. *Perimeter of a sign.* The perimeter of a rectangular sign is not to exceed 50 ft. The length is to be twice the width. What widths will meet these conditions?

35. *Well drilling.* All Seasons Well Drilling offers two plans. Under the "pay-as-you-go" plan, they charge $500 plus $8 a foot for a well of any depth. Under their "guaranteed-water" plan, they charge a flat fee of $4000 for a well that is guaranteed to provide adequate water for a household. For what depths would it save a customer money to use the pay-as-you-go plan?

36. *Cost of road service.* Rick's Automotive charges $50 plus $15 for each (15-min) unit of time when making a road call. Twin City Repair charges $70 plus $10 for each unit of time. Under what circumstances would it be more economical for a motorist to call Rick's?

37. *Insurance-covered repairs.* Most insurance companies will replace a vehicle if an estimated repair exceeds 80% of the "blue-book" value of the vehicle. Michele's insurance company paid $8500 for repairs to her Subaru after an accident. What can be concluded about the blue-book value of the car?

38. *Insurance-covered repairs.* Following an accident, Jeff's Ford pickup was replaced by his insurance company because the damage was so extensive. Before the damage, the blue-book value of the truck was $21,000. How much would it have cost to repair the truck? (See Exercise 37.)

39. *Sizes of packages.* The U.S. Postal Service defines a "package" as a parcel for which the sum of the length and the girth is less than 84 in. (Length is the longest side of a package and girth is the distance around the other two sides of the package.) A box has a fixed girth of 29 in. Determine (in terms of an inequality) those lengths for which the box is considered a "package."

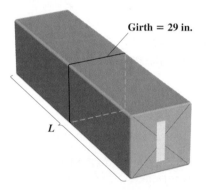

Girth = 29 in.

L

40. *Sizes of envelopes.* Rhetoric Advertising is a direct-mail company. It determines that for a particular campaign, it can use any envelope with a fixed width of $3\frac{1}{2}$ in. and an area of at least $17\frac{1}{2}$ in². Determine (in terms of an inequality) those lengths that will satisfy the company constraints.

41. *Body temperature.* A person is considered to be feverish when his or her temperature is higher than 98.6°F. The formula $F = \frac{9}{5}C + 32$ can be used to convert Celsius temperatures C to Fahrenheit temperatures F. For which Celsius temperatures is a person considered feverish?

42. *Gold temperatures.* Gold stays solid at Fahrenheit temperatures below 1945.4°. Determine (in terms of an inequality) those Celsius temperatures for which gold stays solid. Use the formula given in Exercise 41.

43. *Area of a triangular sign.* Zoning laws in Harrington prohibit displaying signs with areas exceeding 12 ft². If Flo's Marina is ordering a triangular sign with an 8-ft base, how tall can the sign be?

44. *Area of a triangular flag.* As part of an outdoor education course, Trisha needs to make a bright-colored triangular flag with an area of at least 3 ft². What heights can the triangle be if the base is $1\frac{1}{2}$ ft?

45. *Fat content in foods.* Reduced Fat Skippy® peanut butter contains 12 g of fat per serving. In order for a food to be labeled "reduced fat," it must have at least 25% less fat than the regular item. What can you conclude about the number of grams of fat in a serving of the regular Skippy peanut butter?
Source: Best Foods

46. *Fat content in foods.* Reduced Fat Chips Ahoy!® cookies contain 5 g of fat per serving. What can you conclude about the number of grams of fat in regular Chips Ahoy! cookies (see Exercise 45)?
Source: Nabisco Brands, Inc.

47. *Weight gain.* In the last weeks before the yearly Topsfield Weigh In, heavyweight pumpkins gain about 26 lb per day. Charlotte's heaviest pumpkin weighs 532 lb on September 5. For what dates will its weight exceed 818 lb?
Source: Based on a story in the *Burlington Free Press*

48. *Pond depth.* On July 1, Garrett's Pond was 25 ft deep. Since that date, the water level has dropped $\frac{2}{3}$ ft per week. For what dates will the water level not exceed 21 ft?

49. *Cell-phone budget.* Liam has budgeted $60 a month for his cell phone. For his service, he pays a monthly fee of $39.95, plus taxes of $6.65, plus 10¢ for each text message sent or received. How many text messages can he send or receive and not exceed his budget?

50. *Banquet costs.* The women's volleyball team can spend at most $700 for its awards banquet at a local restaurant. If the restaurant charges a $100 setup fee plus $24 per person, at most how many can attend?

51. *World records in the mile run.* The formula
$$R = -0.0065t + 4.3259$$
can be used to predict the world record, in minutes, for the 1-mi run t years after 1900. Determine (in terms of an inequality) those years for which the world record will be less than 3.6 min.
Source: Based on information from Information Please Database 2007, Pearson Education, Inc.

52. *Women's records in the women's 1500-m run.* The formula
$$R = -0.0026t + 4.0807$$
can be used to predict the world record, in minutes, for the 1500-m run t years after 1900. Determine (in terms of an inequality) those years for which the world record will be less than 3.8 min.
Source: Based on information from *Track and Field*

53. *Toll charges.* The equation

$$y = 0.06x + 0.50$$

can be used to determine the approximate cost y, in dollars, of driving x miles on the Pennsylvania Turnpike. For what mileages x will the cost be at most $14?

54. *Price of a movie ticket.* The average price of a movie ticket can be estimated by the equation

$$P = 0.169Y - 333.04,$$

where Y is the year and P is the average price, in dollars. For what years will the average price of a movie ticket be at least $7? (Include the year in which the $7 ticket first occurs.)
Source: National Association of Theatre Owners

55. If f represents Fran's age and t represents Todd's age, write a sentence that would translate to $t + 3 < f$.

56. Explain how the meanings of "Five more than a number" and "Five is more than a number" differ.

Skill Review

Review operations with real numbers (Sections 1.5–1.8).

Simplify.

57. $-2 + (-5) - 7$ [1.6]

58. $\dfrac{1}{2} \div \left(-\dfrac{3}{4}\right)$ [1.7]

59. $3 \cdot (-10) \cdot (-1) \cdot (-2)$ [1.7]

60. $-6.3 + (-4.8)$ [1.5]

61. $(3 - 7) - (4 - 8)$ [1.8]

62. $3 - 2 + 5 \cdot 10 \div 5^2 \cdot 2$ [1.8]

63. $\dfrac{-2 - (-6)}{8 - 10}$ [1.8]

64. $\dfrac{1 - (-7)}{-3 - 5}$ [1.8]

Synthesis

65. Write a problem for a classmate to solve. Devise the problem so the answer is "At most 18 passengers can go on the boat." Design the problem so that at least one number in the solution must be rounded down.

66. Write a problem for a classmate to solve. Devise the problem so the answer is "The Rothmans can drive 90 mi without exceeding their truck rental budget."

67. *Ski wax.* Green ski wax works best between 5° and 15° Fahrenheit. Determine those Celsius temperatures for which green ski wax works best. (See Exercise 41.)

68. *Parking fees.* Mack's Parking Garage charges $4.00 for the first hour and $2.50 for each additional hour. For how long has a car been parked when the charge exceeds $16.50?

Aha! **69.** The area of a square can be no more than 64 cm². What lengths of a side will allow this?

Aha! **70.** The sum of two consecutive odd integers is less than 100. What is the largest pair of such integers?

71. *Nutritional standards.* In order for a food to be labeled "lowfat," it must have fewer than 3 g of fat per serving. Reduced-fat tortilla chips contain 60% less fat than regular nacho cheese tortilla chips, but still cannot be labeled lowfat. What can you conclude about the fat content of a serving of nacho cheese tortilla chips?

72. *Parking fees.* When asked how much the parking charge is for a certain car (see Exercise 68), Mack replies, "between 14 and 24 dollars." For how long has the car been parked?

73. *Frequent buyer bonus.* Alice's Books allows customers to select one free book for every 10 books purchased. The price of that book cannot exceed the average cost of the 10 books. Neoma has bought 9 books whose average cost is $12 per book. How much should her tenth book cost if she wants to select a $15 book for free?

74. *Grading.* After 9 quizzes, Blythe's average is 84. Is it possible for Blythe to improve her average by two points with the next quiz? Why or why not?

75. *Discount card.* Barnes & Noble offers a member card for $25 a year. This card entitles a customer to a 40% discount off list price on hardcover bestsellers, a 20% discount on adult hardcovers, and a 10% discount on other purchases. Describe two sets of circumstances for which an individual would save money by becoming a member.
Source: Barnes & Noble

Study Summary

KEY TERMS AND CONCEPTS	EXAMPLES

SECTION 2.1: SOLVING EQUATIONS

Equivalent equations share the same solution.

$3x - 1 = 10$, $3x = 11$, and $x = \dfrac{11}{3}$ are equivalent equations.

The Addition Principle for Equations

$a = b$ is equivalent to
$a + c = b + c$.

$$x + 5 = -2 \quad \text{is equivalent to}$$
$$x + 5 + (-5) = -2 + (-5) \quad \text{and to}$$
$$x = -7.$$

The Multiplication Principle for Equations

$a = b$ is equivalent to $ac = bc$,
for $c \neq 0$.

$$-\tfrac{1}{3}x = 7 \quad \text{is equivalent to}$$
$$(-3)(-\tfrac{1}{3}x) = (-3)(7) \quad \text{and to}$$
$$x = -21.$$

SECTION 2.2: USING THE PRINCIPLES TOGETHER

We can **clear fractions** by multiplying both sides of an equation by the least common multiple of the denominators in the equation.

Solve: $\tfrac{1}{2}x - \tfrac{1}{3} = \tfrac{1}{6}x + \tfrac{2}{3}$.

$6\left(\tfrac{1}{2}x - \tfrac{1}{3}\right) = 6\left(\tfrac{1}{6}x + \tfrac{2}{3}\right)$	Multiplying by 6, the least common denominator
$6 \cdot \tfrac{1}{2}x - 6 \cdot \tfrac{1}{3} = 6 \cdot \tfrac{1}{6}x + 6 \cdot \tfrac{2}{3}$	Using the distributive law
$3x - 2 = x + 4$	Simplifying
$2x = 6$	Subtracting x from and adding 2 to both sides
$x = 3$	

We can **clear decimals** by multiplying both sides by a power of 10. If there is at most one decimal place in any one number, multiply by 10. If there are at most two decimal places, multiply by 100, and so on.

Solve: $3.6t - 1.5 = 2 - 0.8t$.

$10(3.6t - 1.5) = 10(2 - 0.8t)$	Multiplying both sides by 10 because the greatest number of decimal places is 1.
$36t - 15 = 20 - 8t$	Using the distributive law
$44t = 35$	Adding $8t$ and 15 to both sides
$t = \dfrac{35}{44}$	Dividing both sides by 44

SECTION 2.3: FORMULAS

A **formula** uses letters to show a relationship among two or more quantities. Formulas can be solved for a given letter using the addition and multiplication principles.

Solve: $x = \tfrac{2}{5}y + 7$ *for* y.

$x = \tfrac{2}{5}y + 7$	We are solving for y.
$x - 7 = \tfrac{2}{5}y$	Isolating the term containing y
$\tfrac{5}{2}(x - 7) = \tfrac{5}{2} \cdot \tfrac{2}{5}y$	Multiplying both sides by $\tfrac{5}{2}$
$\tfrac{5}{2}x - \tfrac{5}{2} \cdot 7 = 1 \cdot y$	Using the distributive law
$\tfrac{5}{2}x - \tfrac{35}{2} = y$	We have solved for y.

SECTION 2.4: APPLICATIONS WITH PERCENT

Percent Notation

$n\%$ means $\dfrac{n}{100}$, or $n \times \dfrac{1}{100}$, or $n \times 0.01$

$31\% = 0.31;$ $\dfrac{1}{8} = 0.125 = 12.5\%;$

$2.9\% = 0.029;$ $2.94 = 294\%$

Key Words in Percent Translations

"Of" translates to "$\cdot$" or "$\times$"

"What" translates to a variable

"Is" or "Was" translates to "$=$"

"%" translates to "$\times \frac{1}{100}$" or "$\times 0.01$"

What percent   of   60   is   7.2?

n $\cdot$ 60 $=$ 7.2

$$n = \frac{7.2}{60}$$

$$n = 0.12$$

Thus, 7.2 is 12% of 60.

SECTION 2.5: PROBLEM SOLVING

Five Steps for Problem Solving in Algebra

1. *Familiarize* yourself with the problem.
2. *Translate* to mathematical language. (This often means writing an equation.)
3. *Carry out* some mathematical manipulation. (This often means *solving* an equation.)
4. *Check* your possible answer in the original problem.
5. *State* the answer clearly.

The perimeter of a rectangle is 70 cm. The width is 5 cm longer than half the length. Find the length and the width.

1. **Familiarize.** Look up, if necessary, the formula for the perimeter of a rectangle:

   $$P = 2l + 2w.$$

   We are looking for two values, the length and the width. We can describe the width in terms of the length:

   $$w = \tfrac{1}{2}l + 5.$$

2. **Translate.**

   Rewording: Twice the length   plus   twice the width   is   the perimeter.

   Translating: $2l$ $+$ $2\left(\tfrac{1}{2}l + 5\right)$ $=$ 70

3. **Carry out.** Solve the equation:

   $$2l + 2\left(\tfrac{1}{2}l + 5\right) = 70$$

$2l + l + 10 = 70$	Using the distributive law
$3l + 10 = 70$	Combining like terms
$3l = 60$	Subtracting 10 from both sides
$l = 20.$	Dividing both sides by 3

   If $l = 20$, then $w = \tfrac{1}{2}l + 5 = \tfrac{1}{2} \cdot 20 + 5 = 10 + 5 = 15.$

4. **Check.** The width should be 5 cm longer than half the length. Since half the length is 10 cm, and 15 cm is 5 cm longer, this statement checks. The perimeter should be 70 cm. Since $2l + 2w = 2(20) + 2(15) = 40 + 30 = 70$, this statement also checks.

5. **State.** The length is 20 cm and the width is 15 cm.

SECTION 2.6: SOLVING INEQUALITIES

Solution sets of inequalities are represented using **set-builder notation** and can be **graphed** on a number line.

The solution set of the inequality $x \le 3$ is written $\{x \mid x \le 3\}$ and is graphed as follows.

The Addition Principle for Inequalities

For any real numbers *a*, *b*, and *c*,

$a < b$ is equivalent to $a + c < b + c$;

$a > b$ is equivalent to $a + c > b + c$.

Similar statements hold for $\leq$ and $\geq$.

$x + 3 \leq 5$ is equivalent to

$x + 3 - 3 \leq 5 - 3$ and to

$x \leq 2$.

The Multiplication Principle for Inequalities

For any real numbers *a* and *b*, and for any *positive* number *c*,

$a < b$ is equivalent to $ac < bc$;

$a > b$ is equivalent to $ac > bc$.

For any real numbers *a* and *b*, and for any *negative* number *c*,

$a < b$ is equivalent to $ac > bc$;

$a > b$ is equivalent to $ac < bc$.

Similar statements hold for $\leq$ and $\geq$.

$3x > 9$ is equivalent to

$\frac{1}{3} \cdot 3x > \frac{1}{3} \cdot 9$ The inequality symbol does not change because $\frac{1}{3}$ is positive.

$x > 3$.

$-3x > 9$ is equivalent to

$-\frac{1}{3} \cdot -3x < -\frac{1}{3} \cdot 9$ The inequality symbol is reversed because $-\frac{1}{3}$ is negative.

$x < -3$.

SECTION 2.7: SOLVING APPLICATIONS WITH INEQUALITIES

Many real-world problems can be solved by translating the problem to an inequality and applying the five-step problem-solving strategy.

Translate to an inequality.

The test score must exceed 85.	$s > 85$
At most 15 volunteers greeted visitors.	$v \leq 15$
Ona makes no more than $100 a week.	$w \leq 100$
Herbs need at least 4 hr of sun a day.	$h \geq 4$

Review Exercises: Chapter 2

🖦 **Concept Reinforcement** *Classify each statement as either true or false.*

1. $5x - 4 = 2x$ and $3x = 4$ are equivalent equations. [2.1]

2. $5 - 2t < 9$ and $t > 6$ are equivalent inequalities. [2.6]

3. Some equations have no solution. [2.1]

4. Consecutive odd integers are 2 units apart. [2.5]

5. For any number a, $a \leq a$. [2.6]

6. The addition principle is always used before the multiplication principle. [2.2]

7. A 10% discount results in a sale price that is 90% of the original price. [2.4]

8. Often it is impossible to list all solutions of an inequality number by number. [2.6]

Solve.

9. $x + 9 = -16$ [2.1]

10. $-8x = -56$ [2.1]

11. $-\dfrac{x}{5} = 13$ [2.1]

12. $-8 = n - 11$ [2.1]

13. $\frac{2}{5}t = -8$ [2.1]

14. $x - 0.1 = 1.01$ [2.1]

15. $-\frac{2}{3} + x = -\frac{1}{6}$ [2.1]

16. $4y + 11 = 5$ [2.2]

17. $5 - x = 13$ [2.2]

18. $3t + 7 = t - 1$ [2.2]

19. $7x - 6 = 25x$ [2.2]

20. $\frac{1}{4}x - \frac{5}{8} = \frac{3}{8}$ [2.2]

21. $14y = 23y - 17 - 10$ [2.2]

22. $0.22y - 0.6 = 0.12y + 3 - 0.8y$ [2.2]

23. $\frac{1}{4}x - \frac{1}{8}x = 3 - \frac{1}{16}x$ [2.2]

24. $6(4 - n) = 18$ [2.2]

25. $4(5x - 7) = -56$ [2.2]

26. $8(x - 2) = 4(x - 4)$ [2.2]

27. $-5x + 3(x + 8) = 16$ [2.2]

Solve each formula for the given letter. [2.3]

28. $C = \pi d$, for d

29. $V = \dfrac{1}{3}Bh$, for B

30. $5x - 2y = 10$, for y

31. $tx = ax + b$, for x

32. Find decimal notation: 1.2%. [2.4]

33. Find percent notation: $\frac{11}{25}$. [2.4]

34. What percent of 60 is 42? [2.4]

35. 49 is 35% of what number? [2.4]

Determine whether each number is a solution of $x \leq -5$. [2.6]

36. -3 37. -7 38. 4

Graph on a number line. [2.6]

39. $5x - 6 < 2x + 3$ 40. $-2 < x \leq 5$

41. $t > 0$

Solve. Write the answers in set-builder notation. [2.6]

42. $t + \frac{2}{3} \geq \frac{1}{6}$

43. $9x \geq 63$

44. $2 + 6y > 20$

45. $7 - 3y \geq 27 + 2y$

46. $3x + 5 < 2x - 6$

47. $-4y < 28$

48. $3 - 4x < 27$

49. $4 - 8x < 13 + 3x$

50. $13 \leq -\frac{2}{3}t + 5$

51. $7 \leq 1 - \frac{3}{4}x$

Solve.

52. In 2006, U.S. retailers lost a record $41.6 billion due to theft and fraud. Of this amount, $20 billion was due to employee theft. What percent of the total loss was employee theft? [2.4]
   Source: www.wwaytv3.com

53. A 18-ft beam is cut into two pieces. One piece is 2 ft longer than the other. How long are the pieces? [2.5]

54. In 2004, a total of 103,000 students from China and Japan enrolled in U.S. colleges and universities. The number of Japanese students was 10,000 more than half the number of Chinese students. How many Chinese students and how many Japanese students enrolled in the United States? [2.5]
   Source: Institute of International Education

55. The sum of two consecutive odd integers is 116. Find the integers. [2.5]

56. The perimeter of a rectangle is 56 cm. The width is 6 cm less than the length. Find the width and the length. [2.5]

57. After a 25% reduction, a picnic table is on sale for $120. What was the regular price? [2.4]

58. From 2000 to 2006, the number of U.S. wireless-phone subscribers increased by 114 percent to 233 million. How many subscribers were there in 2000? [2.4]
Source: Cellular Telecommunications and Internet Association

59. The measure of the second angle of a triangle is 50° more than that of the first. The measure of the third angle is 10° less than twice the first. Find the measures of the angles. [2.5]

60. The U.S. Centers for Disease Control recommends that for a typical 2000-calorie daily diet, no more than 65 g of fat be consumed. In the first three days of a four-day vacation, Teresa consumed 55 g, 80 g, and 70 g of fat. Determine how many grams of fat Teresa can consume on the fourth day if she is to average no more than 65 g of fat per day. [2.7]

61. *Blueprints.* To make copies of blueprints, Vantage Reprographics charges a $6 setup fee plus $4 per copy. Myra can spend no more than $65 for the copying. What number of copies will allow her to stay within budget? [2.7]

Synthesis

62. How does the multiplication principle for equations differ from the multiplication principle for inequalities? [2.1], [2.6]

63. Explain how checking the solutions of an equation differs from checking the solutions of an inequality. [2.1], [2.6]

64. A study of sixth- and seventh-graders in Boston revealed that, on average, the students spent 3 hr 20 min per day watching TV or playing video and computer games. This represents 108% more than the average time spent reading or doing homework. How much time each day was spent, on average, reading or doing homework? [2.4]
Source: Harvard School of Public Health

65. In June 2007, a team of Brazilian scientists exploring the Amazon measured its length as 65 mi longer than the Nile. If the combined length of both rivers is 8385 mi, how long is each river? [2.5]
Source: news.nationalgeographic.com

66. Kent purchased a book online at 25% off the retail price. The shipping charges were $4.95. If the amount due was $16.95, what was the retail price of the book? [2.4], [2.5]

Solve.

67. $2|n| + 4 = 50$ [1.4], [2.2]

68. $|3n| - 60$ [1.4], [2.1]

69. $y = 2a - ab + 3$, for a [2.3]

70. The Maryland Heart Center gives the following steps to calculate the number of fat grams needed daily by a moderately active woman. Write the steps as one formula relating the number of fat grams F to a woman's weight w, in pounds. [2.3]

  1. Calculate the total number of calories per day.
    ____ pounds × 12 calories = ____ total calories per day

  2. Take the total number of calories and multiply by 30 percent.
    ____ calories per day × 0.30 = ____ calories from fat per day.

  3. Take the number of calories from fat per day and divide by 9 (there are 9 calories per gram of fat).
    ____ calories from fat per day divided by 9 = ____ fat grams per day

Solve.

1. $t + 7 = 16$

2. $t - 3 = 12$

3. $6x = -18$

4. $-\frac{4}{7}x = -28$

5. $3t + 7 = 2t - 5$

6. $\frac{1}{2}x - \frac{3}{5} = \frac{2}{5}$

7. $8 - y = 16$

8. $4.2x + 3.5 = 1.2 - 2.5x$

9. $4(x + 2) = 36$

10. $9 - 3x = 6(x + 4)$

11. $\frac{5}{6}(3x + 1) = 20$

Solve. Write the answers in set-builder notation.

12. $x + 6 > 1$

13. $14x + 9 > 13x - 4$

14. $-5y \geq 65$

15. $4y \leq -30$

16. $4n + 3 < -17$

17. $3 - 5x > 38$

18. $\frac{1}{2}t - \frac{1}{4} \leq \frac{3}{4}t$

19. $5 - 9x \geq 19 + 5x$

Solve each formula for the given letter.

20. $A = 2\pi rh$, for r

21. $w = \dfrac{P + l}{2}$, for l

22. Find decimal notation: 230%.

23. Find percent notation: 0.003.

24. What number is 18.5% of 80?

25. What percent of 75 is 33?

Graph on a number line.

26. $y < 4$

27. $-2 \leq x \leq 2$

Solve.

28. The perimeter of a rectangular calculator is 36 cm. The length is 4 cm greater than the width. Find the width and the length.

29. In 1948, Earl Shaffer became the first person to hike all 2100 mi of the Appalachian trail—from Springer Mountain, Georgia, to Mt. Katahdin, Maine. Shaffer repeated the feat 50 years later, and at age 79 became the oldest person to hike the entire trail. When Shaffer stood atop Big Walker Mountain, Virginia, he was three times as far from the northern end of the trail as from the southern end. At that point, how far was he from each end of the trail?

30. The perimeter of a triangle is 249 mm. If the sides are consecutive odd integers, find the length of each side.

31. By lowering the temperature of their electric hot-water heater from 140°F to 120°F, the Kellys' average electric bill dropped by 7% to $60.45. What was their electric bill before they lowered the temperature of their hot water?

32. *Mass transit.* Local light rail service in Denver, Colorado, costs $1.50 per trip (one way). A monthly pass costs $54. Gail is a student at Community College of Denver. Express as an inequality the number of trips per month that Gail should make if the pass is to save her money.
Source: rtd-denver.com

Synthesis

Solve.

33. $c = \dfrac{2cd}{a - d}$, for d

34. $3|w| - 8 = 37$

35. Translate to an inequality.

A plant marked "partial sun" needs at least 4 hr but no more than 6 hr of sun each day.
Source: www.yardsmarts.com

36. A concert promoter had a certain number of tickets to give away. Five people got the tickets. The first got one third of the tickets, the second got one fourth of the tickets, and the third got one fifth of the tickets. The fourth person got eight tickets, and there were five tickets left for the fifth person. Find the total number of tickets given away.

Cumulative Review: Chapters 1–2

Simplify.

1. $18 + (-30)$ [1.5]

2. $\frac{1}{2} - \left(-\frac{1}{4}\right)$ [1.6]

3. $-1.2(3.5)$ [1.7]

4. $-5 \div \left(-\frac{1}{2}\right)$ [1.7]

5. $150 - 10^2 \div 25 \cdot 4$ [1.8]

6. $(150 - 10^2) \div (25 \cdot 4)$ [1.8]

Remove parentheses and simplify. [1.8]

7. $5x - (3x - 1)$

8. $2(t + 6) - 12t$

9. $3[4n - 5(2n - 1)] - 3(n - 7)$

10. Graph on a number line: $-\frac{5}{2}$. [1.4]

11. Find the absolute value: $|27|$. [1.4]

12. Factor: $12x + 18y + 30z$. [1.8]

Solve.

13. $12 = -2x$ [2.1]

14. $4x - 7 = 3x + 9$ [2.1]

15. $\frac{2}{3}t + 7 = 13$ [2.2]

16. $9(2a - 1) = 4$ [2.2]

17. $12 - 3(5x - 1) = x - 1$ [2.2]

18. $3(x + 1) - 2 = 8 - 5(x + 7)$ [2.2]

Solve each formula for the given letter. [2.3]

19. $\frac{1}{2}x = 2yz$, for z

20. $4x - 9y = 1$, for y

21. $an = p - rn$, for n

22. Find decimal notation: 183%. [2.4]

23. Find percent notation: $\frac{3}{8}$. [2.4]

24. Graph on a number line: $t > -\frac{5}{2}$. [2.6]

Solve. Write the answer in set-builder notation. [2.6]

25. $4t + 10 \le 2$

26. $8 - t > 5$

27. $4 < 10 - \dfrac{x}{5}$

28. $4(2n - 3) \le 2(5n - 8)$

Solve.

29. The total attendance at NCAA basketball games during the 2006–2007 school year was 33 million. This was 31.25% less than the total attendance at NCAA football games during that year. What was the total attendance at NCAA football games during the 2006–2007 school year? [2.4]
Source: NCAA

30. On an average weekday, a full-time college student spends a total of 7.1 hr in educational activities and in leisure activities. The average student spends 0.7 hr more in leisure activities than in educational activities. On an average weekday, how many hours does a full-time college student spend on educational activities? [2.5]
Source: U.S. Bureau of Labor Statistics

31. The wavelength w, in meters per cycle, of a musical note is given by

$$w = \frac{r}{f},$$

where r is the speed of the sound, in meters per second, and f is the frequency, in cycles per second. The speed of sound in air is 344 m/sec. What is the wavelength of a note whose frequency in air is 24 cycles per second? [2.3]

32. A 24-ft ribbon is cut into two pieces. One piece is 6 ft longer than the other. How long are the pieces? [2.5]

33. Juanita has budgeted an average of $65 a month for entertainment. For the first five months of the year, she has spent $88, $15, $125, $50, and $60. How much can Juanita spend in the sixth month without exceeding her average budget? [2.7]

34. In 2006, about 17 million Americans had diabetes. The U.S. Centers for Disease Control predicts that by 2050, 50 million Americans may have the disease. By what percent would the number of Americans with diabetes increase? [2.4]

35. The second angle of a triangle is twice as large as the first. The third angle is 5° more than four times the first. Find the measure of the largest angle. [2.5]

36. The length of a rectangular frame is 53 cm. For what widths would the perimeter be greater than 160 cm? [2.7]

Synthesis

37. Simplify: $t - \{t - [3t - (2t - t) - t] - 4t\} - t$. [1.8]

38. Solve: $3|n| + 10 = 25$. [2.2]

39. Lindy sold her Fender acoustic guitar on eBay using i-soldit.com. The i-soldit location she used charges 35% of the first $500 of the selling price and 20% of the amount over $500. After these charges were deducted from the selling price, she received $745. For how much did her guitar sell? [2.4], [2.5]
Source: Based on information in *The Wall Street Journal*, 9/11/07

Introduction to Graphing

HEATHER HUTH
DIRECTOR OF VOLUNTEERS
New Orleans, Louisiana

We average about 750 volunteers a month who come to Beacon of Hope to help us clean New Orleans' neighborhoods. We service 22 neighborhoods, so I use math to figure out how many volunteers to send to each location, the quantity of supplies each will need, and how much time is required to complete the task.

AN APPLICATION

An increasing number of college students are donating time and energy in volunteer service. The number of college students volunteering grew from 2.7 million in 2002 to 3.3 million in 2005. Graph the given data and estimate the number of college students who volunteered in 2004, and then predict the number of college students who will volunteer in 2010.

Source: Corporation for National and Community Service

This problem appears as Example 7 in Section 3.7.

We now begin our study of graphing. First we will examine graphs as they commonly appear in newspapers or magazines and develop some terminology. Following that, we will graph certain equations and study the connection between rate and slope. We will also learn how graphs can be used as a problem-solving tool in many applications.

Our work in this chapter centers on equations that contain two variables.

3.1   Reading Graphs, Plotting Points, and Scaling Graphs

Problem Solving with Bar, Circle, and Line Graphs   •   Points and Ordered Pairs   •
Numbering the Axes Appropriately

Today's print and electronic media make almost constant use of graphs. In this section, we consider problem solving with bar graphs, line graphs, and circle graphs. Then we examine graphs that use a coordinate system.

Problem Solving with Bar, Circle, and Line Graphs

A *bar graph* is a convenient way of showing comparisons. In every bar graph, certain categories, such as levels of education in the example below, are paired with certain numbers.

EXAMPLE   1

Lifetime earnings.   Getting a college degree usually means delaying the start of a career. As the bar graph below shows, this loss in earnings is more than made up over a worker's lifetime.

Source: U.S. Census Bureau

Studying Can Pay

Vertical axis: Lifetime earnings estimates, full-time workers ages 25 to 64 (in millions) — $5, 4, 3, 2, 1

Horizontal axis (Levels of education): Less than high school, High school, Associate's degree, Bachelor's degree, Master's degree, Doctoral degree, Professional degree

a) Keagan plans to get an associate's degree. How much can he expect to make in his lifetime?

b) Isabella would like to make at least $2 million in her lifetime. What level of education should she pursue?

SOLUTION

a) Since level of education is shown on the horizontal scale, we go to the top of the bar above the label "associate's degree." Then we move horizontally from the top of the bar to the vertical scale, which shows earnings. We read there that Keagan can expect to make about $1.6 million in his lifetime.

b) By moving up the vertical scale to $2 million and then moving horizontally, we see that the first bar to reach a height of $2 million or higher corresponds to a bachelor's degree. Thus Isabella should pursue a bachelor's, master's, doctoral, or professional degree in order to make at least $2 million in her lifetime.

> **TRY EXERCISE** 5

Circle graphs or *pie charts,* are often used to show what percent of the whole each particular item in a group represents.

EXAMPLE 2

Student aid.   The circle graph below shows the sources for student aid in 2006 and the percentage of aid students received from each source. In that year, the total amount of aid distributed was $134.8 billion. About 5,387,000 students received a federal Pell grant. What was the average amount of the aid per recipient?

Source: Trends in Student Aid 2006, www.collegeboard.com

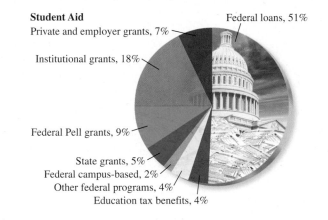

Student Aid

Private and employer grants, 7%
Institutional grants, 18%
Federal loans, 51%
Federal Pell grants, 9%
State grants, 5%
Federal campus-based, 2%
Other federal programs, 4%
Education tax benefits, 4%

SOLUTION

1. **Familiarize.** The problem involves percents, so if we were unsure of how to solve percent problems, we might review Section 2.4.

    The solution of this problem will involve two steps. We are told the total amount of student aid distributed. In order to find the average amount of a Pell grant, we must first calculate the total of all Pell grants and then divide by the number of students.

    We let g = the average amount of a Pell grant in 2006.

2. **Translate.** From the circle graph, we see that federal Pell grants were 9% of the total amount of aid. The total amount distributed was $134.8 billion, or $134,800,000,000, so we have

> Find the value of all Pell grants.

$$\text{the value of all Pell grants} = 0.09(134,800,000,000)$$
$$= 12,132,000,000.$$

Then we reword the problem and translate as follows:

> Calculate the average amount of a Pell grant.

Rewording: The average amount of a Pell grant   is   the value of all Pell grants   divided by   the number of recipients.

Translating: g =   12,132,000,000   ÷   5,387,000

3. **Carry out.** We solve the equation:

$$g = 12,132,000,000 \div 5,387,000$$
$$\approx 2252. \qquad \text{Rounding to the nearest dollar}$$

4. Check. If each student received $2252, the total amount of aid distributed through Pell grants would be $2252 · 5,387,000, or $12,131,524,000. Since this is approximately 9% of the total student aid for 2006, our answer checks.

5. State. In 2006, the average Pell grant was $2252.　　TRY EXERCISE ▶ 9

EXAMPLE ▶ 3

Exercise and pulse rate. The following *line graph* shows the relationship between a person's resting pulse rate and months of regular exercise.* Note that the symbol ⸦ is used to indicate that counting on the vertical scale begins at 50.

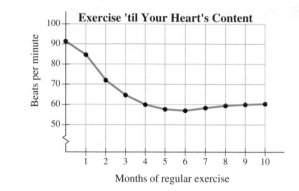

a) How many months of regular exercise are required to lower the pulse rate as much as possible?

b) How many months of regular exercise are needed to achieve a pulse rate of 65 beats per minute?

SOLUTION

a) The lowest point on the graph occurs above the number 6. Thus, after 6 months of regular exercise, the pulse rate is lowered as much as possible.

b) To determine how many months of exercise are needed to lower a person's resting pulse rate to 65, we locate 65 midway between 60 and 70 on the vertical scale. From that location, we move right until the line is reached. At that point, we move down to the horizontal scale and read the number of months required, as shown.

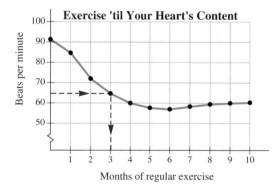

The pulse rate is 65 beats per minute after 3 months of regular exercise.

TRY EXERCISE ▶ 17

---

*Data from *Body Clock* by Dr. Martin Hughes (New York: Facts on File, Inc.), p. 60.

moved, its coordinates appearing at the bottom of the window.

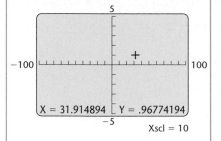

Xscl = 10

Set up the following viewing windows, choosing an appropriate scale for each axis. Then move the cursor and practice reading coordinates.

1. $[-10, 10, -10, 10]$
2. $[-5, 5, 0, 100]$
3. $[-1, 1, -0.1, 0.1]$

Points and Ordered Pairs

The line graph in Example 3 contains a collection of points. Each point pairs up a number of months of exercise with a pulse rate. To create such a graph, we **graph**, or **plot**, pairs of numbers on a plane. This is done using two perpendicular number lines called **axes** (pronounced "ak-sēz"; singular, **axis**). The point at which the axes cross is called the **origin**. Arrows on the axes indicate the positive directions.

Consider the pair $(3, 4)$. The numbers in such a pair are called **coordinates**. The **first coordinate** in this case is 3 and the **second coordinate** is 4.* To plot, or graph, $(3, 4)$, we start at the origin, move horizontally to the 3, move up vertically 4 units, and then make a "dot." Thus, $(3, 4)$ is located above 3 on the first axis and to the right of 4 on the second axis.

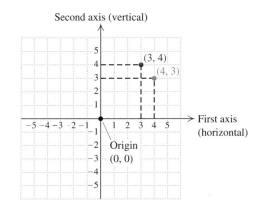

The point $(4, 3)$ is also plotted in the figure above. Note that $(3, 4)$ and $(4, 3)$ are different points. For this reason, coordinate pairs are called **ordered pairs**—the order in which the numbers appear is important.

EXAMPLE   4   Plot the point $(-3, 4)$.

SOLUTION   The first number, -3, is negative. Starting at the origin, we move 3 units in the negative horizontal direction (3 units to the left). The second number, 4, is positive, so we move 4 units in the positive vertical direction (up). The point $(-3, 4)$ is above -3 on the first axis and to the left of 4 on the second axis.

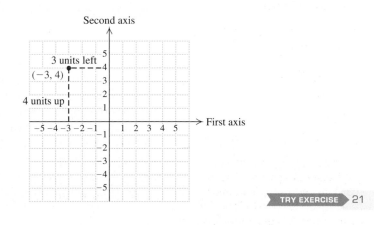

TRY EXERCISE   21

---

*The first coordinate is called the *abscissa* and the second coordinate is called the *ordinate*. The plane is called the *Cartesian coordinate plane* after the French mathematician René Descartes (1595–1650).

To find the coordinates of a point, we see how far to the right or left of the origin the point is and how far above or below the origin it is. Note that the coordinates of the origin itself are $(0, 0)$.

STUDENT NOTES ⎯⎯⎯⎯⎯

It is important that you remember that the first coordinate of an ordered pair is always represented on the horizontal axis. It is better to go slowly and plot points correctly than to go quickly and require more than one attempt.

Find the coordinates of points A, B, C, D, E, F, and G.

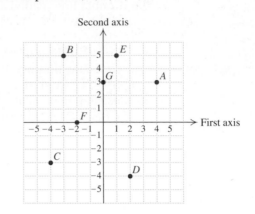

SOLUTION   Point A is 4 units to the right of the origin and 3 units above the origin. Its coordinates are $(4, 3)$. The coordinates of the other points are as follows:

B: $(-3, 5)$; C: $(-4, -3)$; D: $(2, -4)$;

E: $(1, 5)$; F: $(-2, 0)$; G: $(0, 3)$.

TRY EXERCISE   27

The variables x and y are commonly used when graphing on a plane. Coordinates of ordered pairs are often labeled

(x-coordinate, y-coordinate).

The first, or horizontal, axis is labeled the x-axis, and the second, or vertical, axis is labeled the y-axis.

Numbering the Axes Appropriately

In Examples 4 and 5, each square on the grid shown is 1 unit long and 1 unit high: The **scale** of both the x-axis and the y-axis is 1. Often it is necessary to use a different scale on one or both of the axes.

Use a grid 10 squares wide and 10 squares high to plot $(-34, 450)$, $(48, 95)$, and $(10, -200)$.

SOLUTION   Since x-coordinates vary from a low of -34 to a high of 48, the 10 horizontal squares must span $48 - (-34)$, or 82 units. Because 82 is not a multiple of 10, we round *up* to the next multiple of 10, which is 90. Dividing 90 by 10, we find that if each square is 9 units wide (has a scale of 9), we could represent all the x-values. However, since it is more convenient to count by 10's, we will instead use a scale of 10. Starting at 0, we count backward to -40 and forward to 60.

This is how we will arrange the x-axis.

There is more than one correct way to cover the values from -34 to 48 using 10 increments. For instance, we could have counted from -60 to 90, using a scale

This is
how we will
arrange
the *y*-axis.

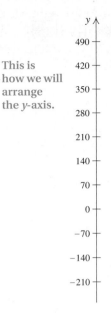

of 15. In general, we try to use the smallest range and scale that will cover the given coordinates. Scales that are multiples of 2, 5, or 10 are especially convenient. It is essential that the numbering always begin at the origin.

Since we must be able to show *y*-values from -200 to 450, the 10 vertical squares must span $450 - (-200)$, or 650 units. For convenience, we round 650 *up* to 700 and then divide by 10: $700 \div 10 = 70$. Using 70 as the scale, we count *down* from 0 until we pass -200 and *up* from 0 until we pass 450, as shown at left.

Next, we combine our work with the *x*-values and the *y*-values to draw a graph in which the *x*-axis extends from -40 to 60 with a scale of 10 and the *y*-axis extends from -210 to 490 with a scale of 70. To correctly locate the axes on the grid, the two 0's must coincide where the axes cross. Finally, once the graph has been numbered, we plot the points as shown below.

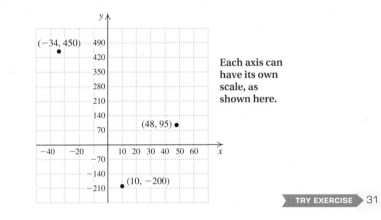

Each axis can
have its own
scale, as
shown here.

TRY EXERCISE ▶ 31

The horizontal and vertical axes divide the plane into four regions, or **quadrants**, as indicated by Roman numerals in the following figure. Note that the point $(-4, 5)$ is in the second quadrant and the point $(5, -5)$ is in the fourth quadrant. The points $(3, 0)$ and $(0, 1)$ are on the axes and are not considered to be in any quadrant.

Second quadrant:
First coordinate negative,
second coordinate positive:
$(-, +)$

First quadrant:
Both coordinates positive:
$(+, +)$

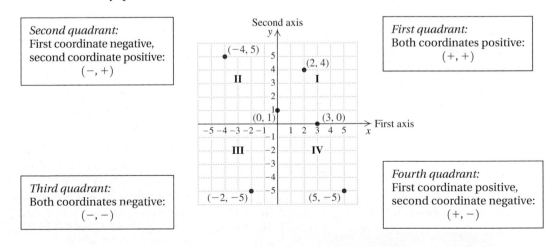

Third quadrant:
Both coordinates negative:
$(-, -)$

Fourth quadrant:
First coordinate positive,
second coordinate negative:
$(+, -)$

| 3.1 | EXERCISE SET |

↩ *Concept Reinforcement*   *In each of Exercises 1–4, match the set of coordinates with the graph on the right that would be the best for plotting the points.*

1. ____ $(-9, 3), (-2, -1), (4, 5)$

2. ____ $(-2, -1), (1, 5), (7, 3)$

3. ____ $(-2, -9), (2, 1), (4, -6)$

4. ____ $(-2, -1), (-9, 3), (-4, -6)$

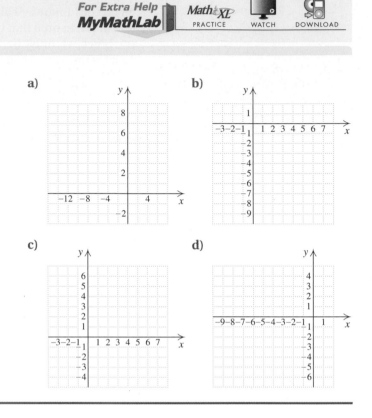

The ▸TRY EXERCISES▸ *for examples are indicated by a shaded block* ▢ *on the exercise number. Complete step-by-step solutions for these exercises appear online at www.pearsonhighered.com/bittingerellenbogen.*

Driving under the influence.   A blood-alcohol level of 0.08% or higher makes driving illegal in the United States. This bar graph shows how many drinks a person of a certain weight would need to consume in 1 hr to achieve a blood-alcohol level of 0.08%. Note that a 12-oz beer, a 5-oz glass of wine, or a cocktail containing $1\frac{1}{2}$ oz of distilled liquor all count as one drink.
Source: Adapted from soberup.com and vsa.vassar.edu/~source/drugs/alcohol.html

Friends Don't Let Friends Drive Drunk

Drinks (y-axis: 1 2 3 4 5 6)
Body weight (in pounds) (x-axis: 100 120 140 160 180 200 220 240)

5. Approximately how many drinks would a 100-lb person have consumed in 1 hr to reach a blood-alcohol level of 0.08%?

6. Approximately how many drinks would a 160-lb person have consumed in 1 hr to reach a blood-alcohol level of 0.08%?

7. What can you conclude about the weight of someone who has consumed 3 drinks in 1 hr without reaching a blood-alcohol level of 0.08%?

8. What can you conclude about the weight of someone who has consumed 4 drinks in 1 hr without reaching a blood-alcohol level of 0.08%?

Student aid.   Use the information in Example 2 to answer Exercises 9–12.

9. In 2006, there were 13,334,170 full-time equivalent students in U.S. colleges and universities. What was the average federal loan per full-time equivalent student?

10. In 2006, there were 13,334,170 full-time equivalent students in U.S. colleges and universities. What was the average education tax benefit received per full-time equivalent student?

11. Approximately 8.6% of campus-based federal student aid is given to students in two-year public institutions. How much campus-based aid did students at two-year public institutions receive in 2006?

12. Approximately 17.7% of Pell grant dollars is given to students in for-profit institutions. How much did students in for-profit institutions receive in Pell grants in 2006?

Sorting solid waste. *Use the following pie chart to answer Exercises 13–16.*

Sorting Solid Waste

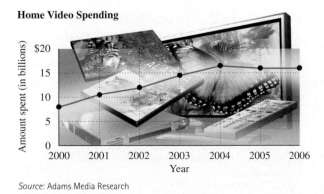

Paper and cardboard, 34.2%
Yard trimmings, 13.1%
Food scraps, 11.7%
Metals, 7.6%
Other, 3.4%
Glass, 5.2%
Rubber, leather, and textiles, 7.3%
Plastics, 11.9%
Wood, 5.7%

Source: Environmental Protection Agency

13. In 2005, Americans generated 245 million tons of waste. How much of the waste was plastic?

14. In 2005, the average American generated 4.5 lb of waste per day. How much of that was paper and cardboard?

15. Americans are recycling about 25.3% of all glass that is in the waste stream. How much glass did Americans recycle in 2005? (See Exercise 13.)

16. Americans are recycling about 61.9% of all yard trimmings. What amount of yard trimmings did the average American recycle per day in 2005? (Use the information in Exercise 14.)

Home video spending. *The line graph below shows U.S. consumer spending on home-video movies.*

Home Video Spending

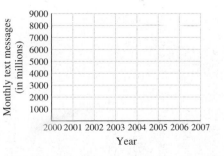

Source: Adams Media Research

17. Approximately how much was spent on home videos in 2002?

18. Approximately how much was spent on home videos in 2006?

19. In what year was approximately $10.5 billion spent on home videos?

20. In what year was approximately $14.5 billion spent on home videos?

Plot each group of points.

21. $(1, 2), (-2, 3), (4, -1), (-5, -3), (4, 0), (0, -2)$

22. $(-2, -4), (4, -3), (5, 4), (-1, 0), (-4, 4), (0, 5)$

23. $(4, 4), (-2, 4), (5, -3), (-5, -5), (0, 4), (0, -4),$ $(-4, 0), (0, 0)$

24. $(2, 5), (-1, 3), (3, -2), (-2, -4), (0, 0), (0, -5),$ $(5, 0), (-5, 0)$

25. *Text messaging.* Listed below are estimates of the number of text messages sent in the United States. Make a line graph of the data.

Year	Monthly Text Messages (in millions)
2000	12
2001	34
2002	931
2003	1221
2004	2862
2005	7253
2006	8000 (estimated)

Source: CSCA

26. *Ozone layer.* Listed below are estimates of the ozone level. Make a line graph of the data, listing years on the horizontal scale.

Year	Ozone Level (in Dobson Units)
2000	287.1
2001	288.2
2002	285.8
2003	285.0
2004	281.2
2005	283.5

Source: johnstonsarchive.net

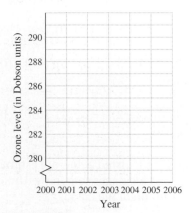

In Exercises 27–30, find the coordinates of points A, B, C, D, and E.

27.

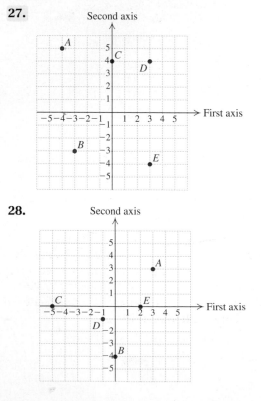

28.

29.

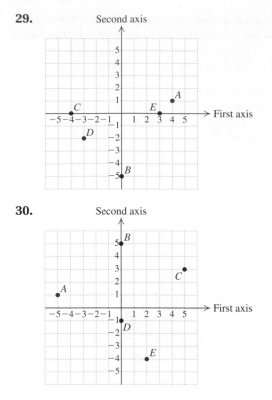

30.

In Exercises 31–40, use a grid 10 squares wide and 10 squares high to plot the given coordinates. Choose your scale carefully. Scales may vary.

31. $(-75, 5), (-18, -2), (9, -4)$

32. $(-13, 3), (48, -1), (62, -4)$

33. $(-1, 83), (-5, -14), (5, 37)$

34. $(2, -79), (4, -25), (-4, 12)$

35. $(-10, -4), (-16, 7), (3, 15)$

36. $(5, -16), (-7, -4), (12, 3)$

37. $(-100, -5), (350, 20), (800, 37)$

38. $(750, -8), (-150, 17), (400, 32)$

39. $(-83, 491), (-124, -95), (54, -238)$

40. $(738, -89), (-49, -6), (-165, 53)$

In which quadrant is each point located?

41. $(7, -2)$ **42.** $(-1, -4)$ **43.** $(-4, -3)$

44. $(1, -5)$ **45.** $(2, 1)$ **46.** $(-4, 6)$

47. $(-4.9, 8.3)$ **48.** $(7.5, 2.9)$

49. In which quadrants are the first coordinates positive?

50. In which quadrants are the second coordinates negative?

51. In which quadrants do both coordinates have the same sign?

52. In which quadrants do the first and second coordinates have opposite signs?

53. The following graph was included in a mailing sent by Agway® to their oil customers in 2000. What information is missing from the graph and why is the graph misleading?

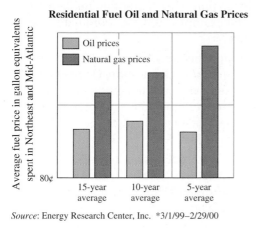

Residential Fuel Oil and Natural Gas Prices

Source: Energy Research Center, Inc. *3/1/99–2/29/00

54. What do all points plotted on the vertical axis of a graph have in common?

Skill Review

To prepare for Section 3.2, review solving for a variable (Section 2.3).

Solve for y. [2.3]

55. $5y = 2x$

56. $2y = -3x$

57. $x - y = 8$

58. $2x + 5y = 10$

59. $2x + 3y = 5$

60. $5x - 8y = 1$

Synthesis

61. In an article about consumer spending on home videos (see the graph used for Exercise 17), the *Wall Street Journal* (9/2/06) stated that "the movie industry has hit a wall." To what were they referring?

62. Describe what the result would be if the first and second coordinates of every point in the following graph of an arrow were interchanged.

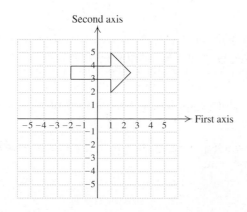

63. In which quadrant(s) could a point be located if its coordinates are opposites of each other?

64. In which quadrant(s) could a point be located if its coordinates are reciprocals of each other?

65. The points $(-1, 1)$, $(4, 1)$, and $(4, -5)$ are three vertices of a rectangle. Find the coordinates of the fourth vertex.

66. The pairs $(-2, -3)$, $(-1, 2)$, and $(4, -3)$ can serve as three (of four) vertices for three different parallelograms. Find the fourth vertex of each parallelogram.

67. Graph eight points such that the sum of the coordinates in each pair is 7. Answers may vary.

68. Find the perimeter of a rectangle if three of its vertices are $(5, -2)$, $(-3, -2)$, and $(-3, 3)$.

69. Find the area of a triangle whose vertices have coordinates $(0, 9)$, $(0, -4)$, and $(5, -4)$.

Coordinates on the globe. *Coordinates can also be used to describe the location on a sphere: 0° latitude is the equator and 0° longitude is a line from the North Pole to the South Pole through France and Algeria. In the figure shown here, hurricane Clara is at a point about 260 mi northwest of Bermuda near latitude 36.0° North, longitude 69.0° West.*

70. Approximate the latitude and the longitude of Bermuda.

71. Approximate the latitude and the longitude of Lake Okeechobee.

72. The graph accompanying Example 3 flattens out. Why do you think this occurs?

73. In the *Star Trek* science-fiction series, a three-dimensional coordinate system is used to locate objects in space. If the center of a planet is used as the origin, how many "quadrants" will exist? Why? If possible, sketch a three-dimensional coordinate system and label each "quadrant."

COLLABORATIVE

CORNER

You Sank My Battleship!

Focus: Graphing points; logical questioning
Time: 15–25 minutes
Group size: 3–5
Materials: Graph paper

In the game Battleship®, a player places a miniature ship on a grid that only that player can see. An opponent guesses at coordinates that might "hit" the "hidden" ship. The following activity is similar to this game.

ACTIVITY

1. Using only integers from −10 to 10 (inclusive), one group member should secretly record the coordinates of a point on a slip of paper. (This point is the hidden "battleship.")

2. The other group members can then ask up to 10 "yes/no" questions in an effort to determine the coordinates of the secret point. Be sure to phrase each question mathematically (for example, "Is the *x*-coordinate negative?")

3. The group member who selected the point should answer each question. On the basis of the answer given, another group member should cross out the points no longer under consideration. All group members should check that this is done correctly.

4. If the hidden point has not been determined after 10 questions have been answered, the secret coordinates should be revealed to all group members.

5. Repeat parts (1)–(4) until each group member has had the opportunity to select the hidden point and answer questions.

3.2 Graphing Linear Equations

Solutions of Equations • Graphing Linear Equations • Applications

We have seen how bar, line, and circle graphs can represent information. Now we begin to learn how graphs can be used to represent solutions of equations.

Solutions of Equations

When an equation contains two variables, solutions are ordered pairs in which each number in the pair replaces a letter in the equation. Unless stated otherwise, the first number in each pair replaces the variable that occurs first alphabetically.

EXAMPLE **1** Determine whether each of the following pairs is a solution of $4b - 3a = 22$: **(a)** $(2, 7)$; **(b)** $(1, 6)$.

SOLUTION

a) We substitute 2 for a and 7 for b (alphabetical order of variables):

$$\begin{array}{c|c} 4b - 3a = 22 \\ \hline 4(7) - 3(2) & 22 \\ 28 - 6 & \\ & 22 \stackrel{?}{=} 22 \quad \text{TRUE} \end{array}$$

Since $22 = 22$ is *true*, the pair $(2, 7)$ *is* a solution.

b) In this case, we replace a with 1 and b with 6:

$$\begin{array}{c|c} 4b - 3a = 22 \\ \hline 4(6) - 3(1) & 22 \\ 24 - 3 & \\ & 21 \stackrel{?}{=} 22 \quad \text{FALSE} \qquad 21 \neq 22 \end{array}$$

Since $21 = 22$ is *false*, the pair $(1, 6)$ is *not* a solution. **TRY EXERCISE** 7

EXAMPLE 2 Show that the pairs $(3, 7)$, $(0, 1)$, and $(-3, -5)$ are solutions of $y = 2x + 1$. Then graph the three points to determine another pair that is a solution.

SOLUTION To show that a pair is a solution, we substitute, replacing x with the first coordinate and y with the second coordinate of each pair:

$$\begin{array}{c|c} y = 2x + 1 \\ \hline 7 & 2 \cdot 3 + 1 \\ & 6 + 1 \\ 7 \stackrel{?}{=} 7 & \text{TRUE} \end{array} \qquad \begin{array}{c|c} y = 2x + 1 \\ \hline 1 & 2 \cdot 0 + 1 \\ & 0 + 1 \\ 1 \stackrel{?}{=} 1 & \text{TRUE} \end{array} \qquad \begin{array}{c|c} y = 2x + 1 \\ \hline -5 & 2(-3) + 1 \\ & -6 + 1 \\ -5 \stackrel{?}{=} -5 & \text{TRUE} \end{array}$$

In each of the three cases, the substitution results in a true equation. Thus the pairs $(3, 7)$, $(0, 1)$, and $(-3, -5)$ are all solutions. We graph them as shown at left.

Note that the three points appear to "line up." Will other points that line up with these points also represent solutions of $y = 2x + 1$? To find out, we use a ruler and draw a line passing through $(-3, -5)$, $(0, 1)$, and $(3, 7)$.

The line appears to pass through $(2, 5)$. Let's check to see if this pair is a solution of $y = 2x + 1$:

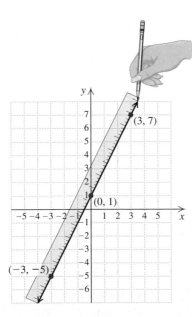

$$\begin{array}{c|c} y = 2x + 1 \\ \hline 5 & 2 \cdot 2 + 1 \\ & 4 + 1 \\ 5 \stackrel{?}{=} 5 & \text{TRUE} \end{array}$$

We see that $(2, 5)$ *is* a solution. You should perform a similar check for at least one other point that appears to be on the line. **TRY EXERCISE** 13

Example 2 leads us to suspect that *any* point on the line passing through $(3, 7)$, $(0, 1)$, and $(-3, -5)$ represents a solution of $y = 2x + 1$. In fact, every solution of $y = 2x + 1$ is represented by a point on this line and every point on this line represents a solution. The line is called the **graph** of the equation.

Graphing Linear Equations

Equations like $y = 2x + 1$ or $4b - 3a = 22$ are said to be **linear** because the graph of each equation is a line. In general, any equation that can be written in the form $y = mx + b$ or $Ax + By = C$ (where $m, b, A, B,$ and C are constants and A and B are not both 0) is linear.

To *graph* an equation is to make a drawing that represents its solutions. Linear equations can be graphed as follows.

> ### To Graph a Linear Equation
> 1. Select a value for one coordinate and calculate the corresponding value of the other coordinate. Form an ordered pair. This pair is one solution of the equation.
> 2. Repeat step (1) to find a second ordered pair. A third ordered pair should be found to use as a check.
> 3. Plot the ordered pairs and draw a straight line passing through the points. The line represents all solutions of the equation.

EXAMPLE   3   Graph: $y = -3x + 1$.

SOLUTION   Since $y = -3x + 1$ is in the form $y = mx + b$, the equation is linear and the graph is a straight line. We select a convenient value for x, compute y, and form an ordered pair. Then we repeat the process for other choices of x.

If $x = 2$,   then $y = -3 \cdot 2 + 1 = -5$,   and $(2, -5)$ is a solution.
If $x = 0$,   then $y = -3 \cdot 0 + 1 = 1$,   and $(0, 1)$ is a solution.
If $x = -1$,   then $y = -3(-1) + 1 = 4$,   and $(-1, 4)$ is a solution.

Results are often listed in a table, as shown below. The points corresponding to each pair are then plotted.

$$y = -3x + 1$$

Calculate ordered pairs.

x	y	(x, y)
2	-5	$(2, -5)$
0	1	$(0, 1)$
-1	4	$(-1, 4)$

(1) Choose x.
(2) Compute y.
(3) Form the pair (x, y).
(4) Plot the points.

Plot the points.

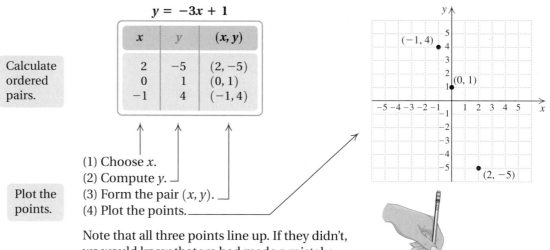

Note that all three points line up. If they didn't, we would know that we had made a mistake, because the equation is linear. When only two points are plotted, an error is more difficult to detect.

Finally, we use a ruler or other straight-edge to draw a line. We add arrowheads to the ends of the line to indicate that it extends indefinitely beyond the edge of the grid drawn. Every point on the line represents a solution of $y = -3x + 1$.

Draw the graph.

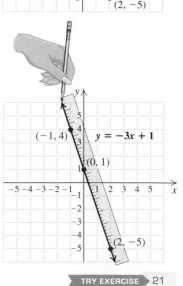

TRY EXERCISE 21

EXAMPLE **4**   Graph: $y = 2x - 3$.

SOLUTION   We select some convenient x-values and compute y-values.

If $x = 0$,   then $y = 2 \cdot 0 - 3 = -3$,   and $(0, -3)$ is a solution.

If $x = 1$,   then $y = 2 \cdot 1 - 3 = -1$,   and $(1, -1)$ is a solution.

If $x = 4$,   then $y = 2 \cdot 4 - 3 = 5$,   and $(4, 5)$ is a solution.

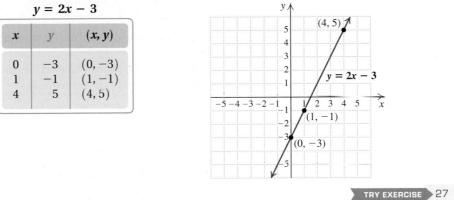

$y = 2x - 3$

x	y	(x, y)
0	-3	$(0, -3)$
1	-1	$(1, -1)$
4	5	$(4, 5)$

TRY EXERCISE 27

EXAMPLE **5**   Graph: $4x + 2y = 12$.

SOLUTION   To form ordered pairs, we can replace either variable with a number and then calculate the other coordinate:

If $y = 0$, we have $4x + 2 \cdot 0 = 12$
$$4x = 12$$
$$x = 3,$$

so $(3, 0)$ is a solution.

If $x = 0$, we have $4 \cdot 0 + 2y = 12$
$$2y = 12$$
$$y = 6,$$

so $(0, 6)$ is a solution.

If $y = 2$, we have $4x + 2 \cdot 2 = 12$
$$4x + 4 = 12$$
$$4x = 8$$
$$x = 2,$$

so $(2, 2)$ is a solution.

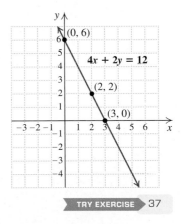

$4x + 2y = 12$

x	y	(x, y)
3	0	$(3, 0)$
0	6	$(0, 6)$
2	2	$(2, 2)$

TRY EXERCISE 37

Note that in Examples 3 and 4 the variable y is isolated on one side of the equation. This generally simplifies calculations, so it is important to be able to solve for y before graphing.

EXAMPLE   6   Graph $3y = 2x$ by first solving for y.

SOLUTION   To isolate y, we divide both sides by 3, or multiply both sides by $\frac{1}{3}$:

$$3y = 2x$$

$$\frac{1}{3} \cdot 3y = \frac{1}{3} \cdot 2x \qquad \text{Using the multiplication principle to multiply both sides by } \frac{1}{3}$$

$$\left.\begin{array}{l} 1y = \frac{2}{3} \cdot x \\ y = \frac{2}{3}x. \end{array}\right\} \quad \text{Simplifying}$$

Because all the equations above are equivalent, we can use $y = \frac{2}{3}x$ to draw the graph of $3y = 2x$.

To graph $y = \frac{2}{3}x$, we can select x-values that are multiples of 3. This will allow us to avoid fractions when the corresponding y-values are computed.

$$\left.\begin{array}{ll} \text{If } x = 3, & \text{then } y = \frac{2}{3} \cdot 3 = 2. \\ \text{If } x = -3, & \text{then } y = \frac{2}{3}(-3) = -2. \\ \text{If } x = 6, & \text{then } y = \frac{2}{3} \cdot 6 = 4. \end{array}\right\} \quad \begin{array}{l} \text{Note that when multiples of 3 are} \\ \text{substituted for } x, \text{ the } y\text{-coordinates} \\ \text{are not fractions.} \end{array}$$

The following table lists these solutions. Next, we plot the points and see that they form a line. Finally, we draw and label the line.

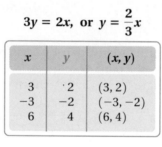

$$3y = 2x, \quad \text{or} \quad y = \frac{2}{3}x$$

x	y	(x, y)
3	2	$(3, 2)$
-3	-2	$(-3, -2)$
6	4	$(6, 4)$

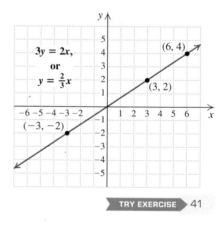

TRY EXERCISE 41

EXAMPLE   7   Graph $x + 5y = -10$ by first solving for y.

SOLUTION   We have

$$x + 5y = -10$$

$$5y = -x - 10 \qquad \text{Adding } -x \text{ to both sides}$$

$$y = \frac{1}{5}(-x - 10) \qquad \text{Multiplying both sides by } \frac{1}{5}$$

$$y = -\frac{1}{5}x - 2. \qquad \text{Using the distributive law}$$

> **CAUTION!**   It is very important to multiply *both* $-x$ and -10 by $\frac{1}{5}$.

Thus, $x + 5y = -10$ is equivalent to $y = -\frac{1}{5}x - 2$. It is important to note that if we now choose x-values that are multiples of 5, we can avoid fractions when calculating the corresponding y-values.

$$\text{If } x = 5, \quad \text{then } y = -\frac{1}{5} \cdot 5 - 2 = -1 - 2 = -3.$$

$$\text{If } x = 0, \quad \text{then } y = -\frac{1}{5} \cdot 0 - 2 = 0 - 2 = -2.$$

$$\text{If } x = -5, \quad \text{then } y = -\frac{1}{5}(-5) - 2 = 1 - 2 = -1.$$

$$x + 5y = -10, \text{ or } y = -\frac{1}{5}x - 2$$

x	y	(x, y)
5	−3	(5, −3)
0	−2	(0, −2)
−5	−1	(−5, −1)

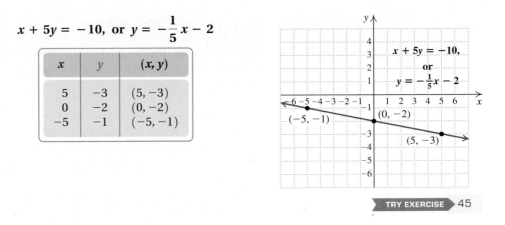

TRY EXERCISE 45

Applications

Linear equations appear in many real-life situations.

EXAMPLE 8 *Fuel efficiency.* A typical tractor-trailer will move 18 tons of air per mile at 55 mph. Air resistance increases with speed, causing fuel efficiency to decrease at higher speeds. At highway speeds, a certain truck's fuel efficiency *t*, in miles per gallon (mpg), can be given by

$$t = -0.1s + 13.1,$$

where *s* is the speed of the truck, in miles per hour (mph). Graph the equation and then use the graph to estimate the fuel efficiency at 66 mph.

Source: Based on data from Kenworth Truck Co.

SOLUTION We graph $t = -0.1s + 13.1$ by first selecting values for *s* and then calculating the associated values *t*. Since the equation is true for highway speeds, we use $s \geq 50$.

s	t
50	8.1
60	7.1
70	6.1

If $s = 50$, then $t = -0.1(50) + 13.1 = 8.1$.
If $s = 60$, then $t = -0.1(60) + 13.1 = 7.1$.
If $s = 70$, then $t = -0.1(70) + 13.1 = 6.1$.

Because we are *selecting* values for *s* and *calculating* values for *t,* we represent *s* on the horizontal axis and *t* on the vertical axis. Counting by 5's horizontally, beginning at 50, and by 0.5 vertically, beginning at 4, will allow us to plot all three pairs, as shown below.

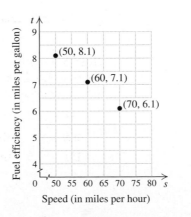

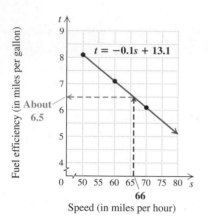

Fuel efficiency (in miles per gallon)

$t = -0.1s + 13.1$

About 6.5

Speed (in miles per hour)

Since the three points line up, our calculations are probably correct. We draw a line, beginning at $(50, 8.1)$. To estimate the fuel efficiency at 66 mph, we locate the point on the line that is above 66 and then find the value on the t-axis that corresponds to that point, as shown at left. The fuel efficiency at 66 mph is about 6.5 mpg.

TRY EXERCISE ▶ 49

CAUTION!   When the coordinates of a point are read from a graph, as in Example 8, values should not be considered exact.

Many equations in two variables have graphs that are not straight lines. Three such graphs are shown below. As before, each graph represents the solutions of the given equation. Graphing calculators are especially helpful when drawing these *nonlinear* graphs. Nonlinear graphs are studied in Chapter 9 and in more advanced courses.

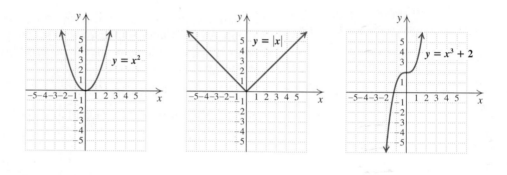

$y = x^2$ $y = |x|$ $y = x^3 + 2$

TECHNOLOGY CONNECTION

Most graphing calculators require that y be alone on one side before the equation is entered. For example, to graph $5y + 4x = 13$, we would first solve for y. The student can check that solving for y yields the equation $y = -\frac{4}{5}x + \frac{13}{5}$.

We press $\boxed{Y=}$, enter $-\frac{4}{5}x + \frac{13}{5}$ as Y1, and press $\boxed{GRAPH}$. The graph is shown here in the standard viewing window $[-10, 10, -10, 10]$.

Using a graphing calculator, graph each of the following. Select the "standard" $[-10, 10, -10, 10]$ window.

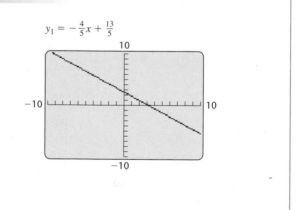

$y_1 = -\frac{4}{5}x + \frac{13}{5}$

1. $y = -5x + 6.5$ 2. $y = 3x + 4.5$
3. $7y - 4x = 22$ 4. $5y + 11x = -20$
5. $2y - x^2 = 0$ 6. $y + x^2 = 8$

3.2 EXERCISE SET

For Extra Help
MyMathLab
Math XL PRACTICE
WATCH
DOWNLOAD

🖐 **Concept Reinforcement** *Classify each statement as either true or false.*

1. A linear equation in two variables has at most one solution.

2. Every solution of $y = 3x - 7$ is an ordered pair.

3. The graph of $y = 3x - 7$ represents all solutions of the equation.

4. If a point is on the graph of $y = 3x - 7$, the corresponding ordered pair is a solution of the equation.

5. To find a solution of $y = 3x - 7$, we can choose any value for x and calculate the corresponding value for y.

6. The graph of every equation is a straight line.

Determine whether each equation has the given ordered pair as a solution.

7. $y = 4x - 7$; $(2, 1)$

8. $y = 5x + 8$; $(0, 8)$

9. $3y + 4x = 19$; $(5, 1)$

10. $5x - 3y = 15$; $(0, 5)$

11. $4m - 5n = 7$; $(3, -1)$

12. $3q - 2p = -8$; $(1, -2)$

In Exercises 13–20, an equation and two ordered pairs are given. Show that each pair is a solution of the equation. Then graph the two pairs to determine another solution. Answers may vary.

13. $y = x + 3$; $(-1, 2), (4, 7)$ $y = 2 + 3$

14. $y = x - 2$; $(3, 1), (-2, -4)$

15. $y = \frac{1}{2}x + 3$; $(4, 5), (-2, 2)$ $0 = \frac{1}{2}5 + 3$

16. $y = \frac{1}{2}x - 1$; $(6, 2), (0, -1)$

17. $y + 3x = 7$; $(2, 1), (4, -5)$

18. $2y + x = 5$; $(-1, 3), (7, -1)$

19. $4x - 2y = 10$; $(0, -5), (4, 3)$

20. $6x - 3y = 3$; $(1, 1), (-1, -3)$

Graph each equation.

21. $y = x + 1$

22. $y = x - 1$

23. $y = -x$

24. $y = x$

25. $y = 2x$

26. $y = -3x$

27. $y = 2x + 2$

28. $y = 3x - 2$

29. $y = -\frac{1}{2}x$

30. $y = \frac{1}{4}x$

31. $y = \frac{1}{3}x - 4$

32. $y = \frac{1}{2}x + 1$

33. $x + y = 4$

34. $x + y = -5$

35. $x - y = -2$

36. $y - x - 3$

37. $x + 2y = -6$

38. $x + 2y = 8$

39. $y = -\frac{2}{3}x + 4$

40. $y = \frac{3}{2}x + 1$

41. $4x = 3y$

42. $2x = 5y$

43. $5x - y = 0$

44. $3x - 5y = 0$

45. $6x - 3y = 9$

46. $8x - 4y = 12$

47. $6y + 2x = 8$

48. $8y + 2x = -4$

49. *Student aid.* The average award a of federal student financial assistance per student is approximated by

$$a = 0.08t + 2.5,$$

where a is in thousands of dollars and t is the number of years since 1994. Graph the equation and use the graph to estimate the average amount of federal student aid per student in 2010.
Source: Based on data from U.S. Department of Education, Office of Postsecondary Education

50. *Value of a color copier.* The value of Dupliographic's color copier is given by

$$v = -0.68t + 3.4,$$

where v is the value, in thousands of dollars, t years from the date of purchase. Graph the equation and use the graph to estimate the value of the copier after $2\frac{1}{2}$ yr.

51. *FedEx mailing costs.* Recently, the cost c, in dollars, of shipping a FedEx Priority Overnight package weighing 1 lb or more a distance of 1001 to 1400 mi was given by

$$c = 3.1w + 29.07,$$

where w is the package's weight, in pounds. Graph the equation and use the graph to estimate the cost of shipping a $6\frac{1}{2}$-lb package.
Source: Based on data from FedEx.com

52. *Increasing life expectancy.* A smoker is 15 times more likely to die of lung cancer than a nonsmoker. An ex-smoker who stopped smoking t years ago is

w times more likely to die of lung cancer than a nonsmoker, where

$$w = 15 - t.$$

Graph the equation and use the graph to estimate how much more likely it is for Sandy to die of lung cancer than Polly, if Polly never smoked and Sandy quit $2\frac{1}{2}$ yr ago.
Source: Data from *Body Clock* by Dr. Martin Hughes, p. 60. New York: Facts on File, Inc.

53. *Scrapbook pricing.* The price *p*, in dollars, of an 8-in. by 8-in. assembled scrapbook is given by

$$p = 3.5n + 9,$$

where *n* is the number of pages in the scrapbook. Graph the equation and use the graph to estimate the price of a scrapbook containing 25 pages.
Source: www.scrapbooksplease.com

54. *Value of computer software.* The value *v* of a shopkeeper's inventory software program, in hundreds of dollars, is given by

$$v = -\tfrac{3}{4}t + 6,$$

where *t* is the number of years since the shopkeeper first bought the program. Graph the equation and use the graph to estimate what the program is worth 4 yr after it was first purchased.

55. *Bottled water.* The number of gallons of bottled water *w* consumed by the average American in one year is given by

$$w = 1.6t + 16.7,$$

where *t* is the number of years since 2000. Graph the equation and use the graph to predict the number of gallons consumed by the average American in 2010.
Source: Based on data from Beverage Marketing Corporation

56. *Record temperature drop.* On January 22, 1943, the temperature *T*, in degrees Fahrenheit, in Spearfish, South Dakota, could be approximated by

$$T = -2m + 54,$$

where *m* is the number of minutes since 9:00 A.M. that morning. Graph the equation and use the graph to estimate the temperature at 9:15 A.M.
Source: Based on information from the National Oceanic Atmospheric Administration

57. *Cost of college.* The cost *T*, in hundreds of dollars, of tuition and fees at many community colleges can be approximated by

$$T = \tfrac{5}{4}c + 2,$$

where *c* is the number of credits for which a student registers. Graph the equation and use the graph to

estimate the cost of tuition and fees when a student registers for 4 three-credit courses.

Bartonville Community College
Ferndell Hall

Tuition $125 / credit
Fees
Registration $85
Transcripts $55
Computer $45
Activities $15
Activities include: Math team, fencing, drama club

58. *Cost of college.* The cost *C*, in thousands of dollars, of a year at a private four-year college (all expenses) can be approximated by

$$C = \tfrac{13}{10}t + 21,$$

where *t* is the number of years since 1995. Graph the equation and use the graph to predict the cost of a year at a private four-year college in 2012.
Source: Based on information in *Statistical Abstract of the United States,* 2007

59. The equations $3x + 4y = 8$ and $y = -\tfrac{3}{4}x + 2$ are equivalent. Which equation would be easier to graph and why?

60. Suppose that a linear equation is graphed by plotting three points and that the three points line up with each other. Does this *guarantee* that the equation is being correctly graphed? Why or why not?

Skill Review

Review solving equations and formulas (Sections 2.2 and 2.3).

Solve and check. [2.2]

61. $5x + 3 \cdot 0 = 12$

62. $3 \cdot 0 - 8y = 6$

63. $5x + 3(2 - x) = 12$

64. $3(y - 5) - 8y = 6$

Solve. [2.3]

65. $A = \dfrac{T + Q}{2}$, for Q

66. $pq + p = w$, for p

67. $Ax + By = C$, for y

68. $\dfrac{y - k}{m} = x - h$, for y

Synthesis

📝 **69.** Janice consistently makes the mistake of plotting the *x*-coordinate of an ordered pair using the *y*-axis, and the *y*-coordinate using the *x*-axis. How will Janice's incorrect graph compare with the appropriate graph?

📝 **70.** Explain how the graph in Example 8 can be used to determine the speed for which the fuel efficiency is 6 mpg.

71. *Bicycling.* Long Beach Island in New Jersey is a long, narrow, flat island. For exercise, Laura routinely bikes to the northern tip of the island and back. Because of the steady wind, she uses one gear going north and another for her return. Laura's bike has 21 gears and the sum of the two gears used on her ride is always 24. Write and graph an equation that represents the different pairings of gears that Laura uses. Note that there are no fraction gears on a bicycle.

In Exercises 72–75, try to find an equation for the graph shown.

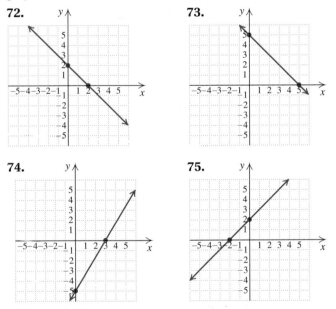

72.

73.

74.

75.

76. Translate to an equation:

    d dimes and *n* nickels total $1.75.

Then graph the equation and use the graph to determine three different combinations of dimes and nickels that total $1.75 (see also Exercise 90).

77. Translate to an equation:

    d $25 dinners and *l* $5 lunches total $225.

Then graph the equation and use the graph to determine three different combinations of lunches and dinners that total $225 (see also Exercise 90).

Use the suggested x-values $-3, -2, -1, 0, 1, 2,$ *and* 3 *to graph each equation.*

78. $y = |x|$

Aha! **79.** $y = -|x|$

Aha! **80.** $y = |x| - 2$

81. $y = x^2$

82. $y = x^2 + 1$

📉 *For Exercises 83–88, use a graphing calculator to graph the equation. Use a* $[-10, 10, -10, 10]$ *window.*

83. $y = -2.8x + 3.5$

84. $y = 4.5x + 2.1$

85. $y = 2.8x - 3.5$

86. $y = -4.5x - 2.1$

87. $y = x^2 + 4x + 1$

88. $y = -x^2 + 4x - 7$

89. Example 8 discusses fuel efficiency. If fuel costs $3.50 a gallon, how much money will a truck driver save on a 500-mi trip by driving at 55 mph instead of 70 mph? How many gallons of fuel will be saved?

📝 **90.** Study the graph of Exercises 76 and 77. Does *every* point on the graph represent a solution of the associated problem? Why or why not?

3.3 Graphing and Intercepts

Intercepts   •   Using Intercepts to Graph   •   Graphing Horizontal or Vertical Lines

Unless a line is horizontal or vertical, it will cross both axes. Often, finding the points where the axes are crossed gives us a quick way of graphing linear equations.

Intercepts

In Example 5 of Section 3.2, we graphed $4x + 2y = 12$ by plotting the points $(3, 0)$, $(0, 6)$, and $(2, 2)$ and then drawing the line.

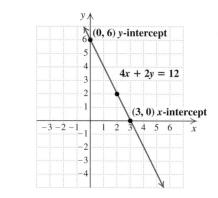

- The point at which a graph crosses the y-axis is called the **y-intercept**. In the figure above, the y-intercept is $(0, 6)$. The x-coordinate of a y-intercept is always 0.

- The point at which a graph crosses the x-axis is called the **x-intercept**. In the figure above, the x-intercept is $(3, 0)$. The y-coordinate of an x-intercept is always 0.

It is possible for the graph of a curve to have more than one y-intercept or more than one x-intercept.

EXAMPLE **1**   For the graph shown below, **(a)** give the coordinates of any x-intercepts and **(b)** give the coordinates of any y-intercepts.

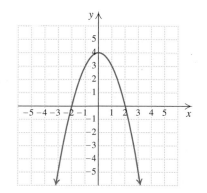

SOLUTION

a) The x-intercepts are the points at which the graph crosses the x-axis. For the graph shown, the x-intercepts are $(-2, 0)$ and $(2, 0)$.

b) The y-intercept is the point at which the graph crosses the y-axis. For the graph shown, the y-intercept is $(0, 4)$.

> TRY EXERCISE ▶ 7

Using Intercepts to Graph

It is important to know how to locate a graph's intercepts from the equation being graphed.

To Find Intercepts

To find the y-intercept(s) of an equation's graph, replace x with 0 and solve for y.

To find the x-intercept(s) of an equation's graph, replace y with 0 and solve for x.

EXAMPLE **2**   Find the y-intercept and the x-intercept of the graph of $2x + 4y = 20$.

SOLUTION   To find the y-intercept, we let $x = 0$ and solve for y:

$$2 \cdot 0 + 4y = 20 \qquad \text{Replacing } x \text{ with } 0$$
$$4y = 20$$
$$y = 5.$$

Thus the y-intercept is $(0, 5)$.

To find the x-intercept, we let $y = 0$ and solve for x:

$$2x + 4 \cdot 0 = 20 \qquad \text{Replacing } y \text{ with } 0$$
$$2x = 20$$
$$x = 10.$$

Thus the x-intercept is $(10, 0)$.

> TRY EXERCISE ▶ 15

Since two points are sufficient to graph a line, intercepts can be used to graph linear equations.

EXAMPLE **3**   Graph $2x + 4y = 20$ using intercepts.

SOLUTION   In Example 2, we showed that the y-intercept is $(0, 5)$ and the x-intercept is $(10, 0)$. Before drawing a line, we plot a third point as a check. We substitute any convenient value for x and solve for y.

If we let $x = 5$, then

$$2 \cdot 5 + 4y = 20 \qquad \text{Substituting 5 for } x$$

$$10 + 4y = 20$$

$$4y = 10 \qquad \text{Subtracting 10 from both sides}$$

$$y = \frac{10}{4}, \text{ or } 2\frac{1}{2}. \qquad \text{Solving for } y$$

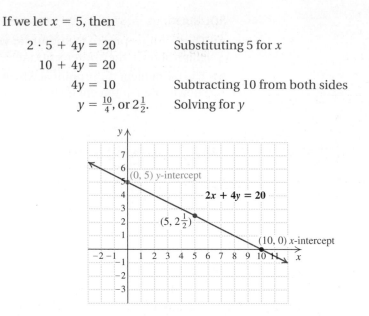

The point $\left(5, 2\frac{1}{2}\right)$ appears to line up with the intercepts, so our work is probably correct. To finish, we draw and label the line.   **TRY EXERCISE** 25

Note that when we solved for the y-intercept, we replaced x with 0 and simplified $2x + 4y = 20$ to $4y = 20$. Thus, to find the y-intercept, we can momentarily ignore the x-term and solve the remaining equation.

In a similar manner, when we solved for the x-intercept, we simplified $2x + 4y = 20$ to $2x = 20$. Thus, to find the x-intercept, we can momentarily ignore the y-term and then solve this remaining equation.

EXAMPLE **4**   Graph $3x - 2y = 60$ using intercepts.

SOLUTION   To find the y-intercept, we let $x = 0$. This amounts to temporarily ignoring the x-term and then solving:

$$-2y = 60 \qquad \text{For } x = 0, \text{ we have } 3 \cdot 0 - 2y, \text{ or simply } -2y.$$

$$y = -30.$$

The y-intercept is $(0, -30)$.

To find the x-intercept, we let $y = 0$. This amounts to temporarily disregarding the y-term and then solving:

$$3x = 60 \qquad \text{For } y = 0, \text{ we have } 3x - 2 \cdot 0, \text{ or simply } 3x.$$

$$x = 20.$$

The x-intercept is $(20, 0)$.

To find a third point, we can replace x with 4 and solve for y:

$$3 \cdot 4 - 2y = 60 \qquad \text{Numbers other than 4 can be used for } x.$$

$$12 - 2y = 60$$

$$-2y = 48$$

$$y = -24. \qquad \text{This means that } (4, -24) \text{ is on the graph.}$$

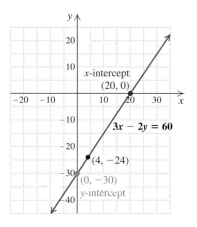

In order for us to graph all three points, the y-axis of our graph must go down to at least -30 and the x-axis must go up to at least 20. Using a scale of 5 units per square allows us to display both intercepts and $(4, -24)$, as well as the origin.

The point $(4, -24)$ appears to line up with the intercepts, so we draw and label the line, as shown at left.

TRY EXERCISE 45

TECHNOLOGY CONNECTION

When an equation has been entered into a graphing calculator, we may not be able to see both intercepts. For example, if $y = -0.8x + 17$ is graphed in the window $[-10, 10, -10, 10]$, neither intercept is visible.

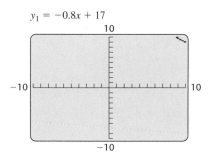

To better view the intercepts, we can change the window dimensions or we can zoom out. The ZOOM feature allows us to reduce or magnify a graph or a portion

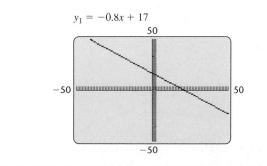

of a graph. Before zooming, the ZOOM *factors* must be set in the memory of the ZOOM key. If we zoom out with factors set at 5, both intercepts are visible but the axes are heavily drawn, as shown in the preceding figure.

This suggests that the *scales* of the axes should be changed. To do this, we use the WINDOW menu and set Xscl to 5 and Yscl to 5. The resulting graph has tick marks 5 units apart and clearly shows both intercepts. Other choices for Xscl and Yscl can also be made.

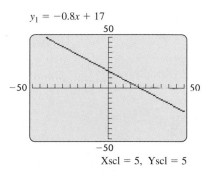

Graph each equation so that both intercepts can be easily viewed. Zoom or adjust the window settings so that tick marks can be clearly seen on both axes.

1. $y = -0.72x - 15$　　**2.** $y - 2.13x = 27$
3. $5x + 6y = 84$　　　　**4.** $2x - 7y = 150$
5. $19x - 17y = 200$　　**6.** $6x + 5y = 159$

Graphing Horizontal or Vertical Lines

The equations graphed in Examples 3 and 4 are both in the form $Ax + By = C$. We have already stated that any equation in the form $Ax + By = C$ is linear, provided A and B are not both zero. What if A or B (but not both) is zero? We will find that when A is zero, there is no x-term and the graph is a horizontal line. We will also find that when B is zero, there is no y-term and the graph is a vertical line.

EXAMPLE 5

Graph: $y = 3$.

SOLUTION We can regard the equation $y = 3$ as $0 \cdot x + y = 3$. No matter what number we choose for x, we find that y must be 3 if the equation is to be solved. Consider the following table.

STUDENT NOTES

Many students draw horizontal lines when they should be drawing vertical lines and vice versa. To avoid this mistake, first locate the correct number on the axis whose label is given. Thus, to graph $x = 2$, we locate 2 on the x-axis and then draw a line perpendicular to that axis at that point. Note that the graph of $x = 2$ on a plane is a line, whereas the graph of $x = 2$ on a number line is a point.

Choose any number for x. →

y must be 3.

$$y = 3$$

x	y	(x, y)
-2	3	$(-2, 3)$
0	3	$(0, 3)$
4	3	$(4, 3)$

All pairs will have 3 as the y-coordinate.

When we plot the ordered pairs $(-2, 3)$, $(0, 3)$, and $(4, 3)$ and connect the points, we obtain a horizontal line. Any ordered pair of the form $(x, 3)$ is a solution, so the line is parallel to the x-axis with y-intercept $(0, 3)$. Note that the graph of $y = 3$ has no x-intercept.

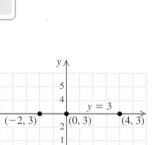

TRY EXERCISE 53

EXAMPLE 6

Graph: $x = -4$.

SOLUTION We can regard the equation $x = -4$ as $x + 0 \cdot y = -4$. We make up a table with all -4's in the x-column.

$$x = -4$$

x must be -4. →

Any number can be used for y.

x	y	(x, y)
-4	-5	$(-4, -5)$
-4	1	$(-4, 1)$
-4	3	$(-4, 3)$

All pairs will have -4 as the x-coordinate.

When we plot the ordered pairs $(-4, -5)$, $(-4, 1)$, and $(-4, 3)$ and connect them, we obtain a vertical line. Any ordered pair of the form $(-4, y)$ is a solution. The line is parallel to the y-axis with x-intercept $(-4, 0)$. Note that the graph of $x = -4$ has no y-intercept.

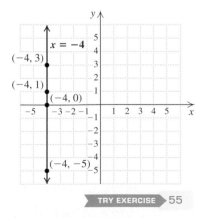

TRY EXERCISE 55

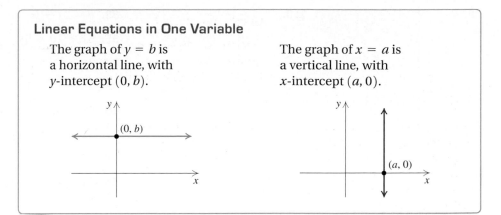

Linear Equations in One Variable

The graph of $y = b$ is a horizontal line, with y-intercept $(0, b)$.

The graph of $x = a$ is a vertical line, with x-intercept $(a, 0)$.

EXAMPLE 7 Write an equation for each graph.

a)

b)

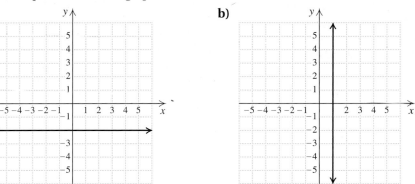

SOLUTION

a) Note that every point on the horizontal line passing through $(0, -2)$ has -2 as the y-coordinate. Thus the equation of the line is $y = -2$.

b) Note that every point on the vertical line passing through $(1, 0)$ has 1 as the x-coordinate. Thus the equation of the line is $x = 1$.

TRY EXERCISE ▸ 71

3.3   EXERCISE SET

For Extra Help

↪ **Concept Reinforcement** *In each of Exercises 1–6, match the phrase with the most appropriate choice from the column on the right.*

1. ____ A vertical line
2. ____ A horizontal line
3. ____ A y-intercept
4. ____ An x-intercept
5. ____ A third point as a check
6. ____ Use a scale of 10 units per square.

a) $2x + 5y = 100$
b) $(3, -2)$
c) $(1, 0)$
d) $(0, 2)$
e) $y = 3$
f) $x = -4$

For Exercises 7–14, list **(a)** *the coordinates of the y-intercept and* **(b)** *the coordinates of all x-intercepts.*

7.

8.

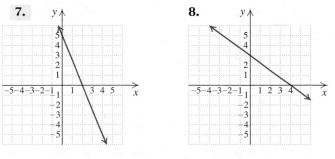

9.
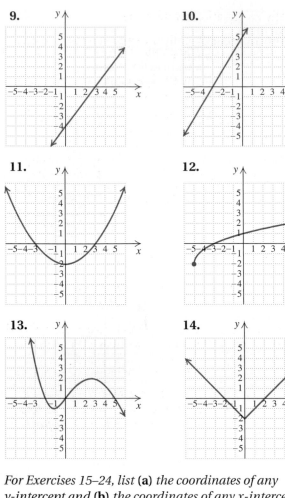

10.
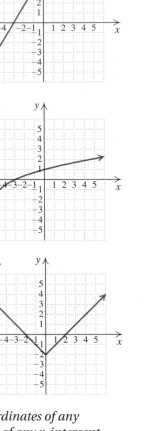

11.

12.

13.

14.

For Exercises 15–24, list **(a)** *the coordinates of any y-intercept and* **(b)** *the coordinates of any x-intercept. Do not graph.*

15. $3x + 5y = 15$

16. $2x + 7y = 14$

17. $9x - 2y = 36$

18. $10x - 3y = 60$

19. $-4x + 5y = 80$

20. $-5x + 6y = 100$

Aha! **21.** $x = 12$

22. $y = 10$

23. $y = -9$

24. $x = -5$

Find the intercepts. Then graph.

25. $3x + 5y = 15$ **26.** $2x + y = 6$

27. $x + 2y = 4$ **28.** $2x + 5y = 10$

29. $-x + 2y = 8$ **30.** $-x + 3y = 9$

31. $3x + y = 9$ **32.** $2x - y = 8$

33. $y = 2x - 6$ **34.** $y = -3x + 6$

35. $5x - 10 = 5y$ **36.** $3x - 9 = 3y$

37. $2x - 5y = 10$ **38.** $2x - 3y = 6$

39. $6x + 2y = 12$

40. $4x + 5y = 20$

41. $4x + 3y = 16$

42. $3x + 2y = 8$

43. $2x + 4y = 1$

44. $3x - 6y = 1$

45. $5x - 3y = 180$

46. $10x + 7y = 210$

47. $y = -30 + 3x$

48. $y = -40 + 5x$

49. $-4x = 20y + 80$

50. $60 = 20x - 3y$

51. $y - 3x = 0$

52. $x + 2y = 0$

Graph.

53. $y = 1$ **54.** $y = 4$

55. $x = 3$ **56.** $x = 6$

57. $y = -2$ **58.** $y = -4$

59. $x = -1$ **60.** $x = -6$

61. $y = -15$ **62.** $x = 20$

63. $y = 0$ **64.** $y = \frac{3}{2}$

65. $x = -\frac{5}{2}$ **66.** $x = 0$

67. $-4x = -100$

68. $12y = -360$

69. $35 + 7y = 0$

70. $-3x - 24 = 0$

Write an equation for each graph.

71.   **72.**

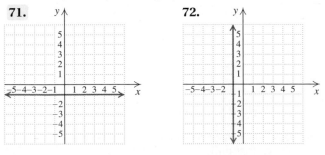

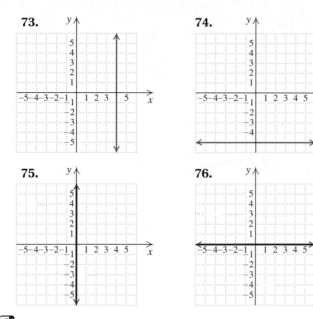

73.

74.

75.

76.

77. Explain in your own words why the graph of $y = 8$ is a horizontal line.

78. Explain in your own words why the graph of $x = -4$ is a vertical line.

Skill Review

Review translating to algebraic expressions (Section 1.1).

Translate to an algebraic expression. [1.1]

79. 7 less than d

80. 5 more than w

81. The sum of 7 and four times a number

82. The product of 3 and a number

83. Twice the sum of two numbers

84. Half of the sum of two numbers

Synthesis

85. Describe what the graph of $x + y = C$ will look like for any choice of C.

86. If the graph of a linear equation has one point that is both the x- and the y-intercepts, what is that point? Why?

87. Write an equation for the x-axis.

88. Write an equation of the line parallel to the x-axis and passing through $(3, 5)$.

89. Write an equation of the line parallel to the y-axis and passing through $(-2, 7)$.

90. Find the coordinates of the point of intersection of the graphs of $y = x$ and $y = 6$.

91. Find the coordinates of the point of intersection of the graphs of the equations $x = -3$ and $y = 4$.

92. Write an equation of the line shown in Exercise 7.

93. Write an equation of the line shown in Exercise 10.

94. Find the value of C such that the graph of $3x + C = 5y$ has an x-intercept of $(-4, 0)$.

95. Find the value of C such that the graph of $4x = C - 3y$ has a y-intercept of $(0, -8)$.

96. For A and B nonzero, the graphs of $Ax + D = C$ and $By + D = C$ will be parallel to an axis. Explain why.

97. Find the x-intercept of the graph of $Ax + D = C$.

In Exercises 98–103, find the intercepts of each equation algebraically. Then adjust the window and scale so that the intercepts can be checked graphically with no further window adjustments.

98. $3x + 2y = 50$

99. $2x - 7y = 80$

100. $y = 1.3x - 15$

101. $y = 0.2x - 9$

102. $25x - 20y = 1$

103. $50x + 25y = 1$

3.4   Rates

Rates of Change   •   Visualizing Rates

Rates of Change

Because graphs make use of two axes, they allow us to visualize how two quantities change with respect to each other. A number accompanied by units is used to represent this type of change and is referred to as a *rate*.

Rate

A *rate* is a ratio that indicates how two quantities change with respect to each other.

Rates occur often in everyday life:

A business whose customer base grows by 1500 customers over a period of 2 yr has an average *growth rate* of $\frac{1500}{2}$, or 750, customers per year.

A vehicle traveling 260 mi in 4 hr is moving at a *rate* of $\frac{260}{4}$, or 65, mph (miles per hour).

A class of 25 students pays a total of $93.75 to visit a museum. The *rate* is $\frac{\$93.75}{25}$, or $3.75, per student.

> *CAUTION!*   To calculate a rate, it is important to keep track of the units being used.

EXAMPLE   1   On January 3, Alisha rented a Ford Focus with a full tank of gas and 9312 mi on the odometer. On January 7, she returned the car with 9630 mi on the odometer.* If the rental agency charged Alisha $108 for the rental and needed 12 gal of gas to fill up the gas tank, find the following rates.

a) The car's rate of gas consumption, in miles per gallon

b) The average cost of the rental, in dollars per day

c) The car's rate of travel, in miles per day

SOLUTION

a) The rate of gas consumption, in miles per gallon, is found by dividing the number of miles traveled by the number of gallons used for that amount of driving:

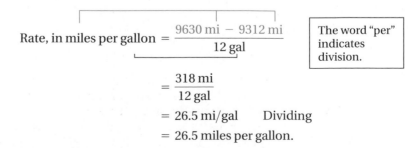

$$\text{Rate, in miles per gallon} = \frac{9630 \text{ mi} - 9312 \text{ mi}}{12 \text{ gal}}$$

The word "per" indicates division.

$$= \frac{318 \text{ mi}}{12 \text{ gal}}$$
$$= 26.5 \text{ mi/gal} \qquad \text{Dividing}$$
$$= 26.5 \text{ miles per gallon.}$$

---

*For all problems concerning rentals, assume that the pickup time was later in the day than the return time so that no late fees were applied.

b) The average cost of the rental, in dollars per day, is found by dividing the cost of the rental by the number of days:

$$\text{Rate, in dollars per day} = \frac{108 \text{ dollars}}{4 \text{ days}}$$

From January 3 to January 7 is $7 - 3 = 4$ days.

$$= 27 \text{ dollars/day}$$
$$= \$27 \text{ per day.}$$

c) The car's rate of travel, in miles per day, is found by dividing the number of miles traveled by the number of days:

$$\text{Rate, in miles per day} = \frac{318 \text{ mi}}{4 \text{ days}}$$

$9630 \text{ mi} - 9312 \text{ mi} = 318 \text{ mi}$;
From January 3 to January 7 is $7 - 3 = 4$ days.

$$= 79.5 \text{ mi/day}$$
$$= 79.5 \text{ mi per day.}$$

TRY EXERCISE 7

CAUTION! Units are a vital part of real-world problems. They must be considered in the translation of a problem and included in the answer to a problem.

Many problems involve a rate of travel, or *speed*. The **speed** of an object is found by dividing the distance traveled by the time required to travel that distance.

EXAMPLE 2

Transportation. An Atlantic City Express bus makes regular trips between Paramus and Atlantic City, New Jersey. At 6:00 P.M., the bus is at mileage marker 40 on the Garden State Parkway, and at 8:00 P.M. it is at marker 170. Find the average speed of the bus.

SOLUTION Speed is the distance traveled divided by the time spent traveling:

$$\text{Bus speed} = \frac{\text{Distance traveled}}{\text{Time spent traveling}}$$
$$= \frac{\text{Change in mileage}}{\text{Change in time}}$$
$$= \frac{130 \text{ mi}}{2 \text{ hr}}$$

$170 \text{ mi} - 40 \text{ mi} = 130 \text{ mi}$;
$8:00 \text{ P.M.} - 6:00 \text{ P.M.} = 2 \text{ hr}$

$$= 65 \frac{\text{mi}}{\text{hr}}$$
$$= 65 \text{ miles per hour.}$$

This *average* speed does not indicate by how much the bus speed may vary along the route.

TRY EXERCISE 13

Visualizing Rates

Graphs allow us to visualize a rate of change. As a rule, the quantity listed in the numerator appears on the vertical axis and the quantity listed in the denominator appears on the horizontal axis.

EXAMPLE **3**

Recycling. Between 1991 and 2006, the amount of paper recycled in the United States increased at a rate of approximately 1.5 million tons per year. In 1991, approximately 31 million tons of paper was recycled. Draw a graph to represent this information.

Source: Based on information from American Forest and Paper Association

SOLUTION To label the axes, note that the rate is given as 1.5 million tons per year, or

$$1.5 \text{ million } \frac{\text{tons}}{\text{yr}}. \qquad \longleftarrow \text{Numerator: vertical axis}$$
$$\longleftarrow \text{Denominator: horizontal axis}$$

We list *Amount of paper recycled (in millions of tons)* on the vertical axis and *Year* on the horizontal axis. (See the figure on the left below.)

Next, we select a scale for each axis that allows us to plot the given information. If we count by increments of 10 million on the vertical axis, we can show 31 million tons for 1991 and increasing amounts for later years. On the horizontal axis, we count by increments of 2 years to make certain that both 1991 and 2006 are included. (See the figure in the middle below.)

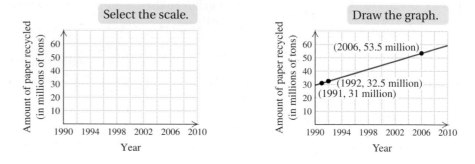

We now plot the point corresponding to (1991, 31 million). Then, to display the rate of growth, we move from that point to a second point that represents 1.5 million more tons 1 year later.

(1991,	31 million)	Beginning point
(1991 + 1,	31 million + 1.5 million)	1.5 million more tons, 1 year later
(1992,	32.5 million)	A second point on the graph

Similarly, we can find the coordinates for 2006. Since 2006 is 15 years after 1991, we add 15 to the year and 15(1.5 million) = 22.5 million to the amount.

(1991,	31 million)	Beginning point
(1991 + 15,	31 million + 22.5 million)	15(1.5) million more tons, 15 years later
(2006,	53.5 million)	A third point on the graph

After plotting the three points, we draw a line through them, as shown in the figure on the right above. This gives us the graph. **TRY EXERCISE** 19

EXAMPLE **4**

Banking. Nadia prepared the following graph from data collected on a recent day at a branch bank.

a) What rate can be determined from the graph?

b) What is that rate?

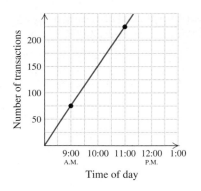

Time of day

SOLUTION

a) Because the vertical axis shows the number of transactions and the horizontal axis lists the time in hour-long increments, we can find the rate *Number of transactions per hour.*

b) The points (9:00, 75) and (11:00, 225) are both on the graph. This tells us that in the 2 hours between 9:00 and 11:00, there were $225 - 75 = 150$ transactions. Thus the rate is

$$\frac{225 \text{ transactions} - 75 \text{ transactions}}{11:00 - 9:00} = \frac{150 \text{ transactions}}{2 \text{ hours}}$$

$$= 75 \text{ transactions per hour.}$$

Note that this is an *average* rate.

TRY EXERCISE 29

3.4 EXERCISE SET

For Extra Help
MyMathLab Math XL WATCH DOWNLOAD
PRACTICE

 Concept Reinforcement   *For Exercises 1–6, fill in the missing units for each rate.*

1. If Eva biked 100 miles in 5 hours, her average rate was 20 _____.

2. If it took Lauren 18 hours to read 6 chapters, her average rate was 3 _____.

3. If Denny's ticket cost $300 for a 150-mile flight, his average rate was 2 _____.

4. If Geoff planted 36 petunias along a 12-ft sidewalk, his average rate was 3 _____.

5. If Christi ran 8 errands in 40 minutes, her average rate was 5 _____.

6. If Ben made 8 cakes using 20 cups of flour, his average rate was $2\frac{1}{2}$ _____.

Solve. For Exercises 7–14, round answers to the nearest cent.

7. *Car rentals.*   Late on June 5, Gaya rented a Ford Focus with a full tank of gas and 13,741 mi on the odometer. On June 8, she returned the car with 14,131 mi on the odometer. The rental agency charged Gaya $118 for the rental and needed 13 gal of gas to fill up the tank.

    a) Find the car's rate of gas consumption, in miles per gallon.

    b) Find the average cost of the rental, in dollars per day.

    c) Find the average rate of travel, in miles per day.

    d) Find the rental rate, in cents per mile.

8. *SUV rentals.* On February 10, Oscar rented a Chevy Trailblazer with a full tank of gas and 13,091 mi on the odometer. On February 12, he returned the vehicle with 13,322 mi on the odometer. The rental agency charged $92 for the rental and needed 14 gal of gas to fill the tank.

   a) Find the SUV's rate of gas consumption, in miles per gallon.

   b) Find the average cost of the rental, in dollars per day.

   c) Find the average rate of travel, in miles per day.

   d) Find the rental rate, in cents per mile.

9. *Bicycle rentals.* At 9:00, Jodi rented a mountain bike from The Bike Rack. She returned the bicycle at 11:00, after cycling 14 mi. Jodi paid $15 for the rental.

   a) Find Jodi's average speed, in miles per hour.

   b) Find the rental rate, in dollars per hour.

   c) Find the rental rate, in dollars per mile.

10. *Bicycle rentals.* At 2:00, Braden rented a mountain bike from The Slick Rock Cyclery. He returned the bike at 5:00, after cycling 18 mi. Braden paid $12 for the rental.

   a) Find Braden's average speed, in miles per hour.

   b) Find the rental rate, in dollars per hour.

   c) Find the rental rate, in dollars per mile.

11. *Proofreading.* Sergei began proofreading at 9:00 A.M., starting at the top of page 93. He worked until 2:00 P.M. that day and finished page 195. He billed the publishers $110 for the day's work.

   a) Find the rate of pay, in dollars per hour.

   b) Find the average proofreading rate, in number of pages per hour.

   c) Find the rate of pay, in dollars per page.

12. *Temporary help.* A typist for Kelly Services reports to 3E's Properties for work at 10:00 A.M. and leaves at 6:00 P.M. after having typed from the end of page 8 to the end of page 50 of a proposal. 3E's pays $120 for the typist's services.

   a) Find the rate of pay, in dollars per hour.

   b) Find the average typing rate, in number of pages per hour.

   c) Find the rate of pay, in dollars per page.

13. *National debt.* The U.S. federal budget debt was $5770 billion in 2001 and $8612 billion in 2006. Find the rate at which the debt was increasing.
Source: U.S. Office of Management and Budget

14. *Four-year-college tuition.* The average tuition at a public four-year college was $3983 in 2001 and $5948 in 2005. Find the rate at which tuition was increasing.
Source: U.S. National Center for Education Statistics

15. *Elevators.* At 2:38, Lara entered an elevator on the 34th floor of the Regency Hotel. At 2:40, she stepped off at the 5th floor.

   a) Find the elevator's average rate of travel, in number of floors per minute.

   b) Find the elevator's average rate of travel, in seconds per floor.

16. *Snow removal.* By 1:00 P.M., Olivia had already shoveled 2 driveways, and by 6:00 P.M. that day, the number was up to 7.

   a) Find Olivia's average shoveling rate, in number of driveways per hour.

   b) Find Olivia's average shoveling rate, in hours per driveway.

17. *Mountaineering.* The fastest ascent of Mt. Everest was accomplished by the Sherpa guide Pemba Dorje of Nepal in 2004. Pemba Dorje climbed from base camp, elevation 17,552 ft, to the summit, elevation 29,028 ft, in 8 hr 10 min.
Source: *Guinness Book of World Records* 2006 Edition

   a) Find Pemba Dorje's average rate of ascent, in feet per minute.

   b) Find Pemba Dorje's average rate of ascent, in minutes per foot.

▤ **18.** *Mountaineering.*   As part of an ill-fated expedition to climb Mt. Everest in 1996, author Jon Krakauer departed "The Balcony," elevation 27,600 ft, at 7:00 A.M. and reached the summit, elevation 29,028 ft, at 1:25 P.M.
Source: Krakauer, Jon, *Into Thin Air, the Illustrated Edition.* New York: Random House, 1998

  a) Find Krakauer's average rate of ascent, in feet per minute.
  b) Find Krakauer's average rate of ascent, in minutes per foot.

In Exercises 19–28, draw a linear graph to represent the given information. Be sure to label and number the axes appropriately (see Example 3).

19. *Landfills.*   In 2006, 35,700,000 tons of paper was deposited in landfills in the United States, and this figure was decreasing by 700,000 tons per year.
Source: Based on data from American Forest and Paper Association

20. *Health insurance.*   In 2005, the average cost for health insurance for a family was about $11,000 and the figure was rising at a rate of about $1100 per year.
Source: Based on data from Kaiser/HRET Survey of Health Benefits

21. *Prescription drug sales.*   In 2006, there were sales of approximately $11 billion of asthma drug products in the United States, and the figure was increasing at a rate of about $1.2 billion per year.
Source: *The Wall Street Journal,* 6/28/2007

22. *Violent crimes.*   In 2004, there were approximately 21.1 violent crimes per 1000 population in the United States, and the figure was dropping at a rate of about 1.2 crimes per 1000 per year.
Source: U.S. Bureau of Justice Statistics

23. *Train travel.*   At 3:00 P.M., the Boston–Washington Metroliner had traveled 230 mi and was cruising at a rate of 90 miles per hour.

24. *Plane travel.*   At 4:00 P.M., the Seattle–Los Angeles shuttle had traveled 400 mi and was cruising at a rate of 300 miles per hour.

25. *Wages.*   By 2:00 P.M., Diane had earned $50. She continued earning money at a rate of $15 per hour.

26. *Wages.*   By 3:00 P.M., Arnie had earned $70. He continued earning money at a rate of $12 per hour.

27. *Telephone bills.*   Roberta's phone bill was already $7.50 when she made a call for which she was charged at a rate of $0.10 per minute.

28. *Telephone bills.*   At 3:00 P.M., Larry's phone bill was $6.50 and increasing at a rate of 7¢ per minute.

In Exercises 29–38, use the graph provided to calculate a rate of change in which the units of the horizontal axis are used in the denominator.

29. *Call center.*   The following graph shows data from a technical assistance call center. At what rate are calls being handled?

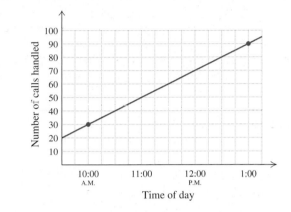

30. *Hairdresser.*   Eve's Custom Cuts has a graph displaying data from a recent day of work. At what rate does Eve work?

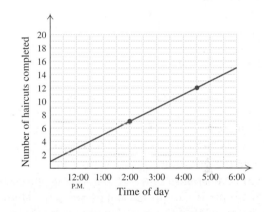

31. *Train travel.*   The following graph shows data from a recent train ride from Chicago to St. Louis. At what rate did the train travel?

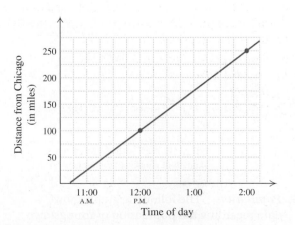

32. *Train travel.* The following graph shows data from a recent train ride from Denver to Kansas City. At what rate did the train travel?

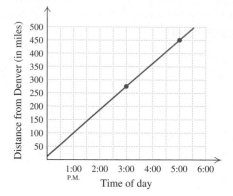

33. *Cost of a telephone call.* The following graph shows data from a recent phone call between the United States and the Netherlands. At what rate was the customer being billed?

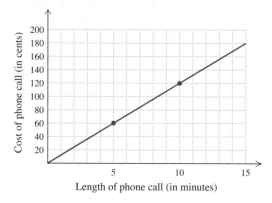

34. *Cost of a telephone call.* The following graph shows data from a recent phone call between the United States and South Korea. At what rate was the customer being billed?

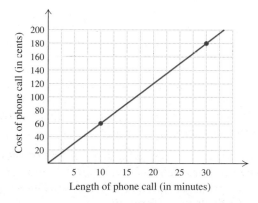

35. *Population.* The following graph shows data regarding the population of Youngstown,

Ohio. At what rate was the population changing?

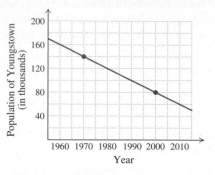

36. *Depreciation of an office machine.* Data regarding the value of a particular color copier is represented in the following graph. At what rate is the value changing?

37. *Gas mileage.* The following graph shows data for a 2008 Toyota Prius driven on interstate highways. At what rate was the vehicle consuming gas?
Source: www.fueleconomy.gov

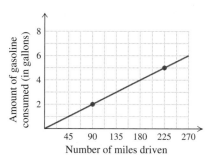

38. *Gas mileage.* The following graph shows data for a 2008 Chevy Malibu driven on city streets. At what rate was the vehicle consuming gas?
Source: Chevrolet

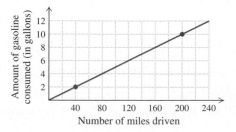

In each of Exercises 39–44, match the description with the most appropriate graph from the choices below. Scales are intentionally omitted. Assume that of the three sports listed, swimming is the slowest and biking is the fastest.

39. ____ Robin trains for triathlons by running, biking, and then swimming every Saturday.

40. ____ Gene trains for triathlons by biking, running, and then swimming every Sunday.

41. ____ Shirley trains for triathlons by swimming, biking, and then running every Sunday.

42. ____ Evan trains for triathlons by swimming, running, and then biking every Saturday.

43. ____ Angie trains for triathlons by biking, swimming, and then running every Sunday.

44. ____ Mick trains for triathlons by running, swimming, and then biking every Saturday.

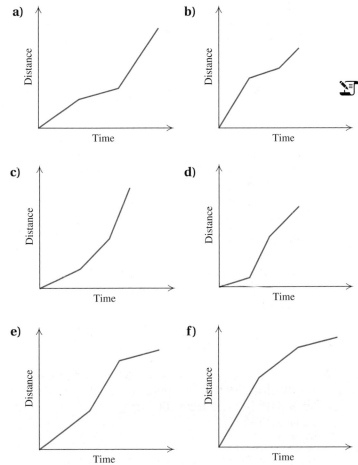

a)

b)

c)

d)

e)

f)

45. What does a negative rate of travel indicate? Explain.

46. Explain how to convert from kilometers per hour to meters per second.

Skill Review

To prepare for Section 3.5, review subtraction and order of operations (Sections 1.6 and 1.8).

Simplify.

47. $-2 - (-7)$ [1.6]

48. $-9 - (-3)$ [1.6]

49. $\dfrac{5 - (-4)}{-2 - 7}$ [1.8]

50. $\dfrac{8 - (-4)}{2 - 11}$ [1.8]

51. $\dfrac{-4 - 8}{11 - 2}$ [1.8]

52. $\dfrac{-5 - (-3)}{4 - 6}$ [1.8]

53. $\dfrac{-6 - (-6)}{-2 - 7}$ [1.8]

54. $\dfrac{-3 - 5}{-1 - (-1)}$ [1.8]

Synthesis

55. How would the graphs of Jon's and Jenny's total earnings compare in each of the following situations?

   a) Jon earns twice as much per hour as Jenny.
   b) Jon and Jenny earn the same hourly rate, but Jenny received a bonus for a cost-saving suggestion.
   c) Jon is paid by the hour, and Jenny is paid a weekly salary.

56. Write an exercise similar to those in Exercises 7–18 for a classmate to solve. Design the problem so that the solution is "The motorcycle's rate of gas consumption was 65 miles per gallon."

57. *Aviation.* A Boeing 737 climbs from sea level to a cruising altitude of 31,500 ft at a rate of 6300 ft/min. After cruising for 3 min, the jet is forced to land, descending at a rate of 3500 ft/min. Represent the flight with a graph in which altitude is measured on the vertical axis and time on the horizontal axis.

58. *Wages with commissions.* Each salesperson at Mike's Bikes is paid $140 a week plus 13% of all sales up to $2000, and then 20% on any sales in excess of $2000. Draw a graph in which sales are measured on the horizontal axis and wages on the vertical axis. Then use the graph to estimate the wages paid when a salesperson sells $2700 in merchandise in one week.

59. *Taxi fares.* The driver of a New York City Yellow Cab recently charged $2 plus 50¢ for each fifth of a mile traveled. Draw a graph that could be used to determine the cost of a fare.

60. *Gas mileage.* Suppose that a Kawasaki motorcycle travels three times as far as a Chevy Malibu on the same amount of gas (see Exercise 38). Draw a graph that reflects this information.

61. *Aviation.*   Tim's F-16 jet is moving forward at a deck speed of 95 mph aboard an aircraft carrier that is traveling 39 mph in the same direction. How fast is the jet traveling, in minutes per mile, with respect to the sea?

62. *Navigation.*   In 3 sec, Penny walks 24 ft, to the bow (front) of a tugboat. The boat is cruising at a rate of 5 ft/sec. What is Penny's rate of travel with respect to land?

63. *Running.*   Anne ran from the 4-km mark to the 7-km mark of a 10-km race in 15.5 min. At this rate, how long would it take Anne to run a 5-mi race?

64. *Running.*   Jerod ran from the 2-mi marker to the finish line of a 5-mi race in 25 min. At this rate, how long would it take Jerod to run a 10-km race?

65. Alex picks apples twice as fast as Ryan. By 4:30, Ryan had already picked 4 bushels of apples. Fifty minutes later, his total reached $5\frac{1}{2}$ bushels. Find Alex's picking rate. Give your answer in number of bushels per hour.

66. At 3:00 P.M., Catanya and Chad had already made 46 candles. By 5:00 P.M., the total reached 100 candles. Assuming a constant production rate, at what time did they make their 82nd candle?

CORNER

COLLABORATIVE

Determining Depreciation Rates

Focus: Modeling, graphing, and rates

Time: 30 minutes

Group size: 3

Materials: Graph paper and straightedges

From the minute a new car is driven out of the dealership, it *depreciates*, or drops in value with the passing of time. The N.A.D.A. Official Used Car Guide is a periodic listing of the trade-in values of used cars. The data below are taken from two such reports from 2007.

ACTIVITY

1. Each group member should select a different one of the cars listed in the table below as his or her own. Assuming that the values are dropping linearly, each student should draw a line representing the trade-in value of his or her car. Draw all three lines on the same graph. Let the horizontal axis represent the time, in months, since January 2007, and let the vertical axis represent the trade-in value of each car. Decide as a group how many months or dollars each square should represent. Make the drawings as neat as possible.

2. At what *rate* is each car depreciating and how are the different rates illustrated in the graph of part (1)?

3. If one of the three cars had to be sold in January 2009, which one would your group sell and why? Compare answers with other groups.

Car	Trade-in Value in January 2007	Trade-in Value in June 2007
2005 Mustang V6 Coupe	$13,625	$13,125
2005 Nissan Sentra SE-R	$11,825	$11,000
2005 Volkswagen Jetta Sedan GL	$12,600	$11,425

3.5 Slope

Rate and Slope ▪ Horizontal and Vertical Lines ▪ Applications

In Section 3.4, we introduced *rate* as a method of measuring how two quantities change with respect to each other. In this section, we will discuss how rate can be related to the slope of a line.

Rate and Slope

Automated digitization machines use robotic arms to carefully turn pages of books so that they can take pictures of each page. Suppose that a large university library purchased a DL-1500 and an APT 1200. The DL-1500 digitizes 3 volumes of an encyclopedia every 2 hr. The APT 1200 digitizes 6 volumes of an encyclopedia every 5 hr. The following tables list the number of books digitized after various amounts of time for each machine.

Source: Based on information from Kirtas and 4DigitalBooks

DL-1500	
Hours Elapsed	**Books Digitized**
0	0
2	3
4	6
6	9
8	12

APT 1200	
Hours Elapsed	**Books Digitized**
0	0
5	6
10	12
15	18
20	24

We now graph the pairs of numbers listed in the tables, using the horizontal axis for the number of hours elapsed and the vertical axis for the number of books digitized.

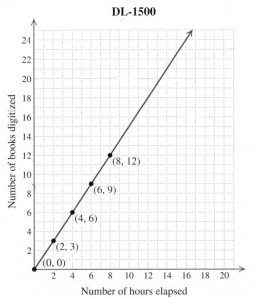

DL-1500

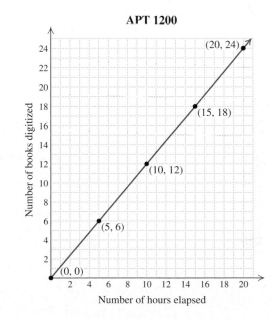

APT 1200

By comparing the number of books digitized by each machine over a specified period of time, we can compare the two rates. For example, the DL-1500 digitizes 3 books every 2 hr, so its *rate* is $3 \div 2 = \frac{3}{2}$ books per hour. Since the APT 1200 digitizes 6 books every 5 hr, its rate is $6 \div 5 = \frac{6}{5}$ books per hour. Note that the rate of the DL-1500 is greater so its graph is steeper.

The rates $\frac{3}{2}$ and $\frac{6}{5}$ can also be found using the coordinates of any two points that are on the line. For example, we can use the points $(6, 9)$ and $(8, 12)$ to find the digitization rate for the DL-1500. To do so, remember that these coordinates tell us that after 6 hr, 9 books have been digitized, and after 8 hr, 12 books have been digitized. In the 2 hr between the 6-hr and 8-hr points, $12 - 9$, or 3, books were digitized. Thus we have

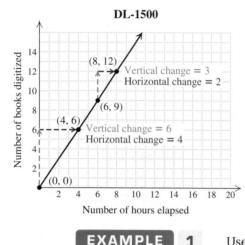

$$\text{DL-1500 digitization rate} = \frac{\text{change in number of books digitized}}{\text{corresponding change in time}}$$

$$= \frac{12 - 9 \text{ books}}{8 - 6 \text{ hr}}$$

$$= \frac{3 \text{ books}}{2 \text{ hr}} = \frac{3}{2} \text{ books per hour.}$$

Because the line is straight, the same rate is found using *any* pair of points on the line. For example, using $(0, 0)$ and $(4, 6)$, we have

$$\text{DL-1500 digitization rate} = \frac{6 - 0 \text{ books}}{4 - 0 \text{ hr}} = \frac{6 \text{ books}}{4 \text{ hr}} = \frac{3}{2} \text{ books per hour.}$$

Note that the rate is always the vertical change divided by the corresponding horizontal change.

EXAMPLE   1   Use the graph of book digitization by the APT 1200 to find the rate at which books are digitized.

SOLUTION   We can use any two points on the line, such as $(15, 18)$ and $(20, 24)$:

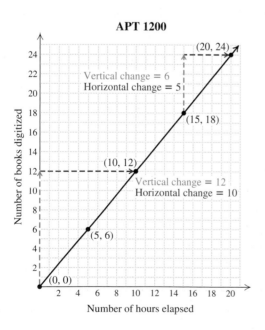

$$\text{APT 1200 digitization rate} = \frac{\text{change in number of books digitized}}{\text{corresponding change in time}}$$

$$= \frac{24 - 18 \text{ books}}{20 - 15 \text{ hr}}$$

$$= \frac{6 \text{ books}}{5 \text{ hr}}$$

$$= \frac{6}{5} \text{ books per hour.}$$

As a check, we can use another pair of points, like $(0, 0)$ and $(10, 12)$:

$$\text{APT 1200 digitization rate} = \frac{12 - 0 \text{ books}}{10 - 0 \text{ hr}}$$

$$= \frac{12 \text{ books}}{10 \text{ hr}}$$

$$= \frac{6}{5} \text{ books per hour.}$$

TRY EXERCISE   11

When the axes of a graph are simply labeled x and y, the ratio of vertical change to horizontal change is the rate at which y is changing with respect to x. This ratio is a measure of a line's slant, or **slope**.

Consider a line passing through $(2, 3)$ and $(6, 5)$, as shown below. We find the ratio of vertical change, or *rise*, to horizontal change, or *run*, as follows:

$$\frac{\text{Ratio of vertical change}}{\text{to horizontal change}} = \frac{\text{change in } y}{\text{change in } x} = \frac{\text{rise}}{\text{run}}$$

$$= \frac{5 - 3}{6 - 2}$$

$$= \frac{2}{4}, \text{ or } \frac{1}{2}.$$

Note that these calculations can be performed without viewing a graph.

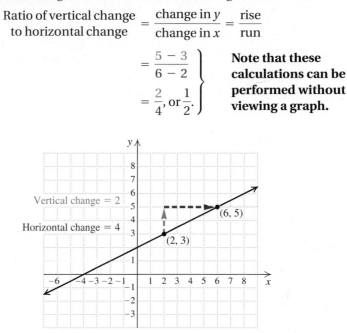

Thus the y-coordinates of points on this line increase at a rate of 2 units for every 4-unit increase in x, which is 1 unit for every 2-unit increase in x, or $\frac{1}{2}$ unit for every 1-unit increase in x. The slope of the line is $\frac{1}{2}$.

In the box below, the *subscripts* 1 and 2 are used to distinguish two arbitrary points, point 1 and point 2, from each other. The slightly lowered 1's and 2's are not exponents but are used to denote x-values (or y-values) that may differ from each other.

Slope

The *slope* of the line containing points (x_1, y_1) and (x_2, y_2) is given by

$$m = \frac{\text{change in } y}{\text{change in } x} = \frac{\text{rise}}{\text{run}} = \frac{y_2 - y_1}{x_2 - x_1}.$$

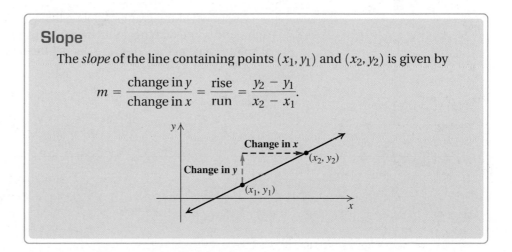

EXAMPLE 2   Graph the line containing the points $(-4, 3)$ and $(2, -6)$ and find the slope.

SOLUTION   The graph is shown below. From $(-4, 3)$ to $(2, -6)$, the change in y, or rise, is $-6 - 3$, or -9. The change in x, or run, is $2 - (-4)$, or 6. Thus,

$$\text{Slope} = \frac{\text{change in } y}{\text{change in } x}$$

$$= \frac{\text{rise}}{\text{run}}$$

$$= \frac{-6 - 3}{2 - (-4)}$$

$$= \frac{-9}{6}$$

$$= -\frac{9}{6}, \text{ or } -\frac{3}{2}.$$

STUDENT NOTES

You may wonder which point should be regarded as (x_1, y_1) and which should be (x_2, y_2). To see that the math works out the same either way, perform both calculations on your own.

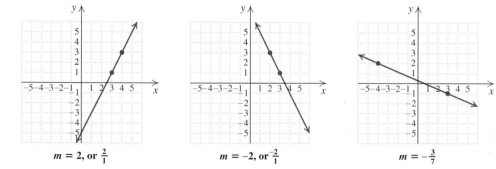

Change in $y = -9$

Change in $x = 6$

TRY EXERCISE 39

CAUTION!   When we use the formula

$$m = \frac{y_2 - y_1}{x_2 - x_1},$$

it makes no difference which point is considered (x_1, y_1). What matters is that we subtract the y-coordinates in the same order that we subtract the x-coordinates.

To illustrate, we reverse *both* of the subtractions in Example 2. The slope is still $-\frac{3}{2}$:

$$\text{Slope} = \frac{\text{change in } y}{\text{change in } x} = \frac{3 - (-6)}{-4 - 2} = \frac{9}{-6} = -\frac{3}{2}.$$

As shown in the graphs below, a line with positive slope slants up from left to right, and a line with negative slope slants down from left to right. The larger the absolute value of the slope, the steeper the line.

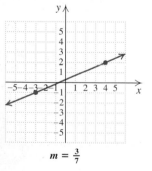

$m = \frac{3}{7}$ $m = 2, \text{ or } \frac{2}{1}$ $m = -2, \text{ or } \frac{-2}{1}$ $m = -\frac{3}{7}$

Horizontal and Vertical Lines

What about the slope of a horizontal line or a vertical line?

EXAMPLE 3   Find the slope of the line $y = 4$.

SOLUTION   Consider the points $(2, 4)$ and $(-3, 4)$, which are on the line. The change in y, or the rise, is $4 - 4$, or 0. The change in x, or the run, is $-3 - 2$, or -5. Thus,

$$m = \frac{4 - 4}{-3 - 2}$$

$$= \frac{0}{-5}$$

$$= 0.$$

Any two points on a horizontal line have the same y-coordinate. Thus the change in y is 0, so the slope is 0.

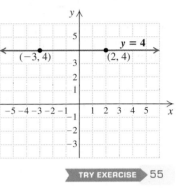

TRY EXERCISE 55

A horizontal line has slope 0.

EXAMPLE 4   Find the slope of the line $x = -3$.

SOLUTION   Consider the points $(-3, 4)$ and $(-3, -2)$, which are on the line. The change in y, or the rise, is $-2 - 4$, or -6. The change in x, or the run, is $-3 - (-3)$, or 0. Thus,

$$m = \frac{-2 - 4}{-3 - (-3)}$$

$$= \frac{-6}{0}. \quad \text{(undefined)}$$

Since division by 0 is not defined, the slope of this line is not defined. The answer to a problem of this type is "The slope of this line is undefined."

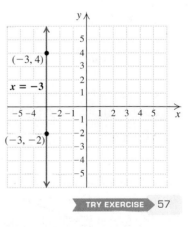

TRY EXERCISE 57

The slope of a vertical line is undefined.

Applications

We have seen that slope has many real-world applications, ranging from car speed to production rate. Some applications use slope to measure steepness. For example, numbers like 2%, 3%, and 6% are often used to represent the **grade** of a road, a measure of a road's steepness. That is, since $3\% = \frac{3}{100}$, a 3% grade means that for every horizontal distance of 100 ft, the road rises or drops 3 ft. The concept of grade also occurs in skiing or snowboarding, where a 7% grade is considered very tame, but a 70% grade is considered steep.

EXAMPLE   5   *Skiing.*   Among the steepest skiable terrain in North America, the Headwall on Mount Washington, in New Hampshire, drops 720 ft over a horizontal distance of 900 ft. Find the grade of the Headwall.

SOLUTION   The grade of the Headwall is its slope, expressed as a percent:

$$m = \frac{720}{900}$$
$$= \frac{8}{10}$$
$$= 80\%.$$

Grade is slope expressed as a percent.

TRY EXERCISE   63

Carpenters use slope when designing stairs, ramps, or roof pitches. Another application occurs in the engineering of a dam—the force or strength of a river depends on how much the river drops over a specified distance.

3.5   EXERCISE SET

🖐 *Concept Reinforcement*   *State whether each of the following rates is positive, negative, or zero.*

1. The rate at which a teenager's height changes

2. The rate at which an elderly person's height changes

3. The rate at which a pond's water level changes during a drought

4. The rate at which a pond's water level changes during the rainy season

5. The rate at which a runner's distance from the starting point changes during a race

6. The rate at which a runner's distance from the finish line changes during a race

7. The rate at which the number of U.S. Senators changes

8. The rate at which the number of people in attendance at a basketball game changes in the moments before the opening tipoff

9. The rate at which the number of people in attendance at a basketball game changes in the moments after the final buzzer sounds

10. The rate at which a person's I.Q. changes during his or her sleep

11. *Blogging.*   Find the rate at which a career blogger is paid.

Source: Based on information from Wired

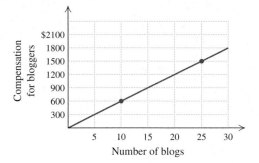

Compensation for bloggers / Number of blogs

12. *Fitness.*   Find the rate at which a runner burns calories.

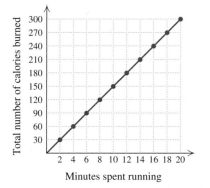

Total number of calories burned / Minutes spent running

13. *Cell-phone prices.*   Find the rate of change in the average price of a new cell phone.

Source: Based on information from Market Reporter at PriceGrabber.com 2006

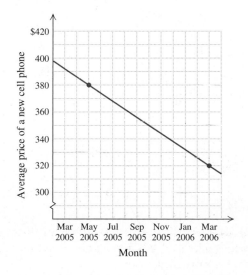

Average price of a new cell phone / Month

14. *Retail sales.*   Find the rate of change in the percentage of department stores' share of total retail sales.

Source: Based on information from the National Retail Federation

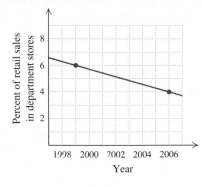

Percent of retail sales in department stores / Year

15. *College admission tests.*   Find the rate of change in SAT verbal scores with respect to family income.

Source: Based on 2004–2005 data from the National Center for Education Statistics

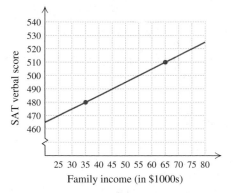

SAT verbal score / Family income (in $1000s)

16. *Long-term care.*   Find the rate of change in Medicaid spending on long-term care.

Source: Based on data from Thomson Medstat, prepared by AARP Public Policy Institute

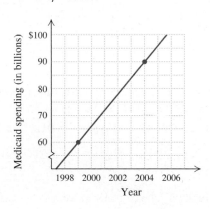

Medicaid spending (in billions) / Year

17. *Meteorology.*   Find the rate of change in the temperature in Spearfish, Montana, on January 22, 1943, as shown below.

Source: National Oceanic Atmospheric Administration

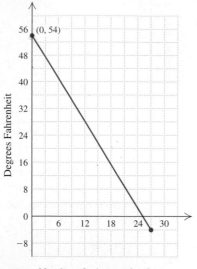

Number of minutes after 9 A.M.

18. Find the rate of change in the number of union-represented Ford employees.

Source: Ford Motor Co.

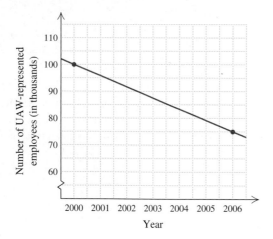

Find the slope, if it is defined, of each line. If the slope is undefined, state this.

19.

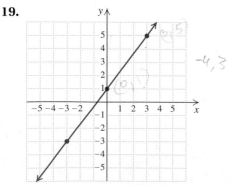

20.

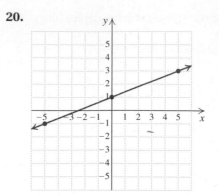

21.

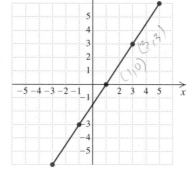

22.

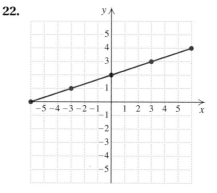

23.

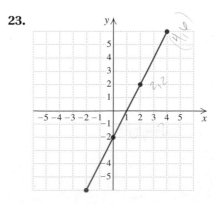

24.

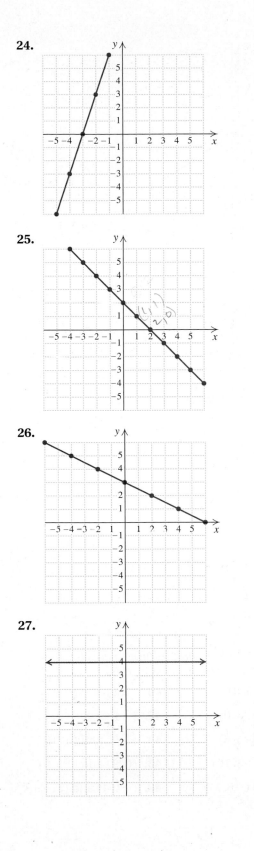

28.

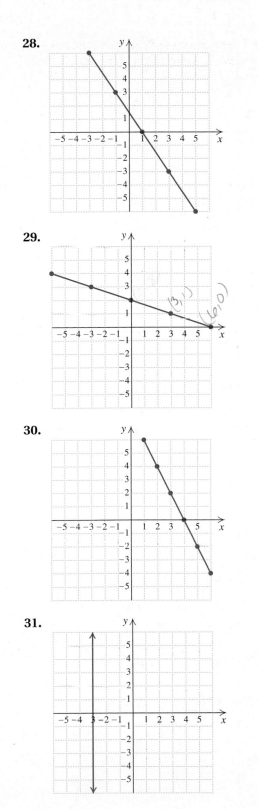

25.

29.

26.

30.

27.

31.

32.

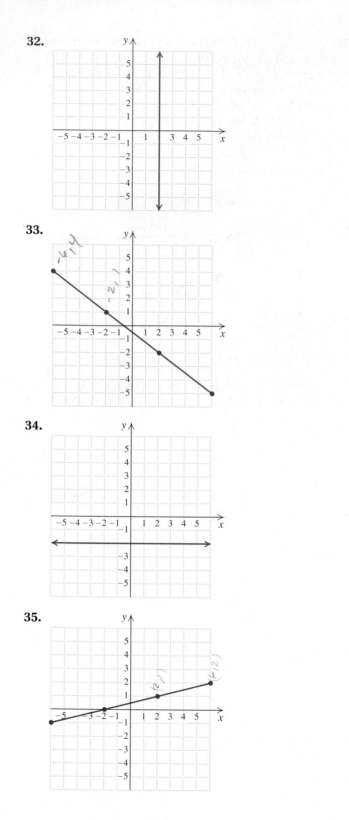

33.

34.

35.

36.

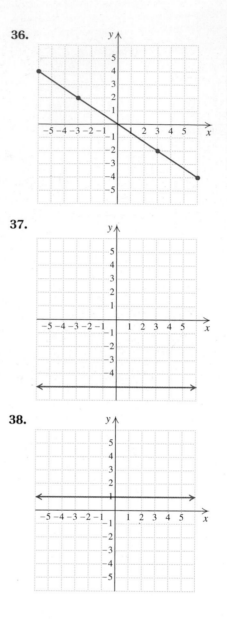

37.

38.

Find the slope of the line containing each given pair of points. If the slope is undefined, state this.

39. $(1, 3)$ and $(5, 8)$ **40.** $(1, 8)$ and $(6, 9)$

41. $(-2, 4)$ and $(3, 0)$ **42.** $(-4, 2)$ and $(2, -3)$

43. $(-4, 0)$ and $(5, 6)$ **44.** $(3, 0)$ and $(6, 9)$

45. $(0, 7)$ and $(-3, 10)$ **46.** $(0, 9)$ and $(-5, 0)$

47. $(-2, 3)$ and $(-6, 5)$ **48.** $(-1, 4)$ and $(5, -8)$

Aha! 49. $\left(-2, \frac{1}{2}\right)$ and $\left(-5, \frac{1}{2}\right)$

50. $(-5, -1)$ and $(2, -3)$

51. $(5, -4)$ and $(2, -7)$

52. $(-10, 3)$ and $(-10, 4)$

53. $(6, -4)$ and $(6, 5)$

54. $(5, -2)$ and $(-4, -2)$

Find the slope of each line whose equation is given. If the slope is undefined, state this.

55. $y = 5$

56. $y = 13$

57. $x = -8$

58. $x = 18$

59. $x = 9$

60. $x = -7$

61. $y = -10$

62. $y = -4$

63. *Surveying.* Lick Skillet Road, near Boulder, Colorado, climbs 792 ft over a horizontal distance of 5280 ft. What is the grade of the road?

64. *Navigation.* Capital Rapids drops 54 ft vertically over a horizontal distance of 1080 ft. What is the slope of the rapids?

65. *Construction.* Part of New Valley rises 28 ft over a horizontal distance of 80 ft, and is too steep to build on. What is the slope of the land?

66. *Engineering.* At one point, Yellowstone's Beartooth Highway rises 315 ft over a horizontal distance of 4500 ft. Find the grade of the road.

67. *Carpentry.* Find the slope (or pitch) of the roof.

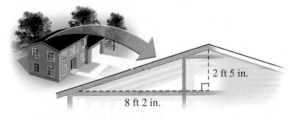

2 ft 5 in.

8 ft 2 in.

68. *Exercise.* Find the slope (or grade) of the treadmill.

0.4 ft

5 ft

69. *Bicycling.* To qualify as a rated climb on the Tour de France, a grade must average at least 4%. The ascent of Dooley Mountain, Oregon, part of the Elkhorn Classic, begins at 3500 ft and climbs to 5400 ft over a horizontal distance of 37,000 ft. What is the grade of the road? Would it qualify as a rated climb if it were part of the Tour de France?
Source: barkercityherald.com

70. *Construction.* Public buildings regularly include steps with 7-in. risers and 11-in. treads. Find the grade of such a stairway.

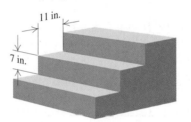

11 in.

7 in.

71. Explain why the order in which coordinates are subtracted to find slope does not matter so long as *y*-coordinates and *x*-coordinates are subtracted in the same order.

72. If one line has a slope of -3 and another has a slope of 2, which line is steeper? Why?

Skill Review

To prepare for Section 3.6, review solving a formula for a variable and graphing linear equations (Sections 2.3 and 3.2).

Solve. [2.3]

73. $ax + by = c$, for y

74. $rx - mn = p$, for r

75. $ax - by = c$, for y

76. $rs + nt = q$, for t

Graph. [3.2]

77. $8x + 6y = 24$

78. $3y = 4$

Synthesis

79. The points $(-4, -3)$, $(1, 4)$, $(4, 2)$, and $(-1, -5)$ are vertices of a quadrilateral. Use slopes to explain why the quadrilateral is a parallelogram.

80. Which is steeper and why: a ski slope that is 50° or one with a grade of 100%?

81. The plans below are for a skateboard "Fun Box". For the ramps labeled A, find the slope or grade.
Source: www.heckler.com

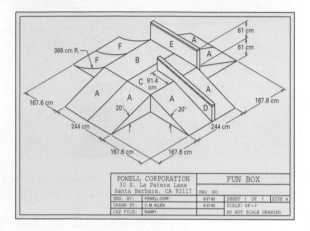

82. A line passes through $(4, -7)$ and never enters the first quadrant. What numbers could the line have for its slope?

83. A line passes through $(2, 5)$ and never enters the second quadrant. What numbers could the line have for its slope?

84. *Architecture.* Architects often use the equation $x + y = 18$ to determine the height y, in inches, of the riser of a step when the tread is x inches wide. Express the slope of stairs designed with this equation without using the variable y.

In Exercises 85 and 86, the slope of the line is $-\frac{2}{3}$, but the numbering on one axis is missing. How many units should each tick mark on that unnumbered axis represent?

85.

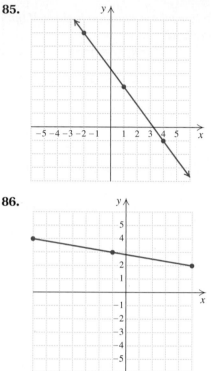

86.

3.6 ## Slope–Intercept Form

Using the *y*-intercept and the Slope to Graph a Line ▪ Equations in Slope-Intercept Form ▪ Graphing and Slope-Intercept Form

If we know the slope and the *y*-intercept of a line, it is possible to graph the line. In this section, we will discover that a line's slope and *y*-intercept can be determined directly from the line's equation, provided the equation is written in a certain form.

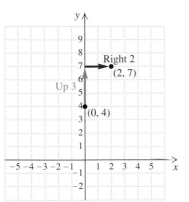

DL -1500

Using the *y*–intercept and the Slope to Graph a Line

Let's modify the book-digitization situation that first appeared in Section 3.5. Suppose that as the information technologist arrives, 4 books had already been digitized by the DL-1500. If the rate of $\frac{3}{2}$ books per hour remains in effect, the table and graph shown here can be made.

DL-1500	
Hours Elapsed	**Books Digitized**
0	4
2	7
4	10
6	13
8	16

To confirm that the digitization rate is still $\frac{3}{2}$, we calculate the slope. Recall that

$$\text{Slope} = \frac{\text{change in } y}{\text{change in } x} = \frac{\text{rise}}{\text{run}} = \frac{y_2 - y_1}{x_2 - x_1},$$

where (x_1, y_1) and (x_2, y_2) are any two points on the graphed line. Here we select $(0, 4)$ and $(2, 7)$:

$$\text{Slope} = \frac{\text{change in } y}{\text{change in } x} = \frac{7 - 4}{2 - 0} = \frac{3}{2}.$$

Knowing that the slope is $\frac{3}{2}$, we could have drawn the graph by plotting $(0, 4)$ and from there moving *up* 3 units and *to the right* 2 units. This would have located the point $(2, 7)$. Using $(0, 4)$ and $(2, 7)$, we can then draw the line. This is the method used in the next example.

EXAMPLE 1 Draw a line that has slope $\frac{1}{4}$ and *y*-intercept $(0, 2)$.

SOLUTION We plot $(0, 2)$ and from there move *up* 1 unit and *to the right* 4 units. This locates the point $(4, 3)$. We plot $(4, 3)$ and draw a line passing through $(0, 2)$ and $(4, 3)$, as shown on the right below.

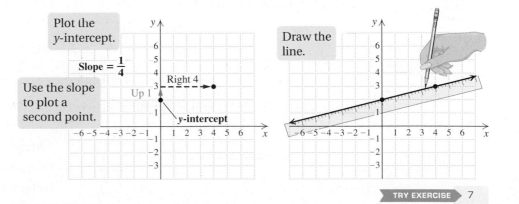

Plot the *y*-intercept.

Use the slope to plot a second point.

Slope = $\frac{1}{4}$

Draw the line.

TRY EXERCISE 7

Equations in Slope–Intercept Form

It is possible to read the slope and the y-intercept of a line directly from its equation. Recall from Section 3.3 that to find the y-intercept of an equation's graph, we replace x with 0 and solve the resulting equation for y. For example, to find the y-intercept of the graph of $y = 2x + 3$, we replace x with 0 and solve as follows:

$$y = 2x + 3$$
$$= 2 \cdot 0 + 3 = 0 + 3 = 3. \qquad \text{The } y\text{-intercept is } (0, 3).$$

The y-intercept of the graph of $y = 2x + 3$ is $(0, 3)$. It can be similarly shown that the graph of $y = mx + b$ has the y-intercept $(0, b)$.

To calculate the slope of the graph of $y = 2x + 3$, we need two ordered pairs that are solutions of the equation. The y-intercept $(0, 3)$ is one pair; a second pair, $(1, 5)$, can be found by substituting 1 for x. We then have

$$\text{Slope} = \frac{\text{change in } y}{\text{change in } x} = \frac{5 - 3}{1 - 0} = \frac{2}{1} = 2.$$

Note that the slope, 2, is also the x-coefficient in $y = 2x + 3$. It can be similarly shown that the graph of any equation of the form $y = mx + b$ has slope m (see Exercise 79).

STUDENT NOTES

An equation for a given line can be written in many different forms. Note that in the slope–intercept form, the equation is solved for y.

The Slope–Intercept Equation

The equation $y = mx + b$ is called the *slope–intercept equation*. The equation represents a line of slope m with y-intercept $(0, b)$.

The equation of any nonvertical line can be written in this form.

EXAMPLE   2   Find the slope and the y-intercept of each line whose equation is given.

a) $y = \frac{4}{5}x - 8$ **b)** $2x + y = 5$ **c)** $3x - 4y = 7$

SOLUTION

a) We rewrite $y = \frac{4}{5}x - 8$ as $y = \frac{4}{5}x + (-8)$. Now we simply read the slope and the y-intercept from the equation:

$$y = \frac{4}{5}x + (-8).$$

The slope is $\frac{4}{5}$. The y-intercept is $(0, -8)$.

b) We first solve for y to find an equivalent equation in the form $y = mx + b$:

$$2x + y = 5$$
$$y = -2x + 5. \qquad \text{Adding } -2x \text{ to both sides}$$

The slope is -2. The y-intercept is $(0, 5)$.

c) We rewrite the equation in the form $y = mx + b$:

$$3x - 4y = 7$$
$$-4y = -3x + 7 \qquad \text{Adding } -3x \text{ to both sides}$$
$$y = -\frac{1}{4}(-3x + 7) \qquad \text{Multiplying both sides by } -\frac{1}{4}$$
$$y = \frac{3}{4}x - \frac{7}{4}. \qquad \text{Using the distributive law}$$

The slope is $\frac{3}{4}$. The y-intercept is $\left(0, -\frac{7}{4}\right)$.

TRY EXERCISE   19

EXAMPLE   3   A line has slope $-\frac{12}{5}$ and y-intercept $(0, 11)$. Find an equation of the line.

SOLUTION   We use the slope–intercept equation, substituting $-\frac{12}{5}$ for m and 11 for b:

$$y = mx + b = -\frac{12}{5}x + 11.$$

The desired equation is $y = -\frac{12}{5}x + 11$.

> TRY EXERCISE 35

EXAMPLE   4   *Threatened species.*   A threatened species is a species that is likely to become endangered in the future. Determine an equation for the graph of threatened species in the Blue Mountains.

Source: Based on information from Sustainable Blue Mountains

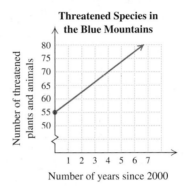

Threatened Species in the Blue Mountains

SOLUTION   To write an equation for a line, we can use slope–intercept form, provided the slope and the y-intercept are known. From the graph, we see that $(0, 55)$ is the y-intercept. Looking closely, we see that the line passes through $(4, 70)$. We can either count squares on the graph or use the formula to calculate the slope:

$$m = \frac{\text{change in } y}{\text{change in } x} = \frac{70 - 55}{4 - 0} = \frac{15}{4}.$$

The desired equation is

$$y = \frac{15}{4}x + 55, \qquad \text{Using } \tfrac{15}{4} \text{ for } m \text{ and 55 for } b$$

where y is the number of threatened species in the Blue Mountains x years after 2000.

> TRY EXERCISE 43

Graphing and Slope–Intercept Form

In Example 1, we drew a graph, knowing only the slope and the y-intercept. In Example 2, we determined the slope and the y-intercept of a line by examining its equation. We now combine the two procedures to develop a quick way to graph a linear equation.

EXAMPLE   5   Graph: **(a)** $y = \frac{3}{4}x + 5$; **(b)** $2x + 3y = 3$.

SOLUTION

a)   From the equation $y = \frac{3}{4}x + 5$, we see that the slope of the graph is $\frac{3}{4}$ and the y-intercept is $(0, 5)$. We plot $(0, 5)$ and then consider the slope, $\frac{3}{4}$. Starting at $(0, 5)$, we plot a second point by moving *up* 3 units (since the numerator is *positive* and corresponds to the change in y) and *to the right* 4 units (since the denominator is *positive* and corresponds to the change in x). We reach a new point, $(4, 8)$.

STUDENT NOTES ———

Remember that

$$\frac{3}{4} = \frac{-3}{-4}.$$

Similarly,

$$-\frac{3}{4} = \frac{-3}{4} = \frac{3}{-4}.$$

Also note that

$$2 = \frac{2}{1} \quad \text{and} \quad -2 = \frac{-2}{1}.$$

Since $\frac{3}{4} = \frac{-3}{-4}$, we can again start at the *y*-intercept, $(0, 5)$, but move *down* 3 units (since the numerator is *negative* and corresponds to the change in *y*) and *to the left* 4 units (since the denominator is *negative* and corresponds to the change in *x*). We reach another point, $(-4, 2)$. Once two or three points have been plotted, the line representing all solutions of $y = \frac{3}{4}x + 5$ can be drawn.

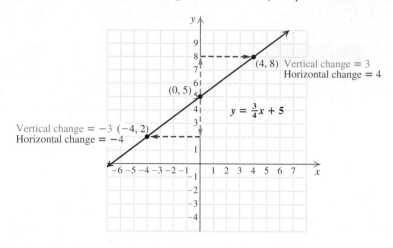

STUDENT NOTES ———

The signs of the numerator and the denominator of the slope indicate whether to move up, down, left, or right. Compare the following slopes.

$\dfrac{1}{2}$ ← 1 unit up
    ← 2 units right

$\dfrac{-1}{-2}$ ← 1 unit down
    ← 2 units left

$\dfrac{-1}{2}$ ← 1 unit down
    ← 2 units right

$\dfrac{1}{-2}$ ← 1 unit up
    ← 2 units left

b) To graph $2x + 3y = 3$, we first rewrite it in slope–intercept form:

$$2x + 3y = 3$$
$$3y = -2x + 3 \qquad \text{Adding } -2x \text{ to both sides}$$
$$y = \tfrac{1}{3}(-2x + 3) \qquad \text{Multiplying both sides by } \tfrac{1}{3}$$
$$y = -\tfrac{2}{3}x + 1. \qquad \text{Using the distributive law}$$

To graph $y = -\frac{2}{3}x + 1$, we first plot the *y*-intercept, $(0, 1)$. We can think of the slope as $\frac{-2}{3}$. Starting at $(0, 1)$ and using the slope, we find a second point by moving *down* 2 units (since the numerator is *negative*) and *to the right* 3 units (since the denominator is *positive*). We plot the new point, $(3, -1)$. In a similar manner, we can move from the point $(3, -1)$ to locate a third point, $(6, -3)$. The line can then be drawn.

Since $-\frac{2}{3} = \frac{2}{-3}$, an alternative approach is to again plot $(0, 1)$, but this time move *up* 2 units (since the numerator is *positive*) and *to the left* 3 units (since the denominator is *negative*). This leads to another point on the graph, $(-3, 3)$.

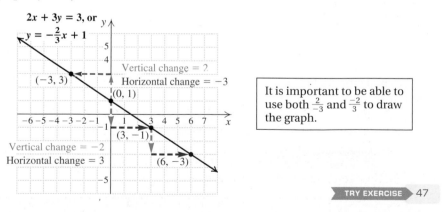

> It is important to be able to use both $\frac{2}{-3}$ and $\frac{-2}{3}$ to draw the graph.

TRY EXERCISE 47

Recall that *parallel lines* are lines in the same plane that do not intersect. Thus, when two lines have the same slope but different *y*-intercepts, they are parallel.

EXAMPLE **6**   Determine whether the graphs of $y = -3x + 4$ and $6x + 2y = -10$ are parallel.

SOLUTION   We compare the slopes of the two lines to determine whether the graphs are parallel.

One of the two equations given is in slope–intercept form:

$y = -3x + 4.$ The slope is -3 and the y-intercept is $(0, 4)$.

To find the slope of the other line, we need to rewrite the other equation in slope–intercept form:

$$6x + 2y = -10$$
$$2y = -6x - 10$$ Adding $-6x$ to both sides
$$y = -3x - 5.$$ The slope is -3 and the y-intercept is $(0, -5)$.

Since both lines have slope -3 but different y-intercepts, the graphs are parallel. There is no need for us to actually graph either equation.   **TRY EXERCISE** 63

TECHNOLOGY CONNECTION

Using a standard $[-10, 10, -10, 10]$ window, graph the equations $y_1 = \frac{2}{3}x + 1$, $y_2 = \frac{3}{8}x + 1$, $y_3 = \frac{2}{3}x + 5$, and $y_4 = \frac{3}{8}x + 5$. If you can, use your graphing calculator in the MODE that graphs equations *simultaneously*. Once all lines have been drawn, try to decide which equation corresponds to each line. After matching equations with lines, you can check your matches by using TRACE and the up and down arrow keys to move from one line to the

next. The number of the equation will appear in a corner of the screen.

1. Graph $y_1 = -\frac{3}{4}x - 2$, $y_2 = -\frac{1}{5}x - 2$, $y_3 = -\frac{3}{4}x - 5$, and $y_4 = -\frac{1}{5}x - 5$ using the SIMULTANEOUS mode. Then match each line with the corresponding equation. Check using TRACE.

3.6   EXERCISE SET

For Extra Help   **MyMathLab**   Math XL PRACTICE   WATCH   DOWNLOAD

↬ *Concept Reinforcement   In each of Exercises 1–6, match the phrase with the most appropriate choice from the column on the right.*

1. ____ The slope of the graph of $y = 3x - 2$

2. ____ The slope of the graph of $y = 2x - 3$

3. ____ The slope of the graph of $y = \frac{2}{3}x + 3$

4. ____ The y-intercept of the graph of $y = 2x - 3$

5. ____ The y-intercept of the graph of $y = 3x - 2$

6. ____ The y-intercept of the graph of $y = \frac{2}{3}x + \frac{3}{4}$

a) $\left(0, \frac{3}{4}\right)$

b) 2

c) $(0, -3)$

d) $\frac{2}{3}$

e) $(0, -2)$

f) 3

Draw a line that has the given slope and y-intercept.

7. Slope $\frac{2}{3}$; y-intercept $(0, 1)$

8. Slope $\frac{3}{5}$; y-intercept $(0, -1)$

9. Slope $\frac{5}{3}$; y-intercept $(0, -2)$

10. Slope $\frac{1}{2}$; y-intercept $(0, 0)$

11. Slope $-\frac{1}{3}$; y-intercept $(0, 5)$

12. Slope $-\frac{4}{5}$; y-intercept $(0, 6)$

13. Slope 2; y-intercept $(0, 0)$

14. Slope -2; y-intercept $(0, -3)$

15. Slope -3; y-intercept $(0, 2)$

16. Slope 3; y-intercept $(0, 4)$

Aha! **17.** Slope 0; y-intercept $(0, -5)$

18. Slope 0; y-intercept $(0, 1)$

Find the slope and the y-intercept of each line whose equation is given.

19. $y = -\frac{2}{7}x + 5$ **20.** $y = -\frac{3}{8}x + 4$

21. $y = \frac{1}{3}x + 7$ **22.** $y = \frac{4}{5}x + 1$

23. $y = \frac{9}{5}x - 4$ **24.** $y = -\frac{9}{10}x - 5$

25. $-3x + y = 7$ **26.** $-4x + y = 7$

27. $4x + 2y = 8$ **28.** $3x + 4y = 12$

Aha! **29.** $y = 3$ **30.** $y - 3 = 5$

31. $2x - 5y = -8$ **32.** $12x - 6y = 9$

33. $9x - 8y = 0$ **34.** $7x = 5y$

Find the slope–intercept equation for the line with the indicated slope and y-intercept.

35. Slope 5; y-intercept $(0, 7)$

36. Slope -4; y-intercept $\left(0, -\frac{3}{5}\right)$

37. Slope $\frac{7}{8}$; y-intercept $(0, -1)$

38. Slope $\frac{5}{7}$; y-intercept $(0, 4)$

39. Slope $-\frac{5}{3}$; y-intercept $(0, -8)$

40. Slope $\frac{3}{4}$; y-intercept $(0, -35)$

Aha! **41.** Slope 0; y-intercept $\left(0, \frac{1}{3}\right)$

42. Slope 7; y-intercept $(0, 0)$

Determine an equation for each graph shown.

43.

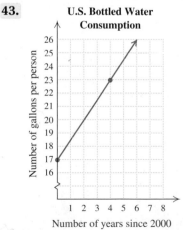

U.S. Bottled Water Consumption

Based on information from the International Bottled Water Association

44.

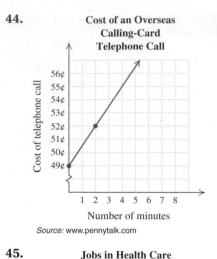

Cost of an Overseas Calling-Card Telephone Call

Source: www.pennytalk.com

45.

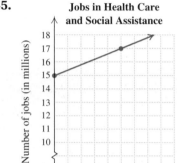

Jobs in Health Care and Social Assistance

Source: U.S. Bureau of Labor Statistics

46.

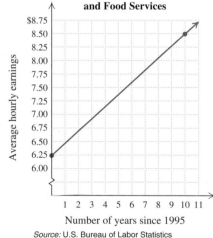

Average Hourly Earnings in Accommodations and Food Services

Source: U.S. Bureau of Labor Statistics

Graph.

47. $y = \frac{2}{3}x + 2$ **48.** $y = -\frac{2}{3}x - 3$

49. $y = -\frac{2}{3}x + 3$ **50.** $y = \frac{2}{3}x - 2$

51. $y = \frac{3}{2}x + 3$ **52.** $y = \frac{3}{2}x - 2$

53. $y = -\frac{4}{3}x + 3$ **54.** $y = -\frac{3}{2}x - 2$

55. $2x + y = 1$ **56.** $3x + y = 2$

57. $3x + y = 0$ **58.** $2x + y = 0$

59. $4x + 5y = 15$

60. $2x + 3y = 9$

61. $x - 4y = 12$

62. $x + 5y = 20$

Determine whether each pair of equations represents parallel lines.

63. $y = \frac{3}{4}x + 6$,
  $y = \frac{3}{4}x - 2$

64. $y = \frac{1}{3}x - 2$,
  $y = -\frac{1}{3}x + 1$

65. $y = 2x - 5$,
  $4x + 2y = 9$

66. $y = -3x + 1$,
  $6x + 2y = 8$

67. $3x + 4y = 8$,
  $7 - 12y = 9x$

68. $3x = 5y - 2$,
  $10y = 4 - 6x$

69. Can a horizontal line be graphed using the method of Example 5? Why or why not?

70. Can a vertical line be graphed using the method of Example 5? Why or why not?

Skill Review

To prepare for Section 3.7, review solving a formula for a variable and subtracting real numbers (Sections 1.6 and 2.3).

Solve. [2.3]

71. $y - k = m(x - h)$, for y

72. $y - 9 = -2(x + 4)$, for y

Simplify. [1.6]

73. $-10 - (-3)$

74. $8 - (-5)$

75. $-4 - 5$

76. $-6 - 5$

Synthesis

77. Explain how it is possible for an incorrect graph to be drawn, even after plotting three points that line up.

78. Which would you prefer, and why: graphing an equation of the form $y = mx + b$ or graphing an equation of the form $Ax + By = C$?

79. Show that the slope of the line given by $y = mx + b$ is m. (*Hint*: Substitute both 0 and 1 for x to find two pairs of coordinates. Then use the formula, Slope = change in y/change in x.)

80. Write an equation of the line with the same slope as the line given by $5x + 2y = 8$ and the same y-intercept as the line given by $3x - 7y = 10$.

81. Write an equation of the line parallel to the line given by $2x - 6y = 10$ and having the same y-intercept as the line given by $9x + 6y = 18$.

82. Write an equation of the line parallel to the line given by $3x - 2y = 8$ and having the same y-intercept as the line given by $2y + 3x = -4$.

83. Write an equation of the line parallel to the line given by $4x + 5y = 9$ and having the same y-intercept as the line given by $2x + 3y = 12$.

84. Write an equation of the line parallel to the line given by $2x + 7y = 10$ and having the same y-intercept as the line given by $3x - 5y = 4$.

85. Write an equation of the line parallel to the line given by $-4x + 8y = 5$ and having the same y-intercept as the line given by $4x - 3y = 0$.

Two lines are perpendicular if either the product of their slopes is -1, or one line is vertical and the other horizontal. For Exercises 86–91, determine whether each pair of equations represents perpendicular lines.

86. $3y = 5x - 3$,
  $3x + 5y = 10$

87. $y + 3x = 10$,
  $2x - 6y = 18$

88. $3x + 5y = 10$,
  $15x + 9y = 18$

89. $10 - 4y = 7x$,
  $7y + 21 = 4x$

90. $x = 5$,
  $y = \frac{1}{2}$

91. $y = -2x$,
  $x = \frac{1}{2}$

92. Write an equation of the line perpendicular to the line given by $2x + 3y = 7$ (see Exercises 86–91) and having the same y-intercept as the line given by $5x + 2y = 10$.

93. Write an equation of the line perpendicular to the line given by $3x - 5y = 8$ (see Exercises 86–91) and having the same y-intercept as the line given by $2x + 4y = 12$.

94. Write an equation of the line perpendicular to the line given by $3x - 2y = 9$ (see Exercises 86–91) and having the same y-intercept as the line given by $2x + 5y = 0$.

95. Write an equation of the line perpendicular to the line given by $2x + 5y = 6$ (see Exercises 86–91) that passes through $(2, 6)$. (*Hint*: Draw a graph.)

CONNECTING the CONCEPTS

If any two points are plotted on a plane, there is only one line that will go through both points. Thus, if we know that the graph of an equation is a straight line, we need only find two points that are on that line. Then we can plot those points and draw the line that goes through them.

   The different graphing methods discussed in this chapter present efficient ways of finding two points that are on the graph of the equation. Following is a general strategy for graphing linear equations.

1. Make sure that the equation is linear. Linear equations can always be written in the form $Ax + By = C$. (A or B may be 0.)
2. Graph the line. Use substitution to find two points on the line, or use a more efficient method based on the form of the equation.
3. Check the graph by finding another point that should be on the line and determining whether it does actually fall on the line.

Form of Equation	Graph
$x = a$	Draw a vertical line through $(a, 0)$.
$y = b$	Draw a horizontal line through $(0, b)$.
$y = mx + b$	Plot the y-intercept $(0, b)$. Start at the y-intercept and count off the rise and run using the slope m to find another point. Draw a line through the two points.
$Ax + By = C$	Determine the x- and y-intercepts $(a, 0)$ and $(0, b)$. Draw a line through the two points.

MIXED REVIEW

For each equation, **(a)** *determine whether it is linear and* **(b)** *if it is linear, graph the line.*

1. $x = 3$

2. $y = -2$

3. $y = \frac{1}{2}x + 3$

4. $4x + 3y = 12$

5. $y - 5 = x$

6. $x + y = -2$

7. $3xy = 6$

8. $2x = 3y$

9. $3 - y = 4$

10. $y = x^2 - 4$

11. $2y = 9x - 10$

12. $x + 8 = 7$

13. $2x - 6 = 3y$

14. $y - 2x = 4$

15. $2y - x = 4$

16. $y = \frac{1}{x}$

17. $x - 2y = 0$

18. $y = 4 - x$

19. $y = 4 + x$

20. $4x - 5y = 20$

3.7   Point–Slope Form

Writing Equations in Point–Slope Form   ● Graphing and Point–Slope Form   ●
Estimations and Predictions Using Two Points

There are many applications in which a slope—or a rate of change—and an ordered pair are known. When the ordered pair is the y-intercept, an equation in slope–intercept form can be easily produced. When the ordered pair represents a point other than the y-intercept, a different form, known as *point–slope form*, is more convenient.

Writing Equations in Point–Slope Form

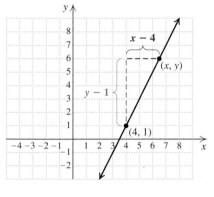

Consider a line with slope 2 passing through the point $(4, 1)$, as shown in the figure. In order for a point (x, y) to be on the line, the coordinates x and y must be solutions of the slope equation

$$\frac{y - 1}{x - 4} = 2. \qquad \text{If } (x, y) \text{ is not on the line, this equation will not be true.}$$

Take a moment to examine this equation. Pairs like $(5, 3)$ and $(3, -1)$ are on the line and are solutions, since

$$\frac{3 - 1}{5 - 4} = 2 \quad \text{and} \quad \frac{-1 - 1}{3 - 4} = 2.$$

When $x \neq 4$, then $x - 4 \neq 0$, and we can multiply on both sides of the slope equation by $x - 4$:

$$(x - 4) \cdot \frac{y - 1}{x - 4} = 2(x - 4)$$
$$y - 1 = 2(x - 4). \qquad \text{Removing a factor equal to 1: } \frac{x - 4}{x - 4} = 1$$

Every point on the line is a solution of this equation.

This is considered **point–slope form** for the line shown in the figure at left. A point–slope equation can be written any time a line's slope and a point on the line are known.

> ## The Point–Slope Equation
> The equation $y - y_1 = m(x - x_1)$ is called the *point–slope equation* for the line with slope m that contains the point (x_1, y_1).

Point–slope form is especially useful in more advanced mathematics courses, where problems similar to the following often arise.

EXAMPLE **1**   Write a point–slope equation for the line with slope $\frac{1}{5}$ that contains the point $(7, 2)$.

SOLUTION   We substitute $\frac{1}{5}$ for m, 7 for x_1, and 2 for y_1:

$$y - y_1 = m(x - x_1)$$ Using the point–slope equation
$$y - 2 = \tfrac{1}{5}(x - 7).$$ Substituting   **TRY EXERCISE** ▶ 13

EXAMPLE **2**   Write a point–slope equation for the line with slope $-\frac{4}{3}$ that contains the point $(1, -6)$.

SOLUTION   We substitute $-\frac{4}{3}$ for m, 1 for x_1, and -6 for y_1:

$$y - y_1 = m(x - x_1)$$ Using the point–slope equation
$$y - (-6) = -\tfrac{4}{3}(x - 1).$$ Substituting   **TRY EXERCISE** ▶ 19

EXAMPLE **3**   Write the slope–intercept equation for the line with slope 2 that contains the point $(3, 1)$.

SOLUTION   There are two parts to this solution. First, we write an equation in point–slope form:

$$y - y_1 = m(x - x_1)$$
$$y - 1 = 2(x - 3).$$ Substituting   Write in point–slope form.

Next, we find an equivalent equation of the form $y = mx + b$:

$$y - 1 = 2(x - 3)$$
$$y - 1 = 2x - 6$$ Using the distributive law   Write in slope–intercept form.
$$y = 2x - 5.$$ Adding 1 to both sides to get slope–intercept form   **TRY EXERCISE** ▶ 27

STUDENT NOTES

There are several forms in which a line's equation can be written. For instance, as shown in Example 3, $y - 1 = 2(x - 3)$, $y - 1 = 2x - 6$, and $y = 2x - 5$ all are equations for the same line.

Graphing and Point–Slope Form

When we know a line's slope and a point that is on the line, we can draw the graph, much as we did in Section 3.6. For example, the information given in the statement of Example 3 is sufficient for drawing a graph.

EXAMPLE **4**   Graph the line with slope 2 that passes through $(3, 1)$.

SOLUTION   We plot $(3, 1)$, move *up* 2 and *to the right* 1 $\left(\text{since } 2 = \tfrac{2}{1}\right)$, and draw the line.

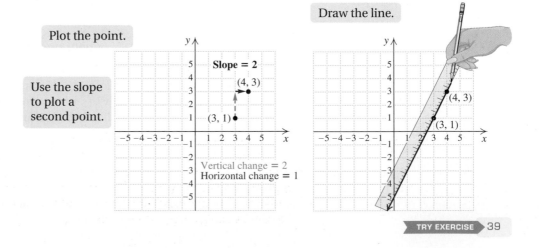

TRY EXERCISE ▶ 39

EXAMPLE 5

Graph: $y - 2 = 3(x - 4)$.

SOLUTION Since $y - 2 = 3(x - 4)$ is in point–slope form, we know that the line has slope 3, or $\frac{3}{1}$, and passes through the point $(4, 2)$. We plot $(4, 2)$ and then find a second point by moving *up* 3 units and *to the right* 1 unit. The line can then be drawn, as shown below.

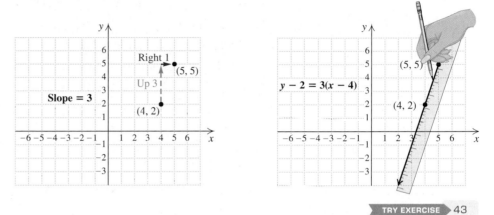

TRY EXERCISE 43

EXAMPLE 6

Graph: $y + 4 = -\frac{5}{2}(x + 3)$.

SOLUTION Once we have written the equation in point–slope form, $y - y_1 = m(x - x_1)$, we can proceed much as we did in Example 5. To find an equivalent equation in point–slope form, we subtract opposites instead of adding:

$$y + 4 = -\frac{5}{2}(x + 3)$$
$$y - (-4) = -\frac{5}{2}(x - (-3)).$$

Subtracting a negative instead of adding a positive. This is now in point–slope form.

From this last equation, $y - (-4) = -\frac{5}{2}(x - (-3))$, we see that the line passes through $(-3, -4)$ and has slope $-\frac{5}{2}$, or $\frac{5}{-2}$.

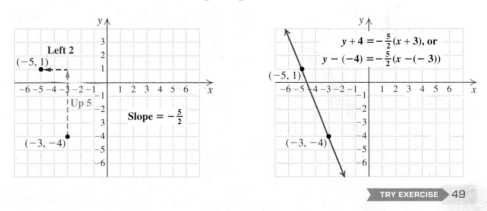

TRY EXERCISE 49

STUDY SKILLS

Understand Your Mistakes

When you receive a graded quiz, test, or assignment back from your instructor, it is important to review and understand what your mistakes were. While the material is still fresh, take advantage of the opportunity to learn from your mistakes.

Estimations and Predictions Using Two Points

We can estimate real-life quantities that are not already known by using two points with known coordinates. When the unknown point is located *between* the two points, this process is called **interpolation**. If a graph passing through the known points is *extended* to predict future values, the process is called **extrapolation**. In statistics, methods exist for using a set of several points to interpolate or extrapolate values using curves other than lines.

EXAMPLE **7**

Student volunteers. An increasing number of college students are donating time and energy in volunteer service. The number of college-student volunteers grew from 2.7 million in 2002 to 3.3 million in 2005.

Source: Corporation for National and Community Service

a) Graph the line passing through the given data points, letting x = the number of years since 2000.

b) Determine an equation for the line and estimate the number of college students who volunteered in 2004.

c) Predict the number of college students who will volunteer in 2010.

SOLUTION

a) We first draw and label a horizontal axis to display the year and a vertical axis to display the number of college-student volunteers, in millions. Next, we number the axes, choosing scales that will include both the given values and the values to be estimated.

Since x = the number of years since 2000, we plot the points $(2, 2.7)$ and $(5, 3.3)$ and draw a line passing through both points.

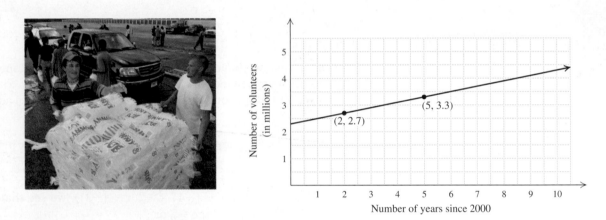

b) To find an equation for the line, we first calculate its slope:

$$m = \frac{\text{change in } y}{\text{change in } x} = \frac{3.3 - 2.7}{5 - 2} = \frac{0.6}{3} = 0.2.$$

The number of college-student volunteers increased at a rate of 0.2 million students per year. We can use either of the given points to write a point–slope equation for the line. Let's use $(2, 2.7)$ and then write an equivalent equation in slope–intercept form:

$y - 2.7 = 0.2(x - 2)$	This is a point–slope equation.
$y - 2.7 = 0.2x - 0.4$	Using the distributive law
$y = 0.2x + 2.3.$	Adding 2.7 to both sides. This is slope–intercept form.

To estimate the number of college-student volunteers in 2004, we substitute 4 for x in the slope–intercept equation:

$$y = 0.2 \cdot 4 + 2.3 = 3.1.$$

As the graph confirms, in 2004 there were about 3.1 million college-student volunteers.

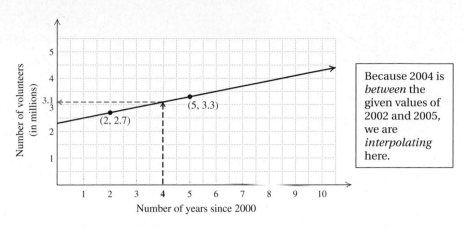

Because 2004 is *between* the given values of 2002 and 2005, we are *interpolating* here.

c) To predict the number of college-student volunteers in 2010, we again substitute for x in the slope–intercept equation:

$$y = 0.2 \cdot 10 + 2.3 = 4.3.$$ 2010 is 10 yr after 2000.

As we can see from the graph, if the trend continues, there will be about 4.3 million college-student volunteers in 2010.

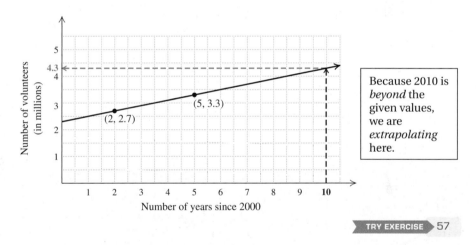

Because 2010 is *beyond* the given values, we are *extrapolating* here.

TRY EXERCISE 57

Visualizing for Success

A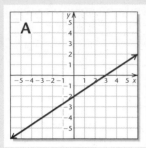

F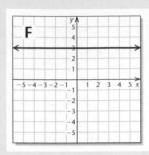

Match each equation with its graph.

1. $y = x + 4$

2. $y = 2x$

B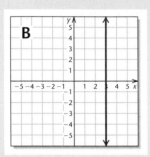

G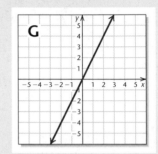

3. $y = 3$

4. $x = 3$

5. $y = -\frac{1}{2}x$

C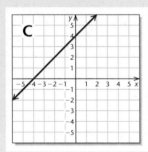

H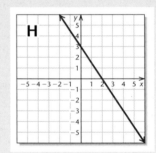

6. $2x - 3y = 6$

7. $y = -3x - 2$

D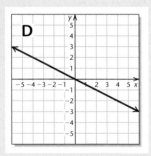

I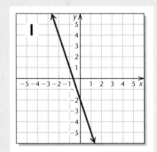

8. $3x + 2y = 6$

9. $y - 3 = 2(x - 1)$

10. $y + 2 = \frac{1}{2}(x + 1)$

E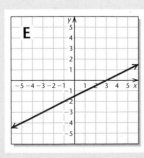

J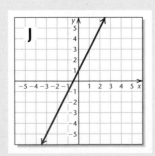

Answers on page A-12

An additional, animated version of this activity appears in MyMathLab. To use MyMathLab, you need a course ID and a student access code. Contact your instructor for more information.

3.7 EXERCISE SET

Concept Reinforcement   *In each of Exercises 1–8, match the given information about a line with the appropriate equation from the column on the right.*

1. _____ Slope 5; includes (2, 3)

2. _____ Slope 5; includes (3, 2)

3. _____ Slope −5; includes (2, 3)

4. _____ Slope −5; includes (3, 2)

5. _____ Slope −5; includes (−2, −3)

6. _____ Slope 5; includes (−2, −3)

7. _____ Slope −5; includes (−3, −2)

8. _____ Slope 5; includes (−3, −2)

a) $y + 3 = 5(x + 2)$

b) $y - 2 = 5(x - 3)$

c) $y + 2 = 5(x + 3)$

d) $y - 3 = -5(x - 2)$

e) $y + 3 = -5(x + 2)$

f) $y + 2 = -5(x + 3)$

g) $y - 3 = 5(x - 2)$

h) $y - 2 = -5(x - 3)$

In each of Exercises 9–12, match the graph with the appropriate equation from the column on the right.

9.

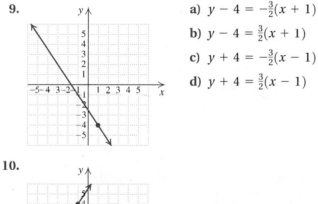

a) $y - 4 = -\frac{3}{2}(x + 1)$

b) $y - 4 = \frac{3}{2}(x + 1)$

c) $y + 4 = -\frac{3}{2}(x - 1)$

d) $y + 4 = \frac{3}{2}(x - 1)$

10.

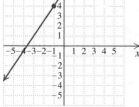

11.

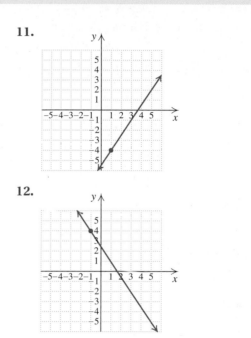

12.

Write a point–slope equation for the line with the given slope that contains the given point.

13. $m = 3$; $(1, 6)$ **14.** $m = 2$; $(3, 7)$

15. $m = \frac{3}{5}$; $(2, 8)$ **16.** $m = \frac{2}{3}$; $(4, 1)$

17. $m = -4$; $(3, 1)$ **18.** $m = -5$; $(6, 2)$

19. $m = \frac{3}{2}$; $(5, -4)$ **20.** $m = -\frac{4}{3}$; $(7, -1)$

21. $m = -\frac{5}{4}$; $(-2, 6)$ **22.** $m = \frac{7}{2}$; $(-3, 4)$

23. $m = -2$; $(-4, -1)$ **24.** $m = -3$; $(-2, -5)$

25. $m = 1$; $(-2, 8)$ **26.** $m = -1$; $(-3, 6)$

Write the slope–intercept equation for the line with the given slope that contains the given point.

27. $m = 4$; $(3, 5)$ **28.** $m = 3$; $(6, 2)$

29. $m = \frac{7}{4}$; $(4, -2)$ **30.** $m = \frac{8}{3}$; $(3, -4)$

31. $m = -2$; $(-3, 7)$ **32.** $m = -3$; $(-2, 1)$

33. $m = -4$; $(-2, -1)$ **34.** $m = -5$; $(-1, -4)$

35. $m = \frac{2}{3}$; $(5, 6)$ **36.** $m = \frac{3}{2}$; $(7, 4)$

Aha! **37.** $m = -\frac{5}{6}$; $(0, 4)$ **38.** $m = -\frac{3}{4}$; $(0, 5)$

39. Graph the line with slope $\frac{4}{3}$ that passes through the point $(1, 2)$.

40. Graph the line with slope $\frac{2}{5}$ that passes through the point $(3, 4)$.

41. Graph the line with slope $-\frac{3}{4}$ that passes through the point $(2, 5)$.

42. Graph the line with slope $-\frac{3}{2}$ that passes through the point $(1, 4)$.

Graph.

43. $y - 5 = \frac{1}{3}(x - 2)$ **44.** $y - 2 = \frac{1}{2}(x - 1)$

45. $y - 1 = -\frac{1}{4}(x - 3)$ **46.** $y - 1 = -\frac{1}{2}(x - 3)$

47. $y + 2 = \frac{2}{3}(x - 1)$ **48.** $y - 1 = \frac{3}{4}(x + 5)$

49. $y + 4 = 3(x + 1)$

50. $y + 3 = 2(x + 1)$

51. $y - 4 = -2(x + 1)$

52. $y + 3 = -1(x - 4)$

53. $y + 4 = 3(x + 2)$

54. $y + 3 = -(x + 2)$

55. $y + 1 = -\frac{3}{5}(x - 2)$

56. $y - 2 = -\frac{2}{3}(x + 1)$.

In Exercises 57–64, graph the given data and determine an equation for the related line (as in Example 7). Then use the equation to answer parts (a) and (b).

57. *Birth rate among teenagers.* The birth rate among teenagers, measured in births per 1000 females age 15–19, fell steadily from 62.1 in 1991 to 41.1 in 2007.
Source: National Center for Health Statistics

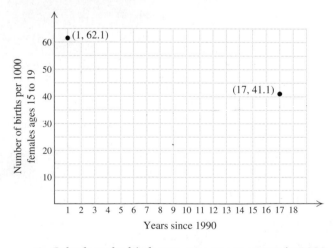

a) Calculate the birth rate among teenagers in 1999.
b) Predict the birth rate among teenagers in 2008.

58. *Food-stamp program participation.* Participation in the U.S. food-stamp program grew from approxi-

mately 17.1 million people in 2000 to approximately 23 million in 2004.
Source: U.S. Department of Agriculture

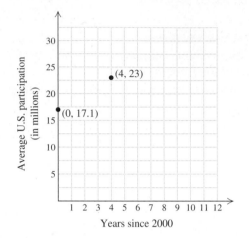

a) Calculate the number of participants in 2001.
b) Predict the number of participants in 2010.

59. *Cigarette smoking.* The percentage of people age 25–44 who smoke has changed from 14.2% in 2001 to 10.8% in 2004.
Source: Office on Smoking and Health, National Center for Chronic Disease Prevention and Health Promotion

a) Calculate the percentage of people age 25–44 who smoked in 2002.
b) Predict the percentage of people age 25–44 who will smoke in 2010.

60. *Cigarette smoking.* The percentage of people age 18–24 who smoke has dropped from 5.2% in 2001 to 3.4% in 2004.
Source: Office on Smoking and Health, National Center for Chronic Disease Prevention and Health Promotion

a) Calculate the percentage of people age 18–24 who smoked in 2003.
b) Estimate the percentage of people age 18–24 who smoked in 2008.

61. *College enrollment.* U.S. college enrollment has grown from approximately 14.3 million in 1995 to 17.4 million in 2005.
Source: National Center for Education Statistics

a) Calculate the U.S. college enrollment for 2002.
b) Predict the U.S. college enrollment for 2010.

62. *High school enrollment.* U.S. high school enrollment has changed from approximately 13.7 million in 1995 to 16.3 million in 2005.
Source: National Center for Education Statistics

a) Calculate the U.S. high school enrollment for 2002.
b) Predict the U.S. high school enrollment for 2010.

63. *Aging population.* The number of U.S. residents over the age of 65 was approximately 31 million in 1990 and 36.3 million in 2004.
Source: U.S. Census Bureau

Aha! **a)** Calculate the number of U.S. residents over the age of 65 in 1997.
b) Predict the number of U.S. residents over the age of 65 in 2010.

64. *Urban population.* The percentage of the U.S. population that resides in metropolitan areas increased from about 78% in 1980 to about 83% in 2006.

a) Calculate the percentage of the U.S. population residing in metropolitan areas in 1992.
b) Predict the percentage of the U.S. population residing in metropolitan areas in 2012.

Write the slope–intercept equation for the line containing the given pair of points.

65. $(2, 3)$ and $(4, 1)$

66. $(6, 8)$ and $(3, 5)$

67. $(-3, 1)$ and $(3, 5)$

68. $(-3, 4)$ and $(3, 1)$

69. $(5, 0)$ and $(0, -2)$

70. $(-2, 0)$ and $(0, 3)$

71. $(-4, -1)$ and $(1, 9)$

72. $(-3, 5)$ and $(-1, -3)$

73. Can equations for horizontal or vertical lines be written in point–slope form? Why or why not?

74. Describe a situation in which it is easier to graph the equation of a line in point–slope form than in slope–intercept form.

Skill Review

To prepare for Chapter 4, review exponential notation and order of operations (Section 1.8).

Simplify. [1.8]

75. $(-5)^3$

76. $(-2)^6$

77. -2^6

78. $3 \cdot 2^4 - 5 \cdot 2^3$

79. $2 - (3 - 2^2) + 10 \div 2 \cdot 5$

80. $(5 - 7)^2(3 - 2 \cdot 2)$

Synthesis

81. Describe a procedure that can be used to write the slope–intercept equation for any nonvertical line passing through two given points.

82. Any nonvertical line has many equations in point–slope form, but only one in slope–intercept form. Why is this?

Graph.

Aha! **83.** $y - 3 = 0(x - 52)$ **84.** $y + 4 = 0(x + 93)$

Write the slope–intercept equation for each line shown.

85.

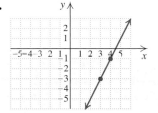

86.

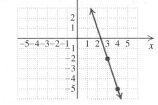

87.

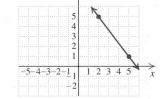

88.

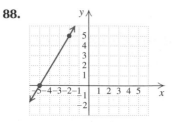

89. Write a point–slope equation of the line passing through $(-4, 7)$ that is parallel to the line given by $2x + 3y = 11$.

90. Write a point–slope equation of the line passing through $(3, -1)$ that is parallel to the line given by $4x - 5y = 9$.

Aha! 91. Write an equation of the line parallel to the line given by $y = 3 - 4x$ that passes through $(0, 7)$.

92. Write the slope–intercept equation of the line that has the same y-intercept as the line $x - 3y = 6$ and contains the point $(5, -1)$.

93. Write the slope–intercept equation of the line that contains the point $(-1, 5)$ and is parallel to the line passing through $(2, 7)$ and $(-1, -3)$.

94. Write the slope–intercept equation of the line that has x-intercept $(-2, 0)$ and is parallel to $4x - 8y = 12$.

Another form of a linear equation is the double-intercept *form:* $\dfrac{x}{a} + \dfrac{y}{b} = 1$. *From this form, we can read the* x-intercept $(a, 0)$ *and the* y-intercept $(0, b)$ *directly.*

95. Find the x-intercept and the y-intercept of the graph of the line given by $\dfrac{x}{2} + \dfrac{y}{5} = 1$.

96. Find the x-intercept and the y-intercept of the graph of the line given by $\dfrac{x}{10} - \dfrac{y}{3} = 1$.

97. Write the equation $4y - 3x = 12$ in double-intercept form and find the intercepts.

98. Write the equation $6x + 5y = 30$ in double-intercept form and find the intercepts.

99. Why is slope–intercept form more useful than point–slope form when using a graphing calculator? How can point–slope form be modified so that it is more easily used with graphing calculators?

CONNECTING the CONCEPTS

Any line can be described by a number of equivalent equations. We write the equation in the form that is most useful for us. For example, all four of the equations below describe the given line.

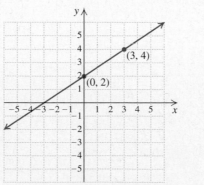

$2x - 3y = -6$;

$y = \dfrac{2}{3}x + 2$;

$y - 4 = \dfrac{2}{3}(x - 3)$;

$2x + 6 = 3y$

Form of a Linear Equation	Example	Uses
Standard form: $Ax + By = C$	$2x - 3y = -6$	Finding x- and y-intercepts; Graphing using intercepts
Slope–intercept form: $y = mx + b$	$y = \dfrac{2}{3}x + 2$	Finding slope and y-intercept; Graphing using slope and y-intercept; Writing an equation given slope and y-intercept
Point–slope form: $y - y_1 = m(x - x_1)$	$y - 4 = \dfrac{2}{3}(x - 3)$	Finding slope and a point on the line; Graphing using slope and a point on the line; Writing an equation given slope and a point on the line

MIXED REVIEW

Tell whether each equation is in standard form, slope–intercept form, point–slope form, or none of these.

1. $y = -\frac{1}{2}x - 7$

2. $5x - 8y = 10$

3. $x = y + 2$

4. $\frac{1}{2}x + \frac{1}{3}y = 5$

5. $y - 2 = 5(x + 1)$

6. $3y + 7 = x$

Write each equation in standard form.

7. $2x = 5y + 10$

8. $x = y + 2$

9. $y = 2x + 7$

10. $y = -\frac{1}{2}x + 3$

11. $y - 2 = 3(x + 7)$

12. $x - 7 = 11$

Write each equation in slope–intercept form.

13. $2x - 7y = 8$

14. $y + 5 = -(x + 3)$

15. $8x = y + 3$

16. $6x + 10y = 30$

17. $9y = 5 - 8x$

18. $x - y = 3x + y$

19. $2 - 3y = 5y + 6$

20. $3(x - 4) = 6(y - 2)$

Study Summary

KEY TERMS AND CONCEPTS	EXAMPLES

SECTION 3.1: READING GRAPHS, PLOTTING POINTS, AND SCALING GRAPHS

Ordered pairs, like $(-3, 2)$ or $(4, 3)$, can be **plotted** or **graphed** using a **coordinate system** that uses two **axes**, which are most often labeled x and y. The axes intersect at the **origin**, $(0, 0)$, and divide a plane into four **quadrants**.

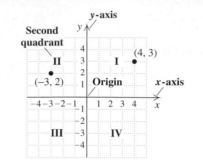

SECTION 3.2: GRAPHING LINEAR EQUATIONS

To **graph** an equation means to make a drawing that represents all of its solutions.

A **linear equation** has a graph that is a straight line.

An equation such as $y = 2x - 7$ or $2x + 3y = 12$ is said to be **linear** because its graph is a straight line.

Any linear equation can be graphed by finding two ordered pairs that are solutions of the equation, plotting the corresponding points, and drawing a line through those points. Plotting a third point serves as a check.

Graph: $3x = y + 1$.

x	y	(x, y)
1	2	$(1, 2)$
0	-1	$(0, -1)$
-1	-4	$(-1, -4)$

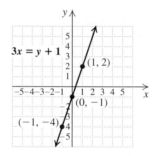

SECTION 3.3: GRAPHING AND INTERCEPTS

To find a y-intercept $(0, b)$, let $x = 0$ and solve for y.

To find an x-intercept $(a, 0)$, let $y = 0$ and solve for x.

Graph $2x - y = 4$ *using intercepts.*

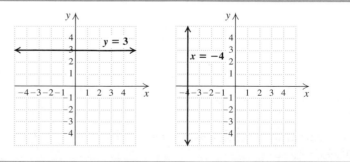

The graph of $y = b$ is a horizontal line, with y-intercept $(0, b)$.

The graph of $x = a$ is a vertical line, with x-intercept $(a, 0)$.

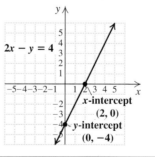

SECTION 3.4: RATES

A **rate** is a ratio that indicates how two quantities change with respect to each other.

Lara had $1500 in her savings account at the beginning of February, and $2400 at the beginning of May. Find the rate at which Lara is saving.

$$\text{Savings rate} = \frac{\text{Amount saved}}{\text{Number of months}}$$

$$= \frac{\$2400 - \$1500}{3 \text{ months}}$$

$$= \frac{\$900}{3 \text{ months}} = \$300 \text{ per month}$$

SECTION 3.5: SLOPE

A line's slant is measured by its **slope**.

$$\text{Slope} = m = \frac{\text{change in } y}{\text{change in } x}$$

$$= \frac{\text{rise}}{\text{run}}$$

$$= \frac{y_2 - y_1}{x_2 - x_1}$$

The slope of the line containing the points $(-1, -4)$ and $(2, -6)$ is

$$m = \frac{-6 - (-4)}{2 - (-1)} = \frac{-2}{3} = -\frac{2}{3}.$$

A line with positive slope slants up from left to right.

A line with negative slope slants down from left to right.

The slope of a horizontal line is 0.

The slope of a vertical line is undefined.

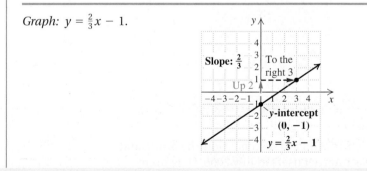

Positive slope   Negative slope   Slope $= 0$ Undefined slope

SECTION 3.6: SLOPE–INTERCEPT FORM

An equation in **slope–intercept form**,

$$y = mx + b,$$

represents a line with slope m and y-intercept $(0, b)$.

For the line given by $y = \frac{2}{3}x - 8$:

The slope is $\frac{2}{3}$ and the y-intercept is $(0, -8)$.

To graph a line written in slope–intercept form, plot the y-intercept and count off the slope.

Graph: $y = \frac{2}{3}x - 1$.

Slope: $\frac{2}{3}$ To the right 3

Up 2

y-intercept $(0, -1)$

$y = \frac{2}{3}x - 1$

SECTION 3.7: POINT–SLOPE FORM

An equation in **point–slope form**,

$$y - y_1 = m(x - x_1),$$

represents a line with slope m passing through (x_1, y_1).

Write a point–slope equation for the line with slope -2 that contains the point $(3, -5)$.

$$y - y_1 = m(x - x_1)$$

$$y - (-5) = -2(x - 3)$$

Review Exercises: Chapter 3

🖎 Concept Reinforcement *Classify each statement as either true or false.*

1. Not every ordered pair lies in one of the four quadrants. [3.1]

2. The equation of a vertical line cannot be written in slope–intercept form. [3.6]

3. Equations for lines written in slope–intercept form appear in the form $Ax + By = C$. [3.6]

4. Every horizontal line has an x-intercept. [3.3]

5. A line's slope is a measure of rate. [3.5]

6. A positive rate of ascent means that an airplane is flying increasingly higher above the earth. [3.4]

7. Any two points on a line can be used to determine the line's slope. [3.5]

8. Knowing a line's slope is enough to write the equation of the line. [3.6]

9. Knowing two points on a line is enough to write the equation of the line. [3.7]

10. Parallel lines that are not vertical have the same slope. [3.6]

The following circle graph shows the percentage of online searches done in July 2006 that were performed by a particular search engine. [3.1]
Source: NetRatings for SearchEngineWatch.com

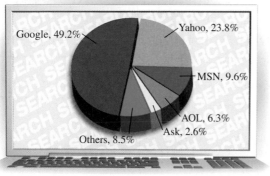

Online Searches

Google, 49.2%
Yahoo, 23.8%
MSN, 9.6%
AOL, 6.3%
Ask, 2.6%
Others, 8.5%

11. There were 5.6 billion searches done by home and business Internet users in July 2006. How many searches were done using Yahoo?

12. About 55% of the online searches done by Waterworks Graphics are image searches. In July 2006, Waterworks employees did 4200 online searches. If their search engine use is typical, how many image searches did they do using Google?

Plot each point. [3.1]

13. $(5, -1)$ 14. $(2, 3)$ 15. $(-4, 0)$

In which quadrant is each point located? [3.1]

16. $(-8, -7)$ 17. $(15.3, -13.8)$ 18. $\left(-\frac{1}{2}, \frac{1}{10}\right)$

Find the coordinates of each point in the figure. [3.1]

19. *A*   20. *B*   21. *C*

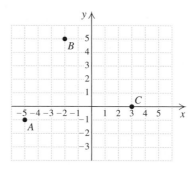

22. Use a grid 10 squares wide and 10 squares high to plot $(-65, -2), (-10, 6)$, and $(25, 7)$. Choose the scale carefully. [3.1]

23. Determine whether the equation $y = 2x + 7$ has *each* ordered pair as a solution: **(a)** $(3, 1)$; **(b)** $(-3, 1)$. [3.2]

24. Show that the ordered pairs $(0, -3)$ and $(2, 1)$ are solutions of the equation $2x - y = 3$. Then use the graph of the two points to determine another solution. Answers may vary. [3.2]

Graph.

25. $y = x - 5$ [3.2]   26. $y = -\frac{1}{4}x$ [3.2]

27. $y = -x + 4$ [3.2]   28. $4x + y = 3$ [3.2]

29. $4x + 5 = 3$ [3.3]   30. $5x - 2y = 10$ [3.3]

31. *TV viewing.* The average number of daily viewers v, in millions, of ABC's soap opera "General Hospital" is given by $v = -\frac{1}{4}t + 9$, where t is the number of years since 2000. Graph the equation and use the graph to predict the average number of daily viewers of "General Hospital" in 2008. [3.2]
Source: Nielsen Media Research

32. At 4:00 P.M., Jesse's Honda Civic was at mile marker 17 of Interstate 290 in Chicago. At 4:45 P.M., Jesse was at mile marker 23. [3.4]

   a) Find Jesse's driving rate, in number of miles per minute.

   b) Find Jesse's driving rate, in number of minutes per mile.

33. *Gas mileage.* The following graph shows data for a Ford Explorer driven on city streets. At what rate was the vehicle consuming gas? [3.4]

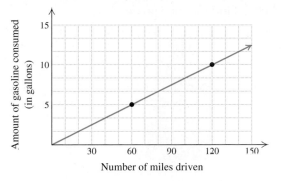

Find the slope of each line. [3.5]

34.

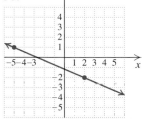

35.

36.

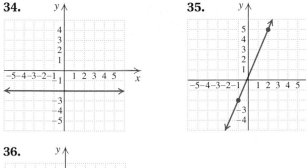

Find the slope of the line containing the given pair of points. If it is undefined, state this. [3.5]

37. $(-2, 5)$ and $(3, -1)$

38. $(6, 5)$ and $(-2, 5)$

39. $(-3, 0)$ and $(-3, 5)$

40. $(-8.3, 4.6)$ and $(-9.9, 1.4)$

41. *Architecture.* To meet federal standards, a wheelchair ramp cannot rise more than 1 ft over a horizontal distance of 12 ft. Express this slope as a grade. [3.5]

42. Find the x-intercept and the y-intercept of the line given by $5x - 8y = 80$. [3.3]

43. Find the slope and the y-intercept of the line given by $3x + 5y = 45$. [3.6]

44. Write the slope–intercept equation of the line with slope $\frac{3}{8}$ and y-intercept $(0, 7)$. [3.6]

45. Write a point–slope equation for the line with slope $-\frac{1}{3}$ that contains the point $(-2, 9)$. [3.7]

46. The average tuition at a public two-year college was \$1359 in 2001 and \$1847 in 2005. Graph the data and determine an equation for the related line. Then **(a)** calculate the average tuition at a public two-year college in 2004 and **(b)** predict the average tuition at a public two-year college in 2012. [3.7]
Source: U.S. National Center for Education Statistics

47. Write the slope–intercept equation for the line with slope 5 that contains the point $(3, -10)$. [3.7]

Graph.

48. $y = \frac{2}{3}x - 5$ [3.6]

49. $2x + y = 4$ [3.3]

50. $y = 6$ [3.3]

51. $x = -2$ [3.3]

52. $y + 2 = -\frac{1}{2}(x - 3)$ [3.7]

Synthesis

53. Can two perpendicular lines share the same y-intercept? Why or why not? [3.3]

54. Is it possible for a graph to have only one intercept? Why or why not? [3.3]

55. Find the value of m in $y = mx + 3$ such that $(-2, 5)$ is on the graph. [3.2]

56. Find the value of b in $y = -5x + b$ such that $(3, 4)$ is on the graph. [3.2]

57. Find the area and the perimeter of a rectangle for which $(-2, 2)$, $(7, 2)$, and $(7, -3)$ are three of the vertices. [3.1]

58. Find three solutions of $y = 4 - |x|$. [3.2]

Volunteering. *The following pie chart shows the types of organizations in which college students volunteer.*
Source: Corporation for National and Community Service

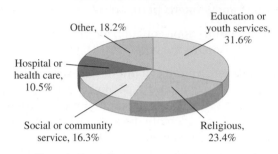

Volunteering by College Students, 2005

Education or youth services, 31.6%

Other, 18.2%

Hospital or health care, 10.5%

Social or community service, 16.3%

Religious, 23.4%

1. At Rolling Hills College, 25% of the 1200 students volunteer. If their choice of organizations is typical, how many students will volunteer in education or youth services?

2. At Valley University, $\frac{1}{3}$ of the 3900 students volunteer. If their choice of organizations is typical, how many students will volunteer in hospital or health-care services?

In which quadrant is each point located?

3. $(-2, -10)$ 4. $(-1.6, 2.3)$

Find the coordinates of each point in the figure.

5. A

6. B

7. C

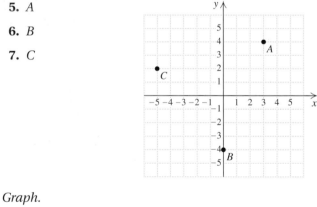

Graph.

8. $y = 2x - 1$ 9. $2x - 4y = -8$

10. $y + 1 = 6$ 11. $y = \frac{3}{4}x$

12. $2x - y = 3$ 13. $x = -1$

Find the slope of the line containing each pair of points. If it is undefined, state this.

14. $(3, -2)$ and $(4, 3)$ 15. $(-5, 6)$ and $(-1, -3)$

16. $(4, 7)$ and $(4, -8)$

17. *Running.* Jon reached the 3-km mark of a race at 2:15 P.M. and the 6-km mark at 2:24 P.M. What is his running rate?

18. At one point Filbert Street, the steepest street in San Francisco, drops 63 ft over a horizontal distance of 200 ft. Find the road grade.

19. Find the x-intercept and the y-intercept of the line given by $5x - y = 30$.

20. Find the slope and the y-intercept of the line given by $y - 8x = 10$.

21. Write the slope–intercept equation of the line with slope $-\frac{1}{3}$ and y-intercept $(0, -11)$.

22. *Aerobic exercise.* A person's target heart rate is the number of beats per minute that brings the most aerobic benefit to his or her heart. The target heart rate for a 20-year-old is 150 beats per minute; for a 60-year-old, it is 120 beats per minute.

   a) Graph the data and determine an equation for the related line. Let a = age and r = target heart rate, in number of beats per minute.

   b) Calculate the target heart rate for a 36-year-old.

Graph.

23. $y = \frac{1}{4}x - 2$ 24. $y + 4 = -\frac{1}{2}(x - 3)$

Synthesis

25. Write an equation of the line that is parallel to the graph of $2x - 5y = 6$ and has the same y-intercept as the graph of $3x + y = 9$.

26. A diagonal of a square connects the points $(-3, -1)$ and $(2, 4)$. Find the area and the perimeter of the square.

27. List the coordinates of three other points that are on the same line as $(-2, 14)$ and $(17, -5)$. Answers may vary.

Cumulative Review: Chapters 1–3

1. Evaluate $\dfrac{x}{5y}$ for $x = 70$ and $y = 2$. [1.1]

2. Multiply: $6(2a - b + 3)$. [1.2]

3. Factor: $8x - 4y + 4$. [1.2]

4. Find the prime factorization of 54. [1.3]

5. Find decimal notation: $-\frac{3}{20}$. [1.4]

6. Find the absolute value: $|-37|$. [1.4]

7. Find the opposite of $-\frac{1}{10}$. [1.6]

8. Find the reciprocal of $-\frac{1}{10}$. [1.7]

9. Find decimal notation: 36.7%. [2.4]

Simplify.

10. $\frac{3}{5} - \frac{5}{12}$ [1.3]

11. $3.4 + (-0.8)$ [1.5]

12. $(-2)(-1.4)(2.6)$ [1.7]

13. $\frac{3}{8} \div \left(-\frac{9}{10}\right)$ [1.7]

14. $1 - [32 \div (4 + 2^2)]$ [1.8]

15. $-5 + 16 \div 2 \cdot 4$ [1.8]

16. $y - (3y + 7)$ [1.8]

17. $3(x - 1) - 2[x - (2x + 7)]$ [1.8]

Solve.

18. $2.7 = 5.3 + x$ [2.1]

19. $\frac{5}{3}x = -45$ [2.1]

20. $3x - 7 = 41$ [2.2]

21. $\dfrac{3}{4} = \dfrac{-n}{8}$ [2.1]

22. $14 - 5x = 2x$ [2.2]

23. $3(5 - x) = 2(3x + 4)$ [2.2]

24. $\frac{1}{4}x - \frac{2}{3} = \frac{3}{4} + \frac{1}{3}x$ [2.2]

25. $y + 5 - 3y = 5y - 9$ [2.2]

26. $x - 28 < 20 - 2x$ [2.6]

27. $2(x + 2) \geq 5(2x + 3)$ [2.6]

28. Solve $A = 2\pi rh + \pi r^2$ for h. [2.3]

29. In which quadrant is the point $(3, -1)$ located? [3.1]

30. Graph on a number line: $-1 < x \leq 2$. [2.6]

31. Use a grid 10 squares wide and 10 squares high to plot $(-150, -40)$, $(40, -7)$, and $(0, 6)$. Choose the scale carefully. [3.1]

Graph.

32. $x = 3$ [3.3]

33. $2x - 5y = 10$ [3.3]

34. $y = -2x + 1$ [3.2]

35. $y = \frac{2}{3}x$ [3.2]

36. $y = -\frac{3}{4}x + 2$ [3.6]

37. $2y - 5 = 3$ [3.3]

Find the coordinates of the x- and y-intercepts. Do not graph.

38. $2x - 7y = 21$ [3.3]

39. $y = 4x + 5$ [3.3]

40. Find the slope and the y-intercept of the line given by $3x - y = 2$. [3.6]

41. Find the slope of the line containing the points $(-4, 1)$ and $(2, -1)$. [3.5]

42. Write an equation of the line with slope $\frac{2}{7}$ and y-intercept $(0, -4)$. [3.6]

43. Write a point–slope equation of the line with slope $-\frac{3}{8}$ that contains the point $(-6, 4)$. [3.7]

44. Write the slope–intercept form of the equation in Exercise 43. [3.6]

45. Determine an equation for the following graph. [3.6], [3.7]

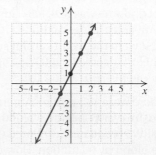

Solve.

46. U.S. bicycle sales rose from 15 million in 1995 to 20 million in 2005. Find the rate of change of bicycle sales. [3.4]
Sources: National Bicycle Dealers Association; U.S. Department of Transportation

47. A 150-lb person will burn 240 calories per hour when riding a bicycle at 6 mph. The same person will burn 410 calories per hour when cycling at 12 mph. [3.7]
Source: American Heart Association

a) Graph the data and determine an equation for the related line. Let r = the rate at which the person is cycling and c = the number of calories burned per hour.
b) Use the equation of part (a) to estimate the number of calories burned per hour by a 150-lb person cycling at 10 mph.

48. Americans spent an estimated $238 billion on home remodeling in 2006. This was $\frac{17}{15}$ of the amount spent on remodeling in 2005. How much was spent on remodeling in 2005? [2.5]
Source: National Association of Home Builders' Remodelers Council

49. In 2005, the mean earnings of individuals with a high school diploma was $29,448. This was about 54% of the mean earnings of those with a bachelor's degree. What were the mean earnings of individuals with a bachelor's degree in 2005? [2.4]
Source: U.S. Census Bureau

50. Recently there were 132 million Americans with either O-positive or O-negative blood. Those with O-positive blood outnumbered those with O-negative blood by 90 million. How many Americans had O-negative blood? [2.5]
Source: American Red Cross

51. Tina paid $126 for a cordless drill, including a 5% sales tax. How much did the drill itself cost? [2.4]

52. A 143-m wire is cut into three pieces. The second is 3 m longer than the first. The third is four fifths as long as the first. How long is each piece? [2.5]

53. In order to qualify for availability pay, a criminal investigator must average at least 2 hr of unscheduled duty per workday. For the first four days of one week, Alayna worked 1, 0, 3, and 2 extra hours. How many extra hours must she work on Friday in order to qualify for availability pay? [2.7]
Source: U.S. Department of Justice

Synthesis

54. Anya's salary at the end of a year is $26,780. This reflects a 4% salary increase in February and then a 3% cost-of-living adjustment in June. What was her salary at the beginning of the year? [2.4]

Solve. If no solution exists, state this.

55. $4|x| - 13 = 3$ [1.4], [2.2]

56. $4(x + 2) = 9(x - 2) + 16$ [2.2]

57. $2(x + 3) + 4 = 0$ [2.2]

58. $\dfrac{2 + 5x}{4} = \dfrac{11}{28} + \dfrac{8x + 3}{7}$ [2.2]

59. $5(7 + x) = (x + 6)5$ [2.2]

60. Solve $p = \dfrac{2}{m + Q}$ for Q. [2.3]

61. The points $(-3, 0)$, $(0, 7)$, $(3, 0)$, and $(0, -7)$ are vertices of a parallelogram. Find four equations of lines that intersect to form the parallelogram. [3.6]

Polynomials

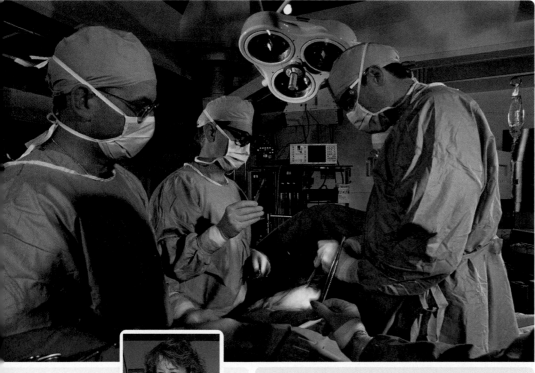

SHELLEY ZOMAK
TRANSPLANT COORDINATOR
(NURSE)
Pittsburgh, Pennsylvania

It is very important that a transplant recipient take the medications prescribed to maintain a determined blood concentration of the drug. As a transplant coordinator, I use math to educate patients on the correct dosing of medications, since the doctor's orders do not always match the pills available from the pharmacy. Some medications need to be adjusted more frequently on the basis of the patient's weight.

AN APPLICATION

Often a patient needing a kidney transplant has a willing kidney donor who does not match the patient medically. A kidney-paired donation matches donor–recipient pairs. Two kidney transplants are performed simultaneously, with each patient receiving the kidney of a stranger. The number k of such "kidney swaps" t years after 2003 can be approximated by

$$k = 14.3t^3 - 56t^2 + 57.7t + 19.$$

Estimate the number of kidney-paired donations in 2006.

Source: Based on data from United Network for Organ Sharing

This problem appears as Exercise 69 in Section 4.2.

Algebraic expressions such as $16t^2$, $5a^2 - 3ab$, and $3x^2 - 7x + 5$ are called polynomials. Polynomials occur frequently in applications and appear in most branches of mathematics. Thus learning to add, subtract, multiply, and divide polynomials is an important part of nearly every course in elementary algebra. The focus of this chapter is finding equivalent expressions, not solving equations.

4.1 Exponents and Their Properties

Multiplying Powers with Like Bases ▪ Dividing Powers with Like Bases ▪ Zero as an Exponent ▪ Raising a Power to a Power ▪ Raising a Product or a Quotient to a Power

In Section 4.2, we begin our study of polynomials. Before doing so, however, we must develop some rules for working with exponents.

Multiplying Powers with Like Bases

Recall from Section 1.8 that an expression like a^3 means $a \cdot a \cdot a$. We can use this fact to find the product of two expressions that have the same base:

$a^3 \cdot a^2 = (a \cdot a \cdot a)(a \cdot a)$ There are three factors in a^3 and two factors in a^2.

$a^3 \cdot a^2 = a \cdot a \cdot a \cdot a \cdot a$ Using an associative law

$a^3 \cdot a^2 = a^5$.

Note that the exponent in a^5 is the sum of the exponents in $a^3 \cdot a^2$. That is, $3 + 2 = 5$. Similarly,

$b^4 \cdot b^3 = (b \cdot b \cdot b \cdot b)(b \cdot b \cdot b)$

$b^4 \cdot b^3 = b^7$,   where $4 + 3 = 7$.

Adding the exponents gives the correct result.

STUDENT NOTES

There are several rules for manipulating exponents in this section. One way to remember them all is to replace variables with numbers (other than 1) and see what the results suggest. For example, multiplying $2^2 \cdot 2^3$ and examining the result is a fine way of reminding yourself that $a^m \cdot a^n = a^{m+n}$.

> **The Product Rule**
>
> For any number a and any positive integers m and n,
>
> $a^m \cdot a^n = a^{m+n}$.
>
> (To multiply powers with the same base, keep the base and add the exponents.)

EXAMPLE 1   Multiply and simplify each of the following. (Here "simplify" means express the product as one base to a power whenever possible.)

a) $2^3 \cdot 2^8$

b) $5^3 \cdot 5^8 \cdot 5^1$

c) $(r + s)^7 (r + s)^6$

d) $(a^3 b^2)(a^3 b^5)$

SOLUTION

a) $2^3 \cdot 2^8 = 2^{3+8}$ Adding exponents: $a^m \cdot a^n = a^{m+n}$

$\qquad\;\; = 2^{11}$ **CAUTION!** The base is unchanged: $2^3 \cdot 2^8 \neq 4^{11}$.

b) $5^3 \cdot 5^8 \cdot 5^1 = 5^{3+8+1}$ Adding exponents

$\qquad\qquad\;\; = 5^{12}$ **CAUTION!** $5^{12} \neq 5 \cdot 12$.

c) $(r + s)^7(r + s)^6 = (r + s)^{7+6}$ The base here is $r + s$.

$\qquad\qquad\qquad = (r + s)^{13}$ **CAUTION!** $(r + s)^{13} \neq r^{13} + s^{13}$.

d) $(a^3b^2)(a^3b^5) = a^3b^2a^3b^5$ Using an associative law

$\qquad\qquad\quad = a^3a^3b^2b^5$ Using a commutative law

$\qquad\qquad\quad = a^6b^7$ Adding exponents   **TRY EXERCISE** ▸ 15

Dividing Powers with Like Bases

Recall that any expression that is divided or multiplied by 1 is unchanged. This, together with the fact that anything (besides 0) divided by itself is 1, can lead to a rule for division:

$$\frac{a^5}{a^2} = \frac{a \cdot a \cdot a \cdot a \cdot a}{a \cdot a}$$

$$\frac{a^5}{a^2} = \frac{a \cdot a \cdot a}{1} \cdot \frac{a \cdot a}{a \cdot a}$$

$$\frac{a^5}{a^2} = \frac{a \cdot a \cdot a}{1} \cdot 1$$

$$\frac{a^5}{a^2} = a \cdot a \cdot a = a^3.$$

Note that the exponent in a^3 is the difference of the exponents in a^5/a^2. Similarly,

$$\frac{x^4}{x^3} = \frac{x \cdot x \cdot x \cdot x}{x \cdot x \cdot x} = \frac{x}{1} \cdot \frac{x \cdot x \cdot x}{x \cdot x \cdot x} = \frac{x}{1} \cdot 1 = x = x^1.$$

Subtracting the exponents gives the correct result.

> ### The Quotient Rule
>
> For any nonzero number a and any positive integers m and n for which $m > n$,
>
> $$\frac{a^m}{a^n} = a^{m-n}.$$
>
> (To divide powers with the same base, subtract the exponent of the denominator from the exponent of the numerator.)

EXAMPLE   2   Divide and simplify. (Here "simplify" means express the quotient as one base to a power whenever possible.)

a) $\dfrac{x^8}{x^2}$ **b)** $\dfrac{7^9}{7^4}$ **c)** $\dfrac{(5a)^{12}}{(5a)^4}$ **d)** $\dfrac{4p^5q^7}{6p^2q}$

SOLUTION

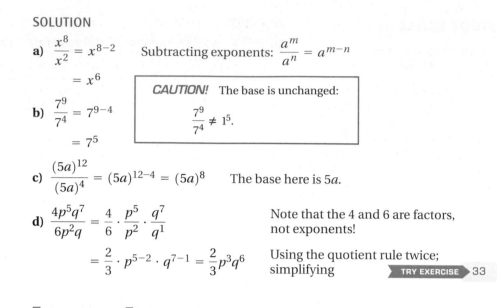

a) $\dfrac{x^8}{x^2} = x^{8-2}$ Subtracting exponents: $\dfrac{a^m}{a^n} = a^{m-n}$

$= x^6$

> **CAUTION!**   The base is unchanged:
>
> $$\dfrac{7^9}{7^4} \neq 1^5.$$

b) $\dfrac{7^9}{7^4} = 7^{9-4}$

$= 7^5$

c) $\dfrac{(5a)^{12}}{(5a)^4} = (5a)^{12-4} = (5a)^8$ The base here is $5a$.

d) $\dfrac{4p^5q^7}{6p^2q} = \dfrac{4}{6} \cdot \dfrac{p^5}{p^2} \cdot \dfrac{q^7}{q^1}$ Note that the 4 and 6 are factors, not exponents!

$= \dfrac{2}{3} \cdot p^{5-2} \cdot q^{7-1} = \dfrac{2}{3}p^3q^6$ Using the quotient rule twice; simplifying   **TRY EXERCISE** 33

Zero as an Exponent

The quotient rule can be used to help determine what 0 should mean when it appears as an exponent. Consider a^4/a^4, where a is nonzero. Since the numerator and the denominator are the same,

$$\dfrac{a^4}{a^4} = 1.$$

On the other hand, using the quotient rule would give us

$$\dfrac{a^4}{a^4} = a^{4-4} = a^0.$$ Subtracting exponents

Since $a^0 = a^4/a^4 = 1$, this suggests that $a^0 = 1$ for any nonzero value of a.

> **The Exponent Zero**
>
> For any real number a, with $a \neq 0$,
>
> $$a^0 = 1.$$
>
> (Any nonzero number raised to the 0 power is 1.)

Note that in the above box, 0^0 is not defined. For this text, we will assume that expressions like a^m do not represent 0^0.

EXAMPLE   3   Simplify: **(a)** 1948^0; **(b)** $(-9)^0$; **(c)** $(3x)^0$; **(d)** $3x^0$; **(e)** $(-1)9^0$; **(f)** -9^0.

SOLUTION

a) $1948^0 = 1$ Any nonzero number raised to the 0 power is 1.

b) $(-9)^0 = 1$ Any nonzero number raised to the 0 power is 1. The base here is -9.

c) $(3x)^0 = 1$, for any $x \neq 0$.   The parentheses indicate that the base is $3x$.

d) Since $3x^0$ means $3 \cdot x^0$, the base is x. Recall that simplifying exponential expressions is done before multiplication in the rules for order of operations:

$$3x^0 = 3 \cdot 1 = 3, \quad \text{for any } x \neq 0.$$

e) $(-1)9^0 = (-1)1 = -1$ The base here is 9.

f) -9^0 is read "the opposite of 9^0" and is equivalent to $(-1)9^0$:

$$-9^0 = (-1)9^0 = (-1)1 = -1.$$

Note from parts (b), (e), and (f) that $-9^0 = (-1)9^0$ and $-9^0 \neq (-9)^0$.

TRY EXERCISE 49

CAUTION! $-9^0 \neq (-9)^0$, and, in general, $-a^n \neq (-a)^n$.

Raising a Power to a Power

Consider an expression like $(7^2)^4$:

$(7^2)^4 = (7^2)(7^2)(7^2)(7^2)$ There are four factors of 7^2.

$(7^2)^4 = (7 \cdot 7)(7 \cdot 7)(7 \cdot 7)(7 \cdot 7)$ We could also use the product rule.

$(7^2)^4 = 7 \cdot 7 \cdot 7 \cdot 7 \cdot 7 \cdot 7 \cdot 7 \cdot 7$ Using an associative law

$(7^2)^4 = 7^8.$

Note that the exponent in 7^8 is the product of the exponents in $(7^2)^4$. Similarly,

$(y^5)^3 = y^5 \cdot y^5 \cdot y^5$ There are three factors of y^5.

$(y^5)^3 = (y \cdot y \cdot y \cdot y \cdot y)(y \cdot y \cdot y \cdot y \cdot y)(y \cdot y \cdot y \cdot y \cdot y)$

$(y^5)^3 = y^{15}.$

Once again, we get the same result if we multiply exponents:

$$(y^5)^3 = y^{5 \cdot 3} = y^{15}.$$

The Power Rule

For any number a and any whole numbers m and n,

$$(a^m)^n = a^{mn}.$$

(To raise a power to a power, multiply the exponents and leave the base unchanged.)

Remember that for this text we assume that 0^0 is not considered.

EXAMPLE 4   Simplify: **(a)** $(3^5)^4$; **(b)** $(m^2)^5$.

SOLUTION

a) $(3^5)^4 = 3^{5 \cdot 4}$ Multiplying exponents: $(a^m)^n = a^{mn}$

$\qquad = 3^{20}$

b) $(m^2)^5 = m^{2 \cdot 5}$

$\qquad = m^{10}$

TRY EXERCISE 57

Raising a Product or a Quotient to a Power

When an expression inside parentheses is raised to a power, the inside expression is the base. Let's compare $2a^3$ and $(2a)^3$:

$2a^3 = 2 \cdot a \cdot a \cdot a;$ The base is a.

$(2a)^3 = (2a)(2a)(2a)$ The base is $2a$.

$(2a)^3 = (2 \cdot 2 \cdot 2)(a \cdot a \cdot a)$

$(2a)^3 = 2^3 a^3$

$(2a)^3 = 8a^3.$

We see that $2a^3$ and $(2a)^3$ are *not* equivalent. Note too that $(2a)^3$ can be simplified by cubing each factor in $2a$. This leads to the following rule for raising a product to a power.

> ### Raising a Product to a Power
>
> For any numbers a and b and any whole number n,
>
> $$(ab)^n = a^n b^n.$$
>
> (To raise a product to a power, raise each factor to that power.)

EXAMPLE 5   Simplify: **(a)** $(4a)^3$; **(b)** $(-5x^4)^2$; **(c)** $(a^7 b)^2 (a^3 b^4)$.

SOLUTION

a) $(4a)^3 = 4^3 a^3 = 64a^3$ Raising each factor to the third power and simplifying

b) $(-5x^4)^2 = (-5)^2 (x^4)^2$ Raising each factor to the second power. Parentheses are important here.

$\qquad\qquad = 25x^8$ Simplifying $(-5)^2$ and using the power rule

c) $(a^7 b)^2 (a^3 b^4) = (a^7)^2 b^2 a^3 b^4$ Raising a product to a power

$\qquad\qquad = a^{14} b^2 a^3 b^4$ Multiplying exponents

$\qquad\qquad = a^{17} b^6$ Adding exponents   **TRY EXERCISE** 63

CAUTION!   The rule $(ab)^n = a^n b^n$ applies only to *products* raised to a power, not to sums or differences. For example, $(3 + 4)^2 \neq 3^2 + 4^2$ since $49 \neq 9 + 16$. Similarly, $(5x)^2 = 5^2 \cdot x^2$, but $(5 + x)^2 \neq 5^2 + x^2$.

There is a similar rule for raising a quotient to a power.

> ## Raising a Quotient to a Power
>
> For any numbers a and b, $b \neq 0$, and any whole number n,
>
> $$\left(\frac{a}{b}\right)^n = \frac{a^n}{b^n}.$$
>
> (To raise a quotient to a power, raise the numerator to the power and divide by the denominator to the power.)

EXAMPLE   6   Simplify: **(a)** $\left(\dfrac{x}{5}\right)^2$; **(b)** $\left(\dfrac{5}{a^4}\right)^3$; **(c)** $\left(\dfrac{3a^4}{b^3}\right)^2$.

SOLUTION

a) $\left(\dfrac{x}{5}\right)^2 = \dfrac{x^2}{5^2} = \dfrac{x^2}{25}$ Squaring the numerator and the denominator

b) $\left(\dfrac{5}{a^4}\right)^3 = \dfrac{5^3}{(a^4)^3}$ Raising a quotient to a power

$\qquad = \dfrac{125}{a^{4 \cdot 3}} = \dfrac{125}{a^{12}}$ Using the power rule and simplifying

c) $\left(\dfrac{3a^4}{b^3}\right)^2 = \dfrac{(3a^4)^2}{(b^3)^2}$ Raising a quotient to a power

$\qquad = \dfrac{3^2(a^4)^2}{b^{3 \cdot 2}} = \dfrac{9a^8}{b^6}$ Raising a product to a power and using the power rule

TRY EXERCISE ❯ 75

In the following summary of definitions and rules, we assume that no denominators are 0 and that 0^0 is not considered.

> ## Definitions and Properties of Exponents
>
> For any whole numbers m and n,
>
> | | |
> |---|---|
> | 1 as an exponent: | $a^1 = a$ |
> | 0 as an exponent: | $a^0 = 1$ |
> | The Product Rule: | $a^m \cdot a^n = a^{m+n}$ |
> | The Quotient Rule: | $\dfrac{a^m}{a^n} = a^{m-n}$ |
> | The Power Rule: | $(a^m)^n = a^{mn}$ |
> | Raising a product to a power: | $(ab)^n = a^n b^n$ |
> | Raising a quotient to a power: | $\left(\dfrac{a}{b}\right)^n = \dfrac{a^n}{b^n}$ |

4.1 EXERCISE SET

For Extra Help
MyMathLab   Math XL
PRACTICE   WATCH   DOWNLOAD

Concept Reinforcement   *In each of Exercises 1–8, complete the sentence using the most appropriate phrase from the column on the right.*

1. To raise a product to a power, ____

2. To raise a quotient to a power, ____

3. To raise a power to a power, ____

4. To divide powers with the same base, ____

5. Any nonzero number raised to the 0 power ____

6. To multiply powers with the same base, ____

7. To square a fraction, ____

8. To square a product, ____

a) keep the base and add the exponents.

b) multiply the exponents and leave the base unchanged.

c) square the numerator and square the denominator.

d) square each factor.

e) raise each factor to that power.

f) raise the numerator to the power and divide by the denominator to the power.

g) is one.

h) subtract the exponent of the denominator from the exponent of the numerator.

Identify the base and the exponent in each expression.

9. $(2x)^5$

10. $(x + 1)^0$

11. $2x^3$

12. $-y^6$

13. $\left(\dfrac{4}{y}\right)^7$

14. $(-5x)^4$

Simplify. Assume that no denominator is 0 and that 0^0 is not considered.

15. $d^3 \cdot d^{10}$

16. $8^4 \cdot 8^3$

17. $a^6 \cdot a$

18. $y^7 \cdot y^9$

19. $6^5 \cdot 6^{10}$

20. $t^0 \cdot t^{16}$

21. $(3y)^4(3y)^8$

22. $(2t)^8(2t)^{17}$

23. $(8n)(8n)^9$

24. $(5p)^0(5p)^1$

25. $(a^2b^7)(a^3b^2)$

26. $(m - 3)^4(m - 3)^5$

27. $(x + 3)^5(x + 3)^8$

28. $(a^8b^3)(a^4b)$

29. $r^3 \cdot r^7 \cdot r^0$

30. $s^4 \cdot s^5 \cdot s^2$

31. $(mn^5)(m^3n^4)$

32. $(a^3b)(ab)^4$

33. $\dfrac{7^5}{7^2}$

34. $\dfrac{4^7}{4^3}$

35. $\dfrac{t^8}{t}$

36. $\dfrac{x^7}{x}$

37. $\dfrac{(5a)^7}{(5a)^6}$

38. $\dfrac{(3m)^9}{(3m)^8}$

Aha! 39. $\dfrac{(x + y)^8}{(x + y)^8}$

40. $\dfrac{(9x)^{10}}{(9x)^2}$

41. $\dfrac{(r + s)^{12}}{(r + s)^4}$

42. $\dfrac{(a - b)^4}{(a - b)^3}$

43. $\dfrac{12d^9}{15d^2}$

44. $\dfrac{10n^7}{15n^3}$

45. $\dfrac{8a^9b^7}{2a^2b}$

46. $\dfrac{12r^{10}s^7}{4r^2s}$

47. $\dfrac{x^{12}y^9}{x^0y^2}$

48. $\dfrac{a^{10}b^{12}}{a^2b^0}$

Simplify.

49. t^0 when $t = 15$

50. y^0 when $y = 38$

51. $5x^0$ when $x = -22$

52. $7m^0$ when $m = 1.7$

53. $7^0 + 4^0$

54. $(8 + 5)^0$

55. $(-3)^1 - (-3)^0$

56. $(-4)^0 - (-4)^1$

Simplify. Assume that no denominator is 0 and that 0^0 is not considered.

57. $(x^3)^{11}$

58. $(a^5)^8$

59. $(5^8)^4$

60. $(2^5)^2$

61. $(t^{20})^4$

62. $(x^{25})^6$

63. $(10x)^2$

64. $(5a)^2$

65. $(-2a)^3$

66. $(-3x)^3$

67. $(-5n^7)^2$

68. $(-4m^4)^2$

69. $(a^2b)^7$

70. $(xy^4)^9$

71. $(r^5t)^3(r^2t^8)$

72. $(a^4b^6)(a^2b)^5$

73. $(2x^5)^3(3x^4)$

74. $(5x^3)^2(2x^7)$

75. $\left(\dfrac{x}{5}\right)^3$

76. $\left(\dfrac{2}{a}\right)^4$

77. $\left(\dfrac{7}{6n}\right)^2$

78. $\left(\dfrac{4x}{3}\right)^3$

79. $\left(\dfrac{a^3}{b^8}\right)^6$

80. $\left(\dfrac{x^5}{y^2}\right)^7$

81. $\left(\dfrac{x^2y}{z^3}\right)^4$

82. $\left(\dfrac{a^4}{b^2c}\right)^5$

83. $\left(\dfrac{a^3}{-2b^5}\right)^4$

84. $\left(\dfrac{x^5}{-3y^3}\right)^4$

85. $\left(\dfrac{5x^7y}{-2z^4}\right)^3$

86. $\left(\dfrac{-4p^5}{3m^2n^3}\right)^3$

Aha! **87.** $\left(\dfrac{4x^3y^5}{3z^7}\right)^0$

88. $\left(\dfrac{5a^7}{2b^5c}\right)^0$

89. Explain in your own words why $-5^2 \neq (-5)^2$.

90. Under what circumstances should exponents be added?

Skill Review

To prepare for Section 4.2, review combining like terms and evaluating expressions (Sections 1.6 and 1.8).

Combine like terms. [1.6]

91. $9x + 2y - x - 2y$

92. $5a - 7b - 8a + b$

93. $-3x + (-2) - 5 - (-x)$

94. $2 - t - 3t - r - 7$

Evaluate. [1.8]

95. $4 + x^3$, for $x = 10$

96. $-x^2 - 5x + 3$, for $x = -2$

Synthesis

97. Under what conditions does a^n represent a negative number? Why?

98. Using the quotient rule, explain why 9^0 is 1.

99. Suppose that the width of a square is three times the width of a second square. How do the areas of the squares compare? Why?

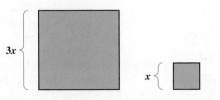

100. Suppose that the width of a cube is twice the width of a second cube. How do the volumes of the cubes compare? Why?

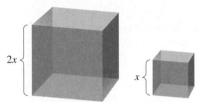

Find a value of the variable that shows that the two expressions are not *equivalent. Answers may vary.*

101. $3x^2$; $(3x)^2$

102. $(a + 5)^2$; $a^2 + 5^2$

103. $\dfrac{t^6}{t^2}$; t^3

104. $\dfrac{a + 7}{7}$; a

Simplify.

105. $y^{4x} \cdot y^{2x}$

106. $a^{10k} \div a^{2k}$

107. $\dfrac{x^{5t}(x^t)^2}{(x^{3t})^2}$

108. $\dfrac{\left(\frac{1}{2}\right)^3\left(\frac{2}{3}\right)^4}{\left(\frac{5}{6}\right)^3}$

109. Solve for x:
$$\dfrac{t^{26}}{t^x} = t^x.$$

Replace ■ *with* $>$, $<$, *or* $=$ *to write a true sentence.*

110. 3^5 ■ 3^4

111. 4^2 ■ 4^3

112. 4^3 ■ 5^3

113. 4^3 ■ 3^4

114. 9^7 ■ 3^{13}

115. 25^8 ■ 125^5

Use the fact that $10^3 \approx 2^{10}$ *to estimate each of the following powers of 2. Then compute the power of 2 with a calculator and find the difference between the exact value and the approximation.*

116. 2^{14}

117. 2^{22}

118. 2^{26}

119. 2^{31}

In computer science, 1 KB of memory refers to 1 kilobyte, or 1×10^3 *bytes, of memory. This is really an approximation of* 1×2^{10} *bytes (since computer memory actually uses powers of 2).*

120. The TI-84 Plus graphing calculator has 480 KB of "FLASH ROM." How many bytes is this?

121. The TI-84 Plus Silver Edition graphing calculator has 1.5 MB (megabytes) of FLASH ROM, where 1 MB is 1000 KB (see Exercise 120). How many bytes of FLASH ROM does this calculator have?

4.2   Polynomials

Terms • Types of Polynomials • Degree and Coefficients • Combining Like Terms •
Evaluating Polynomials and Applications

We now examine an important algebraic expression known as a *polynomial*. Certain polynomials have appeared earlier in this text so you already have some experience working with them.

Terms

At this point, we have seen a variety of algebraic expressions like

$$3a^2b^4, \quad 2l + 2w, \quad \text{and} \quad 5x^2 + x - 2.$$

Within these expressions, $3a^2b^4$, $2l$, $2w$, $5x^2$, x, and -2 are examples of *terms*. A **term** (see p. 17) can be a number (like -2), a variable (like x), a product of numbers and/or variables (like $3a^2b^4$, $2l$, $2w$, or $5x^2$), or a quotient of numbers and/or variables (like $7/t$ or $(a^2b^3)/(4c)$).*

Types of Polynomials

If a term is a product of constants and/or variables, it is called a **monomial**. Note that a term, but not a monomial, can include division by a variable. A **polynomial** is a monomial or a sum of monomials.

Examples of monomials: $7, \quad t, \quad 2l, \quad 2w, \quad 5x^3y, \quad \frac{3}{7}a^5$

Examples of polynomials: $4x + 7, \quad \frac{2}{3}t^2, \quad 6a + 7, \quad -5n^2 + m - 1, \quad 42r^5,$
$\qquad\qquad\qquad\qquad\quad x, \quad 0$

When a polynomial is written as a sum of monomials, each monomial is called a *term of the polynomial.*

EXAMPLE   1   Identify the terms of the polynomial $3t^4 - 5t^6 - 4t + 2$.

SOLUTION   The terms are $3t^4$, $-5t^6$, $-4t$, and 2. We can see this by rewriting all subtractions as additions of opposites:

$$3t^4 - 5t^6 - 4t + 2 = 3t^4 + (-5t^6) + (-4t) + 2.$$
$$\qquad\qquad\qquad\quad \uparrow \qquad\quad \uparrow \qquad\quad \uparrow \quad \uparrow$$

These are the terms of the polynomial.

TRY EXERCISE   9

A polynomial that is composed of two terms is called a **binomial**, whereas those composed of three terms are called **trinomials**. Polynomials with four or more terms have no special name.

---

*Later in this text, expressions like $5x^{3/2}$ and $2a^{-7}b$ will be discussed. Such expressions are also considered terms.

Monomials	Binomials	Trinomials	No Special Name
$4x^2$	$2x + 4$	$3t^3 + 4t + 7$	$4x^3 - 5x^2 + xy - 8$
9	$3a^5 + 6bc$	$6x^7 - 8z^2 + 4$	$z^5 + 2z^4 - z^3 + 7z + 3$
$-7a^{19}b^5$	$-9x^7 - 6$	$4x^2 - 6x - \frac{1}{2}$	$4x^6 - 3x^5 + x^4 - x^3 + 2x - 1$

The following algebraic expressions are *not* polynomials:

$$\textbf{(1)} \ \frac{x + 3}{x - 4}, \qquad \textbf{(2)} \ 5x^3 - 2x^2 + \frac{1}{x}, \qquad \textbf{(3)} \ \frac{1}{x^3 - 2}.$$

Expressions (1) and (3) are not polynomials because they represent quotients, not sums. Expression (2) is not a polynomial because $1/x$ is not a monomial.

Degree and Coefficients

The **degree of a term of a polynomial** is the number of variable factors in that term. Thus the degree of $7t^2$ is 2 because $7t^2$ has two variable factors: $7t^2 = 7 \cdot t \cdot t$. We will revisit the meaning of degree in Section 4.6 when polynomials in several variables are examined.

EXAMPLE 2   Determine the degree of each term: **(a)** $8x^4$; **(b)** $3x$; **(c)** 7.

SOLUTION

a) The degree of $8x^4$ is 4. x^4 represents 4 variable factors: $x \cdot x \cdot x \cdot x$.

b) The degree of $3x$ is 1.   There is 1 variable factor.

c) The degree of 7 is 0.   There is no variable factor.

The degree of a constant polynomial, as in Example 2(c), is 0 since there are no variable factors. There is an exception to this statement, however. Since $0 = 0x = 0x^2 = 0x^3$ and so on, we say that the polynomial 0 has *no* degree.

The part of a term that is a constant factor is the **coefficient** of that term. Thus the coefficient of $3x$ is 3, and the coefficient for the term 7 is simply 7.

EXAMPLE 3   Identify the coefficient of each term in the polynomial

$$4x^3 - 7x^2y + x - 8.$$

SOLUTION

The coefficient of $4x^3$ is 4.

The coefficient of $-7x^2y$ is -7.

The coefficient of the third term is 1, since $x = 1x$.

The coefficient of -8 is simply -8.    **TRY EXERCISE** 13

The **leading term** of a polynomial is the term of highest degree. Its coefficient is called the **leading coefficient** and its degree is referred to as the **degree of the polynomial**. To see how this terminology is used, consider the polynomial

$$3x^2 - 8x^3 + 5x^4 + 7x - 6.$$

The *terms* are $3x^2$, $-8x^3$, $5x^4$, $7x$,   and -6.

The *coefficients* are   3, -8,   5,   7,   and -6.

The *degree of each term* is   2,   3,   4,   1,   and   0.

The *leading term* is $5x^4$ and the *leading coefficient* is 5.

The *degree of the polynomial* is 4.

Combining Like Terms

Recall from Section 1.8 that *like*, or *similar*, *terms* are either constant terms or terms containing the same variable(s) raised to the same power(s). To simplify certain polynomials, we can often *combine*, or *collect*, like terms.

EXAMPLE 4   Identify the like terms in $4x^3 + 5x - 7x^2 + 2x^3 + x^2$.

SOLUTION

Like terms: $4x^3$ and $2x^3$ Same variable and exponent

Like terms: $-7x^2$ and x^2 Same variable and exponent

EXAMPLE 5   Write an equivalent expression by combining like terms.

a) $2x^3 + 6x^3$

b) $5x^2 + 7 + 2x^4 - 6x^2 - 11 - 2x^4$

c) $7a^3 - 5a^2 + 9a^3 + a^2$

d) $\frac{2}{3}x^4 - x^3 - \frac{1}{6}x^4 + \frac{2}{5}x^3 - \frac{3}{10}x^3$

STUDENT NOTES

Remember that when we combine like terms, we are not solving equations, but are forming equivalent expressions.

SOLUTION

a) $2x^3 + 6x^3 = (2 + 6)x^3$ Using the distributive law

$\qquad\qquad\quad = 8x^3$

b) $5x^2 + 7 + 2x^4 - 6x^2 - 11 - 2x^4$

$\qquad = 5x^2 - 6x^2 + 2x^4 - 2x^4 + 7 - 11$

$\qquad = (5 - 6)x^2 + (2 - 2)x^4 + (7 - 11)$ ⎫

$\qquad = -1x^2 + 0x^4 + (-4)$　　　　　　⎬ These steps are often done mentally.

$\qquad = -x^2 - 4$　　　　　　　　　　⎭

c) $7a^3 - 5a^2 + 9a^3 + a^2 = 7a^3 - 5a^2 + 9a^3 + 1a^2$ $a^2 = 1 \cdot a^2 = 1a^2$

$\qquad\qquad\qquad\qquad\qquad = 16a^3 - 4a^2$

d) $\frac{2}{3}x^4 - x^3 - \frac{1}{6}x^4 + \frac{2}{5}x^3 - \frac{3}{10}x^3 = \left(\frac{2}{3} - \frac{1}{6}\right)x^4 + \left(-1 + \frac{2}{5} - \frac{3}{10}\right)x^3$

$\qquad\qquad\qquad\qquad\qquad\qquad = \left(\frac{4}{6} - \frac{1}{6}\right)x^4 + \left(-\frac{10}{10} + \frac{4}{10} - \frac{3}{10}\right)x^3$

$\qquad\qquad\qquad\qquad\qquad\qquad = \frac{3}{6}x^4 - \frac{9}{10}x^3$

$\qquad\qquad\qquad\qquad\qquad\qquad = \frac{1}{2}x^4 - \frac{9}{10}x^3$ There are no similar terms, so we are done.

TRY EXERCISE 41

Note in Example 5 that the solutions are written so that the term of highest degree appears first, followed by the term of next highest degree, and so on. This is known as **descending order** and is the form in which answers will normally appear.

Evaluating Polynomials and Applications

When each variable in a polynomial is replaced with a number, the polynomial then represents a number, or *value*, that can be calculated using the rules for order of operations.

EXAMPLE 6   Evaluate $-x^2 + 3x + 9$ for $x = -2$.

SOLUTION   For $x = -2$, we have

Substitute. $-x^2 + 3x + 9 = -(-2)^2 + 3(-2) + 9$ The negative sign in front of x^2 remains.

$\qquad\qquad\qquad\qquad = -(4) + 3(-2) + 9$

Simplify. $\qquad\qquad\qquad = -4 + (-6) + 9$

$\qquad\qquad\qquad\qquad = -10 + 9 = -1.$

TRY EXERCISE 51

EXAMPLE **7**

Games in a sports league.   In a sports league of n teams in which each team plays every other team twice, the total number of games to be played is given by the polynomial

$$n^2 - n.$$

A girls' soccer league has 10 teams. How many games are played if each team plays every other team twice?

SOLUTION   We evaluate the polynomial for $n = 10$:

$$n^2 - n = 10^2 - 10$$
$$= 100 - 10 = 90.$$

The league plays 90 games.

TRY EXERCISE 61

EXAMPLE **8**

Vehicle miles traveled.   The average annual number of vehicle miles traveled per vehicle (VMT), in thousands, for a driver of age a can be approximated by the polynomial

$$-0.003a^2 + 0.2a + 8.6.$$

Find the VMT per vehicle for a 20-year-old driver.

Source: Based on information from the Energy Information Administration

SOLUTION   To find the VMT per vehicle for a 20-year-old driver, we evaluate the polynomial for $a = 20$:

$$-0.003a^2 + 0.2a + 8.6 = -0.003(20)^2 + 0.2(20) + 8.6$$
$$= -0.003 \cdot 400 + 4 + 8.6$$
$$= -1.2 + 4 + 8.6$$
$$= 11.4.$$

The average annual number of VMT per vehicle by a 20-year-old driver is 11.4 thousand, or 11,400.

TRY EXERCISE 65

Sometimes, a graph can be used to estimate the value of a polynomial visually.

EXAMPLE **9**

Vehicle miles traveled.   In the following graph, the polynomial from Example 8 has been graphed by evaluating it for several choices of a. Use the graph to estimate the number of vehicle miles traveled each year, per vehicle, by a 30-year-old driver.

TECHNOLOGY CONNECTION

One way to evaluate a polynomial is to use the TRACE key. For example, to evaluate $-0.003a^2 + 0.2a + 8.6$ in Example 9 for $a = 30$, we can enter the polynomial as $y = -0.003x^2 + 0.2x + 8.6$. We then use TRACE and enter an x-value of 30.

(continued)

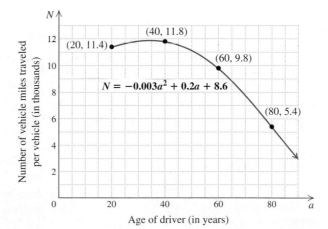

The value of the polynomial appears as y, and the cursor automatically appears at $(30, 11.9)$. The Value option of the CALC menu works in a similar way.

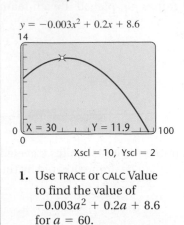

$y = -0.003x^2 + 0.2x + 8.6$

Xscl = 10, Yscl = 2

1. Use TRACE or CALC Value to find the value of $-0.003a^2 + 0.2a + 8.6$ for $a = 60$.

SOLUTION   To estimate the number of vehicle miles traveled by a 30-year-old driver, we locate 30 on the horizontal axis. From there, we move vertically until we meet the curve at some point. From that point, we move horizontally to the N-axis.

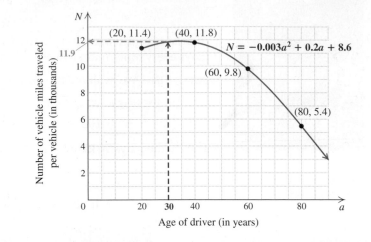

The average number of vehicle miles traveled each year, per vehicle, by a 30-year-old driver is 11.9 thousand, or 11,900. (For $a = 30$, the value of $-0.003a^2 + 0.2a + 8.6$ is approximately 11.9.)

TRY EXERCISE ▶ 71

4.2 EXERCISE SET

For Extra Help
MyMathLab Math XL PRACTICE WATCH DOWNLOAD

↝ *Concept Reinforcement*   *In each of Exercises 1–8, match the description with the most appropriate algebraic expression from the column on the right.*

1. ____ A polynomial with four terms

2. ____ A polynomial with 7 as its leading coefficient

3. ____ A trinomial written in descending order

4. ____ A polynomial with degree 5

5. ____ A binomial with degree 7

6. ____ A monomial of degree 0

7. ____ An expression with two terms that is not a binomial

8. ____ An expression with three terms that is not a trinomial

a) $8x^3 + \dfrac{2}{x^2}$

b) $5x^4 + 3x^3 - 4x + 7$

c) $\dfrac{3}{x} - 6x^2 + 9$

d) $8t - 4t^5$

e) 5

f) $6x^2 + 7x^4 - 2x^3$

g) $4t - 2t^7$

h) $3t^2 + 4t + 7$

Identify the terms of each polynomial.

9. $8x^3 - 11x^2 + 6x + 1$ 10. $5a^3 + 4a^2 - a - 7$

11. $-t^6 - 3t^3 + 9t - 4$ 12. $n^5 - 4n^3 + 2n - 8$

Determine the coefficient and the degree of each term in each polynomial.

13. $8x^4 + 2x$

14. $9a^3 - 4a^2$

15. $9t^2 - 3t + 4$

16. $7x^4 + 5x - 3$

17. $6a^5 + 9a + a^3$

18. $4t^8 - t + 6t^5$

19. $x^4 - x^3 + 4x - 3$

20. $2a^5 + a^2 + 8a + 10$

For each of the following polynomials, (a) list the degree of each term; (b) determine the leading term and the leading coefficient; and (c) determine the degree of the polynomial.

21. $5t + t^3 + 8t^4$

22. $1 + 6n + 4n^2$

23. $3a^2 - 7 + 2a^4$

24. $9x^4 + x^2 + x^7 - 12$

25. $8 + 6x^2 - 3x - x^5$

26. $9a - a^4 + 3 + 2a^3$

27. Complete the following table for the polynomial
$7x^2 + 8x^5 - 4x^3 + 6 - \frac{1}{2}x^4$.

Term	Coefficient	Degree of the Term	Degree of the Polynomial
		5	
$-\frac{1}{2}x^4$			
	-4		
		2	
	6		

28. Complete the following table for the polynomial
$-3x^4 + 6x^3 - 2x^2 + 8x + 7$.

Term	Coefficient	Degree of the Term	Degree of the Polynomial
	-3		
$6x^3$			
		2	
		1	
	7		

Classify each polynomial as a monomial, a binomial, a trinomial, or a polynomial with no special name.

29. $x^2 - 23x + 17$

30. $-9x^2$

31. $x^3 - 7x + 2x^2 - 4$

32. $t^3 + 4$

33. $y + 8$

34. $3x^8 + 12x^3 - 9$

35. 17

36. $2x^4 - 7x^3 + x^2 + x - 6$

Combine like terms. Write all answers in descending order.

37. $5n^2 + n + 6n^2$

38. $5a + 7a^2 + 3a$

39. $3a^4 - 2a + 2a + a^4$

40. $9b^5 + 3b^2 - 2b^5 - 3b^2$

41. $7x^3 - 11x + 5x + x^2$

42. $3x^4 - 7x + x^4 - 2x$

43. $4b^3 + 5b + 7b^3 + b^2 - 6b$

44. $6x^2 + 2x^4 - 2x^2 - x^4 - 4x^2 + x$

45. $10x^2 + 2x^3 - 3x^3 - 4x^2 - 6x^2 - x^4$

46. $12t^6 - t^3 + 8t^6 + 4t^3 - t^7 - 3t^3$

47. $\frac{1}{5}x^4 + 7 - 2x^2 + 3 - \frac{2}{15}x^4 + 2x^2$

48. $\frac{1}{6}x^3 + 3x^2 - \frac{1}{3}x^3 + 7 + x^2 - 10$

49. $8.3a^2 + 3.7a - 8 - 9.4a^2 + 1.6a + 0.5$

50. $1.4y^3 + 2.9 - 7.7y - 1.3y - 4.1 + 9.6y^3$

Evaluate each polynomial for $x = 3$ and for $x = -3$.

51. $-4x + 9$

52. $-6x + 5$

53. $2x^2 - 3x + 7$

54. $4x^2 - 6x + 9$

55. $-3x^3 + 7x^2 - 4x - 8$

56. $-2x^3 - 3x^2 + 4x + 2$

57. $2x^4 - \frac{1}{9}x^3$

58. $\frac{1}{3}x^4 - 2x^3$

59. $-x - x^2 - x^3$

60. $-x^2 - 3x^3 - x^4$

Back-to-college expenses. *The amount of money, in billions of dollars, spent on shoes for college can be estimated by the polynomial*

$$0.4t + 1.13,$$

where t is the number of years since 2004.
Source: Based on data from the National Retail Federation

61. Estimate the amount spent on shoes for college in 2006.

62. Estimate the amount spent on shoes for college in 2010.

63. *Skydiving.* During the first 13 sec of a jump, the number of feet that a skydiver falls in t seconds is approximated by the polynomial

$$11.12t^2.$$

Approximately how far has a skydiver fallen 10 sec after having jumped from a plane?

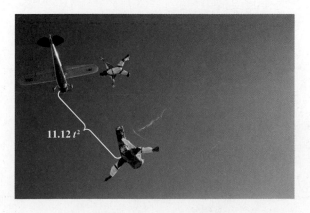
$11.12\,t^2$

64. *Skydiving.* For jumps that exceed 13 sec, the polynomial $173t - 369$ can be used to approximate the distance, in feet, that a skydiver has fallen in t seconds. Approximately how far has a skydiver fallen 20 sec after having jumped from a plane?

Circumference.   The circumference of a circle of radius r is given by the polynomial $2\pi r$, where π is an irrational number. For an approximation of π, use 3.14.

65. Find the circumference of a circle with radius 10 cm.

66. Find the circumference of a circle with radius 5 ft.

Area of a circle.   The area of a circle of radius r is given by the polynomial πr^2. Use 3.14 as an approximation for π.

67. Find the area of a circle with radius 7 m.

68. Find the area of a circle with radius 6 ft.

Kidney transplants.   Often a patient needing a kidney transplant has a willing kidney donor who does not match the patient medically. A kidney-paired donation matches donor–recipient pairs. Two kidney transplants are performed simultaneously, with each patient receiving the kidney of a stranger. The number k of such "kidney swaps" t years after 2003 can be approximated by

$$k = 14.3t^3 - 56t^2 + 57.7t + 19.$$

Use the following graph for Exercises 69 and 70.
Source: Based on data from United Network for Organ Sharing

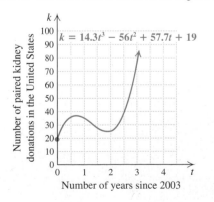

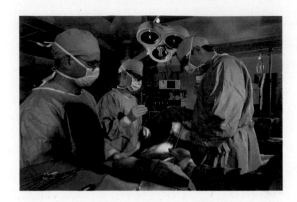

69. Estimate the number of kidney-paired donations in 2006.

70. Estimate the number of kidney-paired donations in 2004.

Memorizing words.   Participants in a psychology experiment were able to memorize an average of M words in t minutes, where $M = -0.001t^3 + 0.1t^2$. Use the following graph for Exercises 71–74.

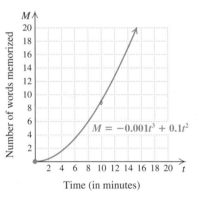

71. Estimate the number of words memorized after 10 min.

72. Estimate the number of words memorized after 14 min.

73. Find the approximate value of M for $t = 8$.

74. Find the approximate value of M for $t = 12$.

Body mass index.   The body mass index, or BMI, is one measure of a person's health. The average BMI B for males of age x, where x is between 2 and 20, is approximated by

$$B = -0.003x^3 + 0.13x^2 - 1.2x + 18.6.$$

Use the following graph for Exercises 75 and 76.
Source: Based on information from the National Center for Health Statistics

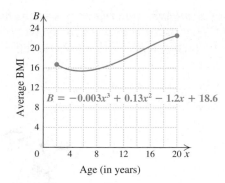

Age (in years)

75. Approximate the average BMI for 4-year-old males; for 14-year-old males.

76. Approximate the average BMI for 10-year-old males; for 16-year-old males.

77. Explain how it is possible for a term to not be a monomial.

78. Is it possible to evaluate polynomials without understanding the rules for order of operations? Why or why not?

Skill Review

To prepare for Section 4.3, review simplifying expressions containing parentheses (Section 1.8).

Simplify. [1.8]

79. $2x + 5 - (x + 8)$

80. $3x - 7 - (5x - 1)$

81. $4a + 3 - (-2a + 6)$

82. $\frac{1}{2}t - \frac{1}{4} - \left(\frac{3}{2}t + \frac{3}{4}\right)$

83. $4t^4 + 8t - (5t^4 - 9t)$

84. $0.1a^2 + 5 - (-0.3a^2 + a - 6)$

Synthesis

85. Suppose that the coefficients of a polynomial are all integers and the polynomial is evaluated for some integer. Must the value of the polynomial then also be an integer? Why or why not?

86. Is it easier to evaluate a polynomial before or after like terms have been combined? Why?

87. Construct a polynomial in x (meaning that x is the variable) of degree 5 with four terms, with coefficients that are consecutive even integers. Write in descending order.

Revenue, cost, and profit.   *Gigabytes Electronics is selling a new type of computer monitor.* Total revenue *is the total amount of money taken in and* total cost *is the total amount paid for producing the items. The firm estimates that for the monitor's first year, revenue from the sale of x monitors is*

$$250x - 0.5x^2 \text{ dollars,}$$

and the total cost is given by

$$4000 + 0.6x^2 \text{ dollars.}$$

Profit *is the difference between revenue and cost.*

88. Find the profit when 20 monitors are produced and sold.

89. Find the profit when 30 monitors are produced and sold.

Simplify.

90. $\frac{9}{2}x^8 + \frac{1}{9}x^2 + \frac{1}{2}x^9 + \frac{9}{2}x + \frac{9}{2}x^9 + \frac{8}{9}x^2 + \frac{1}{2}x - \frac{1}{2}x^8$

91. $(3x^2)^3 + 4x^2 \cdot 4x^4 - x^4(2x)^2 + ((2x)^2)^3 - 100x^2(x^2)^2$

92. A polynomial in x has degree 3. The coefficient of x^2 is 3 less than the coefficient of x^3. The coefficient of x is three times the coefficient of x^2. The remaining constant is 2 more than the coefficient of x^3. The sum of the coefficients is -4. Find the polynomial.

93. Use the graph for Exercises 75 and 76 to determine the ages for which the average BMI is 16.

94. *Path of the Olympic arrow.*   The Olympic flame at the 1992 Summer Olympics was lit by a flaming arrow. As the arrow moved d meters horizontally from the archer, its height h, in meters, was approximated by the polynomial

$$-0.0064d^2 + 0.8d + 2.$$

Complete the table for the choices of d given. Then plot the points and draw a graph representing the path of the arrow.

d	$-0.0064d^2 + 0.8d + 2$
0	
30	
60	
90	
120	

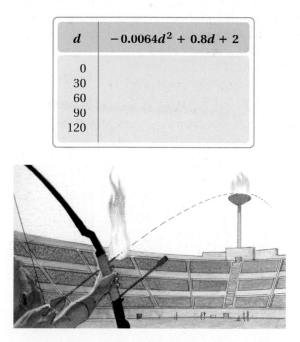

▦ *Semester averages.* *Professor Sakima calculates a student's average for her course using*

$$A = 0.3q + 0.4t + 0.2f + 0.1h,$$

with q, t, f, and h representing a student's quiz average, test average, final exam score, and homework average, respectively. In Exercises 95 and 96, find the given student's course average rounded to the nearest tenth.

95. Galina: quizzes: 60, 85, 72, 91; final exam: 84; tests: 89, 93, 90; homework: 88

96. Nigel: quizzes: 95, 99, 72, 79; final exam: 91; tests: 68, 76, 92; homework: 86

In Exercises 97 and 98, complete the table for the given choices of t. Then plot the points and connect them with a smooth curve representing the graph of the polynomial.

97.

t	$-t^2 + 10t - 18$
3	
4	
5	
6	
7	

98.

t	$-t^2 + 6t - 4$
1	
2	
3	
4	
5	

4.3 Addition and Subtraction of Polynomials

Addition of Polynomials ▪ Opposites of Polynomials ▪ Subtraction of Polynomials ▪ Problem Solving

Addition of Polynomials

To add two polynomials, we write a plus sign between them and combine like terms.

EXAMPLE **1** Write an equivalent expression by adding.

a) $(-5x^3 + 6x - 1) + (4x^3 + 3x^2 + 2)$
b) $\left(\frac{2}{3}x^4 + 3x^2 - 7x + \frac{1}{2}\right) + \left(-\frac{1}{3}x^4 + 5x^3 - 3x^2 + 3x - \frac{1}{2}\right)$

SOLUTION

a) $(-5x^3 + 6x - 1) + (4x^3 + 3x^2 + 2)$

$\quad = -5x^3 + 6x - 1 + 4x^3 + 3x^2 + 2$ Writing without parentheses

$\quad = -5x^3 + 4x^3 + 3x^2 + 6x - 1 + 2$ Using the commutative and associative laws to write like terms together

$\quad = (-5 + 4)x^3 + 3x^2 + 6x + (-1 + 2)$ Combining like terms; using the distributive law

$\quad = -x^3 + 3x^2 + 6x + 1$ Note that $-1x^3 = -x^3$.

b) $\left(\frac{2}{3}x^4 + 3x^2 - 7x + \frac{1}{2}\right) + \left(-\frac{1}{3}x^4 + 5x^3 - 3x^2 + 3x - \frac{1}{2}\right)$

$= \left(\frac{2}{3} - \frac{1}{3}\right)x^4 + 5x^3 + (3 - 3)x^2 + (-7 + 3)x + \left(\frac{1}{2} - \frac{1}{2}\right)$ Combining like terms

$= \frac{1}{3}x^4 + 5x^3 - 4x$ **TRY EXERCISE** ▸ 9

After some practice, polynomial addition is often performed mentally.

EXAMPLE 2

Add: $(2 - 3x + x^2) + (-5 + 7x - 3x^2 + x^3)$.

SOLUTION   We have

$(2 - 3x + x^2) + (-5 + 7x - 3x^2 + x^3)$

$- (2 - 5) + (-3 + 7)x + (1 - 3)x^2 + x^3$ You might do this step mentally.

$= -3 + 4x - 2x^2 + x^3.$ Then you would write only this.

TRY EXERCISE ▸ 17

In the polynomials of the last example, the terms are arranged according to degree, from least to greatest. Such an arrangement is called *ascending order*. As a rule, answers are written in ascending order when the polynomials in the original problem are given in ascending order. If the polynomials in the original problem are given in descending order, the answer is usually written in descending order.

We can also add polynomials by writing like terms in columns. Sometimes this makes like terms easier to see.

EXAMPLE 3

Add: $9x^5 - 2x^3 + 6x^2 + 3$ and $5x^4 - 7x^2 + 6$ and $3x^6 - 5x^5 + x^2 + 5$.

SOLUTION   We arrange the polynomials with like terms in columns.

$$
\begin{array}{llll}
9x^5 & -2x^3 + 6x^2 + & 3 & \\
& 5x^4 & -7x^2 + & 6 & \text{We leave spaces for missing terms.} \\
3x^6 - 5x^5 & & + 1x^2 + & 5 & \text{Writing } x^2 \text{ as } 1x^2 \\
\hline
3x^6 + 4x^5 + 5x^4 - 2x^3 & & + 14 & \text{Adding}
\end{array}
$$

The answer is $3x^6 + 4x^5 + 5x^4 - 2x^3 + 14$.   **TRY EXERCISE** ▸ 23

Opposites of Polynomials

In Section 1.8, we used the property of -1 to show that the opposite of a sum is the sum of the opposites. This idea can be extended.

> ### The Opposite of a Polynomial
> To find an equivalent polynomial for the *opposite*, or *additive inverse*, of a polynomial, change the sign of every term. This is the same as multiplying the polynomial by -1.

EXAMPLE 4

Write two equivalent expressions for the opposite of $4x^5 - 7x^3 - 8x + \frac{5}{6}$.

SOLUTION

i) $-\left(4x^5 - 7x^3 - 8x + \frac{5}{6}\right)$ This is one representation of the opposite of $4x^5 - 7x^3 - 8x + \frac{5}{6}$.

ii) $-4x^5 + 7x^3 + 8x - \frac{5}{6}$ Changing the sign of every term

Thus, $-\left(4x^5 - 7x^3 - 8x + \frac{5}{6}\right)$ and $-4x^5 + 7x^3 + 8x - \frac{5}{6}$ are equivalent. Both expressions represent the opposite of $4x^5 - 7x^3 - 8x + \frac{5}{6}$.   **TRY EXERCISE** 27

EXAMPLE 5   Simplify: $-\left(-7x^4 - \frac{5}{9}x^3 + 8x^2 - x + 67\right)$.

SOLUTION   We have

$$-\left(-7x^4 - \frac{5}{9}x^3 + 8x^2 - x + 67\right) = 7x^4 + \frac{5}{9}x^3 - 8x^2 + x - 67.$$

The same result can be found by multiplying by -1:

$$-\left(-7x^4 - \frac{5}{9}x^3 + 8x^2 - x + 67\right)$$
$$= -1(-7x^4) + (-1)\left(-\frac{5}{9}x^3\right) + (-1)(8x^2) + (-1)(-x) + (-1)67$$
$$= 7x^4 + \frac{5}{9}x^3 - 8x^2 + x - 67.$$ **TRY EXERCISE** 31

Subtraction of Polynomials

We can now subtract one polynomial from another by adding the opposite of the polynomial being subtracted.

EXAMPLE 6   Write an equivalent expression by subtracting.

a) $(9x^5 + x^3 - 2x^2 + 4) - (-2x^5 + x^4 - 4x^3 - 3x^2)$

b) $(7x^5 + x^3 - 9x) - (3x^5 - 4x^3 + 5)$

SOLUTION

a) $(9x^5 + x^3 - 2x^2 + 4) - (-2x^5 + x^4 - 4x^3 - 3x^2)$
$$= 9x^5 + x^3 - 2x^2 + 4 + 2x^5 - x^4 + 4x^3 + 3x^2 \quad \text{Adding the opposite}$$
$$= 11x^5 - x^4 + 5x^3 + x^2 + 4 \quad \text{Combining like terms}$$

b) $(7x^5 + x^3 - 9x) - (3x^5 - 4x^3 + 5)$
$$= 7x^5 + x^3 - 9x + (-3x^5) + 4x^3 - 5 \quad \text{Adding the opposite}$$
$$= 7x^5 + x^3 - 9x - 3x^5 + 4x^3 - 5 \quad \text{Try to go directly to this step.}$$
$$= 4x^5 + 5x^3 - 9x - 5 \quad \text{Combining like terms}$$

TRY EXERCISE 39

To subtract using columns, we first replace the coefficients in the polynomial being subtracted with their opposites. We then add as before.

EXAMPLE 7   Write in columns and subtract: $(5x^2 - 3x + 6) - (9x^2 - 5x - 3)$.

SOLUTION

i) $\begin{aligned} 5x^2 - 3x + 6 \\ -(9x^2 - 5x - 3) \end{aligned}$ Writing similar terms in columns

ii) $\begin{aligned} 5x^2 - 3x + 6 \\ -9x^2 + 5x + 3 \end{aligned}$ Changing signs and removing parentheses

iii) $\begin{aligned} 5x^2 - 3x + 6 \\ \underline{-9x^2 + 5x + 3} \\ -4x^2 + 2x + 9 \end{aligned}$ Adding   **TRY EXERCISE** 53

If you can do so without error, you can arrange the polynomials in columns, mentally find the opposite of each term being subtracted, and write the answer. Lining up like terms is important and may require leaving some blank space.

EXAMPLE 8   Write in columns and subtract: $(x^3 + x^2 - 12) - (-2x^3 + x^2 - 3x + 6)$.

SOLUTION   We have

$$
\begin{array}{l}
x^3 + x^2 - 12 \qquad \text{Leaving a blank space for the missing term} \\
\underline{-(-2x^3 + x^2 - 3x + 6)} \\
3x^3 + 3x - 18
\end{array}
$$

TRY EXERCISE 55

CAUTION!   Be sure to subtract every term of the polynomial being subtracted when using columns.

Problem Solving

EXAMPLE 9   Find a polynomial for the sum of the areas of rectangles A, B, C, and D.

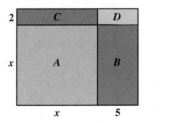

SOLUTION

1. **Familiarize.** Recall that the area of a rectangle is the product of its length and width.

2. **Translate.** We translate the problem to mathematical language. The sum of the areas is a sum of products. We find each product and then add:

$$
\underbrace{\text{Area of } A}_{x \cdot x} \ \text{plus} \ \underbrace{\text{area of } B}_{5x} \ \text{plus} \ \underbrace{\text{area of } C}_{2x} \ \text{plus} \ \underbrace{\text{area of } D}_{2 \cdot 5.}
$$

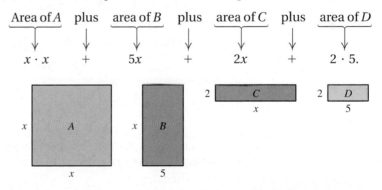

3. **Carry out.** We simplify $x \cdot x$ and $2 \cdot 5$ and combine like terms:

$$
x^2 + 5x + 2x + 10 = x^2 + 7x + 10.
$$

4. **Check.** A partial check is to replace x with a number, say 3. Then we evaluate $x^2 + 7x + 10$ and compare that result with an alternative calculation:

$$
3^2 + 7 \cdot 3 + 10 = 9 + 21 + 10 = 40.
$$

When we substitute 3 for x and calculate the total area by regarding the figure as one large rectangle, we should also get 40:

$$
\text{Total area} = (x + 5)(x + 2) = (3 + 5)(3 + 2) = 8 \cdot 5 = 40.
$$

Our check is only partial, since it is possible for an incorrect answer to equal 40 when evaluated for $x = 3$. This would be unlikely, especially if a second choice of x, say $x = 5$, also checks. We leave that check to the student.

5. State. A polynomial for the sum of the areas is $x^2 + 7x + 10$.

TRY EXERCISE ▸ 61

EXAMPLE   10   A 16-ft wide round fountain is built in a square city park that measures x ft by x ft. Find a polynomial for the remaining area of the park.

SOLUTION

1. Familiarize. We make a drawing of the square park and the circular fountain, and let x represent the length of a side of the park.

The area of a square is given by $A = s^2$, and the area of a circle is given by $A = \pi r^2$. Note that a circle with a diameter of 16 ft has a radius of 8 ft.

2. Translate. We reword the problem and translate as follows.

Rewording:	Area of park	minus	area of fountain	is	area left over.
	↓	↓	↓	↓	↓
Translating:	x ft · x ft	−	$\pi \cdot 8$ ft · 8 ft	=	Area left over

3. Carry out. We carry out the multiplication:

$$x^2 \text{ ft}^2 - 64\pi \text{ ft}^2 = \text{Area left over.}$$

4. Check. As a partial check, note that the units in the answer are square feet (ft^2), a measure of area, as expected.

5. State. The remaining area of the park is $(x^2 - 64\pi)$ ft^2.   TRY EXERCISE ▸ 69

TECHNOLOGY CONNECTION

To check polynomial addition or subtraction, we can let $y_1 =$ the expression before the addition or subtraction has been performed and $y_2 =$ the simplified sum or difference. If the addition or subtraction is correct, y_1 will equal y_2 and $y_2 - y_1$ will be 0. We enter $y_2 - y_1$ as y_3, using **VARS**. Below is a check of Example 6(b) in which

$$y_1 = (7x^5 + x^3 - 9x) - (3x^5 - 4x^3 + 5),$$
$$y_2 = 4x^5 + 5x^3 - 9x - 5,$$

and

$$y_3 = y_2 - y_1.$$

We graph only y_3. If indeed y_1 and y_2 are equivalent, then y_3 should equal 0. This means its graph should coincide with the x-axis. The TRACE or TABLE features can confirm that y_3 is always 0, or we can select y_3 to be drawn bold at the Y= window.

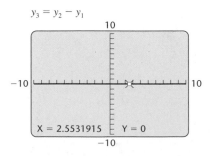

1. Use a graphing calculator to check Examples 1, 2, and 6.

| **4.3** | **EXERCISE SET** | For Extra Help MyMathLab | Math XL PRACTICE | WATCH | DOWNLOAD |

↩ *Concept Reinforcement*   *For Exercises 1–4, replace ▪ with the correct expression or operation sign.*

1. $(3x^2 + 2) + (6x^2 + 7) = (3 + 6)\ \blacksquare + (2 + 7)$

2. $(5t - 6) + (4t + 3) = (5 + 4)t + (\blacksquare + 3)$

3. $(9x^3 - x^2) - (3x^3 + x^2) = 9x^3 - x^2 - 3x^3\ \blacksquare\ x^2$

4. $(-2n^3 + 5) - (n^2 - 2) = -2n^3 + 5 - n^2\ \blacksquare\ 2$

Add.

5. $(3x + 2) + (x + 7)$

6. $(x + 1) + (12x + 10)$

7. $(2t + 7) + (-8t + 1)$

8. $(4t - 3) + (-11t + 2)$

9. $(x^2 + 6x + 3) + (-4x^2 - 5)$

10. $(x^2 - 5x + 4) + (8x - 9)$

11. $(7t^2 - 3t - 6) + (2t^2 + 4t + 9)$

12. $(8a^2 + 4a - 7) + (6a^2 - 3a - 1)$

13. $(4m^3 - 7m^2 + m - 5) + (4m^3 + 7m^2 - 4m - 2)$

14. $(5n^3 - n^2 + 4n + 11) + (2n^3 - 4n^2 + n - 11)$

15. $(3 + 6a + 7a^2 + a^3) + (4 + 7a - 8a^2 + 6a^3)$

16. $(7 + 4t - 5t^2 + 6t^3) + (2 + t + 6t^2 - 4t^3)$

17. $(3x^6 + 2x^4 - x^3 + 5x) + (-x^6 + 3x^3 - 4x^2 + 7x^4)$

18. $(4x^5 - 6x^3 - 9x + 1) + (3x^4 + 6x^3 + 9x^2 + x)$

19. $\left(\frac{3}{5}x^4 + \frac{1}{2}x^3 - \frac{2}{3}x + 3\right) + \left(\frac{2}{5}x^4 - \frac{1}{4}x^3 - \frac{3}{4}x^2 - \frac{1}{6}x\right)$

20. $\left(\frac{1}{3}x^9 + \frac{1}{5}x^5 - \frac{1}{2}x^2 + 7\right) + \left(-\frac{1}{5}x^9 + \frac{1}{4}x^4 - \frac{3}{5}x^5\right)$

21. $(5.3t^2 - 6.4t - 9.1) + (4.2t^3 - 1.8t^2 + 7.3)$

22. $(4.9a^3 + 3.2a^2 - 5.1a) + (2.1a^2 - 3.7a + 4.6)$

23. $\begin{array}{r} -4x^3 + 8x^2 + 3x - 2 \\ -4x^2 + 3x + 2 \\ \hline \end{array}$

24. $\begin{array}{r} -3x^4 + 6x^2 + 2x - 4 \\ -3x^2 + 2x + 4 \\ \hline \end{array}$

25. $\begin{array}{r} 0.05x^4 + 0.12x^3 - 0.5x^2 \\ -0.02x^3 + 0.02x^2 + 2x \\ 1.5x^4 \quad\quad + 0.01x^2 \quad\quad + 0.15 \\ 0.25x^3 \quad\quad\quad\quad + 0.85 \\ -0.25x^4 \quad\quad + 10x^2 \quad - 0.04 \\ \hline \end{array}$

26. $\begin{array}{r} 0.15x^4 + 0.10x^3 - 0.9x^2 \\ -0.01x^3 + 0.01x^2 + x \\ 1.25x^4 \quad\quad + 0.11x^2 \quad + 0.01 \\ 0.27x^3 \quad\quad\quad\quad + 0.99 \\ -0.35x^4 \quad\quad + 15x^2 \quad - 0.03 \\ \hline \end{array}$

Write two equivalent expressions for the opposite of each polynomial, as in Example 4.

27. $-3t^3 + 4t^2 - 7$

28. $-x^3 - 5x^2 + 2x$

29. $x^4 - 8x^3 + 6x$

30. $5a^3 + 2a - 17$

Simplify.

31. $-(9x - 10)$

32. $-(-5x + 8)$

33. $-(3a^4 - 5a^2 + 1.2)$

34. $-(-6a^3 + 0.2a^2 - 7)$

35. $-\left(-4x^4 + 6x^2 + \frac{3}{4}x - 8\right)$

36. $-\left(3x^5 - 2x^3 - \frac{3}{5}x^2 + 16\right)$

Subtract.

37. $(3x + 1) - (5x + 8)$

38. $(7x + 3) - (3x + 2)$

39. $(-9t + 12) - (t^2 + 3t - 1)$

40. $(a^2 - 3a - 2) - (2a^2 - 6a - 2)$

41. $(4a^2 + a - 7) - (3 - 8a^3 - 4a^2)$

42. $(-4x^2 + 2x) - (-5x^2 + 2x^3 + 3)$

43. $(1.2x^3 + 4.5x^2 - 3.8x) - (-3.4x^3 - 4.7x^2 + 23)$

44. $(0.5x^4 - 0.6x^2 + 0.7) - (2.3x^4 + 1.8x - 3.9)$

Aha! **45.** $(7x^3 - 2x^2 + 6) - (6 - 2x^2 + 7x^3)$

46. $(8x^5 + 3x^4 + x - 1) - (8x^5 + 3x^4 - 1)$

47. $(3 + 5a + 3a^2 - a^3) - (2 + 4a - 9a^2 + 2a^3)$

48. $(7 + t - 5t^2 + 2t^3) - (1 + 2t - 4t^2 + 5t^3)$

49. $\left(\frac{5}{8}x^3 - \frac{1}{4}x - \frac{1}{3}\right) - \left(-\frac{1}{2}x^3 + \frac{1}{4}x - \frac{1}{3}\right)$

50. $\left(\frac{1}{5}x^3 + 2x^2 - \frac{3}{10}\right) - \left(-\frac{2}{5}x^3 + 2x^2 + \frac{7}{1000}\right)$

51. $(0.07t^3 - 0.03t^2 + 0.01t) - (0.02t^3 + 0.04t^2 - 1)$

52. $(0.9a^3 + 0.2a - 5) - (0.7a^4 - 0.3a - 0.1)$

53. $\quad x^3 + 3x^2 + 1$
$\quad \underline{-(x^3 + x^2 - 5)}$

54. $\quad x^2 + 5x + 6$
$\quad \underline{-(x^2 + 2x + 1)}$

55. $\quad 4x^4 - 2x^3$
$\quad \underline{-(7x^4 + 6x^3 + 7x^2)}$

56. $\quad 5x^4 + 6x^3 - 9x^2$
$\quad \underline{-(-6x^4 + x^2)}$

57. Solve.

   a) Find a polynomial for the sum of the areas of the rectangles shown in the figure.

   b) Find the sum of the areas when $x = 5$ and $x = 7$.

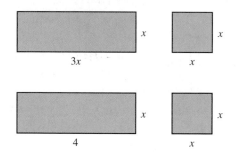

58. Solve. Leave the answers in terms of π.

   a) Find a polynomial for the sum of the areas of the circles shown in the figure.

   b) Find the sum of the areas when $r = 5$ and $r = 11.3$.

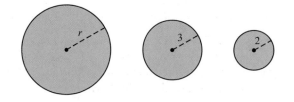

Find a polynomial for the perimeter of each figure in Exercises 59 and 60.

59.

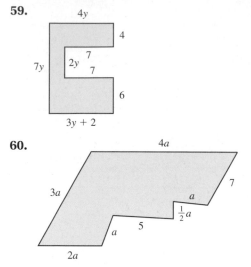

60.

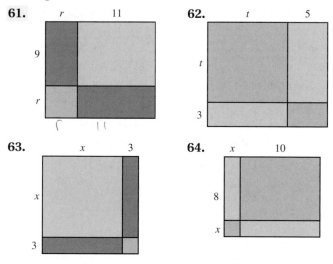

Find two algebraic expressions for the area of each figure. First, regard the figure as one large rectangle, and then regard the figure as a sum of four smaller rectangles.

61.

62.

63.

64.

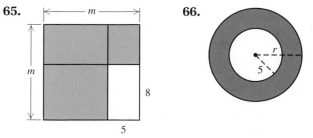

Find a polynomial for the shaded area of each figure.

65.

66.

67.   **68.**

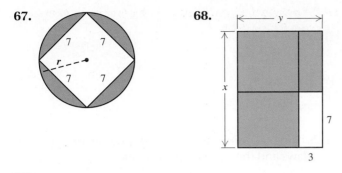

69. A 2-ft by 6-ft bath enclosure is installed in a new bathroom measuring x ft by x ft. Find a polynomial for the remaining floor area.

70. A 5-ft by 7-ft Jacuzzi™ is installed on an outdoor deck measuring y ft by y ft. Find a polynomial for the remaining area of the deck.

71. A 12-ft wide round patio is laid in a garden measuring z ft by z ft. Find a polynomial for the remaining area of the garden.

72. A 10-ft wide round water trampoline is floating in a pool measuring x ft by x ft. Find a polynomial for the remaining surface area of the pool.

73. A 12-m by 12-m mat includes a circle of diameter d meters for wrestling. Find a polynomial for the area of the mat outside the wrestling circle.

74. A 2-m by 3-m rug is spread inside a tepee that has a diameter of x meters. Find a polynomial for the area of the tepee's floor that is not covered.

75. Explain why parentheses are used in the statement of the solution of Example 10: $(x^2 - 64\pi)$ ft².

76. Is the sum of two trinomials always a trinomial? Why or why not?

Skill Review

To prepare for Section 4.4, review multiplying using the distributive law and multiplying with exponential notation (Sections 1.8 and 4.1).

Simplify.

77. $2(x^2 - x + 3)$ [1.8] **78.** $-5(3x^2 - 2x - 7)$ [1.8]

79. $x^2 \cdot x^6$ [4.1] **80.** $y^6 \cdot y$ [4.1]

81. $2n \cdot n^2$ [4.1] **82.** $-6n^4 \cdot n^8$ [4.1]

Synthesis

83. What can be concluded about two polynomials whose sum is zero?

84. Which, if any, of the commutative, associative, and distributive laws are needed for adding polynomials? Why?

Simplify.

85. $(6t^2 - 7t) + (3t^2 - 4t + 5) - (9t - 6)$

86. $(3x^2 - 4x + 6) - (-2x^2 + 4) + (-5x - 3)$

87. $4(x^2 - x + 3) - 2(2x^2 + x - 1)$

88. $3(2y^2 - y - 1) - (6y^2 - 3y - 3)$

89. $(345.099x^3 - 6.178x) - (94.508x^3 - 8.99x)$

Find a polynomial for the surface area of the right rectangular solid.

90.   **91.**

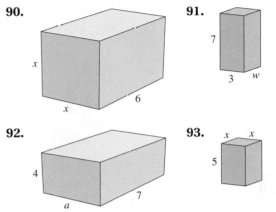

92.   **93.**

94. Find a polynomial for the total length of all edges in the figure appearing in Exercise 93.

95. Find a polynomial for the total length of all edges in the figure appearing in Exercise 90.

96. *Total profit.* Hadley Electronics is marketing a new digital camera. Total revenue is the total amount of money taken in. The firm determines that when it sells x cameras, its total revenue is given by
$$R = 175x - 0.4x^2.$$
Total cost is the total cost of producing x cameras. Hadley Electronics determines that the total cost of producing x cameras is given by
$$C = 5000 + 0.6x^2.$$
The total profit P is
$$(\text{Total Revenue}) - (\text{Total Cost}) = R - C.$$
a) Find a polynomial for total profit.
b) What is the total profit on the production and sale of 75 cameras?
c) What is the total profit on the production and sale of 120 cameras?

97. Does replacing each occurrence of the variable x in $4x^7 - 6x^3 + 2x$ with its opposite result in the opposite of the polynomial? Why or why not?

4.4   ## Multiplication of Polynomials

Multiplying Monomials   •   Multiplying a Monomial and a Polynomial   •
Multiplying Any Two Polynomials   •   Checking by Evaluating

We now multiply polynomials using techniques based largely on the distributive, associative, and commutative laws and the rules for exponents.

Multiplying Monomials

Consider $(3x)(4x)$. We multiply as follows:

$$
\begin{aligned}
(3x)(4x) &= 3 \cdot x \cdot 4 \cdot x && \text{Using an associative law} \\
&= 3 \cdot 4 \cdot x \cdot x && \text{Using a commutative law} \\
&= (3 \cdot 4) \cdot x \cdot x && \text{Using an associative law} \\
&= 12x^2.
\end{aligned}
$$

To Multiply Monomials

To find an equivalent expression for the product of two monomials, multiply the coefficients and then multiply the variables using the product rule for exponents.

EXAMPLE   **1**   Multiply to form an equivalent expression.

a) $(5x)(6x)$

b) $(-a)(3a)$

c) $(7x^5)(-4x^3)$

SOLUTION

a) $(5x)(6x) = (5 \cdot 6)(x \cdot x)$ Multiplying the coefficients; multiplying the variables

$\qquad\qquad = 30x^2$ Simplifying

b) $(-a)(3a) = (-1a)(3a)$ Writing $-a$ as $-1a$ can ease calculations.

$\qquad\qquad = (-1 \cdot 3)(a \cdot a)$ Using an associative law and a commutative law

$\qquad\qquad = -3a^2$

c) $(7x^5)(-4x^3) = 7(-4)(x^5 \cdot x^3)$

$\qquad\qquad\quad = -28x^{5+3}$ ⎱

$\qquad\qquad\quad = -28x^8$ ⎰ Using the product rule for exponents

TRY EXERCISE ▶ 13

After some practice, you can try writing only the answer.

Multiplying a Monomial and a Polynomial

To find an equivalent expression for the product of a monomial, such as $5x$, and a polynomial, such as $2x^2 - 3x + 4$, we use the distributive law.

EXAMPLE 2   Multiply: **(a)** $x(x + 3)$; **(b)** $5x(2x^2 - 3x + 4)$.

SOLUTION

a) $x(x + 3) = x \cdot x + x \cdot 3$ Using the distributive law

$\qquad\qquad = x^2 + 3x$

b) $5x(2x^2 - 3x + 4) = (5x)(2x^2) - (5x)(3x) + (5x)(4)$ Using the distributive law

$\qquad\qquad\qquad\qquad = 10x^3 - 15x^2 + 20x$ Performing the three multiplications

TRY EXERCISE ▸ 29

The product in Example 2(a) can be visualized as the area of a rectangle with width x and length $x + 3$.

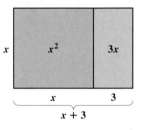

Note that the total area can be expressed as $x(x + 3)$ or, by adding the two smaller areas, $x^2 + 3x$.

> ### The Product of a Monomial and a Polynomial
> To multiply a monomial and a polynomial, multiply each term of the polynomial by the monomial.

Try to do this mentally, when possible. Remember that we multiply coefficients and, when the bases match, add exponents.

EXAMPLE 3   Multiply: $2x^2(x^3 - 7x^2 + 10x - 4)$.

SOLUTION

$$\textit{Think:} \quad \underbrace{2x^2 \cdot x^3} - \underbrace{2x^2 \cdot 7x^2} + \underbrace{2x^2 \cdot 10x} - \underbrace{2x^2 \cdot 4}$$

$$2x^2(x^3 - 7x^2 + 10x - 4) = 2x^5 \quad - \quad 14x^4 \quad + \quad 20x^3 \quad - \quad 8x^2$$

TRY EXERCISE ▸ 31

Multiplying Any Two Polynomials

Before considering the product of *any* two polynomials, let's look at products when both polynomials are binomials.

To find an equivalent expression for the product of two binomials, we again begin by using the distributive law. This time, however, it is a *binomial* rather than a monomial that is being distributed.

EXAMPLE **4**   Multiply each pair of binomials.

a) $x + 5$ and $x + 4$ **b)** $4x - 3$ and $x - 2$

SOLUTION

a) $(x + 5)(x + 4) = (x + 5)x + (x + 5)4$ Using the distributive law

$= x(x + 5) + 4(x + 5)$ Using the commutative law for multiplication

$= x \cdot x + x \cdot 5 + 4 \cdot x + 4 \cdot 5$ Using the distributive law (twice)

$= x^2 + 5x + 4x + 20$ Multiplying the monomials

$= x^2 + 9x + 20$ Combining like terms

b) $(4x - 3)(x - 2) = (4x - 3)x - (4x - 3)2$ Using the distributive law

$= x(4x - 3) - 2(4x - 3)$ Using the commutative law for multiplication. This step is often omitted.

$= x \cdot 4x - x \cdot 3 - 2 \cdot 4x - 2(-3)$ Using the distributive law (twice)

$= 4x^2 - 3x - 8x + 6$ Multiplying the monomials

$= 4x^2 - 11x + 6$ Combining like terms

TRY EXERCISE 37

To visualize the product in Example 4(a), consider a rectangle of length $x + 5$ and width $x + 4$, as shown here.

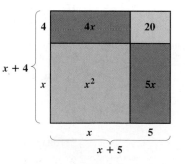

The total area can be expressed as $(x + 5)(x + 4)$ or, by adding the four smaller areas, $x^2 + 5x + 4x + 20$.

Let's consider the product of a binomial and a trinomial. Again we make repeated use of the distributive law.

EXAMPLE 5   Multiply: $(x^2 + 2x - 3)(x + 4)$.

SOLUTION

$$(x^2 + 2x - 3)\ (x + 4)$$

$$= (x^2 + 2x - 3)\ x + (x^2 + 2x - 3)\ 4 \qquad \text{Using the distributive law}$$

$$= x(x^2 + 2x - 3) + 4(x^2 + 2x - 3) \qquad \text{Using the commutative law}$$

$$= x \cdot x^2 + x \cdot 2x - x \cdot 3 + 4 \cdot x^2 + 4 \cdot 2x - 4 \cdot 3 \qquad \text{Using the distributive law (twice)}$$

$$= x^3 + 2x^2 - 3x + 4x^2 + 8x - 12 \qquad \text{Multiplying the monomials}$$

$$= x^3 + 6x^2 + 5x - 12 \qquad \text{Combining like terms}$$

TRY EXERCISE 57

Perhaps you have discovered the following in the preceding examples.

The Product of Two Polynomials

To multiply two polynomials P and Q, select one of the polynomials, say P. Then multiply each term of P by every term of Q and combine like terms.

To use columns for long multiplication, multiply each term in the top row by every term in the bottom row. We write like terms in columns, and then add the results. Such multiplication is like multiplying with whole numbers.

$$\begin{array}{r} 321 \\ \times\ 12 \\ \hline 642 \\ 321 \\ \hline 3852 \end{array} \qquad \begin{array}{r} 300 + 20 + 1 \\ \times \qquad\quad 10 + 2 \\ \hline 600 + 40 + 2 \\ 3000 + 200 + 10 \\ \hline 3000 + 800 + 50 + 2 \end{array} \qquad \begin{array}{l} \\ \\ \text{Multiplying the top row by 2} \\ \text{Multiplying the top row by 10} \\ \text{Adding} \end{array}$$

EXAMPLE 6   Multiply: $(5x^4 - 2x^2 + 3x)(x^2 + 2x)$.

SOLUTION

$$\left.\begin{array}{r} 5x^4 \qquad\quad - 2x^2 + 3x \\ x^2 + 2x \end{array}\right\} \qquad \text{Note that each polynomial is written in descending order, and space is left for missing terms.}$$

$$\begin{array}{r} 10x^5 \qquad - 4x^3 + 6x^2 \\ 5x^6 \qquad\quad - 2x^4 + 3x^3 \qquad\qquad \\ \hline 5x^6 + 10x^5 - 2x^4 - x^3 + 6x^2 \end{array}$$

Multiplying the top row by $2x$

Multiplying the top row by x^2

Combining like terms

Line up like terms in columns.

TRY EXERCISE 61

With practice, you will be able to skip some steps. Sometimes we multiply horizontally, while still aligning like terms as we write the product.

EXAMPLE 7   Multiply: $(2x^3 + 3x^2 - 4x + 6)(3x + 5)$.

SOLUTION

$$
(2x^3 + 3x^2 - 4x + 6)(3x + 5) = \overbrace{6x^4 + 9x^3 - 12x^2 + 18x}^{\text{Multiplying by } 3x}
$$
$$
+ \underbrace{10x^3 + 15x^2 - 20x + 30}_{\text{Multiplying by } 5}
$$
$$
= 6x^4 + 19x^3 + 3x^2 - 2x + 30
$$

TRY EXERCISE ▶ 65

Checking by Evaluating

How can we be certain that our multiplication (or addition or subtraction) of polynomials is correct? One check is to simply review our calculations. A different type of check, used in Example 9 of Section 4.3, makes use of the fact that equivalent expressions have the same value when evaluated for the same replacement. Thus a quick, partial, check of Example 7 can be made by selecting a convenient replacement for x (say, 1) and comparing the values of the expressions $(2x^3 + 3x^2 - 4x + 6)(3x + 5)$ and $6x^4 + 19x^3 + 3x^2 - 2x + 30$:

$$
(2x^3 + 3x^2 - 4x + 6)(3x + 5) = (2 \cdot 1^3 + 3 \cdot 1^2 - 4 \cdot 1 + 6)(3 \cdot 1 + 5)
$$
$$
= (2 + 3 - 4 + 6)(3 + 5)
$$
$$
= 7 \cdot 8 = 56;
$$

$$
6x^4 + 19x^3 + 3x^2 - 2x + 30 = 6 \cdot 1^4 + 19 \cdot 1^3 + 3 \cdot 1^2 - 2 \cdot 1 + 30
$$
$$
= 6 + 19 + 3 - 2 + 30
$$
$$
= 28 - 2 + 30 = 56.
$$

Since the value of both expressions is 56, the multiplication in Example 7 is very likely correct.

It is possible, by chance, for two expressions that are not equivalent to share the same value when evaluated. For this reason, checking by evaluating is only a partial check. Consult your instructor for the checking approach that he or she prefers.

TECHNOLOGY CONNECTION

Tables can also be used to check polynomial multiplication. To illustrate, we can check Example 7 by entering $y_1 = (2x^3 + 3x^2 - 4x + 6)(3x + 5)$ and $y_2 = 6x^4 + 19x^3 + 3x^2 - 2x + 30$.

When (TABLE) is then pressed, we are shown two columns of values—one for y_1 and one for y_2. If our multiplication was correct, the columns of values will match.

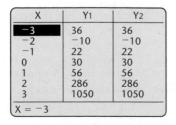

X	Y₁	Y₂
−3	36	36
−2	−10	−10
−1	22	22
0	30	30
1	56	56
2	286	286
3	1050	1050

X = −3

1. Form a table and scroll up and down to check Example 6.
2. Check Example 7 using the method discussed in Section 4.3: Let
$$
y_1 = (2x^3 + 3x^2 - 4x + 6)(3x + 5),
$$
$$
y_2 = 6x^4 + 19x^3 + 3x^2 - 2x + 30,
$$

and

$$
y_3 = y_2 - y_1.
$$

Then check that y_3 is always 0.

4.4 EXERCISE SET

Concept Reinforcement *In each of Exercises 1–6, match the expression with the correct result from the column on the right. Choices may be used more than once.*

1. ___ $3x^2 \cdot 2x^4$
2. ___ $3x^8 + 5x^8$
3. ___ $4x^3 \cdot 2x^5$
4. ___ $3x^5 \cdot 2x^3$
5. ___ $4x^6 + 2x^6$
6. ___ $4x^4 \cdot 2x^2$

a) $6x^8$
b) $8x^6$
c) $6x^6$
d) $8x^8$

Multiply.

7. $(3x^5)7$
8. $2x^3 \cdot 11$
9. $(-x^3)(x^4)$
10. $(-x^2)(-x)$
11. $(-x^6)(-x^2)$
12. $(-x^5)(x^3)$
13. $4t^2(9t^2)$
14. $(6a^8)(3a^2)$
15. $(0.3x^3)(-0.4x^6)$
16. $(-0.1x^6)(0.2x^4)$
17. $\left(-\frac{1}{4}x^4\right)\left(\frac{1}{5}x^8\right)$
18. $\left(-\frac{1}{5}x^3\right)\left(-\frac{1}{3}x\right)$
19. $(-5n^3)(-1)$
20. $19t^2 \cdot 0$
21. $11x^5(-4x^5)$
22. $12x^3(-5x^3)$
23. $(-4y^5)(6y^2)(-3y^3)$
24. $7x^2(-2x^3)(2x^6)$
25. $5x(4x + 1)$
26. $3x(2x - 7)$
27. $(a - 9)3a$
28. $(a - 7)4a$
29. $x^2(x^3 + 1)$
30. $-2x^3(x^2 - 1)$
31. $-3n(2n^2 - 8n + 1)$
32. $4n(3n^3 - 4n^2 \quad 5n + 10)$
33. $-5t^2(3t + 6)$
34. $7t^2(2t + 1)$
35. $\frac{2}{3}a^4\left(6a^5 - 12a^3 - \frac{5}{8}\right)$
36. $\frac{3}{4}t^5\left(8t^6 - 12t^4 + \frac{12}{7}\right)$
37. $(x + 3)(x + 4)$
38. $(x + 7)(x + 3)$
39. $(t + 7)(t - 3)$
40. $(t - 4)(t + 3)$
41. $(a - 0.6)(a - 0.7)$
42. $(a - 0.4)(a - 0.8)$
43. $(x + 3)(x - 3)$
44. $(x + 6)(x - 6)$
45. $(4 - x)(7 - 2x)$
46. $(5 + x)(5 + 2x)$
47. $\left(t + \frac{3}{2}\right)\left(t + \frac{4}{3}\right)$
48. $\left(a - \frac{2}{5}\right)\left(a + \frac{5}{2}\right)$
49. $\left(\frac{1}{4}a + 2\right)\left(\frac{3}{4}a - 1\right)$
50. $\left(\frac{2}{5}t - 1\right)\left(\frac{3}{5}t + 1\right)$

Draw and label rectangles similar to those following Examples 2 and 4 to illustrate each product.

51. $x(x + 5)$
52. $x(x + 2)$
53. $(x + 1)(x + 2)$
54. $(x + 3)(x + 1)$
55. $(x + 5)(x + 3)$
56. $(x + 4)(x + 6)$

Multiply and check.

57. $(x^2 - x + 3)(x + 1)$
58. $(x^2 + x - 7)(x + 2)$
59. $(2a + 5)(a^2 - 3a + 2)$
60. $(3t - 4)(t^2 - 5t + 1)$
61. $(y^2 - 7)(3y^4 + y + 2)$
62. $(a^2 + 4)(5a^3 - 3a - 1)$

Aha! 63. $(3x + 2)(7x + 4x + 1)$
64. $(4x - 5x - 3)(1 + 2x^2)$
65. $(x^2 + 5x - 1)(x^2 - x + 3)$
66. $(x^2 - 3x + 2)(x^2 + x + 1)$
67. $\left(5t^2 - t + \frac{1}{2}\right)(2t^2 + t - 4)$
68. $(2t^2 - 5t - 4)\left(3t^2 - t + \frac{1}{2}\right)$
69. $(x + 1)(x^3 + 7x^2 + 5x + 4)$
70. $(x + 2)(x^3 + 5x^2 + 9x + 3)$

71. Is it possible to understand polynomial multiplication without understanding the distributive law? Why or why not?

72. The polynomials

$$(a + b + c + d) \quad \text{and} \quad (r + s + m + p)$$

are multiplied. Without performing the multiplication, determine how many terms the product will contain. Provide a justification for your answer.

Skill Review

Review simplifying expressions using the rules for order of operations (Section 1.8).

Simplify. [1.8]

73. $(9 - 3)(9 + 3) + 3^2 - 9^2$
74. $(7 + 2)(7 - 2) + 2^2 - 7^2$
75. $5 + \dfrac{7 + 4 + 2 \cdot 5}{7}$

76. $11 - \dfrac{2 + 6 \cdot 3 + 4}{6}$

77. $(4 + 3 \cdot 5 + 5) \div 3 \cdot 4$

78. $(2 + 2 \cdot 7 + 4) \div 2 \cdot 5$

Synthesis

79. Under what conditions will the product of two binomials be a trinomial?

80. How can the following figure be used to show that $(x + 3)^2 \neq x^2 + 9$?

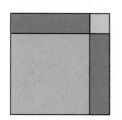

Find a polynomial for the shaded area of each figure.

81.

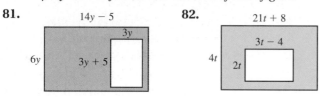

14y − 5

3y

6y   3y + 5

82.

21t + 8

3t − 4

4t   2t

For each figure, determine what the missing number must be in order for the figure to have the given area.

83. Area is $x^2 + 8x + 15$

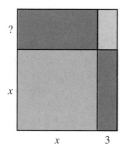

?

x

x   3

84. Area is $x^2 + 7x + 10$

2

x

x   ?

85. A box with a square bottom and no top is to be made from a 12-in.–square piece of cardboard. Squares with side x are cut out of the corners and the sides are folded up. Find the polynomials for the volume and the outside surface area of the box.

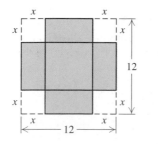

x   x

x   x

12

x   x

x   ── 12 ──

86. Find a polynomial for the volume of the solid shown below.

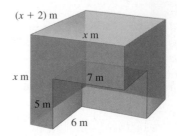

$(x + 2)$ m

x m

x m   7 m

5 m

6 m

87. An open wooden box is a cube with side x cm. The box, including its bottom, is made of wood that is 1 cm thick. Find a polynomial for the interior volume of the cube.

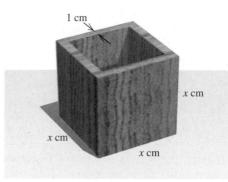

1 cm

x cm

x cm   x cm

88. A side of a cube is $(x + 2)$ cm long. Find a polynomial for the volume of the cube.

89. A rectangular garden is twice as long as it is wide and is surrounded by a sidewalk that is 4 ft wide (see the figure below). The area of the sidewalk is 256 ft². Find the dimensions of the garden.

← 4 ft →

Compute and simplify.

90. $(x + 3)(x + 6) + (x + 3)(x + 6)$

Aha! **91.** $(x - 2)(x - 7) - (x - 7)(x - 2)$

92. $(x + 5)^2 - (x - 3)^2$

93. $(x + 2)(x + 4)(x - 5)$

94. $(x - 3)^3$

Aha! **95.** Extend the pattern and simplify
$$(x - a)(x - b)(x - c)(x - d) \cdots (x - z).$$

96. Use a graphing calculator to check your answers to Exercises 25, 45, and 57. Use graphs, tables, or both, as directed by your instructor.

COLLABORATIVE CORNER

Slick Tricks with Algebra

Focus: Polynomial multiplication

Time: 15 minutes

Group size: 2

Consider the following dialogue.

Jinny: Cal, let me do a number trick with you. Think of a number between 1 and 7. I'll have you perform some manipulations to this number, you'll tell me the result, and I'll tell me your number.

Cal: OK. I've thought of a number.

Jinny: Good. Write it down so I can't see it. Now double it, and then subtract x from the result.

Cal: Hey, this is algebra!

Jinny: I know. Now square your binomial. After you're through squaring, subtract x^2.

Cal: How did you know I had an x^2? I *thought* this was rigged!

Jinny: It is. Now divide each of the remaining terms by 4 and tell me either your constant term or your x-term. I'll tell you the other term and the number you chose.

Cal: OK. The constant term is 16.

Jinny: Then the other term is $-4x$ and the number you chose was 4.

Cal: You're right! How did you do it?

ACTIVITY

1. Each group member should follow Jinny's instructions. Then determine how Jinny determined Cal's number and the other term.

2. Suppose that, at the end, Cal told Jinny the x-term. How would Jinny have determined Cal's number and the other term?

3. Would Jinny's "trick" work with *any* real number? Why do you think she specified numbers between 1 and 7?

4.5   Special Products

Products of Two Binomials   •   Multiplying Sums and Differences of Two Terms   •
Squaring Binomials   •   Multiplications of Various Types

We can observe patterns in the products of two binomials. These patterns allow us to compute such products quickly.

Products of Two Binomials

In Section 4.4, we found the product $(x + 5)(x + 4)$ by using the distributive law a total of three times (see p. 248). Note that each term in $x + 5$ is multiplied by each term in $x + 4$. To shorten our work, we can go right to this step:

$$\begin{aligned} (x + 5)(x + 4) &= x \cdot x + x \cdot 4 + 5 \cdot x + 5 \cdot 4 \\ &= x^2 + 4x + 5x + 20 \\ &= x^2 + 9x + 20. \end{aligned}$$

Note that $x \cdot x$ is found by multiplying the *First* terms of each binomial, $x \cdot 4$ is found by multiplying the *Outer* terms of the two binomials, $5 \cdot x$ is the product of the *Inner* terms of the two binomials, and $5 \cdot 4$ is the product of the *Last* terms of each binomial:

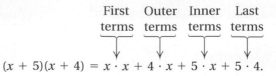

$$(x + 5)(x + 4) = x \cdot x + 4 \cdot x + 5 \cdot x + 5 \cdot 4.$$

To remember this shortcut for multiplying, we use the initials **FOIL**.

The FOIL Method

To multiply two binomials, $A + B$ and $C + D$, multiply the First terms AC, the Outer terms AD, the Inner terms BC, and then the Last terms BD. Then combine like terms, if possible.

$$(A + B)(C + D) = AC + AD + BC + BD$$

1. Multiply First terms: AC.
2. Multiply Outer terms: AD.
3. Multiply Inner terms: BC.
4. Multiply Last terms: BD.

$$\downarrow$$
FOIL

$$\overset{F \qquad L}{(A + B)(C + D)}$$
$$I$$
$$O$$

Because addition is commutative, the individual multiplications can be performed in any order. Both FLOI and FIOL yield the same result as FOIL, but FOIL is most easily remembered and most widely used.

EXAMPLE **1** Form an equivalent expression by multiplying: $(x + 8)(x^2 + 5)$.

SOLUTION

$$\overset{F \qquad L}{(x + 8)(x^2 + 5)} = \overset{F}{x^3} + \overset{O}{5x} + \overset{I}{8x^2} + \overset{L}{40} \qquad \text{There are no like terms.}$$
$$= x^3 + 8x^2 + 5x + 40 \qquad \text{Writing in descending order}$$

TRY EXERCISE 5

After multiplying, remember to combine any like terms.

EXAMPLE **2** Multiply to form an equivalent expression.

a) $(x + 7)(x + 4)$ **b)** $(y + 3)(y - 2)$
c) $(4t^3 + 5t)(3t^2 - 2)$ **d)** $(3 - 4x)(7 - 5x^3)$

SOLUTION
a) $(x + 7)(x + 4) = x^2 + 4x + 7x + 28$ Using FOIL
$$= x^2 + 11x + 28 \qquad \text{Combining like terms}$$

b) $(y + 3)(y - 2) = y^2 - 2y + 3y - 6$
$$= y^2 + y - 6$$

c) $(4t^3 + 5t)(3t^2 - 2) = 12t^5 - 8t^3 + 15t^3 - 10t$ Remember to add exponents when multiplying terms with the same base.

$$= 12t^5 + 7t^3 - 10t$$

d) $(3 - 4x)(7 - 5x^3) = 21 - 15x^3 - 28x + 20x^4$

$$= 21 - 28x - 15x^3 + 20x^4$$ In general, if the original binomials are written in *ascending* order, the answer is also written that way.

TRY EXERCISE ▶ 9

Multiplying Sums and Differences of Two Terms

Consider the product of the sum and the difference of the same two terms, such as

$$(x + 5)(x - 5).$$

Since this is the product of two binomials, we can use FOIL. In doing so, we find that the "outer" and "inner" products are opposites:

a) $(x + 5)(x - 5) = x^2 - 5x + 5x - 25$
$$= x^2 - 25;$$

b) $(3a - 2)(3a + 2) = 9a^2 + 6a - 6a - 4$ The "outer" and "inner" terms
$$= 9a^2 - 4;$$ "drop out." Their sum is zero.

c) $\left(x^3 + \frac{2}{7}\right)\left(x^3 - \frac{2}{7}\right) = x^6 - \frac{2}{7}x^3 + \frac{2}{7}x^3 - \frac{4}{49}$
$$= x^6 - \frac{4}{49}.$$

Because opposites always add to zero, for products like $(x + 5)(x - 5)$ we can use a shortcut that is faster than FOIL.

The Product of a Sum and a Difference

The product of the sum and the difference of the same two terms is the square of the first term minus the square of the second term:

$$(A + B)(A - B) = \underbrace{A^2 - B^2}.$$

This is called a *difference of squares*.

EXAMPLE 3   Multiply.

a) $(x + 4)(x - 4)$

b) $(5 + 2w)(5 - 2w)$

c) $(3a^4 - 5)(3a^4 + 5)$

SOLUTION

$$(A + B)(A - B) = A^2 - B^2$$
$$\downarrow\ \ \downarrow\ \ \downarrow\ \ \ \downarrow\ \ \ \downarrow\ \ \ \ \downarrow$$
a) $(x + 4)(x - 4) = x^2 - 4^2$ Saying the words can help: "The square of the first term, x^2, minus the square of the second, 4^2"

$$= x^2 - 16$$ Simplifying

b) $(5 + 2w)(5 - 2w) = 5^2 - (2w)^2$

$\qquad\qquad\qquad\quad = 25 - 4w^2$ Squaring both 5 and $2w$

c) $(3a^4 - 5)(3a^4 + 5) = (3a^4)^2 - 5^2$

$\qquad\qquad\qquad\qquad\quad = 9a^8 - 25$ Remember to multiply exponents when raising a power to a power.

TRY EXERCISE 41

Squaring Binomials

Consider the square of a binomial, such as $(x + 3)^2$. This can be expressed as $(x + 3)(x + 3)$. Since this is the product of two binomials, we can use FOIL. But again, this product occurs so often that a faster method has been developed. Look for a pattern in the following:

a) $(x + 3)^2 = (x + 3)(x + 3)$

$\qquad\qquad\quad = x^2 + 3x + 3x + 9$

$\qquad\qquad\quad = x^2 + 6x + 9;$

b) $(5 - 3p)^2 = (5 - 3p)(5 - 3p)$

$\qquad\qquad\quad\; = 25 - 15p - 15p + 9p^2$

$\qquad\qquad\quad\; = 25 - 30p + 9p^2;$

c) $(a^3 - 7)^2 = (a^3 - 7)(a^3 - 7)$

$\qquad\qquad\quad\; = a^6 - 7a^3 - 7a^3 + 49$

$\qquad\qquad\quad\; = a^6 - 14a^3 + 49.$

Perhaps you noticed that in each product the "outer" product and the "inner" product are identical. The other two terms, the "first" product and the "last" product, are squares.

The Square of a Binomial

The square of a binomial is the square of the first term, plus twice the product of the two terms, plus the square of the last term:

$$(A + B)^2 = A^2 + 2AB + B^2;$$
$$(A - B)^2 = A^2 - 2AB + B^2.$$

These are called *perfect-square trinomials.**

EXAMPLE 4   Write an equivalent expression for each square of a binomial.

a) $(x + 7)^2$ **b)** $(t - 5)^2$

c) $(3a + 0.4)^2$ **d)** $(5x - 3x^4)^2$

SOLUTION

$$(A + B)^2 = A^2 + 2 \cdot A \cdot B + B^2$$

a) $(x + 7)^2 = x^2 + 2 \cdot x \cdot 7 + 7^2$ Saying the words can help: "The square of the first term, x^2, plus twice the product of the terms, $2 \cdot 7x$, plus the square of the second term, 7^2"

$\qquad\qquad\quad = x^2 + 14x + 49$

---

*Another name for these is *trinomial squares.*

b) $(t - 5)^2 = t^2 - 2 \cdot t \cdot 5 + 5^2$

$\qquad\qquad = t^2 - 10t + 25$

c) $(3a + 0.4)^2 = (3a)^2 + 2 \cdot 3a \cdot 0.4 + 0.4^2$

$\qquad\qquad\quad = 9a^2 + 2.4a + 0.16$

d) $(5x - 3x^4)^2 = (5x)^2 - 2 \cdot 5x \cdot 3x^4 + (3x^4)^2$

$\qquad\qquad\quad = 25x^2 - 30x^5 + 9x^8$ Using the rules for exponents

TRY EXERCISE ▶ 57

CAUTION! Although the square of a product is the product of the squares, the square of a sum is *not* the sum of the squares. That is, $(AB)^2 = A^2B^2$, but

The term $2AB$ is missing.

$(A + B)^2 \neq A^2 + B^2.$

To confirm this inequality, note that

$(7 + 5)^2 = 12^2 = 144,$

whereas

$7^2 + 5^2 = 49 + 25 = 74,$ and $74 \neq 144.$

Geometrically, $(A + B)^2$ can be viewed as the area of a square with sides of length $A + B$:

$(A + B)(A + B) = (A + B)^2.$

This is equal to the sum of the areas of the four smaller regions:

$A^2 + AB + AB + B^2 = A^2 + 2AB + B^2.$

Thus,

$(A + B)^2 = A^2 + 2AB + B^2.$

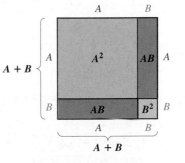

Note that the areas A^2 and B^2 do not fill the area $(A + B)^2$. Two additional areas, AB and AB, are needed.

Multiplications of Various Types

Recognizing patterns often helps when new problems are encountered. To simplify a new multiplication problem, always examine what type of product it is so that the best method for finding that product can be used. To do this, ask yourself questions similar to the following.

> **Multiplying Two Polynomials**
>
> 1. Is the multiplication the product of a monomial and a polynomial? If so, multiply each term of the polynomial by the monomial.
> 2. Is the multiplication the product of two binomials? If so:
>
>   a) Is it the product of the sum and the difference of the *same* two terms? If so, use the pattern
> $$(A + B)(A - B) = A^2 - B^2.$$
>
>   b) Is the product the square of a binomial? If so, use the pattern
> $$(A + B)(A + B) = (A + B)^2 = A^2 + 2AB + B^2,$$
>   or
> $$(A - B)(A - B) = (A - B)^2 = A^2 - 2AB + B^2.$$
>
>   c) If neither (a) nor (b) applies, use FOIL.
>
> 3. Is the multiplication the product of two polynomials other than those above? If so, multiply each term of one by every term of the other. Use columns if you wish.

EXAMPLE 5   Multiply.

a) $(x + 3)(x - 3)$ b) $(t + 7)(t - 5)$

c) $(x + 7)(x + 7)$ d) $2x^3(9x^2 + x - 7)$

e) $(p + 3)(p^2 + 2p - 1)$ f) $\left(3x - \frac{1}{4}\right)^2$

SOLUTION

a) $(x + 3)(x - 3) = x^2 - 9$ This is the product of the sum and the difference of the same two terms.

b) $(t + 7)(t - 5) = t^2 - 5t + 7t - 35$ Using FOIL
$$= t^2 + 2t - 35$$

c) $(x + 7)(x + 7) = x^2 + 14x + 49$ This is the square of a binomial, $(x + 7)^2$.

d) $2x^3(9x^2 + x - 7) = 18x^5 + 2x^4 - 14x^3$ Multiplying each term of the trinomial by the monomial

e) We multiply each term of $p^2 + 2p - 1$ by every term of $p + 3$:
$$(p + 3)(p^2 + 2p - 1) = p^3 + 2p^2 - p \qquad \text{Multiplying by } p$$
$$+ 3p^2 + 6p - 3 \qquad \text{Multiplying by 3}$$
$$= p^3 + 5p^2 + 5p - 3.$$

f) $\left(3x - \frac{1}{4}\right)^2 = 9x^2 - 2(3x)\left(\frac{1}{4}\right) + \frac{1}{16}$ Squaring a binomial
$$= 9x^2 - \frac{3}{2}x + \frac{1}{16}$$

TRY EXERCISE 69

1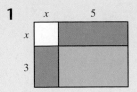

6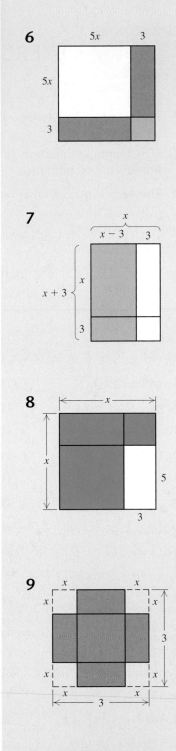

In each of Exercises 1–10, find two algebraic expressions for the shaded area of the figure from the list below.

A. $9 - 4x^2$

B. $x^2 - (x - 6)^2$

C. $(x + 3)(x - 3)$

D. $10^2 + 2^2$

E. $8x + 15$

F. $(x + 5)(x + 3) - x^2$

G. $x^2 - 6x + 9$

H. $(3 - 2x)^2 + 4x(3 - 2x)$

I. $(x + 3)^2$

J. $(5x + 3)^2 - 25x^2$

K. $(5 - 2x)^2 + 4x(5 - 2x)$

L. $x^2 - 9$

M. 104

N. $x^2 - 15$

O. $12x - 36$

P. $30x + 9$

Q. $(x - 5)(x - 3) + 3(x - 5) + 5(x - 3)$

R. $(x - 3)^2$

S. $25 - 4x^2$

T. $x^2 + 6x + 9$

Answers on page A-16

An additional, animated version of this activity appears in MyMathLab. To use MyMathLab, you need a course ID and a student access code. Contact your instructor for more information.

2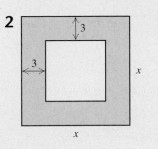

7

3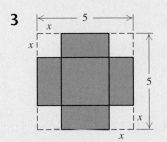

8

4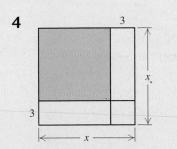

9

5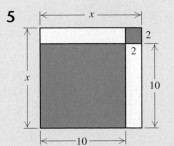

10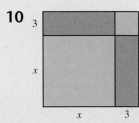

4.5   EXERCISE SET

Concept Reinforcement   *Classify each statement as either true or false.*

1. FOIL is simply a memory device for finding the product of two binomials.

2. The square of a binomial cannot be found using FOIL.

3. Once FOIL is used, it is always possible to combine like terms.

4. The square of $A + B$ is not the sum of the squares of A and B.

Multiply.

5. $(x^2 + 2)(x + 3)$ **6.** $(x - 5)(x^2 - 6)$

7. $(t^4 - 2)(t + 7)$ **8.** $(n^3 + 8)(n - 4)$

9. $(y + 2)(y - 3)$ **10.** $(a + 2)(a + 2)$

11. $(3x + 2)(3x + 5)$ **12.** $(4x + 1)(2x + 7)$

13. $(5x - 3)(x + 4)$ **14.** $(4x - 5)(4x + 5)$

15. $(3 - 2t)(5 - t)$ **16.** $(7 - a)(4 - 3a)$

17. $(x^2 + 3)(x^2 - 7)$ **18.** $(x^2 + 2)(x^2 - 8)$

19. $\left(p - \frac{1}{4}\right)\left(p + \frac{1}{4}\right)$ **20.** $\left(q + \frac{3}{4}\right)\left(q + \frac{3}{4}\right)$

21. $(x - 0.3)(x - 0.3)$ **22.** $(x - 0.1)(x + 0.1)$

23. $(-3n + 2)(n + 7)$ **24.** $(-m + 5)(2m - 9)$

25. $(x + 10)(x + 10)$ **26.** $(x + 12)(x + 12)$

27. $(1 - 3t)(1 + 5t^2)$ **28.** $(1 + 2t)(1 - 3t^2)$

29. $(x^2 + 3)(x^3 - 1)$ **30.** $(x^4 - 3)(2x + 1)$

31. $(3x^2 - 2)(x^4 - 2)$ **32.** $(x^{10} + 3)(x^{10} - 3)$

33. $(2t^3 + 5)(2t^3 + 5)$ **34.** $(5t^2 + 1)(2t^2 + 3)$

35. $(8x^3 + 5)(x^2 + 2)$ **36.** $(5 - 4x^5)(5 + 4x^5)$

37. $(10x^2 + 3)(10x^2 - 3)$ **38.** $(7x - 2)(2x - 7)$

Multiply. Try to recognize the type of product before multiplying.

39. $(x + 8)(x - 8)$ **40.** $(x + 1)(x - 1)$

41. $(2x + 1)(2x - 1)$ **42.** $(4n + 7)(4n - 7)$

43. $(5m^2 + 4)(5m^2 - 4)$ **44.** $(3x^4 + 2)(3x^4 - 2)$

45. $(9a^3 + 1)(9a^3 - 1)$ **46.** $(t^2 - 0.2)(t^2 + 0.2)$

47. $(x^4 + 0.1)(x^4 - 0.1)$ **48.** $(a^3 + 5)(a^3 - 5)$

49. $\left(t - \frac{3}{4}\right)\left(t + \frac{3}{4}\right)$ **50.** $\left(m - \frac{2}{3}\right)\left(m + \frac{2}{3}\right)$

51. $(x + 3)^2$ **52.** $(2x - 1)^2$

53. $(7x^3 - 1)^2$ **54.** $(5x^3 + 2)^2$

55. $\left(a - \frac{2}{5}\right)^2$ **56.** $\left(t - \frac{1}{5}\right)^2$

57. $(t^4 + 3)^2$ **58.** $(a^3 + 6)^2$

59. $(2 - 3x^4)^2$ **60.** $(5 - 2t^3)^2$

61. $(5 + 6t^2)^2$ **62.** $(3p^2 - p)^2$

63. $(7x - 0.3)^2$ **64.** $(4a - 0.6)^2$

65. $7n^3(2n^2 - 1)$ **66.** $5m^3(4 - 3m^2)$

67. $(a - 3)(a^2 + 2a - 4)$ **68.** $(x^2 - 5)(x^2 + x - 1)$

69. $(7 - 3x^4)(7 - 3x^4)$ **70.** $(x - 4x^3)^2$

71. $5x(x^2 + 6x - 2)$ **72.** $6x(-x^5 + 6x^2 + 9)$

73. $(q^5 + 1)(q^5 - 1)$ **74.** $(p^4 + 2)(p^4 - 2)$

75. $3t^2(5t^3 - t^2 + t)$ **76.** $-5x^3(x^2 + 8x - 9)$

77. $(6x^4 - 3x)^2$ **78.** $(8a^3 + 5)(8a^3 - 5)$

79. $(9a + 0.4)(2a^3 + 0.5)$ **80.** $(2a - 0.7)(8a^3 - 0.5)$

81. $\left(\frac{1}{5} - 6x^4\right)\left(\frac{1}{5} + 6x^4\right)$ **82.** $\left(3 + \frac{1}{2}t^5\right)\left(3 + \frac{1}{2}t^5\right)$

83. $(a + 1)(a^2 - a + 1)$

84. $(x - 5)(x^2 + 5x + 25)$

Find the total area of all shaded rectangles.

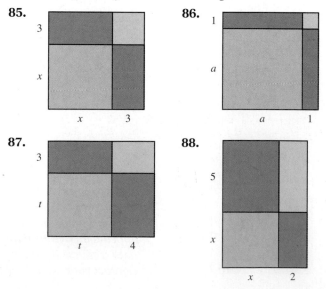

85.

86.

87.

88.

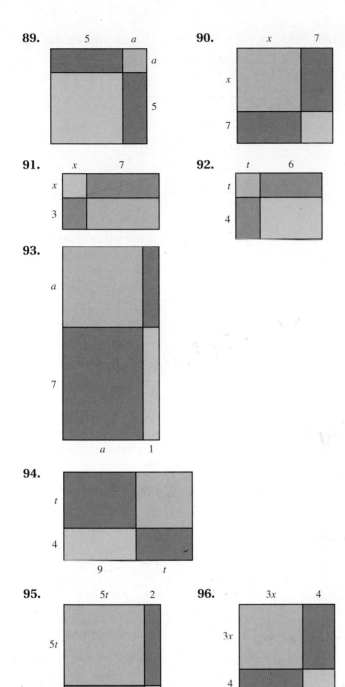

89. **90.**

91. **92.**

93.

94.

95. **96.**

Draw and label rectangles similar to those in Exercises 85–96 to illustrate each of the following.

97. $(x + 5)^2$ **98.** $(x + 8)^2$

99. $(t + 9)^2$ **100.** $(a + 12)^2$

101. $(3 + x)^2$ **102.** $(7 + t)^2$

103. Kristi feels that since she can find the product of any two binomials using FOIL, she needn't study the other special products. What advice would you give her?

104. Under what conditions is the product of two binomials a binomial?

Skill Review

Review problem solving and solving a formula for a variable (Sections 2.3 and 2.5).

Solve. [2.5]

105. *Energy use.*   Under typical use, a refrigerator, a freezer, and a washing machine together use 297 kilowatt-hours per month (kWh/mo). A refrigerator uses 21 times as much energy as a washing machine, and a freezer uses 11 times as much energy as a washing machine. How much energy is used by each appliance?

106. *Advertising.*   North American advertisers spent $9.4 billion on search-engine marketing in 2006. This was a 62% increase over the amount spent in 2005. How much was spent in 2005?
Source: Search Engine Marketing Professional Organization

Solve. [2.3]

107. $5xy = 8$, for y

108. $3ab = c$, for a

109. $ax - by = c$, for x

110. $ax - by = c$, for y

Synthesis

111. By writing $19 \cdot 21$ as $(20 - 1)(20 + 1)$, Justin can find the product mentally. How do you think he does this?

112. The product $(A + B)^2$ can be regarded as the sum of the areas of four regions (as shown following Example 4). How might one visually represent $(A + B)^3$? Why?

Multiply.

Aha! **113.** $(4x^2 + 9)(2x + 3)(2x - 3)$

114. $(9a^2 + 1)(3a - 1)(3a + 1)$

Aha! **115.** $(3t - 2)^2(3t + 2)^2$

116. $(5a + 1)^2(5a - 1)^2$

117. $(t^3 - 1)^4(t^3 + 1)^4$

▤ **118.** $(32.41x + 5.37)^2$

Calculate as the difference of squares.

119. 18×22 [*Hint*: $(20 - 2)(20 + 2)$.]

120. 93×107

Solve.

121. $(x + 2)(x - 5) = (x + 1)(x - 3)$

122. $(2x + 5)(x - 4) = (x + 5)(2x - 4)$

Find a polynomial for the total shaded area in each figure.

123.

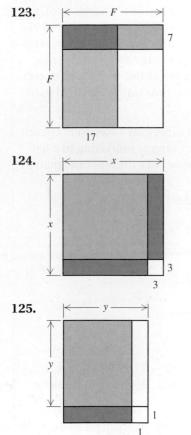

124.

125.

126. Find $(10 - 2x)^2$ by subtracting the white areas from 10^2.

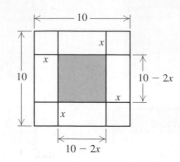

127. Find $(y - 2)^2$ by subtracting the white areas from y^2.

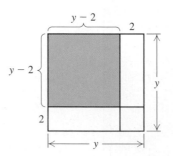

128. Find three consecutive integers for which the sum of the squares is 65 more than three times the square of the smallest integer.

⚞ **129.** Use a graphing calculator and the method developed on p. 242 to check your answers to Exercises 22, 47, and 83.

CONNECTING ↕ the CONCEPTS

When writing equivalent polynomial expressions, look first at the operation that you are asked to perform.

Operation	Procedure	Examples
Addition	Combine like terms.	$(2x^3 + 3x^2 - 5x - 7) + (9x^3 - 11x + 8)$ $= \underline{2x^3 + 9x^3} + 3x^2 \underbrace{\ 5x - 11x\ } \underbrace{-7 + 8}$ $= \ \ 11x^3 \ \ + 3x^2 \ \ \ -16x \ \ \ +1$
Subtraction	Add the opposite of the polynomial being subtracted.	$(9x^4 - 3x^2 + x - 7) - (4x^3 - 8x^2 - 9x + 11)$ $= 9x^4 - 3x^2 + x - 7 + (-4x^3 + 8x^2 + 9x - 11)$ $= 9x^4 - 3x^2 + x - 7 - 4x^3 + 8x^2 + 9x - 11$ $= 9x^4 - 4x^3 + 5x^2 + 10x - 18$
Multiplication	Multiply each term of one polynomial by every term of the other.	$(x^2 - 5x)(3x^4 - 7x^3 + 1)$ $= x^2(3x^4 - 7x^3 + 1) - 5x(3x^4 - 7x^3 + 1)$ $= 3x^6 - 7x^5 + x^2 - 15x^5 + 35x^4 - 5x$ $= 3x^6 - 22x^5 + 35x^4 + x^2 - 5x$
	Special products: $(A + B)(A - B) = A^2 - B^2;$ $(A + B)^2 = A^2 + 2AB + B^2;$ $(A - B)^2 = A^2 - 2AB + B^2;$ $(A + B)(C + D)$ $= AC + AD + BC + BD$	$(x + 5)(x - 5) = x^2 - 25;$ $(2x + 3)^2 = (2x)^2 + 2(2x)(3) + (3)^2 = 4x^2 + 12x + 9;$ $(x^2 - 1)^2 = (x^2)^2 - 2(x^2)(1) + (1)^2 = x^4 - 2x^2 + 1;$ $(x^2 + 3)(x - 2) = x^2(x) + x^2(-2) + 3(x) + 3(-2)$ $= x^3 - 2x^2 + 3x - 6$

MIXED REVIEW

Identify the operation to be performed. Then simplify to form an equivalent expression.

1. $(3x^2 - 2x + 6) + (5x - 3)$
2. $(9x + 6) - (3x - 7)$
3. $6x^3(8x^2 - 7)$
4. $(3x + 2)(2x - 1)$
5. $(9x^3 - 7x + 3) - (5x^2 - 10)$
6. $(2x + 1)(x^2 + x - 3)$
7. $(9x + 1)(9x - 1)$
8. $(8x^3 + 5x) + (9x^4 - 6x^3 - 10x)$

Perform the indicated operation to form an equivalent expression.

9. $(4x^2 - x - 7) - (10x^2 - 3x + 5)$
10. $(3x + 8)(3x + 7)$
11. $8x^5(5x^4 - 6x^3 + 2)$
12. $(t^9 + 3t^6 - 8t^2) + (5t^7 - 3t^6 + 8t^2)$
13. $(2m - 1)^2$
14. $(x - 1)(x^2 + x + 1)$
15. $(5x^3 - 6x^2 - 2x) + (6x^2 + 2x + 3)$
16. $(c + 3)(c - 3)$
17. $(4y^3 + 7)^2$
18. $(3a^4 - 9a^3 - 7) - (4a^3 + 13a^2 - 3)$
19. $(4t^2 - 5)(4t^2 + 5)$
20. $(a^4 + 3)(a^4 - 8)$

4.6   Polynomials in Several Variables

Evaluating Polynomials ▪ Like Terms and Degree ▪ Addition and Subtraction ▪ Multiplication

Thus far, the polynomials that we have studied have had only one variable. Polynomials such as

$$5x + x^2y - 3y + 7, \qquad 9ab^2c - 2a^3b^2 + 8a^2b^3, \quad \text{and} \quad 4m^2 - 9n^2$$

contain two or more variables. In this section, we will add, subtract, multiply, and evaluate such **polynomials in several variables**.

Evaluating Polynomials

To evaluate a polynomial in two or more variables, we substitute numbers for the variables. Then we compute, using the rules for order of operations.

EXAMPLE 1   Evaluate the polynomial $4 + 3x + xy^2 + 8x^3y^3$ for $x = -2$ and $y = 5$.

SOLUTION   We substitute -2 for x and 5 for y:

$$4 + 3x + xy^2 + 8x^3y^3 = 4 + 3(-2) + (-2) \cdot 5^2 + 8(-2)^3 \cdot 5^3$$
$$= 4 - 6 - 50 - 8000 = -8052.$$

> **TRY EXERCISE** 9

EXAMPLE 2   *Surface area of a right circular cylinder.*   The surface area of a right circular cylinder is given by the polynomial

$$2\pi rh + 2\pi r^2,$$

where h is the height and r is the radius of the base. A 12-oz can has a height of 4.7 in. and a radius of 1.2 in. Approximate its surface area to the nearest tenth of a square inch.

SOLUTION   We evaluate the polynomial for $h = 4.7$ in. and $r = 1.2$ in. If 3.14 is used to approximate π, we have

$$2\pi rh + 2\pi r^2 \approx 2(3.14)(1.2\,\text{in.})(4.7\,\text{in.}) + 2(3.14)(1.2\,\text{in.})^2$$
$$\approx 2(3.14)(1.2\,\text{in.})(4.7\,\text{in.}) + 2(3.14)(1.44\,\text{in}^2)$$
$$\approx 35.4192\,\text{in}^2 + 9.0432\,\text{in}^2 \approx 44.4624\,\text{in}^2.$$

If the π key of a calculator is used, we have

$$2\pi rh + 2\pi r^2 \approx 2(3.141592654)(1.2\,\text{in.})(4.7\,\text{in.})$$
$$+ 2(3.141592654)(1.2\,\text{in.})^2$$
$$\approx 44.48495197\,\text{in}^2.$$

Note that the unit in the answer (square inches) is a unit of area. The surface area is about 44.5 in^2 (square inches).

> **TRY EXERCISE** 13

Like Terms and Degree

Recall that the degree of a monomial is the number of variable factors in the term. For example, the degree of $5x^2$ is 2 because there are two variable factors in $5 \cdot x \cdot x$. Similarly, the degree of $5a^2b^4$ is 6 because there are 6 variable factors in $5 \cdot a \cdot a \cdot b \cdot b \cdot b \cdot b$. Note that 6 can be found by adding the exponents 2 and 4.

As we learned in Section 4.2, the degree of a polynomial is the degree of the term of highest degree.

EXAMPLE 3

Identify the coefficient and the degree of each term and the degree of the polynomial

$$9x^2y^3 - 14xy^2z^3 + xy + 4y + 5x^2 + 7.$$

SOLUTION

Term	Coefficient	Degree	Degree of the Polynomial
$9x^2y^3$	9	5	
$-14xy^2z^3$	-14	6	6
xy	1	2	
$4y$	4	1	
$5x^2$	5	2	
7	7	0	

TRY EXERCISE 21

Note in Example 3 that although both xy and $5x^2$ have degree 2, they are *not* like terms. *Like,* or *similar, terms* either have exactly the same variables with exactly the same exponents or are constants. For example,

$$8a^4b^7 \text{ and } 5b^7a^4 \text{ are like terms}$$

and

$$17 \text{ and } 3 \text{ are like terms,}$$

but

$$-2x^2y \text{ and } 9xy^2 \text{ are } not \text{ like terms.}$$

As always, combining like terms is based on the distributive law.

EXAMPLE 4

Combine like terms to form equivalent expressions.

a) $9x^2y + 3xy^2 - 5x^2y - xy^2$
b) $7ab - 5ab^2 + 3ab^2 + 6a^3 + 9ab - 11a^3 + b - 1$

SOLUTION

a) $9x^2y + 3xy^2 - 5x^2y - xy^2 = (9 - 5)x^2y + (3 - 1)xy^2$
$$= 4x^2y + 2xy^2 \qquad \text{Try to go directly to this step.}$$

b) $7ab - 5ab^2 + 3ab^2 + 6a^3 + 9ab - 11a^3 + b - 1$
$$= -5a^3 - 2ab^2 + 16ab + b - 1 \qquad \text{We choose to write descending}$$
powers of a. Other, equivalent, forms can also be used.

TRY EXERCISE 25

Addition and Subtraction

The procedure used for adding polynomials in one variable is used to add polynomials in several variables.

EXAMPLE 5

Add.

a) $(-5x^3 + 3y - 5y^2) + (8x^3 + 4x^2 + 7y^2)$

b) $(5ab^2 - 4a^2b + 5a^3 + 2) + (3ab^2 - 2a^2b + 3a^3b - 5)$

SOLUTION

a) $(-5x^3 + 3y - 5y^2) + (8x^3 + 4x^2 + 7y^2)$

$\qquad = (-5 + 8)x^3 + 4x^2 + 3y + (-5 + 7)y^2$ Try to do this step mentally.

$\qquad = 3x^3 + 4x^2 + 3y + 2y^2$

b) $(5ab^2 - 4a^2b + 5a^3 + 2) + (3ab^2 - 2a^2b + 3a^3b - 5)$

$\qquad = 8ab^2 - 6a^2b + 5a^3 + 3a^3b - 3$ **TRY EXERCISE** 33

When subtracting a polynomial, remember to find the opposite of each term in that polynomial and then add.

EXAMPLE 6

Subtract: $(4x^2y + x^3y^2 + 3x^2y^3 + 6y) - (4x^2y - 6x^3y^2 + x^2y^2 - 5y)$.

SOLUTION

$$(4x^2y + x^3y^2 + 3x^2y^3 + 6y) - (4x^2y - 6x^3y^2 + x^2y^2 - 5y)$$

$\qquad = 4x^2y + x^3y^2 + 3x^2y^3 + 6y - 4x^2y + 6x^3y^2 - x^2y^2 + 5y$

$\qquad = 7x^3y^2 + 3x^2y^3 - x^2y^2 + 11y$ Combining like terms

TRY EXERCISE 35

Multiplication

To multiply polynomials in several variables, multiply each term of one polynomial by every term of the other, just as we did in Sections 4.4 and 4.5.

EXAMPLE 7

Multiply: $(3x^2y - 2xy + 3y)(xy + 2y)$.

SOLUTION

$$\begin{array}{r} 3x^2y - 2xy + 3y \\ xy + 2y \\ \hline 6x^2y^2 - 4xy^2 + 6y^2 \\ 3x^3y^2 - 2x^2y^2 + 3xy^2 \\ \hline 3x^3y^2 + 4x^2y^2 - xy^2 + 6y^2 \end{array}$$

Multiplying by $2y$

Multiplying by xy

Adding   **TRY EXERCISE** 51

The special products discussed in Section 4.5 can speed up our work.

EXAMPLE 8

Multiply.

a) $(p + 5q)(2p - 3q)$ **b)** $(3x + 2y)^2$

c) $(a^3 - 7a^2b)^2$ **d)** $(3x^2y + 2y)(3x^2y - 2y)$

e) $(-2x^3y^2 + 5t)(2x^3y^2 + 5t)$ **f)** $(2x + 3 - 2y)(2x + 3 + 2y)$

SOLUTION

$$\text{F} \qquad \text{O} \qquad \text{I} \qquad \text{L}$$

a) $(p + 5q)(2p - 3q) = 2p^2 - 3pq + 10pq - 15q^2$

$$= 2p^2 + 7pq - 15q^2 \qquad \text{Combining like terms}$$

$$(A + B)^2 = A^2 + 2 \cdot A \cdot B + B^2$$

b) $(3x + 2y)^2 = (3x)^2 + 2(3x)(2y) + (2y)^2 \qquad \text{Using the pattern for squaring a binomial}$

$$= 9x^2 + 12xy + 4y^2$$

$$(A - B)^2 = A^2 - 2 \cdot A \cdot B + B^2$$

c) $(a^3 - 7a^2b)^2 = (a^3)^2 - 2(a^3)(7a^2b) + (7a^2b)^2 \qquad \text{Squaring a binomial}$

$$= a^6 - 14a^5b + 49a^4b^2 \qquad \text{Using the rules for exponents}$$

$$(A + B)(A - B) = A^2 - B^2$$

d) $(3x^2y + 2y)(3x^2y - 2y) = (3x^2y)^2 - (2y)^2 \qquad \begin{array}{l}\text{Using the pattern for} \\ \text{multiplying the sum and the} \\ \text{difference of two terms}\end{array}$

$$= 9x^4y^2 - 4y^2 \qquad \text{Using the rules for exponents}$$

e) $(-2x^3y^2 + 5t)(2x^3y^2 + 5t) = (5t - 2x^3y^2)(5t + 2x^3y^2) \qquad \begin{array}{l}\text{Using the com-} \\ \text{mutative law for} \\ \text{addition twice}\end{array}$

$$= (5t)^2 - (2x^3y^2)^2 \qquad \begin{array}{l}\text{Multiplying the sum and} \\ \text{the difference of the same} \\ \text{two terms}\end{array}$$

$$= 25t^2 - 4x^6y^4$$

$$(A - B)(A + B) = A^2 - B^2$$

f) $(2x + 3 - 2y)(2x + 3 + 2y) = (2x + 3)^2 - (2y)^2 \qquad \begin{array}{l}\text{Multiplying a sum} \\ \text{and a difference}\end{array}$

$$= 4x^2 + 12x + 9 - 4y^2 \qquad \begin{array}{l}\text{Squaring a} \\ \text{binomial}\end{array}$$

TRY EXERCISE 57

TECHNOLOGY CONNECTION

One way to evaluate the polynomial in Example 1 for $x = -2$ and $y = 5$ is to store -2 to X and 5 to Y and enter the polynomial.

```
-2 → X
                        -2
5 → Y
                         5
4+3X+XY²+8X^3Y^3
                     -8052
■
```

Evaluate.

1. $3x^2 - 2y^2 + 4xy + x$, for $x = -6$ and $y = 2.3$

2. $a^2b^2 - 8c^2 + 4abc + 9a$, for $a = 11, b = 15,$ and $c = -7$

In Example 8, we recognized patterns that might elude some students, particularly in parts (e) and (f). In part (e), we *can* use FOIL, and in part (f), we *can* use long multiplication, but doing so is much slower. By carefully inspecting a problem before "jumping in," we can save ourselves considerable work. At least one instructor refers to this as "working smart" instead of "working hard."*

---

*Thanks to Pauline Kirkpatrick of Wharton County Junior College for this language.

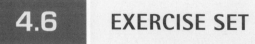

4.6   EXERCISE SET

Concept Reinforcement   *Each of the expressions in Exercises 1–8 can be regarded as either* **(a)** *the square of a binomial,* **(b)** *the product of the sum and the difference of the same two terms, or* **(c)** *neither of the above. Select the appropriate choice for each expression.*

1. $(3x + 5y)^2$

2. $(4x - 9y)(4x + 9y)$

3. $(5a + 6b)(-6b + 5a)$

4. $(4a - 3b)(4a - 3b)$

5. $(r - 3s)(5r + 3s)$

6. $(2x - 7y)(7y - 2x)$

7. $(4x - 9y)(4x - 9y)$

8. $(2r + 9t)^2$

Evaluate each polynomial for $x = 5$ *and* $y = -2$.

9. $x^2 - 2y^2 + 3xy$

10. $x^2 + 5y^2 - 4xy$

Evaluate each polynomial for $x = 2$, $y = -3$, *and* $z = -4$.

11. $xy^2z - z$

12. $xy - x^2z + yz^2$

Lung capacity.   *The polynomial*
$$0.041h - 0.018A - 2.69$$
can be used to estimate the lung capacity, in liters, of a female with height h, in centimeters, and age A, in years.

13. Find the lung capacity of a 20-year-old woman who is 160 cm tall.

14. Find the lung capacity of a 40-year-old woman who is 165 cm tall.

15. *Female caloric needs.*   The number of calories needed each day by a moderately active woman who weighs w pounds, is h inches tall, and is a years old can be estimated by the polynomial
$$917 + 6w + 6h - 6a.$$
Rachel is moderately active, weighs 125 lb, is 64 in. tall, and is 27 yr old. What are her daily caloric needs?
Source: Parker, M., *She Does Math.* Mathematical Association of America

16. *Male caloric needs.*   The number of calories needed each day by a moderately active man who weighs w kilograms, is h centimeters tall, and is a years old can be estimated by the polynomial
$$19.18w + 7h - 9.52a + 92.4.$$
One of the authors of this text is moderately active, weighs 87 kg, is 185 cm tall, and is 59 yr old. What are his daily caloric needs?
Source: Parker, M., *She Does Math.* Mathematical Association of America

Surface area of a silo.   *A silo is a structure that is shaped like a right circular cylinder with a half sphere on top. The surface area of a silo of height h and radius r (including the area of the base) is given by the polynomial*
$$2\pi rh + \pi r^2.$$

17. A coffee grinder is shaped like a silo, with a height of 7 in. and a radius of $1\frac{1}{2}$ in. Find the surface area of the coffee grinder. Use 3.14 for π.

18. A $1\frac{1}{2}$-oz bottle of roll-on deodorant has a height of 4 in. and a radius of $\frac{3}{4}$ in. Find the surface area of the bottle if the bottle is shaped like a silo. Use 3.14 for π.

Altitude of a launched object.   *The altitude of an object, in meters, is given by the polynomial*
$$h + vt - 4.9t^2,$$
where h is the height, in meters, at which the launch occurs, v is the initial upward speed (or velocity), in meters per second, and t is the number of seconds for which the object is airborne.

19. A bocce ball is thrown upward with an initial speed of 18 m/sec by a person atop the Leaning Tower of Pisa, which is 50 m above the ground. How high will the ball be 2 sec after it has been thrown?

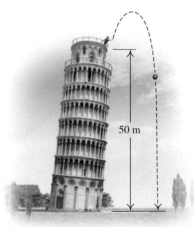

50 m

20. A golf ball is launched upward with an initial speed of 30 m/sec by a golfer atop the Washington Monument, which is 160 m above the ground. How high above the ground will the ball be after 3 sec?

Identify the coefficient and the degree of each term of each polynomial. Then find the degree of each polynomial.

21. $3x^2y - 5xy + 2y^2 - 11$

22. $xy^3 + 7x^3y^2 - 6xy^4 + 2$

23. $7 - abc + a^2b + 9ab^2$

24. $3p - pq - 7p^2q^3 - 8pq^6$

Combine like terms.

25. $3r + s - r - 7s$

26. $9a + b - 8a - 5b$

27. $5xy^2 - 2x^2y + x + 3x^2$

28. $m^3 + 2m^2n - 3m^2 + 3mn^2$

29. $6u^2v - 9uv^2 + 3vu^2 - 2v^2u + 11u^2$

30. $3x^2 + 6xy + 3y^2 - 5x^2 - 10xy$

31. $5a^2c - 2ab^2 + a^2b - 3ab^2 + a^2c - 2ab^2$

32. $3s^2t + r^2t - 9ts^2 - st^2 + 5t^2s - 7tr^2$

Add or subtract, as indicated.

33. $(6x^2 - 2xy + y^2) + (5x^2 - 8xy - 2y^2)$

34. $(7r^3 + rs - 5r^2) - (2r^3 - 3rs + r^2)$

35. $(3a^4 - 5ab + 6ab^2) - (9a^4 + 3ab - ab^2)$

36. $(2r^2t - 5rt + rt^2) - (7r^2t + rt - 5rt^2)$

Aha! **37.** $(5r^2 - 4rt + t^2) + (-6r^2 - 5rt - t^2) + (-5r^2 + 4rt - t^2)$

38. $(2x^2 - 3xy + y^2) + (-4x^2 - 6xy - y^2) + (4x^2 + 6xy + y^2)$

39. $(x^3 - y^3) - (-2x^3 + x^2y - xy^2 + 2y^3)$

40. $(a^3 + b^3) - (-5a^3 + 2a^2b - ab^2 + 3b^3)$

41. $(2y^4x^3 - 3y^3x) + (5y^4x^3 - y^3x) - (9y^4x^3 - y^3x)$

42. $(5a^2b - 7ab^2) - (3a^2b + ab^2) + (a^2b - 2ab^2)$

43. Subtract $7x + 3y$ from the sum of $4x + 5y$ and $-5x + 6y$.

44. Subtract $5a + 2b$ from the sum of $2a + b$ and $3a - 4b$.

Multiply.

45. $(4c - d)(3c + 2d)$

46. $(5x + y)(2x - 3y)$

47. $(xy - 1)(xy + 5)$

48. $(ab + 3)(ab - 5)$

49. $(2a - b)(2a + b)$

50. $(a - 3b)(a + 3b)$

51. $(5rt - 2)(4rt - 3)$

52. $(3xy - 1)(4xy + 2)$

53. $(m^3n + 8)(m^3n - 6)$

54. $(9 - u^2v^2)(2 - u^2v^2)$

55. $(6x - 2y)(5x - 3y)$

56. $(7a - 6b)(5a + 4b)$

57. $(pq + 0.1)(-pq + 0.1)$

58. $(rt + 0.2)(-rt + 0.2)$

59. $(x + h)^2$

60. $(a - r)^2$

61. $(4a - 5b)^2$

62. $(2x + 5y)^2$

63. $(ab + cd^2)(ab - cd^2)$

64. $(p^3 - 5q)(p^3 + 5q)$

65. $(2xy + x^2y + 3)(xy + y^2)$

66. $(5cd - c^2 - d^2)(2c - c^2d)$

Aha! **67.** $(a + b - c)(a + b + c)$

68. $(x + y + 2z)(x + y - 2z)$

69. $[a + b + c][a - (b + c)]$

70. $(a + b + c)(a - b - c)$

Find the total area of each shaded area.

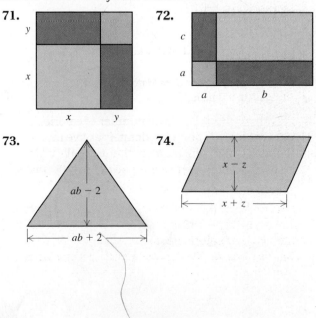

71.

72.

73.

74.

75.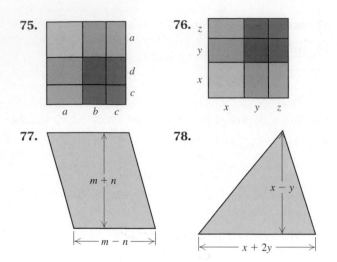

76.

77.

78.

Draw and label rectangles similar to those in Exercises 71, 72, 75, and 76 to illustrate each product.

79. $(r + s)(u + v)$

80. $(m + r)(n + v)$

81. $(a + b + c)(a + d + f)$

82. $(r + s + t)^2$

83. Is it possible for a polynomial in 4 variables to have a degree less than 4? Why or why not?

84. A fourth-degree monomial is multiplied by a third-degree monomial. What is the degree of the product? Explain your reasoning.

Skill Review

To prepare for Section 4.7, review subtraction of polynomials using columns (Section 4.3).

Subtract. [4.3]

85. $x^2 - 3x - 7$
$- (\quad\ 5x - 3)$

86. $2x^3 \quad\quad - x + 3$
$- (\quad\ x^2 \quad\ - 1)$

87. $3x^2 + x + 5$
$- (3x^2 + 3x)$

88. $4x^3 - 3x^2 + x$
$- (4x^3 - 8x^2)$

89. $5x^3 - 2x^2 + 1$
$-(5x^3 - 15x^2)$

90. $2x^2 + 5x - 3$
$- (2x^2 + 6x)$

Synthesis

91. The concept of "leading term" was intentionally not discussed in this section. Why?

92. Explain how it is possible for the sum of two trinomials in several variables to be a binomial in one variable.

Find a polynomial for the shaded area. (Leave results in terms of π where appropriate.)

93. 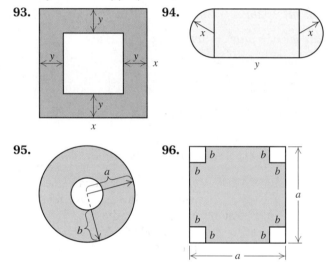 **94.**

95. **96.**

97. Find a polynomial for the total volume of the figure shown.

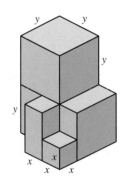

98. Find the shaded area in this figure using each of the approaches given below. Then check that both answers match.

  a) Find the shaded area by subtracting the area of the unshaded square from the total area of the figure.

  b) Find the shaded area by adding the areas of the three shaded rectangles.

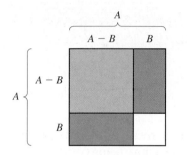

Find a polynomial for the surface area of each solid object shown. (Leave results in terms of π.)

99.   **100.**

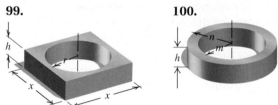

101. The observatory at Danville University is shaped like a silo that is 40 ft high and 30 ft wide (see Exercise 17). The Heavenly Bodies Astronomy Club is to paint the exterior of the observatory using paint that covers 250 ft^2 per gallon. How many gallons should they purchase? Explain your reasoning.

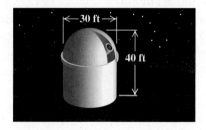

102. Multiply: $(x + a)(x - b)(x - a)(x + b)$.

The computer application Excel allows values for cells in a spreadsheet to be calculated from values in other cells. For example, if the cell C1 contains the formula

$$= A1 + 2*B1,$$

the value in C1 will be the sum of the value in A1 and twice the value in B1. This formula is a polynomial in the two variables A1 and B1.

103. The cell D4 contains the formula

$$= 2*A4 + 3*B4.$$

What is the value in D4 if the value in A4 is 5 and the value in B4 is 10?

104. The cell D6 contains the formula

$$= A1 - 0.2*B1 + 0.3*C1.$$

What is the value in D6 if the value in A1 is 10, the value in B1 is -3, and the value in C1 is 30?

105. *Interest compounded annually.* An amount of money P that is invested at the yearly interest rate r grows to the amount $P(1 + r)^t$ after t years. Find a polynomial that can be used to determine the amount to which P will grow after 2 yr.

106. *Yearly depreciation.* An investment P that drops in value at the yearly rate r drops in value to

$$P(1 - r)^t$$

after t years. Find a polynomial that can be used to determine the value to which P has dropped after 2 yr.

107. Suppose that $10,400 is invested at 8.5% compounded annually. How much is in the account at the end of 5 yr? (See Exercise 105.)

108. A $90,000 investment in computer hardware is depreciating at a yearly rate of 12.5%. How much is the investment worth after 4 yr? (See Exercise 106.)

COLLABORATIVE

CORNER

Finding the Magic Number

Focus: Evaluating polynomials in several variables

Time: 15–25 minutes

Group size: 3

Materials: A coin for each person

When a team nears the end of its schedule in first place, fans begin to discuss the team's "magic number." A team's magic number is the combined number of wins by that team and losses by the second-place team that guarantee the leading team a first-place finish. For example, if the Cubs' magic number is 3 over the Reds, any combination of Cubs wins and Reds losses that totals 3 will guarantee a first-place finish for the Cubs. A team's magic number is computed using the polynomial

$$G - P - L + 1,$$

where G is the length of the season, in games, P is the number of games that the leading team has played, and L is the total number of games that the second-place team has lost minus the total number of games that the leading team has lost.

ACTIVITY

1. The standings below are from a fictitious league. Each group should calculate the Jaguars' magic number with respect to the Catamounts as well as the Jaguars' magic number with respect to the Wildcats.

(Assume that the schedule is 162 games long.)

	W	L
Jaguars	92	64
Catamounts	90	66
Wildcats	89	66

2. Each group member should play the role of one of the teams, using coin tosses to simulate the remaining games. If a group member correctly predicts the side (heads or tails) that comes up, the coin toss represents a win for that team. Should the other side appear, the toss represents a loss. Assume that these games are against other (unlisted) teams in the league. Each group member should perform three coin tosses and then update the standings.

3. Recalculate the two magic numbers, using the updated standings from part (2).

4. Slowly—one coin toss at a time—play out the remainder of the season. Record all wins and losses, update the standings, and recalculate the magic numbers each time all three group members have completed a round of coin tosses.

5. Examine the work in part (4) and explain why a magic number of 0 indicates that a team has been eliminated from contention.

4.7   Division of Polynomials

Dividing by a Monomial   ▪   Dividing by a Binomial

In this section, we study division of polynomials. We will find that polynomial division is similar to division in arithmetic.

Dividing by a Monomial

We first consider division by a monomial. When dividing a monomial by a monomial, we use the quotient rule of Section 4.1 to subtract exponents when bases are the same. For example,

$$\frac{15x^{10}}{3x^4} = 5x^{10-4}$$

$$= 5x^6$$

> **CAUTION!** The coefficients are divided but the exponents are subtracted.

and

$$\frac{42a^2b^5}{-3ab^2} = \frac{42}{-3}a^{2-1}b^{5-2} \qquad \text{Recall that } a^m/a^n = a^{m-n}.$$

$$= -14ab^3.$$

To divide a polynomial by a monomial, we note that since

$$\frac{A}{C} + \frac{B}{C} = \frac{A+B}{C},$$

it follows that

$$\frac{A+B}{C} = \frac{A}{C} + \frac{B}{C}. \qquad \begin{array}{l}\text{Switching the left side and the right side}\\\text{of the equation}\end{array}$$

This is actually how we perform divisions like $86 \div 2$. Although we might simply write

$$\frac{86}{2} = 43,$$

we are really saying

$$\frac{80+6}{2} = \frac{80}{2} + \frac{6}{2} = 40 + 3.$$

Similarly, to divide a polynomial by a monomial, we divide each term by the monomial:

$$\frac{80x^5 + 6x^7}{2x^3} = \frac{80x^5}{2x^3} + \frac{6x^7}{2x^3}$$

$$= \frac{80}{2}x^{5-3} + \frac{6}{2}x^{7-3} \qquad \begin{array}{l}\text{Dividing coefficients and}\\\text{subtracting exponents}\end{array}$$

$$= 40x^2 + 3x^4.$$

EXAMPLE 1 Divide $x^4 + 15x^3 - 6x^2$ by $3x$.

SOLUTION   We divide each term of $x^4 + 15x^3 - 6x^2$ by $3x$:

$$\frac{x^4 + 15x^3 - 6x^2}{3x} = \frac{x^4}{3x} + \frac{15x^3}{3x} - \frac{6x^2}{3x}$$

$$= \frac{1}{3}x^{4-1} + \frac{15}{3}x^{3-1} - \frac{6}{3}x^{2-1} \qquad \begin{array}{l}\text{Dividing coefficients}\\\text{and subtracting}\\\text{exponents}\end{array}$$

$$= \frac{1}{3}x^3 + 5x^2 - 2x. \qquad \text{This is the quotient.}$$

To check, we multiply our answer by $3x$, using the distributive law:

$$3x\left(\frac{1}{3}x^3 + 5x^2 - 2x\right) = 3x \cdot \frac{1}{3}x^3 + 3x \cdot 5x^2 - 3x \cdot 2x$$

$$= x^4 + 15x^3 - 6x^2.$$

This is the polynomial that was being divided, so our answer, $\frac{1}{3}x^3 + 5x^2 - 2x$, checks.

TRY EXERCISE 5

EXAMPLE 2   Divide and check: $(10a^5b^4 - 2a^3b^2 + 6a^2b) \div (-2a^2b)$.

SOLUTION   We have

$$\frac{10a^5b^4 - 2a^3b^2 + 6a^2b}{-2a^2b} = \frac{10a^5b^4}{-2a^2b} - \frac{2a^3b^2}{-2a^2b} + \frac{6a^2b}{-2a^2b}$$

> We divide coefficients and subtract exponents.

$$= -\frac{10}{2}a^{5-2}b^{4-1} - \left(-\frac{2}{2}\right)a^{3-2}b^{2-1} + \left(-\frac{6}{2}\right)$$

$$= -5a^3b^3 + ab - 3.$$

Check: $-2a^2b(-5a^3b^3 + ab - 3)$

$$= -2a^2b\,(-5a^3b^3) + (-2a^2b)(ab) + (-2a^2b)(-3)$$

$$= 10a^5b^4 - 2a^3b^2 + 6a^2b$$

Our answer, $-5a^3b^3 + ab - 3$, checks.

TRY EXERCISE 7

Dividing by a Binomial

The divisors in Examples 1 and 2 have just one term. For divisors with more than one term, we use long division, much as we do in arithmetic. Polynomials are written in descending order and any missing terms in the dividend are written in, using 0 for the coefficients.

EXAMPLE 3   Divide $x^2 + 5x + 6$ by $x + 3$.

SOLUTION   We begin by dividing x^2 by x:

> **Divide** the first term, x^2, by the first term in the divisor: $x^2/x = x$. Ignore the term 3 for the moment.

$$
\begin{array}{r}
x \\
x + 3 \overline{\smash{)}\, x^2 + 5x + 6} \\
-(x^2 + 3x) \\
\hline
2x.
\end{array}
$$

Multiply $x + 3$ by x, using the distributive law.

Subtract both x^2 and $3x$: $x^2 + 5x - (x^2 + 3x) = 2x.$

Now we "bring down" the next term—in this case, 6. The current remainder, $2x + 6$, now becomes the focus of our division. We divide $2x$ by x.

STUDENT NOTES

Long division of polynomials offers many opportunities to make errors. Rather than have one small mistake throw off all your work, we recommend that you double-check each step of your work as you move forward.

$$
\begin{array}{r}
x + 2 \\
x + 3\overline{)x^2 + 5x + 6} \\
-(x^2 + 3x) \\
\hline
2x + 6
\end{array}
$$

Divide 2x by x: $2x/x = 2$.

$-(2x + 6)$ ← **Multiply** 2 by the divisor, $x + 3$, using the distributive law.

0 ← **Subtract** $(2x + 6) - (2x + 6) = 0$.

The quotient is $x + 2$. The notation R 0 indicates a remainder of 0, although a remainder of 0 is generally not listed in an answer.

Check: To check, we multiply the quotient by the divisor and add any remainder to see if we get the dividend:

$$
\underbrace{(x + 3)}_{\text{Divisor}} \underbrace{(x + 2)}_{\text{Quotient}} + \underbrace{0}_{\text{Remainder}} = \underbrace{x^2 + 5x + 6}_{\text{Dividend}}.
$$

Our answer, $x + 2$, checks.

TRY EXERCISE 17

EXAMPLE 4   Divide: $(2x^2 + 5x - 1) \div (2x - 1)$.

SOLUTION   We begin by dividing $2x^2$ by $2x$:

$$
\begin{array}{r}
x \\
2x - 1\overline{)2x^2 + 5x - 1} \\
-(2x^2 - x) \\
\hline
6x
\end{array}
$$

Divide the first term by the first term: $2x^2/(2x) = x$.

Multiply $2x - 1$ by x.

Subtract by changing signs and adding: $(2x^2 + 5x) - (2x^2 - x) = 6x$.

> **CAUTION!** Write the parentheses around the polynomial being subtracted to remind you to subtract all its terms.

Now, we bring down the -1 and divide $6x - 1$ by $2x - 1$.

$$
\begin{array}{r}
x + 3 \\
2x - 1\overline{)2x^2 + 5x - 1} \\
-(2x^2 - x) \\
\hline
6x - 1 \\
-(6x - 3) \\
\hline
2
\end{array}
$$

Divide 6x by 2x: $6x/(2x) = 3$.

Multiply 3 by the divisor, $2x - 1$.

Subtract. Note that $-1 - (-3) = -1 + 3 = 2$.

The answer is $x + 3$ with R 2.

Another way to write $x + 3$ R 2 is as

$$
\underbrace{x + 3}_{\text{Quotient}} + \dfrac{\overset{\text{Remainder}}{2}}{\underset{\text{Divisor}}{2x - 1}}.
$$

(This is the way answers will be given at the back of the book.)

Check: To check, we multiply the divisor by the quotient and add the remainder:

$$
\begin{aligned}
(2x - 1)(x + 3) + 2 &= 2x^2 + 5x - 3 + 2 \\
&= 2x^2 + 5x - 1. \quad \text{Our answer checks.}
\end{aligned}
$$

TRY EXERCISE 29

Our division procedure ends when the degree of the remainder is less than that of the divisor. Check that this was indeed the case in Example 4.

EXAMPLE 5   Divide each of the following.

a) $(x^3 + 1) \div (x + 1)$ 　　　　　　　　**b)** $(x^4 - 3x^2 + 4x - 3) \div (x^2 - 5)$

SOLUTION

a)
$$
\begin{array}{r}
x^2 - x + 1 \\
x + 1 \overline{\smash{)}\ x^3 + 0x^2 + 0x + 1} \\
\underline{-(x^3 + x^2)} \\
-x^2 + 0x \\
\underline{-(-x^2 - x)} \\
x + 1 \\
\underline{-(x + 1)} \\
0
\end{array}
$$

← Writing in the missing terms

← Subtracting $x^3 + x^2$ from $x^3 + 0x^2$ and bringing down the $0x$

← Subtracting $-x^2 - x$ from $-x^2 + 0x$ and bringing down the 1

The answer is $x^2 - x + 1$.

Check:　　$(x + 1)(x^2 - x + 1) = x^3 - x^2 + x + x^2 - x + 1$
$$= x^3 + 1.$$

b)
$$
\begin{array}{r}
x^2 \qquad + 2 \\
x^2 - 5 \overline{\smash{)}\ x^4 + 0x^3 - 3x^2 + 4x - 3} \\
\underline{-(x^4 \qquad - 5x^2)} \\
2x^2 + 4x - 3 \\
\underline{-(2x^2 \qquad - 10)} \\
4x + 7
\end{array}
$$

Writing in the missing term

← Subtracting $x^4 - 5x^2$ from $x^4 - 3x^2$ and bringing down $4x - 3$

← Subtracting $2x^2 - 10$ from $2x^2 + 4x - 3$

Since the remainder, $4x + 7$, is of lower degree than the divisor, the division process stops. The answer is

$$x^2 + 2 + \frac{4x + 7}{x^2 - 5}.$$

Check:　　$(x^2 - 5)(x^2 + 2) + 4x + 7 = x^4 + 2x^2 - 5x^2 - 10 + 4x + 7$
$$= x^4 - 3x^2 + 4x - 3.$$

TRY EXERCISE 35

4.7   EXERCISE SET

For Extra Help *MathXL* **MyMathLab** PRACTICE WATCH DOWNLOAD

Divide and check.

1. $\dfrac{40x^6 - 25x^3}{5}$

2. $\dfrac{16a^5 - 24a^2}{8}$

3. $\dfrac{u - 2u^2 + u^7}{u}$

4. $\dfrac{50x^5 - 7x^4 + 2x}{x}$

5. $(18t^3 - 24t^2 + 6t) \div (3t)$

6. $(20t^3 - 15t^2 + 30t) \div (5t)$

7. $(42x^5 - 36x^3 + 9x^2) \div (6x^2)$

8. $(24x^6 + 18x^4 + 8x^3) \div (4x^3)$

9. $(32t^5 + 16t^4 - 8t^3) \div (-8t^3)$

10. $(36t^6 - 27t^5 - 9t^2) \div (-9t^2)$

11. $\dfrac{8x^2 - 10x + 1}{2x}$

12. $\dfrac{9x^2 + 3x - 2}{3x}$

13. $\dfrac{5x^3y + 10x^5y^2 + 15x^2y}{5x^2y}$

14. $\dfrac{12a^3b^2 + 4a^4b^5 + 16ab^2}{4ab^2}$

15. $\dfrac{9r^2s^2 + 3r^2s - 6rs^2}{-3rs}$

16. $\dfrac{4x^4y - 8x^6y^2 + 12x^8y^6}{4x^4y}$

17. $(x^2 - 8x + 12) \div (x - 2)$

18. $(x^2 + 2x - 15) \div (x + 5)$

19. $(t^2 - 10t - 20) \div (t - 5)$

20. $(t^2 + 8t - 15) \div (t + 4)$

21. $(2x^2 + 11x - 5) \div (x + 6)$

22. $(3x^2 - 2x - 13) \div (x - 2)$

23. $\dfrac{t^3 + 27}{t + 3}$　　　24. $\dfrac{a^3 + 8}{a + 2}$

25. $\dfrac{a^2 - 21}{a - 5}$　　　26. $\dfrac{t^2 - 13}{t - 4}$

27. $(5x^2 - 16x) \div (5x - 1)$

28. $(3x^2 - 7x + 1) \div (3x - 1)$

29. $(6a^2 + 17a + 8) \div (2a + 5)$

30. $(10a^2 + 19a + 9) \div (2a + 3)$

31. $\dfrac{2t^3 - 9t^2 + 11t - 3}{2t - 3}$

32. $\dfrac{8t^3 - 22t^2 - 5t + 12}{4t + 3}$

33. $(x^3 - x^2 + x - 1) \div (x - 1)$

34. $(t^3 - t^2 + t - 1) \div (t + 1)$

35. $(t^4 + 4t^2 + 3t - 6) \div (t^2 + 5)$

36. $(t^4 - 2t^2 + 4t - 5) \div (t^2 - 3)$

37. $(6x^4 - 3x^2 + x - 4) \div (2x^2 + 1)$

38. $(4x^4 - 4x^2 - 3) \div (2x^2 - 3)$

39. How is the distributive law used when dividing a polynomial by a binomial?

40. On an assignment, Emmy Lou *incorrectly* writes

$$\dfrac{12x^3 - 6x}{3x} = 4x^2 - 6x.$$

What mistake do you think she is making and how might you convince her that a mistake has been made?

Skill Review

To prepare for Section 4.8, review the properties of exponents (Section 4.1).

Simplify. [4.1]

41. $y^8 \cdot y^5$　　　42. $(y^8)^5$

43. $\dfrac{y^8}{y^5}$　　　44. $(2p^2q^5)^3$

45. $(a^2b)(a^3b^4)$　　　46. $(2c^2d^3)^0$

47. $\left(\dfrac{3}{8n^3}\right)^2$　　　48. $\left(\dfrac{2x^5y^2}{-3z^3}\right)^3$

Synthesis

49. Explain how to form trinomials for which division by $x - 5$ results in a remainder of 3.

50. Under what circumstances will the quotient of two binomials have more than two terms?

Divide.

51. $(10x^{9k} - 32x^{6k} + 28x^{3k}) \div (2x^{3k})$

52. $(45a^{8k} + 30a^{6k} - 60a^{4k}) \div (3a^{2k})$

53. $(6t^{3h} + 13t^{2h} - 4t^h - 15) \div (2t^h + 3)$

54. $(x^4 + a^2) \div (x + a)$

55. $(5a^3 + 8a^2 - 23a - 1) \div (5a^2 - 7a - 2)$

56. $(15y^3 - 30y + 7 - 19y^2) \div (3y^2 - 2 - 5y)$

57. Divide the sum of $4x^5 - 14x^3 - x^2 + 3$ and $2x^5 + 3x^4 + x^3 - 3x^2 + 5x$ by $3x^3 - 2x - 1$.

58. Divide $5x^7 - 3x^4 + 2x^2 - 10x + 2$ by the sum of $(x - 3)^2$ and $5x - 8$.

If the remainder is 0 when one polynomial is divided by another, the divisor is a factor *of the dividend. Find the value(s) of c for which x − 1 is a factor of each polynomial.*

59. $x^2 - 4x + c$

60. $2x^2 - 3cx - 8$

61. $c^2x^2 + 2cx + 1$

4.8   Negative Exponents and Scientific Notation

Negative Integers as Exponents   •   Scientific Notation   •
Multiplying and Dividing Using Scientific Notation

We now attach a meaning to negative exponents. Once we understand both positive exponents and negative exponents, we can study a method for writing numbers known as *scientific notation*.

Negative Integers as Exponents

Let's define negative exponents so that the rules that apply to whole-number exponents will hold for all integer exponents. To do so, consider a^{-5} and the rule for adding exponents:

$$a^{-5} = a^{-5} \cdot 1 \qquad \text{Using the identity property of 1}$$

$$= \frac{a^{-5}}{1} \cdot \frac{a^5}{a^5} \qquad \text{Writing 1 as } \frac{a^5}{a^5} \text{ and } a^{-5} \text{ as } \frac{a^{-5}}{1}$$

$$= \frac{a^{-5+5}}{a^5} \qquad \text{Adding exponents}$$

$$= \frac{1}{a^5}. \qquad -5 + 5 = 0 \text{ and } a^0 = 1$$

This leads to our definition of negative exponents.

> **Negative Exponents**
>
> For any real number a that is nonzero and any integer n,
>
> $$a^{-n} = \frac{1}{a^n}.$$
>
> (The numbers a^{-n} and a^n are reciprocals of each other.)

EXAMPLE   1   Express using positive exponents and, if possible, simplify.

a) m^{-3} **b)** 4^{-2} **c)** $(-3)^{-2}$ **d)** ab^{-1}

SOLUTION

a) $m^{-3} = \frac{1}{m^3}$ m^{-3} is the reciprocal of m^3.

b) $4^{-2} = \frac{1}{4^2} = \frac{1}{16}$ 4^{-2} is the reciprocal of 4^2. Note that $4^{-2} \neq 4(-2)$.

c) $(-3)^{-2} = \frac{1}{(-3)^2} = \frac{1}{(-3)(-3)} = \frac{1}{9}$ $(-3)^{-2}$ is the reciprocal of $(-3)^2$. Note that $(-3)^{-2} \neq -\frac{1}{3^2}$.

d) $ab^{-1} = a\left(\frac{1}{b^1}\right) = a\left(\frac{1}{b}\right) = \frac{a}{b}$ b^{-1} is the reciprocal of b^1. Note that the base is b, not ab.

TRY EXERCISE   5

> *CAUTION!*   A negative exponent does not, in itself, indicate that an expression is negative. As shown in Example 1,
>
> $$4^{-2} \neq 4(-2) \quad \text{and} \quad (-3)^{-2} \neq -\frac{1}{3^2}.$$

The following is another way to illustrate why negative exponents are defined as they are.

On this side, we divide by 5 at each step.		On this side, the exponents decrease by 1.
	$125 = 5^3$	
	$25 = 5^2$	
	$5 = 5^1$	
	$1 = 5^0$	
	$\dfrac{1}{5} = 5^?$	
	$\dfrac{1}{25} = 5^?$	

To continue the pattern, it follows that

$$\frac{1}{5} = \frac{1}{5^1} = 5^{-1}, \qquad \frac{1}{25} = \frac{1}{5^2} = 5^{-2}, \quad \text{and, in general,} \quad \frac{1}{a^n} = a^{-n}.$$

EXAMPLE **2**   Express $\dfrac{1}{x^7}$ using negative exponents.

SOLUTION   We know that $\dfrac{1}{a^n} = a^{-n}$. Thus,

$$\frac{1}{x^7} = x^{-7}.$$

> **TRY EXERCISE** 25

The rules for powers still hold when exponents are negative.

EXAMPLE **3**   Simplify. Do not use negative exponents in the answer.

a) $t^5 \cdot t^{-2}$ b) $(y^{-5})^{-7}$ c) $(5x^2y^{-3})^4$

d) $\dfrac{x^{-4}}{x^{-5}}$ e) $\dfrac{1}{t^{-5}}$ f) $\dfrac{s^{-3}}{t^{-5}}$

SOLUTION

a) $t^5 \cdot t^{-2} = t^{5+(-2)} = t^3$ Adding exponents

b) $(y^{-5})^{-7} = y^{(-5)(-7)}$ Multiplying exponents
$\qquad\quad = y^{35}$

c) $(5x^2y^{-3})^4 = 5^4(x^2)^4(y^{-3})^4$ Raising each factor to the fourth power
$\qquad\quad = 625x^8y^{-12} = \dfrac{625x^8}{y^{12}}$

d) $\dfrac{x^{-4}}{x^{-5}} = x^{-4-(-5)} = x^1 = x$ We subtract exponents even if the exponent in the denominator is negative.

e) Since $\dfrac{1}{a^n} = a^{-n}$, we have $\dfrac{1}{t^{-5}} = t^{-(-5)} = t^5$.

f) $\dfrac{s^{-3}}{t^{-5}} = s^{-3} \cdot \dfrac{1}{t^{-5}}$

$\qquad = \dfrac{1}{s^3} \cdot t^5 = \dfrac{t^5}{s^3}$ Using the result from part (e) above

> **TRY EXERCISE** 33

The result from Example 3(f) can be generalized.

Factors and Negative Exponents

For any nonzero real numbers a and b and any integers m and n,

$$\frac{a^{-n}}{b^{-m}} = \frac{b^m}{a^n}.$$

(A factor can be moved to the other side of the fraction bar if the sign of the exponent is changed.)

EXAMPLE   4   Simplify: $\dfrac{-15x^{-7}}{5y^2z^{-4}}$.

SOLUTION   We can move the factors x^{-7} and z^{-4} to the other side of the fraction bar if we change the sign of each exponent:

$$\frac{-15x^{-7}}{5y^2z^{-4}} = \frac{-15}{5} \cdot \frac{x^{-7}}{y^2z^{-4}}$$ We can simply divide the constant factors.

$$= -3 \cdot \frac{z^4}{y^2x^7}$$

$$= \frac{-3z^4}{x^7y^2}.$$

> **TRY EXERCISE** 55

Another way to change the sign of the exponent is to take the reciprocal of the base. To understand why this is true, note that

$$\left(\frac{s}{t}\right)^{-5} = \frac{s^{-5}}{t^{-5}} = \frac{t^5}{s^5} = \left(\frac{t}{s}\right)^5.$$

This often provides the easiest way to simplify an expression containing a negative exponent.

Reciprocals and Negative Exponents

For any nonzero real numbers a and b and any integer n,

$$\left(\frac{a}{b}\right)^{-n} = \left(\frac{b}{a}\right)^n.$$

(Any base to a power is equal to the reciprocal of the base raised to the opposite power.)

EXAMPLE 5

Simplify: $\left(\dfrac{x^4}{2y}\right)^{-3}$.

SOLUTION

$$\left(\frac{x^4}{2y}\right)^{-3} = \left(\frac{2y}{x^4}\right)^3 \qquad \text{Taking the reciprocal of the base and changing the sign of the exponent}$$

$$= \frac{(2y)^3}{(x^4)^3} \qquad \text{Raising a quotient to a power by raising both the numerator and the denominator to the power}$$

$$= \frac{2^3 y^3}{x^{12}} \qquad \text{Raising a product to a power; using the power rule in the denominator}$$

$$= \frac{8y^3}{x^{12}} \qquad \text{Cubing 2}$$

TRY EXERCISE 73

Scientific Notation

When we are working with the very large numbers or very small numbers that frequently occur in science, **scientific notation** provides a useful way of writing them. The following are examples of scientific notation.

The mass of the earth:

6.0×10^{24} kilograms (kg) $= 6,000,000,000,000,000,000,000,000$ kg

The mass of a hydrogen atom:

1.7×10^{-24} g $= 0.0000000000000000000000017$ g

> ### Scientific Notation
> *Scientific notation* for a number is an expression of the type
> $$N \times 10^m,$$
> where N is at least 1 but less than 10 (that is, $1 \le N < 10$), N is expressed in decimal notation, and m is an integer.

STUDENT NOTES

Definitions are usually written as concisely as possible, so that every phrase included is important. The definition for scientific notation states that $1 \le N < 10$. Thus, 2.68×10^5 is written in scientific notation, but 26.8×10^5 and 0.268×10^5 are *not* written in scientific notation.

Converting from scientific notation to decimal notation involves multiplying by a power of 10. Consider the following.

Scientific Notation	Multiplication	Decimal Notation
4.52×10^2	4.52×100	452.
4.52×10^1	4.52×10	45.2
4.52×10^0	4.52×1	4.52
4.52×10^{-1}	4.52×0.1	0.452
4.52×10^{-2}	4.52×0.01	0.0452

Note that when m, the power of 10, is positive, the decimal point moves right m places in decimal notation. When m is negative, the decimal point moves left $|m|$ places. We generally try to perform this multiplication mentally.

EXAMPLE 6   Convert to decimal notation: **(a)** 7.893×10^5; **(b)** 4.7×10^{-8}.

SOLUTION

a) Since the exponent is positive, the decimal point moves to the right:

$$7.89300.\quad\quad 7.893 \times 10^5 = 789,300 \quad\quad \text{The decimal point moves to the}$$
$$\overset{\curvearrowright}{\underset{\text{5 places}}{}} \quad\quad\quad\quad\quad\quad\quad\quad\quad\quad \text{right 5 places.}$$

b) Since the exponent is negative, the decimal point moves to the left:

$$0.00000004.7\quad\quad 4.7 \times 10^{-8} = 0.000000047 \quad\quad \text{The decimal point}$$
$$\overset{\curvearrowleft}{\underset{\text{8 places}}{}} \quad\quad\quad\quad\quad\quad\quad\quad\quad\quad\quad \text{moves to the left 8 places.}$$

TRY EXERCISE ▶ 81

To convert from decimal notation to scientific notation, this procedure is reversed.

EXAMPLE 7   Write in scientific notation: **(a)** 83,000; **(b)** 0.0327.

SOLUTION

a) We need to find m such that $83,000 = 8.3 \times 10^m$. To change 8.3 to 83,000 requires moving the decimal point 4 places to the right. This can be accomplished by multiplying by 10^4. Thus,

$$83,000 = 8.3 \times 10^4. \quad\quad \text{This is scientific notation.}$$

b) We need to find m such that $0.0327 = 3.27 \times 10^m$. To change 3.27 to 0.0327 requires moving the decimal point 2 places to the left. This can be accomplished by multiplying by 10^{-2}. Thus,

$$0.0327 = 3.27 \times 10^{-2}. \quad\quad \text{This is scientific notation.}$$

TRY EXERCISE ▶ 91

Conversions to and from scientific notation are often made mentally. Remember that positive exponents are used to represent large numbers and negative exponents are used to represent small numbers between 0 and 1.

Multiplying and Dividing Using Scientific Notation

Products and quotients of numbers written in scientific notation are found using the rules for exponents.

EXAMPLE 8   Simplify.

a) $(1.8 \times 10^9) \cdot (2.3 \times 10^{-4})$

b) $(3.41 \times 10^5) \div (1.1 \times 10^{\ 3})$

SOLUTION

a) $(1.8 \times 10^9) \cdot (2.3 \times 10^{-4})$

$\quad\quad = 1.8 \times 2.3 \times 10^9 \times 10^{-4} \quad\quad$ Using the associative and commutative laws

$\quad\quad = 4.14 \times 10^{9+(-4)} \quad\quad\quad\quad\quad$ Adding exponents

$\quad\quad = 4.14 \times 10^5$

b) $(3.41 \times 10^5) \div (1.1 \times 10^{-3})$

$$= \frac{3.41 \times 10^5}{1.1 \times 10^{-3}}$$

$$= \frac{3.41}{1.1} \times \frac{10^5}{10^{-3}}$$

$$= 3.1 \times 10^{5-(-3)} \qquad \text{Subtracting exponents}$$

$$= 3.1 \times 10^8$$

TRY EXERCISE 103

When a problem is stated using scientific notation, we generally use scientific notation for the answer. This often requires an additional conversion.

EXAMPLE 9 Simplify.

a) $(3.1 \times 10^5) \cdot (4.5 \times 10^{-3})$ **b)** $(7.2 \times 10^{-7}) \div (8.0 \times 10^6)$

SOLUTION

a) We have

$$(3.1 \times 10^5) \cdot (4.5 \times 10^{-3}) = 3.1 \times 4.5 \times 10^5 \times 10^{-3}$$

$$= 13.95 \times 10^2.$$

Our answer is not yet in scientific notation because 13.95 is not between 1 and 10. We convert to scientific notation as follows:

$$13.95 \times 10^2 = 1.395 \times 10^1 \times 10^2 \qquad \text{Substituting } 1.395 \times 10^1 \text{ for } 13.95$$

$$= 1.395 \times 10^3. \qquad \text{Adding exponents}$$

b) $(7.2 \times 10^{-7}) \div (8.0 \times 10^6) = \dfrac{7.2 \times 10^{-7}}{8.0 \times 10^6} = \dfrac{7.2}{8.0} \times \dfrac{10^{-7}}{10^6}$

$$= 0.9 \times 10^{-13}$$

$$= 9.0 \times 10^{-1} \times 10^{-13} \qquad \text{Substituting } 9.0 \times 10^{-1} \text{ for } 0.9$$

$$= 9.0 \times 10^{-14} \qquad \text{Adding exponents}$$

TRY EXERCISE 111

TECHNOLOGY CONNECTION

A key labeled $\boxed{10^x}$, $\boxed{\wedge}$, or $\boxed{\text{EE}}$ is used to enter scientific notation into a calculator. Sometimes this is a secondary function, meaning that another key—often labeled SHIFT or **2ND** —must be pressed first.

To check Example 8(a), we press

1.8 $\boxed{\text{EE}}$ 9 $\boxed{\times}$ 2.3 $\boxed{\text{EE}}$ $\boxed{(-)}$ 4.

When we then press $\boxed{=}$ or **ENTER**, the result 4.14E5 appears. This represents 4.14×10^5. On many calculators, the MODE Sci must be selected in order to display scientific notation.

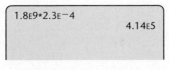

On some calculators, this appears as

4.14 05

or

4.14e+05

Calculate each of the following.

1. $(3.8 \times 10^9) \cdot (4.5 \times 10^7)$
2. $(2.9 \times 10^{-8}) \div (5.4 \times 10^6)$
3. $(9.2 \times 10^7) \div (2.5 \times 10^{-9})$

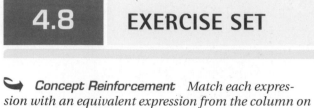

4.8   EXERCISE SET

⤴ Concept Reinforcement   *Match each expression with an equivalent expression from the column on the right.*

1. ____ $\left(\dfrac{x^3}{y^2}\right)^{-2}$ a) $\dfrac{y^6}{x^9}$

2. ____ $\left(\dfrac{y^2}{x^3}\right)^{-2}$ b) $\dfrac{x^9}{y^6}$

3. ____ $\left(\dfrac{y^{-2}}{x^{-3}}\right)^{-3}$ c) $\dfrac{y^4}{x^6}$

4. ____ $\left(\dfrac{x^{-3}}{y^{-2}}\right)^{-3}$ d) $\dfrac{x^6}{y^4}$

Express using positive exponents. Then, if possible, simplify.

5. 2^{-3} 6. 10^{-5} 7. $(-2)^{-6}$

8. $(-3)^{-4}$ 9. t^{-9} 10. x^{-7}

11. xy^{-2} 12. $a^{-3}b$ 13. $r^{-5}t$

14. xy^{-9} 15. $\dfrac{1}{a^{-8}}$ 16. $\dfrac{1}{z^{-6}}$

17. 7^{-1} 18. 3^{-1} 19. $\left(\dfrac{3}{5}\right)^{-2}$

20. $\left(\dfrac{3}{4}\right)^{-2}$ 21. $\left(\dfrac{x}{2}\right)^{-5}$ 22. $\left(\dfrac{a}{2}\right)^{-4}$

23. $\left(\dfrac{s}{t}\right)^{-7}$ 24. $\left(\dfrac{r}{v}\right)^{-5}$

Express using negative exponents.

25. $\dfrac{1}{9^2}$ 26. $\dfrac{1}{5^2}$ 27. $\dfrac{1}{y^3}$

28. $\dfrac{1}{t^4}$ 29. $\dfrac{1}{5}$ 30. $\dfrac{1}{8}$

31. $\dfrac{1}{t}$ 32. $\dfrac{1}{m}$

Simplify. Do not use negative exponents in the answer.

33. $2^{-5} \cdot 2^8$ 34. $5^{-8} \cdot 5^{10}$

35. $x^{-3} \cdot x^{-9}$ 36. $x^{-4} \cdot x^{-7}$

37. $t^{-3} \cdot t$ 38. $y^{-5} \cdot y$

39. $(n^{-5})^3$ 40. $(m^{-5})^{10}$

41. $(t^{-3})^{-6}$ 42. $(a^{-4})^{-7}$

43. $(t^4)^{-3}$ 44. $(x^4)^{-5}$

45. $(mn)^{-7}$ 46. $(ab)^{-9}$

47. $(3x^{-4})^2$ 48. $(2a^{-5})^3$

49. $(5r^{-4}t^3)^2$ 50. $(4x^5y^{-6})^3$

51. $\dfrac{t^{12}}{t^{-2}}$ 52. $\dfrac{x^7}{x^{-2}}$

53. $\dfrac{y^{-7}}{y^{-3}}$ 54. $\dfrac{z^{-6}}{z^{-2}}$

55. $\dfrac{15y^{-7}}{3y^{-10}}$ 56. $\dfrac{-12a^{-5}}{2a^{-8}}$

57. $\dfrac{2x^6}{x}$ 58. $\dfrac{3x}{x^{-1}}$

59. $\dfrac{-15a^{-7}}{10b^{-9}}$ 60. $\dfrac{12x^{-6}}{8y^{-10}}$

Aha! 61. $\dfrac{t^{-7}}{t^{-7}}$ 62. $\dfrac{a^{-5}}{b^{-7}}$

63. $\dfrac{8x^{-3}}{y^{-7}z^{-1}}$ 64. $\dfrac{10a^{-1}}{b^{-7}c^{-3}}$

65. $\dfrac{3t^4}{s^{-2}u^{-4}}$ 66. $\dfrac{5x^{-8}}{y^{-3}z^2}$

67. $(x^4y^5)^{-3}$ 68. $(t^5x^3)^{-4}$

69. $(3m^{-5}n^{-3})^{-2}$ 70. $(2y^{-4}z^{-2})^{-3}$

71. $(a^{-5}b^7c^{-2})(a^{-3}b^{-2}c^6)$

72. $(x^3y^{-4}z^{-5})(x^{-4}y^{-2}z^9)$

73. $\left(\dfrac{a^4}{3}\right)^{-2}$ 74. $\left(\dfrac{y^2}{2}\right)^{-2}$

75. $\left(\dfrac{m^{-1}}{n^{-4}}\right)^3$ 76. $\left(\dfrac{x^2y}{z^{-5}}\right)^3$

77. $\left(\dfrac{2a^2}{3b^4}\right)^{-3}$ 78. $\left(\dfrac{a^2b}{2d^3}\right)^{-5}$

Aha! 79. $\left(\dfrac{5x^{-2}}{3y^{-2}z}\right)^0$ 80. $\left(\dfrac{4a^3b^{-2}}{5c^{-3}}\right)^1$

Convert to decimal notation.

81. 4.92×10^3 82. 8.13×10^4

83. 8.92×10^{-3} 84. 7.26×10^{-4}

85. 9.04×10^8 86. 1.35×10^7

87. 3.497×10^{-6}

88. 9.043×10^{-3}

89. 4.209×10^{7}

90. 5.029×10^{8}

Convert to scientific notation.

91. 36,000,000

92. 27,400

93. 0.00583

94. 0.0814

95. 78,000,000,000

96. 3,700,000,000,000

97. 0.000000527

98. 0.0000506

99. 0.000001032

100. 0.00000008

101. 1,094,000,000,000,000

102. 1,030,200,000,000,000,000

Multiply or divide, and write scientific notation for the result.

103. $(3 \times 10^{5})(2 \times 10^{8})$

104. $(3.1 \times 10^{7})(2.1 \times 10^{-4})$

105. $(3.8 \times 10^{9})(6.5 \times 10^{-2})$

106. $(7.1 \times 10^{-7})(8.6 \times 10^{-5})$

107. $(8.7 \times 10^{-12})(4.5 \times 10^{-5})$

108. $(4.7 \times 10^{5})(6.2 \times 10^{-12})$

109. $\dfrac{8.5 \times 10^{8}}{3.4 \times 10^{-5}}$

110. $\dfrac{5.6 \times 10^{-2}}{2.5 \times 10^{5}}$

111. $(4.0 \times 10^{3}) \div (8.0 \times 10^{8})$

112. $(1.5 \times 10^{-3}) \div (1.6 \times 10^{-6})$

113. $\dfrac{7.5 \times 10^{-9}}{2.5 \times 10^{12}}$

114. $\dfrac{3.0 \times 10^{-2}}{6.0 \times 10^{10}}$

115. Without performing actual computations, explain why 3^{-29} is smaller than 2^{-29}.

116. Explain why each of the following is not scientific notation:

12.6×10^{8};

$4.8 \times 10^{1.7}$;

0.207×10^{-5}.

Skill Review

Review graphing linear equations (Chapter 3).

Graph.

117. $3x - 4y = 12$ [3.3]

118. $y = -\frac{2}{3}x + 4$ [3.6]

119. $3y - 2 = 7$ [3.3]

120. $8x = 4y$ [3.2]

121. Find the slope of the line containing the points $(3, 2)$ and $(-7, 5)$. [3.5]

122. Find the slope and the y-intercept of the line given by $2y = 8x + 7$. [3.6]

123. Find the slope–intercept form of the line with slope -5 and y-intercept $(0, -10)$. [3.6]

124. Find the slope–intercept form of the line containing the points $(6, 3)$ and $(-2, -7)$. [3.7]

Synthesis

125. Explain what requirements must be met in order for x^{-n} to represent a negative integer.

126. Explain why scientific notation cannot be used without an understanding of the rules for exponents.

127. Write the reciprocal of 1.25×10^{-6} in scientific notation.

128. Write the reciprocal of 2.5×10^{9} in scientific notation.

129. Write $8^{-3} \cdot 32 \div 16^{2}$ as a power of 2.

130. Write $81^{3} \div 27 \cdot 9^{2}$ as a power of 3.

Simplify each of the following. Use a calculator only where indicated.

Aha! **131.** $\dfrac{125^{-4}(25^{2})^{4}}{125}$

132. $(13^{-12})^{2} \cdot 13^{25}$

133. $[(5^{-3})^2]^{-1}$

134. $5^0 - 5^{-1}$

135. $3^{-1} + 4^{-1}$

136. $\dfrac{4.2 \times 10^8[(2.5 \times 10^{-5}) \div (5.0 \times 10^{-9})]}{3.0 \times 10^{-12}}$

137. $\dfrac{27^{-2}(81^2)^3}{9^8}$

138. $\dfrac{7.4 \times 10^{29}}{(5.4 \times 10^{-6})(2.8 \times 10^8)}$

139. $\dfrac{5.8 \times 10^{17}}{(4.0 \times 10^{-13})(2.3 \times 10^4)}$

140. $\dfrac{(7.8 \times 10^7)(8.4 \times 10^{23})}{2.1 \times 10^{-12}}$

141. $\dfrac{(2.5 \times 10^{-8})(6.1 \times 10^{-11})}{1.28 \times 10^{-3}}$

142. Determine whether each of the following is true for all pairs of integers m and n and all positive numbers x and y.

   a) $x^m \cdot y^n = (xy)^{mn}$
   b) $x^m \cdot y^m = (xy)^{2m}$
   c) $(x - y)^m = x^m - y^m$

Solve. Write scientific notation for each answer.

143. *Ecology.* In one year, a large tree can remove from the air the same amount of carbon dioxide produced by a car traveling 500 mi. If New York City contains approximately 600,000 trees, how many miles of car traffic can those trees clean in a year?
Sources: Colorado Tree Coalition; New York City Department of Parks and Recreations

144. *Computer technology.* One gigabit is about 1 billion bits of information. In 2007, Intel Corp. began making silicon modulators that can encode data onto a beam of light at a rate of 40 gigabits per second. If 25 of these communication lasers are packed on a single chip, how many bits per second could that chip encode?
Source: *The Wall Street Journal,* 7/25/2007

145. *Hotel management.* The new Four Seasons Hotel in Seattle contains 110,000 ft^2 of condominium space. If these condos sold for about $2100 per ft^2, how much money did the hotel make selling the condominiums?
Source: seattletimes.nwsource.com

146. *Coral reefs.* There are 10 million bacteria per square centimeter of coral in a coral reef. The coral reefs near the Hawaiian Islands cover 14,000 km^2. How many bacteria are there in Hawaii's coral reefs?
Sources: livescience.com; U.S. Geological Survey

147. *Hospital care.* In 2005, 115 million patients visited emergency rooms in the United States. If the average visit lasted 3.3 hr, how many minutes in all did people spend in emergency rooms in 2005?
Source: *The Indianapolis Star,* 7/25/07

CONNECTING the CONCEPTS

The following properties of exponents hold for all integers m and n, assuming that no denominator is 0 and that 0^0 is not considered.

Definitions and Properties of Exponents

The following summary assumes that no denominators are 0 and that 0^0 is not considered. For any integers m and n,

1 as an exponent:	$a^1 = a$
0 as an exponent:	$a^0 = 1$
Negative exponents:	$a^{-n} = \dfrac{1}{a^n},$
	$\dfrac{a^{-n}}{b^{-m}} = \dfrac{b^m}{a^n},$
	$\left(\dfrac{a}{b}\right)^{-n} = \left(\dfrac{b}{a}\right)^n$
The Product Rule:	$a^m \cdot a^n = a^{m+n}$
The Quotient Rule:	$\dfrac{a^m}{a^n} = a^{m-n}$
The Power Rule:	$(a^m)^n = a^{mn}$
Raising a product to a power:	$(ab)^n = a^n b^n$
Raising a quotient to a power:	$\left(\dfrac{a}{b}\right)^n = \dfrac{a^n}{b^n}$

MIXED REVIEW

Simplify. Do not use negative exponents in the answer.

1. $x^4 x^{10}$

2. $x^{-4} x^{-10}$

3. $\dfrac{x^4}{x^{10}}$

4. $\dfrac{x^4}{x^{-10}}$

5. $(x^{-4})^{-10}$

6. $(x^4)^{10}$

7. $\dfrac{1}{c^{-8}}$

8. c^{-8}

9. $(2x^3 y)^4$

10. $(2x^3 y)^{-4}$

11. $(3xy^{-1}z^5)^0$

12. $(a^2 b)(a^3 b^{-1})$

13. $\left(\dfrac{a^3}{b^4}\right)^5$

14. $\left(\dfrac{a^3}{b^4}\right)^{-5}$

15. $\dfrac{30x^4 y^3}{12xy^7}$

16. $\dfrac{12ab^{-8}}{14a^{-1}b^{-3}}$

17. $\dfrac{7p^{-5}}{xt^{-6}}$

18. $\left(\dfrac{3a^{-1}}{4b^{-3}}\right)^{-2}$

19. $(2p^2 q^4)(3pq^5)^2$

20. $(2xy^{-1})^{-1}(3x^2 y^{-3})^2$

Study Summary

KEY TERMS AND CONCEPTS	EXAMPLES

SECTION 4.1: EXPONENTS AND THEIR PROPERTIES

(Assume that no denominators are 0 and that 0^0 is not considered.)

For any integers m and n:

1 as an exponent:	$a^1 = a$	$3^1 = 3$
0 as an exponent:	$a^0 = 1$	$3^0 = 1$
The Product Rule:	$a^m \cdot a^n = a^{m+n}$	$3^5 \cdot 3^9 = 3^{5+9} = 3^{14}$
The Quotient Rule:	$\dfrac{a^m}{a^n} = a^{m-n}$	$\dfrac{3^7}{3} = 3^{7-1} = 3^6$
The Power Rule:	$(a^m)^n = a^{mn}$	$(3^4)^2 = 3^{4 \cdot 2} = 3^8$
Raising a product to a power:	$(ab)^n = a^n b^n$	$(3x^5)^4 = 3^4(x^5)^4 = 81x^{20}$
Raising a quotient to a power:	$\left(\dfrac{a}{b}\right)^n = \dfrac{a^n}{b^n}$	$\left(\dfrac{3}{x}\right)^6 = \dfrac{3^6}{x^6}$

SECTION 4.2: POLYNOMIALS

A **monomial** is a product of constants and/or variables.

$4,\quad x,\quad 4x^5,\quad$ and $\quad 4x^2y^3\quad$ are monomials.

A **polynomial** is a monomial or a sum of monomials.

When a polynomial is written as a sum of monomials, each monomial is a **term** of the polynomial.

The **degree of a term** of a polynomial is the number of variable factors in that term.

The **coefficient** of a term is the part of the term that is a constant factor.

The **leading term** of a polynomial is the term of highest degree.

The **leading coefficient** is the coefficient of the leading term.

The **degree of the polynomial** is the degree of the leading term.

Polynomial: $10x - x^3 - \frac{1}{2}x^2 + 3x^5 + 7$

Term	$10x$	$-x^3$	$-\frac{1}{2}x^2$	$3x^5$	7
Degree of term	1	3	2	5	0
Coefficient of term	10	-1	$-\frac{1}{2}$	3	7
Leading term			$3x^5$		
Leading coefficient			3		
Degree of polynomial			5		

A polynomial is written in **descending order** if it is arranged with terms of higher degree written before terms of lower degree.

Written in descending order:
$$3x^5 - x^3 - \tfrac{1}{2}x^2 + 10x + 7$$

A **monomial** has one term.

A **binomial** has two terms.

A **trinomial** has three terms.

Monomial (one term): $4x^3$

Binomial (two terms): $x^2 - 5$

Trinomial (three terms): $3t^3 + 2t - 10$

Like terms, or **similar terms**, are either constant terms or terms containing the same variable(s) raised to the same power(s). These can be **combined** within a polynomial.

Combine like terms: $3y^4 + 6y^2 - 7 - y^4 - 6y^2 + 8.$

$3y^4 + 6y^2 - 7 - y^4 - 6y^2 + 8 = \underline{3y^4 - y^4} + \underline{6y^2 - 6y^2} \underline{-7 + 8}$

$$= \quad 2y^4 \quad + \quad 0 \quad + 1$$

$$= 2y^4 + 1$$

To **evaluate** a polynomial, replace the variable with a number. The **value** is calculated using the rules for order of operations.

Evaluate $t^3 - 2t^2 - 5t + 1$ *for* $t = -2.$

$$t^3 - 2t^2 - 5t + 1 = (-2)^3 - 2(-2)^2 - 5(-2) + 1$$

$$= -8 - 2(4) - (-10) + 1$$

$$= -8 - 8 + 10 + 1$$

$$= -5$$

SECTION 4.3: ADDITION AND SUBTRACTION OF POLYNOMIALS

Add polynomials by combining like terms.

$(2x^2 - 3x + 7) + (5x^3 + 3x - 9)$

$= 2x^2 + (-3x) + 7 + 5x^3 + 3x + (-9)$

$= 5x^3 + 2x^2 - 2$

Subtract polynomials by adding the opposite of the polynomial being subtracted.

$(2x^2 - 3x + 7) - (5x^3 + 3x - 9)$

$= 2x^2 - 3x + 7 + (-5x^3 - 3x + 9)$

$= 2x^2 - 3x + 7 - 5x^3 - 3x + 9$

$= -5x^3 + 2x^2 - 6x + 16$

SECTION 4.4: MULTIPLICATION OF POLYNOMIALS

Multiply polynomials by multiplying each term of one polynomial by each term of the other.

$(x + 2)(x^2 - x - 1)$

$= x \cdot x^2 - x \cdot x - x \cdot 1 + 2 \cdot x^2 - 2 \cdot x - 2 \cdot 1$

$= x^3 - x^2 - x + 2x^2 - 2x - 2$

$= x^3 + x^2 - 3x - 2$

SECTION 4.5: SPECIAL PRODUCTS

FOIL (First, Outer, Inner, Last):

$(A + B)(C + D) = AC + AD + BC + BD$

$(x + 3)(x - 2) = x^2 - 2x + 3x - 6$

$$= x^2 + x - 6$$

The product of a sum and a difference:

$$(A + B)(A - B) = A^2 - B^2$$

$A^2 - B^2$ is called a **difference of squares.**

$(t^3 + 5)(t^3 - 5) = (t^3)^2 - 5^2$

$$= t^6 - 25$$

The square of a binomial:

$$(A + B)^2 = A^2 + 2AB + B^2;$$
$$(A - B)^2 = A^2 - 2AB + B^2$$

$A^2 + 2AB + B^2$ and $A^2 - 2AB + B^2$ are called **perfect-square trinomials.**

$$(5x + 3)^2 = (5x)^2 + 2(5x)(3) + 3^2 = 25x^2 + 30x + 9;$$
$$(5x - 3)^2 = (5x)^2 - 2(5x)(3) + 3^2 = 25x^2 - 30x + 9$$

SECTION 4.6: POLYNOMIALS IN SEVERAL VARIABLES

To **evaluate** a polynomial, replace each variable with a number and simplify.

Evaluate $4 - 3xy + x^2y$ *for* $x = 5$ *and* $y = -1$.

$$4 - 3xy + x^2y = 4 - 3(5)(-1) + (5)^2(-1)$$
$$= 4 - (-15) + (-25)$$
$$= -6$$

The **degree** of a term is the number of variables in the term or the sum of the exponents of the variables.

The degree of $-19x^3yz^2$ is 6.

Add, subtract, and multiply polynomials in several variables in the same way as polynomials in one variable.

$$(3xy^2 - 4x^2y + 5xy) + (xy - 6x^2y) = 3xy^2 - 10x^2y + 6xy;$$
$$(3xy^2 - 4x^2y + 5xy) - (xy - 6x^2y) = 3xy^2 + 2x^2y + 4xy;$$
$$(2a^2b + 3a)(5a^2b - a) = 10a^4b^2 + 13a^3b - 3a^2$$

SECTION 4.7: DIVISION OF POLYNOMIALS

To divide a polynomial by a monomial, divide each term by the monomial. Divide coefficients and subtract exponents.

$$\frac{3t^5 - 6t^4 + 4t^2 + 9t}{3t} = \frac{3t^5}{3t} - \frac{6t^4}{3t} + \frac{4t^2}{3t} + \frac{9t}{3t}$$
$$= t^4 - 2t^3 + \tfrac{4}{3}t + 3$$

To divide a polynomial by a binomial, use long division.

Divide: $(x^2 + 5x - 2) \div (x - 3)$.

$$\begin{array}{r} x + 8 \\ x - 3 \overline{)\, x^2 + 5x - 2} \\ \underline{-(x^2 - 3x)} \\ 8x - 2 \\ \underline{-(8x - 24)} \\ 22 \end{array}$$

$$(x^2 + 5x - 2) \div (x - 3) = x + 8 + \frac{22}{x - 3}$$

SECTION 4.8: NEGATIVE EXPONENTS AND SCIENTIFIC NOTATION

$$a^{-n} = \frac{1}{a^n};$$

$$\frac{a^{-n}}{b^{-m}} = \frac{b^m}{a^n};$$

$$\left(\frac{a}{b}\right)^{-n} = \left(\frac{b}{a}\right)^n$$

$$3^{-2} = \frac{1}{3^2} = \frac{1}{9};$$

$$\frac{3^{-7}}{x^{-5}} = \frac{x^5}{3^7};$$

$$\left(\frac{3}{x}\right)^{-2} = \left(\frac{x}{3}\right)^2$$

Scientific notation is given by $N \times 10^m$, where m is an integer, N is in decimal notation, and $1 \le N < 10$.

$$4100 = 4.1 \times 10^3;$$
$$5 \times 10^{-3} = 0.005$$

Review Exercises: Chapter 4

Concept Reinforcement *Classify each statement as either true or false.*

1. When two polynomials that are written in descending order are added, the result is generally written in descending order. [4.3]

2. The product of the sum and the difference of the same two terms is a difference of squares. [4.5]

3. When a binomial is squared, the result is a perfect-square trinomial. [4.5]

4. FOIL can be used whenever two polynomials are being multiplied. [4.5]

5. The degree of a polynomial cannot exceed the value of the polynomial's leading coefficient. [4.2]

6. Scientific notation is used only for extremely large numbers. [4.8]

7. FOIL can be used with polynomials in several variables. [4.6]

8. A positive number raised to a negative exponent can never represent a negative number. [4.8]

Simplify. [4.1]

9. $n^3 \cdot n^8 \cdot n$

10. $(7x)^8 \cdot (7x)^2$

11. $t^6 \cdot t^0$

12. $\dfrac{4^5}{4^2}$

13. $\dfrac{(a+b)^4}{(a+b)^4}$

14. $\dfrac{-18c^9 d^3}{2c^5 d}$

15. $(-2xy^2)^3$

16. $(2x^3)(-3x)^2$

17. $(a^2 b)(ab)^5$

18. $\left(\dfrac{2t^5}{3s^4}\right)^2$

Identify the terms of each polynomial. [4.2]

19. $8x^2 - x + \frac{2}{3}$

20. $-4y^5 + 7y^2 - 3y - 2$

List the coefficients of the terms in each polynomial. [4.2]

21. $9x^2 - x + 7$

22. $7n^4 - \frac{5}{6}n^2 - 4n + 10$

For each polynomial, (a) *list the degree of each term;* (b) *determine the leading term and the leading coefficient; and* (c) *determine the degree of the polynomial.* [4.2]

23. $4t^2 + 6 + 15t^5$

24. $-2x^5 + 7 - 3x^2 + x$

Classify each polynomial as a monomial, a binomial, a trinomial, or a polynomial with no special name. [4.2]

25. $4x^3 - 5x + 3$

26. $4 - 9t^3 - 7t^4 + 10t^2$

27. $7y^2$

Combine like terms and write in descending order. [4.2]

28. $3x - x^2 + 4x$

29. $\frac{3}{4}x^3 + 4x^2 - x^3 + 7$

30. $-4t^3 + 2t + 4t^3 + 8 - t - 9$

31. $-a + \frac{1}{3} + 20a^5 - 1 - 6a^5 - 2a^2$

Evaluate each polynomial for $x = -2$. [4.2]

32. $9x - 6$

33. $x^2 - 3x + 6$

Add or subtract. [4.3]

34. $(8x^4 - x^3 + x - 4) + (x^5 + 7x^3 - 3x - 5)$

35. $(5a^5 - 2a^3 - 9a^2) + (2a^5 + a^3) + (-a^5 - 3a^2)$

36. $(y^2 + 8y - 7) - (4y^2 - 10)$

37. $(3x^5 - 4x^4 + 2x^2 + 3) - (2x^5 - 4x^4 + 3x^3 + 4x^2 - 5)$

38.
$$\begin{aligned}
-\tfrac{3}{4}x^4 &+ \tfrac{1}{2}x^3 &&&&+ \tfrac{7}{8}\\
&- \tfrac{1}{4}x^3 &- x^2 &- \tfrac{7}{4}x\\
+\tfrac{3}{2}x^4 &&+ \tfrac{2}{3}x^2 &&&- \tfrac{1}{2}
\end{aligned}$$

39.
$$\begin{aligned}
2x^5 \quad &- x^3 \quad + x + 3\\
-(3x^5 - x^4 &+ 4x^3 + 2x^2 - x + 3)
\end{aligned}$$

40. The length of a rectangle is 3 m greater than its width.

w + 3

w

a) Find a polynomial for the perimeter. [4.3]
b) Find a polynomial for the area. [4.4]

Multiply.

41. $5x^2(-6x^3)$ [4.4]

42. $(7x + 1)^2$ [4.5]

43. $(a - 7)(a + 4)$ [4.5]

44. $(d - 8)(d + 8)$ [4.5]

45. $(4x^2 - 5x + 1)(3x - 2)$ [4.4]

46. $(x - 8)^2$ [4.5]

47. $3t^2(5t^3 - 2t^2 + 4t)$ [4.4]

48. $(2a + 9)(2a - 9)$ [4.5]

49. $(x - 0.8)(x - 0.5)$ [4.5]

50. $(x^4 - 2x + 3)(x^3 + x - 1)$ [4.4]

51. $(4y^3 - 5)^2$ [4.5]

52. $(2t^2 + 3)(t^2 - 7)$ [4.5]

53. $\left(a - \frac{1}{2}\right)\left(a + \frac{2}{3}\right)$ [4.5]

54. $(-7 + 2n)(7 + 2n)$ [4.5]

55. Evaluate $2 - 5xy + y^2 - 4xy^3 + x^6$ for $x = -1$ and $y = 2$. [4.6]

Identify the coefficient and the degree of each term of each polynomial. Then find the degree of each polynomial. [4.6]

56. $x^5y - 7xy + 9x^2 - 8$

57. $a^3b^8c^2 - c^{22} + a^5c^{10}$

Combine like terms. [4.6]

58. $u + 3v - 5u + v - 7$

59. $6m^3 + 3m^2n + 4mn^2 + m^2n - 5mn^2$

Add or subtract. [4.6]

60. $(4a^2 - 10ab - b^2) + (-2a^2 - 6ab + b^2)$

61. $(6x^3y^2 - 4x^2y - 6x) - (-5x^3y^2 + 4x^2y + 6x^2 - 6)$

Multiply. [4.6]

62. $(2x + 5y)(x - 3y)$

63. $(5ab - cd^2)^2$

64. Find a polynomial for the shaded area. [4.6]

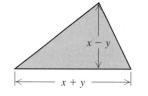

Divide. [4.7]

65. $(3y^5 - y^2 + 12y) \div (3y)$

66. $(6x^3 - 5x^2 - 13x + 13) \div (2x + 3)$

67. $\dfrac{t^4 + t^3 + 2t^2 - t - 3}{t + 1}$

68. Express using a positive exponent: 8^{-6}. [4.8]

69. Express using a negative exponent: $\dfrac{1}{a^9}$. [4.8]

Simplify. Do not use negative exponents in the answer. [4.8]

70. $4^5 \cdot 4^{-7}$

71. $\dfrac{6a^{-5}b}{3a^8b^{-8}}$

72. $(w^3)^{-5}$

73. $(2x^{-3}y)^{-2}$

74. $\left(\dfrac{2x}{y}\right)^{-3}$

75. Convert to decimal notation: 4.7×10^8. [4.8]

76. Convert to scientific notation: 0.0000109. [4.8]

Multiply or divide and write scientific notation for the result. [4.8]

77. $(3.8 \times 10^4)(5.5 \times 10^{-1})$

78. $\dfrac{1.28 \times 10^{-8}}{2.5 \times 10^{-4}}$

Synthesis

79. Explain why $5x^3$ and $(5x)^3$ are not equivalent expressions. [4.1]

80. A binomial is squared and the result, written in descending order, is $x^2 - 6x + 9$. Is it possible to determine what binomial was squared? Why or why not? [4.5]

81. Determine, without performing the multiplications, the degree of each product. [4.4]

   a) $(x^5 - 6x^2 + 3)(x^4 + 3x^3 + 7)$

   b) $(x^7 - 4)^4$

82. Simplify:
$$(-3x^5 \cdot 3x^3 - x^6(2x)^2 + (3x^4)^2 + (2x^2)^4 - 20x^2(x^3)^2)^2. \quad [4.1], [4.2]$$

83. A polynomial has degree 4. The x^2-term is missing. The coefficient of x^4 is two times the coefficient of x^3. The coefficient of x is 3 less than the coefficient of x^4. The remaining coefficient is 7 less than the coefficient of x. The sum of the coefficients is 15. Find the polynomial. [4.2]

84. Multiply: $[(x - 5) - 4x^3][(x - 5) + 4x^3]$. [4.5]

85. Solve: $(x - 7)(x + 10) = (x - 4)(x - 6)$. [2.2], [4.5]

86. *Blood donors.* Every 4–6 weeks, Jordan donates 1.14×10^6 cubic millimeters (two pints) of whole blood, from which platelets are removed and the blood returned to the body. In one cubic millimeter of blood, there are about 2×10^5 platelets. Approximate the number of platelets in Jordan's typical donation. [4.8]

Test: Chapter 4

CHAPTER Test Prep VIDEO CD

Step-by-step test solutions are found on the video CD in the front of this book.

Simplify.

1. $x^7 \cdot x \cdot x^5$

2. $\dfrac{3^8}{3^7}$

3. $\dfrac{(3m)^4}{(3m)^4}$

4. $(t^5)^9$

5. $(-3y^2)^3$

6. $(5x^4y)(-2x^5y)^3$

7. $\dfrac{24a^7b^4}{20a^2b}$

8. $\left(\dfrac{4p}{5q^3}\right)^2$

9. Classify $4x^2y - 7y^3$ as a monomial, a binomial, a trinomial, or a polynomial with no special name.

10. Identify the coefficient of each term of the polynomial:
$$3x^5 - x + \tfrac{1}{9}.$$

11. Determine the degree of each term, the leading term and the leading coefficient, and the degree of the polynomial:
$$2t^3 - t + 7t^5 + 4.$$

12. Evaluate $x^2 + 5x - 1$ for $x = -3$.

Combine like terms and write in descending order.

13. $4a^2 - 6 + a^2$

14. $y^2 - 3y - y + \tfrac{3}{4}y^2$

15. $3 - x^2 + 8x + 5x^2 - 6x - 2x + 4x^3$

Add or subtract.

16. $(3x^5 + 5x^3 - 5x^2 - 3) + (x^5 + x^4 - 3x^2 + 2x - 4)$

17. $\left(x^4 + \tfrac{2}{3}x + 5\right) + \left(4x^4 + 5x^2 + \tfrac{1}{3}x\right)$

18. $(5a^4 + 3a^3 - a^2 - 2a - 1) - (7a^4 - a^2 - a + 6)$

19. $(t^3 - 0.3t^2 - 20) - (t^4 - 1.5t^3 + 0.3t^2 - 11)$

Multiply.

20. $-2x^2(3x^2 - 3x - 5)$

21. $\left(x - \tfrac{1}{3}\right)^2$

22. $(5t - 7)(5t + 7)$

23. $(3b + 5)(2b - 1)$

24. $(x^6 - 4)(x^8 + 4)$

25. $(8 - y)(6 + 5y)$

26. $(2x + 1)(3x^2 - 5x - 3)$

27. $(8a^3 + 3)^2$

28. Evaluate $2x^2y - 3y^2$ for $x = -3$ and $y = 2$.

29. Combine like terms:
$$2x^3y - y^3 + xy^3 + 8 - 6x^3y - x^2y^2 + 11.$$

30. Subtract:
$$(8a^2b^2 - ab + b^3) - (-6ab^2 - 7ab - ab^3 + 5b^3).$$

31. Multiply: $(3x^5 - y)(3x^5 + y)$.

Divide.

32. $(12x^4 + 9x^3 - 15x^2) \div (3x^2)$

33. $(6x^3 - 8x^2 - 14x + 13) \div (3x + 2)$

34. Express using a positive exponent: y^{-7}.

35. Express using a negative exponent: $\dfrac{1}{5^6}$.

Simplify.

36. $t^{-4} \cdot t^{-5}$

37. $\dfrac{9x^3y^2}{3x^8y^{-3}}$

38. $(2a^3b^{-1})^{-4}$

39. $\left(\dfrac{ab}{c}\right)^{-3}$

40. Convert to scientific notation: 3,060,000,000.

41. Convert to decimal notation: 5×10^{-8}.

Multiply or divide and write scientific notation for the result.

42. $\dfrac{5.6 \times 10^6}{3.2 \times 10^{-11}}$

43. $(2.4 \times 10^5)(5.4 \times 10^{16})$

Synthesis

44. The height of a box is 1 less than its length, and the length is 2 more than its width. Express the volume in terms of the length.

45. Solve: $x^2 + (x - 7)(x + 4) = 2(x - 6)^2$.

46. Simplify: $2^{-1} - 4^{-1}$.

47. Every day about 12.4 billion spam e-mails are sent. If each spam e-mail wastes 4 sec, how many hours are wasted each day due to spam?
Source: spam-filter-review.toptenreviews.com

Cumulative Review: Chapters 1–4

1. Evaluate $\dfrac{2x + y}{5}$ for $x = 12$ and $y = 6$. [1.1]

2. Evaluate $5x^2y - xy + y^2$ for $x = -1$ and $y = -2$. [4.6]

Simplify.

3. $\frac{1}{15} - \frac{2}{9}$ [1.3]

4. $2 - [10 - (5 + 12 \div 2^2 \cdot 3)]$ [1.8]

5. $2y - (y - 7) + 3$ [1.8]

6. $t^4 \cdot t^7 \cdot t$ [4.1]

7. $\dfrac{-100x^6y^8}{25xy^5}$ [4.1]

8. $(2a^2b)(5ab^3)^2$ [4.1]

9. Factor: $10a - 6b + 12$. [1.2]

10. Find the absolute value: $\left|\dfrac{11}{16}\right|$. [1.4]

11. Determine the degree of the polynomial
$-x^4 + 5x^3 + 3x^6 - 1$. [4.2]

12. Combine like terms and write in descending order:
$-\frac{1}{2}t^3 + 3t^2 + \frac{3}{4}t^3 + 0.1 - 8t^2 - 0.45$. [4.2]

13. In which quadrant is $(-2, 5)$ located? [3.1]

14. Graph on a number line: $x > -1$. [2.6]

Graph.

15. $3y + 2x = 0$ [3.2]

16. $3y - 2x = 12$ [3.3]

17. $3y = 2$ [3.3]

18. $3y = 2x + 9$ [3.6]

19. Find the slope and the y-intercept of the line given by $y = \frac{1}{10}x + \frac{3}{8}$. [3.6]

20. Find the slope of the line containing the points $(2, 3)$ and $(-6, 8)$. [3.5]

21. Write an equation of the line with slope $-\frac{2}{3}$ and y-intercept $(0, -10)$. [3.6]

22. Find the coordinates of the x- and y-intercepts of the graph of $2x + 5y = 8$. Do not graph. [3.3]

Solve.

23. $\frac{1}{6}n = -\frac{2}{3}$ [2.1]

24. $3 - 5x = 0$ [2.2]

25. $5y + 7 = 8y - 1$ [2.2]

26. $0.4t - 0.5 = 8.3$ [2.2]

27. $5(3 - x) = 2 + 7(x - 1)$ [2.2]

28. $2 - (x - 7) = 8 - 4(x + 5)$ [2.2]

29. $-\frac{1}{2}t \le 4$ [2.6]

30. $3x - 5 > 9x - 8$ [2.6]

31. Solve $c = \dfrac{5pq}{2t}$ for t. [2.3]

Add or subtract.

32. $(2u^2v - uv^2 + uv) + (3u^2 - v^2u + 5vu^2)$ [4.6]

33. $(3x^3 - 2x^2 + 6x) - (x^2 - 6x + 7)$ [4.3]

34. $(2x^5 - x^4 - x) - (x^5 - x^4 + x)$ [4.3]

Multiply.

35. $10(2a - 3b + 7)$ [1.2]

36. $8x^3(-2x^2 - 6x + 7)$ [4.4]

37. $(3a + 7)(2a - 1)$ [4.5]

38. $(x - 2)(x^2 + x - 5)$ [4.4]

39. $(4t^2 + 3)^2$ [4.5]

40. $\left(\frac{1}{2}x + 1\right)\left(\frac{1}{2}x - 1\right)$ [4.5]

41. $(2r^2 + s)(3r^2 - 4s)$ [4.6]

42. Divide: $(x^2 - x + 3) \div (x - 1)$. [4.7]

Simplify. Do not use negative exponents in the answer. [4.8]

43. 7^{-10}

44. $x^{-8} \cdot x$

45. $\left(\dfrac{4s}{3t^{-5}}\right)^{-2}$

46. $(3x^{-7}y^{-2})^{-1}$

Solve.

47. In 2007, Europe and the United States together had installed wind turbines capable of producing about 60 thousand megawatts of electricity. Europe's wind-turbine capacity was four times that of the United States. What was Europe's wind-turbine capacity in 2007? [2.5]
Source: BP Statistical Review of World Energy, 2007

48. In the first four months of 2007, U.S. electric utilities used coal to generate 644 megawatt hours of electricity. This was 49.6% of the total amount of electricity generated. How much electricity was generated in the first four months of 2007? [2.4]
Source: U.S. Energy Information Administration

49. Antonio's energy-efficient washer and dryer use a total of 70 kilowatt hours (kWh) of electricity each month. The dryer uses 10 more than twice as many kilowatt hours as the washer. How many kilowatt hours does each appliance use in a month? [2.5]

50. A typical two-person household will use 195 kWh of electricity each month to heat water. The usage increases to 315 kWh per month for a four-person household. [3.7]
Source: Lee County Electric Cooperative

   a) Graph the data and determine an equation for the related line. Let w represent the number of kilowatt hours used each month and n represent the number of people in a household.

   b) Use the equation of part (a) to estimate the number of kilowatt hours used each month by a five-person household.

51. Abrianna's contract specifies that she cannot work more than 40 hr per week. For the first 4 days of one week, she worked 8, 10, 7, and 9 hr. How many hours can she work on the fifth day without violating her contract? [2.7]

52. In 2006, Brazil owed about \$240 billion in external debt. This amount was $\frac{3}{2}$ of Russia's debt. How much did Russia owe in 2006? [2.5]

53. In 2006, the average GPA of incoming freshmen at the University of South Carolina was 3.73. This was a 3.6% increase over the average GPA in 2001. What was the average GPA in 2001? [2.4]
Source: *The Wall Street Journal*, 11/10/06

54. U.S. retail losses due to crime have increased from \$26 billion in 1998 to \$38 billion in 2005. Find the rate of change of retail losses due to crime. [3.4]
Source: University of Florida

Synthesis

Solve. If no solution exists, state this.

55. $3x - 2(x + 6) = 4(x - 3)$ [2.2]

56. $x - (2x - 1) = 3x - 4(x + 1) + 10$ [2.2]

57. $(x - 2)(x + 3) = (4 - x)^2$ [2.2], [4.5]

58. Find the equation of a line with the same slope as $y = \frac{1}{2}x - 7$ and the same y-intercept as $2y = 5x + 8$. [3.6]

Simplify.

59. $7^{-1} + 8^0$ [4.8]

60. $-2x^5(x^7) + (x^3)^4 - (4x^5)^2(-x^2)$ [4.1], [4.2]

Polynomials and Factoring

CHRIS GJERSVIK
PRINCIPAL NETWORK ENGINEER
Upper Saddle River, New Jersey

The math skills we use in net-working–telecommunications help us in so many ways. For instance, we use math when planning the budgets for a computer network or phone system and when determining capacity of a network. We need to accurately size network circuits so they have enough bandwidth to carry phone calls, Internet access, streaming video, e-mail, and other such applications. Finally, we use math when we need to analyze network traffic and break it down into its most basic form, which uses the binary numeral system.

AN APPLICATION

The number N, in millions, of broadband cable and DSL subscribers in the United States t years after 1998 can be approximated by

$$N = 0.3t^2 + 0.6t.$$

When will there be 36 million broadband cable and DSL subscribers?

Source: Based on information from Leichtman Research Group

This problem appears as Exercise 19 in Section 5.7.

1 n Chapter 1, we learned that *factoring* is multiplying reversed. Thus factoring polynomials requires a solid command of the multiplication methods learned in Chapter 4. Factoring is an important skill that will be used to solve equations and simplify other types of expressions found later in the study of algebra.

In Sections 5.1–5.5, we factor polynomials to find equivalent expressions that are products. In Sections 5.6 and 5.7, we use factoring to solve equations, including those that arise from real-world problems.

5.1   Introduction to Factoring

Factoring Monomials • Factoring When Terms Have a Common Factor •
Factoring by Grouping • Checking by Evaluating

Just as a number like 15 can be factored as $3 \cdot 5$, a polynomial like $x^2 + 7x$ can be factored as $x(x + 7)$. In both cases, we ask ourselves, "What was multiplied to obtain the given result?" The situation is much like a popular television game show in which an "answer" is given and participants must find the "question" to which the answer corresponds.

STUDY SKILLS

You've Got Mail

Many students overlook an excellent opportunity to get questions answered—e-mail. If your instructor makes his or her e-mail address available, consider using it to get help. Often, just the act of writing out your question brings some clarity. If you do use e-mail, allow some time for your instructor to reply.

> ### Factoring
>
> To *factor* a polynomial is to find an equivalent expression that is a product. An equivalent expression of this type is called a *factorization* of the polynomial.

Factoring Monomials

To factor a monomial, we find two monomials whose product is equivalent to the original monomial. For example, $20x^2$ can be factored as $2 \cdot 10x^2$, $4x \cdot 5x$, or $10x \cdot 2x$, as well as several other ways. To check, we multiply.

EXAMPLE   1   Find three factorizations of $15x^3$.

SOLUTION

a) $15x^3 = (3 \cdot 5)(x \cdot x^2)$ Thinking of how 15 and x^3 can be factored

$ = (3x)(5x^2)$ The factors are $3x$ and $5x^2$. *Check:* $3x \cdot 5x^2 = 15x^3$.

b) $15x^3 = (3 \cdot 5)(x^2 \cdot x)$

$ = (3x^2)(5x)$ The factors are $3x^2$ and $5x$. *Check:* $3x^2 \cdot 5x = 15x^3$.

c) $15x^3 = ((-5)(-3))x^3$

$ = (-5)(-3x^3)$ The factors are -5 and $-3x^3$. *Check:*
 $(-5)(-3x^3) = 15x^3$.

$(3x)(5x^2)$, $(3x^2)(5x)$, and $(-5)(-3x^3)$ are all factorizations of $15x^3$. Other factorizations exist as well.

   9

Recall from Section 1.2 that the word "factor" can be a verb or a noun, depending on the context in which it appears.

Factoring When Terms Have a Common Factor

To multiply a polynomial of two or more terms by a monomial, we use the distributive law: $a(b + c) = ab + ac$. To factor a polynomial with two or more terms of the form $ab + ac$, we use the distributive law with the sides of the equation switched: $ab + ac = a(b + c)$.

Multiply   *Factor*

$3(x + 2y - z)$ $3x + 6y - 3z$

$\quad = 3 \cdot x + 3 \cdot 2y - 3 \cdot z$ $\quad = 3 \cdot x + 3 \cdot 2y - 3 \cdot z$

$\quad = 3x + 6y - 3z$ $\quad = 3(x + 2y - z)$

In the factorization on the right, note that since 3 appears as a factor of $3x$, $6y$, and $-3z$, it is a *common factor* for all the terms of the trinomial $3x + 6y - 3z$.

When we factor, we are forming an equivalent expression that is a product.

EXAMPLE **2**   Factor to form an equivalent expression: $10y + 15$.

SOLUTION   We write the prime factorization of both terms to determine any common factors:

> The prime factorization of $10y$ is $2 \cdot 5 \cdot y$; $\Big\}$ 5 is a common factor.
> The prime factorization of 15 is $3 \cdot 5$.

We "factor out" the common factor 5 using the distributive law:

> We can always check a factorization by multiplying.

$10y + 15 = 5 \cdot 2y + 5 \cdot 3$ Try to do this step mentally.

$\quad\quad\quad\quad = 5(2y + 3)$.   Using the distributive law

Check: $5(2y + 3) = 5 \cdot 2y + 5 \cdot 3 = 10y + 15$.

The factorization of $10y + 15$ is $5(2y + 3)$.   ▶ **TRY EXERCISE** 15

We generally factor out the *largest* common factor.

EXAMPLE **3**   Factor: $8a - 12$.

SOLUTION   Lining up common factors in columns can help us determine the largest common factor:

> The prime factorization of $8a$ is $2 \cdot 2 \cdot 2 \cdot \quad a$;
> The prime factorization of 12 is $2 \cdot 2 \cdot \quad 3$.

Since both factorizations include two factors of 2, the largest common factor is $2 \cdot 2$, or 4:

$8a - 12 = 4 \cdot 2a - 4 \cdot 3$ 4 is a factor of $8a$ and of 12.

$8a - 12 = 4(2a - 3)$.   Try to go directly to this step.

Check: $4(2a - 3) = 4 \cdot 2a - 4 \cdot 3 = 8a - 12$, as expected.

The factorization of $8a - 12$ is $4(2a - 3)$.   ▶ **TRY EXERCISE** 17

CAUTION! $2 \cdot 2 \cdot 2a - 2 \cdot 2 \cdot 3$ is a factorization of the *terms* of $8a - 12$ but not of the polynomial itself. The factorization of $8a - 12$ is $4(2a - 3)$.

A common factor may contain a variable.

EXAMPLE **4**   Factor: $24x^5 + 30x^2$.

SOLUTION

The prime factorization of $24x^5$ is $2 \cdot 2 \cdot 2 \cdot 3 \cdot \quad x \cdot x \cdot x \cdot x \cdot x.$
The prime factorization of $30x^2$ is $2 \cdot \quad\quad 3 \cdot 5 \cdot x \cdot x.$

The largest common factor is $2 \cdot 3 \cdot x \cdot x$, or $6x^2$. ←

$$24x^5 + 30x^2 = 6x^2 \cdot 4x^3 + 6x^2 \cdot 5 \qquad \text{Factoring each term}$$
$$= 6x^2(4x^3 + 5) \qquad\qquad \text{Factoring out } 6x^2$$

Check: $6x^2(4x^3 + 5) = 6x^2 \cdot 4x^3 + 6x^2 \cdot 5 = 24x^5 + 30x^2$, as expected.

The factorization of $24x^5 + 30x^2$ is $6x^2(4x^3 + 5)$.   **TRY EXERCISE** 27

The largest common factor of a polynomial is the largest common factor of the coefficients times the largest common factor of the variable(s) in all the terms. Suppose in Example 4 that you did not recognize the *largest* common factor, and removed only part of it, as follows:

$$24x^5 + 30x^2 = 2x^2 \cdot 12x^3 + 2x^2 \cdot 15 \qquad 2x^2 \text{ is a common factor.}$$
$$= 2x^2(12x^3 + 15). \qquad\qquad 12x^3 + 15 \text{ itself contains a}$$
$$\qquad\qquad\qquad\qquad\qquad\qquad \text{common factor.}$$

Note that $12x^3 + 15$ still has a common factor, 3. To find the largest common factor, continue factoring out common factors, as follows, until no more exist:

$$24x^5 + 30x^2 = 2x^2[3(4x^3 + 5)] \qquad \text{Factoring } 12x^3 + 15. \text{ Remember to rewrite}$$
$$\qquad\qquad\qquad\qquad\qquad \text{the first common factor, } 2x^2.$$
$$= 6x^2(4x^3 + 5). \qquad \text{Using an associative law; } 2x^2 \cdot 3 = 6x^2$$

Since $4x^3 + 5$ cannot be factored any further, we say that we have factored *completely.* When we are directed simply to factor, it is understood that we should always factor completely.

EXAMPLE **5**   Factor: $12x^5 - 15x^4 + 27x^3$.

SOLUTION

The prime factorization of $12x^5$ is $2 \cdot 2 \cdot 3 \cdot \quad\quad\quad x \cdot x \cdot x \cdot x \cdot x.$
The prime factorization of $15x^4$ is $\quad\quad 3 \cdot \quad 5 \cdot x \cdot x \cdot x \cdot x.$
The prime factorization of $27x^3$ is $\quad\quad 3 \cdot 3 \cdot 3 \cdot \quad x \cdot x \cdot x.$

The largest common factor is $3 \cdot x \cdot x \cdot x$, or $3x^3$. ←

$$12x^5 - 15x^4 + 27x^3 = 3x^3 \cdot 4x^2 - 3x^3 \cdot 5x + 3x^3 \cdot 9$$
$$= 3x^3(4x^2 - 5x + 9)$$

Since $4x^2 - 5x + 9$ has no common factor, we are done, except for a check:

$$3x^3(4x^2 - 5x + 9) = 3x^3 \cdot 4x^2 - 3x^3 \cdot 5x + 3x^3 \cdot 9$$
$$= 12x^5 - 15x^4 + 27x^3,$$

as expected. The factorization of $12x^5 - 15x^4 + 27x^3$ is $3x^3(4x^2 - 5x + 9)$.

TRY EXERCISE 31

Note in Examples 4 and 5 that the *largest* common variable factor is the *smallest* power of x in the original polynomial.

With practice, we can determine the largest common factor without writing the prime factorization of each term. Then, to factor, we write the largest common factor and parentheses and then fill in the parentheses. It is customary for the leading coefficient of the polynomial inside the parentheses to be positive.

EXAMPLE 6

Factor: **(a)** $8r^3s^2 + 16rs^3$; **(b)** $-3xy + 6xz - 3x$.

SOLUTION

a) $8r^3s^2 + 16rs^3 = 8rs^2(r^2 + 2s)$ Try to go directly to this step.

The largest common factor is $8rs^2$. $\longleftarrow$ $\begin{cases} 8r^3s^2 = 2 \cdot 2 \cdot 2 \cdot \quad r \cdot r^2 \cdot s^2 \\ 16rs^3 = 2 \cdot 2 \cdot 2 \cdot 2 \cdot r \cdot \quad s^2 \cdot s \end{cases}$

Check: $8rs^2(r^2 + 2s) = 8r^3s^2 + 16rs^3$.

b) $-3xy + 6xz - 3x = -3x(y - 2z + 1)$ Note that either $-3x$ or $3x$ can be the largest common factor.

STUDENT NOTES

The 1 in $(y - 2z + 1)$ plays an important role in Example 6(b). If left out of the factorization, the term $-3x$ would not appear in the check.

We generally factor out a negative when the first coefficient is negative. The way we factor can depend on the situation in which we are working. We might also factor as follows:

$$-3xy + 6xz - 3x = 3x(-y + 2z - 1).$$

The checks are left to the student.

TRY EXERCISE 35

In some texts, the largest common factor is referred to as the *greatest* common factor. We have avoided this language because, as shown in Example 6(b), the largest common factor may represent a negative value that is actually *less* than other common factors.

> **Tips for Factoring**
> 1. Factor out the largest common factor, if one exists.
> 2. The common factor multiplies a polynomial with the same number of terms as the original polynomial.
> 3. Factoring can always be checked by multiplying. Multiplication should yield the original polynomial.

Factoring by Grouping

Sometimes algebraic expressions contain a common factor with two or more terms.

EXAMPLE 7

Factor: $x^2(x + 1) + 2(x + 1)$.

SOLUTION   The binomial $x + 1$ is a factor of both $x^2(x + 1)$ and $2(x + 1)$. Thus, $x + 1$ is a common factor:

$$x^2(x + 1) + 2(x + 1) = (x + 1)x^2 + (x + 1)2$$ Using a commutative law twice. Try to do this step mentally.

$$= (x + 1)(x^2 + 2).$$ Factoring out the common factor, $x + 1$

To check, we could simply reverse the above steps.
    The factorization is $(x + 1)(x^2 + 2)$.

TRY EXERCISE 37

In Example 7, the common binomial factor was clearly visible. How do we find such a factor in a polynomial like $5x^3 - x^2 + 15x - 3$? Although there is no factor, other than 1, common to all four terms, $5x^3 - x^2$ and $15x - 3$ can be grouped and factored separately:

$$5x^3 - x^2 = x^2(5x - 1) \quad \text{and} \quad 15x - 3 = 3(5x - 1).$$

Note that $5x^3 - x^2$ and $15x - 3$ share a common factor, $5x - 1$. This means that the original polynomial, $5x^3 - x^2 + 15x - 3$, can be factored:

$$5x^3 - x^2 + 15x - 3 = (5x^3 - x^2) + (15x - 3) \qquad \text{Each binomial has a common factor.}$$

$$= x^2(5x - 1) + 3(5x - 1) \qquad \text{Factoring each binomial}$$

$$= (5x - 1)(x^2 + 3). \qquad \text{Factoring out the common factor, } 5x - 1$$

Check: $(5x - 1)(x^2 + 3) = 5x \cdot x^2 + 5x \cdot 3 - 1 \cdot x^2 - 1 \cdot 3$

$$= 5x^3 - x^2 + 15x - 3.$$

If a polynomial can be split into groups of terms and the groups share a common factor, then the original polynomial can be factored. This method, known as **factoring by grouping**, can be tried on any polynomial with four or more terms.

EXAMPLE   8   Factor by grouping.

a) $2x^3 + 8x^2 + x + 4$ **b)** $8x^4 + 6x - 28x^3 - 21$

SOLUTION

a) $2x^3 + 8x^2 + x + 4 = (2x^3 + 8x^2) + (x + 4)$

$$= 2x^2(x + 4) + 1(x + 4) \qquad \text{Factoring } 2x^3 + 8x^2 \text{ to find a common binomial factor. Writing the 1 helps with the next step.}$$

$$= (x + 4)(2x^2 + 1) \qquad \text{Factoring out the common factor, } x + 4. \text{ The 1 is essential in the factor } 2x^2 + 1.$$

> *CAUTION!*   Be sure to include the term 1. The check below shows why it is essential.

Check: $(x + 4)(2x^2 + 1) = x \cdot 2x^2 + x \cdot 1 + 4 \cdot 2x^2 + 4 \cdot 1$ Using FOIL

$$= 2x^3 + x + 8x^2 + 4$$

$$= 2x^3 + 8x^2 + x + 4. \qquad \text{Using a commutative law}$$

The factorization is $(x + 4)(2x^2 + 1)$.

b) We have a choice of either

$$8x^4 + 6x - 28x^3 - 21 = (8x^4 + 6x) + (-28x^3 - 21)$$
$$= 2x(4x^3 + 3) + 7(-4x^3 - 3) \longleftarrow \text{No common factor}$$

or

$$8x^4 + 6x - 28x^3 - 21 = (8x^4 + 6x) + (-28x^3 - 21)$$
$$= 2x(4x^3 + 3) + (-7)(4x^3 + 3). \longleftarrow \text{Common factor}$$

Because of the common factor $4x^3 + 3$, we choose the latter:

$$8x^4 + 6x - 28x^3 - 21 = 2x(4x^3 + 3) + (-7)(4x^3 + 3)$$
$$= (4x^3 + 3)(2x + (-7)) \qquad \text{Try to do this step mentally.}$$
$$= (4x^3 + 3)(2x - 7). \qquad \text{The common factor } 4x^3 + 3 \text{ was factored out.}$$

Check: $(4x^3 + 3)(2x - 7) = 8x^4 - 28x^3 + 6x - 21$
$$= 8x^4 + 6x - 28x^3 - 21. \qquad \text{This is the original polynomial.}$$

The factorization is $(4x^3 + 3)(2x - 7)$.   ▸ **TRY EXERCISE** ▸ 43

Although factoring by grouping can be useful, some polynomials, like $x^3 + x^2 + 2x - 2$, cannot be factored this way. Factoring polynomials of this type is beyond the scope of this text.

Checking by Evaluating

One way to check a factorization is to multiply. A second type of check, discussed toward the end of Section 4.4, uses the fact that equivalent expressions have the same value when evaluated for the same replacement. Thus a quick, partial check of Example 8(a) can be made by using a convenient replacement for x (say, 1) and evaluating both $2x^3 + 8x^2 + x + 4$ and $(x + 4)(2x^2 + 1)$:

$$2 \cdot 1^3 + 8 \cdot 1^2 + 1 + 4 = 2 + 8 + 1 + 4$$
$$= 15;$$
$$(1 + 4)(2 \cdot 1^2 + 1) = 5 \cdot 3$$
$$= 15.$$

Since the value of both expressions is the same, the factorization is probably correct.

Keep in mind the possibility that two expressions that are not equivalent may share the same value when evaluated at a certain value. Because of this, unless several values are used (at least one more than the degree of the polynomial, it turns out), evaluating offers only a partial check. Consult with your instructor before making extensive use of this type of check.

TECHNOLOGY CONNECTION

A partial check of a factorization can be performed using a table or a graph. To check Example 8(a), we let

$$y_1 = 2x^3 + 8x^2 + x + 4 \quad \text{and} \quad y_2 = (x + 4)(2x^2 + 1).$$

Then we set up a table in AUTO mode (see p. 88). If the factorization is correct, the values of y_1 and y_2 will be the same regardless of the table settings used.

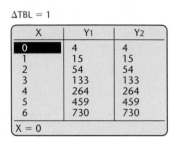

ΔTBL = 1

X	Y₁	Y₂
0	4	4
1	15	15
2	54	54
3	133	133
4	264	264
5	459	459
6	730	730

X = 0

We can also graph $y_1 = 2x^3 + 8x^2 + x + 4$ and $y_2 = (x + 4)(2x^2 + 1)$. If the graphs appear to coincide, the factorization is probably correct. The TRACE feature can be used to confirm this.

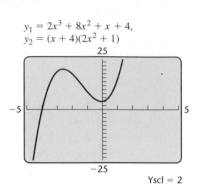

$y_1 = 2x^3 + 8x^2 + x + 4,$
$y_2 = (x + 4)(2x^2 + 1)$

Yscl = 2

Use a table or a graph to determine whether each factorization is correct.

1. $x^2 - 7x - 8 = (x - 8)(x + 1)$
2. $4x^2 - 5x - 6 = (4x + 3)(x - 2)$
3. $5x^2 + 17x - 12 = (5x + 3)(x - 4)$
4. $10x^2 + 37x + 7 = (5x - 1)(2x + 7)$
5. $12x^2 - 17x - 5 = (6x + 1)(2x - 5)$
6. $12x^2 - 17x - 5 = (4x + 1)(3x - 5)$
7. $x^2 - 4 = (x - 2)(x - 2)$
8. $x^2 - 4 = (x + 2)(x - 2)$

5.1 EXERCISE SET

For Extra Help MyMathLab | Math XL PRACTICE | WATCH | DOWNLOAD

Concept Reinforcement *In each of Exercises 1–8, match the phrase with the most appropriate choice from the column on the right.*

1. ____ A factorization of $35a^2b$
2. ____ A factor of $35a^2b$
3. ____ A common factor of $5x + 10$ and $4x + 8$
4. ____ A factorization of $3x^4 - 9x^2$
5. ____ A factorization of $9x^4 - 3x^2$
6. ____ A common factor of $2x + 10$ and $4x + 8$
7. ____ A factor of $3a + 6a^2$
8. ____ A factorization of $3a + 6a^2$

a) $3a(1 + 2a)$
b) $x + 2$
c) $3x^2(3x^2 - 1)$
d) $1 + 2a$
e) $3x^2(x^2 - 3)$
f) $5a^2$
g) 2
h) $7a \cdot 5ab$

Find three factorizations for each monomial. Answers may vary.

9. $14x^3$
10. $22x^3$
11. $-15a^4$
12. $-8t^5$
13. $25t^5$
14. $9a^4$

Factor. Remember to use the largest common factor and to check by multiplying. Factor out a negative factor if the first coefficient is negative.

15. $8x + 24$
16. $10x + 50$
17. $6x - 30$
18. $7x - 21$
19. $2x^2 + 2x - 8$
20. $6x^2 + 3x - 15$
21. $3t^2 + t$
22. $2t^2 + t$
23. $-5y^2 - 10y$
24. $-4y^2 - 12y$
25. $x^3 + 6x^2$
26. $5x^4 - x^2$

27. $16a^4 - 24a^2$

28. $25a^5 + 10a^3$

29. $-6t^6 + 9t^4 - 4t^2$

30. $-10t^5 + 15t^4 + 9t^3$

31. $6x^8 + 12x^6 - 24x^4 + 30x^2$

32. $10x^4 - 30x^3 - 50x - 20$

33. $x^5y^5 + x^4y^3 + x^3y^3 - x^2y^2$

34. $x^9y^6 - x^7y^5 + x^4y^4 + x^3y^3$

35. $-35a^3b^4 + 10a^2b^3 - 15a^3b^2$

36. $-21r^5t^4 - 14r^4t^6 + 21r^3t^6$

Factor.

37. $n(n - 6) + 3(n - 6)$

38. $b(b + 5) + 3(b + 5)$

39. $x^2(x + 3) - 7(x + 3)$

40. $3z^2(2z + 9) + (2z + 9)$

41. $y^2(2y - 9) + (2y - 9)$

42. $x^2(x - 7) - 3(x - 7)$

Factor by grouping, if possible, and check.

43. $x^3 + 2x^2 + 5x + 10$

44. $z^3 + 3z^2 + 7z + 21$

45. $5a^3 + 15a^2 + 2a + 6$

46. $3a^3 + 2a^2 + 6a + 4$

47. $9n^3 - 6n^2 + 3n - 2$

48. $10x^3 - 25x^2 + 2x - 5$

49. $4t^3 - 20t^2 + 3t - 15$

50. $8a^3 - 2a^2 + 12a - 3$

51. $7x^3 + 5x^2 - 21x - 15$

52. $5x^3 + 4x^2 - 10x - 8$

53. $6a^3 + 7a^2 + 6a + 7$

54. $7t^3 - 5t^2 + 7t - 5$

55. $2x^3 + 12x^2 - 5x - 30$

56. $x^3 - x^2 - 2x + 5$

57. $p^3 + p^2 - 3p + 10$

58. $w^3 + 7w^2 + 4w + 28$

59. $y^3 + 8y^2 - 2y - 16$

60. $3x^3 + 18x^2 - 5x - 25$

61. $2x^3 - 8x^2 - 9x + 36$

62. $20g^3 - 4g^2 - 25g + 5$

63. In answering a factoring problem, Taylor says the largest common factor is $-5x^2$ and Madison says the largest common factor is $5x^2$. Can they both be correct? Why or why not?

64. Write a two-sentence paragraph in which the word "factor" is used at least once as a noun and once as a verb.

Skill Review

To prepare for Section 5.2, review multiplying binomials using FOIL (Section 4.5).

Multiply. [4.5]

65. $(x + 2)(x + 7)$

66. $(x - 2)(x - 7)$

67. $(x + 2)(x - 7)$

68. $(x - 2)(x + 7)$

69. $(a - 1)(a - 3)$

70. $(t + 3)(t + 5)$

71. $(t - 5)(t + 10)$

72. $(a + 4)(a - 6)$

Synthesis

73. Azrah recognizes that evaluating provides only a partial check of her factoring. Because of this, she often performs a second check with a different replacement value. Is this a good idea? Why or why not?

74. Holly factors $12x^2y - 18xy^2$ as $6xy \cdot 2x - 6xy \cdot 3y$. Is this the factorization of the polynomial? Why or why not?

Factor, if possible.

75. $4x^5 + 6x^2 + 6x^3 + 9$

76. $x^6 + x^2 + x^4 + 1$

77. $2x^4 + 2x^3 - 4x^2 - 4x$

78. $x^3 + x^2 - 2x + 2$

Aha! **79.** $5x^5 - 5x^4 + x^3 - x^2 + 3x - 3$

Aha! **80.** $ax^2 + 2ax + 3a + x^2 + 2x + 3$

81. Write a trinomial of degree 7 for which $8x^2y^3$ is the largest common factor. Answers may vary.

5.2   Factoring Trinomials of the Type $x^2 + bx + c$

When the Constant Term Is Positive   •   When the Constant Term Is Negative   •   Prime Polynomials

We now learn how to factor trinomials like

$$x^2 + 5x + 4 \quad \text{or} \quad x^2 + 3x - 10,$$

for which no common factor exists and the leading coefficient is 1. Recall that when factoring, we are writing an equivalent expression that is a product. For these trinomials, the factors will be binomials.

As preparation for the factoring that follows, compare the following multiplications:

$$
\begin{array}{c}
\quad\quad\quad\quad\quad F \quad\ O \quad\ I \quad\quad L \\
\quad\quad\quad\quad\quad \downarrow \quad \downarrow \quad \downarrow \quad\quad \downarrow \\
(x + 2)(x + 5) = x^2 + 5x + 2x + 2 \cdot 5 \\
\quad\quad\quad\quad\quad = x^2 + \quad 7x \quad + \quad 10;
\end{array}
$$

$$
\begin{array}{c}
(x - 2)(x - 5) = x^2 - 5x - 2x + (-2)(-5) \\
\quad\quad\quad\quad\quad = x^2 - \quad 7x \quad + \quad 10;
\end{array}
$$

$$
\begin{array}{c}
(x + 3)(x - 7) = x^2 - 7x + 3x + 3(-7) \\
\quad\quad\quad\quad\quad = x^2 - \quad 4x \quad - \quad 21;
\end{array}
$$

$$
\begin{array}{c}
(x - 3)(x + 7) = x^2 + 7x - 3x + (-3)7 \\
\quad\quad\quad\quad\quad = x^2 + \quad 4x \quad - \quad 21.
\end{array}
$$

Note that for all four products:

- The product of the two binomials is a trinomial.
- The coefficient of x in the trinomial is the sum of the constant terms in the binomials.
- The constant term in the trinomial is the product of the constant terms in the binomials.

These observations lead to a method for factoring certain trinomials. We first consider trinomials that have a positive constant term, just as in the first two multiplications above.

When the Constant Term Is Positive

STUDENT NOTES ──────

One way to approach finding the numbers p and q is to think of filling in a diamond:

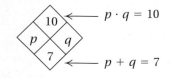

Since $2 \cdot 5 = 10$ and $2 + 5 = 7$, the numbers we want are 2 and 5.

To factor a polynomial like $x^2 + 7x + 10$, we think of FOIL in reverse. The x^2 resulted from x times x, which suggests that the first term of each binomial factor is x. Next, we look for numbers p and q such that

$$x^2 + 7x + 10 = (x + p)(x + q).$$

To get the middle term and the last term of the trinomial, we need two numbers, p and q, whose product is 10 and whose sum is 7. Those numbers are 2 and 5. Thus the factorization is

$$(x + 2)(x + 5).$$

Check: $(x + 2)(x + 5) = x^2 + 5x + 2x + 10$
$$= x^2 + 7x + 10.$$

EXAMPLE **1**

A GEOMETRIC APPROACH TO EXAMPLE 1

In Section 4.4, we saw that the product of two binomials can be regarded as the sum of the areas of four rectangles (see p. 248). Thus we can regard the factoring of $x^2 + 5x + 6$ as a search for p and q so that the sum of areas A, B, C, and D is $x^2 + 5x + 6$.

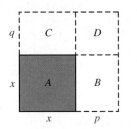

Note that area D is simply the product of p and q. In order for area D to be 6, p and q must be either 1 and 6 or 2 and 3. We illustrate both below.

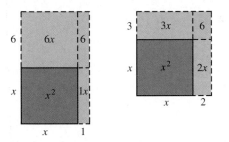

When p and q are 1 and 6, the total area is $x^2 + 7x + 6$, but when p and q are 2 and 3, as shown on the right, the total area is $x^2 + 5x + 6$, as desired. Thus the factorization of $x^2 + 5x + 6$ is $(x + 2)(x + 3)$.

Factor to form an equivalent expression: $x^2 + 5x + 6$.

SOLUTION   Think of FOIL in reverse. The first term of each factor is x:

$$(x + \quad)(x + \quad).$$

To complete the factorization, we need a constant term for each binomial. The constants must have a product of 6 and a sum of 5. We list some pairs of numbers that multiply to 6 and then check the sum of each pair of factors.

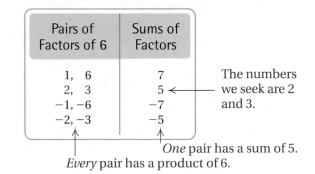

Pairs of Factors of 6	Sums of Factors
1, 6	7
2, 3	5 ←
−1, −6	−7
−2, −3	−5

The numbers we seek are 2 and 3.

One pair has a sum of 5. *Every* pair has a product of 6.

Since

$$2 \cdot 3 = 6 \quad \text{and} \quad 2 + 3 = 5,$$

the factorization of $x^2 + 5x + 6$ is $(x + 2)(x + 3)$.

Check: $(x + 2)(x + 3) = x^2 + 3x + 2x + 6$
$$= x^2 + 5x + 6.$$

Thus, $(x + 2)(x + 3)$ is a product that is equivalent to $x^2 + 5x + 6$.

Note that since 5 and 6 are both positive, when factoring $x^2 + 5x + 6$ we need not consider negative factors of 6. Note too that changing the signs of the factors changes only the sign of the sum (see the table above).

TRY EXERCISE **7**

At the beginning of this section, we considered the multiplication $(x - 2)(x - 5)$. For this product, the resulting trinomial, $x^2 - 7x + 10$, has a positive constant term but a negative coefficient of x. This is because the *product* of two negative numbers is always positive, whereas the *sum* of two negative numbers is always negative.

To Factor $x^2 + bx + c$ When c Is Positive

When the constant term c of a trinomial is positive, look for two numbers with the same sign. Select pairs of numbers with the sign of b, the coefficient of the middle term.

$$x^2 - 7x + 10 = (x - 2)(x - 5);$$

$$x^2 + 7x + 10 = (x + 2)(x + 5)$$

EXAMPLE **2** Factor: $y^2 - 8y + 12$.

SOLUTION Since the constant term is positive and the coefficient of the middle term is negative, we look for a factorization of 12 in which both factors are negative. Their sum must be -8.

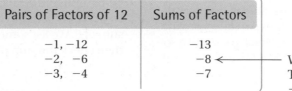

Pairs of Factors of 12	Sums of Factors
$-1, -12$	-13
$-2, \ -6$	-8 ←
$-3, \ -4$	-7

— We need a sum of -8. The numbers we need are -2 and -6.

STUDENT NOTES

It is important to be able to list *all* the pairs of factors of a number. See Example 1 on p. 21 for an organized approach for listing pairs of factors.

The factorization of $y^2 - 8y + 12$ is $(y - 2)(y - 6)$. The check is left to the student.

> **TRY EXERCISE** 13

When the Constant Term Is Negative

As we saw in two of the multiplications at the start of this section, the product of two binomials can have a negative constant term:

$$(x + 3)(x - 7) = x^2 - 4x - 21$$

and

$$(x - 3)(x + 7) = x^2 + 4x - 21.$$

It is important to note that when the signs of the constants in the binomials are reversed, only the sign of the middle term of the trinomial changes.

EXAMPLE **3** Factor: $x^2 - 8x - 20$.

SOLUTION The constant term, -20, must be expressed as the product of a negative number and a positive number. Since the sum of these two numbers must be negative (specifically, -8), the negative number must have the greater absolute value.

Pairs of Factors of -20	Sums of Factors
$1, -20$	-19
$2, -10$	-8 ←
$4, \ -5$	-1
$5, \ -4$	1
$10, \ -2$	8
$20, \ -1$	19

The numbers we need are 2 and -10.

Because in these three pairs, the positive number has the greater absolute value, these sums are all positive. For this problem, these pairs can be eliminated even before calculating the sum.

The numbers that we are looking for are 2 and -10.

Check: $(x + 2)(x - 10) = x^2 - 10x + 2x - 20$
$$= x^2 - 8x - 20.$$

The factorization of $x^2 - 8x - 20$ is $(x + 2)(x - 10)$.

> **TRY EXERCISE** 21

> ### To Factor $x^2 + bx + c$ When c Is Negative
>
> When the constant term c of a trinomial is negative, look for a positive number and a negative number that multiply to c. Select pairs of numbers for which the number with the larger absolute value has the same sign as b, the coefficient of the middle term.
>
> $$x^2 - 4x - 21 = (x + 3)(x - 7);$$
>
> $$x^2 + 4x - 21 = (x - 3)(x + 7)$$

EXAMPLE 4

Factor: $t^2 - 24 + 5t$.

SOLUTION It helps to first write the trinomial in descending order: $t^2 + 5t - 24$. The factorization of the constant term, -24, must have one factor positive and one factor negative. The sum must be 5, so the positive factor must have the larger absolute value. Thus we consider only pairs of factors in which the positive factor has the larger absolute value.

Pairs of Factors of -24	Sums of Factors
$-1, 24$	23
$-2, 12$	10
$-3, 8$	5 ← ──── The numbers we need are -3 and 8.
$-4, 6$	2

The factorization is $(t - 3)(t + 8)$. The check is left to the student.

TRY EXERCISE 23

Polynomials in two or more variables, such as $a^2 + 4ab - 21b^2$, are factored in a similar manner.

EXAMPLE 5

Factor: $a^2 + 4ab - 21b^2$.

SOLUTION It may help to write the trinomial in the equivalent form

$$a^2 + 4ba - 21b^2.$$

We now regard $-21b^2$ as the "constant" term and $4b$ as the "coefficient" of a. Then we try to express $-21b^2$ as a product of two factors whose sum is $4b$. Those factors are $-3b$ and $7b$.

Check: $(a - 3b)(a + 7b) = a^2 + 7ab - 3ba - 21b^2$
$$= a^2 + 4ab - 21b^2.$$

The factorization of $a^2 + 4ab - 21b^2$ is $(a - 3b)(a + 7b)$.

TRY EXERCISE 55

Prime Polynomials

EXAMPLE **6**   Factor: $x^2 - x + 5$.

SOLUTION   Since 5 has very few factors, we can easily check all possibilities.

Pairs of Factors of 5	Sums of Factors
5, 1	6
−5, −1	−6

Since there are no factors whose sum is −1, the polynomial is *not* factorable into binomials.

> **TRY EXERCISE** 37

In this text, a polynomial like $x^2 - x + 5$ that cannot be factored further is said to be **prime**. In more advanced courses, other types of numbers are considered. There, polynomials like $x^2 - x + 5$ can be factored and are not considered prime.

Often factoring requires two or more steps. Remember, when told to factor, we should *factor completely*. This means that the final factorization should contain only prime polynomials.

EXAMPLE **7**   Factor: $-2x^3 + 20x^2 - 50x$.

SOLUTION   *Always* look first for a common factor. Since the leading coefficient is negative, we begin by factoring out $-2x$:

$$-2x^3 + 20x^2 - 50x = -2x(x^2 - 10x + 25).$$

STUDENT NOTES

Whenever a new set of parentheses is created while factoring, check the expression inside the parentheses to see if it can be factored further.

Now consider $x^2 - 10x + 25$. Since the constant term is positive and the coefficient of the middle term is negative, we look for a factorization of 25 in which both factors are negative. Their sum must be -10.

Pairs of Factors of 25	Sums of Factors	
−25, −1	−26	
−5, −5	−10 ←	The numbers we need are −5 and −5.

The factorization of $x^2 - 10x + 25$ is $(x - 5)(x - 5)$, or $(x - 5)^2$.

> ***CAUTION!***   When factoring involves more than one step, be careful to write out the *entire* factorization.

Check: $-2x(x - 5)(x - 5) = -2x[x^2 - 10x + 25]$ Multiplying binomials
$$= -2x^3 + 20x^2 - 50x.$$ Using the distributive law

The factorization of $-2x^3 + 20x^2 - 50x$ is $-2x(x - 5)(x - 5)$, or $-2x(x - 5)^2$.

> **TRY EXERCISE** 27

Once any common factors have been factored out, the following summary can be used to factor $x^2 + bx + c$.

> **To Factor $x^2 + bx + c$**
>
> 1. Find a pair of factors that have c as their product and b as their sum.
>
>   a) If c is positive, both factors will have the same sign as b.
>   b) If c is negative, one factor will be positive and the other will be negative. Select the factors such that the factor with the larger absolute value has the same sign as b.
>
> 2. Check by multiplying.

Note that each polynomial has a unique factorization (except for the order in which the factors are written).

5.2 EXERCISE SET

For Extra Help MyMathLab | Math XP PRACTICE | WATCH | DOWNLOAD

🦢 **Concept Reinforcement** *For Exercises 1–6, assume that a polynomial of the form $x^2 + bx + c$ can be factored as $(x + p)(x + q)$. Complete each sentence by replacing each blank with either "positive" or "negative."*

1. If b is positive and c is positive, then p will be _____ and q will be _____.

2. If b is negative and c is positive, then p will be _____ and q will be _____.

3. If p is negative and q is negative, then b must be _____ and c must be _____.

4. If p is positive and q is positive, then b must be _____ and c must be _____.

5. If b, c, and p are all negative, then q must be _____.

6. If b and c are negative and p is positive, then q must be _____.

Factor completely. Remember to look first for a common factor. Check by multiplying. If a polynomial is prime, state this.

7. $x^2 + 8x + 16$
8. $x^2 + 9x + 20$
9. $x^2 + 11x + 10$
10. $y^2 + 8y + 7$
11. $x^2 + 10x + 21$
12. $x^2 + 6x + 9$
13. $t^2 - 9t + 14$
14. $a^2 - 9a + 20$
15. $b^2 - 5b + 4$
16. $x^2 - 10x + 25$
17. $a^2 - 7a + 12$
18. $z^2 - 8z + 7$
19. $d^2 - 7d + 10$
20. $x^2 - 8x + 15$
21. $x^2 - 2x - 15$
22. $x^2 - x - 42$

23. $x^2 + 2x - 15$
24. $x^2 + x - 42$
25. $2x^2 - 14x - 36$
26. $3y^2 - 9y - 84$
27. $-x^3 + 6x^2 + 16x$
28. $-x^3 + x^2 + 42x$
29. $4y - 45 + y^2$
30. $7x - 60 + x^2$
31. $x^2 - 72 + 6x$
32. $-2x - 99 + x^2$
33. $-5b^2 - 35b + 150$
34. $-c^4 - c^3 + 56c^2$
35. $x^5 - x^4 - 2x^3$
36. $2a^2 - 4a - 70$
37. $x^2 + 5x + 10$
38. $x^2 + 11x + 18$
39. $32 + 12t + t^2$
40. $y^2 - y + 1$
41. $x^2 + 20x + 99$
42. $x^2 + 20x + 100$
43. $3x^3 - 63x^2 - 300x$
44. $2x^3 - 40x^2 + 192x$
45. $-2x^2 + 42x + 144$
46. $-4x^2 - 40x - 100$
47. $y^2 - 20y + 96$
48. $144 - 25t + t^2$
49. $-a^6 - 9a^5 + 90a^4$
50. $-a^4 - a^3 + 132a^2$
51. $t^2 + \frac{2}{3}t + \frac{1}{9}$
52. $x^2 - \frac{2}{5}x + \frac{1}{25}$
53. $11 + w^2 - 4w$
54. $6 + p^2 + 2p$
55. $p^2 - 7pq + 10q^2$
56. $a^2 - 2ab - 3b^2$
57. $m^2 + 5mn + 5n^2$
58. $x^2 - 11xy + 24y^2$
59. $s^2 - 4st - 12t^2$
60. $b^2 + 8bc - 20c^2$
61. $6a^{10} + 30a^9 - 84a^8$
62. $5a^8 - 20a^7 - 25a^6$

📝 63. Without multiplying $(x - 17)(x - 18)$, explain why it cannot possibly be a factorization of $x^2 + 35x + 306$.

📝 64. Shari factors $x^3 - 8x^2 + 15x$ as $(x^2 - 5x)(x - 3)$. Is she wrong? Why or why not? What advice would you offer?

Skill Review

To prepare for Section 5.3, review multiplying binomials using FOIL (Section 4.5).

Multiply. [4.5]

65. $(2x + 3)(3x + 4)$ **66.** $(2x + 3)(3x - 4)$

67. $(2x - 3)(3x + 4)$ **68.** $(2x - 3)(3x - 4)$

69. $(5x - 1)(x - 7)$ **70.** $(x + 6)(3x - 5)$

Synthesis

71. When searching for a factorization, why do we list pairs of numbers with the correct *product* instead of pairs of numbers with the correct *sum*?

72. When factoring $x^2 + bx + c$ with a large value of c, Riley begins by writing out the prime factorization of c. What is the advantage of doing this?

73. Find all integers b for which $a^2 + ba - 50$ can be factored.

74. Find all integers m for which $y^2 + my + 50$ can be factored.

Factor completely.

75. $y^2 - 0.2y - 0.08$ **76.** $x^2 + \frac{1}{2}x - \frac{3}{16}$

77. $-\frac{1}{3}a^3 + \frac{1}{3}a^2 + 2a$ **78.** $-a^7 + \frac{25}{7}a^5 + \frac{30}{7}a^6$

79. $x^{2m} + 11x^m + 28$ **80.** $-t^{2n} + 7t^n - 10$

Aha! **81.** $(a + 1)x^2 + (a + 1)3x + (a + 1)2$

82. $ax^2 - 5x^2 + 8ax - 40x - (a - 5)9$ (*Hint*: See Exercise 81.)

83. Find the volume of a cube if its surface area is $6x^2 + 36x + 54$ square meters.

Find a polynomial in factored form for the shaded area in each figure. (Use π in your answers where appropriate.)

84.
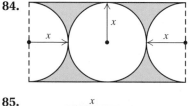

85.
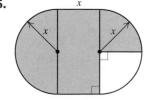

86.
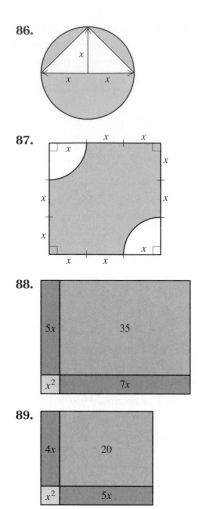

87.

88.

89.

90. A census taker asks a woman, "How many children do you have?"

"Three," she answers.

"What are their ages?"

She responds, "The product of their ages is 36. The sum of their ages is the house number next door."

The math-savvy census taker walks next door, reads the house number, appears puzzled, and returns to the woman, asking, "Is there something you forgot to tell me?"

"Oh yes," says the woman. "I'm sorry. The oldest child is at the park."

The census taker records the three ages, thanks the woman for her time, and leaves.

How old is each child? Explain how you reached this conclusion. (*Hint*: Consider factorizations.)

Source: Adapted from Anita Harnadek, *Classroom Quickies*. Pacific Grove, CA: Critical Thinking Press and Software

CORNER

Visualizing Factoring

Focus: Visualizing factoring

Time: 20–30 minutes

Group size: 3

Materials: Graph paper and scissors

The product $(x + 2)(x + 3)$ can be regarded as the area of a rectangle with width $x + 2$ and length $x + 3$. Similarly, factoring a polynomial like $x^2 + 5x + 6$ can be thought of as determining the length and the width of a rectangle that has area $x^2 + 5x + 6$. This is the approach used below.

ACTIVITY

1. **a)** To factor $x^2 + 11x + 10$ geometrically, the group needs to cut out shapes like those below to represent x^2, $11x$, and 10. This can be done by either tracing the figures below or by selecting a value for x, say 4, and using the squares on the graph paper to cut out the following:

   x^2:   Using the value selected for x, cut out a square that is x units on each side.
   $11x$:   Using the value selected for x, cut out a rectangle that is 1 unit wide and x units long. Repeat this to form 11 such strips.
   10:   Cut out two rectangles with whole-number dimensions and an area of 10. One should be 2 units by 5 units and the other 1 unit by 10 units.

   b) The group, working together, should then attempt to use one of the two rectangles with area 10, along with all of the other shapes, to piece together one large rectangle. Only one of the rectangles with area 10 will work.

   c) From the large rectangle formed in part (b), use the length and the width to determine the factorization of $x^2 + 11x + 10$. Where do the dimensions of the rectangle representing 10 appear in the factorization?

2. Repeat step (1) above, but this time use the other rectangle with area 10, and use only 7 of the 11 strips, along with the x^2-shape. Piece together the shapes to form one large rectangle. What factorization do the dimensions of this rectangle suggest?

3. Cut out rectangles with area 12 and use the above approach to factor $x^2 + 8x + 12$. What dimensions should be used for the rectangle with area 12?

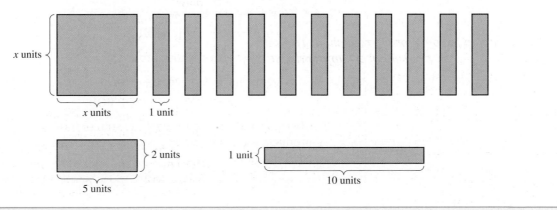

5.3 Factoring Trinomials of the Type $ax^2 + bx + c$

Factoring with FOIL ● The Grouping Method

In Section 5.2, we learned a FOIL-based method for factoring trinomials of the type $x^2 + bx + c$. Now we learn to factor trinomials in which the leading, or x^2, coefficient is not 1. First we will use another FOIL-based method and then we will use an alternative method that involves factoring by grouping. Use the method that you prefer or the one recommended by your instructor.

Factoring with FOIL

Before factoring trinomials of the type $ax^2 + bx + c$, consider the following:

$$\begin{array}{cccc} \text{F} & \text{O} & \text{I} & \text{L} \end{array}$$
$$(2x + 5)(3x + 4) = 6x^2 + 8x + 15x + 20$$
$$= 6x^2 + \quad 23x \quad + 20.$$

To factor $6x^2 + 23x + 20$, we could reverse the multiplication and look for two binomials whose product is this trinomial. We see from the multiplication above that:

- the product of the First terms must be $6x^2$;
- the product of the Outer terms plus the product of the Inner terms must be $23x$; and
- the product of the Last terms must be 20.

How can such a factorization be found without first seeing the corresponding multiplication? Our first approach relies on trial and error and FOIL.

To Factor $ax^2 + bx + c$ Using FOIL

1. Make certain that all common factors have been removed. If any remain, factor out the largest common factor.
2. Find two First terms whose product is ax^2:

$$(\blacksquare x + \quad)(\blacksquare x + \quad) = ax^2 + bx + c.$$
$$\underline{\qquad\qquad}\text{FOIL}$$

3. Find two Last terms whose product is c:

$$(\quad x + \blacksquare)(\quad x + \blacksquare) = ax^2 + bx + c.$$
$$\underline{\qquad\qquad}\text{FOIL}$$

4. Check by multiplying to see if the sum of the Outer and Inner products is bx. If necessary, repeat steps 2 and 3 until the correct combination is found.

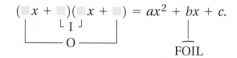

If no correct combination exists, state that the polynomial is prime.

EXAMPLE **1**   Factor: $3x^2 - 10x - 8$.

SOLUTION

1. First, check for a common factor. In this case, there is none (other than 1 or -1).
2. Find two **First** terms whose product is $3x^2$.
   The only possibilities for the **First** terms are $3x$ and x:

   $$(3x +)(x +).$$

3. Find two **Last** terms whose product is -8. There are four pairs of factors of -8 and each can be listed in two ways:

   $$
   \begin{array}{ll}
   -1,\ \ 8 & \quad 8,\ -1 \\
   1,\ -8 & \quad -8,\ \ 1 \\
   -2,\ \ 4 \quad \text{and} & \quad 4,\ -2 \\
   2,\ -4 & \quad -4,\ \ 2.
   \end{array}
   $$

   Important! Since the First terms are not identical, we must consider the pairs of factors in both orders.

4. Knowing that all **First** and **Last** products will check, systematically inspect the **O**uter and **I**nner products resulting from steps (2) and (3). Look for the combination in which the sum of the products is the middle term, $-10x$. Our search ends as soon as the correct combination is found. If none exists, we state that the polynomial is prime.

Pair of Factors	*Corresponding Trial*	*Product*	
$-1,\ \ 8$	$(3x - 1)(x + 8)$	$3x^2 + 24x - x - 8$ $= 3x^2 + 23x - 8$	Wrong middle term
$1,\ -8$	$(3x + 1)(x - 8)$	$3x^2 - 24x + x - 8$ $= 3x^2 - 23x - 8$	Wrong middle term
$-2,\ \ 4$	$(3x - 2)(x + 4)$	$3x^2 + 12x - 2x - 8$ $= 3x^2 + 10x - 8$	Wrong middle term
$2,\ -4$	$(3x + 2)(x - 4)$	$3x^2 - 12x + 2x - 8$ $= 3x^2 - 10x - 8$	Correct middle term!
$8,\ -1$	$(3x + 8)(x - 1)$	$3x^2 - 3x + 8x - 8$ $= 3x^2 + 5x - 8$	Wrong middle term
$-8,\ \ 1$	$(3x - 8)(x + 1)$	$3x^2 + 3x - 8x - 8$ $= 3x^2 - 5x - 8$	Wrong middle term
$4,\ -2$	$(3x + 4)(x - 2)$	$3x^2 - 6x + 4x - 8$ $= 3x^2 - 2x - 8$	Wrong middle term
$-4,\ \ 2$	$(3x - 4)(x + 2)$	$3x^2 + 6x - 4x - 8$ $= 3x^2 + 2x - 8$	Wrong middle term

The correct factorization is $(3x + 2)(x - 4)$.

TRY EXERCISE 5

Two observations can be made from Example 1. First, we listed all possible trials even though we generally stop after finding the correct factorization. We did this to show that **each trial differs only in the middle term of the product**. Second, note that as in Section 5.2, **only the sign of the middle term changes when the signs in the binomials are reversed**.

EXAMPLE **2**   Factor: $10x^2 + 37x + 7$.

SOLUTION

1. There is no factor (other than 1 or -1) common to all three terms.

2. Because $10x^2$ factors as $10x \cdot x$ or $5x \cdot 2x$, we have two possibilities:

$$(10x + \quad)(x + \quad) \quad \text{or} \quad (5x + \quad)(2x + \quad).$$

3. There are two pairs of factors of 7 and each can be listed in two ways:

$$\begin{matrix} 1, & 7 \\ -1, & -7 \end{matrix} \quad \text{and} \quad \begin{matrix} 7, & 1 \\ -7, & -1. \end{matrix}$$

4. Look for **O**uter and **I**nner products for which the sum is the middle term. Because all coefficients in $10x^2 + 37x + 7$ are positive, we need consider only those combinations involving positive factors of 7.

Trial	*Product*	
$(10x + 1)(x + 7)$	$10x^2 + 70x + 1x + 7$	
	$= 10x^2 + 71x + 7$	Wrong middle term
$(10x + 7)(x + 1)$	$10x^2 + 10x + 7x + 7$	
	$= 10x^2 + 17x + 7$	Wrong middle term
$(5x + 7)(2x + 1)$	$10x^2 + 5x + 14x + 7$	
	$= 10x^2 + 19x + 7$	Wrong middle term
$(5x + 1)(2x + 7)$	$10x^2 + 35x + 2x + 7$	
	$= 10x^2 + 37x + 7$	Correct middle term!

The correct factorization is $(5x + 1)(2x + 7)$.

TRY EXERCISE 9

A GEOMETRIC APPROACH TO EXAMPLE 2

The factoring of $10x^2 + 37x + 7$ can be regarded as a search for r and s so that the sum of areas A, B, C, and D is $10x^2 + 37x + 7$.

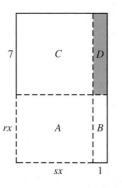

Because A must be $10x^2$, the product rs must be 10. Only when r is 2 and s is 5 will the sum of areas B and C be $37x$ (see below).

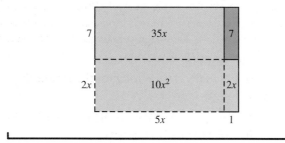

EXAMPLE **3**   Factor: $24x^3 - 76x^2 + 40x$.

SOLUTION

1. First, we factor out the largest common factor, $4x$:

$$4x(6x^2 - 19x + 10).$$

2. Next, we factor $6x^2 - 19x + 10$. Since $6x^2$ can be factored as $3x \cdot 2x$ or $6x \cdot x$, we have two possibilities:

$$(3x + \quad)(2x + \quad) \quad \text{or} \quad (6x + \quad)(x + \quad).$$

3. There are four pairs of factors of 10 and each can be listed in two ways:

$$\begin{matrix} 1, & 10 \\ -1, & -10 \\ 2, & 5 \\ -2, & -5 \end{matrix} \quad \text{and} \quad \begin{matrix} 10, & 1 \\ -10, & -1 \\ 5, & 2 \\ -5, & -2. \end{matrix}$$

4. The two possibilities from step (2) and the eight possibilities from step (3) give $2 \cdot 8$, or 16 possibilities for factorizations. With careful consideration,

we can eliminate some possibilities without multiplying. Since the sign of the middle term, $-19x$, is negative, but the sign of the last term, 10, is positive, the two factors of 10 must both be negative. This means only four pairings from step (3) need be considered. We first try these factors with $(3x + \ \ )(2x + \ \ )$, looking for **O**uter and **I**nner products for which the sum is $-19x$. If none gives the correct factorization, then we will consider $(6x + \ \ )(x + \ \ )$.

Trial	*Product*	
$(3x - 1)(2x - 10)$	$6x^2 - 30x - 2x + 10$	
	$= 6x^2 - 32x + 10$	Wrong middle term
$(3x - 10)(2x - 1)$	$6x^2 - 3x - 20x + 10$	
	$= 6x^2 - 23x + 10$	Wrong middle term
$(3x - 2)(2x - 5)$	$6x^2 - 15x - 4x + 10$	
	$- 6x^2 - 19x + 10$	Correct middle term!
$(3x - 5)(2x - 2)$	$6x^2 - 6x - 10x + 10$	
	$= 6x^2 - 16x + 10$	Wrong middle term

Since we have a correct factorization, we need not consider

$$(6x + \ \ )(x + \ \ ).$$

Look again at the possibility $(3x - 5)(2x - 2)$. Without multiplying, we can reject such a possibility. To see why, note that

$$(3x - 5)(2x - 2) = (3x - 5)2(x - 1).$$

The expression $2x - 2$ has a common factor, 2. But we removed the *largest* common factor in step (1). If $2x - 2$ were one of the factors, then 2 would be *another* common factor in addition to the original, $4x$. Thus, $(2x - 2)$ cannot be part of the factorization of $6x^2 - 19x + 10$. Similar reasoning can be used to reject $(3x - 1)(2x - 10)$ as a possible factorization.

Once the largest common factor is factored out, none of the remaining factors can have a common factor.

The factorization of $6x^2 - 19x + 10$ is $(3x - 2)(2x - 5)$, but do not forget the common factor! The factorization of $24x^3 - 76x^2 + 40x$ is

$$4x(3x - 2)(2x - 5).$$

> **TRY EXERCISE** 15

STUDENT NOTES

Keep your work organized so that you can see what you have already considered. For example, when factoring $6x^2 - 19x + 10$, we can list all possibilities and cross out those in which a common factor appears:

$$\cancel{(3x - 1)(2x - 10)}$$
$$(3x - 10)(2x - 1)$$
$$(3x - 2)(2x - 5)$$
$$\cancel{(3x - 5)(2x - 2)}$$
$$(6x - 1)(x - 10)$$
$$\cancel{(6x - 10)(x - 1)}$$
$$\cancel{(6x - 2)(x - 5)}$$
$$(6x - 5)(x - 2)$$

By being organized and not erasing, we can see that there are only four possible factorizations.

Tips for Factoring $ax^2 + bx + c$

To factor $ax^2 + bx + c \ (a > 0)$:

- Make sure that any common factor has been factored out.
- Once the largest common factor has been factored out of the original trinomial, no binomial factor can contain a common factor (other than 1 or -1).
- If c is positive, then the signs in both binomial factors must match the sign of b.
- Reversing the signs in the binomials reverses the sign of the middle term of their product.
- Organize your work so that you can keep track of which possibilities you have checked.
- Remember to include the largest common factor—if there is one—in the final factorization.
- Always check by multiplying.

EXAMPLE 4

Factor: $10x + 8 - 3x^2$.

SOLUTION   An important problem-solving strategy is to find a way to make new problems look like problems we already know how to solve. The factoring tips above apply only to trinomials of the form $ax^2 + bx + c$, with $a > 0$. This leads us to rewrite $10x + 8 - 3x^2$ in descending order:

$$10x + 8 - 3x^2 = -3x^2 + 10x + 8.\qquad \text{Using the commutative law to write descending order}$$

Although $-3x^2 + 10x + 8$ looks similar to the trinomials we have factored, the tips above require a positive leading coefficient. This can be found by factoring out -1:

$$-3x^2 + 10x + 8 = -1(3x^2 - 10x - 8)\qquad \text{Factoring out } -1 \text{ changes the signs of the coefficients.}$$
$$= -1(3x + 2)(x - 4).\qquad \text{Using the result from Example 1}$$

The factorization of $10x + 8 - 3x^2$ is $-1(3x + 2)(x - 4)$.   **TRY EXERCISE** 31

EXAMPLE 5

Factor: $6r^2 - 13rs - 28s^2$.

SOLUTION   In order for the product of the first terms to be $6r^2$ and the product of the last terms to be $-28s^2$, the binomial factors will be of the form

$$(\blacksquare r + \blacksquare s)(\blacksquare r + \blacksquare s).$$

We verify that no common factor exists and then examine the first term, $6r^2$. There are two possibilities:

$$(2r +\ )(3r +\ )\quad \text{or}\quad (6r +\ )(r +\ ).$$

The last term, $-28s^2$, has the following pairs of factors:

$$
\begin{array}{ccc}
s, -28s & & -28s,\ \ s \\
-s,\ \ 28s & & 28s,\ -s \\
2s, -14s & \text{and} & -14s,\ \ 2s \\
-2s,\ \ 14s & & 14s, -2s \\
4s,\ -7s & & -7s,\ \ 4s \\
-4s,\ \ 7s & & 7s, -4s.
\end{array}
$$

Note that listing the pairs of factors of $-28s^2$ is just like listing the pairs of factors of -28, except that each factor also contains a factor of s.

Some trials, like $(2r + 28s)(3r - s)$ and $(2r + 14s)(3r - 2s)$, cannot be correct because both $(2r + 28s)$ and $(2r + 14s)$ contain a common factor, 2. We try $(2r + 7s)(3r - 4s)$:

$$(2r + 7s)(3r - 4s) = 6r^2 - 8rs + 21rs - 28s^2$$
$$= 6r^2 + 13rs - 28s^2.$$

Our trial is incorrect, but only because of the sign of the middle term. To correctly factor $6r^2 - 13rs - 28s^2$, we simply change the signs in the binomials:

$$(2r - 7s)(3r + 4s) = 6r^2 + 8rs - 21rs - 28s^2$$
$$= 6r^2 - 13rs - 28s^2.$$

The correct factorization of $6r^2 - 13rs - 28s^2$ is $(2r - 7s)(3r + 4s)$.

TRY EXERCISE 67

The Grouping Method

Another method of factoring trinomials of the type $ax^2 + bx + c$ is known as the *grouping method*. The grouping method relies on rewriting $ax^2 + bx + c$ in the form $ax^2 + px + qx + c$ and then factoring by grouping. To develop this method, consider the following*:

$$
\begin{aligned}
(2x + 5)(3x + 4) &= 2x \cdot 3x + 2x \cdot 4 + 5 \cdot 3x + 5 \cdot 4 \qquad \text{Using FOIL} \\
&= 2 \cdot 3 \cdot x^2 + 2 \cdot 4x + 5 \cdot 3x + 5 \cdot 4 \\
&= 2 \cdot 3 \cdot x^2 + (2 \cdot 4 + 5 \cdot 3)x + 5 \cdot 4
\end{aligned}
$$

$$
\begin{array}{ccc}
\uparrow & \uparrow & \uparrow \\
a & b & c \\
\downarrow & \downarrow & \downarrow \\
= \quad 6x^2 \quad + & 23x \quad + & 20.
\end{array}
$$

Note that reversing these steps shows that $6x^2 + 23x + 20$ can be rewritten as $6x^2 + 8x + 15x + 20$ and then factored by grouping. Note that the numbers that add to b (in this case, $2 \cdot 4$ and $5 \cdot 3$) also multiply to ac (in this case, $2 \cdot 3 \cdot 5 \cdot 4$).

To Factor $ax^2 + bx + c$, Using the Grouping Method

1. Factor out the largest common factor, if one exists.
2. Multiply the leading coefficient a and the constant c.
3. Find a pair of factors of ac whose sum is b.
4. Rewrite the middle term, bx, as a sum or a difference using the factors found in step (3).
5. Factor by grouping.
6. Include any common factor from step (1) and check by multiplying.

EXAMPLE 6   Factor: $3x^2 - 10x - 8$.

SOLUTION

1. First, we note that there is no common factor (other than 1 or -1).
2. We multiply the leading coefficient, 3, and the constant, -8:

$$3(-8) = -24.$$

3. We next look for a factorization of -24 in which the sum of the factors is the coefficient of the middle term, -10.

Pairs of Factors of -24	Sums of Factors
1, -24	-23
-1, 24	23
2, -12	-10 ← 2 + (−12) = −10
-2, 12	10
3, -8	-5
-3, 8	5
4, -6	-2
-4, 6	2

We normally stop listing pairs of factors once we have found the one we are after.

---

*This discussion was inspired by a lecture given by Irene Doo at Austin Community College.

4. Next, we express the middle term as a sum or a difference using the factors found in step (3):

$$-10x = 2x - 12x.$$

5. We now factor by grouping as follows:

$$3x^2 - 10x - 8 = 3x^2 + 2x - 12x - 8$$ Substituting $2x - 12x$ for $-10x$. We could also use $-12x + 2x$.

$$= x(3x + 2) - 4(3x + 2)$$ Factoring by grouping; see Section 5.1

$$= (3x + 2)(x - 4).$$ Factoring out the common factor, $3x + 2$

6. *Check:* $(3x + 2)(x - 4) = 3x^2 - 12x + 2x - 8 = 3x^2 - 10x - 8.$

The factorization of $3x^2 - 10x - 8$ is $(3x + 2)(x - 4)$.   TRY EXERCISE ▶ 51

EXAMPLE **7**   Factor: $8x^3 + 22x^2 - 6x$.

SOLUTION

1. We factor out the largest common factor, $2x$:

$$8x^3 + 22x^2 - 6x = 2x(4x^2 + 11x - 3).$$

2. To factor $4x^2 + 11x - 3$ by grouping, we multiply the leading coefficient, 4, and the constant term, -3:

$$4(-3) = -12.$$

3. We next look for factors of -12 that add to 11.

Pairs of Factors of -12	Sums of Factors
1, -12	-11
-1, 12	11 ←
.	.
.	.
.	.

Since $-1 + 12 = 11$, there is no need to list other pairs of factors.

4. We then rewrite the $11x$ in $4x^2 + 11x - 3$ using

$$11x = -1x + 12x, \quad \text{or} \quad 11x = 12x - 1x.$$

5. Next, we factor by grouping:

$$4x^2 + 11x - 3 = 4x^2 - 1x + 12x - 3$$ Rewriting the middle term; $12x - 1x$ could also be used.

$$= x(4x - 1) + 3(4x - 1)$$ Factoring by grouping. Note the common factor, $4x - 1$.

$$= (4x - 1)(x + 3).$$ Factoring out the common factor

6. The factorization of $4x^2 + 11x - 3$ is $(4x - 1)(x + 3)$. But don't forget the common factor, $2x$. The factorization of the original trinomial is

$$2x(4x - 1)(x + 3).$$ TRY EXERCISE ▶ 57

5.3 EXERCISE SET

Concept Reinforcement *In each of Exercises 1–4, match the polynomial with the correct factorization from the column on the right.*

1. ____ $12x^2 + 16x - 3$ a) $(7x - 1)(2x + 3)$

2. ____ $14x^2 + 19x - 3$ b) $(6x + 1)(2x - 3)$

3. ____ $14x^2 - 19x - 3$ c) $(6x - 1)(2x + 3)$

4. ____ $12x^2 - 16x - 3$ d) $(7x + 1)(2x - 3)$

Factor completely. If a polynomial is prime, state this.

5. $2x^2 + 7x - 4$ 6. $3x^2 + x - 4$

7. $3x^2 - 17x - 6$ 8. $5x^2 - 19x - 4$

9. $4t^2 + 12t + 5$ 10. $6t^2 + 17t + 7$

11. $15a^2 - 14a + 3$ 12. $10a^2 - 11a + 3$

13. $6x^2 + 17x + 12$ 14. $6x^2 + 19x + 10$

15. $6x^2 - 10x - 4$ 16. $5t^3 - 21t^2 + 18t$

17. $7t^3 + 15t^2 + 2t$ 18. $15t^2 + 20t - 75$

19. $10 - 23x + 12x^2$ 20. $-20 + 31x - 12x^2$

21. $-35x^2 - 34x - 8$ 22. $28x^2 + 38x - 6$

23. $4 + 6t^2 - 13t$ 24. $9 + 8t^2 - 18t$

25. $25x^2 + 40x + 16$ 26. $49t^2 + 42t + 9$

27. $20y^2 + 59y - 3$ 28. $25a^2 - 23a - 2$

29. $14x^2 + 73x + 45$ 30. $35x^2 - 57x - 44$

31. $-2x^2 + 15 + x$ 32. $2t^2 - 19 - 6t$

33. $-6x^2 - 33x - 15$ 34. $-12x^2 - 28x + 24$

35. $10a^2 - 8a - 18$ 36. $20y^2 - 25y + 5$

37. $12x^2 + 68x - 24$ 38. $6x^2 + 21x + 15$

39. $4x + 1 + 3x^2$ 40. $-9 + 18x^2 + 21x$

Factor. Use factoring by grouping even though it would seem reasonable to first combine like terms.

41. $x^2 + 3x - 2x - 6$ 42. $x^2 + 4x - 2x - 8$

43. $8t^2 - 6t - 28t + 21$

44. $35t^2 - 40t + 21t - 24$

45. $6x^2 + 4x + 15x + 10$ 46. $3x^2 - 2x + 3x - 2$

47. $2y^2 + 8y - y - 4$ 48. $7n^2 + 35n - n - 5$

49. $6a^2 - 8a - 3a + 4$ 50. $10a^2 - 4a - 5a + 2$

Factor completely. If a polynomial is prime, state this.

51. $16t^2 + 23t + 7$ 52. $9t^2 + 14t + 5$

53. $-9x^2 - 18x - 5$ 54. $-16x^2 - 32x - 7$

55. $10x^2 + 30x - 70$ 56. $10a^2 + 25a - 15$

57. $18x^3 + 21x^2 - 9x$ 58. $6x^3 - 4x^2 - 10x$

59. $89x + 64 + 25x^2$ 60. $47 - 42y + 9y^2$

61. $168x^3 + 45x^2 + 3x$

62. $144x^5 - 168x^4 + 48x^3$

63. $-14t^4 + 19t^3 + 3t^2$

64. $-70a^4 + 68a^3 - 16a^2$

65. $132y + 32y^2 - 54$ 66. $220y + 60y^2 - 225$

67. $2a^2 - 5ab + 2b^2$ 68. $3p^2 - 16pq - 12q^2$

69. $8s^2 + 22st + 14t^2$ 70. $10s^2 + 4st - 6t^2$

71. $27x^2 - 72xy + 48y^2$

72. $-30a^2 - 87ab - 30b^2$

73. $-24a^2 + 34ab - 12b^2$ 74. $15a^2 - 5ab - 20b^2$

75. $19x^3 - 3x^2 + 14x^4$ 76. $10x^5 - 2x^4 + 22x^3$

77. $18a^7 + 8a^6 + 9a^8$ 78. $40a^8 + 16a^7 + 25a^9$

79. Asked to factor $2x^2 - 18x + 36$, Kay *incorrectly* answers
$$2x^2 - 18x + 36 = 2(x^2 + 9x + 18)$$
$$= 2(x + 3)(x + 6).$$
If this were a 10-point quiz question, how many points would you take off? Why?

80. Asked to factor $4x^2 + 28x + 48$, Herb *incorrectly* answers
$$4x^2 + 28x + 48 = (2x + 6)(2x + 8)$$
$$= 2(x + 3)(x + 4).$$
If this were a 10-point quiz question, how many points would you take off? Why?

Skill Review

To prepare for Section 5.4, review the special products in Section 4.5.

Multiply. [4.5]

81. $(x - 2)^2$ 82. $(x + 2)^2$

83. $(x + 2)(x - 2)$ 84. $(5t - 3)^2$

85. $(4a + 1)^2$

86. $(2n + 7)(2n - 7)$

87. $(3c - 10)^2$

88. $(1 - 5a)^2$

89. $(8n + 3)(8n - 3)$

90. $(9 - y)(9 + y)$

95. $9a^2b^3 + 25ab^2 + 16$

96. $-9t^{10} - 12t^5 - 4$

97. $16t^{10} - 8t^5 + 1$

98. $9a^2b^2 - 15ab - 2$

99. $-15x^{2m} + 26x^m - 8$

100. $-20x^{2n} - 16x^n - 3$

101. $3a^{6n} - 2a^{3n} - 1$

102. $a^{2n+1} - 2a^{n+1} + a$

103. $7(t - 3)^{2n} + 5(t - 3)^n - 2$

104. $3(a + 1)^{n+1}(a + 3)^2 - 5(a + 1)^n(a + 3)^3$

Synthesis

91. Explain how you would prove to a fellow student that a given trinomial is prime.

92. For the trinomial $ax^2 + bx + c$, suppose that a is the product of three different prime factors and c is the product of another two prime factors. How many possible factorizations (like those in Example 1) exist? Explain how you determined your answer.

Factor. If a polynomial is prime, state this.

93. $18x^2y^2 - 3xy - 10$

94. $8x^2y^3 + 10xy^2 + 2y$

CONNECTING ↕ the CONCEPTS

In Sections 5.1–5.3, we have considered factoring out a common factor, factoring by grouping, and factoring with FOIL. The following is a good strategy to follow when you encounter a mixed set of factoring problems.

1. Factor out any common factor.
2. Try factoring by grouping for polynomials with four terms.
3. Try factoring with FOIL for trinomials. If the leading coefficient of the trinomial is not 1, you may instead try factoring by grouping.

Polynomial	Number of Terms	Factorization
$12y^5 - 6y^4 + 30y^2$	3	There is a common factor: $6y^2$. $12y^5 - 6y^4 + 30y^2 = 6y^2(2y^3 - y^2 + 5)$. The trinomial in the parentheses cannot be factored further.
$t^4 - 5t^3 - 3t + 15$	4	There is no common factor. We factor by grouping. $t^4 - 5t^3 - 3t + 15 = t^3(t - 5) - 3(t - 5)$ $= (t - 5)(t^3 - 3)$
$-4x^4 + 4x^3 + 80x^2$	3	There is a common factor: $-4x^2$. $-4x^4 + 4x^3 + 80x^2 = -4x^2(x^2 - x - 20)$ The trinomial in the parentheses can be factored: $-4x^4 + 4x^3 + 80x^2 = -4x^2(x^2 - x - 20)$ $= -4x^2(x + 4)(x - 5)$.
$10n^2 - 17n + 3$	3	There is no common factor. We factor with FOIL or by grouping. $10n^2 - 17n + 3 = (2n - 3)(5n - 1)$

MIXED REVIEW

Factor completely. If a polynomial is prime, state this.

1. $6x^5 - 18x^2$

2. $x^2 + 10x + 16$

3. $2x^2 + 13x - 7$

4. $x^3 + 3x^2 + 2x + 6$

5. $5x^2 + 40x - 100$

6. $x^2 - 2x - 5$

7. $7x^2y - 21xy - 28y$

8. $15a^4 - 27a^2b^2 + 21a^2b$

9. $b^2 - 14b + 49$

10. $12x^2 - x - 1$

11. $c^3 + c^2 - 4c - 4$

12. $2x^2 + 30x - 200$

13. $t^2 + t - 10$

14. $15d^2 - 30d + 75$

15. $15p^2 + 16pq + 4q^2$

16. $-2t^3 - 10t^2 - 12t$

17. $x^2 + 4x - 77$

18. $10c^2 + 20c + 10$

19. $5 + 3x - 2x^2$

20. $2m^3n - 10m^2n - 6mn + 30n$

5.4   Factoring Perfect-Square Trinomials and Differences of Squares

Recognizing Perfect-Square Trinomials ▪ Factoring Perfect-Square Trinomials ▪
Recognizing Differences of Squares ▪ Factoring Differences of Squares ▪ Factoring Completely

In Section 4.5, we studied special products of certain binomials. Reversing these rules provides shortcuts for factoring certain polynomials.

Recognizing Perfect-Square Trinomials

Some trinomials are squares of binomials. For example, $x^2 + 10x + 25$ is the square of the binomial $x + 5$, because

$$(x + 5)^2 = x^2 + 2 \cdot x \cdot 5 + 5^2 = x^2 + 10x + 25.$$

A trinomial that is the square of a binomial is called a **perfect-square trinomial**.

In Section 4.5, we considered squaring binomials as a special-product rule:

$$(A + B)^2 = A^2 + 2AB + B^2;$$
$$(A - B)^2 = A^2 - 2AB + B^2.$$

Reading the right-hand sides first, we can use these equations to factor perfect-square trinomials. Note that in order for a trinomial to be the square of a binomial, it must have the following:

1. Two terms, A^2 and B^2, must be squares, such as

   $$4, \quad x^2, \quad 81m^2, \quad 16t^2.$$

2. Neither A^2 nor B^2 is being subtracted.

3. The remaining term is either $2 \cdot A \cdot B$ or $-2 \cdot A \cdot B$, where A and B are the square roots of A^2 and B^2.

Determine whether each of the following is a perfect-square trinomial.

a) $x^2 + 6x + 9$ **b)** $t^2 - 8t - 9$ **c)** $16x^2 + 49 - 56x$

SOLUTION

a) To see if $x^2 + 6x + 9$ is a perfect-square trinomial, note that:

   1. Two terms, x^2 and 9, are squares.

   2. Neither x^2 nor 9 is being subtracted.

   3. The remaining term, $6x$, is $2 \cdot x \cdot 3$, where x and 3 are the square roots of x^2 and 9.

Thus, $x^2 + 6x + 9$ *is* a perfect-square trinomial.

b) To see if $t^2 - 8t - 9$ is a perfect-square trinomial, note that:

   1. Both t^2 and 9, are squares. But:

   2. Since 9 is being subtracted, $t^2 - 8t - 9$ *is not* a perfect-square trinomial.

c) To see if $16x^2 + 49 - 56x$ is a perfect-square trinomial, it helps to first write it in descending order:

$$16x^2 - 56x + 49.$$

Next, note that:

   1. Two terms, $16x^2$ and 49, are squares.

   2. There is no minus sign before $16x^2$ or 49.

   3. Twice the product of the square roots, $2 \cdot 4x \cdot 7$, is $56x$. The remaining term, $-56x$, is the opposite of this product.

Thus, $16x^2 + 49 - 56x$ *is* a perfect-square trinomial.   **TRY EXERCISE** ▸ 11

Factoring Perfect-Square Trinomials

Either of the factoring methods discussed in Section 5.3 can be used to factor perfect-square trinomials, but a faster method is to recognize the following patterns.

> **Factoring a Perfect-Square Trinomial**
> $$A^2 + 2AB + B^2 = (A + B)^2; \qquad A^2 - 2AB + B^2 = (A - B)^2$$

Each factorization uses the square roots of the squared terms and the sign of the remaining term. To verify these equations, you should compute $(A + B)(A + B)$ and $(A - B)(A - B)$.

EXAMPLE **2**   Factor: **(a)** $x^2 + 6x + 9$; **(b)** $t^2 + 49 - 14t$; **(c)** $16x^2 - 40x + 25$.

SOLUTION

a) $x^2 + 6x + 9 = x^2 + 2 \cdot x \cdot 3 + 3^2 = (x + 3)^2$ The sign of the middle term is positive.

$$A^2 + 2 \quad A \quad B + B^2 = (A + B)^2$$

b) $t^2 + 49 - 14t = t^2 - 14t + 49$ Using a commutative law to write in descending order

$$= t^2 - 2 \cdot t \cdot 7 + 7^2 = (t - 7)^2$$

$$A^2 - 2 \quad A \quad B + B^2 = (A - B)^2$$

c) $16x^2 - 40x + 25 = (4x)^2 - 2 \cdot 4x \cdot 5 + 5^2 = (4x - 5)^2$ Recall that
$(4x)^2 = 16x^2$.

$$A^2 \quad - 2 \quad A \quad B + B^2 = (A - B)^2$$ **TRY EXERCISE** 19

Polynomials in more than one variable can also be perfect-square trinomials.

EXAMPLE 3

Factor: $4p^2 - 12pq + 9q^2$.

SOLUTION   We have

$$4p^2 - 12pq + 9q^2 = (2p)^2 - 2(2p)(3q) + (3q)^2$$ Recognizing the
perfect-square
trinomial

$$= (2p - 3q)^2.$$ The sign of the middle term
is negative.

Check: $(2p - 3q)(2p - 3q) = 4p^2 - 12pq + 9q^2$.

The factorization is $(2p - 3q)^2$.   **TRY EXERCISE** 43

EXAMPLE 4

Factor: $-75m^3 - 60m^2 - 12m$.

SOLUTION   *Always* look first for a common factor. This time there is one. We factor out $-3m$ so that the leading coefficient of the polynomial inside the parentheses is positive:

Factor out the common factor.

$$-75m^3 - 60m^2 - 12m = -3m[25m^2 + 20m + 4]$$

$$= -3m[(5m)^2 + 2(5m)(2) + 2^2]$$ Recognizing the
perfect-square
trinomial. Try to
do this mentally.

Factor the perfect-square trinomial.

$$= -3m(5m + 2)^2.$$

Check: $-3m(5m + 2)^2 = -3m(5m + 2)(5m + 2)$

$$= -3m(25m^2 + 20m + 4)$$

$$= -75m^3 - 60m^2 - 12m.$$

The factorization is $-3m(5m + 2)^2$.   **TRY EXERCISE** 31

Recognizing Differences of Squares

Some binomials represent the difference of two squares. For example, the binomial $16x^2 - 9$ is a difference of two expressions, $16x^2$ and 9, that are squares. To see this, note that $16x^2 = (4x)^2$ and $9 = 3^2$.

Any expression, like $16x^2 - 9$, that can be written in the form $A^2 - B^2$ is called a **difference of squares**. Note that in order for a binomial to be a difference of squares, it must have the following.

1. There must be two expressions, both squares, such as

$$25, \quad t^2, \quad 4x^2, \quad 1, \quad x^6, \quad 49y^8, \quad 100x^2y^2.$$

2. The terms in the binomial must have different signs.

Note that in order for an expression to be a square, its coefficient must be a perfect square and the power(s) of the variable(s) must be even.

EXAMPLE 5   Determine whether each of the following is a difference of squares.

a) $9x^2 - 64$ **b)** $25 - t^3$ **c)** $-4x^{10} + 36$

SOLUTION

a) To see if $9x^2 - 64$ is a difference of squares, note that:

 1. The first expression is a square: $9x^2 = (3x)^2$.
    The second expression is a square: $64 = 8^2$.
 2. The terms have different signs.

Thus, $9x^2 - 64$ is a difference of squares, $(3x)^2 - 8^2$.

b) To see if $25 - t^3$ is a difference of squares, note that:

 1. The expression t^3 is not a square.

Thus, $25 - t^3$ is not a difference of squares.

c) To see if $-4x^{10} + 36$ is a difference of squares, note that:

 1. The expressions $4x^{10}$ and 36 are squares: $4x^{10} = (2x^5)^2$ and $36 = 6^2$.
 2. The terms have different signs.

Thus, $-4x^{10} + 36$ is a difference of squares, $6^2 - (2x^5)^2$. It is often useful to rewrite $-4x^{10} + 36$ in the equivalent form $36 - 4x^{10}$.   **TRY EXERCISE** 51

Factoring Differences of Squares

To factor a difference of squares, we reverse a pattern from Section 4.5.

> ### Factoring a Difference of Squares
> $$A^2 - B^2 = (A + B)(A - B)$$

Once we have identified the expressions that are playing the roles of A and B, the factorization can be written directly. To verify this equation, simply multiply $(A + B)(A - B)$.

EXAMPLE 6   Factor: **(a)** $x^2 - 4$; **(b)** $1 - 9p^2$; **(c)** $s^6 - 16t^{10}$; **(d)** $50x^2 - 8x^8$.

SOLUTION

a) $x^2 - 4 = x^2 - 2^2 = (x + 2)(x - 2)$
$$A^2 - B^2 = (A + B)(A - B)$$

b) $1 - 9p^2 = 1^2 - (3p)^2 = (1 + 3p)(1 - 3p)$
$$A^2 - \quad B^2 = (A + B)(A - B)$$

c) $s^6 - 16t^{10} = (s^3)^2 - (4t^5)^2$ Using the rules for powers
$$A^2 \quad - \quad B^2$$

$$= (s^3 + 4t^5)(s^3 - 4t^5)$$ Try to go directly to this step.
$$(A + B) \ (A - B)$$

d) *Always* look first for a common factor. This time there is one, $2x^2$:

Factor out the common factor.

$$50x^2 - 8x^8 = 2x^2(25 - 4x^6)$$
$$= 2x^2[5^2 - (2x^3)^2] \qquad \text{Recognizing } A^2 - B^2.$$
$$\text{Try to do this mentally.}$$

Factor the difference of squares.

$$= 2x^2(5 + 2x^3)(5 - 2x^3).$$

Check: $2x^2(5 + 2x^3)(5 - 2x^3) = 2x^2(25 - 4x^6)$
$$= 50x^2 - 8x^8.$$

The factorization of $50x^2 - 8x^8$ is $2x^2(5 + 2x^3)(5 - 2x^3)$.

TRY EXERCISE 57

CAUTION!   Note in Example 6 that a difference of squares is *not* the square of the difference; that is,

$$A^2 - B^2 \neq (A - B)^2. \qquad \text{To see this, note that}$$
$$(A - B)^2 = A^2 - 2AB + B^2.$$

Factoring Completely

Sometimes, as in Examples 4 and 6(d), a *complete* factorization requires two or more steps. Factoring is complete when no factor can be factored further.

EXAMPLE 7   Factor: $y^4 - 16$.

SOLUTION   We have

Factor a difference of squares.

$$y^4 - 16 = (y^2)^2 - 4^2 \qquad\qquad \text{Recognizing } A^2 - B^2$$
$$= (y^2 + 4)(y^2 - 4) \qquad \text{Note that } y^2 - 4 \text{ is not prime.}$$

Factor another difference of squares.

$$= (y^2 + 4)(y + 2)(y - 2). \qquad \text{Note that } y^2 - 4 \text{ is itself a}$$
$$\text{difference of squares.}$$

Check: $(y^2 + 4)(y + 2)(y - 2) = (y^2 + 4)(y^2 - 4)$
$$= y^4 - 16.$$

The factorization is $(y^2 + 4)(y + 2)(y - 2)$.

TRY EXERCISE 79

Note in Example 7 that the factor $y^2 + 4$ is a *sum* of squares that cannot be factored further.

CAUTION!   There is no general formula for factoring a sum of squares. In particular,

$$A^2 + B^2 \neq (A + B)^2.$$

As you proceed through the exercises, these suggestions may prove helpful.

> **Tips for Factoring**
> 1. Always look first for a common factor! If there is one, factor it out.
> 2. Be alert for perfect-square trinomials and for binomials that are differences of squares. Once recognized, they can be factored without trial and error.
> 3. Always factor completely.
> 4. Check by multiplying.

5.4 EXERCISE SET

↪ *Concept Reinforcement   Identify each of the following as a perfect-square trinomial, a difference of squares, a prime polynomial, or none of these.*

1. $4x^2 + 49$

2. $x^2 - 64$

3. $t^2 - 100$

4. $x^2 - 5x + 4$

5. $9x^2 + 6x + 1$

6. $a^2 - 8a + 16$

7. $2t^2 + 10t + 6$

8. $-25x^2 - 9$

9. $16t^2 - 25$

10. $4r^2 + 20r + 25$

Determine whether each of the following is a perfect-square trinomial.

11. $x^2 + 18x + 81$

12. $x^2 - 16x + 64$

13. $x^2 - 10x - 25$

14. $x^2 - 14x - 49$

15. $x^2 - 3x + 9$

16. $x^2 + 4x + 4$

17. $9x^2 + 25 - 30x$

18. $36x^2 + 16 - 24x$

Factor completely. Remember to look first for a common factor and to check by multiplying. If a polynomial is prime, state this.

19. $x^2 + 16x + 64$

20. $x^2 + 10x + 25$

21. $x^2 - 10x + 25$

22. $x^2 - 16x + 64$

23. $5p^2 + 20p + 20$

24. $3p^2 - 12p + 12$

25. $1 - 2t + t^2$

26. $1 + t^2 + 2t$

27. $18x^2 + 12x + 2$

28. $25x^2 + 10x + 1$

29. $49 - 56y + 16y^2$

30. $75 - 60m + 12m^2$

31. $-x^5 + 18x^4 - 81x^3$

32. $-2x^2 + 40x - 200$

33. $2n^3 + 40n^2 + 200n$

34. $x^3 + 24x^2 + 144x$

35. $20x^2 + 100x + 125$

36. $27m^2 - 36m + 12$

37. $49 - 42x + 9x^2$

38. $64 - 112x + 49x^2$

39. $16x^2 + 24x + 9$

40. $2a^2 + 28a + 98$

41. $2 + 20x + 50x^2$

42. $9x^2 + 30x + 25$

43. $9p^2 + 12pq + 4q^2$

44. $x^2 - 3xy + 9y^2$

45. $a^2 - 12ab + 49b^2$

46. $25m^2 - 20mn + 4n^2$

47. $-64m^2 - 16mn - n^2$

48. $-81p^2 + 18pq - q^2$

49. $-32s^2 + 80st - 50t^2$

50. $-36a^2 - 96ab - 64b^2$

Determine whether each of the following is a difference of squares.

51. $x^2 - 100$

52. $x^2 + 49$

53. $n^4 + 1$

54. $n^4 - 81$

55. $-1 + 64t^2$

56. $-12 + 25t^2$

Factor completely. Remember to look first for a common factor. If a polynomial is prime, state this.

57. $x^2 - 25$

58. $x^2 - 36$

59. $p^2 - 9$

60. $q^2 + 1$

61. $-49 + t^2$

62. $-64 + m^2$

63. $6a^2 - 24$

64. $x^2 - 8x + 16$

65. $49x^2 - 14x + 1$

66. $3t^2 - 3$

67. $200 - 2t^2$

68. $98 - 8w^2$

69. $-80a^2 + 45$

70. $25x^2 - 4$

71. $5t^2 - 80$

72. $-4t^2 + 64$

73. $8x^2 - 162$

74. $24x^2 - 54$

75. $36x - 49x^3$

76. $16x - 81x^3$

77. $49a^4 - 20$

78. $25a^4 - 9$

79. $t^4 - 1$

80. $x^4 - 16$

81. $-3x^3 + 24x^2 - 48x$

82. $-2a^4 + 36a^3 - 162a^2$

83. $75t^3 - 27t$

84. $80s^4 - 45s^2$

85. $a^8 \quad 2a^7 + a^6$

86. $x^8 - 8x^7 + 16x^6$

87. $10a^2 - 10b^2$

88. $6p^2 - 6q^2$

89. $16x^4 - y^4$

90. $98x^2 - 32y^2$

91. $18t^2 - 8s^2$

92. $a^4 - 81b^4$

93. Explain in your own words how to determine whether a polynomial is a perfect-square trinomial.

94. Explain in your own words how to determine whether a polynomial is a difference of squares.

Skill Review

Review graphing linear equations (Chapter 3).

95. Find the slope of the line containing the points $(-2, -5)$ and $(3, -6)$. [3.5]

96. Find the slope of the line given by $y - 3 = \frac{1}{4}x$. [3.6]

Graph.

97. $2x - 5y = 10$ [3.3]

98. $-5x = 10$ [3.3]

99. $y = \frac{2}{3}x - 1$ [3.6]

100. $y - 2 = -2(x + 4)$ [3.7]

Synthesis

101. Leon concludes that since $x^2 - 9 = (x - 3)(x + 3)$, it must follow that $x^2 + 9 = (x + 3)(x - 3)$. What mistake(s) is he making?

102. Write directions that would enable someone to construct a polynomial that contains a perfect-square trinomial, a difference of squares, and a common factor.

Factor completely. If a polynomial is prime, state this.

103. $x^8 - 2^8$

104. $3x^2 - \frac{1}{3}$

105. $18x^3 - \frac{8}{25}x$

106. $0.81t - t^3$

107. $(y - 5)^4 - z^8$

108. $x^2 - \left(\dfrac{1}{x}\right)^2$

109. $-x^4 + 8x^2 + 9$

110. $-16x^4 + 96x^2 - 144$

Aha! **111.** $(y + 3)^2 + 2(y + 3) + 1$

112. $49(x + 1)^2 - 42(x + 1) + 9$

113. $27p^3 - 45p^2 - 75p + 125$

114. $a^{2n} - 49b^{2n}$

115. $81 - b^{4k}$

116. $9b^{2n} + 12b^n + 4$

117. Subtract $(x^2 + 1)^2$ from $x^2(x + 1)^2$.

Factor by grouping. Look for a grouping of three terms that is a perfect-square trinomial.

118. $t^2 + 4t + 4 - 25$

119. $y^2 + 6y + 9 - x^2 - 8x - 16$

Find c such that each polynomial is the square of a binomial.

120. $cy^2 + 6y + 1$

121. $cy^2 - 24y + 9$

122. Find the value of a if $x^2 + a^2x + a^2$ factors into $(x + a)^2$.

123. Show that the difference of the squares of two consecutive integers is the sum of the integers. (*Hint*: Use x for the smaller number.)

5.5 Factoring: A General Strategy

Choosing the Right Method

Thus far, each section in this chapter has examined one or two different methods for factoring polynomials. In practice, when the need for factoring a polynomial arises, we must decide on our own which method to use. As preparation for such a situation, we now encounter polynomials of various types, in random order. Regardless of the polynomial with which we are faced, the guidelines listed below can always be used.

To Factor a Polynomial

A. Always look for a common factor first. If there is one, factor out the largest common factor. Be sure to include it in your final answer.

B. Then look at the number of terms.

Two terms: If you have a difference of squares, factor accordingly: $A^2 - B^2 = (A + B)(A - B)$.

Three terms: If the trinomial is a perfect-square trinomial, factor accordingly: $A^2 + 2AB + B^2 = (A + B)^2$ or $A^2 - 2AB + B^2 = (A - B)^2$. If it is not a perfect-square trinomial, try using FOIL or grouping.

Four terms: Try factoring by grouping.

C. Always *factor completely*. When a factor can itself be factored, be sure to factor it. Remember that some polynomials, like $x^2 + 9$, are prime.

D. Check by multiplying.

Choosing the Right Method

EXAMPLE 1 Factor: $5t^4 - 80$.

SOLUTION

A. We look for a common factor:

$$5t^4 - 80 = 5(t^4 - 16). \qquad \text{5 is the largest common factor.}$$

B. The factor $t^4 - 16$ is a difference of squares: $(t^2)^2 - 4^2$. We factor it, being careful to rewrite the 5 from step (A):

$$5t^4 - 80 = 5(t^2 + 4)(t^2 - 4). \qquad t^4 - 16 = (t^2 + 4)(t^2 - 4)$$

C. Since $t^2 - 4$ is a difference of squares, we continue factoring:

$$5t^4 - 80 = 5(t^2 + 4)(t^2 - 4) = 5(t^2 + 4)(t - 2)(t + 2).$$

This is a sum of squares, which cannot be factored.

D. *Check:* $\quad 5(t^2 + 4)(t - 2)(t + 2) = 5(t^2 + 4)(t^2 - 4)$
$$= 5(t^4 - 16) = 5t^4 - 80.$$

The factorization is $5(t^2 + 4)(t - 2)(t + 2)$. ▸ TRY EXERCISE 5

EXAMPLE 2

Factor: $2x^3 + 10x^2 + x + 5$.

SOLUTION

A. We look for a common factor. There is none.

B. Because there are four terms, we try factoring by grouping:

$$2x^3 + 10x^2 + x + 5$$

$$= (2x^3 + 10x^2) + (x + 5) \qquad \text{Separating into two binomials}$$

$$= 2x^2(x + 5) + 1(x + 5) \qquad \text{Factoring out the largest common factor from each binomial. The 1 serves as an aid.}$$

$$= (x + 5)(2x^2 + 1). \qquad \text{Factoring out the common factor, } x + 5$$

C. Nothing can be factored further, so we have factored completely.

D. *Check:* $(x + 5)(2x^2 + 1) = 2x^3 + x + 10x^2 + 5$

$$= 2x^3 + 10x^2 + x + 5.$$

The factorization is $(x + 5)(2x^2 + 1)$.

TRY EXERCISE 13

EXAMPLE 3

Factor: $-n^5 + 2n^4 + 35n^3$.

SOLUTION

A. We note that there is a common factor, $-n^3$:

$$-n^5 + 2n^4 + 35n^3 = -n^3(n^2 - 2n - 35).$$

B. The factor $n^2 - 2n - 35$ is not a perfect-square trinomial. We factor it using trial and error:

$$-n^5 + 2n^4 + 35n^3 = -n^3(n^2 - 2n - 35)$$

$$= -n^3(n - 7)(n + 5).$$

C. Nothing can be factored further, so we have factored completely.

D. *Check:* $-n^3(n - 7)(n + 5) = -n^3(n^2 - 2n - 35)$

$$= -n^5 + 2n^4 + 35n^3.$$

The factorization is $-n^3(n - 7)(n + 5)$.

TRY EXERCISE 21

EXAMPLE 4

STUDENT NOTES ———————

Quickly checking the leading and constant terms of a trinomial to see if they are squares can save you time. If they aren't both squares, the trinomial can't possibly be a perfect-square trinomial.

Factor: $x^2 - 20x + 100$.

SOLUTION

A. We look first for a common factor. There is none.

B. This polynomial is a perfect-square trinomial. We factor it accordingly:

$$x^2 - 20x + 100 = x^2 - 2 \cdot x \cdot 10 + 10^2 \qquad \text{Try to do this step mentally.}$$

$$= (x - 10)^2.$$

C. Nothing can be factored further, so we have factored completely.

D. *Check:* $(x - 10)(x - 10) = x^2 - 20x + 100.$

The factorization is $(x - 10)(x - 10)$, or $(x - 10)^2$.

TRY EXERCISE 7

EXAMPLE 5

Factor: $6x^2y^4 - 21x^3y^5 + 3x^2y^6$.

SOLUTION

A. We first factor out the largest common factor, $3x^2y^4$:

$$6x^2y^4 - 21x^3y^5 + 3x^2y^6 = 3x^2y^4(2 - 7xy + y^2).$$

B. The constant term in $2 - 7xy + y^2$ is not a square, so we do not have a perfect-square trinomial. Note that x appears only in $-7xy$. The product of a form like $(1 - y)(2 - y)$ has no x in the middle term. Thus, $2 - 7xy + y^2$ cannot be factored.

C. Nothing can be factored further, so we have factored completely.

D. *Check:* $3x^2y^4(2 - 7xy + y^2) = 6x^2y^4 - 21x^3y^5 + 3x^2y^6$.

The factorization is $3x^2y^4(2 - 7xy + y^2)$.

 **TRY EXERCISE** 33

EXAMPLE 6

Factor: $ax + ay + cx + cy$.

SOLUTION

A. We look first for a common factor. There is none.

B. There are four terms. We try factoring by grouping:

$$ax + ay + cx + cy = a(x + y) + c(x + y) = (x + y)(a + c).$$

C. Nothing can be factored further, so we have factored completely.

D. *Check:* $(x + y)(a + c) = xa + xc + ya + yc = ax + ay + cx + cy$.

The factorization is $(x + y)(a + c)$.

TRY EXERCISE 39

EXAMPLE 7

Factor: $-25m^2 - 20mn - 4n^2$.

SOLUTION

A. We look first for a common factor. Since all the terms are negative, we factor out a -1:

$$-25m^2 - 20mn - 4n^2 = -1(25m^2 + 20mn + 4n^2).$$

B. There are three terms in the parentheses. Note that the first term and the last term are squares: $25m^2 = (5m)^2$ and $4n^2 = (2n)^2$. We see that twice the product of $5m$ and $2n$ is the middle term,

$$2 \cdot 5m \cdot 2n = 20mn,$$

so the trinomial is a perfect square. To factor, we write a binomial squared:

$$-25m^2 - 20mn - 4n^2 = -1(25m^2 + 20mn + 4n^2)$$
$$= -1(5m + 2n)^2.$$

C. Nothing can be factored further, so we have factored completely.

D. *Check:* $-1(5m + 2n)^2 = -1(25m^2 + 20mn + 4n^2)$
$$= -25m^2 - 20mn - 4n^2.$$

The factorization is $-1(5m + 2n)^2$, or $-(5m + 2n)^2$.

TRY EXERCISE 59

EXAMPLE **8**   Factor: $x^2y^2 + 7xy + 12$.

SOLUTION

A. We look first for a common factor. There is none.

B. Since only one term is a square, we do not have a perfect-square trinomial. We use trial and error, thinking of the product xy as a single variable:

$(xy + \quad)(xy + \quad)$.

We factor the last term, 12. All the signs are positive, so we consider only positive factors. Possibilities are 1, 12 and 2, 6 and 3, 4. The pair 3, 4 gives a sum of 7 for the coefficient of the middle term. Thus,

$x^2y^2 + 7xy + 12 = (xy + 3)(xy + 4)$.

C. Nothing can be factored further, so we have factored completely.

D. *Check:* $(xy + 3)(xy + 4) = x^2y^2 + 7xy + 12$.

The factorization is $(xy + 3)(xy + 4)$.   **TRY EXERCISE** 61

Compare the variables appearing in Example 7 with those in Example 8. Note that if the leading term contains one variable and a different variable is in the last term, as in Example 7, each binomial contains two variable terms. When two variables appear in the leading term and no variables appear in the last term, as in Example 8, each binomial contains one term that has two variables and one term that is a constant.

EXAMPLE **9**   Factor: $a^4 - 16b^4$.

SOLUTION

A. We look first for a common factor. There is none.

B. There are two terms. Since $a^4 = (a^2)^2$ and $16b^4 = (4b^2)^2$, we see that we have a difference of squares. Thus,

$a^4 - 16b^4 = (a^2 + 4b^2)(a^2 - 4b^2)$.

C. The factor $(a^2 - 4b^2)$ is itself a difference of squares. Thus,

$a^4 - 16b^4 = (a^2 + 4b^2)(a + 2b)(a - 2b)$.   Factoring $a^2 - 4b^2$

D. *Check:* $(a^2 + 4b^2)(a + 2b)(a - 2b) = (a^2 + 4b^2)(a^2 - 4b^2)$
$= a^4 - 16b^4$.

The factorization is $(a^2 + 4b^2)(a + 2b)(a - 2b)$.   **TRY EXERCISE** 53

5.5   EXERCISE SET

For Extra Help   MyMathLab   Math XL PRACTICE   WATCH   DOWNLOAD

🖐 ***Concept Reinforcement***   *In each of Exercises 1–4, complete the sentence.*

1. As a first step when factoring polynomials, always check for a _____.

2. When factoring a trinomial, if two terms are not squares, it cannot be a _____.

3. If a polynomial has four terms and no common factor, it may be possible to factor by _____.

4. It is always possible to check a factorization by _____.

Factor completely. If a polynomial is prime, state this.

5. $5a^2 - 125$

6. $10c^2 - 810$

7. $y^2 + 49 - 14y$

8. $a^2 + 25 + 10a$

9. $3t^2 + 16t + 21$

10. $8t^2 + 31t - 4$

11. $x^3 + 18x^2 + 81x$

12. $x^3 - 24x^2 + 144x$

13. $x^3 - 5x^2 - 25x + 125$

14. $x^3 + 3x^2 - 4x - 12$

15. $27t^3 - 3t$

16. $98t^2 - 18$

17. $9x^3 + 12x^2 - 45x$

18. $20x^3 - 4x^2 - 72x$

19. $t^2 + 25$

20. $4x^2 + 20x - 144$

21. $6y^2 + 18y - 240$

22. $4n^2 + 81$

23. $-2a^6 + 8a^5 - 8a^4$

24. $-x^5 - 14x^4 - 49x^3$

25. $5x^5 - 80x$

26. $4x^4 - 64$

27. $t^4 - 9$

28. $9 + t^8$

29. $-x^6 + 2x^5 - 7x^4$

30. $-x^5 + 4x^4 - 3x^3$

31. $p^2 - q^2$

32. $a^2b^2 - c^2$

33. $ax^2 + ay^2$

34. $12n^2 + 24n^3$

35. $2\pi rh + 2\pi r^2$

36. $4\pi r^2 + 2\pi r$

Aha! **37.** $(a + b)5a + (a + b)3b$

38. $5c(a^3 + b) - (a^3 + b)$

39. $x^2 + x + xy + y$

40. $n^2 + 2n + np + 2p$

41. $a^2 - 2a - ay + 2y$

42. $2x^2 - 4x + xz - 2z$

43. $3x^2 + 13xy - 10y^2$

44. $-x^2 - y^2 - 2xy$

45. $8m^3n - 32m^2n^2 + 24mn$

46. $a^2 - 7a - 6$

47. $4b^2 + a^2 - 4ab$

48. $7p^4 - 7q^4$

49. $16x^2 + 24xy + 9y^2$

50. $6a^2b^3 + 12a^3b^2 - 3a^4b^2$

51. $m^2 - 5m + 8$

52. $25z^2 + 10zy + y^2$

53. $a^4b^4 - 16$

54. $a^5 - 4a^4b - 5a^3b^2$

55. $80cd^2 - 36c^2d + 4c^3$

56. $2p^2 + pq + q^2$

57. $3b^2 + 17ab - 6a^2$

58. $2mn - 360n^2 + m^2$

59. $-12 - x^2y^2 - 8xy$

60. $m^2n^2 - 4mn - 32$

61. $5p^2q^2 + 25pq - 30$

62. $a^4b^3 + 2a^3b^2 - 15a^2b$

63. $4ab^5 - 32b^4 + a^2b^6$

64. $-60 + 52x - 8x^2$

65. $x^6 + x^5y - 2x^4y^2$

66. $2s^6t^2 + 10s^3t^3 + 12t^4$

67. $36a^2 - 15a + \frac{25}{16}$

68. $a^2 + 2a^2bc + a^2b^2c^2$

69. $\frac{1}{81}x^2 - \frac{8}{27}x + \frac{16}{9}$

70. $\frac{1}{4}a^2 + \frac{1}{3}ab + \frac{1}{9}b^2$

71. $1 - 16x^{12}y^{12}$

72. $b^4a - 81a^5$

73. $4a^2b^2 + 12ab + 9$

74. $9c^2 + 6cd + d^2$

75. $z^4 + 6z^3 - 6z^2 - 36z$

76. $t^5 - 2t^4 + 5t^3 - 10t^2$

77. Kelly factored $16 - 8x + x^2$ as $(x - 4)^2$, while Tony factored it as $(4 - x)^2$. Are they both correct? Why or why not?

78. Describe in your own words or draw a diagram representing a strategy for factoring polynomials.

Skill Review

To prepare for Section 5.6, review solving equations (Section 2.2).

Solve. [2.2]

79. $8x - 9 = 0$

80. $3x + 5 = 0$

81. $2x + 7 = 0$

82. $4x - 1 = 0$

83. $3 - x = 0$

84. $22 - 2x = 0$

85. $2x - 5 = 8x + 1$

86. $3(x - 1) = 9 - x$

Synthesis

87. There are third-degree polynomials in x that we are not yet able to factor, despite the fact that they are not prime. Explain how such a polynomial could be created.

88. Describe a method that could be used to find a binomial of degree 16 that can be expressed as the product of prime binomial factors.

Factor.

89. $-(x^5 + 7x^3 - 18x)$

90. $18 + a^3 - 9a - 2a^2$

91. $-x^4 + 7x^2 + 18$

92. $-3a^4 + 15a^2 - 12$

Aha! **93.** $y^2(y + 1) - 4y(y + 1) - 21(y + 1)$

94. $y^2(y - 1) - 2y(y - 1) + (y - 1)$

95. $(y + 4)^2 + 2x(y + 4) + x^2$

96. $6(x - 1)^2 + 7y(x - 1) - 3y^2$

97. $2(a + 3)^4 - (a + 3)^3(b - 2) - (a + 3)^2(b - 2)^2$

98. $5(t - 1)^5 - 6(t - 1)^4(s - 1) + (t - 1)^3(s - 1)^2$

99. $49x^4 + 14x^2 + 1 - 25x^6$

COLLABORATIVE CORNER

Matching Factorizations*

Focus: Factoring

Time: 20 minutes

Group size: Begin with the entire class. The end result is pairs of students. If there is an odd number of students, the instructor should participate.

Materials: Prepared sheets of paper, pins or tape. On half of the sheets, the instructor writes a polynomial. On the remaining sheets, the instructor writes the factorization of those polynomials. The activity is more interesting if the polynomials and factorizations are similar; for example,

$$x^2 - 2x - 8, \quad (x - 2)(x - 4),$$
$$x^2 - 6x + 8, \quad (x - 1)(x - 8),$$
$$x^2 - 9x + 8, \quad (x + 2)(x - 4).$$

ACTIVITY

1. As class members enter the room, the instructor pins or tapes either a polynomial or a factorization to the back of each student. Class members are told only whether their sheet of paper contains a polynomial or a factorization. All students should remain quiet and not tell others what is on their backs.

2. After all students are wearing a sheet of paper, they should mingle with one another, attempting to match up their factorization with the appropriate polynomial or vice versa. They may ask questions of one another that relate to factoring and polynomials. Answers to the questions should be yes or no. For example, a legitimate question might be "Is my last term negative?" or "Do my factors have opposite signs?"

3. The game is over when all factorization/polynomial pairs have "found" one another.

_____

*Thanks to Jann MacInnes of Florida Community College at Jacksonville–Kent Campus for suggesting this activity.

5.6   Solving Quadratic Equations by Factoring

The Principle of Zero Products   •   Factoring to Solve Equations

When we factor a polynomial, we are forming an *equivalent expression*. We now use our factoring skills to *solve equations*. We already know how to solve linear equations like $x + 2 = 7$ and $2x = 9$. The equations we will learn to solve in this section contain a variable raised to a power greater than 1 and will usually have more than one solution.

Second-degree equations like $4t^2 - 9 = 0$ and $x^2 + 6x + 5 = 0$ are called **quadratic equations**.

Quadratic Equation

A *quadratic equation* is an equation equivalent to one of the form

$$ax^2 + bx + c = 0,$$

where a, b, and c are constants, with $a \neq 0$.

In order to solve quadratic equations, we need to develop a new principle.

The Principle of Zero Products

Suppose we are told that the product of two numbers is 6. On the basis of this information, it is impossible to know the value of either number—the product could be $2 \cdot 3, 6 \cdot 1, 12 \cdot \frac{1}{2}$, and so on. However, if we are told that the product of two numbers is 0, we know that at least one of the two factors must itself be 0. For example, if $(x + 3)(x - 2) = 0$, we can conclude that either $x + 3$ is 0 or $x - 2$ is 0.

> ### The Principle of Zero Products
> An equation $AB = 0$ is true if and only if $A = 0$ or $B = 0$, or both.
> (A product is 0 if and only if at least one factor is 0.)

EXAMPLE 1   Solve: $(x + 3)(x - 2) = 0$.

SOLUTION   We are looking for all values of x that will make the equation true. The equation tells us that the product of $x + 3$ and $x - 2$ is 0. In order for the product to be 0, at least one factor must be 0. Thus we look for any value of x for which $x + 3 = 0$, as well as any value of x for which $x - 2 = 0$, that is, either

$x + 3 = 0 \quad or \quad x - 2 = 0.$ Using the principle of zero products. There are two equations to solve.

We solve each equation:

$$x + 3 = 0 \quad or \quad x - 2 = 0$$
$$x = -3 \quad or \qquad x = 2.$$

Both -3 and 2 should be checked in the original equation.

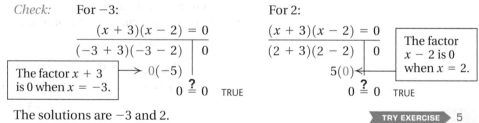

Check: For -3:

$$\begin{array}{c|c} (x + 3)(x - 2) = 0 \\ \hline (-3 + 3)(-3 - 2) & 0 \\ 0(-5) \\ 0 \stackrel{?}{=} 0 & \text{TRUE} \end{array}$$

The factor $x + 3$ is 0 when $x = -3$.

For 2:

$$\begin{array}{c|c} (x + 3)(x - 2) = 0 \\ \hline (2 + 3)(2 - 2) & 0 \\ 5(0) \\ 0 \stackrel{?}{=} 0 & \text{TRUE} \end{array}$$

The factor $x - 2$ is 0 when $x = 2$.

The solutions are -3 and 2.

▶ **TRY EXERCISE** 5

When we are using the principle of zero products, the word "or" is meant to emphasize that any one of the factors could be the one that represents 0.

EXAMPLE 2   Solve: $3(5x + 1)(x - 7) = 0$.

SOLUTION   The factors in this equation are 3, $5x + 1$, and $x - 7$. Since the factor 3 is constant, the only way in which $3(5x + 1)(x - 7)$ can be 0 is for one of the other factors to be 0, that is,

$$5x + 1 = 0 \quad or \quad x - 7 = 0$$ Using the principle of zero products
$$5x = -1 \quad or \qquad x = 7$$ Solving the two equations separately
$$x = -\tfrac{1}{5} \quad or \qquad x = 7.$$ $5x + 1 = 0$ when $x = -\tfrac{1}{5}$; $x - 7 = 0$ when $x = 7$

Check: For $-\frac{1}{5}$:

$$\begin{array}{c|c}
3(5x + 1)(x - 7) = 0 \\ \hline
3\big(5(-\frac{1}{5}) + 1\big)\big(-\frac{1}{5} - 7\big) & 0 \\
3(-1 + 1)\big(-7\frac{1}{5}\big) & \\
3(0)\big(-7\frac{1}{5}\big) & \\
& 0 \stackrel{?}{=} 0 \quad \text{TRUE}
\end{array}$$

For 7:

$$\begin{array}{c|c}
3(5x + 1)(x - 7) = 0 \\ \hline
3(5(7) + 1)(7 - 7) & 0 \\
3(35 + 1)0 & \\
& 0 \stackrel{?}{=} 0 \quad \text{TRUE}
\end{array}$$

The solutions are $-\frac{1}{5}$ and 7.

TRY EXERCISE 11

The constant factor 3 in Example 2 is never 0 and is not a solution of the equation. However, a variable factor such as x or t *can* equal 0, and must be considered when using the principle of zero products.

EXAMPLE 3   Solve: $7t(t - 5) = 0$.

SOLUTION   We have

$$\begin{array}{lll}
7 \cdot t(t - 5) = 0 & & \text{The factors are 7, } t\text{, and } t - 5. \\
t = 0 \quad or \quad t - 5 = 0 & & \text{Using the principle of zero products} \\
t = 0 \quad or \qquad t = 5. & & \text{Solving. Note that the constant factor, 7,} \\
& & \text{is never 0.}
\end{array}$$

The solutions are 0 and 5. The check is left to the student.

TRY EXERCISE 17

Factoring to Solve Equations

By factoring and using the principle of zero products, we can now solve a variety of quadratic equations.

EXAMPLE 4   Solve: $x^2 + 5x + 6 = 0$.

SOLUTION   This equation differs from those solved in Chapter 2. There are no like terms to combine, and there is a squared term. We first factor the polynomial. Then we use the principle of zero products:

$$\begin{array}{lll}
x^2 + 5x + 6 = 0 & & \\
(x + 2)(x + 3) = 0 & & \text{Factoring} \\
x + 2 = 0 \quad or \quad x + 3 = 0 & & \text{Using the principle of zero products} \\
x = -2 \quad or \qquad x = -3. & &
\end{array}$$

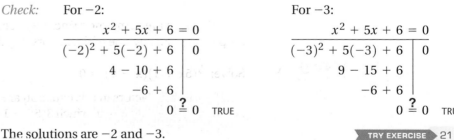

Check: For -2:

$$\begin{array}{c|c}
x^2 + 5x + 6 = 0 \\ \hline
(-2)^2 + 5(-2) + 6 & 0 \\
4 - 10 + 6 & \\
-6 + 6 & \\
& 0 \stackrel{?}{=} 0 \quad \text{TRUE}
\end{array}$$

For -3:

$$\begin{array}{c|c}
x^2 + 5x + 6 = 0 \\ \hline
(-3)^2 + 5(-3) + 6 & 0 \\
9 - 15 + 6 & \\
-6 + 6 & \\
& 0 \stackrel{?}{=} 0 \quad \text{TRUE}
\end{array}$$

The solutions are -2 and -3.

TRY EXERCISE 21

The principle of zero products applies even when there is a common factor.

EXAMPLE 5

Solve: $x^2 + 7x = 0$.

SOLUTION Although there is no constant term, because of the x^2-term, the equation is still quadratic. The methods of Chapter 2 are not sufficient, so we try factoring:

$$x^2 + 7x = 0$$
$$x(x + 7) = 0 \qquad \text{Factoring out the largest common factor, } x$$
$$x = 0 \quad or \quad x + 7 = 0 \qquad \text{Using the principle of zero products}$$
$$x = 0 \quad or \qquad x = -7.$$

The solutions are 0 and −7. The check is left to the student.

STUDENT NOTES ─────────

Checking for a common factor is an important step that is often overlooked. In Example 5, the equation must be factored. If we "divide both sides by x," we will not find the solution 0.

TRY EXERCISE 27

> *CAUTION!* We *must* have 0 on one side of the equation before the principle of zero products can be used. Get all nonzero terms on one side and 0 on the other.

EXAMPLE 6

Solve: **(a)** $x^2 - 8x = -16$; **(b)** $4t^2 = 25$.

SOLUTION

a) We first add 16 to get 0 on one side:

$$x^2 - 8x = -16$$
$$x^2 - 8x + 16 = 0 \qquad \text{Adding 16 to both sides to get 0 on one side}$$
$$(x - 4)(x - 4) = 0 \qquad \text{Factoring}$$
$$x - 4 = 0 \quad or \quad x - 4 = 0 \qquad \text{Using the principle of zero products}$$
$$x = 4 \quad or \qquad x = 4.$$

There is only one solution, 4. The check is left to the student.

b) We have

$$4t^2 = 25$$
$$4t^2 - 25 = 0 \qquad \text{Subtracting 25 from both sides to get 0 on one side}$$
$$(2t - 5)(2t + 5) = 0 \qquad \text{Factoring a difference of squares}$$
$$\left. \begin{array}{lll} 2t - 5 = 0 & or & 2t + 5 = 0 \\ 2t = 5 & or & 2t = -5 \\ t = \frac{5}{2} & or & t = -\frac{5}{2}. \end{array} \right\} \text{Solving the two equations separately}$$

The solutions are $\frac{5}{2}$ and $-\frac{5}{2}$. The check is left to the student.

TRY EXERCISE 41

When solving quadratic equations by factoring, remember that a factorization is not useful unless 0 is on the other side of the equation.

EXAMPLE 7

Solve: $(x + 3)(2x - 1) = 9$.

SOLUTION Be careful with an equation like this! Since we need 0 on one side, we first multiply out the product on the left and then subtract 9 from both sides.

$(x + 3)(2x - 1) = 9$	This is not a product equal to 0.
$2x^2 + 5x - 3 = 9$	Multiplying on the left
$2x^2 + 5x - 3 - 9 = 9 - 9$	Subtracting 9 from both sides to get 0 on one side
$2x^2 + 5x - 12 = 0$	Combining like terms
$(2x - 3)(x + 4) = 0$	Factoring. Now we have a product equal to 0.
$2x - 3 = 0$ *or* $x + 4 = 0$	Using the principle of zero products
$2x = 3$ *or* $x = -4$	
$x = \frac{3}{2}$ *or* $x = -4$	

Check: For $\frac{3}{2}$:

$$\frac{(x + 3)(2x - 1) = 9}{\left(\frac{3}{2} + 3\right)\left(2 \cdot \frac{3}{2} - 1\right) \mid 9}$$
$$\left(\frac{9}{2}\right)(2) \mid$$
$$9 \overset{?}{=} 9 \quad \text{TRUE}$$

For -4:

$$\frac{(x + 3)(2x - 1) = 9}{(-4 + 3)(2(-4)-1) \mid 9}$$
$$(-1)(-9) \mid$$
$$9 \overset{?}{=} 9 \quad \text{TRUE}$$

The solutions are $\frac{3}{2}$ and -4.

▶ **TRY EXERCISE** ▶ 43

ALGEBRAIC–GRAPHICAL CONNECTION

When graphing equations in Chapter 3, we found the x-intercept by replacing y with 0 and solving for x. This procedure is also used to find the x-intercepts when graphing equations of the form $y = ax^2 + bx + c$. Although the details of creating such graphs is left for Chapter 9, we consider them briefly here from the standpoint of finding x-intercepts. The graphs are shaped as shown. Note that each x-intercept represents a solution of $ax^2 + bx + c = 0$.

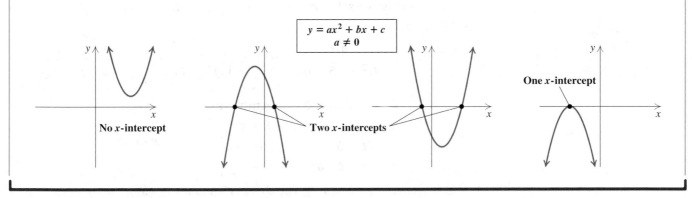

$$y = ax^2 + bx + c$$
$$a \neq 0$$

No x-intercept

Two x-intercepts

One x-intercept

EXAMPLE 8   Find the x-intercepts for the graph of the equation shown. (The grid is intentionally not included.)

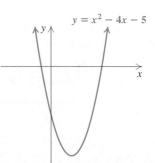

$y = x^2 - 4x - 5$

SOLUTION   To find the x-intercepts, we let $y = 0$ and solve for x:

$$0 = x^2 - 4x - 5 \qquad \text{Substituting 0 for } y$$
$$0 = (x - 5)(x + 1) \qquad \text{Factoring}$$
$$x - 5 = 0 \quad or \quad x + 1 = 0 \qquad \text{Using the principle of zero products}$$
$$x = 5 \quad or \qquad x = -1. \qquad \text{Solving for } x$$

The x-intercepts are $(5, 0)$ and $(-1, 0)$.

> **TRY EXERCISE** ▶ 59

TECHNOLOGY CONNECTION

A graphing calculator allows us to solve quadratic equations even when an equation cannot be solved by factoring. For example, to solve $x^2 - 3x - 5 = 0$, we can let $y_1 = x^2 - 3x - 5$ and $y_2 = 0$. Selecting a bold line type to the left of y_2 in the ⟨ Y= ⟩ window makes the line easier to see. Using the INTERSECT option of the CALC menu, we select the two graphs in which we are interested, along with a guess. The graphing calculator displays the point of intersection.

An alternative method uses only y_1 and the ZERO option of the CALC menu. This option requires you to enter an x-value to the left of each x-intercept as a LEFT BOUND. An x-value to the right of the x-intercept is then entered as a RIGHT BOUND. Finally, a GUESS value between the two bounds is entered and the x-intercept, or ZERO, is displayed.

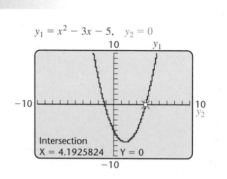

$y_1 = x^2 - 3x - 5, \quad y_2 = 0$

Use a graphing calculator to find the solutions, if they exist, accurate to two decimal places.

1. $x^2 + 4x - 3 = 0$
2. $x^2 - 5x - 2 = 0$
3. $x^2 + 13.54x + 40.95 = 0$
4. $x^2 - 4.43x + 6.32 = 0$
5. $1.235x^2 - 3.409x = 0$

5.6   EXERCISE SET

For Extra Help   MyMathLab   Math XL PRACTICE   WATCH   DOWNLOAD

↬ *Concept Reinforcement*   For each of Exercises 1–4, match the phrase with the most appropriate choice from the column on the right.

1. ____ The name of equations of the type $ax^2 + bx + c = 0$, with $a \neq 0$

2. ____ The maximum number of solutions of quadratic equations

3. ____ The idea that $A \cdot B = 0$ if and only if $A = 0$ or $B = 0$

4. ____ The number that a product must equal before the principle of zero products is used

a) 2

b) 0

c) Quadratic

d) The principle of zero products

Solve using the principle of zero products.

5. $(x + 2)(x + 9) = 0$

6. $(x + 3)(x + 10) = 0$

7. $(x + 1)(x - 8) = 0$

8. $(x + 5)(x - 4) = 0$

9. $(2t - 3)(t + 6) = 0$

10. $(5t - 8)(t - 1) = 0$

11. $4(7x - 1)(10x - 3) = 0$

12. $6(4x - 3)(2x + 9) = 0$

13. $x(x - 7) = 0$

14. $x(x + 2) = 0$

15. $\left(\frac{2}{3}x - \frac{12}{11}\right)\left(\frac{7}{4}x - \frac{1}{12}\right) = 0$

16. $\left(\frac{1}{9} - 3x\right)\left(\frac{1}{5} + 2x\right) = 0$

17. $6n(3n + 8) = 0$

18. $10n(4n - 5) = 0$

19. $(20 - 0.4x)(7 - 0.1x) = 0$

20. $(1 - 0.05x)(1 - 0.3x) = 0$

Solve by factoring and using the principle of zero products.

21. $x^2 - 7x + 6 = 0$

22. $x^2 - 6x + 5 = 0$

23. $x^2 + 4x - 21 = 0$

24. $x^2 - 7x - 18 = 0$

25. $n^2 + 11n + 18 = 0$

26. $n^2 + 8n + 15 = 0$

27. $x^2 - 10x = 0$

28. $x^2 + 8x = 0$

29. $6t + t^2 = 0$

30. $3t - t^2 = 0$

31. $x^2 - 36 = 0$

32. $x^2 - 100 = 0$

33. $4t^2 = 49$

34. $9t^2 = 25$

35. $0 = 25 + x^2 + 10x$

36. $0 = 6x + x^2 + 9$

37. $64 + x^2 = 16x$

38. $x^2 + 1 = 2x$

39. $4t^2 = 8t$

40. $12t = 3t^2$

41. $4y^2 = 7y + 15$

42. $12y^2 - 5y = 2$

43. $(x - 7)(x + 1) = -16$

44. $(x + 2)(x - 7) = -18$

45. $15z^2 + 7 = 20z + 7$

46. $14z^2 - 3 = 21z - 3$

47. $36m^2 - 9 = 40$

48. $81x^2 - 5 = 20$

49. $(x + 3)(3x + 5) = 7$

50. $(x - 1)(5x + 4) = 2$

51. $3x^2 - 2x = 9 - 8x$

52. $x^2 - 2x = 18 + 5x$

53. $(6a + 1)(a + 1) = 21$

54. $(2t + 1)(4t - 1) = 14$

55. Use this graph to solve $x^2 - 3x - 4 = 0$.

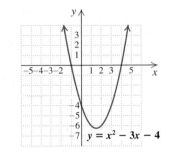

56. Use this graph to solve $x^2 + x - 6 = 0$.

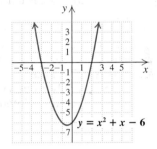

57. Use this graph to solve $-x^2 - x + 6 = 0$.

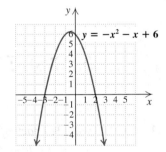

58. Use this graph to solve $-x^2 + 2x + 3 = 0$.

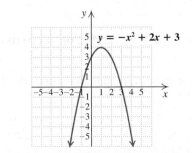

Find the x-intercepts for the graph of each equation. Grids are intentionally not included.

59. $y = x^2 - x - 6$

60. $y = x^2 + 3x - 4$

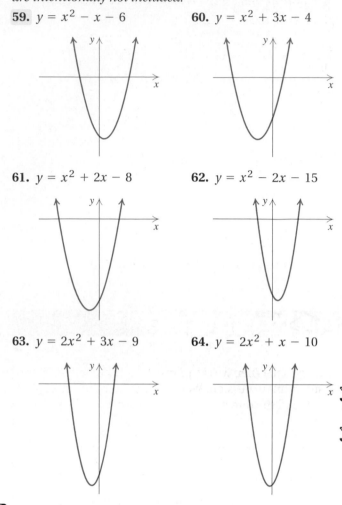

61. $y = x^2 + 2x - 8$

62. $y = x^2 - 2x - 15$

63. $y = 2x^2 + 3x - 9$

64. $y = 2x^2 + x - 10$

65. The equation $x^2 + 1 = 0$ has no real-number solutions. What implications does this have for the graph of $y = x^2 + 1$?

66. What is the difference between a quadratic polynomial and a quadratic equation?

Skill Review

To prepare for Section 5.7, review solving problems using the five-step strategy (Section 2.5).

Translate to an algebraic expression. [1.1]

67. The square of the sum of two numbers

68. The sum of the squares of two numbers

69. The product of two consecutive integers

Solve. [2.5]

70. In 2005, shoppers spent $22.8 billion on gifts for Mother's Day and for Father's Day combined. They spent $4.8 billion more for Mother's Day than for Father's Day. How much did shoppers spend for each holiday?

Source: National Retail Federation

71. The first angle of a triangle is four times as large as the second. The measure of the third angle is 30° less than that of the second. How large are the angles?

72. A rectangular table top is twice as long as it is wide. The perimeter of the table is 192 in. What are the dimensions of the table?

Synthesis

73. What is wrong with solving $x^2 = 3x$ by dividing both sides of the equation by x?

74. When the principle of zero products is used to solve a quadratic equation, will there always be two different solutions? Why or why not?

Solve.

75. $(2x - 11)(3x^2 + 29x + 56) = 0$

76. $(4x + 1)(15x^2 - 7x - 2) = 0$

77. Find an equation with integer coefficients that has the given numbers as solutions. For example, 3 and -2 are solutions to $x^2 - x - 6 = 0$.

    a) $-4, 5$ **b)** $-1, 7$ **c)** $\frac{1}{4}, 3$

    d) $\frac{1}{2}, \frac{1}{3}$ **e)** $\frac{2}{3}, \frac{3}{4}$ **f)** $-1, 2, 3$

Solve.

78. $16(x - 1) = x(x + 8)$

79. $a(9 + a) = 4(2a + 5)$

80. $(t - 5)^2 = 2(5 - t)$

81. $-x^2 + \frac{9}{25} = 0$

82. $a^2 = \frac{49}{100}$

Aha! **83.** $(t + 1)^2 = 9$

84. $\frac{27}{25}x^2 = \frac{1}{3}$

85. For each equation on the left, find an equivalent equation on the right.

a) $x^2 + 10x - 2 = 0$ $4x^2 + 8x + 36 = 0$
b) $(x - 6)(x + 3) = 0$ $(2x + 8)(2x - 5) = 0$
c) $5x^2 - 5 = 0$ $9x^2 - 12x + 24 = 0$
d) $(2x - 5)(x + 4) = 0$ $(x + 1)(5x - 5) = 0$
e) $x^2 + 2x + 9 = 0$ $x^2 - 3x - 18 = 0$
f) $3x^2 - 4x + 8 = 0$ $2x^2 + 20x - 4 = 0$

86. Explain how to construct an equation that has seven solutions.

87. Explain how the graph in Exercise 57 can be used to visualize the solutions of

$$-x^2 - x + 6 = 4.$$

Use a graphing calculator to find the solutions of each equation. Round solutions to the nearest hundredth.

88. $-x^2 + 0.63x + 0.22 = 0$

89. $x^2 - 9.10x + 15.77 = 0$

90. $6.4x^2 - 8.45x - 94.06 = 0$

91. $x^2 + 13.74x + 42.00 = 0$

92. $0.84x^2 - 2.30x = 0$

93. $1.23x^2 + 4.63x = 0$

94. $x^2 + 1.80x - 5.69 = 0$

CONNECTING the CONCEPTS

Recall that an *equation* is a statement that two *expressions* are equal. When we simplify expressions, combine expressions, and form equivalent expressions, each result is an expression. When we are asked to solve an equation, the result is one or more numbers. Remember to read the directions to an exercise carefully so you do not attempt to "solve" an expression.

MIXED REVIEW

For Exercises 1–6, tell whether each is an example of an expression or an equation.

1. $x^2 - 25$

2. $x^2 - 25 = 0$

3. $x^2 + 2x = 5$

4. $(x + 3)(2x - 1)$

5. $x(x + 3) - 2(2x - 7) - (x - 5)$

6. $x = 10$

7. Add the expressions:
$$(2x^3 - 5x + 1) + (x^2 - 3x - 1).$$

8. Subtract the expressions:
$$(x^2 - x - 5) - (3x^2 - x + 6).$$

9. Solve the equation: $t^2 - 100 = 0$.

10. Multiply the expressions: $(3a - 2)(2a - 5)$.

11. Factor the expression: $n^2 - 10n + 9$.

12. Solve the equation: $x^2 + 16 = 10x$.

13. Solve: $4t^2 + 20t + 25 = 0$.

14. Add: $(3x^3 - 5x + 1) + (4x^3 + 7x - 8)$.

15. Factor: $16x^2 - 81$.

16. Solve: $y^2 - 5y - 24 = 0$.

17. Subtract: $(a^2 - 2) - (5a^2 + a + 9)$.

18. Factor: $18x^4 - 24x^3 + 20x^2$.

19. Solve: $3x^2 + 5x + 2 = 0$.

20. Multiply: $4x^2(2x^3 - 5x^2 + 3)$.

5.7   Solving Applications

Applications   •   The Pythagorean Theorem

Applications

We can use the five-step problem-solving process and our new methods of solving quadratic equations to solve new types of problems.

EXAMPLE   1

Race numbers.   Terry and Jody each entered a boat in the Lakeport Race. The racing numbers of their boats were consecutive numbers, the product of which was 156. Find the numbers.

SOLUTION

1. **Familiarize.** Consecutive numbers are one apart, like 49 and 50. Let x = the first boat number; then $x + 1$ = the next boat number.

2. **Translate.** We reword the problem before translating:

Rewording:	The first boat number	times	the next boat number	is	156.
Translating:	x	$\cdot$	$(x + 1)$	$=$	156

3. **Carry out.** We solve the equation as follows:

$$x(x + 1) = 156$$
$$x^2 + x = 156 \qquad \text{Multiplying}$$
$$x^2 + x - 156 = 0 \qquad \text{Subtracting 156 to get 0 on one side}$$
$$(x - 12)(x + 13) = 0 \qquad \text{Factoring}$$
$$x - 12 = 0 \quad or \quad x + 13 = 0 \qquad \text{Using the principle of zero products}$$
$$x = 12 \quad or \qquad x = -13. \qquad \text{Solving each equation}$$

4. **Check.** The solutions of the equation are 12 and -13. Since race numbers are not negative, -13 must be rejected. On the other hand, if x is 12, then $x + 1$ is 13 and $12 \cdot 13 = 156$. Thus the solution 12 checks.

5. **State.** The boat numbers for Terry and Jody were 12 and 13.

TRY EXERCISE ▶ 5

EXAMPLE 2   *Manufacturing.*   Wooden Work, Ltd., builds cutting boards that are twice as long as they are wide. The most popular board that Wooden Work makes has an area of 800 cm². What are the dimensions of the board?

SOLUTION

1. **Familiarize.** We first make a drawing. Recall that the area of any rectangle is Length · Width. We let x = the width of the board, in centimeters. The length is then $2x$, since the board is twice as long as it is wide.

$2x$ x

2. **Translate.** We reword and translate as follows:

Rewording: The area of the rectangle   is   800 cm².

Translating: $2x \cdot x$ =   800

3. **Carry out.** We solve the equation as follows:

$$2x \cdot x = 800$$
$$2x^2 = 800$$

$2x^2 - 800 = 0$	Subtracting 800 to get 0 on one side of the equation
$2(x^2 - 400) = 0$	Factoring out a common factor of 2
$2(x - 20)(x + 20) = 0$	Factoring a difference of squares
$(x - 20)(x + 20) = 0$	Dividing both sides by 2
$x - 20 = 0$ *or* $x + 20 = 0$	Using the principle of zero products
$x = 20$ *or* $x = -20.$	Solving each equation

4. **Check.** The solutions of the equation are 20 and -20. Since the width must be positive, -20 cannot be a solution. To check 20 cm, we note that if the width is 20 cm, then the length is $2 \cdot 20$ cm = 40 cm and the area is 20 cm · 40 cm = 800 cm². Thus the solution 20 checks.

5. **State.** The cutting board is 20 cm wide and 40 cm long.   **TRY EXERCISE** 9

EXAMPLE 3   *Dimensions of a leaf.*   Each leaf of one particular *Philodendron* species is approximately a triangle. A typical leaf has an area of 320 in². If the leaf is 12 in. longer than it is wide, find the length and the width of the leaf.

SOLUTION

1. **Familiarize.** The formula for the area of a triangle is Area = $\frac{1}{2} \cdot$ (base) · (height). We let b = the width, in inches, of the triangle's base and $b + 12$ = the height, in inches.

2. **Translate.** We reword and translate as follows:

Rewording: The area of the leaf   is   320 in².

Translating: $\frac{1}{2} \cdot b(b + 12)$ =   320

3. Carry out. We solve the equation as follows:

$$\frac{1}{2} \cdot b \cdot (b + 12) = 320$$

$$\frac{1}{2}(b^2 + 12b) = 320 \qquad \text{Multiplying}$$

$$b^2 + 12b = 640 \qquad \text{Multiplying by 2 to clear fractions}$$

$$b^2 + 12b - 640 = 0 \qquad \text{Subtracting 640 to get 0 on one side}$$

$$(b + 32)(b - 20) = 0 \qquad \text{Factoring}$$

$$b + 32 = 0 \quad or \quad b - 20 = 0 \qquad \text{Using the principle of zero products}$$

$$b = -32 \quad or \qquad b = 20.$$

4. Check. The width must be positive, so -32 cannot be a solution. Suppose the base is 20 in. The height would be $20 + 12$, or 32 in., and the area $\frac{1}{2}(20)(32)$, or 320 in^2. These numbers check in the original problem.

5. State. The leaf is 32 in. long and 20 in. wide.　　　▶ TRY EXERCISE ▶ 13

EXAMPLE 4

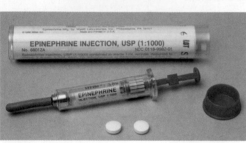

Medicine. For certain people suffering an extreme allergic reaction, the drug epinephrine (adrenaline) is sometimes prescribed. The number of micrograms N of epinephrine in an adult's bloodstream t minutes after 250 micrograms have been injected can be approximated by

$$-10t^2 + 100t = N.$$

How long after an injection will there be about 210 micrograms of epinephrine in the bloodstream?

Source: Based on information in Chohan, Naina, Rita M. Doyle, and Patricia Nayle (eds.), *Nursing Handbook*, 21st ed. Springhouse, PA: Springhouse Corporation, 2001

SOLUTION

1. Familiarize. To familiarize ourselves with this problem, we could calculate N for different choices of t. We leave this for the student. Note that there may be two solutions, one on each side of the time at which the drug's effect peaks.

2. Translate. To find the length of time after injection when 210 micrograms are in the bloodstream, we replace N with 210 in the formula above:

$$-10t^2 + 100t = 210. \qquad \text{Substituting 210 for } N. \text{ This is now an equation in one variable.}$$

3. Carry out. We solve the equation as follows:

$$-10t^2 + 100t = 210$$

$$-10t^2 + 100t - 210 = 0 \qquad \text{Subtracting 210 from both sides to get 0 on one side}$$

$$-10(t^2 - 10t + 21) = 0 \qquad \text{Factoring out the largest common factor, } -10$$

$$-10(t - 3)(t - 7) = 0 \qquad \text{Factoring}$$

$$t - 3 = 0 \quad or \quad t - 7 = 0 \qquad \text{Using the principle of zero products}$$

$$t = 3 \quad or \qquad t = 7.$$

As a visual check for Example 4, we can either let
$y_1 = -10x^2 + 100x$ and
$y_2 = 210$, or let
$y_1 = -10x^2 + 100x - 210$ and
$y_2 = 0$. In either case, the points of intersection occur at
$x = 3$ and $x = 7$, as shown.

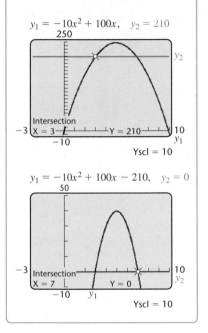

4. **Check.** Since $-10 \cdot 3^2 + 100 \cdot 3 = -90 + 300 = 210$, the number 3 checks. Since $-10 \cdot 7^2 + 100 \cdot 7 = -490 + 700 = 210$, the number 7 also checks.

5. **State.** There will be 210 micrograms of epinephrine in the bloodstream approximately 3 minutes and 7 minutes after injection.

> TRY EXERCISE 17

The Pythagorean Theorem

The following problems involve the Pythagorean theorem, which relates the lengths of the sides of a *right* triangle. A triangle is a **right triangle** if it has a 90°, or *right*, angle. The side opposite the 90° angle is called the **hypotenuse**. The other sides are called **legs**.

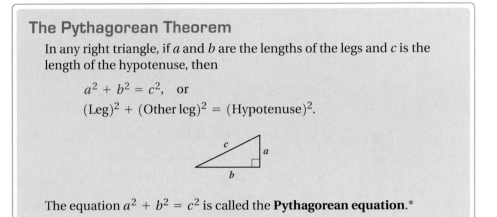

The Pythagorean theorem is named for the Greek mathematician Pythagoras (569?–500? B.C.). We can think of this relationship as adding areas.

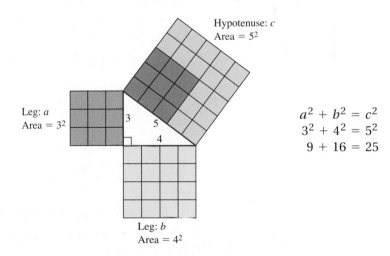

If we know the lengths of any two sides of a right triangle, we can use the Pythagorean equation to determine the length of the third side.

---

*The *converse* of the Pythagorean theorem is also true. That is, if $a^2 + b^2 = c^2$, then the triangle is a right triangle.

EXAMPLE   5   *Travel.*   A zipline canopy tour in Alaska includes a cable that slopes downward from a height of 135 ft to a height of 100 ft. The trees that the cable connects are 120 ft apart. Find the minimum length of the cable.

SOLUTION

1. **Familiarize.** We first make a drawing or visualize the situation. The difference in height between the platforms is 35 ft. Note that the cable must have some extra length to allow for the rider's movement, but we will approximate its length as the hypotenuse of a right triangle, as shown.

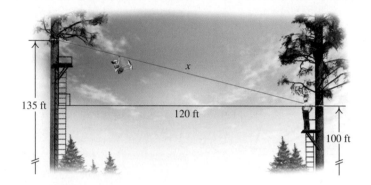

2. **Translate.** Since a right triangle is formed, we can use the Pythagorean theorem:

$$a^2 + b^2 = c^2$$
$$35^2 + 120^2 = x^2. \qquad \text{Substituting}$$

3. **Carry out.** We solve the equation as follows:

$$1225 + 14{,}400 = x^2 \qquad\qquad \text{Squaring 35 and 120}$$
$$15{,}625 = x^2 \qquad\qquad \text{Adding}$$
$$0 = x^2 - 15{,}625 \qquad\qquad \text{Subtracting 15,625 from both sides}$$
$$0 = (x + 125)(x - 125) \qquad\qquad \text{Note that } 15{,}625 = 125^2. \text{ A calculator would be helpful here.}$$
$$x + 125 = 0 \quad or \quad x - 125 = 0 \qquad \text{Using the principle of zero products}$$
$$x = -125 \quad or \qquad\qquad x = 125.$$

4. **Check.** Since the length of the cable must be positive, -125 is not a solution. If the length is 125 ft, we have $35^2 + 120^2 = 1225 + 14{,}400 = 15{,}625$, which is 125^2. Thus the solution 125 checks.

5. **State.** The minimum length of the cable is 125 ft.   **TRY EXERCISE** 27

EXAMPLE   **6**   *Bridge design.*   A 50-ft diagonal brace on a bridge connects a support at the center of the bridge to a side support on the bridge. The horizontal distance that it spans is 10 ft longer than the height that it reaches on the side of the bridge. Find both distances.

SOLUTION

1. **Familiarize.** We first make a drawing. The diagonal brace and the missing distances form the hypotenuse and the legs of a right triangle. We let x = the length of the vertical leg. Then $x + 10$ = the length of the horizontal leg. The hypotenuse has length 50 ft.

2. **Translate.** Since the triangle is a right triangle, we can use the Pythagorean theorem:

$$a^2 + b^2 = c^2$$
$$x^2 + (x + 10)^2 = 50^2. \qquad \text{Substituting}$$

3. **Carry out.** We solve the equation as follows:

$x^2 + (x^2 + 20x + 100) = 2500$	Squaring
$2x^2 + 20x + 100 = 2500$	Combining like terms
$2x^2 + 20x - 2400 = 0$	Subtracting 2500 to get 0 on one side
$2(x^2 + 10x - 1200) = 0$	Factoring out a common factor
$2(x + 40)(x - 30) = 0$	Factoring. A calculator would be helpful here.
$x + 40 = 0 \quad or \quad x - 30 = 0$	Using the principle of zero products
$x = -40 \quad or \qquad x = 30.$	

4. **Check.** The integer -40 cannot be a length of a side because it is negative. If the length is 30 ft, $x + 10 = 40$, and $30^2 + 40^2 = 900 + 1600 = 2500$, which is 50^2. So the solution 30 checks.

5. **State.** The height that the brace reaches on the side of the bridge is 30 ft, and the distance that it reaches to the middle of the bridge is 40 ft.

TRY EXERCISE   31

Translating for Success

1. **Angle measures.**
The measures of the angles of a triangle are three consecutive integers. Find the measures of the angles.

2. **Rectangle dimensions.** The area of a rectangle is 3604 ft^2. The length is 15 ft longer than the width. Find the dimensions of the rectangle.

3. **Sales tax.** Claire paid $3604 for a used pickup truck. This included 6% for sales tax. How much did the truck cost before tax?

4. **Wire cutting.** A 180-m wire is cut into three pieces. The third piece is 2 m longer than the first. The second is two-thirds as long as the first. How long is each piece?

5. **Perimeter.** The perimeter of a rectangle is 240 ft. The length is 2 ft greater than the width. Find the length and the width.

Translate each word problem to an equation and select a correct translation from equations A–O.

A. $2x \cdot x = 288$

B. $x(x + 60) = 7021$

C. $59 = x \cdot 60$

D. $x^2 + (x + 15)^2 = 3604$

E. $x^2 + (x + 70)^2 = 130^2$

F. $0.06x = 3604$

G. $2(x + 2) + 2x = 240$

H. $\frac{1}{2}x(x - 1) = 1770$

I. $x + \frac{2}{3}x + (x + 2) = 180$

J. $0.59x = 60$

K. $x + 0.06x = 3604$

L. $2x^2 + x = 288$

M. $x(x + 15) = 3604$

N. $x^2 + 60 = 7021$

O. $x + (x + 1) + (x + 2) = 180$

Answers on page A-21

An additional, animated version of this activity appears in MyMathLab. To use MyMathLab, you need a course ID and a student access code. Contact your instructor for more information.

6. **Cell-phone tower.** A guy wire on a cell-phone tower is 130 ft long and is attached to the top of the tower. The height of the tower is 70 ft longer than the distance from the point on the ground where the wire is attached to the bottom of the tower. Find the height of the tower.

7. **Sales meeting attendance.** PTQ Corporation holds a sales meeting in Tucson. Of the 60 employees, 59 of them attend the meeting. What percent attend the meeting?

8. **Dimensions of a pool.** A rectangular swimming pool is twice as long as it is wide. The area of the surface is 288 ft^2. Find the dimensions of the pool.

9. **Dimensions of a triangle.** The height of a triangle is 1 cm less than the length of the base. The area of the triangle is 1770 cm^2. Find the height and the length of the base.

10. **Width of a rectangle.** The length of a rectangle is 60 ft longer than the width. Find the width if the area of the rectangle is 7021 ft^2.

5.7 EXERCISE SET

Solve. Use the five-step problem-solving approach.

1. A number is 6 less than its square. Find all such numbers.

2. A number is 30 less than its square. Find all such numbers.

3. One leg of a right triangle is 2 m longer than the other leg. The length of the hypotenuse is 10 m. Find the length of each side.

4. One leg of a right triangle is 7 cm shorter than the other leg. The length of the hypotenuse is 13 cm. Find the length of each side.

5. *Parking-space numbers.* The product of two consecutive parking spaces is 132. Find the parking-space numbers.

6. *Page numbers.* The product of the page numbers on two facing pages of a book is 420. Find the page numbers.

7. The product of two consecutive even integers is 168. Find the integers.

8. The product of two consecutive odd integers is 195. Find the integers.

9. *Construction.* The front porch on Trent's new home is five times as long as it is wide. If the area of the porch is 180 ft^2, find the dimensions.

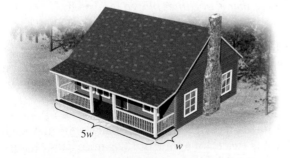

5w
w

10. *Furnishings.* The work surface of Anita's desk is a rectangle that is twice as long as it is wide. If the area of the desktop is 18 ft^2, find the length and the width of the desk.

w
2w

11. *Design.* The screen of the TI-84 Plus graphing calculator is nearly rectangular. The length of the rectangle is 2 cm more than the width. If the area of the rectangle is 24 cm^2, find the length and the width.

w + 2
w

12. *Area of a garden.* The length of a rectangular garden is 4 m greater than the width. The area of the garden is 96 m^2. Find the length and the width.

w
w + 4

13. *Dimensions of a triangle.* The height of a triangle is 3 in. less than the length of the base. If the area of the triangle is 54 in^2, find the height and the length of the base.

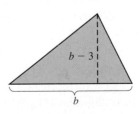

b − 3
b

14. *Dimensions of a triangle.* A triangle is 10 cm wider than it is tall. The area is 48 cm^2. Find the height and the base.

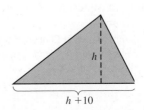

h
h + 10

15. *Dimensions of a sail.*   The height of the jib sail on a Lightning sailboat is 5 ft greater than the length of its "foot." If the area of the sail is 42 ft², find the length of the foot and the height of the sail.

16. *Road design.*   A triangular traffic island has a base half as long as its height. Find the base and the height if the island has an area of 64 m².

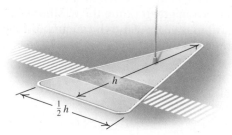

17. *Medicine.*   For many people suffering from constricted bronchial muscles, the drug Albuterol is prescribed. The number of micrograms A of Albuterol in a person's bloodstream t minutes after 200 micrograms have been inhaled can be approximated by

$$A = -50t^2 + 200t.$$

How long after an inhalation will there be about 150 micrograms of Albuterol in the bloodstream?
Source: Based on information in Chohan, Naina, Rita M. Doyle, and Patricia Nayle (eds.), *Nursing Handbook*, 21st ed. Springhouse, PA: Springhouse Corporation, 2001

18. *Medicine.*   For adults with certain heart conditions, the drug Primacor (milrinone lactate) is prescribed. The number of milligrams M of Primacor in the bloodstream of a 132-lb patient t hours after a 3-mg dose has been injected can be approximated by

$$M = -\frac{1}{2}t^2 + \frac{5}{2}t.$$

How long after an injection will there be about 2 mg in the bloodstream?
Source: Based on information in Chohan, Naina, Rita M. Doyle, and Patricia Nayle (eds.), *Nursing Handbook*, 21st ed. Springhouse, PA: Springhouse Corporation, 2001

19. *High-speed Internet.*   The number N, in millions, of broadband cable and DSL subscribers in the United States t years after 1998 can be approximated by

$$N = 0.3t^2 + 0.6t.$$

When will there be 36 million broadband cable and DSL subscribers?
Source: Based on information from Leichtman Research Group

20. *Wave height.*   The height of waves in a storm depends on the speed of the wind. Assuming the wind has no obstructions for a long distance, the maximum wave height H for a wind speed x can be approximated by

$$H = 0.006x^2 + 0.6x.$$

Here H is in feet and x is in knots (nautical miles per hour). For what wind speed would the maximum wave height be 6.6 ft?
Source: Based on information from cimss.ssec.wisc.edu

Games in a league's schedule.   *In a sports league of x teams in which all teams play each other twice, the total number N of games played is given by*

$$x^2 - x = N.$$

Use this formula for Exercises 21 and 22.

21. The Colchester Youth Soccer League plays a total of 240 games, with all teams playing each other twice. How many teams are in the league?

22. The teams in a women's softball league play each other twice, for a total of 132 games. How many teams are in the league?

Number of handshakes.   *The number of possible handshakes H within a group of n people is given by $H = \frac{1}{2}(n^2 - n)$. Use this formula for Exercises 23–26.*

23. At a meeting, there are 12 people. How many handshakes are possible?

24. At a party, there are 25 people. How many handshakes are possible?

25. *High-fives.* After winning the championship, all San Antonio Spurs teammates exchanged "high-fives." Altogether there were 66 high-fives. How many players were there?

26. *Toasting.* During a toast at a party, there were 105 "clicks" of glasses. How many people took part in the toast?

27. *Construction.* The diagonal braces in a lookout tower are 15 ft long and span a horizontal distance of 12 ft. How high does each brace reach vertically?

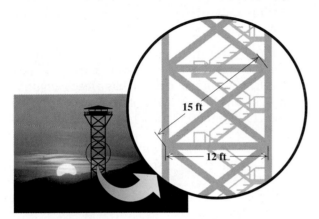

28. *Reach of a ladder.* Twyla has a 26-ft ladder leaning against her house. If the bottom of the ladder is 10 ft from the base of the house, how high does the ladder reach?

29. *Roadway design.* Elliott Street is 24 ft wide when it ends at Main Street in Brattleboro, Vermont. A 40-ft long diagonal crosswalk allows pedestrians to cross Main Street to or from either corner of Elliott Street (see the figure). Determine the width of Main Street.

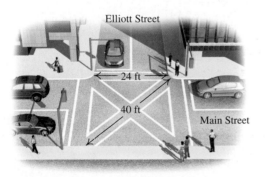

30. *Aviation.* Engine failure forced Robbin to pilot her Cessna 150 to an emergency landing. To land, Robbin's plane glided 17,000 ft over a 15,000-ft stretch of deserted highway. From what altitude did the descent begin?

31. *Archaeology.* Archaeologists have discovered that the 18th-century garden of the Charles Carroll House in Annapolis, Maryland, was a right triangle. One leg of the triangle was formed by a 400-ft long sea wall. The hypotenuse of the triangle was 200 ft longer than the other leg. What were the dimensions of the garden?
Source: www.bsos.umd.edu

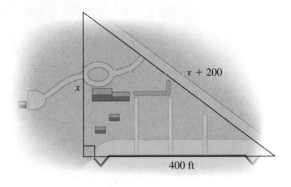

32. *Guy wire.* The guy wire on a TV antenna is 1 m longer than the height of the antenna. If the guy wire is anchored 3 m from the foot of the antenna, how tall is the antenna?

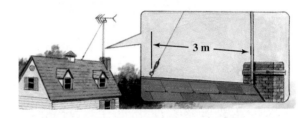

33. *Architecture.* An architect has allocated a rectangular space of 264 ft^2 for a square dining room and a 10-ft wide kitchen, as shown in the figure. Find the dimensions of each room.

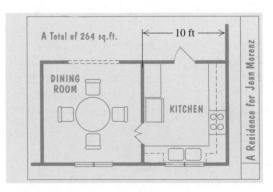

34. *Design.* A window panel for a sun porch consists of a 7-ft high rectangular window stacked above a square window. The windows have the same width. If the total area of the window panel is 18 ft², find the dimensions of each window.

7 ft

Height of a rocket. For Exercises 35–38, assume that a water rocket is launched upward with an initial velocity of 48 ft/sec. Its height h, in feet, after t seconds, is given by $h = 48t - 16t^2$.

35. Determine the height of the rocket $\frac{1}{2}$ sec after it has been launched.

36. Determine the height of the rocket 2.5 sec after it has been launched.

37. When will the rocket be exactly 32 ft above the ground?

38. When will the rocket crash into the ground?

39. Do we now have the ability to solve *any* problem that translates to a quadratic equation? Why or why not?

40. Write a problem for a classmate to solve such that only one of two solutions of a quadratic equation can be used as an answer.

Skill Review

To prepare for Chapter 6, review addition, subtraction, multiplication, and division using fraction notation (Sections 1.3, 1.5, 1.6, and 1.7).

Simplify.

41. $-\dfrac{3}{5} \cdot \dfrac{4}{7}$ [1.7]

42. $-\dfrac{3}{5} \div \dfrac{4}{7}$ [1.7]

43. $-\dfrac{5}{6} - \dfrac{1}{6}$ [1.6]

44. $\dfrac{3}{4} + \left(-\dfrac{5}{2}\right)$ [1.5]

45. $-\dfrac{3}{8} \cdot \left(-\dfrac{10}{15}\right)$ [1.7]

46. $\dfrac{-\dfrac{8}{15}}{-\dfrac{2}{3}}$ [1.7]

47. $\dfrac{5}{24} + \dfrac{3}{28}$ [1.3]

48. $\dfrac{5}{6} - \left(-\dfrac{2}{9}\right)$ [1.6]

Synthesis

The converse of the Pythagorean theorem is also true. That is, if $a^2 + b^2 = c^2$, then the triangle is a right triangle (where a and b are the lengths of the legs and c is the length of the hypotenuse). Use this result to answer Exercises 49 and 50.

49. An archaeologist has measuring sticks of 3 ft, 4 ft, and 5 ft. Explain how she could draw a 7-ft by 9-ft rectangle on a piece of land being excavated.

50. Explain how measuring sticks of 5 cm, 12 cm, and 13 cm can be used to draw a right triangle that has two 45° angles.

51. *Sailing.* The mainsail of a Lightning sailboat is a right triangle in which the hypotenuse is called the leech. If a 24-ft tall mainsail has a leech length of 26 ft and if Dacron® sailcloth costs $1.50 per square foot, find the cost of the fabric for a new mainsail.

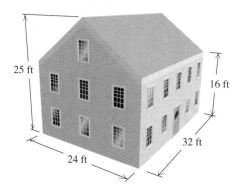

26 ft

24 ft

52. *Roofing.* A *square* of shingles covers 100 ft² of surface area. How many squares will be needed to re-shingle the house shown?

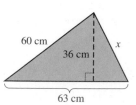

25 ft

16 ft

24 ft

32 ft

53. Solve for x.

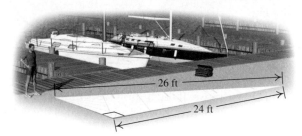

60 cm

36 cm

x

63 cm

54. *Pool sidewalk.*   A cement walk of uniform width is built around a 20-ft by 40-ft rectangular pool. The total area of the pool and the walk is 1500 ft². Find the width of the walk.

55. *Folding sheet metal.*   An open rectangular gutter is made by turning up the sides of a piece of metal 20 in. wide, as shown. The area of the cross-section of the gutter is 48 in². Find the possible depths of the gutter.

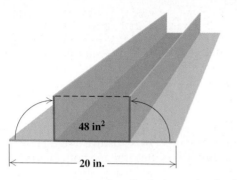

48 in²

20 in.

56. Find a polynomial for the shaded area in the figure below.

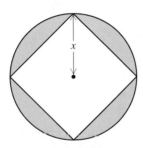

x

57. *Telephone service.*   Use the information in the figure below to determine the height of the telephone pole.

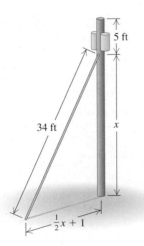

5 ft

34 ft

x

$\frac{1}{2}x + 1$

58. *Dimensions of a closed box.*   The total surface area of a closed box is 350 m². The box is 9 m high and has a square base and lid. Find the length of a side of the base.

Medicine.   *For certain people with acid reflux, the drug Pepcid (famotidine) is used. The number of milligrams N of Pepcid in an adult's bloodstream t hours after a 20-mg tablet has been swallowed can be approximated by*

$$N = -0.009t\,(t - 12)^3.$$

Use a graphing calculator with the window $[-1, 13, -1, 25]$ *and the* TRACE *feature to answer Exercises 59–61.*

Source: Based on information in Chohan, Naina, Rita M. Doyle, and Patricia Nayle (eds.), *Nursing Handbook*, 21st ed. Springhouse, PA: Springhouse Corporation, 2001

59. Approximately how long after a tablet has been swallowed will there be 18 mg in the bloodstream?

60. Approximately how long after a tablet has been swallowed will there be 10 mg in the bloodstream?

61. Approximately how long after a tablet has been swallowed will the peak dosage in the bloodstream occur?

Study Summary

KEY TERMS AND CONCEPTS	EXAMPLES

SECTION 5.1: INTRODUCTION TO FACTORING

To **factor** a polynomial means to write it as a product. Always begin by factoring out the **largest common factor**.

Factor: $12x^4 - 30x^3$.
$$12x^4 - 30x^3 = 6x^3(2x - 5)$$

Some polynomials with four terms can be **factored by grouping**.

Factor: $3x^3 - x^2 - 6x + 2$.
$$3x^3 - x^2 - 6x + 2 = x^2(3x - 1) - 2(3x - 1)$$
$$= (3x - 1)(x^2 - 2)$$

SECTION 5.2: FACTORING TRINOMIALS OF THE TYPE $x^2 + bx + c$

Some trinomials of the type $x^2 + bx + c$ can be factored by reversing the steps of FOIL.

Factor: $x^2 - 11x + 18$.

Pairs of Factors of 18	Sums of Factors
$-1, -18$	-19
$-2, -9$	-11
$-3, -6$	-9

18 is positive and -11 is negative, so both factors will be negative.

←The numbers we need are -2 and -9.

The factorization is $(x - 2)(x - 9)$.

SECTION 5.3: FACTORING TRINOMIALS OF THE TYPE $ax^2 + bx + c$

One method for factoring trinomials of the type $ax^2 + bx + c$ is a FOIL-based method.

Factor: $6x^2 - 5x - 6$.

The factors will be in the form
$$(3x + \;)(2x + \;) \quad \text{or} \quad (6x + \;)(x + \;).$$

We list all pairs of factors of -6, and check possible products by multiplying any possibilities that do not contain a common factor.

$(3x - 2)(2x + 3) = 6x^2 + 5x - 6,$
$(3x + 2)(2x - 3) = 6x^2 - 5x - 6.$ ←This is the correct product, so we stop here.

The factorization is $(3x + 2)(2x - 3)$.

Another method for factoring trinomials of the type $ax^2 + bx + c$ involves factoring by grouping.

Factor: $6x^2 - 5x - 6$.

Multiply the leading coefficient and the constant term:
$$6(-6) = -36.$$

Look for factors of -36 that add to -5.

Pairs of Factors of -36	Sums of Factors
$1, -36$	-35
$2, -18$	-16
$3, -12$	-9
$4, -9$	-5
$6, -6$	0

-5 is negative, so the negative factor must have the greater absolute value.

←The numbers we want are 4 and -9.

Rewrite $-5x$ as $4x - 9x$ and factor by grouping:

$$6x^2 - 5x - 6 = 6x^2 + 4x - 9x - 6$$
$$= 2x(3x + 2) - 3(3x + 2)$$
$$= (3x + 2)(2x - 3).$$

SECTION 5.4: FACTORING PERFECT-SQUARE TRINOMIALS AND DIFFERENCES OF SQUARES

Factoring a perfect-square trinomial

$$A^2 + 2AB + B^2 = (A + B)^2;$$
$$A^2 - 2AB + B^2 = (A - B)^2$$

Factor: $y^2 + 100 - 20y$.

$$A^2 - 2AB + B^2 = (A - B)^2$$

$$y^2 + 100 - 20y = y^2 - 20y + 100 = (y - 10)^2$$

Factoring a difference of squares

$$A^2 - B^2 = (A + B)(A - B)$$

Factor: $9t^2 - 1$.

$$A^2 - B^2 = (A + B)(A - B)$$

$$9t^2 - 1 = (3t + 1)(3t - 1)$$

SECTION 5.5: FACTORING: A GENERAL STRATEGY

To factor a polynomial:

A. Factor out the largest common factor.

B. Look at the number of terms.

Two terms: If a difference of squares, use $A^2 - B^2 = (A + B)(A - B)$.

Three terms: If a trinomial square, use $A^2 + 2AB + B^2 = (A + B)^2$ or $A^2 - 2AB + B^2 = (A - B)^2$. Otherwise, try FOIL or grouping.

Four terms: Try factoring by grouping.

C. Factor completely.

D. Check by multiplying.

Factor: $5x^5 - 80x$.

$$5x^5 - 80x = 5x(x^4 - 16)$$ $5x$ is the largest common factor.

$$= 5x(x^2 + 4)(x^2 - 4)$$ $x^4 - 16$ is a difference of squares.

$$= 5x(x^2 + 4)(x + 2)(x - 2)$$ $x^2 - 4$ is also a difference of squares.

Check: $5x(x^2 + 4)(x + 2)(x - 2) = 5x(x^2 + 4)(x^2 - 4)$
$$= 5x(x^4 - 16) = 5x^5 - 80x.$$

Factor: $-x^2y^2 - 3xy + 10$.

$$-x^2y^2 - 3xy + 10 = -(x^2y^2 + 3xy - 10)$$ Factor out -1 to make the leading coefficient positive.

$$= -(xy + 5)(xy - 2)$$

Check: $-(xy + 5)(xy - 2) = -(x^2y^2 + 3xy - 10)$
$$= -x^2y^2 - 3xy + 10.$$

SECTION 5.6: SOLVING QUADRATIC EQUATIONS BY FACTORING

The Principle of Zero Products

An equation $AB = 0$ is true if and only if $A = 0$ or $B = 0$, or both.

Solve: $x^2 + 7x = 30$.

$$x^2 + 7x = 30$$
$$x^2 + 7x - 30 = 0$$ Getting 0 on one side
$$(x + 10)(x - 3) = 0$$ Factoring
$$x + 10 = 0 \quad or \quad x - 3 = 0$$ Using the principle of zero products

$$x = -10 \quad or \quad x = 3$$

The solutions are -10 and 3.

SECTION 5.7: SOLVING APPLICATIONS

The Pythagorean Theorem

In any right triangle, if a and b are the lengths of the legs and c is the length of the hypotenuse, then

$a^2 + b^2 = c^2,$ or

$(\text{Leg})^2 + (\text{Other leg})^2 = (\text{Hypotenuse})^2.$

Aaron has a 25-ft ladder leaning against his house. The height that the ladder reaches on the house is 17 ft more than the distance that the bottom of the ladder is from the house. Find both distances.

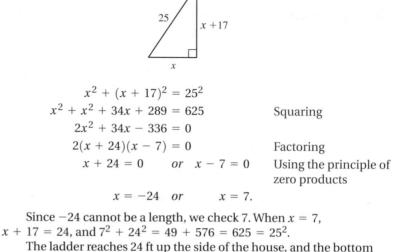

$$x^2 + (x + 17)^2 = 25^2$$
$$x^2 + x^2 + 34x + 289 = 625 \qquad \text{Squaring}$$
$$2x^2 + 34x - 336 = 0$$
$$2(x + 24)(x - 7) = 0 \qquad \text{Factoring}$$
$$x + 24 = 0 \quad or \quad x - 7 = 0 \qquad \text{Using the principle of zero products}$$
$$x = -24 \quad or \qquad x = 7.$$

Since -24 cannot be a length, we check 7. When $x = 7$, $x + 17 = 24$, and $7^2 + 24^2 = 49 + 576 = 625 = 25^2$. The ladder reaches 24 ft up the side of the house, and the bottom of the ladder is 7 ft from the house.

Review Exercises: Chapter 5

🖐 *Concept Reinforcement* *Classify each statement as either true or false.*

1. The largest common variable factor is the largest power of the variable in the polynomial. [5.1]

2. A prime polynomial has no common factor other than 1 or -1. [5.2]

3. Every perfect-square trinomial can be expressed as a binomial squared. [5.4]

4. Every binomial can be regarded as a difference of squares. [5.4]

5. Every quadratic equation has two different solutions. [5.6]

6. The principle of zero products can be applied whenever a product equals 0. [5.6]

7. In a right triangle, the hypotenuse is always longer than either leg. [5.7]

8. The Pythagorean theorem can be applied to any triangle that has an angle measuring at least 90°. [5.7]

Find three factorizations of each monomial. [5.1]

9. $20x^3$

10. $-18x^5$

Factor completely. If a polynomial is prime, state this.

11. $12x^4 - 18x^3$ [5.1]

12. $8a^2 - 12a$ [5.1]

13. $100t^2 - 1$ [5.4]

14. $x^2 + x - 12$ [5.2]

15. $x^2 + 14x + 49$ [5.4]

16. $12x^3 + 12x^2 + 3x$ [5.4]

17. $6x^3 + 9x^2 + 2x + 3$ [5.1]

18. $6a^2 + a - 5$ [5.3]

19. $25t^2 + 9 - 30t$ [5.4]

20. $48t^2 - 28t + 6$ [5.1]

21. $81a^4 - 1$ [5.4]

22. $9x^3 + 12x^2 - 45x$ [5.3]

23. $3x^2 - 27$ [5.4]

24. $x^4 + 4x^3 - 2x - 8$ [5.1]

25. $a^2b^4 - 64$ [5.4]

26. $-8x^6 + 32x^5 - 4x^4$ [5.1]

27. $75 + 12x^2 - 60x$ [5.4]

28. $y^2 + 9$ [5.4]

29. $-t^3 + t^2 + 42t$ [5.2]

30. $4x^2 - 25$ [5.4]

31. $n^2 - 60 - 4n$ [5.2]

32. $5z^2 - 30z + 10$ [5.1]

33. $4t^2 + 13t + 10$ [5.3]

34. $2t^2 - 7t - 4$ [5.3]

35. $7x^3 + 35x^2 + 28x$ [5.2]

36. $5x^3 + 35x^2 + 50x$ [5.2]

37. $20x^2 - 20x + 5$ [5.4]

38. $-6x^3 + 150x$ [5.4]

39. $15 - 8x + x^2$ [5.2]

40. $3x + x^2 + 5$ [5.2]

41. $x^2y^2 + 6xy - 16$ [5.2]

42. $12a^2 + 84ab + 147b^2$ [5.4]

43. $m^2 + 5m + mt + 5t$ [5.1]

44. $32x^4 - 128y^4z^4$ [5.4]

45. $6m^2 + 2mn + n^2 + 3mn$ [5.1], [5.3]

46. $6r^2 + rs - 15s^2$ [5.3]

Solve. [5.6]

47. $(x - 9)(x + 11) = 0$

48. $x^2 + 2x - 35 = 0$

49. $16x^2 = 9$

50. $3x^2 + 2 = 5x$

51. $2x^2 - 7x = 30$

52. $(x + 1)(x - 2) = 4$

53. $9t - 15t^2 = 0$

54. $3x^2 + 3 = 6x$

55. The square of a number is 12 more than the number. Find all such numbers. [5.7]

56. The formula $x^2 - x = N$ can be used to determine the total number of games played, N, in a league of x teams in which all teams play each other twice. Serena referees for a soccer league in which all teams play each other twice and a total of 90 games is played. How many teams are in the league? [5.7]

57. Find the x-intercepts for the graph of $y = 2x^2 - 3x - 5$. [5.6]

58. The front of a house is a triangle that is as wide as it is tall. Its area is 98 ft². Find the height and the base. [5.7]

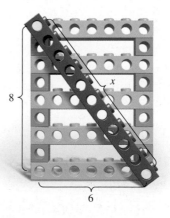

59. Josh needs to add a diagonal brace to his LEGO® robot. The brace must span a height of 8 holes and a width of 6 holes. How long should the brace be? [5.7]

Synthesis

60. On a quiz, Celia writes the factorization of $4x^2 - 100$ as $(2x - 10)(2x + 10)$. If this were a 10-point question, how many points would you give Celia? Why? [5.4]

61. How do the equations solved in this chapter differ from those solved in previous chapters? [5.6]

Solve.

62. The pages of a book measure 15 cm by 20 cm. Margins of equal width surround the printing on each page and constitute one half of the area of the page. Find the width of the margins. [5.7]

63. The cube of a number is the same as twice the square of the number. Find the number. [5.7]

64. The length of a rectangle is two times its width. When the length is increased by 20 cm and the width is decreased by 1 cm, the area is 160 cm². Find the original length and width. [5.7]

65. The length of each side of a square is increased by 5 cm to form a new square. The area of the new square is $2\frac{1}{4}$ times the area of the original square. Find the area of each square. [5.7]

Solve. [5.6]

66. $(x - 2)2x^2 + x(x - 2) - (x - 2)15 = 0$

Aha! **67.** $x^2 + 25 = 0$

Test: Chapter 5

Step-by-step test solutions are found on the video CD in the front of this book.

1. Find three factorizations of $12x^4$.

Factor completely. If a polynomial is prime, state this.

2. $x^2 - 13x + 36$

3. $x^2 + 25 - 10x$

4. $6y^2 - 8y^3 + 4y^4$

5. $x^3 + x^2 + 2x + 2$

6. $t^7 - 3t^5$

7. $a^3 + 3a^2 - 4a$

8. $28x - 48 + 10x^2$

9. $4t^2 - 25$

10. $x^2 - x - 6$

11. $-6m^3 - 9m^2 - 3m$

12. $3w^2 - 75$

13. $45r^2 + 60r + 20$

14. $3x^4 - 48$

15. $49t^2 + 36 + 84t$

16. $x^4 + 2x^3 - 3x - 6$

17. $x^2 + 3x + 6$

18. $4x^2 - 4x - 15$

19. $6t^3 + 9t^2 - 15t$

20. $3m^2 - 9mn - 30n^2$

Solve.

21. $x^2 - 6x + 5 = 0$

22. $2x^2 - 7x = 15$

23. $4t - 10t^2 = 0$

24. $25t^2 = 1$

25. $x(x - 1) = 20$

26. Find the x-intercepts for the graph of $y = 3x^2 - 5x - 8$.

27. The length of a rectangle is 6 m more than the width. The area of the rectangle is 40 m². Find the length and the width.

28. The number of possible handshakes H within a group of n people is given by $H = \frac{1}{2}(n^2 - n)$. At a meeting, everyone shook hands once with everyone else. If there were 45 handshakes, how many people were at the meeting?

29. A mason wants to be sure she has a right corner in a building's foundation. She marks a point 3 ft from the corner along one wall and another point 4 ft from the corner along the other wall. If the corner is a right angle, what should the distance be between the two marked points?

Synthesis

30. *Dimensions of an open box.* A rectangular piece of cardboard is twice as long as it is wide. A 4-cm square is cut out of each corner, and the sides are turned up to make a box with an open top. The volume of the box is 616 cm³. Find the original dimensions of the cardboard.

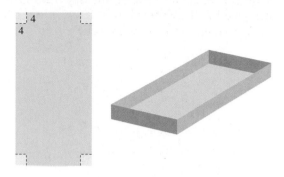

31. Factor: $(a + 3)^2 - 2(a + 3) - 35$.

32. Solve: $20x(x + 2)(x - 1) = 5x^3 - 24x - 14x^2$.

Cumulative Review: Chapters 1–5

Simplify. Do not use negative exponents in the answer.

1. $\frac{3}{8} \div \frac{3}{4}$ [1.3]

2. $\frac{3}{8} \cdot \frac{3}{4}$ [1.3]

3. $\frac{3}{8} + \frac{3}{4}$ [1.3]

4. $-2 + (20 \div 4)^2 - 6 \cdot (-1)^3$ [1.8]

5. $(3x^2y^3)^{-2}$ [4.8]

6. $(t^2)^3 \cdot t^4$ [4.1]

7. $(3x^4 - 2x^2 + x - 7) + (5x^3 + 2x^2 - 3)$ [4.3]

8. $(a^2b - 2ab^2 + 3b^3) - (4a^2b - ab^2 + b^3)$ [4.6]

9. $\dfrac{3t^3s^{-1}}{12t^{-5}s}$ [4.8]

10. $\left(\dfrac{-2x^2y}{3z^4}\right)^3$ [4.1]

11. Evaluate $-x$ for $x = -8$. [1.6]

12. Evaluate $-(-x)$ for $x = -8$. [1.6]

13. Determine the leading term of the polynomial
$$4x^3 - 6x^2 - x^4 + 7. \quad [4.2]$$

14. Divide: $(8x^4 - 20x^3 + 2x^2 - 4x) \div (4x)$. [4.7]

Multiply.

15. $-4t^8(t^3 - 2t - 5)$ [4.4]

16. $(3x - 5)^2$ [4.5]

17. $(10x^5 + y)(10x^5 - y)$ [4.6]

18. $(x - 1)(x^2 - x - 1)$ [4.4]

Factor completely.

19. $c^2 - 1$ [5.4]

20. $5x + 5y + 10x^2 + 10xy$ [5.1]

21. $x^2 - 14x + 24$ [5.2]

22. $4r^2 - 4rt + t^2$ [5.4]

23. $6x^2 - 19x + 10$ [5.3]

24. $10y^2 + 40$ [5.1]

25. $x^2y - 3xy + 2y$ [5.2]

26. $12x^2 - 5xy - 2y^2$ [5.3]

Solve.

27. $\frac{1}{3} + 2x = \frac{1}{2}$ [2.2]

28. $3(t - 1) = 2 - (t + 1)$ [2.2]

29. $8y - 6(y - 2) = 3(2y + 7)$ [2.2]

30. $3x - 7 \geq 4 - 8x$ [2.6]

31. $(x - 1)(x + 3) = 0$ [5.6]

32. $x^2 + x = 12$ [5.6]

33. $3x^2 = 12$ [5.6]

34. $3x^2 = 12x$ [5.6]

35. Solve $a = bc + dc$ for c. [2.3]

36. Find the slope of the line containing the points $(6, 7)$ and $(-2, 7)$. [3.5]

37. Find the slope and the y-intercept of the line given by $2x + y = 5$. [3.6]

38. Write the slope–intercept equation for the line with slope 5 and y-intercept $\left(0, -\frac{1}{3}\right)$. [3.6]

39. Write the slope–intercept equation for the line with slope 5 that contains the point $\left(-\frac{1}{3}, 0\right)$. [3.7]

Graph.

40. $4(x + 1) = 8$ [3.3]

41. $x + y = 5$ [3.3]

42. $y = \frac{3}{2}x - 2$ [3.6]

43. $3x + 5y = 10$ [3.6]

44. Use a grid 10 squares wide and 10 squares high to plot $(5, 40)$, $(18, -60)$, and $(30, -22)$. Choose the scale carefully. [3.1]

Solve.

45. On average, men talk 97 min more per month on cell phones than do women. The sum of men's average minutes and women's average minutes is 647 min. What is the average number of minutes per month that men talk on cell phones? [2.5]
Source: *International Communications Research for Cingular Wireless*

46. The number of cell-phone subscribers increased from 680,000 in 1986 to 233,000,000 in 2006. What was the average rate of increase? [3.4]
Source: CTIA – The Wireless Association

47. In 2007, there were 1.2 billion Internet users worldwide. Of these, 5% spoke French. How many Internet users spoke French? [2.4]
Source: Internetworldstats.com

48. The number of people in the United States, in thousands, who are on a waiting list for an organ transplant can be approximated by the polynomial $2.38t + 77.38$, where t is the number of years since 2000. Estimate the number of people on a waiting list for an organ transplant in 2010. [4.2]
Source: Based on information from The Organ Procurement and Transplantation Network

49. A 13-ft ladder is placed against a building in such a way that the distance from the top of the ladder to the ground is 7 ft more than the distance from the bottom of the ladder to the building. Find both distances. [5.7]

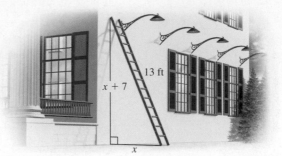

50. A rectangular table in Arlo's House of Tunes is six times as long as it is wide. If the area of the table is 24 ft^2, find the length and the width of the table. [5.7]

51. Donna's quiz grades are 8, 3, 5, and 10. What scores on the fifth quiz will make her average quiz grade at least 7? [2.7]

52. The average amount of sodium in a serving of Chef Boyardee foods dropped from 1100 mg in 2003 to 900 mg in 2007. [3.7]
Source: *The Indianapolis Star*, 11/25/07

  a) Graph the data and determine an equation for the related line. Let s represent the average amount of sodium per serving and t the number of years after 2000.

  b) Use the equation of part (a) to estimate the average amount of sodium in a serving of Chef Boyardee foods in 2006.

Synthesis

53. Solve $x = \dfrac{abx}{2 - b}$ for b. [2.3]

54. Write an equation of the line parallel to the x-axis and passing through $(-6, -8)$. [3.3]

55. a) Multiply: $(3y + 2 + x)(3y + 2 - x)$. [4.5]
  b) Factor: $9y^2 + 12y + 4 - x^2$. [5.4]

56. Solve: $6x^3 + 4x^2 = 2x$. [5.6]

Rational Expressions and Equations

FRED JENKINS
TECHNICAL SERVICES
MANAGER
Goose Creek, South Carolina

As a chemist involved in the quality control of all chemicals used and manufactured where I work, I use mathematics to calculate purities, density, refractive index, acidity, and other parameters that determine whether a material meets intended specifications. Even though many calculations are performed by instrument software, it is important for me as a chemist and manager to know how to verify results and troubleshoot with math when problems occur.

AN APPLICATION

As one alternative to gasoline-powered vehicles, flex fuel vehicles can use both regular gasoline and E85, a fuel containing 85% ethanol. Because using E85 results in a lower fuel economy, its price per gallon must be lower than the price of gasoline in order to make it an economical fuel option. A 2007 Chevrolet Tahoe gets 21 mpg on the highway using gasoline and only 15 mpg on the highway using E85. If the price of gasoline is $3.36 a gallon, what must the price of E85 be in order for the fuel cost per mile to be the same?

Source: www.caranddriver.com

This problem appears as Example 5 in Section 6.7.

R ational expressions are similar to fractions in arithmetic, in that both are ratios of two expressions. We now learn how to simplify, add, subtract, multiply, and divide rational expressions. These skills will then be used to solve the equations that arise from real-life problems like the one on the preceding page.

6.1   Rational Expressions

Simplifying Rational Expressions   ▪   Factors That Are Opposites

Just as a rational number is any number that can be written as a quotient of two integers, a **rational expression** is any expression that can be written as a quotient of two polynomials. The following are examples of rational expressions:

$$\frac{7}{3}, \quad \frac{5}{x + 6}, \quad \frac{t^2 - 5t + 6}{4t^2 - 7}.$$

Rational expressions are examples of *algebraic fractions*. They are also examples of *fraction expressions*.

Because rational expressions indicate division, we must be careful to avoid denominators that are 0. For example, in the expression

$$\frac{x + 3}{x - 7},$$

when x is replaced with 7, the denominator is 0, and the expression is undefined:

$$\frac{x + 3}{x - 7} = \frac{7 + 3}{7 - 7} = \frac{10}{0}. \xleftarrow{} \text{As explained in Chapter 1, division by 0 is undefined.}$$

When x is replaced with a number other than 7—say, 6—the expression *is* defined because the denominator is not 0:

$$\frac{x + 3}{x - 7} = \frac{6 + 3}{6 - 7} = \frac{9}{-1} = -9.$$

The expression is also defined when $x = -3$:

$$\frac{x + 3}{x - 7} = \frac{-3 + 3}{-3 - 7} = \frac{0}{-10} = 0. \qquad \text{0 divided by a nonzero number is 0.}$$

Any replacement for the variable that makes the *denominator* 0 will cause an expression to be undefined.

EXAMPLE 1   Find all numbers for which the rational expression

$$\frac{x + 4}{x^2 - 3x - 10}$$

is undefined.

SOLUTION   The value of the numerator has no bearing on whether or not a rational expression is defined. To determine which numbers make the rational expression undefined, we set the *denominator* equal to 0 and solve:

$$x^2 - 3x - 10 = 0 \qquad \text{We set the denominator equal to 0.}$$
$$(x - 5)(x + 2) = 0 \qquad \text{Factoring}$$
$$x - 5 = 0 \quad or \quad x + 2 = 0 \qquad \text{Using the principle of zero products}$$
$$x = 5 \quad or \qquad x = -2. \qquad \text{Solving each equation}$$

Check:

For $x = 5$:

$$\frac{x + 4}{x^2 - 3x - 10} = \frac{5 + 4}{5^2 - 3 \cdot 5 - 10} \qquad \text{There are no restrictions on the numerator.}$$

$$= \frac{9}{25 - 15 - 10} = \frac{9}{0}. \qquad \text{This expression is undefined, as expected.}$$

For $x = -2$:

$$\frac{x + 4}{x^2 - 3x - 10} = \frac{-2 + 4}{(-2)^2 - 3(-2) - 10}$$

$$= \frac{2}{4 + 6 - 10} = \frac{2}{0}. \qquad \text{This expression is undefined, as expected.}$$

Thus, $\dfrac{x + 4}{x^2 - 3x - 10}$ is undefined for $x = 5$ and $x = -2$.   ▶ **TRY EXERCISE** ▶ 7

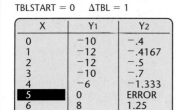
Simplifying Rational Expressions

A rational expression is said to be *simplified* when the numerator and the denominator have no factors (other than 1) in common. To simplify a rational expression, we first factor the numerator and the denominator. We then identify factors common to the numerator and the denominator, rewrite the expression as a product of two rational expressions (one of which is equal to 1), and then remove the factor equal to 1. The process is identical to that used in Section 1.3 to simplify $\frac{15}{40}$:

$$\frac{15}{40} = \frac{3 \cdot 5}{8 \cdot 5} \qquad \text{Factoring the numerator and the denominator. Note the common factor, 5.}$$

$$= \frac{3}{8} \cdot \frac{5}{5} \qquad \text{Rewriting as a product of two fractions}$$

$$= \frac{3}{8} \cdot 1 \qquad \frac{5}{5} = 1$$

$$= \frac{3}{8}. \qquad \text{Using the identity property of 1 to remove the factor 1}$$

Similar steps are followed when simplifying rational expressions: We factor and remove a factor equal to 1, using the fact that

$$\frac{ab}{cb} = \frac{a}{c} \cdot \frac{b}{b}.$$

EXAMPLE 2   Simplify: $\dfrac{8x^2}{24x}$.

SOLUTION

$$\frac{8x^2}{24x} = \frac{8 \cdot x \cdot x}{3 \cdot 8 \cdot x} \qquad \text{Factoring the numerator and the denominator.}$$
Note the common factor of $8 \cdot x$.

$$= \frac{x}{3} \cdot \frac{8x}{8x} \qquad \text{Rewriting as a product of two rational expressions}$$

$$= \frac{x}{3} \cdot 1 \qquad \frac{8x}{8x} = 1$$

$$= \frac{x}{3} \qquad \text{Removing the factor 1} \qquad \boxed{\text{TRY EXERCISE} \; 17}$$

We say that $\dfrac{8x^2}{24x}$ *simplifies* to $\dfrac{x}{3}$.* In the work that follows, we assume that all denominators are nonzero.

EXAMPLE 3   Simplify: $\dfrac{5a + 15}{10}$.

SOLUTION

$$\frac{5a + 15}{10} = \frac{5(a + 3)}{5 \cdot 2} \qquad \text{Factoring the numerator and the denominator.}$$
Note the common factor of 5.

$$= \frac{5}{5} \cdot \frac{a + 3}{2} \qquad \text{Rewriting as a product of two rational expressions}$$

$$= 1 \cdot \frac{a + 3}{2} \qquad \frac{5}{5} = 1$$

$$= \frac{a + 3}{2} \qquad \text{Removing the factor 1} \qquad \boxed{\text{TRY EXERCISE} \; 19}$$

The result in Example 3 can be partially checked using a replacement for a—say, $a = 2$.

Original expression:

$$\frac{5a + 15}{10} = \frac{5 \cdot 2 + 15}{10}$$

$$= \frac{25}{10} = \frac{5}{2} \qquad \text{The results are the same.}$$

Simplified expression:

$$\frac{a + 3}{2} = \frac{2 + 3}{2}$$

$$= \frac{5}{2}$$

| To see why this check is not foolproof, see Exercise 65. |

If we do not get the same result when evaluating both expressions, we know that a mistake has been made. For example, if $(5a + 15)/10$ is *incorrectly* simplified as $(a + 15)/2$ and we evaluate using $a = 2$, we have the following.

Original expression:

$$\frac{5a + 15}{10} = \frac{5 \cdot 2 + 15}{10}$$

$$= \frac{5}{2} \qquad \text{The results are different.}$$

Incorrectly simplified expression:

$$\frac{a + 15}{2} = \frac{2 + 15}{2}$$

$$= \frac{17}{2}$$

This demonstrates that a mistake has been made.

---

*In more advanced courses, we would *not* say that $8x^2/(24x)$ simplifies to $x/3$, but would instead say that $8x^2/(24x)$ simplifies to $x/3$ *with the restriction that $x \neq 0$.*

Sometimes the common factor has two or more terms.

EXAMPLE **4**

Simplify.

a) $\dfrac{6x - 12}{7x - 14}$ **b)** $\dfrac{18t^2 + 6t}{6t^2 + 15t}$ **c)** $\dfrac{x^2 + 3x + 2}{x^2 - 1}$

SOLUTION

a) $\dfrac{6x - 12}{7x - 14} = \dfrac{6(x - 2)}{7(x - 2)}$ Factoring the numerator and the denominator. Note the common factor of $x - 2$.

$= \dfrac{6}{7} \cdot \dfrac{x - 2}{x - 2}$ Rewriting as a product of two rational expressions

$= \dfrac{6}{7} \cdot 1$ $\dfrac{x - 2}{x - 2} = 1$

$= \dfrac{6}{7}$ Removing the factor 1

b) $\dfrac{18t^2 + 6t}{6t^2 + 15t} = \dfrac{3t \cdot 2(3t + 1)}{3t(2t + 5)}$ Factoring the numerator and the denominator. Note the common factor of $3t$.

$= \dfrac{3t}{3t} \cdot \dfrac{2(3t + 1)}{2t + 5}$ Rewriting as a product of two rational expressions

$= 1 \cdot \dfrac{2(3t + 1)}{2t + 5}$ $\dfrac{3t}{3t} = 1$

$= \dfrac{2(3t + 1)}{2t + 5}$ Removing the factor 1. The numerator and the denominator have no common factor so the simplification is complete.

c) $\dfrac{x^2 + 3x + 2}{x^2 - 1} = \dfrac{(x + 1)(x + 2)}{(x + 1)(x - 1)}$ Factoring; $x + 1$ is the common factor.

$= \dfrac{x + 1}{x + 1} \cdot \dfrac{x + 2}{x - 1}$ Rewriting as a product of two rational expressions

$= 1 \cdot \dfrac{x + 2}{x - 1}$ $\dfrac{x + 1}{x + 1} = 1$

$= \dfrac{x + 2}{x - 1}$ Removing the factor 1

> **TRY EXERCISE** 23

Canceling is a shortcut that can be used—and easily *misused*—to simplify rational expressions. As stated in Section 1.3, canceling must be done with care and understanding. Essentially, canceling streamlines the process of removing a factor equal to 1. Example 4(c) could have been streamlined as follows:

$$\dfrac{x^2 + 3x + 2}{x^2 - 1} = \dfrac{\cancel{(x + 1)}(x + 2)}{\cancel{(x + 1)}(x - 1)}$$ When a factor equal to 1 is noted, it is "canceled": $\dfrac{x + 1}{x + 1} = 1$.

$$= \dfrac{x + 2}{x - 1}.$$ Simplifying

TECHNOLOGY CONNECTION

We can use the TABLE feature as a partial check that rational expressions have been simplified correctly. To check the simplification in Example 4(c),

$$\dfrac{x^2 + 3x + 2}{x^2 - 1} = \dfrac{x + 2}{x - 1},$$

we enter $y_1 = (x^2 + 3x + 2)/(x^2 - 1)$ and $y_2 = (x + 2)/(x - 1)$ and select the mode AUTO to look at a table of values of y_1 and y_2. The values should match for all allowable replacements.

X	Y₁	Y₂
−4	.4	.4
−3	.25	.25
−2	0	0
−1	ERROR	−.5
0	−2	−2
1	ERROR	−2
2	4	4
X = −4		

The ERROR messages indicate that −1 and 1 are not allowable replacements in y_1 and 1 is not an allowable replacement in y_2. For all other numbers, y_1 and y_2 are the same, so the simplification appears to be correct.

Use the TABLE feature to determine whether each of the following appears to be correct.

1. $\dfrac{8x^2}{24x} = \dfrac{x}{3}$

2. $\dfrac{5x + 15}{10} = \dfrac{x + 3}{2}$

3. $\dfrac{x + 3}{x} = 3$

4. $\dfrac{x^2 + 3x - 4}{x^2 - 16} = \dfrac{x - 1}{x + 4}$

> *CAUTION!* Canceling is often used incorrectly. The following cancellations are *incorrect*:
>
> $$\frac{\cancel{x} + 7}{\cancel{x} + 3}, \qquad \frac{a^2 - \cancel{8}}{\cancel{8}}, \qquad \frac{6\cancel{x}^2 + 5\cancel{x} + 1}{4\cancel{x}^2 - 3\cancel{x}}.$$
>
> Wrong!   Wrong!   Wrong!
>
> None of the above cancellations removes a factor equal to 1. Factors are parts of products. For example, in $x \cdot 7$, x and 7 are factors, but in $x + 7$, x and 7 are terms, *not* factors. Only factors can be canceled.

EXAMPLE **5**   Simplify: $\dfrac{3x^2 - 2x - 1}{x^2 - 3x + 2}$.

SOLUTION   We factor the numerator and the denominator and look for common factors:

$$\frac{3x^2 - 2x - 1}{x^2 - 3x + 2} = \frac{(3x + 1)\cancel{(x - 1)}}{(x - 2)\cancel{(x - 1)}} \qquad \text{Try to visualize this as } \frac{3x + 1}{x - 2} \cdot \frac{x - 1}{x - 1}.$$

$$= \frac{3x + 1}{x - 2}. \qquad \text{Removing a factor equal to 1: } \frac{x - 1}{x - 1} = 1$$

> **TRY EXERCISE** 33

Factors That Are Opposites

Consider

$$\frac{x - 4}{8 - 2x}, \quad \text{or, equivalently,} \quad \frac{x - 4}{2(4 - x)}.$$

At first glance, the numerator and the denominator do not appear to have any common factors. But $x - 4$ and $4 - x$ are opposites, or additive inverses, of each other. Thus we can find a common factor by factoring out -1 in one expression.

EXAMPLE **6**   Simplify $\dfrac{x - 4}{8 - 2x}$ and check by evaluating.

SOLUTION   We have

$$\frac{x - 4}{8 - 2x} = \frac{x - 4}{2(4 - x)} \qquad \text{Factoring}$$

$$= \frac{x - 4}{2(-1)(x - 4)} \qquad \text{Note that } 4 - x = -x + 4 = -1(x - 4).$$

$$= \frac{x - 4}{-2(x - 4)} \qquad \begin{array}{l}\text{Had we originally factored out } -2, \text{ we} \\ \text{could have gone directly to this step.}\end{array}$$

$$= \frac{1}{-2} \cdot \frac{x - 4}{x - 4} \qquad \begin{array}{l}\text{Rewriting as a product. It is important} \\ \text{to write the 1 in the numerator.}\end{array}$$

$$= -\frac{1}{2}. \qquad \begin{array}{l}\text{Removing a factor equal to 1:} \\ (x - 4)/(x - 4) = 1\end{array}$$

As a partial check, note that for any choice of x other than 4, the value of the rational expression is $-\frac{1}{2}$. For example, if $x = 5$, then

$$\frac{x - 4}{8 - 2x} = \frac{5 - 4}{8 - 2 \cdot 5}$$

$$= \frac{1}{-2} = -\frac{1}{2}.$$

TRY EXERCISE   47

6.1   EXERCISE SET

For Extra Help

Math XL   WATCH   DOWNLOAD
PRACTICE

👀 **Concept Reinforcement**   *In each of Exercises 1–6, match the rational expression with the list of numbers in the column on the right for which the rational expression is undefined.*

1. ___ $\dfrac{x - 5}{(x - 2)(x + 3)}$

2. ___ $\dfrac{3t}{(t + 1)(t - 4)}$

3. ___ $\dfrac{a + 7}{a^2 - a - 12}$

4. ___ $\dfrac{m - 3}{m^2 - 2m - 15}$

5. ___ $\dfrac{2t + 7}{(2t - 1)(3t + 4)}$

6. ___ $\dfrac{4x - 1}{(3x - 1)(3x - 2)}$

a) $-1, 4$

b) $-3, 5$

c) $-\dfrac{4}{3}, \dfrac{1}{2}$

d) $-3, 4$

e) $-3, 2$

f) $\dfrac{1}{3}, \dfrac{2}{3}$

List all numbers for which each rational expression is undefined.

7. $\dfrac{18}{-11x}$

8. $\dfrac{13}{-5t}$

9. $\dfrac{y - 3}{y + 5}$

10. $\dfrac{a + 6}{a - 10}$

11. $\dfrac{t - 5}{3t - 15}$

12. $\dfrac{x^2 - 4}{5x + 10}$

13. $\dfrac{x^2 - 25}{x^2 - 3x - 28}$

14. $\dfrac{p^2 - 9}{p^2 - 7p + 10}$

15. $\dfrac{t^2 + t - 20}{2t^2 + 11t - 6}$

16. $\dfrac{x^2 + 2x + 1}{3x^2 - x - 14}$

Simplify by removing a factor equal to 1. Show all steps.

17. $\dfrac{50a^2b}{40ab^3}$

18. $\dfrac{-24x^4y^3}{6x^7y}$

19. $\dfrac{6t + 12}{6t - 18}$

20. $\dfrac{5n - 30}{5n + 5}$

21. $\dfrac{21t - 7}{24t - 8}$

22. $\dfrac{10n + 25}{8n + 20}$

23. $\dfrac{a^2 - 9}{a^2 + 4a + 3}$

24. $\dfrac{a^2 + 5a + 6}{a^2 - 9}$

Simplify, if possible. Then check by evaluating, as in Example 6.

25. $\dfrac{-36x^8}{54x^5}$

26. $\dfrac{45a^4}{30a^6}$

27. $\dfrac{-2y + 6}{-8y}$

28. $\dfrac{4x - 12}{6x}$

29. $\dfrac{6a^2 + 3a}{7a^2 + 7a}$

30. $\dfrac{-4m^2 + 4m}{-8m^2 + 12m}$

31. $\dfrac{t^2 - 16}{t^2 - t - 20}$

32. $\dfrac{a^2 - 4}{a^2 + 5a + 6}$

33. $\dfrac{3a^2 + 9a - 12}{6a^2 - 30a + 24}$

34. $\dfrac{2t^2 - 6t + 4}{4t^2 + 12t - 16}$

35. $\dfrac{x^2 - 8x + 16}{x^2 - 16}$

36. $\dfrac{x^2 - 25}{x^2 + 10x + 25}$

37. $\dfrac{t^2 - 1}{t + 1}$

38. $\dfrac{a^2 - 1}{a - 1}$

39. $\dfrac{y^2 + 4}{y + 2}$

40. $\dfrac{m^2 + 9}{m + 3}$

41. $\dfrac{5x^2 + 20}{10x^2 + 40}$

42. $\dfrac{6x^2 + 54}{4x^2 + 36}$

43. $\dfrac{y^2 + 6y}{2y^2 + 13y + 6}$

44. $\dfrac{t^2 + 2t}{2t^2 + t - 6}$

45. $\dfrac{4x^2 - 12x + 9}{10x^2 - 11x - 6}$

46. $\dfrac{4x^2 - 4x + 1}{6x^2 + 5x - 4}$

47. $\dfrac{10 - x}{x - 10}$

48. $\dfrac{x - 8}{8 - x}$

49. $\dfrac{7t - 14}{2 - t}$

50. $\dfrac{3 - n}{5n - 15}$

51. $\dfrac{a - b}{4b - 4a}$

52. $\dfrac{2p - 2q}{q - p}$

53. $\dfrac{3x^2 - 3y^2}{2y^2 - 2x^2}$

54. $\dfrac{7a^2 - 7b^2}{3b^2 - 3a^2}$

Aha! 55. $\dfrac{7s^2 - 28t^2}{28t^2 - 7s^2}$

56. $\dfrac{9m^2 - 4n^2}{4n^2 - 9m^2}$

57. Explain how simplifying is related to the identity property of 1.

58. If a rational expression is undefined for $x = 5$ and $x = -3$, what is the degree of the denominator? Why?

Skill Review

To prepare for Section 6.2, review multiplication and division using fraction notation (Section 1.7).

Simplify.

59. $-\dfrac{2}{15} \cdot \dfrac{10}{7}$ [1.7]

60. $\left(\dfrac{3}{4}\right)\left(\dfrac{-20}{9}\right)$ [1.7]

61. $\dfrac{5}{8} \div \left(-\dfrac{1}{6}\right)$ [1.7]

62. $\dfrac{7}{10} \div \left(-\dfrac{8}{15}\right)$ [1.7]

63. $\dfrac{7}{9} - \dfrac{2}{3} \cdot \dfrac{6}{7}$ [1.8]

64. $\dfrac{2}{3} - \left(\dfrac{3}{4}\right)^2$ [1.8]

Synthesis

65. Keith *incorrectly* simplifies

$$\dfrac{x^2 + x - 2}{x^2 + 3x + 2} \quad \text{as} \quad \dfrac{x - 1}{x + 2}.$$

He then checks his simplification by evaluating both expressions for $x = 1$. Use this situation to explain why evaluating is not a foolproof check.

66. How could you convince someone that $a - b$ and $b - a$ are opposites of each other?

Simplify.

67. $\dfrac{16y^4 - x^4}{(x^2 + 4y^2)(x - 2y)}$

68. $\dfrac{(x - 1)(x^4 - 1)(x^2 - 1)}{(x^2 + 1)(x - 1)^2(x^4 - 2x^2 + 1)}$

69. $\dfrac{x^5 - 2x^3 + 4x^2 - 8}{x^7 + 2x^4 - 4x^3 - 8}$

70. $\dfrac{10t^4 - 8t^3 + 15t - 12}{8 - 10t + 12t^2 - 15t^3}$

71. $\dfrac{(t^4 - 1)(t^2 - 9)(t - 9)^2}{(t^4 - 81)(t^2 + 1)(t + 1)^2}$

72. $\dfrac{(t + 2)^3(t^2 + 2t + 1)(t + 1)}{(t + 1)^3(t^2 + 4t + 4)(t + 2)}$

73. $\dfrac{(x^2 - y^2)(x^2 - 2xy + y^2)}{(x + y)^2(x^2 - 4xy - 5y^2)}$

74. $\dfrac{x^4 - y^4}{(y - x)^4}$

75. Select any number x, multiply by 2, add 5, multiply by 5, subtract 25, and divide by 10. What do you get? Explain how this procedure can be used for a number trick.

6.2	Multiplication and Division

Multiplication • Division

Multiplication and division of rational expressions are similar to multiplication and division with fractions. In this section, we again assume that all denominators are nonzero.

Multiplication

Recall that to multiply fractions, we multiply numerator times numerator and denominator times denominator. Rational expressions are multiplied in a similar way.

> **The Product of Two Rational Expressions**
>
> To multiply rational expressions, multiply numerators and multiply denominators:
> $$\frac{A}{B} \cdot \frac{C}{D} = \frac{AC}{BD}.$$
> Then factor and, if possible, simplify the result.

For example,

$$\frac{3}{5} \cdot \frac{8}{11} = \frac{3 \cdot 8}{5 \cdot 11} \quad \text{and} \quad \frac{x}{3} \cdot \frac{x+2}{y} = \frac{x(x+2)}{3y}.$$

Fraction bars are grouping symbols, so parentheses are needed when writing some products. Because we generally simplify, we often leave products involving variables in factored form. There is no need to multiply further.

EXAMPLE 1 Multiply and, if possible, simplify.

a) $\dfrac{5a^3}{4} \cdot \dfrac{2}{5a}$

b) $(x^2 - 3x - 10) \cdot \dfrac{x+4}{x^2 - 10x + 25}$

c) $\dfrac{10x + 20}{2x^2 - 3x + 1} \cdot \dfrac{x^2 - 1}{5x + 10}$

SOLUTION

a) $\dfrac{5a^3}{4} \cdot \dfrac{2}{5a} = \dfrac{5a^3(2)}{4(5a)}$ Forming the product of the numerators and the product of the denominators

$= \dfrac{5 \cdot a \cdot a \cdot a \cdot 2}{2 \cdot 2 \cdot 5 \cdot a}$ Factoring the numerator and the denominator

$= \dfrac{\cancel{5} \cdot a \cdot \cancel{a} \cdot a \cdot \cancel{2}}{\cancel{2} \cdot 2 \cdot \cancel{5} \cdot \cancel{a}}$ ⎤
                                                         ⎥ Removing a factor equal to 1: $\dfrac{2 \cdot 5 \cdot a}{2 \cdot 5 \cdot a} = 1$
$= \dfrac{a^2}{2}$ ⎦

b) $(x^2 - 3x - 10) \cdot \dfrac{x+4}{x^2 - 10x + 25}$

$$= \dfrac{x^2 - 3x - 10}{1} \cdot \dfrac{x+4}{x^2 - 10x + 25} \qquad \text{Writing } x^2 - 3x - 10 \text{ as a rational expression}$$

$$= \dfrac{(x^2 - 3x - 10)(x+4)}{1(x^2 - 10x + 25)} \qquad \text{Multiplying the numerators and the denominators}$$

$$= \dfrac{(x-5)(x+2)(x+4)}{(x-5)(x-5)} \qquad \text{Factoring the numerator and the denominator}$$

$$= \dfrac{\cancel{(x-5)}(x+2)(x+4)}{\cancel{(x-5)}(x-5)} \qquad \text{Removing a factor equal to 1:} \quad \dfrac{x-5}{x-5} = 1$$

$$= \dfrac{(x+2)(x+4)}{x-5}$$

c) $\dfrac{10x + 20}{2x^2 - 3x + 1} \cdot \dfrac{x^2 - 1}{5x + 10}$

$$= \dfrac{(10x + 20)(x^2 - 1)}{(2x^2 - 3x + 1)(5x + 10)} \qquad \boxed{\text{Multiply.}}$$

$$= \dfrac{5(2)(x+2)(x+1)(x-1)}{(x-1)(2x-1)5(x+2)} \qquad \boxed{\text{Factor.}} \;\; \text{Try to go directly to this step.}$$

$$= \dfrac{\cancel{5}(2)\cancel{(x+2)}(x+1)\cancel{(x-1)}}{\cancel{(x-1)}(2x-1)\cancel{5}\cancel{(x+2)}} \qquad \boxed{\text{Simplify.}} \;\; \dfrac{5(x+2)(x-1)}{5(x+2)(x-1)} = 1$$

$$= \dfrac{2(x+1)}{2x-1}$$

TRY EXERCISE ▸ 19

STUDENT NOTES

Neatness is important for work with rational expressions. Consider using the lines on your notebook paper to indicate where the fraction bars should be drawn in each example. Write the equals sign at the same height as the fraction bars.

Because our results are often used in problems that require factored form, there is no need to multiply out the numerator or the denominator.

Division

As with fractions, reciprocals of rational expressions are found by interchanging the numerator and the denominator. For example,

the reciprocal of $\dfrac{2}{7}$ is $\dfrac{7}{2}$,   and   the reciprocal of $\dfrac{3x}{x+5}$ is $\dfrac{x+5}{3x}$.

The Quotient of Two Rational Expressions

To divide by a rational expression, multiply by its reciprocal:

$$\dfrac{A}{B} \div \dfrac{C}{D} = \dfrac{A}{B} \cdot \dfrac{D}{C} = \dfrac{AD}{BC}.$$

Then factor and, if possible, simplify.

For an explanation of why we divide this way, see Exercise 55 in Section 6.5.

EXAMPLE **2**   Divide: **(a)** $\dfrac{x}{5} \div \dfrac{7}{y}$; **(b)** $(x + 2) \div \dfrac{x - 1}{x + 3}$.

SOLUTION

a) $\dfrac{x}{5} \div \dfrac{7}{y} = \dfrac{x}{5} \cdot \dfrac{y}{7}$ Multiplying by the reciprocal of the divisor

$\qquad = \dfrac{xy}{35}$ Multiplying rational expressions

b) $(x + 2) \div \dfrac{x - 1}{x + 3} = \dfrac{x + 2}{1} \cdot \dfrac{x + 3}{x - 1}$ Multiplying by the reciprocal of the divisor. Writing $x + 2$ as $\dfrac{x + 2}{1}$ can be helpful.

$\qquad\qquad = \dfrac{(x + 2)(x + 3)}{x - 1}$

TRY EXERCISE 41

As usual, we should simplify when possible. Often that requires us to factor one or more polynomials, hoping to discover a common factor that appears in both the numerator and the denominator.

EXAMPLE **3**   Divide and, if possible, simplify: $\dfrac{x + 1}{x^2 - 1} \div \dfrac{x + 1}{x^2 - 2x + 1}$.

SOLUTION

$$\dfrac{x + 1}{x^2 - 1} \div \dfrac{x + 1}{x^2 - 2x + 1} = \dfrac{x + 1}{x^2 - 1} \cdot \dfrac{x^2 - 2x + 1}{x + 1} \qquad \text{Rewrite as multiplication.}$$

$$= \dfrac{(x + 1)(x - 1)(x - 1)}{(x + 1)(x - 1)(x + 1)} \qquad \begin{array}{l}\text{Multiply.}\\ \text{Factor.}\end{array}$$

$$= \dfrac{\cancel{(x + 1)}\cancel{(x - 1)}(x - 1)}{\cancel{(x + 1)}\cancel{(x - 1)}(x + 1)} \qquad \begin{array}{l}\text{Simplify.}\\ \dfrac{(x + 1)(x - 1)}{(x + 1)(x - 1)} = 1\end{array}$$

$$= \dfrac{x - 1}{x + 1}$$

TRY EXERCISE 47

EXAMPLE **4**   Divide and, if possible, simplify.

a) $\dfrac{a^2 + 3a + 2}{a^2 + 4} \div (5a^2 + 10a)$ **b)** $\dfrac{x^2 - 2x - 3}{x^2 - 4} \div \dfrac{x + 1}{x + 5}$

SOLUTION

a) $\dfrac{a^2 + 3a + 2}{a^2 + 4} \div (5a^2 + 10a)$

$\quad = \dfrac{a^2 + 3a + 2}{a^2 + 4} \cdot \dfrac{1}{5a^2 + 10a}$ Multiplying by the reciprocal of the divisor

$\quad = \dfrac{(a + 2)(a + 1)}{(a^2 + 4)5a(a + 2)}$ Multiplying rational expressions and factoring

$\quad = \dfrac{\cancel{(a + 2)}(a + 1)}{(a^2 + 4)5a\cancel{(a + 2)}}$ Removing a factor equal to 1: $\dfrac{a + 2}{a + 2} = 1$

$\quad = \dfrac{a + 1}{(a^2 + 4)5a}$

TECHNOLOGY CONNECTION

In performing a partial check of Example 4(b), we must be careful placing parentheses. We enter the original expression as $y_1 = ((x^2 - 2x - 3)/(x^2 - 4))/((x + 1)/(x + 5))$ and the simplified expression as $y_2 = ((x - 3)(x + 5))/((x - 2)(x + 2))$. Comparing values of y_1 and y_2, we see that

(continued)

the simplification is probably correct.

X	Y₁	Y₂
−5	ERROR	0
−4	−.5833	−.5833
−3	−2.4	−2.4
−2	ERROR	ERROR
−1	ERROR	5.3333
0	3.75	3.75
1	4	4

X = −5

1. Check Example 4(a).
2. Why are there 3 ERROR messages shown for y_1 on the screen above, and only 1 for y_2?

b) $\dfrac{x^2 - 2x - 3}{x^2 - 4} \div \dfrac{x + 1}{x + 5}$

$= \dfrac{x^2 - 2x - 3}{x^2 - 4} \cdot \dfrac{x + 5}{x + 1}$ Multiplying by the reciprocal of the divisor

$= \dfrac{(x - 3)(x + 1)(x + 5)}{(x - 2)(x + 2)(x + 1)}$ Multiplying rational expressions and factoring

$= \dfrac{(x - 3)\cancel{(x + 1)}(x + 5)}{(x - 2)(x + 2)\cancel{(x + 1)}}$

$= \dfrac{(x - 3)(x + 5)}{(x - 2)(x + 2)}$ Removing a factor equal to 1: $\dfrac{x + 1}{x + 1} = 1$

TRY EXERCISE ▶ 51

6.2   EXERCISE SET

Multiply. Leave each answer in factored form.

1. $\dfrac{3x}{8} \cdot \dfrac{x + 2}{5x - 1}$

2. $\dfrac{2x}{7} \cdot \dfrac{3x + 5}{x - 1}$

3. $\dfrac{a - 4}{a + 6} \cdot \dfrac{a + 2}{a + 6}$

4. $\dfrac{a + 3}{a + 6} \cdot \dfrac{a + 3}{a - 1}$

5. $\dfrac{2x + 3}{4} \cdot \dfrac{x + 1}{x - 5}$

6. $\dfrac{x + 2}{3x - 4} \cdot \dfrac{4}{5x + 6}$

7. $\dfrac{n - 4}{n^2 + 4} \cdot \dfrac{n + 4}{n^2 - 4}$

8. $\dfrac{t + 3}{t^2 - 2} \cdot \dfrac{t + 3}{t^2 - 4}$

9. $\dfrac{y + 6}{1 + y} \cdot \dfrac{y - 3}{y + 3}$

10. $\dfrac{m + 4}{m + 8} \cdot \dfrac{2 + m}{m + 5}$

Multiply and, if possible, simplify.

11. $\dfrac{8t^3}{5t} \cdot \dfrac{3}{4t}$

12. $\dfrac{18}{a^5} \cdot \dfrac{2a^2}{3a}$

13. $\dfrac{3c}{d^2} \cdot \dfrac{8d}{6c^3}$

14. $\dfrac{3x^2 y}{2} \cdot \dfrac{4}{xy^3}$

15. $\dfrac{x^2 - 3x - 10}{(x - 2)^2} \cdot (x - 2)$

16. $t + 2 \cdot \dfrac{t^2 - 5t + 6}{(t + 2)^2}$

17. $\dfrac{n^2 - 6n + 5}{n + 6} \cdot \dfrac{n - 6}{n^2 + 36}$

18. $\dfrac{a + 2}{a - 2} \cdot \dfrac{a^2 + 4}{a^2 + 5a + 4}$

19. $\dfrac{a^2 - 9}{a^2} \cdot \dfrac{7a}{a^2 + a - 12}$

20. $\dfrac{x^2 + 10x - 11}{9x} \cdot \dfrac{x^3}{x + 11}$

21. $\dfrac{4v - 8}{5v} \cdot \dfrac{15v^2}{4v^2 - 16v + 16}$

22. $\dfrac{m - 2}{3m + 9} \cdot \dfrac{m^2 + 6m + 9}{2m^2 - 8}$

23. $\dfrac{t^2 + 2t - 3}{t^2 + 4t - 5} \cdot \dfrac{t^2 - 3t - 10}{t^2 + 5t + 6}$

24. $\dfrac{x^2 + 5x + 4}{x^2 - 6x + 8} \cdot \dfrac{x^2 + 5x - 14}{x^2 + 8x + 7}$

25. $\dfrac{12y + 12}{5y + 25} \cdot \dfrac{3y^2 - 75}{8y^2 - 8}$

26. $\dfrac{9t^2 - 900}{5t^2 - 20} \cdot \dfrac{5t + 10}{3t - 30}$

Aha! 27. $\dfrac{x^2 + 4x + 4}{(x - 1)^2} \cdot \dfrac{x^2 - 2x + 1}{(x + 2)^2}$

28. $\dfrac{x^2 + 7x + 12}{x^2 + 6x + 8} \cdot \dfrac{4 - x^2}{x^2 + x - 6}$

29. $\dfrac{t^2 - 4t + 4}{2t^2 - 7t + 6} \cdot \dfrac{2t^2 + 7t - 15}{t^2 - 10t + 25}$

30. $\dfrac{5y^2 - 4y - 1}{3y^2 + 5y - 12} \cdot \dfrac{y^2 + 6y + 9}{y^2 - 2y + 1}$

31. $(10x^2 - x - 2) \cdot \dfrac{4x^2 - 8x + 3}{10x^2 - 11x - 6}$

32. $\dfrac{2x^2 - 5x + 3}{6x^2 - 5x - 1} \cdot (6x^2 + 13x + 2)$

Find the reciprocal of each expression.

33. $\dfrac{2x}{9}$

34. $\dfrac{3 - x}{x^2 + 4}$

35. $a^4 + 3a$

36. $\dfrac{1}{a^2 - b^2}$

37. $\dfrac{x^2 + 2x - 5}{x^2 - 4x + 7}$

38. $\dfrac{x^2 - 3xy + y^2}{x^2 + 7xy - y^2}$

Divide and, if possible, simplify.

39. $\dfrac{5}{9} \div \dfrac{3}{4}$

40. $\dfrac{3}{8} \div \dfrac{4}{7}$

41. $\dfrac{x}{4} \div \dfrac{5}{x}$

42. $\dfrac{5}{x} \div \dfrac{x}{12}$

43. $\dfrac{a^5}{b^4} \div \dfrac{a^2}{b}$

44. $\dfrac{x^5}{y^2} \div \dfrac{x^2}{y}$

45. $\dfrac{t - 3}{6} \div \dfrac{t + 1}{8}$

46. $\dfrac{10}{a + 3} \div \dfrac{15}{a}$

47. $\dfrac{4y - 8}{y + 2} \div \dfrac{y - 2}{y^2 - 4}$

48. $\dfrac{x^2 - 1}{x} \div \dfrac{x + 1}{2x - 2}$

49. $\dfrac{a}{a - b} \div \dfrac{b}{b - a}$

50. $\dfrac{x - y}{6} \div \dfrac{y - x}{3}$

51. $(n^2 + 5n + 6) \div \dfrac{n^2 - 4}{n + 3}$

52. $(v^2 - 1) \div \dfrac{(v + 1)(v - 3)}{v^2 + 9}$

53. $\dfrac{-3 + 3x}{16} \div \dfrac{x - 1}{5}$

54. $\dfrac{-4 + 2x}{15} \div \dfrac{x - 2}{3}$

55. $\dfrac{x - 1}{x + 2} \div \dfrac{1 - x}{4 + x^2}$

56. $\dfrac{-12 + 4x}{12} \div \dfrac{6 - 2x}{6}$

57. $\dfrac{a + 2}{a - 1} \div \dfrac{3a + 6}{a - 5}$

58. $\dfrac{t - 3}{t + 2} \div \dfrac{4t - 12}{t + 1}$

59. $(2x - 1) \div \dfrac{2x^2 - 11x + 5}{4x^2 - 1}$

60. $(a + 7) \div \dfrac{3a^2 + 14a - 49}{a^2 + 8a + 7}$

61. $\dfrac{w^2 - 14w + 49}{2w^2 - 3w - 14} \div \dfrac{3w^2 - 20w - 7}{w^2 - 6w - 16}$

62. $\dfrac{2m^2 + 59m - 30}{m^2 - 10m + 25} \div \dfrac{2m^2 - 21m + 10}{m^2 + m - 30}$

63. $\dfrac{c^2 + 10c + 21}{c^2 - 2c - 15} \div (5c^2 + 32c - 21)$

64. $\dfrac{z^2 - 2z + 1}{z^2 - 1} \div (4z^2 - z - 3)$

65. $\dfrac{x - y}{x^2 + 2xy + y^2} \div \dfrac{x^2 - y^2}{x^2 - 5xy + 4y^2}$

66. $\dfrac{a^2 - b^2}{a^2 - 4ab + 4b^2} \div \dfrac{a^2 - 3ab + 2b^2}{a - 2b}$

67. Why is it important to insert parentheses when multiplying rational expressions such as

$$\dfrac{x + 2}{5x - 7} \cdot \dfrac{3x - 1}{x + 4}?$$

68. As a first step in dividing $\dfrac{x}{3}$ by $\dfrac{7}{x}$, Jan canceled the x's. Explain why this was incorrect, and show the correct division.

Skill Review

To prepare for Section 6.3, review addition and subtraction with fraction notation (Sections 1.3 and 1.6) and subtraction of polynomials (Section 4.3).

Simplify.

69. $\dfrac{3}{4} + \dfrac{5}{6}$ [1.3]

70. $\dfrac{7}{8} + \dfrac{5}{6}$ [1.3]

71. $\dfrac{2}{9} - \dfrac{1}{6}$ [1.3]

72. $\dfrac{3}{10} - \dfrac{7}{15}$ [1.6]

73. $2x^2 - x + 1 - (x^2 - x - 2)$ [4.3]

74. $3x^2 + x - 7 - (5x^2 + 5x - 8)$ [4.3]

Synthesis

75. Is the reciprocal of a product the product of the two reciprocals? Why or why not?

76. Explain why the quotient

$$\dfrac{x + 3}{x - 5} \div \dfrac{x - 7}{x + 1}$$

is undefined for $x = 5$, $x = -1$, and $x = 7$, but *is* defined for $x = -3$.

77. Find the reciprocal of $2\frac{1}{3}x$.

78. Find the reciprocal of $7.25x$.

Simplify.

79. $(x - 2a) \div \dfrac{a^2x^2 - 4a^4}{a^2x + 2a^3}$

80. $\dfrac{2a^2 - 5ab}{c - 3d} \div (4a^2 - 25b^2)$

81. $\dfrac{3x^2 - 2xy - y^2}{x^2 - y^2} \div (3x^2 + 4xy + y^2)^2$

82. $\dfrac{3a^2 - 5ab - 12b^2}{3ab + 4b^2} \div (3b^2 - ab)^2$

Aha! **83.** $\dfrac{a^2 - 3b}{a^2 + 2b} \cdot \dfrac{a^2 - 2b}{a^2 + 3b} \cdot \dfrac{a^2 + 2b}{a^2 - 3b}$

84. $\dfrac{y^2 - 4xy}{y - x} \div \dfrac{16x^2y^2 - y^4}{4x^2 - 3xy - y^2} \div \dfrac{4}{x^3y^3}$

85. $\dfrac{z^2 - 8z + 16}{z^2 + 8z + 16} \div \dfrac{(z - 4)^5}{(z + 4)^5} \div \dfrac{3z + 12}{z^2 - 16}$

86. $\dfrac{x^2 - x + xy - y}{x^2 + 6x - 7} \div \dfrac{x^2 + 2xy + y^2}{4x + 4y}$

87. $\dfrac{3x + 3y + 3}{9x} \div \dfrac{x^2 + 2xy + y^2 - 1}{x^4 + x^2}$

88. $\dfrac{(t + 2)^3}{(t + 1)^3} \div \dfrac{t^2 + 4t + 4}{t^2 + 2t + 1} \cdot \dfrac{t + 1}{t + 2}$

89. $\dfrac{a^4 - 81b^4}{a^2c - 6abc + 9b^2c} \cdot \dfrac{a + 3b}{a^2 + 9b^2} \div \dfrac{a^2 + 6ab + 9b^2}{(a - 3b)^2}$

90. $\dfrac{3y^3 + 6y^2}{y^2 - y - 12} \div \dfrac{y^2 - y}{y^2 - 2y - 8} \cdot \dfrac{y^2 + 5y + 6}{y^2}$

91. Use a graphing calculator to check that
$$\frac{x - 1}{x^2 + 2x + 1} \div \frac{x^2 - 1}{x^2 - 5x + 4}$$
is equivalent to
$$\frac{x^2 - 5x + 4}{(x + 1)^3}.$$

COLLABORATIVE CORNER

Currency Exchange

Focus: Least common multiples and proportions
Time: 20 minutes
Group size: 2

International travelers usually exchange currencies. Recently one New Zealand dollar was worth 76 U.S. cents. Use this exchange rate for the activity that follows.

ACTIVITY

1. Within each group of two students, one student should play the role of a U.S. citizen visiting New Zealand. The other student should play the role of a New Zealander visiting the United States. Use the exchange rate of one New Zealand dollar for 76 U.S. cents.

2. The "U.S." student should exchange $76 U.S. and $100 U.S. for New Zealand money. The "New Zealand" student should exchange $76 New Zealand and $100 New Zealand for U.S. money.

3. The exchanges in part (2) should indicate that coins smaller than a dollar are needed to exchange $76 New Zealand for U.S. funds, or

to exchange $100 U.S. for New Zealand money. What is the smallest amount of New Zealand dollars that can be exchanged for a whole-number amount of U.S. dollars? What is the smallest amount of U.S. dollars that can be exchanged for a whole-number amount of New Zealand dollars? (*Hint*: See part 2.)

4. Use the results from part (3) to find two other amounts of U.S. dollars that can be exchanged for a whole-number amount of New Zealand dollars. Answers may vary.

5. Find the smallest number a for which both conversions—from a New Zealand dollars to U.S. funds and from a U.S. dollars to New Zealand funds—use only whole numbers. (*Hint*: Use LCMs and the results of part 2 above.)

6. At one time in 2008, one Polish zloty was worth about 40 U.S. cents. Find the smallest number a for which both conversions—from a Polish zlotys to U.S. funds and from a U.S. dollars to Polish funds—use only whole numbers. (*Hint*: See part 5.)

6.3 Addition, Subtraction, and Least Common Denominators

Addition When Denominators Are the Same ● Subtraction When Denominators Are the Same ●
Least Common Multiples and Denominators

Addition When Denominators Are the Same

Recall that to add fractions having the same denominator, like $\frac{2}{7}$ and $\frac{3}{7}$, we add the numerators and keep the common denominator: $\frac{2}{7} + \frac{3}{7} = \frac{5}{7}$. The same procedure is used when rational expressions share a common denominator.

> ### The Sum of Two Rational Expressions
>
> To add when the denominators are the same, add the numerators and keep the common denominator:
>
> $$\frac{A}{B} + \frac{C}{B} = \frac{A + C}{B}.$$

In this section, we again assume that all denominators are nonzero.

EXAMPLE 1 Add. Simplify the result, if possible.

a) $\dfrac{4}{a} + \dfrac{3 + a}{a}$ b) $\dfrac{3x}{x - 5} + \dfrac{2x + 1}{x - 5}$

c) $\dfrac{2x^2 + 3x - 7}{2x + 1} + \dfrac{x^2 + x - 8}{2x + 1}$ d) $\dfrac{x - 5}{x^2 - 9} + \dfrac{2}{x^2 - 9}$

SOLUTION

a) $\dfrac{4}{a} + \dfrac{3 + a}{a} = \dfrac{7 + a}{a}$ When the denominators are alike, add the numerators and keep the common denominator.

b) $\dfrac{3x}{x - 5} + \dfrac{2x + 1}{x - 5} = \dfrac{5x + 1}{x - 5}$ The denominators are alike, so we add the numerators.

c) $\dfrac{2x^2 + 3x - 7}{2x + 1} + \dfrac{x^2 + x - 8}{2x + 1} = \dfrac{(2x^2 + 3x - 7) + (x^2 + x - 8)}{2x + 1}$

$= \dfrac{3x^2 + 4x - 15}{2x + 1}$ Combining like terms

$= \dfrac{(3x - 5)(x + 3)}{2x + 1}$ Factoring. There are no common factors, so we cannot simplify further.

d) $\dfrac{x - 5}{x^2 - 9} + \dfrac{2}{x^2 - 9} = \dfrac{x - 3}{x^2 - 9}$ Combining like terms in the numerator: $x - 5 + 2 = x - 3$

$= \dfrac{x - 3}{(x - 3)(x + 3)}$ Factoring

$= \dfrac{1 \cdot \cancel{(x - 3)}}{\cancel{(x - 3)}(x + 3)}$ Removing a factor equal to 1: $\dfrac{x - 3}{x - 3} = 1$

$= \dfrac{1}{x + 3}$

TRY EXERCISE 7

STUDY SKILLS

Visualize the Steps

If you have completed all assignments and are studying for a quiz or test, don't feel that you need to redo every assigned problem. A more productive use of your time would be to work through one problem of each type. Then read through the other problems, visualizing the steps that lead to a solution. When you are unsure of how to solve a problem, work that problem in its entirety, seeking outside help as needed.

Subtraction When Denominators Are the Same

When two fractions have the same denominator, we subtract one numerator from the other and keep the common denominator: $\frac{5}{7} - \frac{2}{7} = \frac{3}{7}$. The same procedure is used with rational expressions.

> ### The Difference of Two Rational Expressions
> To subtract when the denominators are the same, subtract the second numerator from the first numerator and keep the common denominator:
>
> $$\frac{A}{B} - \frac{C}{B} = \frac{A - C}{B}.$$

> *CAUTION!*   The fraction bar under a numerator is a grouping symbol, just like parentheses. Thus, when a numerator is subtracted, it is important to subtract *every* term in that numerator.

EXAMPLE   2   Subtract and, if possible, simplify: **(a)** $\dfrac{3x}{x + 2} - \dfrac{x - 5}{x + 2}$; **(b)** $\dfrac{x^2}{x - 4} - \dfrac{x + 12}{x - 4}$.

SOLUTION

a) $\dfrac{3x}{x + 2} - \dfrac{x - 5}{x + 2} = \dfrac{3x - (x - 5)}{x + 2}$ The parentheses are needed to make sure that we subtract both terms.

$= \dfrac{3x - x + 5}{x + 2}$ Removing the parentheses and changing signs (using the distributive law)

$= \dfrac{2x + 5}{x + 2}$ Combining like terms

b) $\dfrac{x^2}{x - 4} - \dfrac{x + 12}{x - 4} = \dfrac{x^2 - (x + 12)}{x - 4}$ Remember the parentheses!

$= \dfrac{x^2 - x - 12}{x - 4}$ Removing parentheses (using the distributive law)

$= \dfrac{(x - 4)(x + 3)}{x - 4}$ Factoring, in hopes of simplifying

$= \dfrac{\cancel{(x - 4)}(x + 3)}{\cancel{x - 4}}$ Removing a factor equal to 1: $\dfrac{x - 4}{x - 4} = 1$

$= x + 3$

TRY EXERCISE 21

Least Common Multiples and Denominators

Thus far, every pair of rational expressions that we have added or subtracted shared a common denominator. To add or subtract rational expressions that have different denominators, we must first find equivalent rational expressions that *do* have a common denominator.

In algebra, we find a common denominator much as we do in arithmetic. Recall that to add $\frac{1}{12}$ and $\frac{7}{30}$, we first identify the smallest number that contains both 12 and 30 as factors. Such a number, the **least common multiple (LCM)** of the denominators, is then used as the **least common denominator (LCD)**.

Let's find the LCM of 12 and 30 using a method that can also be used with polynomials. We begin by writing the prime factorizations of 12 and 30:

$$12 = 2 \cdot 2 \cdot 3;$$
$$30 = 2 \cdot 3 \cdot 5.$$

The LCM must include the factors of each number, so it must include each prime factor the greatest number of times that it appears in either of the factorizations. To find the LCM for 12 and 30, we select one factorization, say

$$2 \cdot 2 \cdot 3,$$

and note that because it lacks a factor of 5, it does not contain the entire factorization of 30. If we multiply $2 \cdot 2 \cdot 3$ by 5, every prime factor occurs just often enough to contain both 12 and 30 as factors.

> ┌─── 12 is a factor of the LCM.
>
> LCM = $2 \cdot 2 \cdot 3 \cdot 5$
>
> └─── 30 is a factor of the LCM.

Note that each prime factor—2, 3, and 5—is used the greatest number of times that it appears in either of the individual factorizations. The factor 2 occurs twice and the factors 3 and 5 once each.

> ## To Find the Least Common Denominator (LCD)
>
> 1. Write the prime factorization of each denominator.
> 2. Select one of the factorizations and inspect it to see if it completely contains the other factorization.
>
>   a) If it does, it represents the LCM of the denominators.
>   b) If it does not, multiply that factorization by any factors of the other denominator that it lacks. The final product is the LCM of the denominators.
>
> The LCD is the LCM of the denominators. It should contain each factor the greatest number of times that it occurs in any of the individual factorizations.

EXAMPLE 3   Find the LCD of $\dfrac{5}{36x^2}$ and $\dfrac{7}{24x}$.

SOLUTION

1. We begin by writing the prime factorizations of $36x^2$ and $24x$:

$$36x^2 = 2 \cdot 2 \cdot 3 \cdot 3 \cdot x \cdot x;$$
$$24x = 2 \cdot 2 \cdot 2 \cdot 3 \cdot x.$$

2. We select the factorization of $36x^2$. Except for a third factor of 2, this factorization contains the entire factorization of $24x$. Thus we multiply $36x^2$ by a third factor of 2.

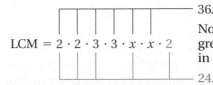

LCM = $2 \cdot 2 \cdot 3 \cdot 3 \cdot x \cdot x \cdot 2$

┌─── $36x^2$ is a factor of the LCM.

Note that each factor appears the greatest number of times that it occurs in either of the above factorizations.

└─── $24x$ is a factor of the LCM.

Since $2^3 \cdot 3^2 \cdot x^2$, or $72x^2$, is the smallest multiple of both $36x^2$ and $24x$, the LCM of the denominators is $72x^2$. The LCD of the expressions is $72x^2$.

Let's add $\dfrac{1}{12}$ and $\dfrac{7}{30}$:

$$\frac{1}{12} + \frac{7}{30} = \frac{1}{2 \cdot 2 \cdot 3} + \frac{7}{2 \cdot 3 \cdot 5}.$$ The least common denominator (LCD) is $2 \cdot 2 \cdot 3 \cdot 5$.

We found above that the LCD is $2 \cdot 2 \cdot 3 \cdot 5$, or 60. To get the LCD, we see that the first denominator needs a factor of 5, and the second denominator needs another factor of 2. Therefore, we multiply $\frac{1}{12}$ by 1, using $\frac{5}{5}$, and we multiply $\frac{7}{30}$ by 1, using $\frac{2}{2}$. Since $a \cdot 1 = a$, for any number a, the values of the fractions are not changed.

$$\frac{1}{12} + \frac{7}{30} = \frac{1}{2 \cdot 2 \cdot 3} \cdot \frac{5}{5} + \frac{7}{2 \cdot 3 \cdot 5} \cdot \frac{2}{2} \qquad \frac{5}{5} = 1 \text{ and } \frac{2}{2} = 1$$

$$= \frac{5}{60} + \frac{14}{60} \qquad \text{Both denominators are now the LCD.}$$

$$= \frac{19}{60} \qquad \text{Adding the numerators and keeping the LCD}$$

Expressions like $\dfrac{5}{36x^2}$ and $\dfrac{7}{24x}$ are added in much the same manner. In Example 3, we found that the LCD is $2 \cdot 2 \cdot 2 \cdot 3 \cdot 3 \cdot x \cdot x$, or $72x^2$. To obtain equivalent expressions with this LCD, we multiply each expression by 1, using the missing factors of the LCD to write 1:

$$\frac{5}{36x^2} + \frac{7}{24x} = \frac{5}{2 \cdot 2 \cdot 3 \cdot 3 \cdot x \cdot x} + \frac{7}{2 \cdot 2 \cdot 2 \cdot 3 \cdot x}$$

$$= \frac{5}{2 \cdot 2 \cdot 3 \cdot 3 \cdot x \cdot x} \cdot \frac{2}{2} + \frac{7}{2 \cdot 2 \cdot 2 \cdot 3 \cdot x} \cdot \frac{3 \cdot x}{3 \cdot x}$$

The LCD requires another factor of 2.   The LCD requires additional factors of 3 and x.

$$= \frac{10}{72x^2} + \frac{21x}{72x^2} \qquad \text{Both denominators are now the LCD.}$$

$$= \frac{21x + 10}{72x^2}.$$

You now have the "big picture" of why LCMs are needed when adding rational expressions. For the remainder of this section, we will practice finding LCMs and rewriting rational expressions so that they have the LCD as the denominator. In Section 6.4, we will return to the addition and subtraction of rational expressions.

EXAMPLE **4**   For each pair of polynomials, find the least common multiple.

a) $15a$ and $35b$

b) $21x^3y^6$ and $7x^5y^2$

c) $x^2 + 5x - 6$ and $x^2 - 1$

SOLUTION

a) We write the prime factorizations and then construct the LCM, starting with the factorization of $15a$.

$$15a = 3 \cdot 5 \cdot a$$
$$35b = 5 \cdot 7 \cdot b$$

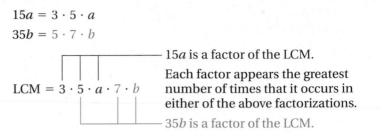

$$\text{LCM} = 3 \cdot 5 \cdot a \cdot 7 \cdot b$$

15a is a factor of the LCM.

Each factor appears the greatest number of times that it occurs in either of the above factorizations.

35b is a factor of the LCM.

The LCM is $3 \cdot 5 \cdot a \cdot 7 \cdot b$, or $105ab$.

b) $21x^3y^6 = 3 \cdot 7 \cdot x \cdot x \cdot x \cdot y \cdot y \cdot y \cdot y \cdot y \cdot y$ Try to visualize the factors of x and y mentally.

$$7x^5y^2 = 7 \cdot x \cdot x \cdot x \cdot x \cdot x \cdot y \cdot y$$

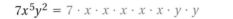

$$\text{LCM} = 3 \cdot 7 \cdot x \cdot x \cdot x \cdot y \cdot y \cdot y \cdot y \cdot y \cdot y \cdot x \cdot x$$

We start with the prime factorization of $21x^3y^6$.

We multiply by the factors of $7x^5y^2$ that are lacking.

Note that we used the highest power of each factor in $21x^3y^6$ and $7x^5y^2$. The LCM is $21x^5y^6$.

c) $x^2 + 5x - 6 = (x - 1)(x + 6)$
$x^2 - 1 = (x - 1)(x + 1)$

$$\text{LCM} = (x - 1)(x + 6)(x + 1)$$

We start with the factorization of $x^2 + 5x - 6$.

We multiply by the factor of $x^2 - 1$ that is missing.

The LCM is $(x - 1)(x + 6)(x + 1)$. There is no need to multiply this out.

TRY EXERCISE 43

The procedure above can be used to find the LCM of three or more polynomials as well. We factor each polynomial and then construct the LCM using each factor the greatest number of times that it appears in any one factorization.

EXAMPLE   5   For each group of polynomials, find the LCM.

a) $12x$, $16y$, and $8xyz$

b) $x^2 + 4$, $x + 1$, and 5

STUDENT NOTES —————

If you prefer, the LCM for a group of three polynomials can be found by finding the LCM of two of them and then finding the LCM of that result and the remaining polynomial.

SOLUTION

a) $12x = 2 \cdot 2 \cdot 3 \cdot x$

$16y = 2 \cdot 2 \cdot 2 \cdot 2 \cdot y$

$8xyz = 2 \cdot 2 \cdot 2 \cdot x \cdot y \cdot z$

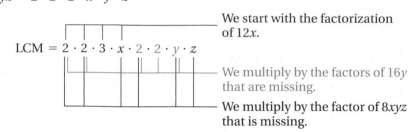

$\text{LCM} = 2 \cdot 2 \cdot 3 \cdot x \cdot 2 \cdot 2 \cdot y \cdot z$

We start with the factorization of $12x$.

We multiply by the factors of $16y$ that are missing.

We multiply by the factor of $8xyz$ that is missing.

The LCM is $2^4 \cdot 3 \cdot xyz$, or $48xyz$.

b) Since $x^2 + 4$, $x + 1$, and 5 are not factorable, the LCM is their product: $5(x^2 + 4)(x + 1)$.

TRY EXERCISE ▸ 51

To add or subtract rational expressions with different denominators, we first write equivalent expressions that have the LCD. To do this, we multiply each rational expression by a carefully constructed form of 1.

EXAMPLE **6**   Find equivalent expressions that have the LCD:

$$\frac{x + 3}{x^2 + 5x - 6}, \qquad \frac{x + 7}{x^2 - 1}.$$

SOLUTION   From Example 4(c), we know that the LCD is

$$(x + 6)(x - 1)(x + 1).$$

Since

$$x^2 + 5x - 6 = (x + 6)(x - 1),$$

the factor of the LCD that is missing from the first denominator is $x + 1$. We multiply by 1 using $(x + 1)/(x + 1)$:

$$\left.\begin{aligned}\frac{x + 3}{x^2 + 5x - 6} &= \frac{x + 3}{(x + 6)(x - 1)} \cdot \frac{x + 1}{x + 1} \\ &= \frac{(x + 3)(x + 1)}{(x + 6)(x - 1)(x + 1)}.\end{aligned}\right\}$$ Finding an equivalent expression that has the least common denominator

For the second expression, we have

$$x^2 - 1 = (x + 1)(x - 1).$$

The factor of the LCD that is missing is $x + 6$. We multiply by 1 using $(x + 6)/(x + 6)$:

$$\left.\begin{aligned}\frac{x + 7}{x^2 - 1} &= \frac{x + 7}{(x + 1)(x - 1)} \cdot \frac{x + 6}{x + 6} \\ &= \frac{(x + 7)(x + 6)}{(x + 1)(x - 1)(x + 6)}.\end{aligned}\right\}$$ Finding an equivalent expression that has the least common denominator

We leave the results in factored form. In Section 6.4, we will carry out the actual addition and subtraction of such rational expressions.

TRY EXERCISE ▸ 59

| **6.3** | **EXERCISE SET** |

🖐 *Concept Reinforcement* *Use one or more words to complete each of the following sentences.*

1. To add two rational expressions when the denominators are the same, add _____ and keep the common _____.

2. When a numerator is being subtracted, use parentheses to make sure to subtract every _____ in that numerator.

3. The least common multiple of two denominators is usually referred to as the _____ and is abbreviated _____.

4. The least common denominator of two fractions must contain the prime _____ of both _____.

Perform the indicated operation. Simplify, if possible.

5. $\dfrac{3}{t} + \dfrac{5}{t}$

6. $\dfrac{8}{y^2} + \dfrac{2}{y^2}$

7. $\dfrac{x}{12} + \dfrac{2x + 5}{12}$

8. $\dfrac{a}{7} + \dfrac{3a - 4}{7}$

9. $\dfrac{4}{a + 3} + \dfrac{5}{a + 3}$

10. $\dfrac{5}{x + 2} + \dfrac{8}{x + 2}$

11. $\dfrac{11}{4x - 7} - \dfrac{3}{4x - 7}$

12. $\dfrac{9}{2x + 3} - \dfrac{5}{2x + 3}$

13. $\dfrac{3y + 8}{2y} - \dfrac{y + 1}{2y}$

14. $\dfrac{5 + 3t}{4t} - \dfrac{2t + 1}{4t}$

15. $\dfrac{5x + 7}{x + 3} + \dfrac{x + 11}{x + 3}$

16. $\dfrac{3x + 4}{x - 1} + \dfrac{2x - 9}{x - 1}$

17. $\dfrac{5x + 7}{x + 3} - \dfrac{x + 11}{x + 3}$

18. $\dfrac{3x + 4}{x - 1} - \dfrac{2x - 9}{x - 1}$

19. $\dfrac{a^2}{a - 4} + \dfrac{a - 20}{a - 4}$

20. $\dfrac{x^2}{x + 5} + \dfrac{7x + 10}{x + 5}$

21. $\dfrac{y^2}{y + 2} - \dfrac{5y + 14}{y + 2}$

22. $\dfrac{t^2}{t - 3} - \dfrac{8t - 15}{t - 3}$

Aha! **23.** $\dfrac{t^2 - 5t}{t - 1} + \dfrac{5t - t^2}{t - 1}$

24. $\dfrac{y^2 + 6y}{y + 2} + \dfrac{2y + 12}{y + 2}$

25. $\dfrac{x - 6}{x^2 + 5x + 6} + \dfrac{9}{x^2 + 5x + 6}$

26. $\dfrac{x - 5}{x^2 - 4x + 3} + \dfrac{2}{x^2 - 4x + 3}$

27. $\dfrac{t^2 - 5t}{t^2 + 6t + 9} + \dfrac{4t - 12}{t^2 + 6t + 9}$

28. $\dfrac{y^2 - 7y}{y^2 + 8y + 16} + \dfrac{6y - 20}{y^2 + 8y + 16}$

29. $\dfrac{2y^2 + 3y}{y^2 - 7y + 12} - \dfrac{y^2 + 4y + 6}{y^2 - 7y + 12}$

30. $\dfrac{3a^2 + 7}{a^2 - 2a - 8} - \dfrac{7 + 3a^2}{a^2 - 2a - 8}$

31. $\dfrac{3 - 2x}{x^2 - 6x + 8} + \dfrac{7 - 3x}{x^2 - 6x + 8}$

32. $\dfrac{1 - 2t}{t^2 - 5t + 4} + \dfrac{4 - 3t}{t^2 - 5t + 4}$

33. $\dfrac{x - 9}{x^2 + 3x - 4} - \dfrac{2x - 5}{x^2 + 3x - 4}$

34. $\dfrac{5 - 3x}{x^2 - 2x + 1} - \dfrac{x + 1}{x^2 - 2x + 1}$

Find the LCM.

35. 15, 36 **36.** 18, 30 **37.** 8, 9

38. 12, 15 **39.** 6, 12, 15 **40.** 8, 32, 50

Find the LCM.

41. $18t^2$, $6t^5$ **42.** $8x^5$, $24x^2$

43. $15a^4b^7$, $10a^2b^8$ **44.** $6a^2b^7$, $9a^5b^2$

45. $2(y - 3)$, $6(y - 3)$ **46.** $4(x - 1)$, $8(x - 1)$

47. $x^2 - 2x - 15$, $x^2 - 9$ **48.** $t^2 - 4$, $t^2 + 7t + 10$

49. $t^3 + 4t^2 + 4t$, $t^2 - 4t$ **50.** $y^3 - y^2$, $y^4 - y^2$

51. $6xz^2$, $8x^2y$, $15y^3z$ **52.** $12s^3t$, $15sv^2$, $6t^4v$

53. $a + 1$, $(a - 1)^2$, $a^2 - 1$

54. $x - 2$, $(x + 2)^2$, $x^2 - 4$

55. $2n^2 + n - 1$, $2n^2 + 3n - 2$

56. $m^2 - 2m - 3$, $2m^2 + 3m + 1$

57. $6x^3 - 24x^2 + 18x$, $4x^5 - 24x^4 + 20x^3$

58. $9x^3 - 9x^2 - 18x$, $6x^5 - 24x^4 + 24x^3$

Find equivalent expressions that have the LCD.

59. $\dfrac{5}{6t^4}, \dfrac{s}{18t^2}$

60. $\dfrac{7}{10y^2}, \dfrac{x}{5y^6}$

61. $\dfrac{7}{3x^4y^2}, \dfrac{4}{9xy^3}$

62. $\dfrac{3}{2a^2b}, \dfrac{7}{8ab^2}$

63. $\dfrac{2x}{x^2 - 4}, \dfrac{4x}{x^2 + 5x + 6}$

64. $\dfrac{5x}{x^2 - 9}, \dfrac{2x}{x^2 + 11x + 24}$

65. Explain why the product of two numbers is not always their least common multiple.

66. If the LCM of two numbers is their product, what can you conclude about the two numbers?

Skill Review

To prepare for Section 6.4, review opposites (Sections 1.7 and 1.8).

Write each number in two equivalent forms. [1.7]

67. $-\dfrac{5}{8}$

68. $\dfrac{4}{-11}$

Write an equivalent expression without parentheses. [1.8]

69. $-(x - y)$

70. $-(3 - a)$

Multiply and simplify. [1.8]

71. $-1(2x - 7)$

72. $-1(a - b)$

Synthesis

73. If the LCM of two third-degree polynomials is a sixth-degree polynomial, what can be concluded about the two polynomials?

74. If the LCM of a binomial and a trinomial is the trinomial, what relationship exists between the two expressions?

Perform the indicated operations. Simplify, if possible.

75. $\dfrac{6x - 1}{x - 1} + \dfrac{3(2x + 5)}{x - 1} + \dfrac{3(2x - 3)}{x - 1}$

76. $\dfrac{2x + 11}{x - 3} \cdot \dfrac{3}{x + 4} + \dfrac{-1}{4 + x} \cdot \dfrac{6x + 3}{x - 3}$

77. $\dfrac{x^2}{3x^2 - 5x - 2} - \dfrac{2x}{3x + 1} \cdot \dfrac{1}{x - 2}$

78. $\dfrac{x + y}{x^2 - y^2} + \dfrac{x - y}{x^2 - y^2} - \dfrac{2x}{x^2 - y^2}$

African artistry. In Southeast Mozambique, the design of every woven handbag, or gipatsi (plural, sipatsi) is created by repeating two or more geometric patterns.

Each pattern encircles the bag, sharing the strands of fabric with any pattern above or below. The length, or period, of each pattern is the number of strands required to construct the pattern. For a gipatsi to be considered beautiful, each individual pattern must fit a whole number of times around the bag.
Source: Gerdes, Paulus, *Women, Art and Geometry in Southern Africa.* Asmara, Eritrea: Africa World Press, Inc., p. 5

79. A weaver is using two patterns to create a gipatsi. Pattern A is 10 strands long, and pattern B is 3 strands long. What is the smallest number of strands that can be used to complete the gipatsi?

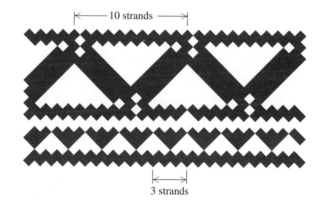

80. A weaver is using a four-strand pattern, a six-strand pattern, and an eight-strand pattern. What is the smallest number of strands that can be used to complete the gipatsi?

81. For technical reasons, the number of strands is generally a multiple of 4. Answer Exercise 79 with this additional requirement in mind.

Find the LCM.

82. 80, 96, 108

83. $4x^2 - 25, 6x^2 - 7x - 20, (9x^2 + 24x + 16)^2$

84. $9n^2 - 9$, $(5n^2 - 10n + 5)^2$, $15n - 15$

85. *Copiers.*   The Brother® MFC240C copier can print 20 color pages per minute. The Lexmark X5470 copier can print 18 color pages per minute. If both machines begin printing at the same instant, how long will it be until they again begin printing a page at exactly the same time?

86. *Running.*   Kim and Trey leave the starting point of a fitness loop at the same time. Kim jogs a lap in 6 min and Trey jogs one in 8 min. Assuming they continue to run at the same pace, when will they next meet at the starting place?

87. *Bus schedules.*   Beginning at 5:00 A.M., a hotel shuttle bus leaves Salton Airport every 25 min, and the downtown shuttle bus leaves the airport every 35 min. What time will it be when both shuttles again leave at the same time?

88. *Appliances.*   Smoke detectors last an average of 10 yr, water heaters an average of 12 yr, and refrigerators an average of 15 yr. If an apartment house is equipped with new smoke detectors, water heaters, and refrigerators in 2010, in what year will all three appliances need to be replaced at once?
Source: Demesne.info

89. Explain how evaluating can be used to perform a partial check on the result of Example 1(d):

$$\frac{x-5}{x^2-9} + \frac{2}{x^2-9} = \frac{1}{x+3}.$$

90. On p. 379, the second step in finding an LCD is to select one of the factorizations of the denominators. Does it matter which one is selected? Why or why not?

6.4   Addition and Subtraction with Unlike Denominators

Adding and Subtracting with LCDs   ■   When Factors Are Opposites

Adding and Subtracting with LCDs

We now know how to rewrite two rational expressions in equivalent forms that use the LCD. Once rational expressions share a common denominator, they can be added or subtracted just as in Section 6.3.

> **To Add or Subtract Rational Expressions Having Different Denominators**
> 1. Find the LCD.
> 2. Multiply each rational expression by a form of 1 made up of the factors of the LCD missing from that expression's denominator.
> 3. Add or subtract the numerators, as indicated. Write the sum or the difference over the LCD.
> 4. Simplify, if possible.

 **1**   Add: $\dfrac{5x^2}{8} + \dfrac{7x}{12}$.

SOLUTION

1. First, we find the LCD:

$$\left.\begin{array}{l} 8 = 2 \cdot 2 \cdot 2 \\ 12 = 2 \cdot 2 \cdot 3 \end{array}\right\} \quad \text{LCD} = 2 \cdot 2 \cdot 2 \cdot 3, \text{ or } 24.$$

2. The denominator 8 must be multiplied by 3 in order to obtain the LCD. The denominator 12 must be multiplied by 2 in order to obtain the LCD. Thus we multiply the first expression by $\frac{3}{3}$ and the second expression by $\frac{2}{2}$ to get the LCD:

$$\begin{aligned} \frac{5x^2}{8} + \frac{7x}{12} &= \frac{5x^2}{2 \cdot 2 \cdot 2} + \frac{7x}{2 \cdot 2 \cdot 3} \\ &= \frac{5x^2}{2 \cdot 2 \cdot 2} \cdot \frac{3}{3} + \frac{7x}{2 \cdot 2 \cdot 3} \cdot \frac{2}{2} \qquad \text{Multiplying each expression by a form of 1 to get the LCD} \\ &= \frac{15x^2}{24} + \frac{14x}{24}. \end{aligned}$$

3. Next, we add the numerators:

$$\frac{15x^2}{24} + \frac{14x}{24} = \frac{15x^2 + 14x}{24}.$$

4. Since $15x^2 + 14x$ and 24 have no common factor,

$$\frac{15x^2 + 14x}{24}$$

cannot be simplified any further.   **TRY EXERCISE** ▶ 13

Subtraction is performed in much the same way.

EXAMPLE **2**   Subtract: $\dfrac{7}{8x} - \dfrac{5}{12x^2}$.

SOLUTION   We follow the four steps shown above. First, we find the LCD:

$$\left.\begin{array}{l} 8x = 2 \cdot 2 \cdot 2 \cdot x \\ 12x^2 = 2 \cdot 2 \cdot 3 \cdot x \cdot x \end{array}\right\} \quad \text{LCD} = 2 \cdot 2 \cdot 3 \cdot x \cdot x \cdot 2, \text{ or } 24x^2.$$

The denominator $8x$ must be multiplied by $3x$ in order to obtain the LCD. The denominator $12x^2$ must be multiplied by 2 in order to obtain the LCD. Thus we multiply by $\dfrac{3x}{3x}$ and $\dfrac{2}{2}$ to get the LCD. Then we subtract and, if possible, simplify.

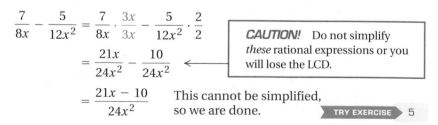

$$\begin{aligned} \frac{7}{8x} - \frac{5}{12x^2} &= \frac{7}{8x} \cdot \frac{3x}{3x} - \frac{5}{12x^2} \cdot \frac{2}{2} \\ &= \frac{21x}{24x^2} - \frac{10}{24x^2} \\ &= \frac{21x - 10}{24x^2} \end{aligned}$$

CAUTION! Do not simplify *these* rational expressions or you will lose the LCD.

This cannot be simplified, so we are done.   **TRY EXERCISE** ▶ 5

When denominators contain polynomials with two or more terms, the same steps are used.

EXAMPLE **3**   Add: $\dfrac{2a}{a^2 - 1} + \dfrac{1}{a^2 + a}$.

SOLUTION   First, we find the LCD:

Find the LCD.
$$\left.\begin{array}{l} a^2 - 1 = (a - 1)(a + 1) \\ a^2 + a = a(a + 1). \end{array}\right\} \quad \text{LCD} = (a - 1)(a + 1)a$$

We multiply by a form of 1 to get the LCD in each expression:

Write each expression with the LCD.
$$\dfrac{2a}{a^2 - 1} + \dfrac{1}{a^2 + a} = \dfrac{2a}{(a - 1)(a + 1)} \cdot \dfrac{a}{a} + \dfrac{1}{a(a + 1)} \cdot \dfrac{a - 1}{a - 1}$$

Multiplying by $\dfrac{a}{a}$ and $\dfrac{a - 1}{a - 1}$ to get the LCD

$$= \dfrac{2a^2}{(a - 1)(a + 1)a} + \dfrac{a - 1}{a(a + 1)(a - 1)}$$

Add numerators.
$$= \dfrac{2a^2 + a - 1}{a(a - 1)(a + 1)}$$ Adding numerators

Simplify.
$$= \dfrac{(2a - 1)\cancel{(a + 1)}}{a(a - 1)\cancel{(a + 1)}}$$
$$= \dfrac{2a - 1}{a(a - 1)}.$$

Simplifying by factoring and removing a factor equal to 1:
$$\dfrac{a + 1}{a + 1} = 1$$

TRY EXERCISE 33

EXAMPLE **4**   Perform the indicated operations.

a) $\dfrac{x + 4}{x - 2} - \dfrac{x - 7}{x + 5}$

b) $\dfrac{t}{t^2 + 11t + 30} + \dfrac{-5}{t^2 + 9t + 20}$

c) $\dfrac{x}{x^2 + 5x + 6} - \dfrac{2}{x^2 + 3x + 2}$

STUDENT NOTES

As you can see, adding or subtracting rational expressions can involve many steps. Therefore, it is important to double-check each step of your work as you work through each problem. Waiting to inspect your work at the end of each problem is usually a less efficient use of your time.

SOLUTION

a) First, we find the LCD. It is just the product of the denominators:
$$\text{LCD} = (x - 2)(x + 5).$$

We multiply by a form of 1 to get the LCD in each expression. Then we subtract and try to simplify.

$$\dfrac{x + 4}{x - 2} - \dfrac{x - 7}{x + 5} = \dfrac{x + 4}{x - 2} \cdot \dfrac{x + 5}{x + 5} - \dfrac{x - 7}{x + 5} \cdot \dfrac{x - 2}{x - 2}$$

$$= \dfrac{x^2 + 9x + 20}{(x - 2)(x + 5)} - \dfrac{x^2 - 9x + 14}{(x - 2)(x + 5)}$$

Multiplying out numerators (but not denominators)

$$= \dfrac{x^2 + 9x + 20 - \overbrace{(x^2 - 9x + 14)}}{(x - 2)(x + 5)}$$

When subtracting a numerator with more than one term, parentheses are important.

$$= \dfrac{x^2 + 9x + 20 - x^2 + 9x - 14}{(x - 2)(x + 5)}$$

Removing parentheses and subtracting every term

$$= \dfrac{18x + 6}{(x - 2)(x + 5)}$$

$$= \dfrac{6(3x + 1)}{(x - 2)(x + 5)}$$

We cannot simplify.

b) $\dfrac{t}{t^2 + 11t + 30} + \dfrac{-5}{t^2 + 9t + 20}$

$= \dfrac{t}{(t + 5)(t + 6)} + \dfrac{-5}{(t + 5)(t + 4)}$ Factoring the denominators in order to find the LCD. The LCD is $(t + 5)(t + 6)(t + 4)$.

$= \dfrac{t}{(t + 5)(t + 6)} \cdot \dfrac{t + 4}{t + 4} + \dfrac{-5}{(t + 5)(t + 4)} \cdot \dfrac{t + 6}{t + 6}$ Multiplying to get the LCD

$= \dfrac{t^2 + 4t}{(t + 5)(t + 6)(t + 4)} + \dfrac{-5t - 30}{(t + 5)(t + 6)(t + 4)}$ Multiplying in each numerator

$= \dfrac{t^2 + 4t - 5t - 30}{(t + 5)(t + 6)(t + 4)}$ Adding numerators

$= \dfrac{t^2 - t - 30}{(t + 5)(t + 6)(t + 4)}$ Combining like terms in the numerator

$= \dfrac{\cancel{(t + 5)}(t - 6)}{\cancel{(t + 5)}(t + 6)(t + 4)}$

$= \dfrac{t - 6}{(t + 6)(t + 4)}$ Always simplify the result, if possible, by removing a factor equal to 1; here $\dfrac{t + 5}{t + 5} = 1$.

c) $\dfrac{x}{x^2 + 5x + 6} - \dfrac{2}{x^2 + 3x + 2}$

Find the LCD.
$= \dfrac{x}{(x + 2)(x + 3)} - \dfrac{2}{(x + 2)(x + 1)}$ Factoring denominators. The LCD is $(x + 2)(x + 3)(x + 1)$.

Write each expression with the LCD.
$= \dfrac{x}{(x + 2)(x + 3)} \cdot \dfrac{x + 1}{x + 1} - \dfrac{2}{(x + 2)(x + 1)} \cdot \dfrac{x + 3}{x + 3}$

$= \dfrac{x^2 + x}{(x + 2)(x + 3)(x + 1)} - \dfrac{2x + 6}{(x + 2)(x + 3)(x + 1)}$

Subtract numerators.
$= \dfrac{x^2 + x - (2x + 6)}{(x + 2)(x + 3)(x + 1)}$ Don't forget the parentheses!

$= \dfrac{x^2 + x - 2x - 6}{(x + 2)(x + 3)(x + 1)}$ Remember to subtract each term in $2x + 6$.

$= \dfrac{x^2 - x - 6}{(x + 2)(x + 3)(x + 1)}$ Combining like terms in the numerator

Simplify.
$= \dfrac{\cancel{(x + 2)}(x - 3)}{\cancel{(x + 2)}(x + 3)(x + 1)}$

$= \dfrac{x - 3}{(x + 3)(x + 1)}$ Factoring and simplifying; $\dfrac{x + 2}{x + 2} = 1$

TRY EXERCISE 45

When Factors Are Opposites

When one denominator is the opposite of the other, we can first multiply either expression by 1 using $-1/-1$.

EXAMPLE 5   Add: **(a)** $\dfrac{t}{2} + \dfrac{3}{-2}$; **(b)** $\dfrac{x}{x-5} + \dfrac{7}{5-x}$.

SOLUTION

a) $\dfrac{t}{2} + \dfrac{3}{-2} = \dfrac{t}{2} + \dfrac{3}{-2} \cdot \dfrac{-1}{-1}$ Multiplying by 1 using $\dfrac{-1}{-1}$

$\qquad\qquad = \dfrac{t}{2} + \dfrac{-3}{2}$ The denominators are now the same.

$\qquad\qquad = \dfrac{t + (-3)}{2}$

$\qquad\qquad = \dfrac{t - 3}{2}$

b) Recall that when an expression of the form $a - b$ is multiplied by -1, the subtraction is reversed: $-1(a - b) = -a + b = b + (-a) = b - a$. Since $x - 5$ and $5 - x$ are opposites, we can find a common denominator by multiplying one of the rational expressions by $-1/-1$. Because polynomials are usually written in descending order, we choose to reverse the subtraction in the second denominator:

$\dfrac{x}{x-5} + \dfrac{7}{5-x} = \dfrac{x}{x-5} + \dfrac{7}{5-x} \cdot \dfrac{-1}{-1}$ Multiplying by 1, where $1 = \dfrac{-1}{-1}$

$\qquad\qquad\qquad\quad = \dfrac{x}{x-5} + \dfrac{-7}{-5+x}$

$\qquad\qquad\qquad\quad = \dfrac{x}{x-5} + \dfrac{-7}{x-5}$ Note that $-5 + x = x + (-5) = x - 5$.

$\qquad\qquad\qquad\quad = \dfrac{x - 7}{x-5}.$

> **TRY EXERCISE** 53

Sometimes, after factoring to find the LCD, we find a factor in one denominator that is the opposite of a factor in the other denominator. When this happens, multiplication by $-1/-1$ can again be helpful.

EXAMPLE 6   Perform the indicated operations and simplify.

a) $\dfrac{x}{x^2 - 25} + \dfrac{3}{5 - x}$ **b)** $\dfrac{x + 9}{x^2 - 4} + \dfrac{6 - x}{4 - x^2} - \dfrac{1 + x}{x^2 - 4}$

SOLUTION

a) $\dfrac{x}{x^2 - 25} + \dfrac{3}{5 - x} = \dfrac{x}{(x-5)(x+5)} + \dfrac{3}{5 - x}$ Factoring

$\qquad\qquad\qquad\qquad = \dfrac{x}{(x-5)(x+5)} + \dfrac{3}{5 - x} \cdot \dfrac{-1}{-1}$ Multiplication by $-1/-1$ changes $5 - x$ to $x - 5$.

$\qquad\qquad\qquad\qquad = \dfrac{x}{(x-5)(x+5)} + \dfrac{-3}{x - 5}$ $(5 - x)(-1) = x - 5$

$\qquad\qquad\qquad\qquad = \dfrac{x}{(x-5)(x+5)} + \dfrac{-3}{(x-5)} \cdot \dfrac{x + 5}{x + 5}$ The LCD is $(x - 5)(x + 5)$.

$\qquad\qquad\qquad\qquad = \dfrac{x}{(x-5)(x+5)} + \dfrac{-3x - 15}{(x-5)(x+5)}$

$\qquad\qquad\qquad\qquad = \dfrac{-2x - 15}{(x-5)(x+5)}$

b) Since $4 - x^2$ is the opposite of $x^2 - 4$, multiplying the second rational expression by $-1/-1$ will lead to a common denominator:

$$\frac{x + 9}{x^2 - 4} + \frac{6 - x}{4 - x^2} - \frac{1 + x}{x^2 - 4} = \frac{x + 9}{x^2 - 4} + \frac{6 - x}{4 - x^2} \cdot \frac{-1}{-1} - \frac{1 + x}{x^2 - 4}$$

$$= \frac{x + 9}{x^2 - 4} + \frac{x - 6}{x^2 - 4} - \frac{1 + x}{x^2 - 4}$$

$$= \frac{x + 9 + x - 6 - 1 - x}{x^2 - 4} \quad \begin{array}{l}\text{Adding and} \\ \text{subtracting} \\ \text{numerators}\end{array}$$

$$= \frac{x + 2}{x^2 - 4}$$

$$= \frac{\cancel{(x + 2)} \cdot 1}{\cancel{(x + 2)}(x - 2)} \quad \left.\begin{array}{l} \\ \\ \end{array}\right\} \text{Simplifying}$$

$$= \frac{1}{x - 2}.$$

TRY EXERCISE ▸ 59

6.4 EXERCISE SET

For Extra Help

↪ *Concept Reinforcement* In Exercises 1–4, the four steps for adding rational expressions with different denominators are listed. Fill in the missing word or words for each step.

1. To add or subtract when the denominators are different, first find the _____.

2. Multiply each rational expression by a form of 1 made up of the factors of the LCD that are _____ from that expression's _____.

3. Add or subtract the _____, as indicated. Write the sum or the difference over the _____.

4. _____, if possible.

Perform the indicated operation. Simplify, if possible.

5. $\dfrac{3}{x^2} + \dfrac{5}{x}$

6. $\dfrac{6}{x} + \dfrac{7}{x^2}$

7. $\dfrac{1}{6r} - \dfrac{3}{8r}$

8. $\dfrac{4}{9t} - \dfrac{7}{6t}$

9. $\dfrac{3}{uv^2} + \dfrac{4}{u^3v}$

10. $\dfrac{8}{cd^2} + \dfrac{1}{c^2d}$

11. $\dfrac{-2}{3xy^2} - \dfrac{6}{x^2y^3}$

12. $\dfrac{8}{9t^3} - \dfrac{5}{6t^2}$

13. $\dfrac{x + 3}{8} + \dfrac{x - 2}{6}$

14. $\dfrac{x - 4}{9} + \dfrac{x + 5}{12}$

15. $\dfrac{x - 2}{6} - \dfrac{x + 1}{3}$

16. $\dfrac{a + 2}{2} - \dfrac{a - 4}{4}$

17. $\dfrac{a + 3}{15a} + \dfrac{2a - 1}{3a^2}$

18. $\dfrac{5a + 1}{2a^2} + \dfrac{a + 2}{6a}$

19. $\dfrac{4z - 9}{3z} - \dfrac{3z - 8}{4z}$

20. $\dfrac{x - 1}{4x} - \dfrac{2x + 3}{x}$

21. $\dfrac{3c + d}{cd^2} + \dfrac{c - d}{c^2d}$

22. $\dfrac{u + v}{u^2v} + \dfrac{2u + v}{uv^2}$

23. $\dfrac{4x + 2t}{3xt^2} - \dfrac{5x - 3t}{x^2t}$

24. $\dfrac{5x + 3y}{2x^2y} - \dfrac{3x + 4y}{xy^2}$

25. $\dfrac{3}{x - 2} + \dfrac{3}{x + 2}$

26. $\dfrac{5}{x - 1} + \dfrac{5}{x + 1}$

27. $\dfrac{t}{t + 3} - \dfrac{1}{t - 1}$

28. $\dfrac{y}{y - 3} + \dfrac{12}{y + 4}$

29. $\dfrac{3}{x + 1} + \dfrac{2}{3x}$

30. $\dfrac{2}{x + 5} + \dfrac{3}{4x}$

31. $\dfrac{3}{2t^2 - 2t} - \dfrac{5}{2t - 2}$

32. $\dfrac{8}{3t^2 - 15t} - \dfrac{3}{2t - 10}$

33. $\dfrac{3a}{a^2 - 9} + \dfrac{a}{a + 3}$

34. $\dfrac{5p}{p^2 - 16} + \dfrac{p}{p - 4}$

35. $\dfrac{6}{z + 4} - \dfrac{2}{3z + 12}$

36. $\dfrac{t}{t - 3} - \dfrac{5}{4t - 12}$

37. $\dfrac{5}{q - 1} + \dfrac{2}{(q - 1)^2}$

38. $\dfrac{3}{w + 2} + \dfrac{7}{(w + 2)^2}$

39. $\dfrac{3}{x + 2} - \dfrac{8}{x^2 - 4}$

40. $\dfrac{2t}{t^2 - 9} - \dfrac{3}{t - 3}$

41. $\dfrac{3a}{4a - 20} + \dfrac{9a}{6a - 30}$

42. $\dfrac{4a}{5a - 10} + \dfrac{3a}{10a - 20}$

Aha! **43.** $\dfrac{x}{x - 5} + \dfrac{x}{5 - x}$

44. $\dfrac{x + 4}{x} + \dfrac{x}{x + 4}$

45. $\dfrac{6}{a^2 + a - 2} + \dfrac{4}{a^2 - 4a + 3}$

46. $\dfrac{x}{x^2 + 2x + 1} + \dfrac{1}{x^2 + 5x + 4}$

47. $\dfrac{x}{x^2 + 9x + 20} - \dfrac{4}{x^2 + 7x + 12}$

48. $\dfrac{x}{x^2 + 5x + 6} - \dfrac{2}{x^2 + 3x + 2}$

49. $\dfrac{3z}{z^2 - 4z + 4} + \dfrac{10}{z^2 + z - 6}$

50. $\dfrac{3}{x^2 - 9} + \dfrac{2}{x^2 - x - 6}$

Aha! **51.** $\dfrac{-7}{x^2 + 25x + 24} - \dfrac{0}{x^2 + 11x + 10}$

52. $\dfrac{x}{x^2 + 17x + 72} - \dfrac{1}{x^2 + 15x + 56}$

53. $\dfrac{5x}{4} - \dfrac{x - 2}{-4}$

54. $\dfrac{x}{6} - \dfrac{2x - 3}{-6}$

55. $-\dfrac{y^2}{y - 3} + \dfrac{9}{3 - y}$

56. $\dfrac{t^2}{t - 2} + \dfrac{4}{2 - t}$

57. $\dfrac{c - 5}{c^2 - 64} - \dfrac{5 - c}{64 - c^2}$

58. $\dfrac{b - 4}{b^2 - 49} + \dfrac{b - 4}{49 - b^2}$

59. $\dfrac{4 - p}{25 - p^2} + \dfrac{p + 1}{p - 5}$

60. $\dfrac{y + 2}{y - 7} + \dfrac{3 - y}{49 - y^2}$

61. $\dfrac{x}{x - 4} - \dfrac{3}{16 - x^2}$

62. $\dfrac{x}{3 - x} - \dfrac{2}{x^2 - 9}$

63. $\dfrac{a}{a^2 - 1} + \dfrac{2a}{a - a^2}$

64. $\dfrac{3x + 2}{3x + 6} + \dfrac{x}{4 - x^2}$

65. $\dfrac{4x}{x^2 - y^2} - \dfrac{6}{y - x}$

66. $\dfrac{4 - a^2}{a^2 - 9} - \dfrac{a - 2}{3 - a}$

Perform the indicated operations. Simplify, if possible.

67. $\dfrac{x - 3}{2 - x} - \dfrac{x + 3}{x + 2} + \dfrac{x + 6}{4 - x^2}$

68. $\dfrac{t - 5}{1 - t} - \dfrac{t + 4}{t + 1} + \dfrac{t + 2}{t^2 - 1}$

69. $\dfrac{2x + 5}{x + 1} + \dfrac{x + 7}{x + 5} - \dfrac{5x + 17}{(x + 1)(x + 5)}$

70. $\dfrac{x + 5}{x + 3} + \dfrac{x + 7}{x + 2} - \dfrac{7x + 19}{(x + 3)(x + 2)}$

71. $\dfrac{1}{x + y} + \dfrac{1}{x - y} - \dfrac{2x}{x^2 - y^2}$

72. $\dfrac{2r}{r^2 - s^2} + \dfrac{1}{r + s} - \dfrac{1}{r - s}$

73. What is the advantage of using the *least* common denominator—rather than just *any* common denominator—when adding or subtracting rational expressions?

74. Describe a procedure that can be used to add any two rational expressions.

Skill Review

To prepare for Section 6.5, review division of fractions and rational expressions (Sections 1.3, 1.7, and 6.2).

Simplify.

75. $-\dfrac{3}{8} \div \dfrac{11}{4}$ [1.7]

76. $-\dfrac{7}{12} \div \left(-\dfrac{3}{4}\right)$ [1.7]

77. $\dfrac{\frac{3}{4}}{\frac{5}{6}}$ [1.3]

78. $\dfrac{\frac{8}{15}}{\frac{9}{10}}$ [1.3]

79. $\dfrac{2x + 6}{x - 1} \div \dfrac{3x + 9}{x - 1}$ [6.2]

80. $\dfrac{x^2 - 9}{x^2 - 4} \div \dfrac{x^2 + 6x + 9}{x^2 + 4x + 4}$ [6.2]

Synthesis

81. How could you convince someone that

$$\dfrac{1}{3 - x} \quad \text{and} \quad \dfrac{1}{x - 3}$$

are opposites of each other?

82. Are parentheses as important for adding rational expressions as they are for subtracting rational expressions? Why or why not?

Write expressions for the perimeter and the area of each rectangle.

83.

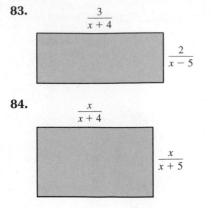

$\dfrac{3}{x + 4}$

$\dfrac{2}{x - 5}$

84.

$\dfrac{x}{x + 4}$

$\dfrac{x}{x + 5}$

Perform the indicated operations.

85. $\dfrac{x^2}{3x^2 - 5x - 2} - \dfrac{2x}{3x + 1} \cdot \dfrac{1}{x - 2}$

86. $\dfrac{2x + 11}{x - 3} \cdot \dfrac{3}{x + 4} + \dfrac{2x + 1}{4 + x} \cdot \dfrac{3}{3 - x}$

Aha! **87.** $\left(\dfrac{x}{x + 7} - \dfrac{3}{x + 2} \right)\left(\dfrac{x}{x + 7} + \dfrac{3}{x + 2} \right)$

88. $\dfrac{1}{ay - 3a + 2xy - 6x} - \dfrac{xy + ay}{a^2 - 4x^2}\left(\dfrac{1}{y - 3} \right)^2$

89. $\left(\dfrac{a}{a - b} + \dfrac{b}{a + b} \right)\left(\dfrac{1}{3a + b} + \dfrac{2a + 6b}{9a^2 - b^2} \right)$

90. $\dfrac{2x^2 + 5x - 3}{2x^2 - 9x + 9} + \dfrac{x + 1}{3 - 2x} + \dfrac{4x^2 + 8x + 3}{x - 3} \cdot \dfrac{x + 3}{9 - 4x^2}$

91. Express

$$\dfrac{a - 3b}{a - b}$$

as a sum of two rational expressions with denominators that are opposites of each other. Answers may vary.

92. Use a graphing calculator to check the answer to Exercise 29.

93. Why does the word ERROR appear in the table displayed on p. 390?

CONNECTING the CONCEPTS

The process of adding and subtracting rational expressions is significantly different from multiplying and dividing them. The first thing you should take note of when combining rational expressions is the operation sign.

Operation	Need Common Denominator?	Procedure	Tips and Cautions
Addition	Yes	Write with a common denominator. Add numerators. Keep denominator.	Do not simplify after writing with the LCD. Instead, simplify after adding the numerators.
Subtraction	Yes	Write with a common denominator. Subtract numerators. Keep denominator.	Use parentheses around the numerator being subtracted. Simplify after subtracting the numerators.
Multiplication	No	Multiply numerators. Multiply denominators.	Do not carry out the multiplications. Instead, factor and try to simplify.
Division	No	Multiply by the reciprocal of the divisor.	Begin by rewriting as a multiplication using the reciprocal of the divisor.

MIXED REVIEW

Tell what operation is being performed. Then perform the operation and, if possible, simplify.

1. $\dfrac{3}{5x} + \dfrac{2}{x^2}$

2. $\dfrac{3}{5x} \cdot \dfrac{2}{x^2}$

3. $\dfrac{3}{5x} \div \dfrac{2}{x^2}$

4. $\dfrac{3}{5x} - \dfrac{2}{x^2}$

5. $\dfrac{2x - 6}{5x + 10} \cdot \dfrac{x + 2}{6x - 12}$

6. $\dfrac{2}{x + 3} \cdot \dfrac{3}{x + 4}$

7. $\dfrac{2}{x - 5} \div \dfrac{6}{x - 5}$

8. $\dfrac{x}{x + 2} - \dfrac{1}{x - 1}$

9. $\dfrac{2}{x + 3} + \dfrac{3}{x + 4}$

10. $\dfrac{5}{2x - 1} + \dfrac{10x}{1 - 2x}$

11. $\dfrac{3}{x - 4} - \dfrac{2}{4 - x}$

12. $\dfrac{(x - 2)(2x + 3)}{(x + 1)(x - 5)} \div \dfrac{(x - 2)(x + 1)}{(x - 5)(x + 3)}$

13. $\dfrac{a}{6a - 9b} - \dfrac{b}{4a - 6b}$

14. $\dfrac{x^2 - 16}{x^2 - x} \cdot \dfrac{x^2}{x^2 - 5x + 4}$

15. $\dfrac{x + 1}{x^2 - 7x + 10} + \dfrac{3}{x^2 - x - 2}$

16. $\dfrac{3u^2 - 3}{4} \div \dfrac{4u + 4}{3}$

17. $\dfrac{t + 2}{10} + \dfrac{2t + 1}{15}$

18. $(t^2 + t - 20) \cdot \dfrac{t + 5}{t - 4}$

19. $\dfrac{a^2 - 2a + 1}{a^2 - 4} \div (a^2 - 3a + 2)$

20. $\dfrac{2x - 7}{x} - \dfrac{3x - 5}{2}$

6.5   Complex Rational Expressions

Using Division to Simplify ● Multiplying by the LCD

A **complex rational expression** is a rational expression that has one or more rational expressions within its numerator or denominator. Here are some examples:

$$\dfrac{1 + \dfrac{2}{x}}{3}, \qquad \dfrac{\dfrac{x + y}{7}}{\dfrac{2x}{x + 1}}, \qquad \dfrac{\dfrac{4}{3} + \dfrac{1}{5}}{\dfrac{2}{x} - \dfrac{x}{y}}.$$

These are rational expressions within the complex rational expression.

When we simplify a complex rational expression, we rewrite it so that it is no longer complex. We will consider two methods for simplifying complex rational expressions. Each method offers certain advantages.

Using Division to Simplify (Method 1)

Our first method for simplifying complex rational expressions involves rewriting
the expression as a quotient of two rational expressions.

> **To Simplify a Complex Rational Expression by Dividing**
> 1. Add or subtract, as needed, to get a single rational expression in the numerator.
> 2. Add or subtract, as needed, to get a single rational expression in the denominator.
> 3. Divide the numerator by the denominator (invert and multiply).
> 4. If possible, simplify by removing a factor equal to 1.

The key here is to express a complex rational expression as one rational
expression divided by another. We can then proceed as in Section 6.2.

EXAMPLE 1   Simplify: $\dfrac{\dfrac{x}{x-3}}{\dfrac{4}{5x-15}}$.

SOLUTION   Here the numerator and the denominator are already single rational
expressions. This allows us to start by dividing (step 3), as in Section 6.2:

$$\frac{\dfrac{x}{x-3}}{\dfrac{4}{5x-15}} = \frac{x}{x-3} \div \frac{4}{5x-15} \qquad \text{Rewriting with a division symbol}$$

$$= \frac{x}{x-3} \cdot \frac{5x-15}{4} \qquad \begin{array}{l}\text{Multiplying by the reciprocal of the} \\ \text{divisor (inverting and multiplying)}\end{array}$$

$$= \frac{x}{\cancel{x-3}} \cdot \frac{5\cancel{(x-3)}}{4} \qquad \begin{array}{l}\text{Factoring and removing a factor} \\ \text{equal to 1: } \dfrac{x-3}{x-3} = 1\end{array}$$

$$= \frac{5x}{4}.$$

TRY EXERCISE ▶ 21

All four steps of the division method are used in Example 2(a).

EXAMPLE 2   Simplify.

a) $\dfrac{\dfrac{5}{2a} + \dfrac{1}{a}}{\dfrac{1}{4a} - \dfrac{5}{6}}$

b) $\dfrac{\dfrac{x^2}{y} - \dfrac{5}{x}}{xz}$

SOLUTION

a) $\dfrac{\dfrac{5}{2a} + \dfrac{1}{a}}{\dfrac{1}{4a} - \dfrac{5}{6}} = \dfrac{\dfrac{5}{2a} + \dfrac{1}{a} \cdot \dfrac{2}{2}}{\dfrac{1}{4a} \cdot \dfrac{3}{3} - \dfrac{5}{6} \cdot \dfrac{2a}{2a}} \left.\begin{array}{l} \\ \\ \end{array}\right\}$

$\leftarrow$ Multiplying by 1 to get the LCD, $2a$, for the numerator of the complex rational expression

$\leftarrow$ Multiplying by 1 to get the LCD, $12a$, for the denominator of the complex rational expression

1. Add to get a single rational expression in the numerator.

2. Subtract to get a single rational expression in the denominator.

3. Divide the numerator by the denominator.

4. Simplify.

$$= \frac{\dfrac{5}{2a} + \dfrac{2}{2a}}{\dfrac{3}{12a} - \dfrac{10a}{12a}} = \frac{\dfrac{7}{2a}}{\dfrac{3-10a}{12a}} \quad \begin{array}{l}\longleftarrow \text{Adding} \\ \nwarrow \text{Subtracting}\end{array}$$

$$= \frac{7}{2a} \div \frac{3-10a}{12a} \qquad \text{Rewriting with a division symbol. This is often done mentally.}$$

$$= \frac{7}{2a} \cdot \frac{12a}{3-10a} \qquad \text{Multiplying by the reciprocal of the divisor (inverting and multiplying)}$$

$$= \frac{7}{2\!\!\!\diagup a} \cdot \frac{2\!\!\!\diagup a \cdot 6}{3-10a} \qquad \text{Removing a factor equal to 1: } \frac{2a}{2a} = 1$$

$$= \frac{42}{3-10a}$$

b)

$$\frac{\dfrac{x^2}{y} - \dfrac{5}{x}}{xz} = \frac{\dfrac{x^2}{y} \cdot \dfrac{x}{x} - \dfrac{5}{x} \cdot \dfrac{y}{y}}{xz} \quad \begin{array}{l}\longleftarrow \text{Multiplying by 1 to get the LCD, } xy, \text{ for the} \\ \text{numerator of the complex rational expression}\end{array}$$

$$= \frac{\dfrac{x^3}{xy} - \dfrac{5y}{xy}}{xz}$$

$$= \frac{\dfrac{x^3 - 5y}{xy}}{xz} \quad \begin{array}{l}\longleftarrow \text{Subtracting} \\ \longleftarrow \text{If you prefer, write } xz \text{ as } \dfrac{xz}{1}.\end{array}$$

$$= \frac{x^3 - 5y}{xy} \div (xz) \qquad \text{Rewriting with a division symbol}$$

$$= \frac{x^3 - 5y}{xy} \cdot \frac{1}{xz} \qquad \text{Multiplying by the reciprocal of the divisor (inverting and multiplying)}$$

$$= \frac{x^3 - 5y}{x^2yz}$$

TRY EXERCISE 11

Multiplying by the LCD (Method 2)

A second method for simplifying complex rational expressions relies on multiplying by a carefully chosen expression that is equal to 1. This multiplication by 1 will result in an expression that is no longer complex.

> **To Simplify a Complex Rational Expression by Multiplying by the LCD**
>
> **1.** Find the LCD of *all* rational expressions within the complex rational expression.
> **2.** Multiply the complex rational expression by an expression equal to 1. Write 1 as the LCD over itself (LCD/LCD).
> **3.** Simplify. No fraction expressions should remain within the complex rational expression.
> **4.** Factor and, if possible, simplify.

EXAMPLE 3 Simplify: $\dfrac{\dfrac{1}{2} + \dfrac{3}{4}}{\dfrac{5}{6} - \dfrac{3}{8}}$.

STUDENT NOTES ─────

Pay careful attention to the difference in the ways in which Method 1 and Method 2 use an LCD. In Method 1, separate LCDs are found in the numerator and in the denominator. In Method 2, *all* rational expressions are considered in order to find one LCD.

SOLUTION

1. The LCD of $\frac{1}{2}, \frac{3}{4}, \frac{5}{6},$ and $\frac{3}{8}$ is 24.

2. We multiply by an expression equal to 1:

$$\frac{\frac{1}{2} + \frac{3}{4}}{\frac{5}{6} - \frac{3}{8}} = \frac{\frac{1}{2} + \frac{3}{4}}{\frac{5}{6} - \frac{3}{8}} \cdot \frac{24}{24}.$$

Multiplying by an expression equal to 1, using the LCD: $\frac{24}{24} = 1$

3. Using the distributive law, we perform the multiplication:

$$\frac{\frac{1}{2} + \frac{3}{4}}{\frac{5}{6} - \frac{3}{8}} \cdot \frac{24}{24} = \frac{\left(\frac{1}{2} + \frac{3}{4}\right)24}{\left(\frac{5}{6} - \frac{3}{8}\right)24}$$

← Multiplying the numerator by 24
Don't forget the parentheses!
← Multiplying the denominator by 24

$$= \frac{\frac{1}{2}(24) + \frac{3}{4}(24)}{\frac{5}{6}(24) - \frac{3}{8}(24)}$$

Using the distributive law

$$= \frac{12 + 18}{20 - 9}, \quad \text{or} \quad \frac{30}{11}.$$

Simplifying

4. The result, $\frac{30}{11}$, cannot be factored or simplified, so we are done.

TRY EXERCISE ▶ 5

Multiplying like this effectively clears fractions in both the top and the bottom of the complex rational expression. Compare the steps in Example 4(a) with those in Example 2(a).

EXAMPLE **4** Simplify.

a) $\dfrac{\dfrac{5}{2a} + \dfrac{1}{a}}{\dfrac{1}{4a} - \dfrac{5}{6}}$

b) $\dfrac{1 - \dfrac{1}{x}}{1 - \dfrac{1}{x^2}}$

SOLUTION

1. Find the LCD.

a) The denominators within the complex expression are $2a$, a, $4a$, and 6, so the LCD is $12a$. We multiply by 1 using $(12a)/(12a)$:

2. Multiply by LCD/LCD.

$$\frac{\frac{5}{2a} + \frac{1}{a}}{\frac{1}{4a} - \frac{5}{6}} = \frac{\frac{5}{2a} + \frac{1}{a}}{\frac{1}{4a} - \frac{5}{6}} \cdot \frac{12a}{12a} = \frac{\frac{5}{2a}(12a) + \frac{1}{a}(12a)}{\frac{1}{4a}(12a) - \frac{5}{6}(12a)}.$$

Using the distributive law

When we multiply by $12a$, all fractions in the numerator and the denominator of the complex rational expression are cleared:

3., 4. Simplify.

$$\frac{\frac{5}{2a}(12a) + \frac{1}{a}(12a)}{\frac{1}{4a}(12a) - \frac{5}{6}(12a)} = \frac{30 + 12}{3 - 10a} = \frac{42}{3 - 10a}.$$

All fractions have been cleared.

Be careful to place parentheses properly when entering complex rational expressions into a graphing calculator. Remember to enclose the entire numerator of the complex rational expression in one set of parentheses and the entire denominator in another. For example, we enter the expression in Example 4(a) as

$$y_1 = (5/(2x) + 1/x)$$
$$/(1/(4x) - 5/6).$$

1. Write Example 4(b) as you would to enter it into a graphing calculator.
2. When must the numerator of a rational expression be enclosed in parentheses? the denominator?

b) $\dfrac{1 - \dfrac{1}{x}}{1 - \dfrac{1}{x^2}} = \dfrac{1 - \dfrac{1}{x}}{1 - \dfrac{1}{x^2}} \cdot \dfrac{x^2}{x^2}$ The LCD is x^2 so we multiply by 1 using x^2/x^2.

$$= \dfrac{1 \cdot x^2 - \dfrac{1}{x} \cdot x^2}{1 \cdot x^2 - \dfrac{1}{x^2} \cdot x^2}$$ Using the distributive law

$$= \dfrac{x^2 - x}{x^2 - 1}$$ All fractions have been cleared within the complex rational expression.

$$= \dfrac{x(x-1)}{(x+1)(x-1)}$$ Factoring and simplifying: $\dfrac{x-1}{x-1} = 1$

$$= \dfrac{x}{x+1}$$

> **TRY EXERCISE** 15

It is important to understand both of the methods studied in this section. Sometimes, as in Example 1, the complex rational expression is either given as—or easily written as—a quotient of two rational expressions. In these cases, Method 1 (using division) is probably the easier method to use. Other times, as in Example 4, it is not difficult to find the LCD of all denominators in the complex rational expression. When this occurs, it is usually easier to use Method 2 (multiplying by the LCD). The more practice you get using both methods, the better you will be at selecting the easier method for any given problem.

6.5 EXERCISE SET

Concept Reinforcement *In each of Exercises 1–4, use the method listed to match the accompanying expression with the expression in the column on the right that arises when that method is used.*

1. $\dfrac{\dfrac{5}{x^2} + \dfrac{1}{x}}{\dfrac{7}{2} - \dfrac{3}{4x}}$;

Multiplying by the LCD (Method 2)

2. $\dfrac{\dfrac{5}{x^2} + \dfrac{1}{x}}{\dfrac{7}{2} - \dfrac{3}{4x}}$;

Using division to simplify (Method 1)

3. $\dfrac{\dfrac{4}{5x} - \dfrac{1}{10}}{\dfrac{8}{x^2} + \dfrac{7}{2}}$;

Using division to simplify (Method 1)

4. $\dfrac{\dfrac{4}{5x} - \dfrac{1}{10}}{\dfrac{8}{x^2} + \dfrac{7}{2}}$;

Multiplying by the LCD (Method 2)

a) $\dfrac{\dfrac{5+x}{x^2}}{\dfrac{14x-3}{4x}}$

b) $\dfrac{\dfrac{8-x}{10x}}{\dfrac{16+7x^2}{2x^2}}$

c) $\dfrac{\dfrac{4}{5x} \cdot 10x^2 - \dfrac{1}{10} \cdot 10x^2}{\dfrac{8}{x^2} \cdot 10x^2 + \dfrac{7}{2} \cdot 10x^2}$

d) $\dfrac{\dfrac{5}{x^2} \cdot 4x^2 + \dfrac{1}{x} \cdot 4x^2}{\dfrac{7}{2} \cdot 4x^2 - \dfrac{3}{4x} \cdot 4x^2}$

Simplify. Use either method or the method specified by your instructor.

5. $\dfrac{1 + \dfrac{1}{4}}{2 + \dfrac{3}{4}}$

6. $\dfrac{3 + \dfrac{1}{4}}{1 + \dfrac{1}{2}}$

7. $\dfrac{\dfrac{1}{2} + \dfrac{1}{3}}{\dfrac{1}{4} - \dfrac{1}{6}}$

8. $\dfrac{\dfrac{2}{5} - \dfrac{1}{10}}{\dfrac{7}{20} - \dfrac{4}{15}}$

9. $\dfrac{\dfrac{x}{4} + x}{\dfrac{4}{x} + x}$

10. $\dfrac{\dfrac{1}{c} + 2}{\dfrac{1}{c} - 5}$

11. $\dfrac{\dfrac{10}{t}}{\dfrac{2}{t^2} - \dfrac{5}{t}}$

12. $\dfrac{\dfrac{5}{x} - \dfrac{2}{x^2}}{\dfrac{2}{x^2}}$

13. $\dfrac{\dfrac{2a - 5}{3a}}{\dfrac{a - 7}{6a}}$

14. $\dfrac{\dfrac{a + 5}{a^2}}{\dfrac{a - 2}{3a}}$

15. $\dfrac{\dfrac{x}{6} - \dfrac{3}{x}}{\dfrac{1}{3} + \dfrac{1}{x}}$

16. $\dfrac{\dfrac{2}{x} + \dfrac{x}{4}}{\dfrac{3}{4} - \dfrac{2}{x}}$

17. $\dfrac{\dfrac{1}{s} - \dfrac{1}{5}}{\dfrac{s - 5}{s}}$

18. $\dfrac{\dfrac{1}{9} - \dfrac{1}{n}}{\dfrac{n + 9}{9}}$

19. $\dfrac{\dfrac{1}{t^2} + 1}{\dfrac{1}{t} - 1}$

20. $\dfrac{2 + \dfrac{1}{x}}{2 - \dfrac{1}{x^2}}$

21. $\dfrac{\dfrac{x^2}{x^2 - y^2}}{\dfrac{x}{x + y}}$

22. $\dfrac{\dfrac{a^2}{a - 2}}{\dfrac{3a}{a^2 - 4}}$

23. $\dfrac{\dfrac{7}{c^2} + \dfrac{4}{c}}{\dfrac{6}{c} - \dfrac{3}{c^3}}$

24. $\dfrac{\dfrac{4}{t^3} - \dfrac{1}{t^2}}{\dfrac{3}{t} + \dfrac{5}{t^2}}$

25. $\dfrac{\dfrac{2}{7a^4} - \dfrac{1}{14a}}{\dfrac{3}{5a^2} + \dfrac{2}{15a}}$

26. $\dfrac{\dfrac{5}{4x^3} - \dfrac{3}{8x}}{\dfrac{3}{2x} + \dfrac{3}{4x^3}}$ Aha! **27.** $\dfrac{\dfrac{x}{5y^3} + \dfrac{3}{10y}}{\dfrac{3}{10y} + \dfrac{x}{5y^3}}$

28. $\dfrac{\dfrac{a}{6b^3} + \dfrac{4}{9b^2}}{\dfrac{5}{6b} - \dfrac{1}{9b^3}}$

29. $\dfrac{\dfrac{3}{ab^4} + \dfrac{4}{a^3b}}{\dfrac{5}{a^3b} - \dfrac{3}{ab}}$

30. $\dfrac{\dfrac{2}{x^2y} + \dfrac{3}{xy^2}}{\dfrac{3}{xy^2} + \dfrac{2}{x^2y}}$

31. $\dfrac{2 - \dfrac{3}{x^2}}{4 + \dfrac{9}{x^4}}$

32. $\dfrac{3 - \dfrac{2}{a^4}}{2 + \dfrac{3}{a^3}}$

33. $\dfrac{t - \dfrac{9}{t}}{t + \dfrac{4}{t}}$

34. $\dfrac{s + \dfrac{2}{s}}{s - \dfrac{3}{s}}$

35. $\dfrac{3 + \dfrac{4}{ab^3}}{\dfrac{3 + a}{a^2b}}$

36. $\dfrac{5 + \dfrac{3}{x^2y}}{\dfrac{3 + x}{x^3y}}$

37. $\dfrac{t + 5 + \dfrac{3}{t}}{t + 2 + \dfrac{1}{t}}$

38. $\dfrac{a + 3 + \dfrac{2}{a}}{a + 2 + \dfrac{5}{a}}$

39. $\dfrac{x - 2 - \dfrac{1}{x}}{x - 5 - \dfrac{4}{x}}$

40. $\dfrac{x - 3 - \dfrac{2}{x}}{x - 4 - \dfrac{3}{x}}$

41. Is it possible to simplify complex rational expressions without knowing how to divide rational expressions? Why or why not?

42. Why is the distributive law important when simplifying complex rational expressions?

Skill Review

To prepare for Section 6.6, review solving linear and quadratic equations (Sections 2.2 and 5.6).

Solve.

43. $3x - 5 + 2(4x - 1) = 12x - 3$ [2.2]

44. $(x - 1)7 - (x + 1)9 = 4(x + 2)$ [2.2]

45. $\dfrac{3}{4}x - \dfrac{5}{8} = \dfrac{3}{8}x + \dfrac{7}{4}$ [2.2]

46. $\dfrac{5}{9} - \dfrac{2x}{3} = \dfrac{5x}{6} + \dfrac{4}{3}$ [2.2]

47. $x^2 - 7x + 12 = 0$ [5.6]

48. $x^2 + 13x - 30 = 0$ [5.6]

Synthesis

49. Which of the two methods presented would you use to simplify Exercise 30? Why?

50. Which of the two methods presented would you use to simplify Exercise 22? Why?

In Exercises 51–54, find all x-values for which the given expression is undefined.

51. $\dfrac{\dfrac{x - 5}{x - 6}}{\dfrac{x - 7}{x - 8}}$

52. $\dfrac{\dfrac{x + 1}{x + 2}}{\dfrac{x + 3}{x + 4}}$

53. $\dfrac{\dfrac{2x + 3}{5x + 4}}{\dfrac{3}{7} - \dfrac{x^2}{21}}$

54. $\dfrac{\dfrac{3x - 5}{2x - 7}}{\dfrac{4x}{5} - \dfrac{8}{15}}$

55. Use multiplication by the LCD (Method 2) to show that

$$\frac{A}{B} \div \frac{C}{D} = \frac{A}{B} \cdot \frac{D}{C}.$$

(*Hint*: Begin by forming a complex rational expression.)

56. The formula

$$\dfrac{P\left(1 + \dfrac{i}{12}\right)^2}{\dfrac{\left(1 + \dfrac{i}{12}\right)^2 - 1}{\dfrac{i}{12}}},$$

where P is a loan amount and i is an interest rate, arises in certain business situations. Simplify this expression. (*Hint*: Expand the binomials.)

Simplify.

57. $\dfrac{\dfrac{x}{x + 5} + \dfrac{3}{x + 2}}{\dfrac{2}{x + 2} - \dfrac{x}{x + 5}}$

58. $\dfrac{\dfrac{5}{x + 2} - \dfrac{3}{x - 2}}{\dfrac{x}{x - 1} + \dfrac{x}{x + 1}}$

Aha! **59.** $\left[\dfrac{\dfrac{x - 1}{x - 1} - 1}{\dfrac{x + 1}{x - 1} + 1}\right]^5$

60. $1 + \dfrac{1}{1 + \dfrac{1}{1 + \dfrac{1}{x}}}$

61. $\dfrac{\dfrac{z}{1 - \dfrac{z}{2 + 2z}} - 2z}{\dfrac{2z}{5z - 2} - 3}$

62. Find the simplified form for the reciprocal of

$$\dfrac{2}{x - 1} - \dfrac{1}{3x - 2}.$$

63. Under what circumstance(s) will there be no restrictions on the variable appearing in a complex rational expression?

64. Use a graphing calculator to check Example 2(a).

| 6.6 | **Solving Rational Equations** |

Solving a New Type of Equation   ■   A Visual Interpretation

Our study of rational expressions allows us to solve a type of equation that we could not have solved prior to this chapter.

Solving a New Type of Equation

A **rational**, or **fraction**, **equation** is an equation containing one or more rational expressions, often with the variable in a denominator. Here are some examples:

$$\dfrac{2}{3} + \dfrac{5}{6} = \dfrac{x}{9}, \qquad t + \dfrac{7}{t} = -5, \qquad \dfrac{x^2}{x - 1} = \dfrac{1}{x - 1}.$$

To Solve a Rational Equation

1. List any restrictions that exist. Numbers that make a denominator equal 0 can never be solutions.
2. Clear the equation of fractions by multiplying both sides by the LCM of the denominators.
3. Solve the resulting equation using the addition principle, the multiplication principle, and the principle of zero products, as needed.
4. Check the possible solution(s) in the original equation.

When clearing an equation of fractions, we use the terminology LCM instead of LCD because we are *not* adding or subtracting rational expressions.

EXAMPLE   1

Solve: $\dfrac{x}{6} - \dfrac{x}{8} = \dfrac{1}{12}$.

SOLUTION   Because no variable appears in a denominator, no restrictions exist. The LCM of 6, 8, and 12 is 24, so we multiply both sides by 24:

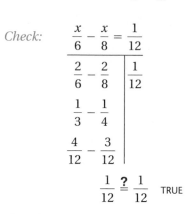

$$24\left(\dfrac{x}{6} - \dfrac{x}{8}\right) = 24 \cdot \dfrac{1}{12}$$

Using the multiplication principle to multiply both sides by the LCM. Parentheses are important!

$$24 \cdot \dfrac{x}{6} - 24 \cdot \dfrac{x}{8} = 24 \cdot \dfrac{1}{12}$$

Using the distributive law

────── Be sure to multiply *each* term by the LCM.

$$\left.\begin{array}{c} \dfrac{24x}{6} - \dfrac{24x}{8} = \dfrac{24}{12} \\[2mm] 4x - 3x = 2 \end{array}\right\}$$

Simplifying. Note that all fractions have been cleared. If fractions remain, we have either made a mistake or have not used the LCM of the denominators.

$$x = 2.$$

Check:

$$\begin{array}{c|c} \dfrac{x}{6} - \dfrac{x}{8} & \dfrac{1}{12} \\[2mm] \dfrac{2}{6} - \dfrac{2}{8} & \dfrac{1}{12} \\[2mm] \dfrac{1}{3} - \dfrac{1}{4} & \\[2mm] \dfrac{4}{12} - \dfrac{3}{12} & \end{array}$$

$$\dfrac{1}{12} \overset{?}{=} \dfrac{1}{12} \quad \text{TRUE}$$

This checks, so the solution is 2.

 **TRY EXERCISE** 5

Recall that the multiplication principle states that $a = b$ is equivalent to $a \cdot c = b \cdot c$, *provided c is not zero.* Because rational equations often have variables in a denominator, clearing fractions will now require us to multiply both sides by a variable expression. Since a variable expression could represent 0, *multiplying both sides of an equation by a variable expression does not always produce an equivalent equation.* Thus checking each solution in the original equation is essential.

EXAMPLE   2

Solve.

a) $\dfrac{2}{3x} + \dfrac{1}{x} = 10$

b) $x + \dfrac{6}{x} = -5$

c) $1 + \dfrac{3x}{x + 2} = \dfrac{-6}{x + 2}$

d) $\dfrac{3}{x - 5} + \dfrac{1}{x + 5} = \dfrac{2}{x^2 - 25}$

e) $\dfrac{x^2}{x - 1} = \dfrac{1}{x - 1}$

SOLUTION

List restrictions.

a) If $x = 0$, both denominators are 0. We list this restriction:

$$x \neq 0.$$

We now clear the equation of fractions and solve:

$$\frac{2}{3x} + \frac{1}{x} = 10 \qquad \text{We cannot have } x = 0. \text{ The LCM is } 3x.$$

Clear fractions.

$$3x\left(\frac{2}{3x} + \frac{1}{x}\right) = 3x \cdot 10 \qquad \begin{array}{l}\text{Using the multiplication principle to} \\ \text{multiply both sides by the LCM.} \\ \textit{Don't forget the parentheses!}\end{array}$$

$$3x \cdot \frac{2}{3x} + 3x \cdot \frac{1}{x} = 3x \cdot 10 \qquad \text{Using the distributive law}$$

$$2 + 3 = 30x \qquad \begin{array}{l}\text{Removing factors equal to 1:} \\ (3x)/(3x) = 1 \text{ and } x/x = 1. \text{ This} \\ \text{clears all fractions.}\end{array}$$

Solve.

$$5 = 30x$$

$$\frac{5}{30} = x, \quad \text{so } x = \frac{1}{6}. \qquad \begin{array}{l}\text{Dividing both sides by 30, or} \\ \text{multiplying both sides by 1/30}\end{array}$$

Since $\frac{1}{6} \neq 0$, and 0 is the only restricted value, $\frac{1}{6}$ *should* check.

Check.

Check:

$$\frac{\dfrac{2}{3x} + \dfrac{1}{x} = 10}{\begin{array}{c|c} \dfrac{2}{3 \cdot \frac{1}{6}} + \dfrac{1}{\frac{1}{6}} & 10 \\[2ex] \dfrac{2}{\frac{1}{2}} + \dfrac{1}{\frac{1}{6}} & \\[2ex] 2 \cdot \dfrac{2}{1} + 1 \cdot \dfrac{6}{1} & \\[1ex] 4 + 6 & \\[1ex] 10 \overset{?}{=} 10 & \text{TRUE} \end{array}}$$

The solution is $\frac{1}{6}$.

b) Again, note that if $x = 0$, the expression $6/x$ is undefined. We list the restriction:

$$x \neq 0.$$

We now clear the equation of fractions and solve:

$$x + \frac{6}{x} = -5 \qquad \text{We cannot have } x = 0. \text{ The LCM is } x.$$

$$x\left(x + \frac{6}{x}\right) = x(-5) \qquad \begin{array}{l}\text{Multiplying both sides by the LCM.} \\ \textit{Don't forget the parentheses!}\end{array}$$

$$x \cdot x + x \cdot \frac{6}{x} = -5x \qquad \text{Using the distributive law}$$

$$x^2 + 6 = -5x \qquad \begin{array}{l}\text{Removing a factor equal to 1: } x/x = 1. \\ \text{We are left with a quadratic equation.}\end{array}$$

$$x^2 + 5x + 6 = 0 \qquad \begin{array}{l}\text{Using the addition principle to add } 5x \text{ to} \\ \text{both sides}\end{array}$$

$$(x + 3)(x + 2) = 0 \qquad \text{Factoring}$$

$$x + 3 = 0 \quad or \quad x + 2 = 0 \qquad \text{Using the principle of zero products}$$

$$x = -3 \quad or \qquad x = -2. \qquad \begin{array}{l}\text{The only restricted value is 0, so} \\ \text{both answers should check.}\end{array}$$

Check: For -3:   For -2:

$$x + \frac{6}{x} = -5$$

$-3 + \dfrac{6}{-3}$	-5
$-3 - 2$	
$-5 \overset{?}{=} -5$ TRUE	

$$x + \frac{6}{x} = -5$$

$-2 + \dfrac{6}{-2}$	-5
$-2 - 3$	
$-5 \overset{?}{=} -5$ TRUE	

Both of these check, so there are two solutions, -3 and -2.

STUDENT NOTES ————

Not all checking is for finding errors in computation. For these equations, the solution process itself can introduce numbers that do not check.

c) The only denominator is $x + 2$. We set this equal to 0 and solve:

$$x + 2 = 0$$
$$x = -2.$$

If $x = -2$, the rational expressions are undefined. We list the restriction:

$$x \neq -2.$$

We clear fractions and solve:

$$1 + \frac{3x}{x + 2} = \frac{-6}{x + 2}$$ We cannot have $x = -2$. The LCM is $x + 2$.

$$(x + 2)\left(1 + \frac{3x}{x + 2}\right) = (x + 2)\frac{-6}{x + 2}$$ Multiplying both sides by the LCM. *Don't forget the parentheses!*

$$(x + 2) \cdot 1 + (x + 2)\frac{3x}{x + 2} = (x + 2)\frac{-6}{x + 2}$$ Using the distributive law; removing a factor equal to 1: $(x + 2)/(x + 2) = 1$

$$x + 2 + 3x = -6$$
$$4x + 2 = -6$$
$$4x = -8$$
$$x = -2.$$ Above, we stated that $x \neq -2$.

Because of the above restriction, -2 must be rejected as a solution. The check below simply confirms this.

Check:

$$1 + \frac{3x}{x + 2} = \frac{-6}{x + 2}$$

$1 + \dfrac{3(-2)}{-2 + 2}$	$\dfrac{-6}{-2 + 2}$
$1 + \dfrac{-6}{0} \overset{?}{=} \dfrac{-6}{0}$	FALSE

The equation has no solution.

d) The denominators are $x - 5$, $x + 5$, and $x^2 - 25$. Setting them equal to 0 and solving, we find that the rational expressions are undefined when $x = 5$ or $x = -5$. We list the restrictions:

$$x \neq 5, \quad x \neq -5.$$

We clear fractions and solve:

$$\frac{3}{x-5} + \frac{1}{x+5} = \frac{2}{x^2-25}$$

We cannot have $x = 5$ or $x = -5$. The LCM is $(x-5)(x+5)$.

$$(x-5)(x+5)\left(\frac{3}{x-5} + \frac{1}{x+5}\right) = (x-5)(x+5)\frac{2}{(x-5)(x+5)}$$

$$\frac{\cancel{(x-5)}(x+5)3}{\cancel{x-5}} + \frac{(x-5)\cancel{(x+5)}}{\cancel{x+5}} = \frac{2\cancel{(x-5)}\cancel{(x+5)}}{\cancel{(x-5)}\cancel{(x+5)}}$$

Using the distributive law

$$(x+5)3 + (x-5) = 2$$

Removing factors equal to 1: $(x-5)/(x-5) = 1$, $(x+5)/(x+5) = 1$, and $\dfrac{(x-5)(x+5)}{(x-5)(x+5)} = 1$

$$3x + 15 + x - 5 = 2$$ Using the distributive law

$$4x + 10 = 2$$

$$4x = -8$$

$$x = -2.$$ $-2 \neq 5$ and $-2 \neq -5$, so -2 *should* check.

The student can check to confirm that -2 is the solution.

e) If $x = 1$, the denominators are 0. We list the restriction:

$x \neq 1$.

We clear fractions and solve:

$$\frac{x^2}{x-1} = \frac{1}{x-1}$$ We cannot have $x = 1$.

$$\cancel{(x-1)} \cdot \frac{x^2}{\cancel{x-1}} = \cancel{(x-1)} \cdot \frac{1}{\cancel{x-1}}$$ Multiplying both sides by $x - 1$, the LCM

$$x^2 = 1$$ Removing a factor equal to 1: $(x-1)/(x-1) = 1$

$$x^2 - 1 = 0$$ Subtracting 1 from both sides

$$(x-1)(x+1) = 0$$ Factoring

$$x - 1 = 0 \quad or \quad x + 1 = 0$$ Using the principle of zero products

$$x = 1 \quad or \quad x = -1.$$ Above, we stated that $x \neq 1$.

Because of the above restriction, 1 must be rejected as a solution. The student should check in the original equation that -1 *does* check. The solution is -1.

TRY EXERCISES ▸ 21 and 29

A Visual Interpretation

ALGEBRAIC–GRAPHICAL CONNECTION

We can obtain a visual check of the solutions of a rational equation by graphing. For example, consider the equation

$$\frac{x}{4} + \frac{x}{2} = 6.$$

We can examine the solution by graphing the equations

$$y = \frac{x}{4} + \frac{x}{2} \quad \text{and} \quad y = 6$$

using the same set of axes, as shown below.

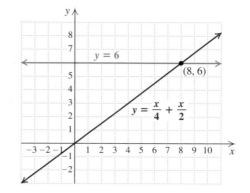

The y-values for each equation will be the same where the graphs intersect. The x-value of that point will yield that value, so it will be a solution of the equation. It appears from the graph that when $x = 8$, the value of $x/4 + x/2$ is 6. We can check by substitution:

$$\frac{x}{4} + \frac{x}{2} = \frac{8}{4} + \frac{8}{2} = 2 + 4 = 6.$$

Thus the solution is 8.

TECHNOLOGY CONNECTION

We can use a table to check possible solutions of rational equations. Consider the equation in Example 2(e),

$$\frac{x^2}{x-1} = \frac{1}{x-1},$$

and the possible solutions that were found, 1 and -1. To check these solutions, we enter $y_1 = x^2/(x-1)$ and $y_2 = 1/(x-1)$. After setting Indpnt to Ask and Depend to Auto in the TBLSET menu, we display the table and enter $x = 1$. The ERROR messages indicate that 1 is not a solution because it is not an allowable replacement for x in the equation. Next, we enter $x = -1$. Since y_1 and y_2 have the same value, we know that the equation is true when $x = -1$, and thus -1 is a solution.

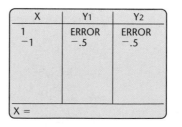

Use a graphing calculator to check the possible solutions of Example 1 and Example 2, parts (a)–(d).

6.6 EXERCISE SET

For Extra Help · MyMathLab · Math XL PRACTICE · WATCH · DOWNLOAD

↪ **Concept Reinforcement** *Classify each statement as either true or false.*

1. Every rational equation has at least one solution.

2. It is possible for a rational equation to have more than one solution.

3. When both sides of an equation are multiplied by a variable expression, the result is not always an equivalent equation.

4. All the equation-solving principles studied thus far may be needed when solving a rational equation.

Solve. If no solution exists, state this.

5. $\dfrac{3}{5} - \dfrac{2}{3} = \dfrac{x}{6}$

6. $\dfrac{5}{8} - \dfrac{3}{5} = \dfrac{x}{10}$

7. $\dfrac{1}{3} + \dfrac{5}{6} = \dfrac{1}{x}$

8. $\dfrac{3}{5} + \dfrac{1}{8} = \dfrac{1}{x}$

9. $\dfrac{1}{8} + \dfrac{1}{12} = \dfrac{1}{t}$

10. $\dfrac{1}{6} + \dfrac{1}{10} = \dfrac{1}{t}$

11. $y + \dfrac{4}{y} = -5$

12. $n + \dfrac{3}{n} = -4$

13. $\dfrac{x}{6} - \dfrac{6}{x} = 0$

14. $\dfrac{x}{7} - \dfrac{7}{x} = 0$

15. $\dfrac{2}{x} = \dfrac{5}{x} - \dfrac{1}{4}$

16. $\dfrac{3}{t} = \dfrac{4}{t} - \dfrac{1}{5}$

17. $\dfrac{5}{3t} + \dfrac{3}{t} = 1$

18. $\dfrac{3}{4x} + \dfrac{5}{x} = 1$

19. $\dfrac{n+2}{n-6} = \dfrac{1}{2}$

20. $\dfrac{a-4}{a+6} = \dfrac{1}{3}$

21. $x + \dfrac{12}{x} = -7$

22. $x + \dfrac{8}{x} = -9$

23. $\dfrac{3}{x-4} = \dfrac{5}{x+1}$

24. $\dfrac{1}{x+3} = \dfrac{4}{x-1}$

25. $\dfrac{a}{6} - \dfrac{a}{10} = \dfrac{1}{6}$

26. $\dfrac{t}{8} - \dfrac{t}{12} = \dfrac{1}{8}$

27. $\dfrac{x+1}{3} - 1 = \dfrac{x-1}{2}$

28. $\dfrac{x+2}{5} - 1 = \dfrac{x-2}{4}$

29. $\dfrac{y+3}{y-3} = \dfrac{6}{y-3}$

30. $\dfrac{3}{a+7} = \dfrac{a+10}{a+7}$

31. $\dfrac{3}{x+4} = \dfrac{5}{x}$

32. $\dfrac{2}{x+3} = \dfrac{7}{x}$

33. $\dfrac{n+1}{n+2} = \dfrac{n-3}{n+1}$

34. $\dfrac{n+2}{n-3} = \dfrac{n+1}{n-2}$

35. $\dfrac{5}{t-2} + \dfrac{3t}{t-2} = \dfrac{4}{t^2-4t+4}$

36. $\dfrac{4}{t-3} + \dfrac{2t}{t-3} = \dfrac{12}{t^2-6t+9}$

37. $\dfrac{x}{x+5} - \dfrac{5}{x-5} = \dfrac{14}{x^2-25}$

38. $\dfrac{5}{x+1} + \dfrac{2x}{x^2-1} = \dfrac{1}{x+1}$

39. $\dfrac{5}{t-3} - \dfrac{30}{t^2-9} = 1$

40. $\dfrac{1}{y+3} + \dfrac{1}{y-3} = \dfrac{1}{y^2-9}$

41. $\dfrac{7}{6-a} = \dfrac{a+1}{a-6}$

42. $\dfrac{t-12}{t-10} = \dfrac{1}{10-t}$

Aha! 43. $\dfrac{-2}{x+2} = \dfrac{x}{x+2}$

44. $\dfrac{3}{2x-6} = \dfrac{x}{2x-6}$

45. $\dfrac{12}{x} = \dfrac{x}{3}$

46. $\dfrac{x}{2} = \dfrac{18}{x}$

47. When solving rational equations, why do we multiply each side by the LCM of the denominators?

48. Explain the difference between adding rational expressions and solving rational equations.

Skill Review

To prepare for Section 6.7, review solving applications and rates of change (Sections 2.5, 3.4, and 5.7).

49. The sum of two consecutive odd numbers is 276. Find the numbers. [2.5]

50. The length of a rectangular picture window is 3 yd greater than the width. The area of the rectangle is 10 yd². Find the perimeter. [5.7]

51. The height of a triangle is 3 cm longer than its base. If the area of the triangle is 54 cm², find the measurements of the base and the height. [5.7]

52. The product of two consecutive even integers is 48. Find the numbers. [5.7]

53. *Human physiology.* Between June 9 and June 24, Seth's beard grew 0.9 cm. Find the rate at which Seth's beard grows. [3.4]

54. *Gardening.* Between July 7 and July 12, Carla's string beans grew 1.4 in. Find the growth rate of the string beans. [3.4]

Synthesis

55. Describe a method that can be used to create rational equations that have no solution.

56. How can a graph be used to determine how many solutions an equation has?

Solve.

57. $1 + \dfrac{x - 1}{x - 3} = \dfrac{2}{x - 3} - x$

58. $\dfrac{4}{y - 2} + \dfrac{3}{y^2 - 4} = \dfrac{5}{y + 2} + \dfrac{2y}{y^2 - 4}$

59. $\dfrac{12 - 6x}{x^2 - 4} = \dfrac{3x}{x + 2} - \dfrac{3 - 2x}{2 - x}$

60. $\dfrac{x}{x^2 + 3x - 4} + \dfrac{x + 1}{x^2 + 6x + 8} = \dfrac{2x}{x^2 + x - 2}$

61. $7 - \dfrac{a - 2}{a + 3} = \dfrac{a^2 - 4}{a + 3} + 5$

62. $\dfrac{x^2}{x^2 - 4} = \dfrac{x}{x + 2} - \dfrac{2x}{2 - x}$

63. $\dfrac{1}{x - 1} + x - 5 = \dfrac{5x - 4}{x - 1} - 6$

64. $\dfrac{5 - 3a}{a^2 + 4a + 3} - \dfrac{2a + 2}{a + 3} = \dfrac{3 - a}{a + 1}$

65. $\dfrac{\dfrac{1}{x} + 1}{x} = \dfrac{\dfrac{1}{x}}{2}$

66. $\dfrac{\dfrac{1}{3}}{x} = \dfrac{1 - \dfrac{1}{x}}{x}$

67. Use a graphing calculator to check your answers to Exercises 13, 21, 31, and 57.

CONNECTING the CONCEPTS

An equation contains an equals sign; an expression does not. Be careful not to confuse simplifying an expression with solving an equation. When expressions are simplified, the result is an equivalent expression. When equations are solved, the result is a solution. Compare the following.

Simplify: $\dfrac{x - 1}{6x} + \dfrac{4}{9}$.

SOLUTION

$\dfrac{x - 1}{6x} + \dfrac{4}{9} = \dfrac{x - 1}{6x} \cdot \dfrac{3}{3} + \dfrac{4}{9} \cdot \dfrac{2x}{2x}$

$\qquad$ The equals signs indicate that all the expressions are equivalent.

$\qquad = \dfrac{3x - 3}{18x} + \dfrac{8x}{18x}$ Writing with the LCD, $18x$

$\qquad = \dfrac{11x - 3}{18x}$ The result is an expression equivalent to $\dfrac{x - 1}{6x} + \dfrac{4}{9}$.

Solve: $\dfrac{x - 1}{6x} = \dfrac{4}{9}$.

SOLUTION

$\qquad \dfrac{x - 1}{6x} = \dfrac{4}{9}$ Each line is an equivalent equation.

$\qquad 18x \cdot \dfrac{x - 1}{6x} = 18x \cdot \dfrac{4}{9}$ Multiplying by the LCM, $18x$

$\qquad 3 \cdot \cancel{6x} \cdot \dfrac{x - 1}{\cancel{6x}} = 2 \cdot \cancel{9} \cdot x \cdot \dfrac{4}{\cancel{9}}$

$\qquad 3(x - 1) = 2x \cdot 4$

$\qquad 3x - 3 = 8x$

$\qquad -3 = 5x$

$\qquad -\dfrac{3}{5} = x$ The result is a solution; $-\dfrac{3}{5}$ is the solution of $\dfrac{x - 1}{6x} = \dfrac{4}{9}$.

MIXED REVIEW

Tell whether each of the following is an expression or an equation. Then simplify the expression or solve the equation.

1. Simplify: $\dfrac{4x^2 - 8x}{4x^2 + 4x}$.

2. Add and, if possible, simplify: $\dfrac{2}{5n} + \dfrac{3}{2n - 1}$.

3. Solve: $\dfrac{3}{y} - \dfrac{1}{4} = \dfrac{1}{y}$.

4. Simplify: $\dfrac{\dfrac{1}{z} + 1}{\dfrac{1}{z^2 - 1}}$.

5. Solve: $\dfrac{5}{x + 3} = \dfrac{3}{x + 2}$.

6. Multiply and, if possible, simplify:

$$\frac{8t + 8}{2t^2 + t - 1} \cdot \frac{t^2 - 1}{t^2 - 2t + 1}.$$

7. Subtract and, if possible, simplify:

$$\frac{2a}{a + 1} - \frac{4a}{1 - a^2}.$$

8. Solve: $\dfrac{15}{x} - \dfrac{15}{x + 2} = 2$.

9. Divide and, if possible, simplify:

$$\frac{18x^2}{25} \div \frac{12x}{5}.$$

10. Solve: $\dfrac{20}{x} = \dfrac{x}{5}$.

6.7 Applications Using Rational Equations and Proportions

Problems Involving Work ● Problems Involving Motion ● Problems Involving Proportions

In many areas of study, applications involving rates, proportions, or reciprocals translate to rational equations. By using the five steps for problem solving and the lessons of Section 6.6, we can now solve such problems.

Problems Involving Work

EXAMPLE 1

Sorting recyclables. Cecilia and Aaron work as volunteers at a town's recycling depot. Cecilia can sort a day's accumulation of recyclables in 4 hr, while Aaron requires 6 hr to do the same job. How long would it take them, working together, to sort the recyclables?

SOLUTION

1. **Familiarize.** We familiarize ourselves with the problem by exploring two common, but *incorrect*, approaches.

   a) One common incorrect approach is to simply add the two times:

   $$4 \,\text{hr} + 6 \,\text{hr} = 10 \,\text{hr}.$$

   Let's think about this. If Cecilia can do the sorting *alone* in 4 hr, then Cecilia and Aaron *together* should take *less* than 4 hr. Thus we reject 10 hr as a solution and reason that the answer must be less than 4 hr.

b) Another incorrect approach is to assume that Cecilia does half the sorting and Aaron does the other half. Then

Cecilia sorts $\frac{1}{2}$ of the accumulation in $\frac{1}{2}$(4 hr), or 2 hr, and

Aaron sorts $\frac{1}{2}$ of the accumulation in $\frac{1}{2}$(6 hr), or 3 hr.

However, since Cecilia would finish 1 hr earlier than Aaron, she would help Aaron after completing her half. She would sort more than half and Aaron would sort less than half of the accumulation. This tells us that the entire job will take them between 2 hr and 3 hr.

A correct approach is to consider how much of the sorting is finished in 1 hr, 2 hr, 3 hr, and so on. It takes Cecilia 4 hr to sort the recyclables alone, so her rate is $\frac{1}{4}$ of the job per hour. It takes Aaron 6 hr to do the sorting alone, so his rate is $\frac{1}{6}$ of the job per hour. Working together, they can complete

$$\frac{1}{4} + \frac{1}{6}, \quad \text{or } \frac{5}{12} \text{ of the sorting in 1 hr.} \qquad \text{We cannot add their } \textit{times,} \text{ but we can add their } \textit{rates.}$$

In 2 hr, Cecilia can do $\frac{1}{4} \cdot 2$ of the sorting and Aaron can do $\frac{1}{6} \cdot 2$ of the sorting. Working together, they can complete

$$\frac{1}{4} \cdot 2 + \frac{1}{6} \cdot 2, \quad \text{or } \frac{5}{6} \text{ of the sorting in 2 hr.} \qquad \text{Note that } \tfrac{5}{12} \cdot 2 = \tfrac{5}{6}.$$

Continuing this reasoning, we can form a table.

Time	Fraction of the Sorting Completed		
	Cecilia	**Aaron**	**Together**
1 hr	$\frac{1}{4}$	$\frac{1}{6}$	$\frac{1}{4} + \frac{1}{6}$, or $\frac{5}{12}$
2 hr	$\frac{1}{4} \cdot 2$	$\frac{1}{6} \cdot 2$	$\left(\frac{1}{4} + \frac{1}{6}\right)2$, or $\frac{5}{12} \cdot 2$, or $\frac{5}{6}$ ⟵ This is too little.
3 hr	$\frac{1}{4} \cdot 3$	$\frac{1}{6} \cdot 3$	$\left(\frac{1}{4} + \frac{1}{6}\right)3$, or $\frac{5}{12} \cdot 3$, or $1\frac{1}{4}$ ⟵ This is too much.
t hr	$\frac{1}{4} \cdot t$	$\frac{1}{6} \cdot t$	$\left(\frac{1}{4} + \frac{1}{6}\right)t$, or $\frac{5}{12} \cdot t$

From the table, we see that if they work 3 hr, the fraction of the sorting that they complete is $1\frac{1}{4}$, which is more of the job than needs to be done. We need to find a number t for which the fraction of the sorting that is completed in t hours is exactly 1, no more and no less.

2. Translate. From the table, we see that the time we want is some number t for which

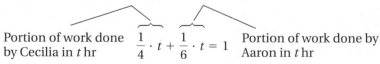

Portion of work done by Cecilia in t hr $\quad \dfrac{1}{4} \cdot t + \dfrac{1}{6} \cdot t = 1 \quad$ Portion of work done by Aaron in t hr

or

$$\underbrace{\left(\frac{1}{4} + \frac{1}{6}\right)t = 1 \quad \text{or} \quad \frac{5}{12} \cdot t = 1.}_{\text{Portion of work done together in } t \text{ hr}}$$

3. **Carry out.** We can choose any one of the above equations to solve:

$$\frac{5}{12} \cdot t = 1$$

$$\frac{12}{5} \cdot \frac{5}{12} \cdot t = \frac{12}{5} \cdot 1 \qquad \text{Multiplying both sides by } \frac{12}{5}$$

$$t = \frac{12}{5}, \quad \text{or} \quad 2\frac{2}{5} \, \text{hr.}$$

4. **Check.** The check can be done following the pattern used in the table of the *Familiarize* step above:

$$\frac{1}{4} \cdot \frac{12}{5} + \frac{1}{6} \cdot \frac{12}{5} = \frac{3}{5} + \frac{2}{5} = \frac{5}{5} = 1.$$

A second, partial, check is that (as we predicted in step 1) the answer is between 2 hr and 3 hr.

5. **State.** Together, it takes Cecilia and Aaron $2\frac{2}{5}$ hr to complete the sorting.

> TRY EXERCISE ▶ 1

The Work Principle

Suppose that A requires a units of time to complete a task and B requires b units of time to complete the same task. Then

A works at a rate of $\dfrac{1}{a}$ tasks per unit of time,

B works at a rate of $\dfrac{1}{b}$ tasks per unit of time, and

A and B together work at a rate of $\dfrac{1}{a} + \dfrac{1}{b}$ tasks per unit of time.

If A and B, working together, require t units of time to complete the task, then their rate is $1/t$ and the following equations hold:

$$\frac{1}{a} \cdot t + \frac{1}{b} \cdot t = 1; \quad \left(\frac{1}{a} + \frac{1}{b}\right)t = 1; \quad \frac{t}{a} + \frac{t}{b} = 1; \quad \frac{1}{a} + \frac{1}{b} = \frac{1}{t}.$$

Problems Involving Motion

Problems that deal with distance, speed (or rate), and time are called **motion problems**. Translation of these problems involves the distance formula, $d = r \cdot t$, and/or the equivalent formulas $r = d/t$ and $t = d/r$.

EXAMPLE 2

Flight speed.   Because of a tailwind, a Delta Boeing 777 is able to fly 200 mph faster than a Northwest 777 that is flying *into* the wind. In the same time that it takes the Delta plane to travel 1800 mi, the Northwest plane flies 1200 mi. How fast is each plane traveling?

SOLUTION

1. **Familiarize.** Suppose that the Northwest plane is traveling 300 mph. Then the Delta plane would be flying $300 + 200$, or 500 mph. Thus if r is the speed of the Northwest plane, in miles per hour, then the Delta plane is traveling $(r + 200)$ mph.

   At 300 mph, the Northwest flight would travel 1200 mi in 1200/300, or 4 hr. At 500 mph, the Delta plane would travel 1800 mi in 1800/500, or $3\frac{3}{5}$ hr. Since both planes spend the same amount of time traveling, and since 4 hr $\neq 3\frac{3}{5}$ hr, we see that our guess of 300 mph is incorrect. Rather than check another guess, we form a table. The columns in the table come from the distance formula: distance, rate, and time. There is one row in the table for each plane.

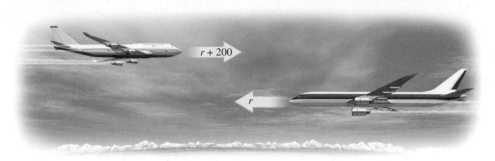

$$Distance \;=\; Rate \;\cdot\; Time$$

	Distance (in miles)	Speed (in miles per hour)	Time (in hours)
Northwest	1200	r	
Delta	1800	$r + 200$	

We do not know the times, but we know that they must be the same.

2. **Translate.** We checked our guess by comparing the two flight times. The times were found by dividing the distances, 1200 mi and 1800 mi, by the rates, 300 mph and 500 mph, respectively. Thus the blanks in the table above can be filled, using *time = distance/rate*. This yields a table that uses only one variable.

	Distance (in miles)	Speed (in miles per hour)	Time (in hours)
Northwest	1200	r	$1200/r$
Delta	1800	$r + 200$	$1800/(r + 200)$

Every variable expression in the table uses the variable r.

The fact that the times must be the same for both planes gives us the translation:

Rewording: Northwest time = Delta time

Translating: $\dfrac{1200}{r} = \dfrac{1800}{r + 200}.$

Note that $\dfrac{mi}{mph} = \dfrac{mi}{mi/hr} = \cancel{mi} \cdot \dfrac{hr}{\cancel{mi}} = hr,$ so we are indeed comparing two times.

3. **Carry out.** To solve the equation, we first multiply both sides by the LCM of the denominators, $r(r + 200)$:

$$r(r + 200) \cdot \frac{1200}{r} = r(r + 200) \cdot \frac{1800}{r + 200}$$

Multiplying both sides by the LCM. Note that we must have $r \neq 0$ and $r \neq -200$.

$$(r + 200)1200 = 1800r \qquad \text{Simplifying}$$

$$1200r + 240{,}000 = 1800r \qquad \text{Using the distributive law}$$

$$240{,}000 = 600r \qquad \text{Subtracting } 1200r \text{ from both sides}$$

$$400 = r. \qquad \text{Dividing both sides by 600}$$

We now have a possible solution. The speed of the Northwest plane is 400 mph, and the speed of the Delta plane is 400 + 200, or 600 mph.

4. **Check.** We first reread the problem to confirm that we were to find the speeds. Note that, at 600 mph, the Delta flight would indeed be traveling 200 mph faster than the Northwest flight. At 600 mph, the Delta plane would cover 1800 mi in 1800/600, or 3 hr. If the Northwest plane flies 1200 mi at 400 mph, it is flying for 1200/400, or 3 hr. Since the times are the same, the speeds check.

5. **State.** The Northwest 777 is flying 400 mph and the Delta 777 is flying 600 mph.

TRY EXERCISE ▶ 11

Problems Involving Proportions

A **ratio** of two quantities is their quotient. For example, 37% is the ratio of 37 to 100, or $\frac{37}{100}$. A **proportion** is an equation stating that two ratios are equal.

> ### Proportion
> An equality of ratios,
> $$\frac{A}{B} = \frac{C}{D},$$
> is called a *proportion*. The numbers within a proportion are said to be *proportional* to each other.

Proportions arise in geometry when we are studying *similar triangles.* If two triangles are **similar**, then their corresponding angles have the same measure and their corresponding sides are proportional. To illustrate, if triangle *ABC* is similar to triangle *RST*, then angles *A* and *R* have the same measure, angles *B* and *S* have the same measure, angles *C* and *T* have the same measure, and

$$\frac{a}{r} = \frac{b}{s} = \frac{c}{t}.$$

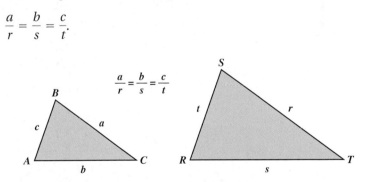

EXAMPLE 3

Similar triangles. Triangles *ABC* and *XYZ* are similar. Solve for *z* if *x* = 10, *a* = 8, and *c* = 5.

SOLUTION We make a drawing, write a proportion, and then solve. Note that side *a* is always opposite angle *A*, side *x* is always opposite angle *X*, and so on.

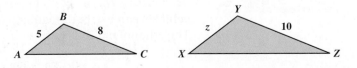

We have

$$\frac{z}{5} = \frac{10}{8}$$ The proportions $\frac{5}{z} = \frac{8}{10}$, $\frac{5}{8} = \frac{z}{10}$, or

$\frac{8}{5} = \frac{10}{z}$ could also be used.

$$40 \cdot \frac{z}{5} = 40 \cdot \frac{10}{8}$$ Multiplying both sides by the LCM, 40

$$8z = 50$$ Simplifying

$$z = \frac{50}{8}, \text{ or } 6.25.$$

TRY EXERCISE 19

EXAMPLE 4

Architecture. A *blueprint* is a scale drawing of a building representing an architect's plans. Ellia is adding 12 ft to the length of an apartment and needs to indicate the addition on an existing blueprint. If a 10-ft long bedroom is represented by $2\frac{1}{2}$ in. on the blueprint, how much longer should Ellia make the drawing in order to represent the addition?

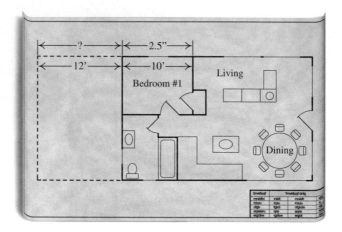

SOLUTION   We let w represent the width, in inches, of the addition that Ellia is drawing. Because the drawing must be to scale, we have

Inches on drawing $\rightarrow \dfrac{w}{12} = \dfrac{2.5}{10}$ $\leftarrow$ Inches on drawing
Feet in real life $\rightarrow$ $\qquad\qquad \leftarrow$ Feet in real life

To solve for w, we multiply both sides by the LCM of the denominators, 60:

$$60 \cdot \frac{w}{12} = 60 \cdot \frac{2.5}{10}$$

$$5w = 6 \cdot 2.5$$ Simplifying

$$w = \frac{15}{5}, \text{ or } 3.$$

Ellia should make the blueprint 3 in. longer.

TRY EXERCISE 23

Proportions can be used to solve a variety of applied problems.

EXAMPLE 5

Alternative fuels. As one alternative to gasoline-powered vehicles, flex fuel vehicles can use both regular gasoline and E85, a fuel containing 85% ethanol. The ethanol in E85 is derived from corn and other plants, a renewable resource, and burning E85 results in less pollution than gasoline. Because using E85 results in a lower fuel economy, its price per gallon must be lower than the price of gasoline in order to make it an economical fuel option. A 2007 Chevrolet Tahoe gets 21 mpg on the highway using gasoline and only 15 mpg on the highway using E85. If the price of gasoline is $3.36 a gallon, what must the price of E85 be in order for the fuel cost per mile to be the same?

Source: www.caranddriver.com

Alternative fuels

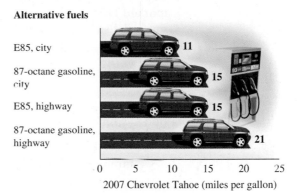

E85, city   11

87-octane gasoline, city   15

E85, highway   15

87-octane gasoline, highway   21

0   5   10   15   20   25

2007 Chevrolet Tahoe (miles per gallon)

Source: www.caranddriver.com

SOLUTION

1. **Familiarize.** We let x = the price per gallon of E85, and organize the given information:

   Gasoline miles per gallon: 21,
   E85 miles per gallon: 15,
   Gasoline price per gallon: $3.36,
   E85 price per gallon: x.

   We need to use the given information to find the fuel cost per mile. If we divide the number of dollars per gallon by the number of miles per gallon, we can find the number of dollars per mile:

   $$\frac{\text{dollars}}{\text{gal}} \div \frac{\text{mi}}{\text{gal}} = \frac{\text{dollars}}{\cancel{\text{gal}}} \cdot \frac{\cancel{\text{gal}}}{\text{mi}} = \frac{\text{dollars}}{\text{mi}}.$$

   Thus we form ratios of the form

   $$\frac{\text{dollars per gallon}}{\text{miles per gallon}}.$$

2. **Translate.** We form a proportion in which the ratio of price per gallon and number of miles per gallon is expressed in two ways:

   $$\begin{array}{c} \text{Gasoline price} \to \\ \text{Gasoline mpg} \to \end{array} \frac{\$3.36}{21} = \frac{x}{15} \begin{array}{c} \leftarrow \text{E85 price} \\ \leftarrow \text{E85 mpg} \end{array}.$$

3. **Carry out.** To solve for x, we multiply both sides of the equation by the LCM, 105:

   $$105 \cdot \frac{3.36}{21} = 105 \cdot \frac{x}{15}$$

   $$5 \cdot \cancel{21} \cdot \frac{3.36}{\cancel{21}} = 7 \cdot \cancel{15} \cdot \frac{x}{\cancel{15}} \qquad \text{Removing factors equal to 1: } 21/21 = 1 \text{ and } 15/15 = 1$$

   $$16.8 = 7x$$

   $$2.4 = x. \qquad\qquad \text{Dividing both sides by 7 and simplifying}$$

4. Check. If E85 costs $2.40 per gallon, the fuel cost per mile will be $2.40/15 = $0.16 per mile. The fuel cost per mile for gasoline is $3.36/21 = $0.16 per mile. The fuel costs are the same.

5. State. The price of E85 must be $2.40 per gallon.   TRY EXERCISE 33

EXAMPLE 6

Wildlife population. To determine the number of brook trout in River Denys, Cape Breton, Nova Scotia, a team of volunteers and professionals caught and marked 1190 brook trout. Later, they captured 915 brook trout, of which 24 were marked. Estimate the number of brook trout in River Denys.

Source: www.gov.ns.ca

SOLUTION We let T = the brook trout population in River Denys. If we assume that the percentage of marked trout in the second group of trout captured is the same as the percentage of marked trout in the entire river, we can form a proportion in which this percentage is expressed in two ways:

Trout originally marked → $\dfrac{1190}{T} = \dfrac{24}{915}$. ← Marked trout in second group
Entire population →   ← Total trout in second group

To solve for T, we multiply by the LCM, $915T$:

$$915T \cdot \frac{1190}{T} = 915T \cdot \frac{24}{915}$$ Multiplying both sides by $915T$

$$915 \cdot 1190 = 24T$$ Removing factors equal to 1: $T/T = 1$ and $915/915 = 1$

$$\frac{915 \cdot 1190}{24} = T \text{ or } T \approx 45{,}369.$$ Dividing both sides by 24

There are about 45,369 brook trout in the river.   TRY EXERCISE 47

Translating for Success

Translate each word problem to an equation and select a correct translation from equations A–O.

A. $2x + 2(x + 1) = 613$

B. $x^2 + (x + 1)^2 = 613$

C. $\dfrac{60}{x + 2} = \dfrac{50}{x}$

D. $x = 62\% \cdot 9.4$

E. $\dfrac{197}{7} = \dfrac{x}{30}$

F. $x + (x + 1) = 613$

G. $\dfrac{7}{197} = \dfrac{x}{30}$

H. $x^2 + (x + 2)^2 = 612$

I. $x^2 + (x + 1)^2 = 612$

J. $\dfrac{50}{x + 2} = \dfrac{60}{x}$

K. $x + 62\% \cdot x = 9.4$

L. $\dfrac{5 + 6}{2} = t$

M. $x^2 + (x + 1)^2 = 452$

N. $\dfrac{1}{5} + \dfrac{1}{6} = \dfrac{1}{t}$

O. $x^2 + (x + 2)^2 = 452$

Answers on page A-25

An additional, animated version of this activity appears in MyMathLab. To use MyMathLab, you need a course ID and a student access code. Contact your instructor for more information.

1. *Search-engine ads.* In 2006, North American advertisers spent \$9.4 billion in marketing through Internet search engines such as Google®. This was a 62% increase over the amount spent in 2005. How much was spent in 2005?
Source: Search Engine Marketing Professional Organization

2. *Bicycling.* The speed of one bicyclist is 2 km/h faster than the speed of another bicyclist. The first bicyclist travels 60 km in the same amount of time that it takes the second to travel 50 km. Find the speed of each bicyclist.

3. *Filling time.* A swimming pool can be filled in 5 hr by hose A alone and in 6 hr by hose B alone. How long would it take to fill the tank if both hoses were working?

4. *Payroll.* In 2007, the total payroll for Kraftside Productions was \$9.4 million. Of this amount, 62% was paid to employees working on an assembly line. How much money was paid to assembly-line workers?

5. *Cycling distance.* A bicyclist traveled 197 mi in 7 days. At this rate, how many miles could the cyclist travel in 30 days?

6. *Sides of a square.* If the sides of a square are increased by 2 ft, the area of the original square plus the area of the enlarged square is 452 ft². Find the length of a side of the original square.

7. *Consecutive integers.* The sum of two consecutive integers is 613. Find the integers.

8. *Sums of squares.* The sum of the squares of two consecutive odd integers is 612. Find the integers.

9. *Sums of squares.* The sum of the squares of two consecutive integers is 613. Find the integers.

10. *Rectangle dimensions.* The length of a rectangle is 1 ft longer than its width. Find the dimensions of the rectangle such that the perimeter of the rectangle is 613 ft.

Solve.

1. *Volunteerism.* As a "Friend of the Library," it takes Kelby 10 hr to price books for a used-book sale. Rosa can do the same job in 8 hr. How long would it take Kelby and Rosa to price the books working together?

2. *Concrete work.* It takes Cara 3 hr to build a form for a concrete porch. Etta can do the same job in 5 hr. How long would it take them, working together, to build the form?

3. *Painting.* Oliver can paint the Gardner's Queen Anne style house in 75 hr. It would take Pat 100 hr to paint the house. How long would it take them to paint the house if they worked together?

4. *Masonry.* By checking work records, a contractor finds that it takes Sonya 8 hr to construct a wall of a certain size. It takes Melissa 6 hr to construct the same wall. How long would it take if they worked together?

5. *Heavy machinery.* By checking work records, a foreman finds that Barry can dig a trench for a water line in 10 hr. Deb can do the same job in 15 hr. How long would it take if they worked together?

6. *Gardening.* Ben can weed his vegetable garden in 50 min, while Natalie can weed the same garden in 40 min. How long would it take if they worked together?

7. *Fruit trees.* It takes Jorell, working alone, 20 hr to prune the Gala apple trees at Tuttle's Orchard. Ferdous can prune the trees in 24 hr. How long would it take the two of them, working together, to prune the trees?

8. *Harvesting.* Bobbi can pick a quart of raspberries in 20 min. Mark can pick a quart in 25 min. How long would it take if Bobbi and Mark worked together?

9. *Multifunction copiers.* The Aficio SP C210SF can copy Mousa's dissertation in 7 min. The MX-3501N can copy the same document in 6 min. If the two machines work together, how long would they take to copy the dissertation?
   Source: Manufacturers' marketing brochures

MX-3501N
600 dpi B/W
600 dpi color

Aficio SP C210SF
600 dpi B/W•600 dpi color

10. *Fax machines.* The FAXPHONE L80 can fax a year-end report in 7 min while the HP 1050 Fax Series can fax the same report in 14 min. How long would it take the two machines, working together, to fax the report? (Assume that the recipient has at least two machines for incoming faxes.)
    Source: Manufacturers' marketing brochures

11. *Train speeds.* A B & M freight train is traveling 14 km/h slower than an AMTRAK passenger train. The B & M train travels 330 km in the same time that it takes the AMTRAK train to travel 400 km. Find their speeds. Complete the following table as part of the familiarization.

Distance	=	Rate	•	Time
	Distance (in km)	Speed (in km/h)		Time (in hours)
B & M	330			
AMTRAK	400	r		$\dfrac{400}{r}$

12. *Speed of travel.* A loaded Roadway truck is moving 40 mph faster than a New York Railways freight train. In the time that it takes the train to travel 150 mi, the truck travels 350 mi. Find their

speeds. Complete the following table as part of the familiarization.

Distance	=	Rate	·	Time

	Distance (in miles)	Speed (in miles per hour)	Time (in hours)
Truck	350	r	$\dfrac{350}{r}$
Train	150		

13. *Driving speed.* Sean's Camaro travels 15 mph faster than Rita's Harley. In the same time that Sean travels 156 mi, Rita travels 120 mi. Find their speeds.

14. *Bicycle speed.* Ada bicycles 5 km/h slower than Elin. In the same time that it takes Ada to ride 48 km, Elin can ride 63 km. How fast does each bicyclist travel?

15. *Walking speed.* Baruti walks 4 km/h faster than Tau. In the time that it takes Tau to walk 7.5 km, Baruti walks 13.5 km. Find their speeds.

16. *Cross-country skiing.* Luca cross-country skis 4 km/h faster than Lea. In the time that it takes Lea to ski 18 km, Luca skis 24 km. Find their speeds.

Aha! 17. *Tractor speed.* Manley's tractor is just as fast as Caledonia's. It takes Manley 1 hr more than it takes Caledonia to drive to town. If Manley is 20 mi from town and Caledonia is 15 mi from town, how long does it take Caledonia to drive to town?

18. *Boat speed.* Tory and Emilio's motorboats travel at the same speed. Tory pilots her boat 40 km before docking. Emilio continues for another 2 hr, traveling a total of 100 km before docking. How long did it take Tory to navigate the 40 km?

Geometry. For each pair of similar triangles, find the value of the indicated letter.

19. *b*

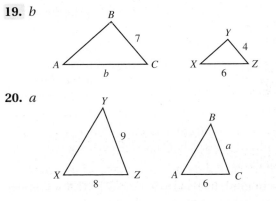

20. *a*

21. *f*

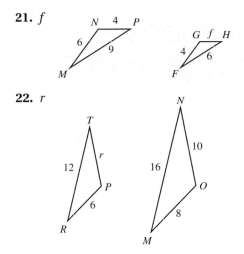

22. *r*

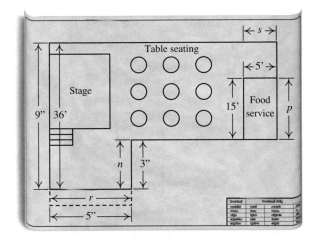

Architecture. Use the blueprint below to find the indicated length.

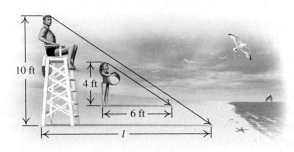

23. *p*, in inches on blueprint

24. *s*, in inches on blueprint

25. *r*, in feet on actual building

26. *n*, in feet on actual building

Find the indicated length.

27. *l*

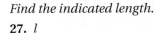

28. *h*

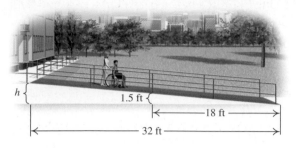

h 1.5 ft 18 ft 32 ft

Geometry. *When three parallel lines are crossed by two or more lines (transversals), the lengths of corresponding segments are proportional (see the following figure).*

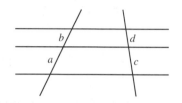

b *d* *a* *c*

29. If *a* is 10 cm when *b* is 6 cm, find *d* when *c* is 5 cm.

30. If *d* is 8 cm when *c* is 9 cm, find *a* when *b* is 10 cm.

Graphing. *Find the indicated length.*

31. *r*

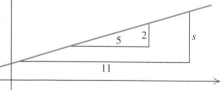

5 9 7 *r*

32. *s*

5 2 *s* 11

33. *Text messaging.* Brett sent or received 384 text messages in 8 days. At this rate, how many text messages would he send or receive in 30 days?

34. *Burning calories.* The average 140-lb adult burns about 380 calories bicycling 10 mi at a moderate rate. How far should the average 140-lb adult ride in order to burn 100 calories?
Source: *ACE Fitness Matters,* Volume 1, Number 4, 1997

35. *Illegal immigration.* Between 2001 and 2006, the Border Patrol caught 12,334 people trying to cross illegally from Canada to the United States along a 295-mi stretch of the border. If this rate were the same for the entire 5525-mi border between the two countries, how many would the Border Patrol catch in those same years along the entire border?
Source: *The Wall Street Journal,* July 10, 2007

Aha! **36.** *Photography.* Aziza snapped 234 photos over a period of 14 days. At this rate, how many would she take in 42 days?

37. *Mileage.* The Honda Civic Hybrid is a gasoline–electric car that travels approximately 180 mi on 4 gal of gas. Find the amount of gas required for an 810-mi trip.
Source: www.greenhybrid.com

38. *Baking.* In a potato bread recipe, the ratio of milk to flour is $\frac{3}{13}$. If 5 cups of milk are used, how many cups of flour are used?

39. *Wing aspect ratio.* The wing aspect ratio for a bird or an airplane is the ratio of the wing span to the wing width. Generally, higher aspect ratios are more efficient during low speed flying. Herons and storks, both waders, have comparable wing aspect ratios. A grey heron has a wing span of 180 cm and a wing width of 24 cm. A white stork has a wing span of 200 cm. What is the wing width of a stork?
Source: birds.ecoport.org

Aha! **40.** *Money.* The ratio of the weight of copper to the weight of zinc in a U.S. penny is $\frac{1}{39}$. If 50 kg of zinc is being turned into pennies, how much copper is needed?
Source: United States Mint

Hat sizes. *U.S. hat sizes typically range between 6 and 8 and are proportional to the circumference of one's head. In Europe, and increasingly in North America, hat size is given in centimeters. Use this information for Exercises 41 and 42.*

41. A felt hat that is 60 cm in circumference is size $7\frac{1}{2}$ in the United States. What U.S. size corresponds to a 56-cm hat?

42. A 58-cm mesh hat is U.S. hat size $7\frac{1}{4}$. What European hat size corresponds to U.S. hat size $6\frac{7}{8}$?

43. *Light bulbs.*   A sample of 220 compact fluorescent light bulbs contained 8 defective bulbs. How many defective bulbs would you expect in a batch of 1430 bulbs?

44. *Flash drives.*   A sample of 150 flash drives contained 7 defective drives. How many defective flash drives would you expect in a batch of 2700 flash drives?

45. *Veterinary science.*   The amount of water needed by a small dog depends on its weight. A moderately active 8-lb Shih Tzu needs approximately 12 oz of water per day. How much water does a moderately active 5-lb Bolognese require each day?
Source: www.smalldogsparadise.com

46. *Miles driven.*   Emmanuel is allowed to drive his leased car for 45,000 mi in 4 yr without penalty. In the first $1\frac{1}{2}$ yr, Emmanuel has driven 16,000 mi. At this rate will he exceed the mileage allowed for 4 yr?

47. *Moose population.*   To determine the size of Pine County's moose population, naturalists catch 69 moose, tag them, and then set them free. Months later, 40 moose are caught, of which 15 have tags. Estimate the size of the moose population.

48. *Deer population.*   To determine the number of deer in the Great Gulf Wilderness, a game warden catches 318 deer, tags them, and lets them loose. Later, 168 deer are caught; 56 of them have tags. Estimate the number of deer in the preserve.

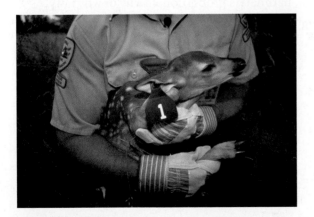

49. *Environmental science.*   To determine the number of humpback whales in a pod, a marine biologist, using tail markings, identifies 27 members of the pod. Several weeks later, 40 whales from the pod are randomly sighted. Of the 40 sighted, 12 are from the 27 originally identified. Estimate the number of whales in the pod.

50. *Fox population.*   To determine the number of foxes in King County, a naturalist catches, tags, and then releases 25 foxes. Later, 36 foxes are caught; 4 of them have tags. Estimate the fox population of the county.

51. *Weight on the moon.*   The ratio of the weight of an object on the moon to the weight of that object on Earth is 0.16 to 1.
   a) How much would a 12-ton rocket weigh on the moon?
   b) How much would a 180-lb astronaut weigh on the moon?

52. *Weight on Mars.*   The ratio of the weight of an object on Mars to the weight of that object on Earth is 0.4 to 1.
   a) How much would a 12-ton rocket weigh on Mars?
   b) How much would a 120-lb astronaut weigh on Mars?

53. Is it correct to assume that two workers will complete a task twice as quickly as one person working alone? Why or why not?

54. If two triangles are exactly the same shape and size, are they similar? Why or why not?

Skill Review

To prepare for Chapter 7, review graphing linear equations (Section 3.2).

Graph. [3.2]

55. $y = 2x - 6$ **56.** $y = -2x + 6$

57. $3x + 2y = 12$ **58.** $x - 3y = 6$

59. $y = -\dfrac{3}{4}x + 2$ **60.** $y = \dfrac{2}{5}x - 4$

Synthesis

61. Write a problem similar to Example 1 for a classmate to solve. Design the problem so that the translation step is

$$\frac{t}{7} + \frac{t}{5} = 1.$$

62. Write a problem similar to Example 2 for a classmate to solve. Design the problem so that the translation step is

$$\frac{30}{r + 4} = \frac{18}{r}.$$

63. Show that the four equations in the box labeled "The Work Principle" on p. 409 are equivalent.

64. *Quilting.*   Ricki and Maura work together and sew a quilt in 4 hr. Working alone, Maura would need 6 hr more than Ricki to sew a quilt. How long would it take each of them working alone?

65. *Car cleaning.*   Together, Michelle, Sal, and Kristen can clean and wax a car in 1 hr 20 min. To complete the job alone, Michelle needs twice the time that Sal needs and 2 hr more than Kristen. How long would it take each to clean and wax the car working alone?

66. *Grading.*   Alma can grade a batch of placement exams in 3 hr. Kevin can grade a batch in 4 hr. If they work together to grade a batch of exams, what percentage of the exams will have been graded by Alma?

Aha! **67.** *Roofing.*   Working alone, Russ can reshingle a roof in 12 hr. When Joan works with Russ, the job takes 6 hr. How long would it take Joan, working alone, to reshingle the roof?

68. *Wiring.*   Janet can wire a house in 28 hr. Linus can wire a house in 34 hr. How long will it take Janet and Linus, working together, to wire *two* houses?

69. According to the U.S. Census Bureau, Population Division, in July 2007, there was one birth every 7 sec, one death every 13 sec, and one new international migrant every 27 sec. How many seconds does it take for a net gain of one person?

70. *Home maintenance.*   Fuel used in many chain saws is made by pouring a 3.2-oz bottle of 2-cycle oil into 160 oz of gasoline. Gus accidentally poured 5.6 oz of 2-cycle oil into 200 oz of gasoline. How much more oil or gasoline should he add in order for the fuel to have the proper ratio of oil to gasoline?

71. *Boating.*   The speed of a boat in still water is 10 mph. It travels 24 mi upstream and 24 mi downstream in a total time of 5 hr. What is the speed of the current?

72. *Commuting.*   To reach an appointment 50 mi away, Dr. Wright allowed 1 hr. After driving 30 mi, she realized that her speed would have to be increased 15 mph for the remainder of the trip. What was her speed for the first 30 mi?

73. *Distances.*   The shadow from a 40-ft cliff just reaches across a water-filled quarry at the same time that a 6-ft tall diver casts a 10-ft shadow. How wide is the quarry?

74. Simplest fraction notation for a rational number is $\frac{9}{17}$. Find an equivalent ratio where the sum of the numerator and the denominator is 104.

75. How soon after 5 o'clock will the hands on a clock first be together?

76. Given that
$$\frac{A}{B} = \frac{C}{D},$$
write three other proportions using A, B, C, and D.

77. If two triangles are similar, are their areas and perimeters proportional? Why or why not?

78. Are the equations
$$\frac{A + B}{B} = \frac{C + D}{D} \quad \text{and} \quad \frac{A}{B} = \frac{C}{D}$$
equivalent? Why or why not?

CORNER

Sharing the Workload

Focus: Modeling, estimation, and work problems

Time: 15–20 minutes

Group size: 3

Materials: Paper, pencils, textbooks, and a watch

Many tasks can be done by two people working together. If both people work at the same rate, each does half the task, and the project is completed in half the time. However, when the work rates differ, the faster worker performs more than half of the task.

ACTIVITY

1. The project is to write down (but not answer) Review Exercises 27–36 from Chapter 6 (p. 425) on a sheet of paper. The problems should be spaced apart and written clearly so that they can be used for studying in the future. Two of the members in each group should write down the exercises, one working slowly and one working quickly. The third group member should record the time required for each to write down all 10 exercises.

2. Using the times from step (1), calculate how long it will take the two workers, working together, to complete the task.

3. Next, have the same workers as in step (1)—working at the same speeds as in step (1)—perform the task together. To do this, one person should begin writing with Exercise 27, while the other worker begins with Exercise 36 and lists the problems counting backward. The third member is again the timekeeper and should observe when the two workers have written all the exercises. To avoid collision, each of the two writers should use a separate sheet of paper.

4. Compare the actual experimental time from part (3) with the time predicted by the model in part (2). List reasons that might account for any discrepancy.

5. Let t_1, t_2, and t_3 represent the times required for the first worker, the second worker, and the two workers together, respectively, to complete a task. Then develop a model that can be used to find t_3 when t_1 and t_2 are known.

Time Required for Worker A, Working Alone	Time Required for Worker B, Working Alone	Estimated Time for the Two Workers, Working Together	Actual Time Required for the Two Workers, Working Together

Study Summary

KEY TERMS AND CONCEPTS	EXAMPLES

SECTION 6.1: RATIONAL EXPRESSIONS

A **rational expression** can be written as a quotient of two polynomials and is undefined when the denominator is 0. We simplify rational expressions by removing a factor equal to 1.

$$\frac{x^2 - 3x - 4}{x^2 - 1} = \frac{(x + 1)(x - 4)}{(x + 1)(x - 1)}$$

Factoring the numerator and the denominator

$$= \frac{x - 4}{x - 1} \qquad \frac{x + 1}{x + 1} = 1$$

SECTION 6.2: MULTIPLICATION AND DIVISION

The Product of Two Rational Expressions

$$\frac{A}{B} \cdot \frac{C}{D} = \frac{AC}{BD}$$

$$\frac{5v + 5}{v - 2} \cdot \frac{2v^2 - 8v + 8}{v^2 - 1}$$

$$= \frac{5(v + 1) \cdot 2(v - 2)(v - 2)}{(v - 2)(v + 1)(v - 1)}$$

Multiplying numerators, multiplying denominators, and factoring

$$= \frac{10(v - 2)}{v - 1} \qquad \frac{(v + 1)(v - 2)}{(v + 1)(v - 2)} = 1$$

The Quotient of Two Rational Expressions

$$\frac{A}{B} \div \frac{C}{D} = \frac{A}{B} \cdot \frac{D}{C} = \frac{AD}{BC}$$

$$(x^2 - 5x - 6) \div \frac{x^2 - 1}{x + 6}$$

$$= \frac{x^2 - 5x - 6}{1} \cdot \frac{x + 6}{x^2 - 1}$$

Multiplying by the reciprocal of the divisor

$$= \frac{(x - 6)(x + 1)(x + 6)}{(x + 1)(x - 1)}$$

Multiplying numerators, multiplying denominators, and factoring

$$= \frac{(x - 6)(x + 6)}{x - 1} \qquad \frac{x + 1}{x + 1} = 1$$

SECTION 6.3: ADDITION, SUBTRACTION, AND LEAST COMMON DENOMINATORS

The Sum of Two Rational Expressions

$$\frac{A}{B} + \frac{C}{B} = \frac{A + C}{B}$$

$$\frac{7x + 8}{x + 1} + \frac{4x + 3}{x + 1} = \frac{7x + 8 + 4x + 3}{x + 1}$$

Adding numerators and keeping the denominator

$$= \frac{11x + 11}{x + 1}$$

$$= \frac{11(x + 1)}{x + 1}$$

Factoring

$$= 11 \qquad \frac{x + 1}{x + 1} = 1$$

The Difference of Two Rational Expressions

$$\frac{A}{B} - \frac{C}{B} = \frac{A - C}{B}$$

$$\frac{7x + 8}{x + 1} - \frac{4x + 3}{x + 1} = \frac{7x + 8 - (4x + 3)}{x + 1}$$

Subtracting numerators and keeping the denominator. The parentheses are necessary.

$$= \frac{7x + 8 - 4x - 3}{x + 1}$$

Removing parentheses

$$= \frac{3x + 5}{x + 1}$$

The **least common denominator (LCD)** of rational expressions is the **least common multiple (LCM)** of the denominators. To find the LCM, write the prime factorizations of the denominators. The LCM contains each factor the greatest number of times that it occurs in any of the individual factorizations.

Find the LCM of $m^2 - 5m + 6$ and $m^2 - 4m + 4$.

$$\left.\begin{array}{l} m^2 - 5m + 6 = (m - 2)(m - 3) \\ m^2 - 4m + 4 = (m - 2)(m - 2) \end{array}\right\}$$ Factoring each expression

LCM $= (m - 2)(m - 2)(m - 3)$ The LCM contains 2 factors of $(m - 2)$ and 1 factor of $(m - 3)$.

SECTION 6.4: ADDITION AND SUBTRACTION WITH UNLIKE DENOMINATORS

To add or subtract rational expressions with different denominators, first rewrite the expressions as equivalent expressions with a common denominator.

$$\frac{2x}{x^2 - 16} + \frac{x}{x - 4}$$

$$= \frac{2x}{(x + 4)(x - 4)} + \frac{x}{x - 4}$$ Factoring denominators: The LCD is $(x + 4)(x - 4)$.

$$= \frac{2x}{(x + 4)(x - 4)} + \frac{x}{x - 4} \cdot \frac{x + 4}{x + 4}$$ Multiplying by 1 to get the LCD in the second expression

$$= \frac{2x}{(x + 4)(x - 4)} + \frac{x^2 + 4x}{(x + 4)(x - 4)}$$

$$= \frac{x^2 + 6x}{(x + 4)(x - 4)}$$ Adding numerators and keeping the denominator. This cannot be simplified.

SECTION 6.5: COMPLEX RATIONAL EXPRESSIONS

Complex rational expressions contain one or more rational expressions within the numerator and/or the denominator. They can be simplified either by using division or by multiplying by a form of 1 to clear the fractions.

Using division to simplify:

$$\frac{\dfrac{1}{6} - \dfrac{1}{x}}{\dfrac{6 - x}{6}} = \frac{\dfrac{1}{6} \cdot \dfrac{x}{x} - \dfrac{1}{x} \cdot \dfrac{6}{6}}{\dfrac{6 - x}{6}} = \frac{\dfrac{x - 6}{6x}}{\dfrac{6 - x}{6}}$$ Subtracting to get a single rational expression in the numerator

$$= \frac{x - 6}{6x} \div \frac{6 - x}{6} = \frac{x - 6}{6x} \cdot \frac{6}{6 - x}$$ Dividing the numerator by the denominator

$$= \frac{6(x - 6)}{6x(-1)(x - 6)} = \frac{1}{-x} = -\frac{1}{x}$$ Factoring and simplifying; $\dfrac{6(x - 6)}{6(x - 6)} = 1$

Multiplying by 1:

$$\frac{\dfrac{4}{x}}{\dfrac{3}{x} + \dfrac{2}{x^2}} = \frac{\dfrac{4}{x}}{\dfrac{3}{x} + \dfrac{2}{x^2}} \cdot \frac{x^2}{x^2}$$

$$= \frac{\dfrac{4}{x} \cdot \dfrac{x^2}{1}}{\left(\dfrac{3}{x} + \dfrac{2}{x^2}\right) \cdot \dfrac{x^2}{1}}$$ The LCD of all the denominators is x^2; multiplying by $\dfrac{x^2}{x^2}$

$$= \frac{\dfrac{4 \cdot x \cdot x}{x}}{\dfrac{3 \cdot x \cdot x}{x} + \dfrac{2 \cdot x^2}{x^2}} = \frac{4x}{3x + 2}$$ The fractions are cleared.

SECTION 6.6: SOLVING RATIONAL EQUATIONS

To Solve a Rational Equation

1. List any restrictions.
2. Clear the equation of fractions.
3. Solve the resulting equation.
4. Check the possible solution(s) in the original equation.

Solve: $\dfrac{2}{x+1} = \dfrac{1}{x-2}$.   The restrictions are $x \neq -1, x \neq 2$.

$$\frac{2}{x+1} = \frac{1}{x-2}$$

$$(x+1)(x-2) \cdot \frac{2}{x+1} = (x+1)(x-2) \cdot \frac{1}{x-2}$$

$$2(x-2) = x+1$$

$$2x - 4 = x + 1$$

$$x = 5$$

Check: Since $\dfrac{2}{5+1} = \dfrac{1}{5-2}$, the solution is 5.

SECTION 6.7: APPLICATIONS USING RATIONAL EQUATIONS AND PROPORTIONS

The Work Principle

If A requires a units of time to complete a task and B requires b units of time to complete the same task, then:

$$\frac{1}{a} \cdot t + \frac{1}{b} \cdot t = 1, \qquad \left(\frac{1}{a} + \frac{1}{b}\right)t = 1,$$

$$\frac{t}{a} + \frac{t}{b} = 1, \qquad \frac{1}{a} + \frac{1}{b} = \frac{1}{t},$$

where t is the time it takes A and B, working together, to complete the task.

Cecilia and Aaron work as volunteers at a town's recycling depot. Cecilia can sort a day's accumulation of recyclables in 4 hr, while Aaron requires 6 hr to do the same job. How long would it take them working together to sort the recyclables?

If $t =$ the time, in hours, that it takes Cecilia and Aaron to do the job working together, then

$$\frac{1}{4} \cdot t + \frac{1}{6} \cdot t = 1 \qquad \text{Using the work principle}$$

$$t = 2\frac{2}{5}\,\text{hr.} \qquad \text{Solving the equation}$$

See Example 1 in Section 6.7 for a complete solution of this problem.

The Motion Formula

$$d = r \cdot t, \quad r = \frac{d}{t}, \quad \text{or} \quad t = \frac{d}{r}$$

Because of a tailwind, a Delta Boeing 777 is able to fly 200 mph faster than a Northwest 777 that is flying *into* the wind. In the same time that it takes the Delta plane to travel 1800 mi, the Northwest plane flies 1200 mi. How fast is each plane traveling?

If $r =$ the speed, in miles per hour, of the Northwest plane, then Northwest time $= \frac{1200}{r}$ and Delta time $= \frac{1800}{r+200}$. Since the times are the same,

$$\frac{1200}{r} = \frac{1800}{r+200}$$

$$r = 400. \qquad \text{Solving the equation}$$

The speed of the Northwest plane is 400 mph, and the speed of the Delta plane is 600 mph.

See Example 2 in Section 6.7 for a complete solution of this problem.

In geometry, proportions arise in the study of **similar triangles**.

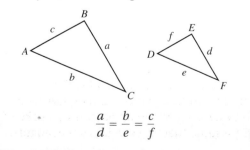

$$\frac{a}{d} = \frac{b}{e} = \frac{c}{f}$$

Triangles DEF and UVW are similar. Solve for u.

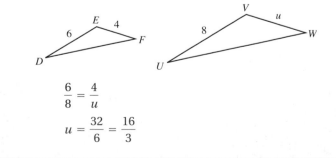

$$\frac{6}{8} = \frac{4}{u}$$

$$u = \frac{32}{6} = \frac{16}{3}$$

Review Exercises: Chapter 6

✎ *Concept Reinforcement* *Classify each statement as either true or false.*

1. Every rational expression can be simplified. [6.1]

2. The expression $(t - 3)/(t^2 - 4)$ is undefined for $t = 2$. [6.1]

3. The expression $(t - 3)/(t^2 - 4)$ is undefined for $t = 3$. [6.1]

4. To multiply rational expressions, a common denominator is never required. [6.2]

5. To divide rational expressions, a common denominator is never required. [6.2]

6. To add rational expressions, a common denominator is never required. [6.3]

7. To subtract rational expressions, a common denominator is never required. [6.3]

8. The number 0 can never be a solution of a rational equation. [6.6]

List all numbers for which each expression is undefined. [6.1]

9. $\dfrac{17}{-x^2}$

10. $\dfrac{9}{2a + 10}$

11. $\dfrac{x - 5}{x^2 - 36}$

12. $\dfrac{x^2 + 3x + 2}{x^2 + x - 30}$

13. $\dfrac{-6}{(t + 2)^2}$

Simplify. [6.1]

14. $\dfrac{3x^2 - 9x}{3x^2 + 15x}$

15. $\dfrac{14x^2 - x - 3}{2x^2 - 7x + 3}$

16. $\dfrac{6y^2 - 36y + 54}{4y^2 - 36}$

17. $\dfrac{5x^2 - 20y^2}{2y - x}$

Multiply or divide and, if possible, simplify. [6.2]

18. $\dfrac{a^2 - 36}{10a} \cdot \dfrac{2a}{a + 6}$

19. $\dfrac{6y - 12}{2y^2 + 3y - 2} \cdot \dfrac{y^2 - 4}{8y - 8}$

20. $\dfrac{16 - 8t}{3} \div \dfrac{t - 2}{12t}$

21. $\dfrac{4x^4}{x^2 - 1} \div \dfrac{2x^3}{x^2 - 2x + 1}$

22. $\dfrac{x^2 + 1}{x - 2} \cdot \dfrac{2x + 1}{x + 1}$

23. $(t^2 + 3t - 4) \div \dfrac{t^2 - 1}{t + 4}$

Find the LCM. [6.3]

24. $10a^3b^8,\ 12a^5b$

25. $x^2 - x,\ x^5 - x^3,\ x^4$

26. $y^2 - y - 2,\ y^2 - 4$

Add or subtract and, if possible, simplify.

27. $\dfrac{x + 6}{x + 3} + \dfrac{9 - 4x}{x + 3}$ [6.3]

28. $\dfrac{6x - 3}{x^2 - x - 12} - \dfrac{2x - 15}{x^2 - x - 12}$ [6.3]

29. $\dfrac{3x - 1}{2x} - \dfrac{x - 3}{x}$ [6.4]

30. $\dfrac{2a + 4b}{5ab^2} - \dfrac{5a - 3b}{a^2b}$ [6.4]

31. $\dfrac{y^2}{y - 2} + \dfrac{6y - 8}{2 - y}$ [6.4]

32. $\dfrac{t}{t + 1} + \dfrac{t}{1 - t^2}$ [6.4]

33. $\dfrac{d^2}{d - 2} + \dfrac{4}{2 - d}$ [6.4]

34. $\dfrac{1}{x^2 - 25} - \dfrac{x - 5}{x^2 - 4x - 5}$ [6.4]

35. $\dfrac{3x}{x + 2} - \dfrac{x}{x - 2} + \dfrac{8}{x^2 - 4}$ [6.4]

36. $\dfrac{3}{4t} + \dfrac{3}{3t + 2}$ [6.4]

Simplify. [6.5]

37. $\dfrac{\dfrac{1}{z} + 1}{\dfrac{1}{z^2} - 1}$

38. $\dfrac{\dfrac{5}{2x^2}}{\dfrac{3}{4x} + \dfrac{4}{x^3}}$

39. $\dfrac{\dfrac{c}{d} - \dfrac{d}{c}}{\dfrac{1}{c} + \dfrac{1}{d}}$

Solve. [6.6]

40. $\dfrac{3}{x} - \dfrac{1}{4} = \dfrac{1}{2}$

41. $\dfrac{3}{x + 4} = \dfrac{1}{x - 1}$

42. $x + \dfrac{6}{x} = -7$

43. Jackson can sand the oak floors and stairs in a two-story home in 12 hr. Charis can do the same job in 9 hr. How long would it take if they worked together? (Assume that two sanders are available.) [6.7]

44. The distance by highway between Richmond and Waterbury is 70 km, and the distance by rail is 60 km. A car and a train leave Richmond at the same time and arrive in Waterbury at the same time, the car having traveled 15 km/h faster than the train. Find the speed of the car and the speed of the train. [6.7]

45. To estimate the harbor seal population in Bristol Bay, scientists radio-tagged 33 seals. Several days later, they collected a sample of 40 seals, and 24 of them were tagged. Estimate the seal population of the bay. [6.7]

46. Triangles *ABC* and *XYZ* are similar. Find the value of *x*. [6.7]

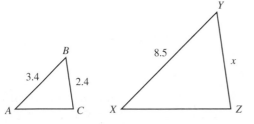

47. A sample of 30 weather-alert radios contained 4 defective ones. How many defective radios would you expect to find in a batch of 540? [6.7]

Synthesis

48. For what procedures in this chapter is the LCM of denominators used to clear fractions? [6.5], [6.6]

49. A student always uses the common denominator found by multiplying the denominators of the expressions being added. How could this approach be improved? [6.3]

Simplify.

50. $\dfrac{2a^2 + 5a - 3}{a^2} \cdot \dfrac{5a^3 + 30a^2}{2a^2 + 7a - 4} \div \dfrac{a^2 + 6a}{a^2 + 7a + 12}$ [6.2]

51. $\dfrac{12a}{(a - b)(b - c)} - \dfrac{2a}{(b - a)(c - b)}$ [6.4]

Aha! **52.** $\dfrac{5(x - y)}{(x - y)(x + 2y)} - \dfrac{5(x - 3y)}{(x + 2y)(x - 3y)}$ [6.3]

53. It has been over 60 yr since a major-league baseball player went an entire season averaging 4 hits in every 10 at-bats. Suppose that Hideki currently has 153 hits after 395 at-bats. If he is assured 125 more at-bats, what percentage of those must be hits if he is to average 4 hits for every 10 at-bats? [6.7]

Test: Chapter 6

CHAPTER
Test Prep
VIDEO CD

Step-by-step test solutions are found on the video CD in the front of this book.

List all numbers for which each expression is undefined.

1. $\dfrac{2 - x}{5x}$

2. $\dfrac{5}{x + 8}$

3. $\dfrac{x - 7}{x^2 - 1}$

4. $\dfrac{x^2 + x - 30}{x^2 - 3x + 2}$

5. Simplify: $\dfrac{6x^2 + 17x + 7}{2x^2 + 7x + 3}$.

Multiply or divide and, if possible, simplify.

6. $\dfrac{t^2 - 9}{12t} \cdot \dfrac{8t^2}{t^2 - 4t + 3}$

7. $\dfrac{25y^2 - 1}{9y^2 - 6y} \div \dfrac{5y^2 + 9y - 2}{3y^2 + y - 2}$

8. $\dfrac{4a^2 + 1}{4a^2 - 1} \div \dfrac{4a^2}{4a^2 + 4a + 1}$

9. $(x^2 + 6x + 9) \cdot \dfrac{(x - 3)^2}{x^2 - 9}$

10. Find the LCM:

$$y^2 - 9, \ y^2 + 10y + 21, \ y^2 + 4y - 21.$$

Add or subtract, and, if possible, simplify.

11. $\dfrac{2 + x}{x^3} + \dfrac{7 - 4x}{x^3}$

12. $\dfrac{5 - t}{t^2 + 1} - \dfrac{t - 3}{t^2 + 1}$

13. $\dfrac{2x - 4}{x - 3} + \dfrac{x - 1}{3 - x}$

14. $\dfrac{2x - 4}{x - 3} - \dfrac{x - 1}{3 - x}$

15. $\dfrac{7}{t-2} + \dfrac{4}{t}$

16. $\dfrac{y}{y^2 + 6y + 9} + \dfrac{1}{y^2 + 2y - 3}$

17. $\dfrac{1}{x-1} + \dfrac{4}{x^2 - 1} - \dfrac{2}{x^2 - 2x + 1}$

Simplify.

18. $\dfrac{9 - \dfrac{1}{y^2}}{3 - \dfrac{1}{y}}$

19. $\dfrac{\dfrac{x}{8} - \dfrac{8}{x}}{\dfrac{1}{8} + \dfrac{1}{x}}$

Solve.

20. $\dfrac{1}{t} + \dfrac{1}{3t} = \dfrac{1}{2}$

21. $\dfrac{15}{x} - \dfrac{15}{x-2} = -2$

22. Kopy Kwik has 2 copiers. One can copy a year-end report in 20 min. The other can copy the same document in 30 min. How long would it take both machines, working together, to copy the report?

23. The average 140-lb adult burns about 320 calories walking 4 mi at a moderate speed. How far should the average 140-lb adult walk in order to burn 100 calories?
Source: www.walking.about.com

24. Ryan drives 20 km/h faster than Alicia. In the same time that Alicia drives 225 km, Ryan drives 325 km. Find the speed of each car.

Synthesis

25. Pe'rez and Rema work together to mulch the flower beds around an office complex in $2\frac{6}{7}$ hr. Working alone, it would take Pe'rez 6 hr more than it would take Rema. How long would it take each of them to complete the landscaping working alone?

26. Simplify: $1 - \dfrac{1}{1 - \dfrac{1}{1 - \dfrac{1}{a}}}$.

27. The square of a number is the opposite of the number's reciprocal. Find the number.

Cumulative Review: Chapters 1–6

1. Use the commutative law of multiplication to write an expression equivalent to $a + bc$. [1.2]

2. Evaluate $-x^2$ for $x = 5$. [1.8]

3. Evaluate $(-x)^2$ for $x = 5$. [1.8]

4. Simplify: $-3[2(x - 3) - (x + 5)]$. [1.8]

Solve.

5. $5(x - 2) = 40$ [2.2]

6. $49 = x^2$ [5.6]

7. $-18n = 30$ [2.1]

8. $4x - 3 = 9x - 11$ [2.2]

9. $4(y - 5) = -2(y - 2)$ [2.2]

10. $x^2 + 11x + 10 = 0$ [5.6]

11. $\dfrac{4}{9}t + \dfrac{2}{3} = \dfrac{1}{3}t - \dfrac{2}{9}$ [2.2]

12. $\dfrac{4}{x} + x = 5$ [6.6]

13. $6 - y \geq 2y + 8$ [2.6]

14. $\dfrac{2}{x-3} = \dfrac{5}{3x+1}$ [6.6]

15. $2x^2 + 7x = 4$ [5.6]

16. $4(x + 7) < 5(x - 3)$ [2.6]

17. $\dfrac{t^2}{t+5} = \dfrac{25}{t+5}$ [6.6]

18. $(2x + 7)(x - 5) = 0$ [5.6]

19. $\dfrac{2}{x^2 - 9} + \dfrac{5}{x - 3} = \dfrac{3}{x + 3}$ [6.6]

Solve each formula. [2.3]

20. $3a - b + 9 = c$, for b

21. $\frac{3}{4}(x + 2y) = z$, for y

Graph. [3.2], [3.3], [3.6]

22. $y = \frac{3}{4}x + 5$

23. $x = -3$

24. $4x + 5y = 20$

25. $y = 6$

26. Find the slope of the line containing the points $(1, 5)$ and $(2, 3)$. [3.5]

27. Find the slope and the y-intercept of the line given by $2x - 4y = 1$. [3.6]

28. Write the slope–intercept equation of the line with slope $-\frac{5}{8}$ and y-intercept $(0, -4)$. [3.6]

Simplify.

29. $\dfrac{x^{-5}}{x^{-3}}$ [4.8]

30. $y \cdot y^{-8}$ [4.8]

31. $-(2a^2b^7)^2$ [4.1]

32. Subtract: [4.3]

$$(-8y^2 - y + 2) - (y^3 - 6y^2 + y - 5).$$

Multiply.

33. $-5(3a - 2b + c)$ [1.2]

34. $(2x^2 - 1)(x^3 + x - 3)$ [4.4]

35. $(6x - 5y)^2$ [4.5]

36. $(3n + 2)(n - 5)$ [4.5]

37. $(2x^3 + 1)(2x^3 - 1)$ [4.5]

Factor.

38. $6x - 2x^2 - 24x^4$ [5.1]

39. $16x^2 - 81$ [5.4]

40. $t^2 - 10t + 24$ [5.2]

41. $8x^2 + 10x + 3$ [5.3]

42. $6x^2 - 28x + 16$ [5.3]

43. $25t^2 + 40t + 16$ [5.4]

44. $x^2y^2 - xy - 20$ [5.2]

45. $x^4 + 2x^3 - 3x - 6$ [5.1]

Simplify.

46. $\dfrac{4t - 20}{t^2 - 16} \cdot \dfrac{t - 4}{t - 5}$ [6.2]

47. $\dfrac{x^2 - 1}{x^2 - x - 2} \div \dfrac{x - 1}{x - 2}$ [6.2]

48. $\dfrac{5ab}{a^2 - b^2} + \dfrac{a + b}{a - b}$ [6.4]

49. $\dfrac{x + 2}{4 - x} - \dfrac{x + 3}{x - 4}$ [6.4]

50. $\dfrac{1 + \dfrac{2}{x}}{1 - \dfrac{4}{x^2}}$ [6.5]

51. $\dfrac{3y + \dfrac{2}{y}}{y - \dfrac{3}{y^2}}$ [6.5]

Divide. [4.7]

52. $\dfrac{18x^4 - 15x^3 + 6x^2 + 12x + 3}{3x^2}$

53. $(15x^4 - 12x^3 + 6x^2 + 2x + 18) \div (x + 3)$

54. For each order, HanBooks.com charges a shipping fee of $5.50 plus $1.99 per book. The shipping cost for Dae's book order was $35.35. How many books did she order? [2.5]
Source: hanbooks.com

55. In 2006, the attendance at movie theaters was 1448.5 billion. This was 425% more than the number of admissions to theme parks. How many theme park admissions were there in 2006? [2.4]
Source: MPAA, PricewaterhouseCoopers

56. A pair of 1976 two-dollar bills with consecutive serial numbers is being sold in an auction. The sum of the serial numbers is 66,679,015. What are the serial numbers of the bills? [2.5]

57. Nikki is laying out two square flower gardens in a client's lawn. Each side of one garden is 2 ft longer than each side of the smaller garden. Together, the area of the gardens is 340 ft². Find the length of a side of the smaller garden. [5.7]

58. It takes Wes 25 min to file a week's worth of receipts. Corey, a new employee, takes 75 min to do the same job. How long would it take if they worked together? [6.7]

59. A game warden catches, tags, and then releases 18 antelope. A month later, a sample of 30 antelope is caught and released and 6 of them have tags. Use this information to estimate the size of the antelope population in that area. [6.7]

60. Rachel burned 450 calories in a workout. She burned twice as many in her aerobics session as she did doing calisthenics. How many calories did she burn doing calisthenics? [2.5]

Synthesis

61. Solve: $\frac{1}{3}|n| + 8 = 56$. [1.4], [2.2]

Aha! **62.** Multiply: $[4y^3 - (y^2 - 3)][4y^3 + (y^2 - 3)]$. [4.5]

63. Solve: $x(x^2 + 3x - 28) - 12(x^2 + 3x - 28) = 0$. [5.6]

64. Solve: $\dfrac{2}{x - 3} \cdot \dfrac{3}{x + 3} - \dfrac{4}{x^2 - 7x + 12} = 0$. [6.6]

Systems and More Graphing

ARUNACHAL SEN
COMPUTER HARDWARE
TECHNICIAN
Pensacola, Florida

Discrete math is fundamental to computer science. During my work as a computer technician, I use math in many different ways—from analyzing system performance to calculating the cost of replacement parts.

AN APPLICATION

The time *t* that it takes to download a file varies directly as the size *s* of the file. Using her broadband Internet connection, La'Neiya can download a 4-MB (megabyte) song in 30 sec. How long would it take her to download a 210-MB television show?

Source: docs.info.apple.com

This problem appears as Example 3 in Section 7.7.

M any problems translate into two or more equations that must all be true in order for the solution found to be correct. Section 7.1 examines how graphing can be used to solve such a *system of equations*, and Sections 7.2 and 7.3 discuss two algebraic methods of solving systems. These three methods are then used in a variety of applications and as an aid in solving systems of inequalities.

7.1   Systems of Equations and Graphing

Solutions of Systems   ▪   Solving Systems of Equations by Graphing

Solutions of Systems

A **system of equations** is a set of two or more equations that are to be solved *simultaneously*. A solution of a system will make all of the equations true. It is often easier to translate real-world situations to a system of two equations that use two variables, or unknowns, than it is to represent the situation with one equation using one variable. To see this, let's see how the following problem from Section 2.5 can be represented by two equations using two unknowns:

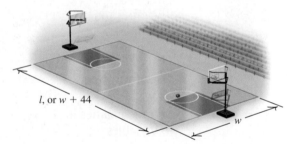

l, or *w* + 44

w

> The perimeter of an NBA basketball court is 288 ft. The length is 44 ft longer than the width. Find the dimensions of the court.
>
> **Source:** National Basketball Association

If we let w = the width of the court, in feet, and l = the length of the court, in feet, the problem translates to the following system of equations:

$$2l + 2w = 288, \qquad \text{The perimeter is 288.}$$
$$l = w + 44. \qquad \text{The length is 44 more than the width.}$$

Both equations must be true. A solution of this system is an ordered pair of the form (l, w) for which $2l + 2w = 288$ *and* $l = w + 44$. Note that variables are listed in alphabetical order within an ordered pair unless stated otherwise.

To solve a system of two equations is to find all ordered pairs (if any exist) for which *both* equations are true.

EXAMPLE   1   Consider the system from above:

$$2l + 2w = 288,$$
$$l = w + 44.$$

Determine whether each pair is a solution of the system: **(a)** (94, 50); **(b)** (90, 46).

SOLUTION

a) We check by substituting (alphabetically) 94 for l and 50 for w:

$$
\begin{array}{c|c}
2l + 2w = 288 & \\
\hline
2 \cdot 94 + 2 \cdot 50 & 288 \\
188 + 100 & \\
288 \overset{?}{=} 288 & \text{TRUE}
\end{array}
\qquad
\begin{array}{c|c}
l = w + 44 & \\
\hline
94 & 50 + 44 \\
94 \overset{?}{=} 94 & \text{TRUE}
\end{array}
$$

Since (94, 50) checks in *both* equations, it is a solution of the system.

b) We substitute 90 for l and 46 for w:

$$
\begin{array}{c|c}
2l + 2w = 288 & l = w + 44 \\
\hline
2 \cdot 90 + 2 \cdot 46 \mid 288 & 90 \mid 46 + 44 \\
180 + 92 \mid & 90 \stackrel{?}{=} 90 \qquad \text{TRUE} \\
272 \stackrel{?}{=} 288 \quad \text{FALSE} &
\end{array}
$$

Since $(90, 46)$ is not a solution of *both* equations, it is not a solution of the system.

TRY EXERCISE 5

In Example 1, we demonstrated that $(94, 50)$ is a solution of the system, but we did not show how the pair $(94, 50)$ was found. One way to find such a solution uses graphs.

Solving Systems of Equations by Graphing

Recall that a graph of an equation is a set of points representing its solution set. Each point on the graph corresponds to an ordered pair that is a solution of the equation. By graphing two equations using one set of axes, we can identify a solution of both equations by looking for a point of intersection.

EXAMPLE **2**

Solve this system of equations by graphing:

$$x + y = 7,$$
$$y = 3x - 1.$$

SOLUTION   We graph the equations using any method studied earlier. The equation $x + y = 7$ can be graphed easily using the intercepts, $(0, 7)$ and $(7, 0)$. The equation $y = 3x - 1$ is in slope–intercept form, so it can be graphed by plotting its y-intercept, $(0, -1)$, and "counting off" a slope of 3.

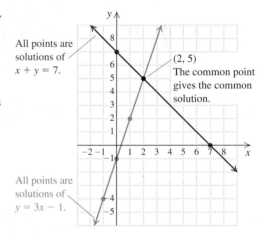

The "apparent" solution of the system, $(2, 5)$, should be checked in both equations.

Check:
$$
\begin{array}{c|c}
x + y = 7 & y = 3x - 1 \\
\hline
2 + 5 \mid 7 & 5 \mid 3 \cdot 2 - 1 \\
7 \stackrel{?}{=} 7 \quad \text{TRUE} & 5 \stackrel{?}{=} 5 \qquad \text{TRUE}
\end{array}
$$

Since it checks in both equations, $(2, 5)$ is a solution of the system.

TRY EXERCISE 11

A system of equations that has at least one solution, like the systems in Examples 1 and 2, is said to be **consistent**. A system for which there is no solution is said to be **inconsistent**.

EXAMPLE **3**

Solve this system of equations by graphing:

$$y = \tfrac{5}{2}x + 4,$$
$$y = \tfrac{5}{2}x - 3.$$

SOLUTION Both equations are in slope–intercept form so it is easy to see that both lines have the same slope, $\tfrac{5}{2}$. The y-intercepts differ so the lines are parallel, as shown in the figure at right.

Because the lines are parallel, there is no point of intersection. Thus the system is inconsistent and has no solution.

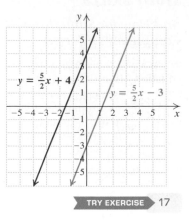

TRY EXERCISE 17

Sometimes both equations in a system have the same graph.

EXAMPLE **4**

Solve this system of equations by graphing:

$$2x + 3y = 6,$$
$$-8x - 12y = -24.$$

SOLUTION Graphing the equations, we see that they both represent the same line. This can also be seen by solving each equation for y, obtaining the equivalent slope–intercept form, $y = -\tfrac{2}{3}x + 2$. Because the equations are equivalent, any solution of one equation is a solution of the other equation as well. We show four such solutions.

We check one solution, $(0, 2)$, in each of the original equations.

$$\frac{2x + 3y = 6}{\begin{array}{c|c} 2(0) + 3(2) & 6 \\ 0 + 6 & \\ & 6 \overset{?}{=} 6 \quad \text{TRUE} \end{array}}$$

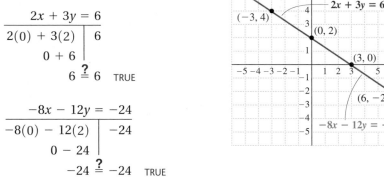

$$\frac{-8x - 12y = -24}{\begin{array}{c|c} -8(0) - 12(2) & -24 \\ 0 - 24 & \\ & -24 \overset{?}{=} -24 \quad \text{TRUE} \end{array}}$$

On your own, check that $(3, 0)$ is also a solution of the system. If two points are solutions, the lines coincide and all points on the line are solutions.

Since a solution exists, the system is consistent. We state that there is an infinite number of solutions.

TRY EXERCISE 25

When one equation can be obtained by multiplying both sides of another equation by a nonzero constant, the two equations are called **dependent**. Thus the equations in Example 4 are dependent, but those in Examples 2 and 3 are **independent**. For systems of two equations, when two equations are dependent, they are equivalent. For systems containing more than two equations, the definition of dependent is slightly different and it is possible for dependent equations to not be equivalent.

When a system of two linear equations in two variables is graphed, one of the following must occur.

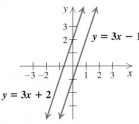

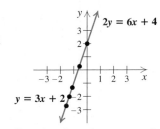

Graphs intersect at one point.
The system is *consistent*
and has one solution.
Since neither equation is
a multiple of the other, the
equations are *independent*.

Graphs are parallel.
The system is *inconsistent*
because there is no solution.
Since neither equation is
a multiple of the other, the
equations are *independent*.

Equations have the same graph.
The system is *consistent* and has
an infinite number of solutions.
Since one equation is a multiple
of the other, the equations are
dependent.

Graphing calculators can be
used to solve systems of equations. Before solving Example 2
on a graphing calculator, we
must solve $x + y = 7$ for y. An
equivalent system is

$$y = -x + 7,$$
$$y = 3x - 1.$$

We then enter the two equations as y_1 and y_2. After graphing the equations, we use the
INTERSECT option of the CALC
menu to identify and display
the point of intersection.

1. Use a graphing calculator
   to check Example 2.
2. Graph the system

$$y_1 = 0.23x + 1.49,$$
$$y_2 = 0.23x + 3.49$$

and then, using the Y-VARS
option of the VARS menu,
let $y_3 = y_1 - y_2$. Because
the graphs of y_1 and y_2 are
parallel, we expect y_3 to be
a constant. Use TRACE or
the TABLE feature to confirm
that this is the case.

ALGEBRAIC–GRAPHICAL CONNECTION

Let's take an algebraic–graphical look at equation solving.
Consider the equation $2x - 5 = 3$. Let's solve it algebraically as we
did in Chapter 2:

$$2x - 5 = 3$$
$$2x = 8 \qquad \text{Adding 5 to both sides}$$
$$x = 4. \qquad \text{Dividing both sides by 2}$$

To solve $2x - 5 = 3$ graphically, we
graph the equations $y = 2x - 5$ and
$y = 3$, as shown at right. The point
of intersection, $(4, 3)$, indicates that
when x is 4, the value of $2x - 5$ is 3.
Thus the solution of $2x - 5 = 3$ is 4.

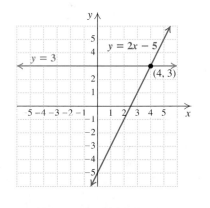

EXAMPLE 5

Solve graphically: $5 - x = x - 1$.

SOLUTION   We graph $y = 5 - x$ and $y = x - 1$,
as shown. The graphs intersect at $(3, 2)$, indicating that for the x-value 3 both $5 - x$ and
$x - 1$ share the same value (in this case, 2). As
a check, note that $5 - 3 = 3 - 1$ is true. The
solution is 3.

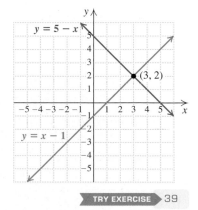

TRY EXERCISE ▸ 39

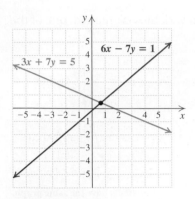

Although graphing lets us "see" the solution of a system, it does not always allow us to find a precise solution. For example, the solution of the system

$$3x + 7y = 5,$$
$$6x - 7y = 1$$

is $\left(\frac{2}{3}, \frac{3}{7}\right)$, but finding that precise solution from a graph—*even with a computer or graphing calculator*—can be difficult. Fortunately, systems like this can be solved accurately with the methods discussed in Sections 7.2 and 7.3.

7.1   EXERCISE SET

For Extra Help   MyMathLab   MathXL PRACTICE   WATCH   DOWNLOAD

✎ **Concept Reinforcement**   *Choose the correct word(s) to complete each sentence.*

1. A solution of a system of two equations is an ordered pair that is a solution of _____ equation(s).   both/at least one

2. A solution of a system of two equations can be found by identifying where the two graphs _____ .
   intersect/cross the *y*-axis

3. A system of equations that has at least one solution is said to be _____ .
   inconsistent/consistent

4. When one equation in a system can be obtained by multiplying both sides of another equation in the system by a nonzero constant, the equations are said to be _____ .
   dependent/independent

Determine whether each ordered pair is a solution of the system of equations. Use alphabetical order of the variables.

5. $(2, 5)$; $2x + 3y = 19$,
    $3x - y = 1$

6. $(1, 4)$; $5x - 2y = -3$,
    $7x - 3y = -5$

7. $(3, 2)$; $3b - 2a = 0$,
    $b + 2a = 15$

8. $(2, -2)$; $b + 2a = 2$,
    $b - a = -4$

9. $(-15, 20)$; $3x + 2y = -5$,
    $4y + 5x = 5$

10. $(-2, -1)$; $r - 3t = 1$,
     $r + 2t = 0$

Solve each system of equations by graphing. If there is no solution or an infinite number of solutions, state this.

11. $x + y = 4$,
     $x - y = 2$

12. $x - y = 3$,
     $x + y = 7$

13. $y = -2x + 5$,
     $x + y = 4$

14. $y = 2x - 5$,
     $x + y = 4$

15. $y = -2x - 1$,
     $y = 2 - x$

16. $y = -3x + 1$,
     $y = 3 - x$

17. $4x - 20 = 5y$,
     $8x - 10y = 12$

18. $6x + 12 = 2y$,
     $6 - y = -3x$

19. $x = 4$,
     $y = -1$

20. $x = -6$,
     $y = 1$

21. $2x + y = 8$,
     $x - y = 7$

22. $3x + y = 4$,
     $x - y = 4$

23. $y - x = 5$,
     $x + 2y = 4$

24. $y - x = 8$,
     $x + 2y = 1$

25. $x + 2y = 7$,
     $3x + 6y = 21$

26. $x + 3y = 6$,
     $4x + 12y = 24$

27. $2x = 3y - 6$,
     $x = 3y$

28. $3y - 9 = 6x$,
     $y = x$

Aha! 29. $y = \frac{1}{5}x + 4$,
     $2y = \frac{2}{5}x + 8$

30. $y = \frac{1}{3}x + 2$,
     $y = \frac{1}{3}x - 7$

31. $4x + y = 2$,
     $x = \frac{1}{2}y + 5$

32. $3x - y = 1$,
     $x = \frac{1}{5}y + 1$

33. $2x - 3y = 5$,
     $x - 2y = 6$

34. $3x + 4y = 8$,
     $x + 2y = 10$

35. $3x + 2y = 1$,
     $2x + 5y = -14$

36. $4x + 2y = -2$,
     $5x + 4y = 5$

37. $x = \frac{1}{3}y$,
     $y = 6$

38. $x = \frac{1}{2}y$,
     $x = 3$

Solve graphically.

39. $2x - 1 = 3$

40. $3x - 1 = 2$

41. $x - 4 = 6 - x$

42. $x - 3 = 1 - x$

43. $2x - 1 = -x + 5$

44. $-x + 4 = 2x - 5$

45. $\frac{1}{2}x + 3 = -\frac{1}{2}x - 1$

46. $\frac{3}{2}x + 5 = -\frac{1}{2}x + 1$

47. Is it possible for a system of two linear equations to have exactly two solutions? Why or why not?

48. Suppose that the graphs of both lines in a system of two equations have the same slope. What must be true of the solution of the system?

Skill Review

To prepare for Section 7.2, review solving linear equations (Sections 2.2 and 2.3).

Solve. [2.2]

49. $3x - (4 - 2x) = 9$

50. $7x - 2(4 + 3x) = 5$

51. $2(8 - 5y) - y = 4$

52. $3(5 - 2y) - 4y = 8$

53. Solve $3x - 4y = 2$ for y. [2.3]

54. Solve $2x + 3y = -1$ for x. [2.3]

Synthesis

55. Suppose that the equations in a system of two linear equations are dependent. Does it follow that the system is consistent? Why or why not?

56. Explain why slope–intercept form can be especially useful when solving systems of equations by graphing.

57. Which of the systems in Exercises 11–38 contain dependent equations?

58. Which of the systems in Exercises 11–38 are consistent?

59. Which of the systems in Exercises 11–38 are inconsistent?

60. Which of the systems in Exercises 11–38 contain independent equations?

61. Write an equation that can be paired with $5x + 2y = 3$ to form a system that has $(-1, 4)$ as the solution. Answers may vary.

62. Write an equation that can be paired with $4x + 3y = 6$ to form a system that has $(3, -2)$ as the solution. Answers may vary.

63. The solution of the following system is $(2, -3)$. Find A and B.
$$Ax - 3y = 13,$$
$$x - By = 8$$

64. Solve by graphing:
$$4x - 8y = -7,$$
$$2x + 3y = 7.$$
(*Hint*: Use four squares per unit on your graph.)

65. *Copying costs.* Shelby occasionally goes to The UPS Store® with small copying jobs. He can purchase a "copy card" for $20 that will entitle him to 500 copies, or he can simply pay 6¢ per page.

   a) Create cost equations for each method of paying for a number (up to 500) of copies.

   b) Graph both cost equations on the same set of axes.

   c) Use the graph to determine how many copies Shelby must make if the card is to be more economical.

Aha! **66.** Solve:
$$3x - 4y = 2,$$
$$-6x + 8y = -6.$$

67. *College faculty.* In 2003, about 632,000 full-time faculty taught in U.S. colleges and universities and that number was increasing at a rate of 8000 faculty members per year. There were 543,000 part-time faculty members in 2003, and that number was growing at a rate of 13,000 faculty members per year.
Source: U.S. National Center for Education Statistics

   a) Write two equations that can be used to predict n, the number of full-time and part-time faculty, in thousands, t years after 2003.

   b) Use a graphing calculator to determine the year in which the numbers of full-time faculty and part-time faculty are the same.

68. Use a graphing calculator to solve the system
$$y = 1.2x - 32.7,$$
$$y = -0.7x + 46.15.$$

69. Use a graphing calculator to solve
$$1.3x - 4.9 = 6.3 - 3.7x.$$

COLLABORATIVE CORNER

Conserving Energy and Money

Focus: System of linear equations (two variables)
Time: 20 minutes
Group size: 2
Materials: Graph paper

Jean, a condo owner, has an old electric water heater that consumes $110 of electricity per month. To replace it with a new gas water heater that consumes only $35 of gas per month will cost $550 plus $250 for installation. Jean wants to know how long it will take for the new gas water heater to "pay for itself," in other words, to produce savings.

ACTIVITY

1. The "break-even" point occurs when the costs of the two water heaters are equal. Determine, by "guessing and checking," the number of months before Jean breaks even.

2. One of the group members should create a cost equation of the form $y = mx + b$ for the electric heater, where y is the cost, in dollars, and x is the time, in months. He or she should also graph the equation on the graph paper provided.

3. The second group member should create a cost equation of the form $y = mx + b$ for the gas heater, where y is the cost, in dollars, and x is the time, in months. This equation should be graphed on the same graph used in part (2).

4. Working together, the group should determine the coordinates of the point of intersection of the two lines. This is the break-even point. Which coordinate indicates the number of months before Jean breaks even? What does the other coordinate indicate? Compare the answer found graphically with the estimate made in part (1).

7.2 Systems of Equations and Substitution

The Substitution Method ▪ Solving for the Variable First ▪ Problem Solving

Near the end of Section 7.1, we mentioned that graphing can be an imprecise method for solving systems. In this section and the next, we develop methods of finding exact solutions using algebra.

The Substitution Method

One method for solving systems is known as the **substitution method**. It uses algebra instead of graphing and is thus considered an *algebraic* method.

EXAMPLE 1 Solve the system

$$x + y = 7, \quad (1)$$
$$y = 3x - 1. \quad (2)$$

We have numbered the equations (1) and (2) for easy reference.

SOLUTION   The second equation says that y and $3x - 1$ represent the same value. Thus, in the first equation, we can substitute $3x - 1$ for y:

$$x + y = 7, \qquad \text{Equation (1)}$$
$$x + 3x - 1 = 7. \qquad \text{Substituting } 3x - 1 \text{ (from equation 2) for } y \text{ (in equation 1)}$$

The equation $x + 3x - 1 = 7$ has only one variable, for which we now solve:

$$4x - 1 = 7 \qquad \text{Combining like terms}$$
$$4x = 8 \qquad \text{Adding 1 to both sides}$$
$$x = 2. \qquad \text{Dividing both sides by 4}$$

We have found the x-value of the solution. To find the y-value, we return to the original pair of equations. Substituting into either equation will give us the y-value. We choose equation (1):

$$x + y = 7 \qquad \text{Equation (1)}$$
$$2 + y = 7 \qquad \text{Substituting 2 for } x$$
$$y = 5. \qquad \text{Subtracting 2 from both sides}$$

The ordered pair $(2, 5)$ appears to be a solution. We check:

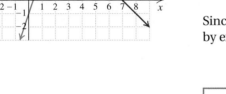

$$\begin{array}{c|c} x + y = 7 \\ \hline 2 + 5 \ | \ 7 \\ 7 \overset{?}{=} 7 \quad \text{TRUE} \end{array} \qquad \begin{array}{c|c} y = 3x - 1 \\ \hline 5 \ | \ 3 \cdot 2 - 1 \\ 5 \overset{?}{=} 5 \qquad \text{TRUE} \end{array}$$

Since $(2, 5)$ checks, it is the solution. For this particular system, we can also check by examining the graph from Example 2 in Section 7.1, as shown at left.

TRY EXERCISE ▶ 5

> *CAUTION!*   A solution of a system of equations in two variables is an ordered *pair* of numbers. Once you have solved for one variable, don't forget the other. A common mistake is to solve for only one variable.

EXAMPLE   2   Solve:

$$x = 3 - 2y, \qquad (1)$$
$$y - 3x = 5. \qquad (2)$$

SOLUTION   We substitute $3 - 2y$ for x in the second equation:

$$y - 3x = 5 \qquad \text{Equation (2)}$$
$$y - 3(3 - 2y) = 5. \qquad \text{Substituting } 3 - 2y \text{ for } x. \text{ The parentheses are very important.}$$

Now we solve for y:

$$y - 9 + 6y = 5 \qquad \text{Using the distributive law}$$
$$\left.\begin{array}{r} 7y - 9 = 5 \\ 7y = 14 \\ y = 2. \end{array}\right\} \quad \text{Solving for } y$$

Next, we substitute 2 for y in equation (1) of the original system:

$$x = 3 - 2y \qquad \text{Equation (1)}$$
$$x = 3 - 2 \cdot 2 \qquad \text{Substituting 2 for } y$$
$$x = -1. \qquad \text{Simplifying}$$

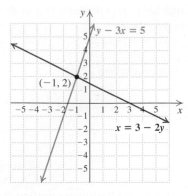

We check the ordered pair $(-1, 2)$.

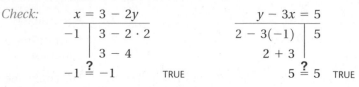

Check:

$x = 3 - 2y$	
-1	$3 - 2 \cdot 2$
	$3 - 4$
$-1 \stackrel{?}{=} -1$	TRUE

$y - 3x = 5$	
$2 - 3(-1)$	5
$2 + 3$	
$5 \stackrel{?}{=} 5$	TRUE

The pair $(-1, 2)$ is the solution. A graph is shown at left as another check.

> TRY EXERCISE ▶ 11

Solving for the Variable First

Sometimes neither equation has a variable alone on one side. In that case, we solve one equation for one of the variables and then proceed as before.

EXAMPLE 3   Solve:

$$x - 2y = 6, \quad (1)$$
$$3x + 2y = 4. \quad (2)$$

SOLUTION   We can solve either equation for either variable. Since the coefficient of x is 1 in equation (1), it is easier to solve that equation for x:

$x - 2y = 6$		Equation (1)
$x = 6 + 2y.$	(3)	Adding $2y$ to both sides

We substitute $6 + 2y$ for x in equation (2) of the original pair and solve for y:

$3x + 2y = 4$	Equation (2)
$3(6 + 2y) + 2y = 4$	Substituting $6 + 2y$ for x

> Remember to use parentheses when you substitute.

$18 + 6y + 2y = 4$	Using the distributive law
$18 + 8y = 4$	Combining like terms
$8y = -14$	Subtracting 18 from both sides
$y = \dfrac{-14}{8} = -\dfrac{7}{4}.$	Dividing both sides by 8 and simplifying

To find x, we can substitute $-\frac{7}{4}$ for y in equation (1), (2), or (3). Because it is generally easier to use an equation that has already been solved for a specific variable, we decide to use equation (3):

$$x = 6 + 2y = 6 + 2\left(-\tfrac{7}{4}\right) = 6 - \tfrac{7}{2} = \tfrac{12}{2} - \tfrac{7}{2} = \tfrac{5}{2}.$$

We check the ordered pair $\left(\frac{5}{2}, -\frac{7}{4}\right)$ in the original equations.

Check:

$x - 2y = 6$	
$\frac{5}{2} - 2\left(-\frac{7}{4}\right)$	6
$\frac{5}{2} + \frac{7}{2}$	
$\frac{12}{2}$	
$6 \stackrel{?}{=} 6$	TRUE

$3x + 2y = 4$	
$3 \cdot \frac{5}{2} + 2\left(-\frac{7}{4}\right)$	4
$\frac{15}{2} - \frac{7}{2}$	
$\frac{8}{2}$	
$4 \stackrel{?}{=} 4$	TRUE

Since $\left(\frac{5}{2}, -\frac{7}{4}\right)$ checks, it is the solution.

> TRY EXERCISE ▶ 17

Some systems have no solution and some have an infinite number of solutions.

TECHNOLOGY CONNECTION

To check Example 3 with a graphing calculator, we must first solve each equation for y. When we do so, equation (1) becomes $y = (6 - x)/(-2)$ and equation (2) becomes $y = (4 - 3x)/2$.

1. Use the INTERSECT option of the CALC menu to determine the solution of the system.
2. What happens when parentheses are deleted from the two equations above?

EXAMPLE 4

Solve each system.

a) $y = \frac{5}{2}x + 4,$ (1)
  $y = \frac{5}{2}x - 3$ (2)

b) $2y = 6x + 4,$ (1)
  $y = 3x + 2$ (2)

SOLUTION

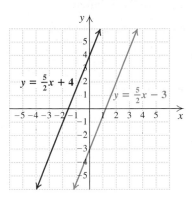

$y = \frac{5}{2}x + 4$

$y = \frac{5}{2}x - 3$

a) We solved this system graphically in Example 3 of Section 7.1. The lines are parallel and the system has no solution. Let's see what happens if we try to solve this system by substituting $\frac{5}{2}x - 3$ for y in the first equation:

$$y = \frac{5}{2}x + 4 \qquad \text{Equation (1)}$$
$$\frac{5}{2}x - 3 = \frac{5}{2}x + 4 \qquad \text{Substituting } \frac{5}{2}x - 3 \text{ for } y$$
$$-3 = 4. \qquad \text{Subtracting } \frac{5}{2}x \text{ from both sides}$$

When we subtract $\frac{5}{2}x$ from both sides, we obtain a *false* equation. In such a case, when solving algebraically leads to a false equation, we state that the system has no solution and thus is inconsistent.

b) A graph of this system first appears on p. 433. The lines coincide, so the system has an infinite number of solutions. If we use substitution to solve the system, we can substitute $3x + 2$ for y in equation (1):

$$2y = 6x + 4 \qquad \text{Equation (1)}$$
$$2(3x + 2) = 6x + 4 \qquad \text{Substituting } 3x + 2 \text{ for } y$$
$$6x + 4 = 6x + 4$$
$$4 = 4. \qquad \text{Subtracting } 6x \text{ from both sides}$$

This last equation is always true. When the algebraic solution of a system of two equations leads to an equation that is true for all real numbers, we state that the system has an infinite number of solutions.

TRY EXERCISE 21

To Solve a System Using Substitution

1. Solve for a variable in either one of the equations if neither equation already has a variable isolated.
2. Using the result of step (1), substitute in the *other* equation for the variable isolated in step (1).
3. Solve the equation from step (2).
4. Substitute the solution from step (3) into one of the other equations to solve for the other variable.
5. Check that the ordered pair resulting from steps (3) and (4) checks in both of the original equations.

Problem Solving

Now let's use the substitution method in problem solving.

EXAMPLE 5

Supplementary angles. Two angles are supplementary. One angle measures 30° more than twice the other. Find the measures of the two angles.

SOLUTION

1. **Familiarize.** Recall that two angles are supplementary if the sum of their measures is 180°. We could try to guess a solution, but instead we make a drawing and translate. Let x and y represent the measures of the two angles.

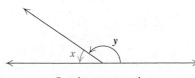

Supplementary angles

2. **Translate.** Since we are told that the angles are supplementary, one equation is

$$x + y = 180. \quad (1)$$

The second sentence can be rephrased and translated as follows:

Rewording: One angle   is   30° more than two times the other.

Translating: y = $2x + 30$ (2)

We now have a system of two equations in two unknowns:

$$x + y = 180, \quad (1)$$
$$y = 2x + 30. \quad (2)$$

3. **Carry out.** We substitute $2x + 30$ for y in equation (1):

$x + y = 180$	Equation (1)
$x + (2x + 30) = 180$	Substituting
$3x + 30 = 180$	
$3x = 150$	Subtracting 30 from both sides
$x = 50.$	Dividing both sides by 3

Substituting 50 for x in equation (1) then gives us

$x + y = 180$	Equation (1)
$50 + y = 180$	Substituting 50 for x
$y = 130.$	

4. **Check.** If one angle is 50° and the other is 130°, then the sum of the measures is 180°. Thus the angles are supplementary. If 30° is added to twice the measure of the smaller angle, we have $2 \cdot 50° + 30°$, or 130°, which is the measure of the other angle. The numbers check.

5. **State.** One angle measures 50° and the other 130°.   ▸ **TRY EXERCISE** ▸ 45

7.2   EXERCISE SET

For Extra Help | MyMathLab | Math XL PRACTICE | WATCH | DOWNLOAD

↩ **Concept Reinforcement**   *Classify each statement as either true or false.*

1. When using the substitution method, we must solve for the variables in the order in which they occur alphabetically.

2. The substitution method often requires us to first solve for a variable, much as we did when solving for a letter in a formula.

3. When solving a system of equations algebraically leads to a false equation, the system has no solution.

4. When solving a system of two equations algebraically leads to an equation that is always true, the system has an infinite number of solutions.

Solve each system using the substitution method. If a system has no solution or an infinite number of solutions, state this.

5. $x + y = 9,$
   $\quad y = x + 1$

6. $x + y = 5,$
   $\quad x = y + 1$

7. $\quad x = y + 1,$
   $x + 2y = 4$

8. $\quad y = x - 3,$
   $3x + y = 5$

9. $\quad y = 5x - 1,$
   $y - 3x = 1$

10. $\quad y = 2x - 5,$
    $2y - x = 2$

11. $\quad a = -4b,$
    $a + 5b = 5$

12. $\quad r = -3s,$
    $r + 4s = 10$

13. $\quad x = y - 5,$
    $2x + 5y = 4$

14. $\quad x = y - 6,$
    $3x + 2y = 2$

15. $x = 2y + 1,$
         $3x - 6y = 2$

16. $y = 3x - 1,$
         $6x - 2y = 2$

17. $s + t = -5,$
         $s - t = 3$

18. $s - t = 2,$
         $s + t = -4$

19. $x - y = 5,$
         $x + 2y = 7$

20. $y - 2x = -6,$
         $2y - x = 5$

21. $x - 2y = 7,$
         $3x - 21 = 6y$

22. $x - 4y = 3,$
         $2x - 6 = 8y$

23. $y = 2x + 5,$
         $-2y = -4x - 10$

24. $y = -2x + 3,$
         $3y = -6x + 9$

25. $4x - y = -3,$
         $2x + 5y = 2$

26. $2x + 3y = -2,$
         $2x - y = 9$

27. $a - b = 6,$
         $3a - 2b = 12$

28. $x - y = -3,$
         $2x + 3y = -6$

29. $s = \frac{1}{2}r,$
         $3r - 4s = 10$

30. $x = \frac{1}{2}y,$
         $2x + y = 12$

31. $x - 3y = 7,$
         $-4x + 12y = 28$

32. $8x + 2y = 6,$
         $y = 3 - 4x$

33. $x - 2y = 5,$
         $2y - 3x = 1$

34. $x - 3y = -1,$
         $5y - 2x = 4$

Aha! **35.** $2x - y = 0,$
             $2x - y = -2$

36. $5x = y - 3,$
         $5x = y + 5$

Solve using a system of equations.

37. The sum of two numbers is 63. One number is 7 more than the other. Find the numbers.

38. The sum of two numbers is 74. One number is 6 more than the other. Find the numbers.

39. Find two numbers for which the sum is 51 and the difference is 13.

40. Find two numbers for which the sum is 51 and the difference is 5.

41. The difference between two numbers is 2. Three times the larger number plus one-half the smaller is 34. What are the numbers?

42. The difference between two numbers is 11. Twice the smaller number plus three times the larger is 93. What are the numbers?

43. *Supplementary angles.* Two angles are supplementary. One angle is 15° more than twice the other. Find the measure of each angle.

44. *Supplementary angles.* Two angles are supplementary. One angle is 8° less than three times the other. Find the measure of each angle.

45. *Complementary angles.* Two angles are complementary. One angle is 3° less than one-half the other. Find the measure of each angle. (*Complementary angles* are pairs of angles for which the sum is 90°.)

Complementary angles

46. *Complementary angles.* Two angles are complementary. One angle is 42° more than one-half the other. Find the measure of each angle.

47. *Billboards.* In January 2004, a record-sized billboard advertising a new line of clothing was unveiled in Times Square, New York City. The perimeter of the billboard was 452 ft, and the length was 118 ft more than the width. Find the length and the width.

48. *Two-by-four.* The perimeter of a cross section of a "two-by-four" piece of lumber is $10\frac{1}{2}$ in. The length is twice the width. Find the actual dimensions of the cross section of a two-by-four.

Two-by-four $P = 10\frac{1}{2}$ in.

49. *Dimensions of Wyoming.* The state of Wyoming is a rectangle with a perimeter of 1280 mi. The width is 90 mi less than the length. Find the length and the width.

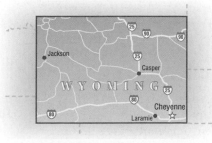

50. *Dimensions of Colorado.* The state of Colorado is roughly in the shape of a rectangle whose perimeter is 1300 mi. The width is 110 mi less than the length. Find the length and the width.

51. *Soccer.* The perimeter of a soccer field is 280 yd. The width is 5 yd more than half the length. Find the length and the width.

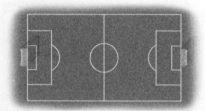

52. *Racquetball.* A regulation racquetball court has a perimeter of 120 ft, with a length that is twice the width. Find the length and the width of a court.

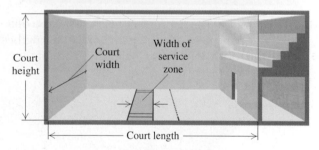

53. *Racquetball.* The height of the front wall of a standard racquetball court is four times the width of the service zone (see the figure). Together, these measurements total 25 ft. Find the height and the width.

54. *Lacrosse.* The perimeter of a lacrosse field is 340 yd. The length is 10 yd less than twice the width. Find the length and the width.

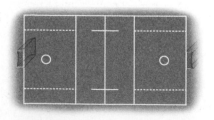

55. Gary solves every system of two equations (in x and y) by first solving for y in the first equation and then substituting into the second equation. Is he using the best approach? Why or why not?

56. Describe two advantages of the substitution method over the graphing method for solving systems of equations.

Skill Review

To prepare for Section 7.3, review simplifying algebraic expressions (Section 1.8).

Simplify. [1.8]

57. $3(4x + 2y) - 5(2x + y)$

58. $5(2x + 3y) - 3(7x + 5y)$

59. $4(5x + 6y) - 5(4x + 7y)$

60. $3(8x + 1 - 2y) - 8(3x + 1)$

61. $2(3x - 4y) + 4(5x + 2y)$

62. $4(2x + 3y) + 3(5x - 4y)$

Synthesis

63. Angenita can tell by inspection that the system
$$x = 2y - 1,$$
$$x = 2y + 3$$
has no solution. How can she tell?

64. Under what circumstances can a system of equations be solved more easily by graphing than by substitution?

Solve by the substitution method.

65. $\frac{1}{6}(a + b) = 1,$
  $\frac{1}{4}(a - b) = 2$

66. $\frac{x}{5} - \frac{y}{2} = 3,$
  $\frac{x}{4} + \frac{3y}{4} = 1$

67. $y + 5.97 = 2.35x,$
  $2.14y - x = 4.88$

68. $a + 4.2b = 25.1,$
  $9a - 1.8b = 39.78$

69. *Age at marriage.* Trudy is 20 yr younger than Dennis. She feels that she needs to be 7 more than half of Dennis's age before they can marry. What is the youngest age at which Trudy can marry Dennis and honor this requirement?

Exercises 70 and 71 contain systems of three equations in three variables. A solution is an ordered triple of the form (x, y, z). Use the substitution method to solve.

70. $x + y + z = 4,$
  $x - 2y - z = 1,$
    $y = -1$

71. $x + y + z = 180,$
    $x = z - 70,$
   $2y - z = 0$

72. *Softball.* The perimeter of a softball diamond is two-thirds of the perimeter of a baseball diamond. Together, the two perimeters measure 200 yd. Find the distance between the bases in each sport.

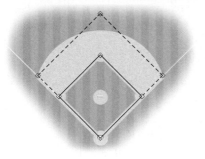

73. Solve Example 3 by first solving for $2y$ in equation (1) and then substituting for $2y$ in equation (2). Is this method easier than the procedure used in Example 3? Why or why not?

74. Write a system of two linear equations that can be solved more quickly—but still precisely—by a graphing calculator than by substitution. Time yourself using both methods to solve the system.

7.3   Systems of Equations and Elimination

Solving by the Elimination Method   ▪   Problem Solving

We have seen that graphing is not always a precise method of solving a system of equations. The substitution method, considered in Section 7.2, is precise but sometimes difficult to use. For example, let's begin to solve the system

$$2x + 3y = 13, \quad (1)$$
$$4x - 3y = 17 \quad (2)$$

by substitution. We need to first solve for a variable in one of the equations. Solving equation (1) for y, we find that $y = \frac{13}{3} - \frac{2}{3}x$. We then use the expression $\frac{13}{3} - \frac{2}{3}x$ in equation (2) as a replacement for y:

$$4x - 3\left(\frac{13}{3} - \frac{2}{3}x\right) = 17.$$

Although substitution can be used to solve this system, another method, *elimination,* allows us to solve this system without using fraction notation.

Solving by the Elimination Method

The **elimination method** for solving systems of equations makes use of the addition principle. To see how it works, we use it to solve the system above.

EXAMPLE 1   Solve the system

$$2x + 3y = 13, \quad (1)$$
$$4x - 3y = 17. \quad (2)$$

SOLUTION   According to equation (2), $4x - 3y$ and 17 are the same number. Thus we can add $4x - 3y$ to the left side of equation (1) and 17 to the right side:

$$2x + 3y = 13 \quad (1)$$
$$\underline{4x - 3y = 17} \quad (2)$$
$$6x + 0y = 30. \quad \text{Adding. Note that } y \text{ has been "eliminated."}$$

The resulting equation has just one variable:

$$6x = 30.$$

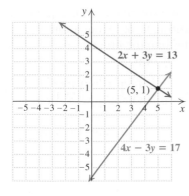

Dividing both sides of this equation by 6, we find that $x = 5$.

Next, we substitute 5 for x in either of the original equations:

$$2x + 3y = 13 \qquad \text{Equation (1)}$$
$$2 \cdot 5 + 3y = 13 \qquad \text{Substituting 5 for } x$$
$$10 + 3y = 13$$
$$3y = 3$$
$$y = 1. \qquad \text{Solving for } y$$

We check the ordered pair $(5, 1)$. The graph shown at left also serves as a check.

Check:

$$\frac{2x + 3y = 13}{\begin{array}{c|c} 2(5) + 3(1) & 13 \\ 10 + 3 & \\ & 13 \overset{?}{=} 13 \quad \text{TRUE} \end{array}} \qquad \frac{4x - 3y = 17}{\begin{array}{c|c} 4(5) - 3(1) & 17 \\ 20 - 3 & \\ & 17 \overset{?}{=} 17 \quad \text{TRUE} \end{array}}$$

Since $(5, 1)$ checks in both equations, it is the solution.   [TRY EXERCISE] 5

Adding in Example 1 eliminated the variable y because two terms, $-3y$ in equation (2) and $3y$ in equation (1), are opposites. Most systems have no pair of terms that are opposites. When this occurs, we can multiply one or both of the equations by appropriate numbers to create a pair of terms that are opposites.

EXAMPLE **2**   Solve:

$$2x + 3y = 8, \qquad (1)$$
$$x + 3y = 7. \qquad (2)$$

SOLUTION   Adding these equations as they now appear will not eliminate a variable. However, if the $3y$ were $-3y$ in one equation, we could eliminate y. We multiply both sides of equation (2) by -1 to find an equivalent equation that contains $-3y$, and then add:

$$2x + 3y = 8 \qquad \text{Equation (1)}$$
$$\underline{-x - 3y = -7} \qquad \text{Multiplying both sides of equation (2) by } -1$$
$$x = 1. \qquad \text{Adding}$$

Next, we substitute 1 for x in either of the original equations:

$$x + 3y = 7 \qquad \text{Equation (2)}$$
$$1 + 3y = 7 \qquad \text{Substituting 1 for } x$$
$$\left.\begin{array}{l} 3y = 6 \\ y = 2. \end{array}\right\} \quad \text{Solving for } y$$

We can check the ordered pair $(1, 2)$. The graph shown at left is also a check.

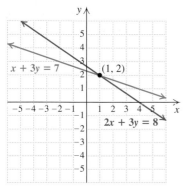

Check:

$$\frac{2x + 3y = 8}{\begin{array}{c|c} 2 \cdot 1 + 3 \cdot 2 & 8 \\ 2 + 6 & \\ & 8 \overset{?}{=} 8 \quad \text{TRUE} \end{array}} \qquad \frac{x + 3y = 7}{\begin{array}{c|c} 1 + 3 \cdot 2 & 7 \\ 1 + 6 & \\ & 7 \overset{?}{=} 7 \quad \text{TRUE} \end{array}}$$

Since $(1, 2)$ checks in both equations, it is the solution.   [TRY EXERCISE] 17

When deciding which variable to eliminate, we inspect the coefficients in both equations. If one coefficient is a multiple of the coefficient of the same variable in the other equation, that is the easiest variable to eliminate.

EXAMPLE 3   Solve:

$$3x + 6y = -6, \quad (1)$$
$$5x - 2y = 14. \quad (2)$$

SOLUTION   No terms are opposites, but if both sides of equation (2) are multiplied by 3 $\left(\text{or if both sides of equation (1) are multiplied by } \frac{1}{3}\right)$, the coefficients of y will be opposites. Note that 6 is the LCM of 2 and 6:

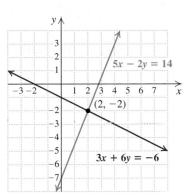

$$3x + 6y = -6 \qquad \text{Equation (1)}$$
$$\underline{15x - 6y = 42} \qquad \text{Multiplying both sides of equation (2) by 3}$$
$$18x \quad\quad = 36 \qquad \text{Adding}$$
$$x = 2. \qquad \text{Solving for } x$$

We then substitute 2 for x in either equation (1) or equation (2):

$$3 \cdot 2 + 6y = -6 \qquad \text{Substituting 2 for } x \text{ in equation (1)}$$
$$\left.\begin{array}{l} 6 + 6y = -6 \\ 6y = -12 \\ y = -2. \end{array}\right\} \quad \text{Solving for } y$$

We leave it to the student to confirm that $(2, -2)$ checks and is the solution. The graph in the margin also serves as a check.

TRY EXERCISE 21

Sometimes both equations must be multiplied to find the least common multiple of two coefficients.

EXAMPLE 4   Solve:

$$3y + 1 + 2x = 0, \quad (1)$$
$$5x = 7 - 4y. \quad (2)$$

SOLUTION   It is often helpful to write both equations in the form $Ax + By = C$ before attempting to eliminate a variable:

$$2x + 3y = -1, \quad (3) \qquad \text{Subtracting 1 from both sides and rearranging the terms of the first equation}$$
$$5x + 4y = 7. \quad (4) \qquad \text{Adding } 4y \text{ to both sides of equation (2)}$$

Since neither coefficient of x is a multiple of the other and neither coefficient of y is a multiple of the other, we use the multiplication principle with *both* equations. Note that we can eliminate the x-term by multiplying both sides of equation (3) by 5 and both sides of equation (4) by -2:

Multiply to get terms that are opposites.

$$10x + 15y = -5 \qquad \text{Multiplying both sides of equation (3) by 5}$$
$$\underline{-10x - 8y = -14} \qquad \text{Multiplying both sides of equation (4) by } -2$$

Solve for one variable.

$$7y = -19 \qquad \text{Adding}$$
$$y = \frac{-19}{7} = -\frac{19}{7}. \qquad \text{Dividing by 7}$$

Substitute.   We substitute $-\frac{19}{7}$ for y in equation (3):

$$2x + 3y = -1 \qquad \text{Equation (3)}$$
$$2x + 3\left(-\frac{19}{7}\right) = -1 \qquad \text{Substituting } -\frac{19}{7} \text{ for } y$$
$$2x - \frac{57}{7} = -1$$
$$2x = -1 + \frac{57}{7} \qquad \text{Adding } \frac{57}{7} \text{ to both sides}$$

Solve for the other variable.

$$2x = -\frac{7}{7} + \frac{57}{7} = \frac{50}{7}$$
$$x = \frac{50}{7} \cdot \frac{1}{2} = \frac{25}{7}. \qquad \text{Solving for } x$$

We check the ordered pair $\left(\frac{25}{7}, -\frac{19}{7}\right)$.

Check in both equations.

Check:

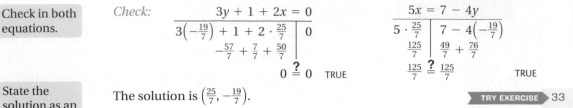

$$\begin{array}{c|c} 3y + 1 + 2x = 0 \\ \hline 3\left(-\frac{19}{7}\right) + 1 + 2 \cdot \frac{25}{7} & 0 \\ -\frac{57}{7} + \frac{7}{7} + \frac{50}{7} & \\ 0 \stackrel{?}{=} 0 & \text{TRUE} \end{array}$$

$$\begin{array}{c|c} 5x = 7 - 4y \\ \hline 5 \cdot \frac{25}{7} & 7 - 4\left(-\frac{19}{7}\right) \\ \frac{125}{7} & \frac{49}{7} + \frac{76}{7} \\ \frac{125}{7} \stackrel{?}{=} \frac{125}{7} & \text{TRUE} \end{array}$$

State the solution as an ordered pair.

The solution is $\left(\frac{25}{7}, -\frac{19}{7}\right)$.

TRY EXERCISE 33

Next, we consider a system with no solution and see what happens when the elimination method is used.

EXAMPLE **5**

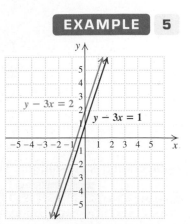

Solve:

$$y - 3x = 2, \quad (1)$$
$$y - 3x = 1. \quad (2)$$

SOLUTION To eliminate y, we multiply both sides of equation (2) by -1. Then we add:

$$\begin{array}{ll} y - 3x = 2 & \\ \underline{-y + 3x = -1} & \text{Multiplying both sides of equation (2) by } -1 \\ 0 = 1. & \text{Adding. Note that this is a } \textit{false} \text{ equation.} \end{array}$$

Note that in eliminating y, we eliminated x as well. The resulting equation, $0 = 1$, is false for any pair (x, y), so there is *no solution*.

TRY EXERCISE 23

Sometimes there is an infinite number of solutions. Consider a system that we graphed in Example 4 of Section 7.1.

EXAMPLE **6**

Solve:

$$2x + 3y = 6, \quad (1)$$
$$-8x - 12y = -24. \quad (2)$$

SOLUTION To eliminate x, we multiply both sides of equation (1) by 4 and then add the two equations:

$$\begin{array}{ll} 8x + 12y = 24 & \text{Multiplying both sides of equation (1) by 4} \\ \underline{-8x - 12y = -24} & \\ 0 = 0. & \text{Adding. Note that this equation is } \textit{always} \text{ true.} \end{array}$$

Again, we have eliminated *both* variables. The resulting equation, $0 = 0$, is always true, indicating that the equations are dependent. Such a system has an infinite number of solutions.

TRY EXERCISE 15

When decimals or fractions appear, we can first multiply to clear them. Then we proceed as before.

EXAMPLE **7**   Solve:

$$\tfrac{1}{2}x + \tfrac{3}{4}y = 2, \quad (1)$$
$$x + 3y = 7. \quad (2)$$

SOLUTION   The number 4 is the LCM of the denominators in equation (1). Thus we multiply both sides of equation (1) by 4 to clear fractions:

$$4\left(\tfrac{1}{2}x + \tfrac{3}{4}y\right) = 4 \cdot 2 \qquad \text{Multiplying both sides of equation (1) by 4}$$
$$4 \cdot \tfrac{1}{2}x + 4 \cdot \tfrac{3}{4}y = 8 \qquad \text{Using the distributive law}$$
$$2x + 3y = 8.$$

The resulting system is

$$2x + 3y = 8, \qquad \text{This equation is equivalent to equation (1).}$$
$$x + 3y = 7.$$

As we saw in Example 2, the solution of this system is $(1, 2)$.   **TRY EXERCISE** 35

Problem Solving

We now use the elimination method to solve a problem.

EXAMPLE **8**   *Printing costs.*   The purchase price of a Canon Pixma Mini 320 printer is $170, and each 4×6 photo printed on the Canon costs 40¢ for ink and paper. An Epson PictureMate Zoom printer costs $200, and each 4×6 photo printed on the Epson costs 25¢. If each printer is used only for 4×6 photos, for what number of photos will the total cost be the same?

Source: www.computershopper.com

SOLUTION

1. **Familiarize.** Some costs are given in dollars and some in cents. Before calculating any costs, we convert the cents to dollars: 40¢ = $0.40, and 25¢ = $0.25.

   To become familiar with the problem, we make and check a guess of 100 photos. If the Canon is purchased, the cost would be $170 + $0.40 · 100, or $210. If the Epson is purchased, the cost would be $200 + $0.25 · 100, or $225. Since 210 ≠ 225, our guess is incorrect. However, from the check, we can see how equations can be written to model the situation. We let p = the number of photos printed and c = the total cost of printing those photos.

2. **Translate.** We reword the problem and translate as follows:

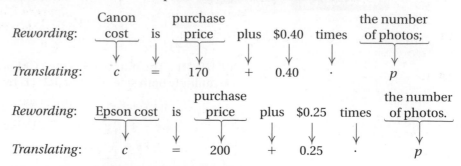

Rewording: $\underbrace{\text{Canon cost}}$ is $\underbrace{\text{purchase price}}$ plus $\$0.40$ times $\underbrace{\text{the number of photos;}}$

Translating: c = 170 + 0.40 · p

Rewording: $\underbrace{\text{Epson cost}}$ is $\underbrace{\text{purchase price}}$ plus $\$0.25$ times $\underbrace{\text{the number of photos.}}$

Translating: c = 200 + 0.25 · p

We now have the system of equations

$$c = 170 + 0.40p,$$
$$c = 200 + 0.25p.$$

3. **Carry out.** To solve the system, we multiply the second equation by -1 and add to eliminate c:

$$
\begin{aligned}
c &= 170 + 0.40p \\
-c &= -200 - 0.25p \\
\hline
0 &= -30 + 0.15p.
\end{aligned}
$$

We can now solve for p:

$30 = 0.15p$ Adding 30 to both sides

$200 = p.$ Dividing both sides by 0.15

4. **Check.** To print 200 photos using the Canon would cost

$170 + 0.40 \cdot 200 = 170 + 80$, or $\$250$.

To print 200 photos using the Epson would cost

$200 + 0.25 \cdot 200 = 200 + 50$, or $\$250$.

The costs are the same for 200 photos.

5. **State.** For 200 photos, the costs are the same.

`TRY EXERCISE` ▸ 39

7.3   EXERCISE SET

For Extra Help   **MyMathLab**   Math XL PRACTICE   WATCH   DOWNLOAD

 Concept Reinforcement   *Classify each statement as either true or false.*

1. The elimination method is never easier to use than the substitution method.

2. The elimination method works especially well when the coefficients of one variable are opposites of each other.

3. When the elimination method yields an equation, like $0 = 0$, that is always true, all real numbers are solutions of the system.

4. When the elimination method yields an equation, like $0 = 3$, that is never true, the system has no solution.

Solve using the elimination method. If a system has no solution or an infinite number of solutions, state this.

5. $x - y = 3,$
    $x + y = 13$

6. $x + y = 5,$
    $x - y = 3$

7. $x + y = 6,$
    $-x + 3y = -2$

8. $x + y = 6,$
    $-x + 2y = 15$

9. $4x + y = 5,$
$2x - y = 7$

10. $2x - y = 3,$
$3x + y = -8$

11. $5a + 4b = 7,$
$-5a + b = 8$

12. $7c + 4d = 16,$
$c - 4d = -4$

13. $8x - 5y = -9,$
$3x + 5y = -2$

14. $3a - 3b = -15,$
$-3a - 3b = -3$

15. $3a - 6b = 8,$
$-3a + 6b = -8$

16. $8x + 3y = 4,$
$-8x - 3y = -4$

17. $-x - y = 3,$
$2x - y = -3$

18. $x - y = 1,$
$3x - y = -5$

19. $x + 3y = 19,$
$x - y = -1$

20. $3x - y = 8,$
$x + 2y = 5$

21. $8x - 3y = -6,$
$5x + 6y = 75$

22. $x - y = 3,$
$2x - 3y = -1$

23. $2w - 3z = -1,$
$-4w + 6z = 5$

24. $7p + 5q = 2,$
$8p - 9q = 17$

25. $4a + 6b = -1,$
$a - 3b = 2$

26. $x + 9y = 1,$
$2x - 6y = 10$

27. $3y = x,$
$5x + 14 = y$

28. $5a = 2b,$
$2a + 11 = 3b$

29. $4x - 10y = 13,$
$-2x + 5y = 8$

30. $2p + 5q = 9,$
$3p - 2q = 4$

31. $2n - 15 - 10m = 40,$
$28 = n - 4m$

32. $30y + 14 + x = 0,$
$41 = 5y - 2x$

33. $3x + 5y = 4,$
$-2x + 3y = 10$

34. $2x + y = 13,$
$4x + 2y = 23$

35. $0.06x + 0.05y = 0.07,$
$0.4x - 0.3y = 1.1$

36. $x - \frac{3}{2}y = 13,$
$\frac{3}{2}x - y = 17$

37. $x + \frac{9}{2}y = \frac{15}{4},$
$\frac{9}{10}x - y = \frac{9}{20}$

38. $1.8x - 2y = 0.9,$
$0.04x + 0.18y = 0.15$

Solve.

39. *Car rentals.* The University of Oklahoma maintains a fleet of vehicles for use by university departments. To transport equipment for a production, the School of Drama needs to rent either a cargo van or a pickup truck for one day. The cargo van costs $27 for a day plus 22¢ per mile. The pickup truck costs $29 for a day plus 17¢ per mile. For what mileage is the cost the same?
Source: fleetservices.ou.edu

40. *RV rentals.* Stoltzfus RV rents a Class A recreational vehicle (RV) for $1545 a week plus 20¢ per mile. Martin RV rents a similar RV for $1183 a week plus 30¢ per mile. For what mileage is the cost the same?
Sources: Stoltzfus RV; Martin RV

41. *Complementary angles.* Two angles are complementary. One angle measures 6° less than twice the measure of the other. Find the measure of each angle.

42. *Complementary angles.* Two angles are complementary. One angle measures 10° more than four times the measure of the other. Find the measure of each angle.

43. *Phone rates.* Recently, AT&T offered the One Rate 10¢ Nationwide Direct Plan for $2.99 per month plus 10¢ per minute. Quest offered the 5 Cent Long Distance Plan for $4.99 per month plus 5¢ per minute. For what number of minutes per month will the two plans cost the same?
Sources: AT&T; Quest

44. *Phone rates.* Daya makes frequent calls from the United States to India. She is choosing between a Quest plan that costs $4 per month plus 31¢ per minute and an AT&T plan that costs $5 per month plus 28¢ per minute. For what number of minutes will the two plans cost the same?
Sources: AT&T; Quest

45. *Supplementary angles.* Two angles are supplementary. One angle measures 5° less than four times the measure of the other. Find the measure of each angle.

46. *Supplementary angles.* Two angles are supplementary. One angle measures 45° more than twice the measure of the other. Find the measure of each angle.

47. *Planting grapes.* South Wind Vineyards uses 820 acres to plant Chardonnay and Riesling grapes. The vintner knows the profits will be greatest by planting 140 more acres of Chardonnay than Riesling. How many acres of each type of grape should be planted?

48. *Baking.* Maple Branch Bakers sells 175 loaves of bread each day—some white and the rest whole-wheat. Because of a regular order from a local sandwich shop, Maple Branch consistently bakes 9 more loaves of white bread than whole-wheat. How many loaves of each type of bread do they bake?

49. *Framing.* Angel has 18 ft of molding from which he needs to make a rectangular frame. Because of the dimensions of the mirror being framed, the frame must be twice as long as it is wide. What should the dimensions of the frame be?

50. *Gardening.* Patrice has 108 ft of fencing for a rectangular garden. If the garden's length is to be $1\frac{1}{2}$ times its width, what should the garden's dimensions be?

51. Describe a method that could be used for writing a system that contains dependent equations.

52. Describe a method that could be used for writing an inconsistent system of equations.

Skill Review

To prepare for Section 7.4, review percent notation (Section 2.4).

Convert to decimal notation. [2.4]

53. 12.2%   **54.** 0.5%

Solve. [2.4]

55. What percent of 65 is 26?

56. What number is 17% of 18?

Translate to an algebraic expression. [1.1]

57. 12% of the number of liters

58. 10.5% of the number of pounds

Synthesis

59. If a system has an infinite number of solutions, does it follow that *any* ordered pair is a solution? Why or why not?

60. Explain how the multiplication and addition principles are used in this section. Then count the number of times that these principles are used in Example 4.

Solve using substitution, elimination, or graphing.

61. $y = 3x + 4$,
$3 + y = 2(y - x)$

62. $x + y = 7$,
$3(y - x) = 9$

63. $0.05x + y = 4$,
$\dfrac{x}{2} + \dfrac{y}{3} = 1\frac{1}{3}$

64. $2(5a - 5b) = 10$,
$-5(2a + 6b) = 10$

Aha! **65.** $y = -\frac{2}{7}x + 3$,
$y = \frac{4}{5}x + 3$

66. $y = \frac{2}{5}x - 7$,
$y = \frac{2}{5}x + 4$

Solve for x and y.

67. $ax + by + c = 0$,
$ax + cy + b = 0$

68. $y = ax + b$,
$y = x + c$

69. *Caged rabbits and pheasants.* Several ancient Chinese books included problems that can be solved by translating to systems of equations. *Arithmetical Rules in Nine Sections* is a book of 246 problems compiled by a Chinese mathematician, Chang Tsang, who died in 152 B.C. One of the problems is: Suppose there are a number of rabbits and pheasants confined in a cage. In all, there are 35 heads and 94 feet. How many rabbits and how many pheasants are there? Solve the problem.

70. *Age.* Patrick's age is 20% of his mother's age. Twenty years from now, Patrick's age will be 52% of his mother's age. How old are Patrick and his mother now?

71. *Age.* If 5 is added to a man's age and the total is divided by 5, the result will be his daughter's age. Five years ago, the man's age was eight times his daughter's age. Find their present ages.

72. *Dimensions of a triangle.* When the base of a triangle is increased by 1 ft and the height is increased by 2 ft, the height changes from being two-thirds of the base to being four-fifths of the base. Find the original dimensions of the triangle.

CONNECTING the CONCEPTS

We now have three distinctly different ways to solve a system. Each method has certain strengths and weaknesses.

Method	Strengths	Weaknesses
Graphical	Solutions are displayed visually. Works with any system that can be graphed.	For some systems, only approximate solutions can be found graphically. The graph drawn may not be large enough to show the solution.
Substitution	Always yields exact solutions. Easy to use when a variable is alone on one side of an equation.	Introduces extensive computations with fractions when solving more complicated systems. Solutions are not graphically displayed.
Elimination	Always yields exact solutions. Easy to use when fractions or decimals appear in the system.	Solutions are not graphically displayed.

When selecting the best method to use for a particular system, consider the strengths and weaknesses listed above. As you gain experience with these methods, it will become easier to choose the best method for any given system.

MIXED REVIEW

Solve using the best method.

1. $x = y,$
$x + y = 4$

2. $x + y = 5,$
$x - y = 3$

3. $y = x + 1,$
$2x + y = 6$

4. $2x + 3y = 6,$
$2x - 3y = 2$

5. $y = 2x + 1,$
$y = \frac{1}{2}x - 2$

6. $2y = x + 7,$
$3x + 2y = 5$

7. $3x - 4y = 1,$
$2x + 2y = 5$

8. $y = \frac{2}{3}x - 5,$
$y = \frac{2}{3}x + 1$

9. $x = -1,$
$y = 5$

10. $3y = 5x + 4,$
$2y = 3x - 1$

11. $y - 3x = 1,$
$y = 3x - 5$

12. $x = 2y + 1,$
$3x + 2y = 4$

13. $x + y = 3,$
$2x + 2y = 6$

14. $6x + 5y = 15,$
$3y - 6x = 1$

15. $8x - 11y = 12,$
$3x + 11y = 10$

16. $y = 3,$
$x = 7$

17. $2x + 7y = 0,$
$9x - 10y = 0$

18. $x - 3y = 6,$
$2y = x - 1$

19. $\frac{1}{2}x + \frac{1}{5}y = 9,$
$\frac{1}{3}x - \frac{4}{15}y = -2$

20. $0.1x - 0.5y = -2.8,$
$0.2x = 0.3y$

7.4 ## More Applications Using Systems

Total-Value Problems ▪ Mixture Problems

The five steps for problem solving and the methods for solving systems of equations can be used in a variety of applications.

Total-Value Problems

EXAMPLE 1

Basketball scores. In one game during the 2007 NBA playoffs, the San Antonio Spurs scored 83 points. Of those, 20 points were from free throws. The remaining 63 points were a result of 29 two- and three-point baskets. How many baskets of each type were made?

Source: National Basketball Association

Spurs 83, Cavaliers 82
SAN ANTONIO (83)
Bowen 1–9 0–0 2, Duncan 4–15 4–10 12, Oberto 3–5 1–1 7, Finley 1–5 2–3 4, Parker 10–14 2–5 24, Ginobili 8–19 8–11 27, Horry 0–2 1–2 1, Elson 1–1 2–2 4, Vaughn 1–1 0–0 2, Barry 0–0 0–0 0. Totals: 29–71 20–34 83.
CLEVELAND (82)
James 10–30 2–6 24, Gooden 5–12 1–1 11, Ilgauskas 3–8 2–2 8, Pavlovic 1–6 0–0 3, Gibson 4–10 0–0 10, Snow 2–4 0–0 4, Varejao 3–3 2–2 8, Marshall 2–6 1–2 5, Jones 2–5 3–3 9, Brown 0–0 0–0 0. Totals: 32–84 11–16 82.
San Antonio 19 20 21 23 — 83
Cleveland 20 14 18 30 — 82

SOLUTION

1. **Familiarize.** Suppose that of the 29 baskets, 20 were two-pointers and 9 were three-pointers. These 29 baskets would then amount to a total of

$$20 \cdot 2 + 9 \cdot 3 = 40 + 27 = 67 \text{ points.}$$

Although our guess is incorrect, checking the guess has familiarized us with the problem. We let $w =$ the number of two-pointers made and $r =$ the number of three-pointers made.

2. **Translate.** Since a total of 29 baskets was made, we must have

$$w + r = 29.$$

To find a second equation, we reword some information and focus on the points scored, just as when we checked our guess above.

Rewording:	The points scored from two-pointers	plus	the points scored from three-pointers	totaled	63.
Translating:	$w \cdot 2$	$+$	$r \cdot 3$	$=$	63

The problem has been translated to the following system of equations:

$$w + r = 29, \quad (1)$$
$$2w + 3r = 63. \quad (2)$$

3. **Carry out.** For purposes of review, we solve by substitution. First we solve equation (1) for w:

$$w + r = 29 \qquad \text{Equation (1)}$$
$$w = 29 - r. \quad (3) \qquad \text{Solving for } w$$

Next, we replace w in equation (2) with $29 - r$:

$$2w + 3r = 63 \qquad \text{Equation (2)}$$
$$2(29 - r) + 3r = 63 \qquad \text{Substituting } 29 - r \text{ for } w$$
$$58 - 2r + 3r = 63$$
$$58 + r = 63 \ \Big\}$$
$$r = 5. \ \Big\} \qquad \text{Solving for } r$$

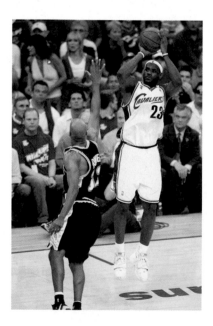

We find w by substituting 5 for r in equation (3):

$$w = 29 - r = 29 - 5 = 24.$$

4. **Check.** If the Spurs made 24 two-pointers and 5 three-pointers, they would have made 29 shots, for a total of

$$24 \cdot 2 + 5 \cdot 3 = 48 + 15 = 63 \text{ points.}$$

The numbers check.

5. **State.** The Spurs made 24 two-pointers and 5 three-pointers.

> TRY EXERCISE ▸ 1

EXAMPLE 2　Recently the Mayfair Clinic purchased 160 stamps for $61.95. If the stamps were a combination of 27¢ postcard stamps and 42¢ first-class stamps, how many of each type were bought?

SOLUTION

1. **Familiarize.** When faced with a new problem, it is often useful to compare it to a similar problem that you have already solved. Here instead of counting two- and three-point baskets, as in Example 1, we are counting postcard stamps and first-class stamps. We let $p =$ the number of postcard stamps and $f =$ the number of first-class stamps purchased.

2. **Translate.** Since a total of 160 stamps was purchased, we have

$$p + f = 160.$$

To find a second equation, we reword some information, focusing on the amount of money paid:

Rewording:　The money paid　　the money paid for
　　　　for postcard stamps　plus　first-class stamps　totaled　$61.95.

Translating:　$p \cdot 0.27$　　+　　$f \cdot 0.42$　　=　　61.95

Note that we changed all the money units to dollars.
Presenting the information in a table can be helpful.

	Postcard Stamps	First-Class Stamps	Total
Cost per Stamp	$0.27	$0.42	
Number of Stamps	p	f	160	→ $p + f = 160$
Amount Paid	$0.27p$	$0.42f$	61.95	→ $0.27p + 0.42f = 61.95$

We have translated to a system of equations:

$$p + f = 160, \qquad (1)$$
$$0.27p + 0.42f = 61.95. \qquad (2)$$

3. Carry out. The system can be solved using elimination:

$$-0.27p - 0.27f = -43.20 \qquad \text{Multiplying both sides}$$
$$\text{of equation (1) by } -0.27$$

$$\underline{0.27p + 0.42f = 61.95} \qquad \text{Equation (2)}$$

$$0.15f = 18.75 \qquad \text{Adding}$$

$$f = 125. \qquad \text{Dividing both sides by } 0.15$$

To solve for p, we substitute 125 for f:

$$p + f = 160 \qquad \text{Using equation (1)}$$
$$p + 125 = 160$$
$$p = 35. \qquad \text{Subtracting 125 from both sides}$$

4. Check. If $p = 35$ and $f = 125$, the total number of stamps is 160. The clinic paid 35($0.27), or $9.45, for the postcard stamps and 125($0.42), or $52.50, for the first-class stamps. The total paid was therefore $9.45 + $52.50, or $61.95. The numbers check.

5. State. The clinic purchased 35 postcard stamps and 125 first-class stamps.

> **TRY EXERCISE** 7

Mixture Problems

EXAMPLE 3

Blending coffees. The Roasted Bean wants to mix fair-trade Ethiopian beans that sell for $11.95 per pound with organic Mexican beans that sell for $9.95 per pound to form a 50-lb batch of Early Riser Blend that sells for $10.75 per pound. How many pounds of each type of bean should go into the blend?

SOLUTION

1. Familiarize. This problem seems similar to Example 2. Instead of two different prices for two types of stamps, we have two different prices per pound for two types of coffee. Instead of knowing the total amount paid, we know the weight and price per pound of the batch of Early Riser Blend being made. Note that we can easily calculate the value of the batch of Early Riser Blend by multiplying 50 lb times $10.75 per pound. We let t = the number of pounds of fair-trade Ethiopian coffee used and m = the number of pounds of organic Mexican coffee used.

2. Translate. Since a 50-lb batch is being made, we must have

$$t + m = 50.$$

To find a second equation, we consider the total value of the 50-lb batch. That value must be the same as the value of the Ethiopian beans plus the value of the Mexican beans that go into the blend.

Rewording:	The value of the Ethiopian beans	plus	the value of the Mexican beans	is	the value of the Early Riser Blend.
	↓	↓	↓	↓	↓
Translating:	$t \cdot 11.95$	$+$	$m \cdot 9.95$	$=$	$50 \cdot 10.75$

This information can be presented in a table.

	Ethiopian Beans	Mexican Beans	Early-Riser Blend	
Number of Pounds	t	m	50	$\longrightarrow$ $t + m = 50$
Price per Pound	11.95	9.95	10.75	
Value of Beans	$11.95t$	$9.95m$	$50 \cdot 10.75$, or 537.50	$\longrightarrow$ $11.95t + 9.95m = 537.50$

We have translated to a system of equations:

$$t + m = 50, \qquad (1)$$
$$11.95t + 9.95m = 537.50. \qquad (2)$$

3. **Carry out.** When equation (1) is solved for t, we have $t = 50 - m$. We then substitute $50 - m$ for t in equation (2):

$11.95(50 - m) + 9.95m = 537.50$	Solving by substitution
$597.50 - 11.95m + 9.95m = 537.50$	Using the distributive law
$-2m = -60$	Combining like terms; subtracting 597.50 from both sides
$m = 30.$	Dividing both sides by 2

If $m = 30$, we see from equation (1) that $t = 20$.

4. **Check.** If 20 lb of fair-trade Ethiopian beans and 30 lb of organic Mexican beans are mixed, a 50-lb blend will result. The value of 20 lb of Ethiopian beans is 20($11.95), or $239. The value of 30 lb of Mexican beans is 30($9.95), or $298.50, so the value of the blend is $239 + $298.50 = $537.50. A 50-lb blend priced at $10.75 a pound is also worth $537.50, so our answer checks.

5. **State.** The Early Riser Blend should be made by combining 20 lb of fair-trade Ethiopian beans with 30 lb of organic Mexican beans. **TRY EXERCISE** 17

EXAMPLE 4

Paint colors. At a local "paint swap," Madison found large supplies of Skylite Pink (12.5% red pigment) and MacIntosh Red (20% red pigment). How many gallons of each color should be mixed in order to create a 15-gal batch of Summer Rose (17% red pigment)?

SOLUTION

1. **Familiarize.** This problem is similar to Example 3. Instead of mixing two types of coffee and keeping an eye on the price of the mixture, we are mixing two types of paint and keeping an eye on the amount of pigment in the mixture.

   To visualize this problem, think of the pigment as a solid that, given time, would settle to the bottom of each can. Let's guess that 3 gal of Skylite Pink and 12 gal of MacIntosh Red are mixed. How much pigment would be in the mixture? The Skylite Pink would contribute 12.5% of 3 gal, or 0.375 gal of pigment, and the MacIntosh Red would contribute 20% of 12 gal, or 2.4 gal of pigment. Thus the 15-gal mixture would contain $0.375 + 2.4 = 2.775$ gal of pigment. Since Madison needs the 15 gal of Summer Rose to be 17% pigment, and since 17% of 15 gal is 2.55 gal, our guess is incorrect. Rather than guess again, we let p = the number of gallons of Skylite Pink used and m = the number of gallons of MacIntosh Red used.

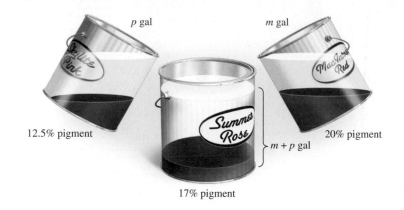

p gal m gal

12.5% pigment   20% pigment

Summer Rose

$m + p$ gal

17% pigment

2. Translate. As in Example 3, the information given can be arranged in a table.

	Skylite Pink	MacIntosh Red	Summer Rose
Amount of Paint (in gallons)	p	m	15
Percent Pigment	12.5%	20%	17%
Amount of Pigment (in gallons)	$0.125p$	$0.2m$	0.17×15, or 2.55

A system of two equations can be formed by reading across the first and third rows of the table. Since Madison needs 15 gal of mixture, we must have

$p + m = 15.$ ← Total amount of paint

Madison also needs 2.55 gal of pigment. Since the amount of pigment in the Summer Rose paint comes from the pigment in the Skylite Pink and the MacIntosh Red paint, we have

$0.125p + 0.2m = 2.55.$ ← Total amount of pigment

We have translated to a system of equations:

$$p + \quad m = 15, \qquad (1)$$
$$0.125p + 0.2m = 2.55. \qquad (2)$$

3. Carry out. Note that if we multiply both sides of equation (2) by -5, we can eliminate m (other approaches will also work):

$$
\begin{array}{ll}
p + m = \quad 15 & \\
\underline{-0.625p - m = -12.75} & \text{We observed that } (-5)(0.2m) = -m. \\
0.375p \quad\quad = \quad 2.25 & \\
\quad\quad\quad p = 6. & \text{Dividing both sides by } 0.375
\end{array}
$$

If $p = 6$, we see from equation (1) that $m = 9$.

4. **Check.** Clearly, 6 gal of Skylite Pink and 9 gal of MacIntosh Red do add up to a total of 15 gal. To determine whether the mixture is the right color, Summer Rose, we calculate the amount of pigment in the mixture: $0.125 \cdot 6 + 0.2 \cdot 9 = 0.75 + 1.8 = 2.55$ gal. Since 2.55 is indeed 17% of 15, the mixture is the correct color.

5. **State.** Madison needs 6 gal of Skylite Pink and 9 gal of MacIntosh Red in order to make 15 gal of Summer Rose. **TRY EXERCISE** 21

Re-examine Examples 1–4, looking for similarities. Examples 3 and 4 are often called *mixture problems*, but they have much in common with Examples 1 and 2.

Problem-Solving Tip

When solving a new problem, see if it is like a problem that you have already solved. You can then modify the earlier approach to fit the new problem.

EXERCISE SET

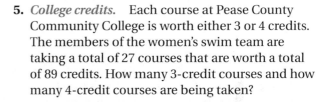

Solve. Use the five steps for problem solving.

1. *Basketball scoring.* In a recent game, Stephen Jackson of the Golden State Warriors scored 27 points on a combination of 10 two- and three-point baskets. How many shots of each type were made?
   Source: National Basketball Association

2. *Basketball scoring.* In a recent game, the Cleveland Cavaliers scored 71 points on a combination of 32 two- and three-point baskets. How many shots of each type were made?
   Source: National Basketball Association

3. *Basketball scoring.* In clinching the Eastern Conference championship, the Miami Heat once scored 84 of their points on a combination of 39 two- and three-point baskets. How many shots of each type were made?
   Source: National Basketball Association

4. *Basketball scoring.* In winning the Western Conference finals, the Dallas Mavericks once scored 78 of their points on a combination of 37 two- and three-pointers. How many shots of each type did they make?
   Source: National Basketball Association

5. *College credits.* Each course at Pease County Community College is worth either 3 or 4 credits. The members of the women's swim team are taking a total of 27 courses that are worth a total of 89 credits. How many 3-credit courses and how many 4-credit courses are being taken?

6. *College credits.* Each course at Mt. Regis College is worth either 3 or 4 credits. The members of the men's soccer team are taking a total of 48 courses that are worth a total of 155 credits. How many 3-credit courses and how many 4-credit courses are being taken?

7. *Returnable bottles.* As part of a fundraiser, the Cobble Hill Daycare collected 430 returnable bottles and cans, some worth 5 cents each and the rest worth 10 cents each. If the total value of the cans and bottles was $26.20, how many 5-cent bottles or cans and how many 10-cent bottles or cans were collected?

8. *Ice cream cones.* A busload of campers stopped at a dairy stand for ice cream. They ordered 40 cones, some soft-serve at $2.25 and the rest hard-pack at $2.75. If the total bill was $96, how many of each type of cone were ordered?

9. *Yellowstone Park admissions.* Entering Yellowstone National Park costs $25 for a car and $20 for a motorcycle. On a typical day, 5950 cars or motorcycles enter and pay a total of $137,650. How many motorcycles enter on a typical day?
Source: National Park Service

10. *Disneyland admissions.* A two-day Park Hopper ticket to Disneyland costs $102 for children ages 3–9 and $122 for adults and children ages 10 and older. A bus full of 23 kindergartners and parents paid a total of $2466 for their tickets. How many children and how many adults were on the bus?

11. *Zoo admissions.* From March 1 through November 30, the Philadelphia Zoo charges an admission fee of $17 for adults and $14 for children. One June day, a total of $9369 was collected from 642 admissions. How many adult admissions were there?
Source: Philadelphia Zoo

12. *Zoo admissions.* The Bronx Zoo charges an admission fee of $14 for adults and $11 for children. One July day, a total of $11,490 was collected from 960 admissions. How many adult admissions were there?
Source: Bronx Zoo

13. *Music lessons.* Jillian charges $25 for a private guitar lesson and $18 for a group guitar lesson. One day in August, Jillian earned $265 from 12 students. How many students of each type did Jillian teach?

14. *Dance lessons.* Dona charges $20 for a private tap lesson and $12 for a group class. One Wednesday, Dona earned $216 from 14 students. How many students of each type did Dona teach?

15. *Video games.* Playtime sells used Xbox 360 games for $9.99 and used PS3 games for $17.99. Jessie recently purchased a total of 11 Xbox 360 and PS3 games for $125.89 (before tax). How many Xbox 360 games and how many PS3 games did she buy?

16. *Pizza.* For a tailgate party, Gil purchased a total of 18 medium and large pizzas for a total of $269.82. If a medium pizza costs $12.99 and a large pizza costs $15.99, how many of each size did he buy?

17. *Coffee blends.* Cafe Europa mixes Brazilian coffee worth $19 per kilogram with Turkish coffee worth $22 per kilogram. The mixture should be worth $20 per kilogram. How much of each type of coffee should be used to make a 300-kg mixture?

18. *Seed mix.* Sunflower seed is worth $1.00 per pound and rolled oats are worth $1.35 per pound. How much of each would you use to make 50 lb of a mixture worth $1.14 per pound?

19. *Mixed nuts.* The Nuthouse has 10 kg of mixed cashews and pecans worth $8.40 per kilogram. Cashews alone sell for $8.00 per kilogram, and pecans sell for $9.00 per kilogram. How many kilograms of each are in the mixture?

20. *Mixed nuts.* A vendor wishes to mix peanuts worth $2.52 per pound with Brazil nuts worth $3.80 per pound to make 480 lb of a mixture worth $3.44 per pound. How much of each should be used?

21. *Acid mixtures.* Jerome's experiment requires him to mix a 50%-acid solution with an 80%-acid solution to create 200 mL of a 68%-acid solution. How much 50%-acid solution and how much 80%-acid solution should he use? Complete the following table as part of the *Translate* step.

Type of Solution	50%-Acid	80%-Acid	68%-Acid Mix
Amount of Solution	x	y	
Percent Acid	50%		68%
Amount of Acid in Solution		$0.8y$	

22. *Production.* Streakfree window cleaner is 12% alcohol and Sunstream window cleaner is 30% alcohol. How much of each should be used to make 90 oz of a cleaner that is 20% alcohol?

23. *Chemistry.* E-Chem Testing has a solution that is 80% base and another that is 30% base. A technician needs 200 L of a solution that is 62% base. The 200 L will be prepared by mixing the two solutions on hand. How much of each should be used?

24. *Horticulture.* A solution containing 28% fungicide is to be mixed with a solution containing 40% fungicide to make 300 L of a solution containing 36% fungicide. How much of each solution should be used?

25. *Octane ratings.* When a tanker delivers gas to a gas station, it brings only two grades of gasoline, the highest and the lowest, filling two large underground tanks. If you purchase a middle grade, the pump's computer mixes the other two grades appropriately. How much 87-octane gas and 93-octane gas should be blended in order to make 12 gal of 91-octane gas?
Source: Champlain Electric and Petroleum Equipment

26. *Octane ratings.* Refer to Exercise 25. If a gas station is supplied with 87-octane gas and 95-octane gas, how much of each should be blended in order to make 10 gal of 93-octane gas?

27. *Printing.* Using some pages that hold 1300 words per page and others that hold 1850 words per page, a typesetter is able to completely fill 12 pages with an 18,350-word document. How many pages of each kind were used?

28. *Coin value.* A collection of quarters and nickels is worth $1.25. There are 13 coins in all. How many of each are there?

29. *Basketball scoring.* Wilt Chamberlain once scored a record 100 points on a combination of 64 foul shots (each worth one point) and two-pointers. How many shots of each type did he make?

30. *Basketball scoring.* Dwayne Wade recently scored 41 points on a combination of 26 foul shots and two-pointers. How many shots of each type did he make?

31. *Suntan lotion.* Lisa has a tube of Kinney's suntan lotion that is rated 15 spf and a second tube of Coppertone that is 30 spf. How many fluid ounces of each type of lotion should be mixed in order to create 50 fluid ounces of sunblock that is rated 20 spf?

32. *Cough syrup.* Dr. Zeke's cough syrup is 2% alcohol. Vitabrite cough syrup is 5% alcohol. How much of each type should be used in order to prepare an 80-oz batch of cough syrup that is 3% alcohol?

Aha! 33. *Nutrition.* New England Natural Bakers Muesli gets 20% of its calories from fat. Breadshop Supernatural granola gets 35% of its calories from fat. How much of each type should be used to create a 40-lb mixture that gets 27.5% of its calories from fat?
Source: Onion River Cooperative, Burlington VT

34. *Textile production.* DRG Outdoor Products uses one insulation that is 20% goose down and another that is 34% goose down. How many pounds of each should be used to create 50 lb of insulation that is 25% goose down?

35. Why might fraction answers be acceptable on problems like Examples 3 and 4, but not on problems like Examples 1 and 2?

36. Write a problem for a classmate to solve by translating to a system of two equations in two unknowns.

Skill Review

To prepare for Section 7.5, review solving and graphing inequalities (Section 2.6).

Graph the solution set on a number line. [2.6]

37. $x + 2 < 5$

38. $2x + 5 \geq 1$

39. $5 - 3x > 20$

40. $13 - 6x \leq 1$

41. $4 \leq -\frac{1}{3}x + 5$

42. $6 > -\frac{1}{2}x + 3$

Synthesis

43. In Exercise 30, suppose that some of Wade's 26 baskets were three-pointers. Could the problem still be solved? Why or why not?

44. In Exercise 28, suppose that some of the 13 coins may have been half-dollars. Could the problem still be solved? Why or why not?

45. *Coffee.* Kona coffee, grown only in Hawaii, is highly desired and quite expensive. To create a blend of beans that is 30% Kona, the Brewtown Beanery is adding pure Kona beans to a 45-lb sack of Colombian beans. How many pounds of Kona should be added to the 45 lb of Colombian?

46. *Chemistry.* A tank contains 8000 L of a solution that is 40% acid. How much water should be added in order to make a solution that is 30% acid?

47. *Automobile maintenance.* The radiator in Candy's Honda Accord contains 6.3 L of antifreeze and water. This mixture is 30% antifreeze. How much should be drained and replaced with pure antifreeze so that the mixture will be 50% antifreeze?

48. *Investing.* One year Shannon made $288 from two investments: $1100 was invested at one yearly rate and $1800 at a rate that was 1.5% higher. Find the two rates of interest.

49. *Octane rating.* Many cars need gasoline with an octane rating of at least 87. After mistakenly putting 5 gal of 85-octane gas in her empty gas tank, Kim plans to add 91-octane gas until the mixture's octane rating is 87. How much 91-octane gas should she add?

50. *Sporting-goods prices.* Together, a bat, ball, and glove cost $99.00. The bat costs $9.95 more than the ball, and the glove costs $65.45 more than the bat. How much does each cost?

51. *Overtime pay.* Juanita is paid "time and a half" for any time she works in excess of 40 hr per week. The week before Christmas she worked 55 hr and was paid $812.50. What was her normal hourly wage?

52. *Investing.* Khalid invested $54,000, part of it at 6% and the rest at 6.5%. The total yield after one year is $3385. How much was invested at each rate?

53. *Payroll.* Ace Engineering pays a total of $325 an hour when employing some workers at $20 an hour and others at $25 an hour. When the number of $20 workers is increased by 50% and the number of $25 workers is decreased by 20%, the cost per hour is $400. How many workers were originally employed at each rate?

54. *Dairy farming.* The Benson Cooperative Creamery has 1000 gal of milk that is 4.6% butterfat. How much skim milk (no butterfat) should be added to make milk that is 2% butterfat?

55. A two-digit number is six times the sum of its digits. The tens digit is 1 more than the ones digit. Find the number.

56. The sum of the digits of a two-digit number is 12. When the digits are reversed, the number is decreased by 18. Find the original number.

57. *Literature.* In Lewis Carroll's *Through the Looking Glass*, Tweedledum says to Tweedledee, "The sum of your weight and twice mine is 361 pounds." Then Tweedledee says to Tweedledum, "Contrariwise, the sum of your weight and twice mine is 362 pounds." Find the weights of Tweedledum and Tweedledee.

COLLABORATIVE CORNER

Sunoco's Custom Blending Pump

Focus: Mixture problems

Time: 30 minutes

Group size: 3

Materials: Calculators

Sunoco® gasoline stations pride themselves on offering customers the "custom blending pump." While most competitors offer just three octane levels—87, 93, and a blend that is 89—Sunoco customers can select an octane level of 86, 87, 89, 92, or 94. The Sunoco supplier brings 86- and 94-octane gasoline to each station and a computerized pump mixes the two to create the blend that a customer desires.

ACTIVITY

1. Assume that your group's gas station has an abundant supply of 86-octane and 94-octane gasoline. Each group member should select a different one of the custom blends: 87, 89, or 92.

2. Each member will be asked to determine the number of gallons of 86-octane gas and 94-octane gas needed to form 100 gal of his or her selected blend. Before doing so, however, the group should outline a series of steps that each member will use to solve his or her problem. Group members should agree on the letters chosen as variables, the variable that will be solved for first, and the sequence of steps that will be followed. For the purposes of this activity, it is best to use substitution to solve the system of equations that will be created.

3. *Following the agreed-upon series of steps,* each group member should determine how many gallons of 86-octane gas and how many gallons of 94-octane gas should be mixed in order to form 100 gal of his or her selected blend. Check that all work is done correctly and consistently.

4. The pumps in use at most gas stations blend mixtures in $\frac{1}{10}$-gal "batches." How can the results of part (3) be adjusted so that $\frac{1}{10}$-gal, not 100-gal, blends are formulated?

7.5   Linear Inequalities in Two Variables

Graphing Linear Inequalities • Linear Inequalities in One Variable

STUDY SKILLS

When Something Seems Easy

Every so often, you may encounter a lesson that you remember from a previous math course. When this occurs, make sure that *all* of that lesson is understood. If anything is amiss, be sure to review all tricky spots.

Just as the solutions of linear equations like $5x + 4y = 13$ or $y = \frac{1}{2}x + 1$ can be graphed, so too can the solutions of *linear inequalities* like $5x + 4y < 13$ or $y > \frac{1}{2}x + 1$ be represented graphically.

Graphing Linear Inequalities

In Section 2.6, we found that solutions of inequalities like $5x + 9 \le 4x + 3$ can be represented by a shaded portion of a number line. When a solution included an endpoint, we drew a solid dot, and when the endpoint was excluded, we drew an open dot. To graph inequalities like $y > \frac{1}{2}x + 1$ or $2x + 3y \le 6$, we will shade a region of a plane. That region will be either above or below the graph of a "boundary line" (in this case, the graph of $y = \frac{1}{2}x + 1$ or $2x + 3y = 6$). If the symbol is $\le$ or $\ge$, we will draw the boundary line solid, since it is part of the solution. When the boundary is excluded—that is, if $<$ or $>$ is used—we will draw a dashed line.

EXAMPLE 1 Graph: $y > \frac{1}{2}x + 1$.

SOLUTION We begin by graphing the boundary line $y = \frac{1}{2}x + 1$. The slope is $\frac{1}{2}$ and the y-intercept is $(0, 1)$. Since the symbol $>$ is used, the solutions of $y = \frac{1}{2}x + 1$ are *not* part of the solutions of the inequality. To indicate this, we draw the line dashed.

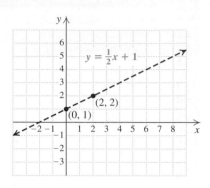

The plane is now split in two. If we consider the coordinates of a few points above the line, we will find that all are solutions of $y > \frac{1}{2}x + 1$.

Here is a check for the points $(2, 3)$ and $(-2, 4)$:

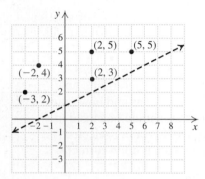

$$\begin{array}{c|c}
y > \frac{1}{2}x + 1 & \\
\hline
3 & \frac{1}{2} \cdot 2 + 1 \\
& 1 + 1 \\
& 3 \overset{?}{>} 2 \quad \text{TRUE}
\end{array}
\qquad
\begin{array}{c|c}
y > \frac{1}{2}x + 1 & \\
\hline
4 & \frac{1}{2}(-2) + 1 \\
& -1 + 1 \\
& 4 \overset{?}{>} 0 \quad \text{TRUE}
\end{array}$$

The student can check that *any* point on the same side of the dashed line as $(2, 3)$ or $(-2, 4)$ is a solution. If one point in a region solves an inequality, then *all* points in that region are solutions. The graph of

$$y > \frac{1}{2}x + 1$$

is shown below. Note that the solution set consists of all points in the shaded region. Furthermore, note that for any inequality of the form $y > mx + b$ or $y \geq mx + b$, we shade the region *above* the boundary line.

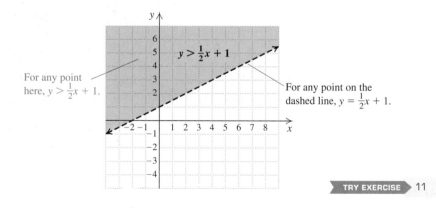

For any point here, $y > \frac{1}{2}x + 1$.

For any point on the dashed line, $y = \frac{1}{2}x + 1$.

TRY EXERCISE 11

EXAMPLE 2 Graph: $2x + 3y \leq 6$.

STUDENT NOTES

A mistake in checking your test point could lead you to shade the wrong region of the plane. For this reason, you may want to use a second test point to assure yourself that the correct region is shaded.

SOLUTION   First, we establish the boundary line by graphing $2x + 3y = 6$. This can be done either by using the intercepts, $(0, 2)$ and $(3, 0)$, or by finding slope–intercept form, $y = -\frac{2}{3}x + 2$. Since the inequality contains the symbol $\leq$, we draw a solid boundary line to include all pairs on the line as part of the solution. The graph of $2x + 3y \leq 6$ also includes either the region above or below the line. By using a "test point" that is clearly above or below the line, we can determine which region to shade. The origin, $(0, 0)$, is often a convenient test point, so long as it does not lie on the boundary line:

$$\frac{2x + 3y \leq 6}{2 \cdot 0 + 3 \cdot 0 \mid 6}$$
$$0 \overset{?}{\leq} 6 \quad \text{TRUE}$$

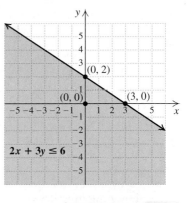

The point $(0, 0)$ is a solution and it is in the region below the boundary line. Thus this region, along with the line itself, represents the solution.

The original inequality is equivalent to $y \leq -\frac{2}{3}x + 2$. Note that for any inequality of the form $y \leq mx + b$ or $y < mx + b$, we shade the region *below* the boundary line.

 17

To Graph a Linear Inequality

1. Draw the boundary line by replacing the inequality symbol with an equals sign and graphing the resulting equation. If the inequality symbol is $<$ or $>$, the line is dashed. If the symbol is $\leq$ or $\geq$, the line is solid.
2. Shade the region on one side of the boundary line. To determine which side, select a point not on the line as a test point. If that point's coordinates are a solution of the inequality, shade the region containing the point. If not, shade the other region.
3. Inequalities of the form $y < mx + b$ or $y \leq mx + b$ are shaded below the boundary line. Inequalities of the form $y > mx + b$ or $y \geq mx + b$ are shaded above the boundary line.

Linear Inequalities in One Variable

EXAMPLE   3

Graph $y \leq -2$ on a plane.

SOLUTION   We graph $y = -2$ as a solid line to indicate that all points on the line are solutions. Again, we select $(0, 0)$ as a test point. It may help to write $y \leq -2$ as $y \leq 0 \cdot x - 2$:

$$\frac{y \leq 0 \cdot x - 2}{0 \mid 0 \cdot 0 - 2}$$
$$0 \overset{?}{\leq} -2 \quad \text{FALSE}$$

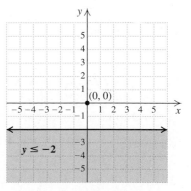

Since $(0, 0)$ is *not* a solution, we do not shade the region in which it appears. Instead, we shade below the boundary line as shown. The solution consists of all ordered pairs with y-coordinates less than or equal to -2.   **TRY EXERCISE** 27

EXAMPLE   4   Graph $x < 3$ on a plane.

SOLUTION   We graph $x = 3$ using a dashed line. To determine which region to shade, we again use the test point $(0, 0)$. It may help to write $x < 3$ as $x + 0 \cdot y < 3$:

$$\frac{x + 0 \cdot y < 3}{0 + 0 \cdot 0 \ \mid \ 3}$$
$$0 \overset{?}{<} 3 \quad \text{TRUE}$$

Since $(0, 0)$ is a solution, we shade to the left. The solution consists of all ordered pairs with first coordinates less than 3.

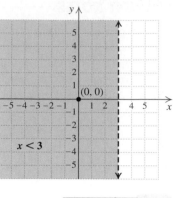

TRY EXERCISE 25

TECHNOLOGY CONNECTION

To graph $2x + 3y \le 6$ on a graphing calculator, we must first solve for y. In Example 2, we found that $y \le -\frac{2}{3}x + 2$. On many graphing calculators, this graph is drawn by entering $(-2/3)x + 2$ as y_1, moving the cursor to the GraphStyle icon just to the left of y_1, pressing **ENTER** until ◤ appears, and then pressing **GRAPH**. The symbol ◤ indicates that the area *below* the line is shaded. Note that the boundary line always appears solid.

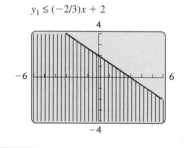

$y_1 \le (-2/3)x + 2$

Many calculators have an Inequalz application that allows us to enter the inequality as written. Graphs drawn using the Inequalz application will include a dashed boundary line if appropriate. Be sure to quit Inequalz when you are finished.

Graph each of the following.

1. $y > \frac{2}{3}x - 2$
2. $y - x \le 5$
3. $2x + 5y \ge 25$

7.5   EXERCISE SET

◆ *Concept Reinforcement   Classify each statement as either true or false.*

1. To graph a linear inequality, we first graph a line.

2. To determine which region to shade, we must test at least two points.

3. Inequalities of the form $y \ge mx + b$ are shaded above the boundary line.

4. Inequalities of the form $y < mx + b$ are shaded above the boundary line.

5. Determine whether $(-2, -6)$ is a solution of $2x + y < -10$.

6. Determine whether $(8, -1)$ is a solution of $2y - x \le -10$.

Aha! 7. Determine whether $\left(\frac{1}{3}, \frac{9}{10}\right)$ is a solution of $2y + 3x \ge -1$.

8. Determine whether $(-2, 3)$ is a solution of $x + 0 \cdot y > -1$.

Graph on a plane.

9. $y \le x - 1$

10. $y \le x + 5$

11. $y < x + 4$

12. $y < x - 2$

13. $y \ge x - 1$

14. $y \ge x - 3$

15. $y \le 3x + 2$

16. $y \le 2x - 1$

17. $x + y \le 4$

18. $x + y \le 6$

19. $x - y > 7$

20. $x - y > 5$

21. $y \ge 1 - 2x$

22. $y > 2 - 3x$

23. $y + 2x > 0$

24. $y + 3x \ge 0$

25. $x \ge 4$

26. $x > -4$

27. $y > -1$

28. $y \le 4$

29. $y < 0$

30. $y \ge -5$

31. $x \le -2$

32. $x < 5$

33. $\frac{1}{3}x - y < -5$

34. $y - \frac{1}{2}x \le -1$

35. $2x + 3y \le 12$

36. $5x + 4y \ge 20$

37. $3x - 2y \ge -6$

38. $2x - 5y \le -10$

39. Examine the solution of Example 2. Why is the point $(4.5, -1)$ *not* a good choice for a test point?

40. Why is $(0, 0)$ such a "convenient" test point to use?

Skill Review

To prepare for Section 7.6, review solving equations (Sections 2.2, 5.6, and 6.6).

Solve.

41. $2x - 7 = 3x + 8$ [2.2]

42. $x^2 - 7x + 6 = 0$ [5.6]

43. $\frac{2}{3}(x - 1) = -12$ [2.2]

44. $5x^2 = 20$ [5.6]

45. $\frac{1}{2} - \frac{1}{4} = \frac{1}{t}$ [6.6]

46. $2(x - 5) - 7 = 5(x + 1) + 3$ [2.2]

47. $x^2 + 64 = 16x$ [5.6]

48. $\frac{1}{x + 1} - \frac{1}{x} = \frac{3}{x + 1}$ [6.6]

Synthesis

49. When graphing a linear inequality, if $(0, 0)$ is on the boundary line, Cyrus chooses a test point that is on an axis. Why is this a convenient choice?

50. Without drawing a graph, explain why the graph of $3x + y < 0$ is shaded *below* the boundary line but the graph of $3x - y < 0$ is shaded *above* the boundary line.

51. *Elevators.*   Many Otis elevators have a capacity of 1000 lb. Suppose that c children, each weighing 75 lb, and a adults, each weighing 150 lb, are on an elevator. Find and graph an inequality that asserts that the elevator is overloaded.

52. *Hockey wins and losses.*   A hockey team needs at least 60 points for the season in order to make the playoffs. Suppose that the Coyotes finish with w wins, each worth 2 points, and t ties, each worth 1 point. Find and graph an inequality that indicates that the Coyotes made the playoffs.

Find an inequality for each graph shown.

53.

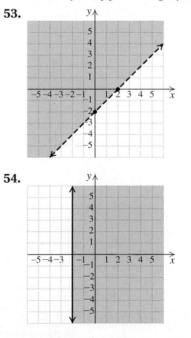

54.

Graph on a plane. (Hint: Use several test points.)

55. $xy \le 0$

56. $xy \ge 0$

57. Graph: $y + 3x \le 4.9$.

58. Graph: $0.7x - y \le 2.3$.

7.6   Systems of Linear Inequalities

Graphing Systems of Inequalities • Locating Solution Sets

Systems of linear equations were graphed in Section 7.1. We now consider **systems of linear inequalities** in two variables, such as

$$x + y \leq 3,$$
$$x - y < 3.$$

A solution of a system of inequalities makes *both* inequalities true.

When systems of equations are solved graphically, we search for points common to both lines. To solve a system of inequalities graphically, we again look for points common to both graphs. This is accomplished by graphing each inequality and determining where the graphs overlap, or intersect.

EXAMPLE 1 Graph the solutions of the system

$$x + y \leq 3,$$
$$x - y < 3.$$

SOLUTION To graph $x + y \leq 3$, we draw the graph of $x + y = 3$ using a solid line. We graph the line using the intercepts $(3, 0)$ and $(0, 3)$ (see the graph on the left below). Since $(0, 0)$ is a solution of $x + y \leq 3$, we shade (in red) all points on that side of the line. The arrows near the ends of the lines also indicate the region that contains solutions.

Next, we superimpose the graph of $x - y < 3$, using a dashed line for $x - y = 3$ and again using $(0, 0)$ as a test point. Since $(0, 0)$ is again a solution, we shade (in blue) the region on the same side of the dashed line as $(0, 0)$.

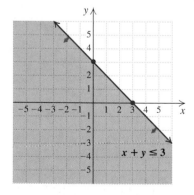

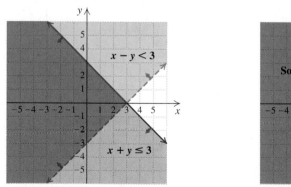

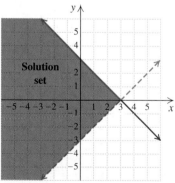

The solution set of the system is the region shaded purple along with the purple portion of the line $x + y = 3$.

TRY EXERCISE 1

EXAMPLE 2 Graph the solutions of the system

$$x \geq 3,$$
$$x - 3y < 6.$$

SOLUTION We graph $x \geq 3$ using blue and $x - 3y < 6$ using red. The solution set is the purple region along with the purple portion of the solid line.

STUDENT NOTES

Unless you have ready access to different colored pencils, using arrows (as in the examples) or patterned shadings to indicate shaded regions is a good idea.

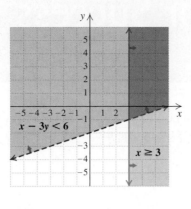

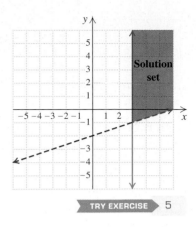

TRY EXERCISE 5

EXAMPLE 3

Graph the solutions of the system

$$x - 2y < 0,$$
$$-2x + y > 2.$$

SOLUTION   We graph $x - 2y < 0$ using red and $-2x + y > 2$ using blue. The region that is purple is the solution set of the system since those points solve both inequalities.

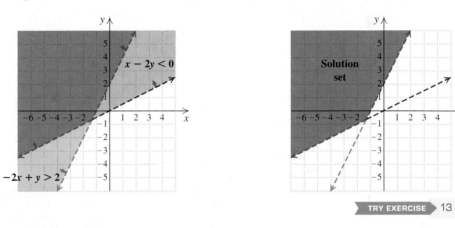

TRY EXERCISE 13

TECHNOLOGY CONNECTION

Many graphing calculators can be used to display systems of linear inequalities. To do so, enter each inequality, indicating the side that should be shaded. The graphing calculator automatically uses a different shading for each region.

- ◣ $y_1 = (-2/3)x + 5,$
- ◥ $y_2 = .5x - 3,$
- ◣ $y_3 = 3x + 4$

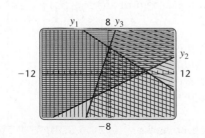

If your calculator has an Inequalz application, these inequalities can be entered as

$$y_1 \leq (-2/3)x + 5,$$
$$y_2 \geq .5x - 3,$$
$$y_3 \leq 3x + 4.$$

You may also be able to choose to shade only the intersection of the inequality.

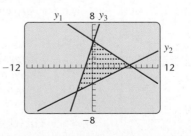

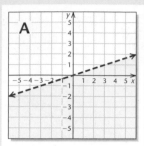

A

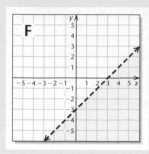

F

Visualizing for Success

Match each equation or inequality or system of equations or inequalities with its graph.

1. $x + y = 3$

2. $x = 3$

3. $y < x - 3$

4. $y = \frac{1}{3}x$

5. $y = 3$

6. $x - y = 3,$
   $x + y = 3$

7. $y < \frac{1}{3}x$

8. $y \geq x - 3,$
   $3x - y > 1$

9. $x - y < 3$

10. $y = \frac{1}{3}x,$
    $y = 3x$

Answers on page A-29

An additional, animated version of this activity appears in MyMathLab. To use MyMathLab, you need a course ID and a student access code. Contact your instructor for more information.

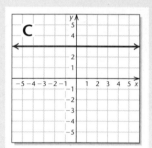

B

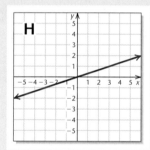

G

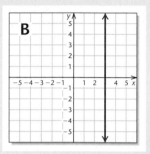

C

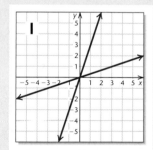

H

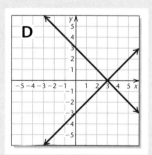

D

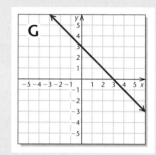

I

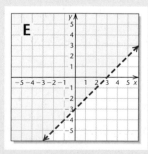

E

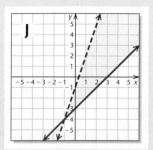

J

7.6 EXERCISE SET

For Extra Help MyMathLab | Math XP PRACTICE | WATCH | DOWNLOAD

Graph the solutions of each system.

1. $x + y \leq 3$,
$x - y \leq 5$

2. $x + y \leq 7$,
$x - y \leq 4$

3. $x + y < 6$,
$x + y > 0$

4. $y - 2x > 1$,
$y - 2x < 3$

5. $x > 3$,
$x + y \leq 4$

6. $y \geq -1$,
$x > 3 + y$

7. $y \geq x$,
$y \leq 1 - x$

8. $y > 3x - 2$,
$y < -x + 4$

9. $x \geq -3$,
$y \leq 2$

10. $x \leq 4$,
$y \geq -1$

11. $x \leq 0$,
$y \leq 0$

12. $x \geq 0$,
$y \geq 0$

13. $2x - 3y \geq 9$,
$2y + x > 6$

14. $3x - 2y \leq 8$,
$2x + y > 6$

15. $y > \frac{1}{2}x - 2$,
$x + y \leq 1$

16. $2y - 3x \leq 4$,
$\frac{2}{3}x + y > 4$

17. $x + y \leq 5$,
$x \geq 0$,
$y \geq 0$,
$y \leq 3$

18. $x + 2y \leq 8$,
$x \leq 6$,
$x \geq 0$,
$y \geq 0$

19. $y - x \geq 1$,
$y - x \leq 3$,
$x \leq 5$,
$x \geq 2$

20. $x - 2y \leq 0$,
$y - 2x \leq 2$,
$x \leq 2$,
$y \leq 2$

21. $y \leq x$,
$x \geq -2$,
$x \leq -y$

22. $y > 0$,
$2y + x \geq -6$,
$x + 2 \leq 2y$

23. Will a system of linear inequalities always have a solution? Why or why not?

24. If shadings are used to represent the inequalities in a system, will the most heavily shaded region always represent the solution set? Why or why not?

Skill Review

To prepare for Section 7.7, review solving equations using the multiplication principle (Section 2.1).

Solve. [2.1]

25. $210 = k \cdot 10$

26. $36 = k \cdot 12$

27. $0.4 = k \cdot 0.5$

28. $3 = k \cdot 7$

29. $5 = \dfrac{k}{8}$

30. $100 = \dfrac{k}{0.16}$

Synthesis

31. Explain how it would be possible for the solution of a system of linear inequalities to be a line.

32. Explain how it would be possible for the solution of a system of linear inequalities to be a single point.

33. Explain how it would be possible for the solution of a system of linear inequalties to be a line segment.

34. Explain how it would be possible for the solution of a system of linear inequalities to be the entire plane except for a line.

Graph the solutions of each system. If no solution exists, state this.

35. $3r + 6t \geq 36$,
$2r + 3t \geq 21$,
$5r + 3t \geq 30$,
$t \geq 0$,
$r \geq 0$

36. $2x + 5y \geq 18$,
$4x + 3y \geq 22$,
$2x + y \geq 8$,
$x \geq 0$,
$y \geq 0$

37. $x + 3y \leq 6$,
$2x + y \geq 4$,
$3x + 9y \geq 18$

38. $2x + 5y \geq 10$,
$x - 3y \leq 6$,
$4x + 10y \leq 20$

Aha! **39.** $2x + 3y \leq 1$,
$4x + 6y > 9$,
$5x - 2y \leq 8$,
$x \leq 12$,
$y \geq -15$

40. $5x - 4y \geq 8$,
$2x + 3y < 9$,
$2y \geq -8$,
$x \leq -5$,
$y < -6$

41. *Quilting.* The International Plowing Match and Country Festival, held in Crosby, Ontario, sponsors several quilting competitions. The maximum perimeter allowed for the Theme Quilt is 200 in. Find and graph a system of inequalities that indicates that a quilt's length and width meet this requirement.
Source: www.ipm2007.ca

42. *Architecture.* Most architects agree that the sum of a step's riser r and tread t, in inches, should not be less than 17 in. In Kennebunk, Maine, the maximum riser height for nonresidential buildings is 7 in., and the minimum tread width is 11 in. Find and graph a system of inequalities that indicates that a stair design has met the architect's and the town's requirements.
Source: www.kennebunkmaine.org

43. Use a graphing calculator to solve Exercise 14.

44. Use a graphing calculator to solve Exercise 13.

7.7   Direct and Inverse Variation

Equations of Direct Variation   •   Problem Solving with Direct Variation   •
Equations of Inverse Variation   •   Problem Solving with Inverse Variation

Many problems lead to equations of the form $y = kx$ or $y = k/x$, where k is a constant. Such equations are called *equations of variation*.

Equations of Direct Variation

A bicycle tour following the Underground Railroad route is traveling at a rate of 12 mph. In 1 hr, it goes 12 mi. In 2 hr, it goes 24 mi. In 3 hr, it goes 36 mi, and so on. In the graph below, we use the number of hours as the first coordinate and the number of miles traveled as the second coordinate:

$$(1, 12), \quad (2, 24), \quad (3, 36), \quad (4, 48), \quad \text{and so on.}$$

Note that the second coordinate is always 12 times the first.

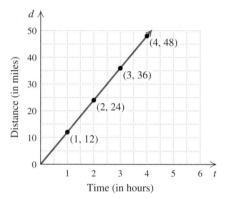

In this example, distance is a constant multiple of time, so we say that there is **direct variation** and that distance **varies directly** as time. The **equation of variation** is $d = 12t$.

> ### Direct Variation
>
> When a situation translates to an equation described by $y = kx$, with k a constant, we say that y *varies directly* as x. The equation $y = kx$ is called an *equation of direct variation*.

Note that for $k > 0$, any equation of the form $y = kx$ indicates that as x increases, y increases as well.

The terminologies

"y varies as x,"

"y is directly proportional to x," and

"y is proportional to x"

also imply direct variation and are used in many situations. The constant k is called the **constant of proportionality** or the **variation constant**. It can be found if one pair of values of x and y is known. Once k is known, other pairs can be determined.

EXAMPLE **1**

If y varies directly as x and $y = 2$ when $x = 5$, find the equation of variation.

SOLUTION   The words "y varies directly as x" indicate an equation of the form $y = kx$:

$$y = kx$$
$$2 = k \cdot 5 \qquad \text{Substituting to solve for } k$$
$$\tfrac{2}{5} = k, \quad \text{or} \quad k = 0.4. \qquad \text{Dividing both sides by 5}$$

Thus the equation of variation is $y = 0.4x$. A visualization of the situation is shown at left.

> TRY EXERCISE ▶ 7

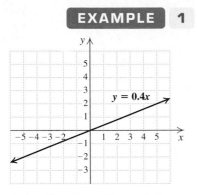

A visualization of Example 1

From the last two graphs, we see that when y varies directly as x, the constant of proportionality is also the slope of the associated graph—the rate at which y changes with respect to x.

EXAMPLE **2**

Find an equation in which s varies directly as t and $s = 10$ when $t = 15$. Then find the value of s when $t = 32$.

SOLUTION   We have

$$s = kt \qquad \text{We are told "} s \text{ varies directly as } t.\text{"}$$
$$10 = k \cdot 15 \qquad \text{Substituting 10 for } s \text{ and 15 for } t$$
$$\tfrac{10}{15} = k, \quad \text{or} \quad k = \tfrac{2}{3}. \qquad \text{Solving for } k$$

Thus the equation of variation is $s = \tfrac{2}{3}t$. When $t = 32$, we have

$$s = \tfrac{2}{3}t$$
$$s = \tfrac{2}{3} \cdot 32 \qquad \text{Substituting 32 for } t \text{ in the equation of variation}$$
$$s = \tfrac{64}{3}, \text{ or } 21\tfrac{1}{3}.$$

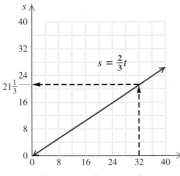

A visualization of Example 2

The value of s is $21\tfrac{1}{3}$ when $t = 32$.

Problem Solving with Direct Variation

In applications, it is often necessary to find an equation of variation and then use it to find other values, much as we did in Example 2.

EXAMPLE **3**

The time t that it takes to download a file varies directly as the size s of the file. Using her broadband Internet connection, La'Neiya can download a 4-MB (megabyte) song in 30 sec. How long would it take her to download a 210-MB television show?

Source: docs.info.apple.com

SOLUTION

1., 2. Familiarize and **Translate.** The problem indicates direct variation between t and s. Thus an equation $t = ks$ applies.

3. Carry out. We find an equation of variation:

$$t = ks$$
$$30 = k(4) \qquad \text{Substituting 30 for } t \text{ and 4 for } s$$
$$\frac{30}{4} = k$$
$$7.5 = k.$$

The equation of variation is $t = 7.5s$. When $s = 210$, we have

$$t = 7.5s$$
$$t = 7.5(210) \qquad \text{Substituting 210 for } s$$
$$t = 1575.$$

4. **Check.** To check, you might note that as the size of the file increased from 4 MB to 210 MB, the download time increased from 30 sec to 1575 sec. Also, the ratios 30/4 and 1575/210 are both 7.5.

5. **State.** La'Neiya will need 1575 sec, or 26 min 15 sec, to download the television show.

TRY EXERCISE ▶ 23

Equations of Inverse Variation

A car is traveling a distance of 20 mi. At a speed of 5 mph, the trip will take 4 hr. At 20 mph, it will take 1 hr. At 40 mph, it will take $\frac{1}{2}$ hr, and so on. This determines a set of pairs of numbers:

$$(5, 4), \qquad (20, 1), \qquad \left(40, \tfrac{1}{2}\right), \quad \text{and so on.}$$

Note that the product of speed and time for each of these pairs is 20. Note too that as the speed *increases*, the time *decreases*.

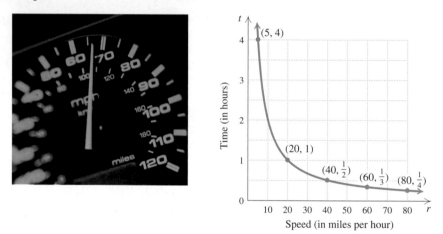

In this case, the product of speed and time is constant so we say that there is **inverse variation** and that time **varies inversely** as speed. The equation of variation is

$$rt = 20 \text{ (a constant)}, \quad \text{or} \quad t = \frac{20}{r}.$$

Inverse Variation

When a situation translates to an equation described by $y = k/x$, with k a constant, we say that y *varies inversely* as x. The equation $y = k/x$ is called an *equation of inverse variation.*

Note that for $k > 0$, any equation of the form $y = k/x$ indicates that as x increases, y decreases.

The terminology

"y is inversely proportional to x"

also implies inverse variation and is used in some situations. The constant k is again called the *constant of proportionality* or the *variation constant.*

EXAMPLE 4 If y varies inversely as x and $y = 145$ when $x = 0.8$, find the equation of variation.

SOLUTION The words "y varies inversely as x" indicate an equation of the form $y = k/x$:

$$y = \frac{k}{x}$$

$$145 = \frac{k}{0.8} \qquad \text{Substituting to solve for } k$$

$$(0.8)145 = k \qquad \text{Multiplying both sides by } 0.8$$

$$116 = k.$$

The equation of variation is $y = \dfrac{116}{x}$.

TRY EXERCISE 15

Problem Solving with Inverse Variation

Often in applications, we must decide what kind of variation, if any, applies.

EXAMPLE 5 *Community service.* Maxwell Elementary School needs 40 volunteers to donate 8 hr per month each to read one-on-one with a young child. If 50 people actually volunteer, how many hours per month would each person need to donate?

SOLUTION

1. **Familiarize.** What kind of variation applies to this situation? It seems reasonable that the greater the number of people volunteering, the less time each will need to donate. The total number of volunteer hours will be constant and inverse variation applies. If $T = $ the number of hours each person must volunteer and $N = $ the number of volunteers, then as N increases, T decreases.

2. **Translate.** Since inverse variation applies, we have

$$T = \frac{k}{N}.$$

3. **Carry out.** We find an equation of variation:

$$T = \frac{k}{N}$$

$$8 = \frac{k}{40} \qquad \text{Substituting 8 for } T \text{ and 40 for } N$$

$$40 \cdot 8 = k \qquad \text{Multiplying both sides by 40}$$

$$320 = k.$$

The equation of variation is $T = \dfrac{320}{N}$. When $N = 50$, we have

$$T = \frac{320}{50} \qquad \text{Substituting 50 for } N$$

$$T = 6.4.$$

4. **Check.** To check, note that 40 volunteers would donate a total of $(40)(8)$, or 320 hr per month. If each of 50 volunteers donated 6.4 hr per month, the total volunteer time would be $(50)(6.4)$, which is also 320 hr. Also, as the number of volunteers increases from 40 to 50, the time each needs to donate per month decreases from 8 hr to 6.4 hr.

5. **State.** Each of the 50 volunteers should donate 6.4 hr each month.

TRY EXERCISE 29

> **To Solve Variation Problems**
> 1. Determine from the language of the problem whether direct variation or inverse variation applies.
> 2. Using an equation of the form $y = kx$ for direct variation or $y = k/x$ for inverse variation, substitute known values and solve for k.
> 3. Write the equation of variation and use it, as needed, to find unknown values.

7.7   EXERCISE SET

↪ **Concept Reinforcement** *Determine whether each situation reflects direct variation or inverse variation.*

1. Two copiers can complete the job in 3 hr, whereas three copiers require only 2 hr.

2. Fionna wrote 5 invitations in 30 min and 8 invitations in 48 min.

3. Tomas found 8 flaws on 5 cabinets and 24 flaws on 15 cabinets.

4. Three neighbors collected roadside debris along a mile of roadway in 40 min while another group of five neighbors collected a mile's worth of debris in 24 min.

5. One heater warmed the room to 65° in 21 min when three heaters would have needed only 7 min.

6. Sara required 4 hr to read 6 term papers and 10 hr to read 15 papers.

For each of the following, find an equation of variation in which y varies directly as x and the following are true.

7. $y = 40$, when $x = 8$

8. $y = 30$, when $x = 3$

9. $y = 1.75$, when $x = 0.25$

10. $y = 3.2$, when $x = 0.08$

11. $y = 0.3$, when $x = 0.5$

12. $y = 1.2$, when $x = 1.1$

13. $y = 200$, when $x = 300$

14. $y = 500$, when $x = 60$

For each of the following, find an equation of variation in which y varies inversely as x and the following are true.

15. $y = 10$, when $x = 12$

16. $y = 8$, when $x = 3$

17. $y = 0.25$, when $x = 4$

18. $y = 0.25$, when $x = 8$

19. $y = 50$, when $x = 0.4$

20. $y = 42$, when $x = 50$

21. $y = 42$, when $x = 5$

22. $y = 0.2$, when $x = 5$

Solve.

23. *Wages.* Chardeé's paycheck P varies directly as the number of hours worked H. For 15 hr of work, the pay is $135. Find the pay for 23 hr of work.

24. *Oatmeal servings.* The number of servings S of oatmeal varies directly as the net weight W of the container purchased. A 42-oz box of oatmeal contains 30 servings. How many servings does a 63-oz box of oatmeal contain?

25. *Energy use.* The number of kilowatt-hours (kWh) of electricity n used by an air conditioner varies directly as its capacity c, measured in tons. A 2-ton air conditioner uses 18.5 kWh in 10 hr. How many kilowatt-hours would a 3-ton air conditioner use in 10 hr?
Source: www.lccc.net

26. *Gas volume.* The volume V of a gas varies inversely as the pressure P on it. The volume of a gas is 200 cm^3 (cubic centimeters) under a pressure of 32 kg/cm^2. What will be its volume under a pressure of 20 kg/cm^2?

27. The karat rating R of a gold object varies directly as the percentage P of gold in the object. A 14-karat gold chain is 58.25% gold. What is the percentage of gold in a 24-karat gold chain?

28. *Lunar weight.* The weight M of an object on the moon varies directly as its weight E on Earth. One of the authors of this book, Marv Bittinger, weighs 192 lb, but would weigh only 32 lb on the moon. The other author, David Ellenbogen, weighs 185 lb. How much would he weigh on the moon?

29. *Job completion.* The number of workers n required to complete a job varies inversely as the time t spent working. It takes 4 hr for 20 people to raise a barn. How long would it take 25 people to complete the job?

30. *Electricity.* A thermistor senses temperature changes by measuring changes in resistance. The resistance R in a PTC thermistor varies directly as the temperature T. A PTC thermistor has a resistance of 5000 ohms at a temperature of 25° C. What is the temperature if the resistance is 6000 ohms?

31. *Musical tones.* The frequency, or pitch P, of a musical tone varies inversely as its wavelength W. The U.S. standard concert A has a pitch of 440 vibrations per second and a wavelength of 2.4 ft. The E above concert A has a frequency of 660 vibrations per second. What is its wavelength?

32. *Filling time.* The time t required to fill a pool varies inversely as the rate r of flow into the pool. A tanker can fill a pool in 90 min at the rate of 1200 L/min. How long would it take to fill the pool at a rate of 2000 L/min?

Aha! 33. *Chartering a boat.* The cost per person c of chartering a boat is inversely proportional to the number of people n who are chartering the boat. If it costs $70 per person when 40 people charter the Sea Otter V, how many people chartered the boat if the cost were $140 per person?

34. *Answering questions.* The number of minutes m that a student should allow for each question on a quiz is inversely proportional to the number of questions n on the quiz. If a 16-question quiz allows students 2.5 min per question, how many questions would appear on a quiz in which students have 4 min per question?

State whether each situation represents direct variation, inverse variation, or neither. Give reasons for your answers.

35. The cost of mailing an overnight package in the United States and the distance that it travels

36. A runner's speed in a race and the time it takes to run the race

37. The weight of a turkey and its cooking time

38. The number of plays it takes to go 80 yd for a touchdown and the average gain per play

Skill Review

To prepare for Chapter 8, review squares (Sections 1.8, 4.1, and 4.5).

Find each square.

39. 10^2 [1.8]

40. $(-10)^2$ [1.8]

41. $(-5x)^2$ [4.1]

42. $(7t)^2$ [4.1]

43. $(ab^2)^2$ [4.1]

44. $(s^3t^4)^2$ [4.1]

45. $(x + y)^2$ [4.5]

46. $(2a - b)^2$ [4.5]

Synthesis

47. If x varies inversely as y and y varies inversely as z, how does x vary with regard to z? Why?

48. If a varies directly as b and b varies inversely as c, how does a vary with regard to c? Why?

Write an equation of variation for each situation. Use k as the variation constant in each equation.

49. *Wind energy.* The power P in a windmill varies directly as the cube of the wind speed v.

50. *Acoustics.* The square of the pitch P of a vibrating string varies directly as the tension t on the string.

51. *Ecology.* In a stream, the amount of salt S carried varies directly as the sixth power of the speed of the stream v.

52. *Lighting.* The intensity of illumination I from a light source varies inversely as the square of the distance d from the source.

53. *Crowd size.*   The number of people attending a First Night celebration N is directly proportional to the temperature T and inversely proportional to the percentage chance of rain or snow, P.

Write an equation of variation for each situation. Include a value for the variation constant in each equation.

54. *Geometry.*   The volume V of a sphere varies directly as the cube of the radius r.

55. *Geometry.*   The perimeter P of an equilateral octagon varies directly as the length S of a side.

56. *Geometry.*   The circumference C of a circle varies directly as the radius r.

57. *Geometry.*   The area A of a circle varies directly as the square of the length of the radius r.

Study Summary

KEY TERMS AND CONCEPTS	EXAMPLES

SECTION 7.1: SYSTEMS OF EQUATIONS AND GRAPHING

A solution of a **system of two equations** is an ordered pair that makes both equations true. The intersection of the graphs of the equations gives the solution of the system.

A **consistent** system has at least one solution; an **inconsistent** system has no solution.

When one equation in a system of two equations is a nonzero multiple of the other, the equations are **dependent** and there is an infinite number of solutions. Otherwise, the equations are **independent.**

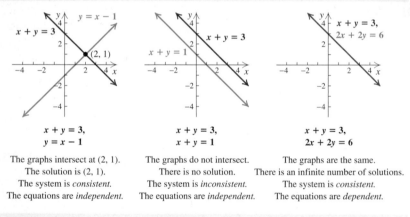

$$x + y = 3,$$
$$y = x - 1$$

The graphs intersect at $(2, 1)$.
The solution is $(2, 1)$.
The system is *consistent*.
The equations are *independent*.

$$x + y = 3,$$
$$x + y = 1$$

The graphs do not intersect.
There is no solution.
The system is *inconsistent*.
The equations are *independent*.

$$x + y = 3,$$
$$2x + 2y = 6$$

The graphs are the same.
There is an infinite number of solutions.
The system is *consistent*.
The equations are *dependent*.

SECTION 7.2: SYSTEMS OF EQUATIONS AND SUBSTITUTION

To use the **substitution method,** we solve one equation for a variable and substitute that expression for that variable in the other equation.

Solve:

$$2x + 3y = 8,$$
$$x = y + 1.$$

Substitute and solve for y:

$$2(y + 1) + 3y = 8$$
$$2y + 2 + 3y = 8$$
$$y = \tfrac{6}{5}.$$

Substitute and solve for x:

$$x = y + 1$$
$$x = \tfrac{6}{5} + 1$$
$$x = \tfrac{11}{5}.$$

The solution is $\left(\tfrac{11}{5}, \tfrac{6}{5}\right)$.

SECTION 7.3: SYSTEMS OF EQUATIONS AND ELIMINATION

To use the **elimination method,** we add to eliminate a variable.

Solve:

$$4x - 2y = 6,$$
$$3x + y = 7.$$

Eliminate y and solve for x:

$$\begin{array}{r} 4x - 2y = 6 \\ 6x + 2y = 14 \\ \hline 10x \phantom{{}+ 2y} = 20 \\ x = 2. \end{array}$$

Substitute and solve for y:

$$3x + y = 7$$
$$3 \cdot 2 + y = 7$$
$$y = 1.$$

The solution is $(2, 1)$.

SECTION 7.4: MORE APPLICATIONS USING SYSTEMS

Total-value and **mixture** problems are two types of problems that are often translated readily into a system of equations.

An industrial cleaning solution that is 40% nitric acid is added to a solution that is 15% nitric acid in order to create 2 L of a solution that is 25% nitric acid. How much 40%-acid and how much 15%-acid should be used?

1. **Familiarize.** Let x = the number of liters of 40%-acid solution and y = the number of liters of 15%-acid solution.

2. **Translate.** We will organize the information in a table.

Type of Solution	40%-Acid	15%-Acid	25%-Acid Mix
Amount of Solution	x	y	2 L
Percent Acid	40%	15%	25%
Amount of Acid in Solution	$0.4x$	$0.15y$	$0.25(2) = 0.5$ L

The amount in the mix must be 2 L, and the amount of acid must total 0.5 L. From the second and fourth lines of the table, we have the system of equations

$$x + y = 2, \quad (1)$$
$$0.4x + 0.15y = 0.5. \quad (2)$$

3. **Carry out.** We solve the system:

$$-0.15x - 0.15y = -0.3 \qquad \text{Multiplying both sides of equation (1) by } -0.15$$

$$\underline{0.4x + 0.15y = 0.5}$$
$$0.25x = 0.2$$
$$x = 0.8. \qquad \text{Dividing both sides by 0.25}$$

If $x = 0.8$, then $y = 1.2$, since $x + y = 2$.

4. **Check.** Combining 0.8 L of the 40%-acid solution and 1.2 L of the 15%-acid solution will result in 2 L of a mixture. There will be $0.4(0.8) + 0.15(1.2)$, or 0.5 L of nitric acid in the mixture, which is 25% of 2 L.

5. **State.** To create the mixture, use 0.8 L of the 40%-acid solution and 1.2 L of the 15%-acid solution.

SECTION 7.5: LINEAR INEQUALITIES IN TWO VARIABLES

The solutions of a **linear inequality** like $y \le \frac{1}{2}x - 1$ are ordered pairs that make the inequality true. The graph of a linear inequality is a region in the plane bounded by a solid line if the inequality symbol is $\le$ or $\ge$ and by a dashed line if the inequality symbol is $<$ or $>$. A test point not on the boundary line is used to determine which half-plane to shade as the graph of the solution.

Graph: $y \le \frac{1}{2}x - 1$.

1. Graph $y = \frac{1}{2}x - 1$, using a solid line.
2. Test $(0, 0)$. Since $0 \le \frac{1}{2}(0) - 1$ is *false*, shade the region that does *not* contain $(0, 0)$.

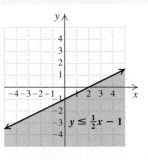

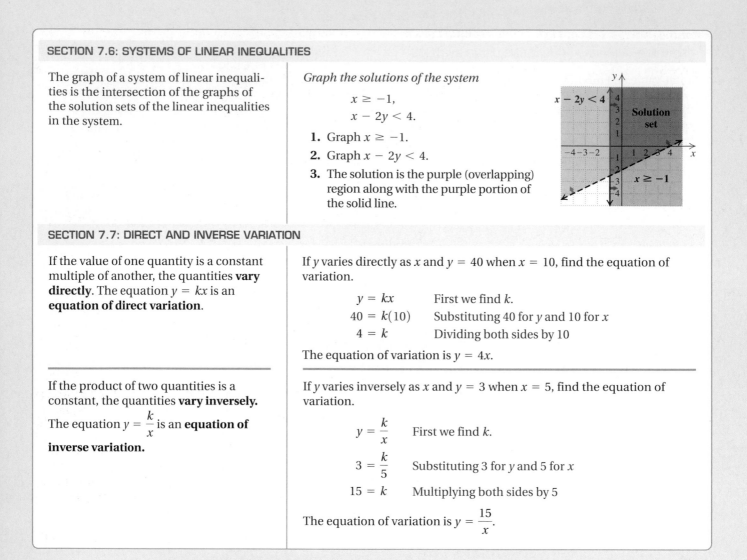

SECTION 7.6: SYSTEMS OF LINEAR INEQUALITIES

The graph of a system of linear inequalities is the intersection of the graphs of the solution sets of the linear inequalities in the system.

Graph the solutions of the system

$$x \geq -1,$$
$$x - 2y < 4.$$

1. Graph $x \geq -1$.
2. Graph $x - 2y < 4$.
3. The solution is the purple (overlapping) region along with the purple portion of the solid line.

SECTION 7.7: DIRECT AND INVERSE VARIATION

If the value of one quantity is a constant multiple of another, the quantities **vary directly**. The equation $y = kx$ is an **equation of direct variation**.

If y varies directly as x and $y = 40$ when $x = 10$, find the equation of variation.

$$y = kx \qquad \text{First we find } k.$$
$$40 = k(10) \qquad \text{Substituting 40 for } y \text{ and 10 for } x$$
$$4 = k \qquad \text{Dividing both sides by 10}$$

The equation of variation is $y = 4x$.

If the product of two quantities is a constant, the quantities **vary inversely.**

The equation $y = \dfrac{k}{x}$ is an **equation of inverse variation.**

If y varies inversely as x and $y = 3$ when $x = 5$, find the equation of variation.

$$y = \dfrac{k}{x} \qquad \text{First we find } k.$$
$$3 = \dfrac{k}{5} \qquad \text{Substituting 3 for } y \text{ and 5 for } x$$
$$15 = k \qquad \text{Multiplying both sides by 5}$$

The equation of variation is $y = \dfrac{15}{x}$.

Review Exercises: Chapter 7

🖐 *Concept Reinforcement* *Complete each sentence.*

1. The solution of a system of two equations in two variables is a(n) _____ pair of numbers. [7.1]

2. When both equations in a system are equivalent to each other, there is a(n) _____ number of solutions. [7.1]

3. When the equations in a system have graphs that are _____, the system has no solution. [7.1]

4. When substituting an expression for a variable, use _____ around the expression being substituted. [7.2]

5. To use the elimination method, the _____ of one of the variables must be made to be (if they are not already) opposites of each other. [7.3]

6. The graph of an inequality of the form $y < mx + b$ is shaded _____ the line $y = mx + b$. [7.5]

7. The graph of the solutions of a system of linear inequalities is the _____ of the graphs of the inequalities. [7.6]

8. If y varies inversely as x, then as x increases, y _____. [7.7]

Determine whether each ordered pair is a solution of the system of equations. [7.1]

9. $(2, -5)$;
$x - 2y = 12,$
$2x - y = 1$

10. $(-3, 1)$;
$3b + a = 0,$
$a + 5b = 2$

Solve by graphing. If there is no solution or an infinite number of solutions, state this. [7.1]

11. $y = 4x - 1,$
$y = x + 2$

12. $x - y = 8,$
$x + y = 4$

13. $3x - 4y = 8,$
$4y - 3x = 6$

14. $2x + y = 3,$
$4x + 2y = 6$

Solve using the substitution method. If there is no solution or an infinite number of solutions, state this. [7.2]

15. $y = 4 - x,$
$3x + 4y = 21$

16. $x + 2y = 6,$
$2x + y = 8$

17. $x + y = 5,$
$y = 2 - x$

18. $x + y = 6,$
$y = 3 - 2x$

19. $x - 2y = 5,$
$3x + 4y = 10$

20. $3x - y = 5,$
$6x = 2y + 10$

Solve using the elimination method. If there is no solution or an infinite number of solutions, state this. [7.3]

21. $3x - 2y = 0,$
$2x + 2y = 50$

22. $x - y = 8,$
$2x + y = 7$

23. $x - \frac{1}{3}y = -\frac{13}{3},$
$3x - y = -13$

24. $4x + 3y = -1,$
$2x + 9y = 2$

25. $5x - 2y = 7,$
$4x - 3y = 14$

26. $-x - y = -5,$
$2x - y = 4$

27. $2x + 5y = 8,$
$3x + 4y = 10$

28. $-4x + 6y = -10,$
$6x - 9y = 12$

Solve. [7.4]

29. *Perimeter of a garden.* Hassam is designing a rectangular garden with a perimeter of 66 ft. The length of the garden is 1 ft longer than three times the width. Find the dimensions of the garden.

30. *Basketball.* In a recent NBA game, Dirk Nowitzki scored 25 points on a combination of 11 two- and three-point baskets. How many shots of each type were made?
Source: National Basketball Association

31. *Lunch costs.* Dramatic Productions ordered a combination of oven-roasted turkey subs at $6.49 each and veggie subs at $5.09 each. The order contained 50 subs and cost a total of $303.50. How many of each type of sub were delivered?

32. *Fat content.* Café Rich instant flavored coffee gets 40% of its calories from fat. Café Light coffee gets 25% of its calories from fat. How much of each brand of coffee should be mixed in order to make 200 g of instant coffee with 30% of its calories from fat?

Graph on a plane. [7.5]

33. $x \leq y$

34. $x - 2y \geq 4$

35. $y > -2$

36. $y \geq \frac{2}{3}x - 5$

37. $2x + y < 1$

38. $x < 4$

Graph the solutions of each system. [7.6]

39. $x \geq 1,$
$y \leq -1$

40. $x - y > 2,$
$x + y < 1$

41. If y varies directly as x and $y = 81$ when $x = 3$, find the equation of variation. [7.7]

42. *Cubicle space.* The size of a cubicle c in an office space varies inversely as the number of workers w in that space. When there are 10 workers in Office A-213, each worker has 180 ft^2 of space. How much space will each worker have if there are 12 workers in that office? [7.7]

Synthesis

43. Explain why any solution of a system of equations is a point of intersection of the graphs of each equation in the system. [7.1]

44. Monroe sketches the boundary lines of a system of two linear inequalities and notes that the lines are parallel. Since there is no point of intersection, he concludes that the solution set is empty. What is wrong with this conclusion? [7.6]

45. The solution of the following system is $(6, 2)$. Find C and D.
$2x - Dy = 6,$
$Cx + 4y = 14$ [7.1]

46. Solve using the substitution method:
$x - y + 2z = -3,$
$2x + y - 3z = 11,$
$z = -2.$ [7.2]

47. Solve:
$3(x - y) = 4 + x,$
$x = 5y + 2.$ [7.2]

48. For a two-digit number, the sum of the ones digit and the tens digit is 6. When the digits are reversed, the new number is 18 more than the original number. Find the original number. [7.4]

49. A sales representative agrees to a compensation package of $42,000 plus a computer. After 7 months, the sales representative leaves the company and receives a prorated compensation package consisting of the computer and $23,750. What was the value of the computer? [7.4]

Test: Chapter 7

1. Determine whether $(3, -2)$ is a solution of the following system of equations:
$$2x + y = 4,$$
$$5x - 6y = 27.$$

Solve by graphing. If there is no solution or an infinite number of solutions, state this.

2. $y = -2x + 5,$
   $y = 4x - 1$

3. $2y - x = 7,$
   $2x - 4y = 4$

Solve using the substitution method. If there is no solution or an infinite number of solutions, state this.

4. $2x + 11 = y,$
   $3x + 2y = 1$

5. $4x + y = 5,$
   $2x + y = 4$

6. $x = 5y - 10,$
   $15y = 3x + 30$

Solve using the elimination method. If there is no solution or an infinite number of solutions, state this.

7. $x - y = 1,$
   $2x + y = 8$

8. $\frac{3}{2}x - y = 24,$
   $2x + \frac{3}{2}y = 15$

9. $4x + 5y = 5,$
   $6x + 7y = 7$

10. $2x + 3y = 8,$
   $3x + 2y = 5$

Solve.

11. *Complementary angles.* Two angles are complementary. One angle is 2° less than three times the other. Find the angles.

12. *Cooking.* A chef has one vinaigrette that is 40% vinegar and another that is 25% vinegar. How much of each is needed in order to make 60 L of a vinaigrette that is 30% vinegar?

13. *Taxi fares.* A New York City taxi recently cost $2.50 plus $1.00 per half mile. A Boston taxi cost $1.75 plus $1.20 per half mile. For what distance will the taxis cost the same?

Graph on a plane.

14. $y > x - 1$

15. $2x - y \leq 4$

16. $y < -2$

Graph the solutions of each system.

17. $y \geq x - 5,$
   $y < \frac{1}{2}x$

18. $x + y \leq 4,$
   $x \geq 0,$
   $y \geq 0$

19. If y varies inversely as x and $y = 9$ when $x = 2$, find the equation of variation.

20. *Hospital costs.* A hospital's linen cost varies directly with the number of patients. Wellview Hospital has a daily linen cost of $1178 when there are 124 patients in the hospital. What is the daily linen cost when there are 140 patients in the hospital?

Synthesis

21. You are in line at a ticket window. There are two more people ahead of you in line than there are behind you. In the entire line, there are three times as many people as there are behind you. How many are in the line?

22. Graph on a plane: $|x| \leq 5$.

23. Find the numbers C and D such that $(-2, 3)$ is a solution of the system
$$Cx - 4y = 7,$$
$$3x + Dy = 8.$$

Cumulative Review: Chapters 1–7

1. Evaluate $9t \div 6t^3$ for $t = -2$. [1.8]

2. Find the opposite of $-\frac{1}{10}$. [1.6]

3. Find the reciprocal of $-\frac{1}{10}$. [1.7]

4. Remove parentheses and simplify:
$$3x^2 - 2(-5x^2 + y) + y. \quad [1.8]$$

Simplify.

5. $-8 - (-15)$ [1.6]

6. $-\frac{3}{10} \div \left(-\frac{9}{20}\right)$ [1.7]

7. $40 - 8^2 \div 4 \cdot 4$ [1.8]

8. $\dfrac{|-2 \cdot 5 - 3 \cdot 4|}{5^2 - 2 \cdot 7}$ [1.8]

Solve.

9. $2x + 1 = 5(2 - x)$ [2.2]

10. $x^2 + 5x + 6 = 0$ [5.6]

11. $t + \dfrac{6}{t} = 5$ [6.6]

12. $\frac{1}{2}t + \frac{1}{6} = \frac{1}{3} - t$ [2.2]

13. $2y + 9 \le 5y + 11$ [2.6]

14. $n^2 = 100$ [5.6]

15. $3x + y = 5$,
$\quad y = x + 1$ [7.2]

16. $\dfrac{4}{x - 1} = \dfrac{3}{x + 2}$ [6.6]

17. $6x^2 = x + 2$ [5.6]

18. $x + y = 10$,
$\quad x - y = 4$ [7.3]

19. $3x + 2y = 4$,
$\quad 5x - 4y = 1$ [7.3]

20. $3(x + 1) - 5(x - 2) = 6(x + 3) - 4x$ [2.2]

Solve each formula. [2.3]

21. $t = \frac{1}{3}pq$, for p

22. $A = \dfrac{r + s}{2}$, for s

Add and, if possible, simplify.

23. $(8a^2 - 6a - 7) + (3a^3 + 6a - 7)$ [4.3]

24. $\dfrac{2x + 1}{x + 2} + \dfrac{x + 5}{x + 2}$ [6.3]

25. $\dfrac{m + n}{2m + n} + \dfrac{n}{m - n}$ [6.4]

Subtract.

26. $(8a^2 - 6a - 7) - (3a^3 + 6a - 7)$ [4.3]

27. $\dfrac{4p - q}{p + q} - \dfrac{10p - q}{p + q}$ [6.3]

28. $\dfrac{x + 5}{x - 2} - \dfrac{x - 1}{2 - x}$ [6.4]

Multiply.

29. $(5a^2 + b)(5a^2 - b)$ [4.6]

30. $-3x^2(5x^3 - 6x^2 - 2x + 1)$ [4.4]

31. $(2n + 5)^2$ [4.5]

32. $(8t^2 + 5)(t^3 + 4)$ [4.5]

33. $\dfrac{x^2 - 2x + 1}{x^2 - 4} \cdot \dfrac{x^2 + 4x + 4}{x^2 - 3x + 2}$ [6.2]

Divide.

34. $(2x^2 - 5x - 3) \div (x - 3)$ [4.7]

35. $\dfrac{x^2 - x}{4x^2 + 8x} \div \dfrac{x^2 - 1}{2x}$ [6.2]

Factor completely.

36. $-5t + 10s - 15$ [1.2]

37. $x^3 + x^2 + 2x + 2$ [5.1]

38. $m^4 - 1$ [5.4]

39. $2x^3 + 18x^2 + 40x$ [5.2]

40. $x^4 - 6x^2 + 9$ [5.4]

41. $4x^2 - 2x - 10$ [5.1]

42. $10x^2 - 29x + 10$ [5.3]

43. $9a^2b^2 - 16a^2b^4$ [5.4]

Graph.

44. $2x + y = 6$ [3.3]

45. $y = -2x + 1$ [3.6]

46. $2x = 10$ [3.3]

47. $x = 2y$ [3.2]

48. $x < 2y$ [7.5]

49. $x + y \le 1$,
$\quad x - y > 2$ [7.6]

50. Find the x-intercept and the y-intercept of the line given by $10x - 15y = 60$. [3.3]

51. Write the slope–intercept equation for the line with slope -2 and y-intercept $\left(0, \frac{4}{7}\right)$. [3.6]

Simplify.

52. $y^6 \cdot y^{-10}$ [4.8]

53. $(-3a^2)^2 \cdot a^5$ [4.1]

54. $\dfrac{3x^{-2}}{6x^{-12}}$ [4.8]

55. $\left(\dfrac{a^2b}{a^3b^4}\right)^{-1}$ [4.8]

Solve.

56. A library was mistakenly charged sales tax on an order of children's books. The invoice, including 5% sales tax, was for $1323. How much should the library have been charged? [2.4]

57. Each nurse practitioner at the Midway Clinic is required to see an average of at least 40 patients per day. During the first 4 workdays of one week, Michael saw 50, 35, 42, and 38 patients. How many patients must he see on the fifth day in order to meet his requirements? [2.7]

58. A snowmobile is traveling 40 mph faster than a dog sled. In the same time that the dog sled travels 24 mi, the snowmobile travels 104 mi. Find the speeds of the sled and the snowmobile. [6.7]

59. Tia and Avery live next door to each other on Meachin Street. Their house numbers are consecutive odd numbers, and the product of their house numbers is 143. Find the house numbers. [5.7]

60. Long-distance calls made using the "Ruby" prepaid calling card cost 1.4¢ per minute plus a maintenance fee of 99¢ per week. The "Sapphire" plan has a maintenance fee of only 69¢ per week, but calls cost 1.7¢ per minute. For what number of minutes per week will the two cards cost the same? [7.3]
Source: www.enjoyprepaid.com

61. The second angle of a triangle is twice as large as the first. The third angle is 15° less than the sum of the first two angles. Find the measure of each angle. [2.5]

62. A newspaper uses self-employed copywriters to write its advertising copy. If they use 12 copywriters, each person works 35 hr per week. How many hours per week would each person work if the newspaper uses 10 copywriters? [7.7]

Synthesis

63. Solve $t = px - qx$ for x. [2.3]

64. Write the slope–intercept equation of the line that contains the point $(-2, 3)$ and is parallel to the line $2x - y = 7$. [3.7]

65. Simplify: $\dfrac{t^2 - 9}{2t + 1} \div \dfrac{t^2 - 4t + 3}{6t^2 + 3t} \cdot \dfrac{2t^3 - t^2 + t}{4t + 12}$. [6.2]

66. Graph on a plane: $y \geq x^2$. [7.5]

67. The maximum length of a postcard that can be mailed with one postcard stamp is $\frac{7}{4}$ in. longer than the maximum width. The maximum area is $\frac{51}{2}$ in². Find the maximum length and width of a postcard. [5.7]
Source: USPS

Radical Expressions and Equations

SARA RAMIREZ
SEA TURTLE BIOLOGIST
St. Kitts and Nevis, West Indies

I use math in many aspects of my work. While in the field collecting data, I take measurements of both the sea turtles and their beach environment, such as their shell length and the degree of inclination of the beach. Often these values must be converted between the metric and standard systems of measurement. When analyzing data, I use math to determine values such as nest hatching success and to graph correlations between nesting variables.

AN APPLICATION

The daily number of calories c needed by a reptile of weight w pounds can be approximated by $c = 10w^{3/4}$. Find the daily calorie requirement of a green iguana weighing 16 lb.

Source: www.anapsid.org

This problem appears as Exercise 87 in Section 8.7.

Y ou may already be familiar with the notion of *square roots.* For example, 3 is a square root of 9 because $3^2 = 9$ In this chapter, we learn how to manipulate square roots of polynomials and rational expressions. Later in this chapter, these *radical expressions* will appear in equations and in problem-solving situations.

8.1   Introduction to Square Roots and Radical Expressions

Square Roots   •   Radicands and Radical Expressions   •   Irrational Numbers   •
Square Roots and Absolute Value   •   Problem Solving

STUDY SKILLS

Professors Are Human

Even the best professors sometimes make mistakes. If your work does not agree with your instructor's, politely ask about it. Your instructor will welcome the opportunity to correct any errors.

We begin our study of radical expressions by examining square roots of numbers, square roots of variable expressions, and an application involving a formula.

Square Roots

To find the square of a number, we multiply that number by itself. When the process is reversed, we say that we are looking for a number's *square root.* For example, since $5^2 = 25$, we say that

25 is the square of 5,   and   5 is a square root of 25.

> ### Square Root
> The number c is a *square root* of a if $c^2 = a$.

We say that 5 is *a* square root of 25 because every positive number has two square roots. The square roots of 25 are 5 and -5 because $5^2 = 25$ and $(-5)^2 = 25$.

EXAMPLE **1**   Find the square roots of 81.

SOLUTION   The square roots of 81 are 9 and -9. To check, note that $9^2 = 81$ and $(-9)^2 = (-9)(-9) = 81$.

> **TRY EXERCISE** 11

We use a **radical sign**, $\sqrt{}$, to indicate the nonnegative square root, or **principal square root**, of a number. Thus the square roots of 81 are 9 and -9, but $\sqrt{81} = 9$. Note that $\sqrt{81} \neq -9$. To represent the negative square root of 81, we write $-\sqrt{81}$.

EXAMPLE **2**   Find each of the following: **(a)** $\sqrt{225}$; **(b)** $-\sqrt{64}$.

SOLUTION

a) The principal square root of 225 is its positive square root, so $\sqrt{225} = 15$.

b) The symbol $-\sqrt{64}$ represents the opposite of $\sqrt{64}$. Since $\sqrt{64} = 8$, we have $-\sqrt{64} = -8$.

> **TRY EXERCISE** 19

Radicands and Radical Expressions

A **radical expression** is an algebraic expression that contains at least one radical sign. Here are some examples:

$$\sqrt{14}, \qquad 8 + \sqrt{2x}, \qquad \sqrt{t^2 + 4}, \qquad \sqrt{\dfrac{x^2 - 5}{2}}.$$

The **radicand** in a radical expression is the expression under the radical sign.

EXAMPLE 3   Identify the radicand in each expression: **(a)** $\sqrt{x}$; **(b)** $3\sqrt{y^2 - 5}$.

SOLUTION

a) In $\sqrt{x}$, the radicand is x.

b) In $3\sqrt{y^2 - 5}$, the radicand is $y^2 - 5$.

TRY EXERCISE 31

The square of any nonzero real number is always positive. For example, $8^2 = 64$ and $(-11)^2 = 121$. No real number, squared, is equal to a negative number. Thus the following expressions are not real numbers:

$$\sqrt{-100}, \qquad \sqrt{-49}, \qquad -\sqrt{-3}.$$

Numbers like $\sqrt{-100}$, $\sqrt{-49}$, and $-\sqrt{-3}$ are discussed in Chapter 9.

Irrational Numbers

In Section 1.4, we learned that $\sqrt{2}$ cannot be written as a ratio of two integers. Numbers like $\sqrt{2}$ are real but not rational and are called *irrational*. A number that is the square of some rational number, like 9 or 64, is called a *perfect square*. The square root of a perfect square is always rational; the square root of a nonnegative number that is not a perfect square is irrational.

EXAMPLE 4   Classify each of the following numbers as rational or irrational.

a) $\sqrt{3}$ **b)** $\sqrt{25}$

c) $\sqrt{35}$ **d)** $-\sqrt{9}$

SOLUTION

a) $\sqrt{3}$ is irrational, since 3 is not a perfect square.

b) $\sqrt{25}$ is rational, since 25 is a perfect square: $\sqrt{25} = 5$.

c) $\sqrt{35}$ is irrational, since 35 is not a perfect square.

d) $-\sqrt{9}$ is rational, since 9 is a perfect square: $-\sqrt{9} = -3$.

TRY EXERCISE 37

For the following list, we have printed the irrational numbers in red: $\sqrt{1}$, $\sqrt{2}$, $\sqrt{3}$, $\sqrt{4}$, $\sqrt{5}$, $\sqrt{6}$, $\sqrt{7}$, $\sqrt{8}$, $\sqrt{9}$, $\sqrt{10}$, $\sqrt{11}$, $\sqrt{12}$, $\sqrt{13}$, $\sqrt{14}$, $\sqrt{15}$, $\sqrt{16}$, $\sqrt{17}$, $\sqrt{18}$, $\sqrt{19}$, $\sqrt{20}$, $\sqrt{21}$, $\sqrt{22}$, $\sqrt{23}$, $\sqrt{24}$, $\sqrt{25}$.

Often, when square roots are irrational, a calculator is used to find decimal approximations.

EXAMPLE 5   Use a calculator to approximate $\sqrt{10}$ to three decimal places.

SOLUTION   Calculators vary in their methods of operation. On some, we press $\boxed{\sqrt{}}$ and then the number; on others, we enter the number and then press $\boxed{\sqrt{}}$.

$$\sqrt{10} \approx 3.162277660 \qquad \text{Using a calculator with a 10-digit display}$$

With most graphing calculators, we press $\boxed{\textbf{2ND}}$ $\boxed{\sqrt{}}$, the radicand (in this case 10), and $\boxed{\textbf{ENTER}}$.

     Decimal representation of an irrational number is nonrepeating and nonending. Rounding to three decimal places, we have $\sqrt{10} \approx 3.162$.

> **TRY EXERCISE** 47

Square Roots and Absolute Value

Let's compare $\sqrt{(-5)^2}$ and $\sqrt{5^2}$:

$$\sqrt{(-5)^2} = \sqrt{25} = 5, \qquad |-5| = 5;$$
$$\sqrt{5^2} = \sqrt{25} = 5, \qquad |5| = 5.$$

Note that squaring a number and then taking the square root is the same as taking the absolute value of the number. In short, the principal square root of the square of A is the absolute value of A:

> **For any real number A, $\sqrt{A^2} = |A|$.**

EXAMPLE 6   Simplify $\sqrt{(3x)^2}$ given that x can represent any real number.

SOLUTION   To simplify $\sqrt{(3x)^2}$, note that if x represents a negative number, say, -2, the result is *not* $3x$. To see this, note that $\sqrt{(3(-2))^2} \neq 3(-2)$. Since the principal square root is always positive, to write $\sqrt{(3x)^2} = 3x$ would be incorrect. Instead, we write

$$\sqrt{(3x)^2} = |3x|, \text{ or } 3|x|. \qquad \text{Note that } 3x \text{ will be negative for } x < 0.$$

> **TRY EXERCISE** 53

     Fortunately, in many cases, it can be assumed that radicands that are variable expressions do not represent the square of a negative number. When this assumption is made, absolute-value symbols are not needed:

> **For $A \geq 0$, $\sqrt{A^2} = A$.**

EXAMPLE 7   Simplify each expression. Assume that all variables represent nonnegative numbers.

a) $\sqrt{(8x)^2}$ **b)** $\sqrt{(t + 2)^2}$

SOLUTION

a) $\sqrt{(8x)^2} = 8x$ Since $8x$ is assumed to be nonnegative, $|8x| = 8x$.

b) $\sqrt{(t + 2)^2} = t + 2$ Since $t + 2$ is assumed to be nonnegative, $|t + 2| = t + 2$.

> **TRY EXERCISE** 61

     In Sections 8.2–8.6, we will often state assumptions that make absolute-value symbols unnecessary.

On most graphing calculators, radicands are enclosed in parentheses. Pressing $\boxed{\sqrt{}}$ often results in both a radical sign and a left parenthesis placed on the home screen.

     To approximate $\sqrt{10}$ with such a graphing calculator, we press $\boxed{\textbf{2ND}}$ $\boxed{\sqrt{}}$ $\boxed{1}$ $\boxed{0}$ $\boxed{)}$ $\boxed{\textbf{ENTER}}$. Note that $\boxed{\sqrt{}}$ is the second operation associated with the $\boxed{x^2}$ key. Use of the $\boxed{)}$ key is optional here, but is essential when more complex calculations arise.

```
√(10)
              3.16227766
```

1. Approximate $\sqrt{8}$, $\sqrt{11}$, and $\sqrt{48}$.

Problem Solving

Radical expressions often appear in applications.

EXAMPLE   8

Parking-lot arrival spaces.   The attendants at a downtown parking lot use spaces to leave cars before they are taken to long-term parking stalls. The required number N of such spaces is approximated by the formula

$$N = 2.5\sqrt{A},$$

where A is the average number of arrivals in peak hours. Find the number of spaces needed when an average of 43 cars arrive during peak hours.

SOLUTION   We substitute 43 into the formula. We use a calculator to find an approximation:

$$N = 2.5\sqrt{43}$$
$$\approx 2.5(6.557) \qquad \text{Rounding } \sqrt{43} \text{ to } 6.557$$
$$\approx 16.393 \approx 17.$$

Note that we round *up* to 17 spaces because rounding down would create some overcrowding. Thus, for an average of 43 arrivals, 17 spaces are needed.

TRY EXERCISE ▸ 69

STUDENT NOTES

In Example 8, the number 16.393 is rounded *up* to 17 because of the real-world situation being modeled. In another situation, a number may be rounded *down*. Always make sure that your answers make sense in the real world.

Calculator Note.   In Example 8, we rounded $\sqrt{43}$. Generally, when using a calculator, we round at the *end* of our work. This can change the answer:

$$N = 2.5\sqrt{43} \approx 2.5(6.557438524) = 16.39359631 \approx 16.394.$$

Note the discrepancy in the third decimal place. When using a calculator for approximation, be aware of possible variations in answers. You may get answers that differ slightly from those at the back of the book. Answers to the exercises have been found by rounding at the end of the calculations.

TECHNOLOGY CONNECTION

Graphing equations that contain radical expressions often involves approximating irrational numbers. Also, since the square root of a negative number is not real, such graphs may not exist for all choices of x. For example, the graph of $y = \sqrt{x - 1}$ does not exist for $x < 1$.

Similarly, the graph of $y = \sqrt{2 - x}$ does not exist for $x > 2$.

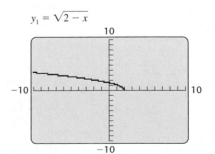

$y_1 = \sqrt{2 - x}$

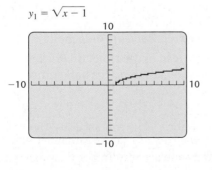

$y_1 = \sqrt{x - 1}$

Graph.

1. $y = \sqrt{x}$ 2. $y = \sqrt{2x}$
3. $y = \sqrt{x^2}$ 4. $y = \sqrt{(2x)^2}$
5. $y = \sqrt{x + 4}$ 6. $y = \sqrt{6 - x}$

8.1 EXERCISE SET

For Extra Help   MyMathLab   Math XL PRACTICE   WATCH   DOWNLOAD

Concept Reinforcement *In each of Exercises 1–6, match the phrase with the most appropriate choice from the column on the right.*

1. _____ The name for an expression written under a radical
2. _____ The name for an algebraic expression that contains at least one radical sign
3. _____ The name for a number that is real but not rational
4. _____ The name for a positive number's positive square root
5. _____ The sign of $3x$ when x represents a negative number
6. _____ The sign of $\sqrt{A^2}$ when A represents a negative number

a) Irrational
b) Negative
c) Radicand
d) Positive
e) Radical expression
f) Principal square root

Classify each statement as either true or false. Do not use a calculator.

7. $\sqrt{37}$ is between 6 and 7.
8. $\sqrt{75}$ is between 7 and 8.
9. $\sqrt{150}$ is between 11 and 12.
10. $\sqrt{103}$ is between 10 and 11.

Find the square roots of each number.

11. 100   12. 4
13. 36   14. 9
15. 1   16. 121
17. 144   18. 169

Simplify.

19. $\sqrt{100}$ 20. $\sqrt{9}$ 21. $-\sqrt{1}$
22. $-\sqrt{49}$ 23. $\sqrt{0}$ 24. $-\sqrt{81}$
25. $-\sqrt{121}$ 26. $\sqrt{400}$ 27. $\sqrt{900}$
28. $\sqrt{441}$ 29. $-\sqrt{144}$ 30. $-\sqrt{625}$

Identify the radicand for each expression.

31. $\sqrt{10a}$ 32. $\sqrt{x^2 y}$
33. $5\sqrt{t^3 - 2}$ 34. $8\sqrt{x^2 + 2}$
35. $x^2 y \sqrt{\dfrac{7}{x+y}}$ 36. $ab^2 \sqrt{\dfrac{a}{a-b}}$

Classify each number as rational or irrational.

37. $\sqrt{4}$ 38. $\sqrt{15}$
39. $\sqrt{11}$ 40. $\sqrt{12}$
41. $\sqrt{32}$ 42. $\sqrt{64}$
43. $-\sqrt{16}$ 44. $-\sqrt{144}$
Aha! 45. $-\sqrt{19^2}$ 46. $-\sqrt{22}$

Use a calculator to approximate each of the following numbers. Round to three decimal places.

47. $\sqrt{8}$ 48. $\sqrt{6}$
49. $\sqrt{15}$ 50. $\sqrt{19}$
51. $\sqrt{83}$ 52. $\sqrt{43}$

Simplify. Assume that x can represent any real number.

53. $\sqrt{x^2}$ 54. $\sqrt{(7x)^2}$
55. $\sqrt{(10x)^2}$ 56. $\sqrt{(x-1)^2}$
57. $\sqrt{(x+7)^2}$ 58. $\sqrt{(4-x)^2}$
59. $\sqrt{(5-2x)^2}$ 60. $\sqrt{(3x+1)^2}$

Simplify. Assume that all variables represent nonnegative numbers.

61. $\sqrt{x^2}$ 62. $\sqrt{t^2}$
63. $\sqrt{(5y)^2}$ 64. $\sqrt{(3a)^2}$
65. $\sqrt{16t^2}$ 66. $\sqrt{25x^2}$
67. $\sqrt{(n+7)^2}$ 68. $\sqrt{(a+2)^2}$

Parking spaces. *Solve. Use the formula $N = 2.5\sqrt{A}$ of Example 8.*

69. Find the number of spaces needed when the average number of arrivals is (a) 36; (b) 29.
70. Find the number of spaces needed when the average number of arrivals is (a) 49; (b) 53.

Hang time. *An athlete's hang time (time airborne for a jump), T, in seconds, is given by $T = 0.144\sqrt{V}$,* where V is the athlete's vertical leap, in inches.*

71. Luther Head of the Houston Rockets can jump 39 in. vertically. Find his hang time.
Source: rockets.com

*Based on an article by Peter Brancazio, "The Mechanics of a Slam Dunk," Popular Mechanics, November 1991. Courtesy of Peter Brancazio, Brooklyn College.

72. Lamar Odom of the Los Angeles Lakers can jump 32 in. vertically. Find his hang time.
Source: vertcoach.com

73. Which is the more exact way to write the square root of 12: 3.464101615 or $\sqrt{12}$? Why?

74. What is the difference between saying "*the* square root of 10" and saying "*a* square root of 10"?

Skill Review

To prepare for Section 8.2, review the product and power rules for exponents (Section 4.1).

Simplify. [4.1]

75. $(t^{10})^2$

76. $(p^5)^2$

77. $2a \cdot 25a^{12}$

78. $5x \cdot 9x^6$

79. $3m \cdot (10m^4)^2$

80. $2y \cdot (5y^7)^2$

Synthesis

81. Explain in your own words why $\sqrt{A^2} \neq A$ when A is negative.

82. One number has only one square root. What is the number and why is it unique in this regard?

Simplify, if possible.

83. $\sqrt{\frac{1}{100}}$

84. $\sqrt{0.01}$

85. $\sqrt{(-7)^2}$

86. $\sqrt{-7^2}$

87. $\sqrt{3^2 + 4^2}$

88. $\sqrt{\sqrt{81}}$

Aha! **89.** Between what two consecutive integers is $-\sqrt{33}$?

90. Find a number that is the square of an integer and the cube of a different integer.

Solve. If no real-number solution exists, state this.

91. $\sqrt{t^2} = 4$

92. $\sqrt{y^2} = -5$

93. $-\sqrt{x^2} = -3$

94. $a^2 = 36$

95. For which values of x is $\sqrt{x - 10}$ not a real number?

96. For which values of x is $\sqrt{10 - x}$ not a real number?

Simplify. Assume that all variables represent positive numbers.

97. $\sqrt{\dfrac{144x^8}{36y^6}}$

98. $\sqrt{\dfrac{y^{12}}{8100}}$

99. $\sqrt{\dfrac{400}{m^{16}}}$

100. $\sqrt{\dfrac{p^2}{3600}}$

101. Use the graph of $y = \sqrt{x}$, shown below, to estimate each of the following to the nearest tenth: **(a)** $\sqrt{3}$; **(b)** $\sqrt{5}$; **(c)** $\sqrt{7}$. Be sure to check by multiplying.

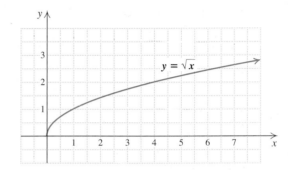

Speed of sound.   The speed V of sound traveling through air, in feet per second, is given by

$$V = \frac{1087\sqrt{273 + t}}{16.52},$$

where t is the temperature, in degrees Celsius. Using a calculator, find the speed of sound through air at each of the following temperatures. Round to the nearest tenth.

102. 28°C

103. 5°C

104. −10°C

105. Use a graphing calculator to draw the graphs of $y_1 = \sqrt{x - 2}$, $y_2 = \sqrt{x + 7}$, $y_3 = 5 + \sqrt{x}$, and $y_4 = -4 + \sqrt{x}$. If possible, graph all four equations using the SIMULTANEOUS mode and a $[-10, 10, -10, 10]$ window. Then determine which equation corresponds to each curve.

106. For which values of t is $\sqrt{(t + 4)^2} = t + 4$ false?

107. For which values of a is $\sqrt{(a - 1)^2} = a - 1$ false?

108. What restrictions on a, if any, are needed in order for
$$\sqrt{64a^{16}} = 8a^8$$
to be true? Explain how you found your answer.

CORNER

Lengths and Cycles of a Pendulum

Focus: Square roots and modeling

Time: 25–35 minutes

Group size: 3

Materials: Rulers, clocks or watches (to measure seconds), pendulums (see below), calculators

A pendulum is simply a string, a rope, or a chain with a weight of some sort attached at one end. When the unweighted end is held, a pendulum can swing freely from side to side. A shoe hanging from a shoelace, a yo-yo, a pendant hanging from a chain, or a hairbrush tied to a length of dental floss are all examples of a pendulum. In this activity, each group will develop a mathematical model (formula) that relates a pendulum's length L to the time T that it takes for one complete swing back and forth (one "cycle").

ACTIVITY

1. The group should design a pendulum with a length that can vary from 1 ft to 4 ft (see above). One group member should hold the pendulum so that its length is 1 ft. A second group member should lift the weight to one side, keeping the string tight, and then release (do not throw) the weight. The third group member should find the average time, in seconds, for one swing (cycle) back and forth, by timing *five* cycles and dividing by five. Repeat this for each pendulum length listed, so that the following table is completed.

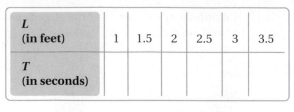

L (in feet)	1	1.5	2	2.5	3	3.5
T (in seconds)						

2. Examine the table your group has created. Can you find one number, a, such that $T \approx aL$ for all pairs of values on the chart?

3. To see if a better model can be found, add a third row to the chart and fill in $\sqrt{L}$ for each value of L listed. Can you find one number, b, such that $T \approx b\sqrt{L}$? Does this appear to be a more accurate model than $T = aL$?

4. Use the model for part (3) to predict T when L is 4 ft. Then check your prediction by measuring T as you did in part (1) above. Was your prediction "acceptable"? Compare your results with those of other groups.

5. In Section 8.3, we use a formula equivalent to

$$T = \frac{2\pi}{\sqrt{32}} \cdot \sqrt{L}.$$

How does your value of b from part (3) compare with $2\pi/\sqrt{32}$?

8.2 Multiplying and Simplifying Radical Expressions

Multiplying • Simplifying and Factoring • Simplifying Square Roots of Powers •
Multiplying and Simplifying

We now learn to multiply and simplify radical expressions. For this section, we assume that all variables represent nonnegative numbers. Because of this assumption, we will not need to use absolute-value symbols when simplifying.

Multiplying

To see how to multiply with radical notation, consider the following:

$$\sqrt{9} \cdot \sqrt{4} = 3 \cdot 2 = 6;$$ This is a product of square roots.
$$\sqrt{9 \cdot 4} = \sqrt{36} = 6.$$ This is the square root of a product.

Note that $\sqrt{9} \cdot \sqrt{4} = \sqrt{9 \cdot 4}$. This is generalized in the following rule.

STUDENT NOTES

Every statement in a definition or a rule is important and should be read carefully for understanding. In the product rule for square roots, the statement "for any real numbers $\sqrt{A}$ and $\sqrt{B}$" means that A and B must be nonnegative. The product rule does not hold, for example, for $\sqrt{-16} \cdot \sqrt{-25}$.

> ### The Product Rule for Square Roots
>
> For any real numbers $\sqrt{A}$ and $\sqrt{B}$,
>
> $$\sqrt{A} \cdot \sqrt{B} = \sqrt{A \cdot B}.$$
>
> (To multiply square roots, multiply the radicands and take the square root.)

EXAMPLE 1   Multiply: **(a)** $\sqrt{5}\,\sqrt{7}$; **(b)** $\sqrt{6}\,\sqrt{6}$; **(c)** $\sqrt{\frac{2}{3}}\,\sqrt{\frac{7}{5}}$; **(d)** $\sqrt{2x}\,\sqrt{3y}$.

SOLUTION

a) $\sqrt{5}\,\sqrt{7} = \sqrt{5 \cdot 7} = \sqrt{35}$

b) $\sqrt{6}\,\sqrt{6} = \sqrt{6 \cdot 6} = \sqrt{36} = 6$ Try to do this one directly: $\sqrt{6}\,\sqrt{6} = 6$.

c) $\sqrt{\dfrac{2}{3}}\,\sqrt{\dfrac{7}{5}} = \sqrt{\dfrac{2}{3} \cdot \dfrac{7}{5}} = \sqrt{\dfrac{14}{15}}$

d) $\sqrt{2x}\,\sqrt{3y} = \sqrt{6xy}$ Note that in order for $\sqrt{2x}$ and $\sqrt{3y}$ to be real numbers, we must have $x, y \geq 0$.   **TRY EXERCISE** 11

Simplifying and Factoring

To factor a square root, we can use the product rule in reverse. That is,

$$\sqrt{AB} = \sqrt{A}\,\sqrt{B}.$$

This property is especially useful when a radicand contains a perfect square as a factor. For example, the radicand in $\sqrt{50}$ is not a perfect square, but one of its factors, 25, *is* a perfect square. Thus,

$$\sqrt{50} = \sqrt{25 \cdot 2}$$ 25 is a perfect-square factor of 50.
$$= \sqrt{25} \cdot \sqrt{2}$$ $\sqrt{25}$ is a rational factor of $\sqrt{50}$.
$$= 5\sqrt{2}.$$ Simplifying $\sqrt{25}$; $\sqrt{2}$ cannot be simplified further.

To determine whether a radicand contains a factor that is a perfect square, it may help to write a prime factorization of the radicand. For example, if we did not easily see that 50 contains a perfect square as a factor, we could write

$$\sqrt{50} = \sqrt{2 \cdot 5 \cdot 5} \qquad \text{Factoring into prime factors}$$
$$= \sqrt{2} \cdot \sqrt{5 \cdot 5} \qquad \begin{array}{l}\text{Grouping pairs of like factors;}\\ 5 \cdot 5 \text{ is a perfect square.}\end{array}$$
$$= \sqrt{2} \cdot 5.$$

To avoid any uncertainty as to what is under the radical sign, it is customary to write the radical factor last. Thus, $\sqrt{50} = 5\sqrt{2}$.

A radical expression, like $\sqrt{26}$, in which the radicand has no perfect-square factors, is considered to be in simplified form.

> ### Simplified Form of a Square Root
> A radical expression for a square root is simplified when its radicand has no factor other than 1 that is a perfect square.

A variable raised to an even power, such as t^2 or x^6, is also a perfect square. Simplifying is always easiest if the *largest* perfect-square factor is identified in the first step.

EXAMPLE 2

Simplify by factoring. Assume that all variables represent nonnegative numbers.

a) $\sqrt{18}$ **b)** $\sqrt{a^2 b}$ **c)** $\sqrt{196 t^2 u}$

SOLUTION

a) $\sqrt{18} = \sqrt{9 \cdot 2}$ 9 is a perfect-square factor of 18.
$\qquad = \sqrt{9} \sqrt{2}$ $\sqrt{9}$ is a rational factor of $\sqrt{18}$.
$\qquad = 3\sqrt{2}$ $\sqrt{2}$ cannot be simplified further.

b) $\sqrt{a^2 b} = \sqrt{a^2} \sqrt{b}$ Identifying a perfect-square factor and factoring into a product of radicals

$\qquad = a\sqrt{b}$ No absolute-value signs are necessary since a is assumed to be nonnegative.

c) We may not recognize that $196 = 14^2$. Let's suppose that we noticed only that 4 and t^2 are perfect-square factors of $196 t^2 u$:

$$\sqrt{196 t^2 u} = \sqrt{4} \cdot \sqrt{t^2} \cdot \sqrt{49u} \qquad \text{Identifying perfect-square factors}$$
$$= 2t\sqrt{49u}. \qquad \text{We assume } t \geq 0.$$

Because the radicand still contains a perfect-square factor, we have not simplified completely. To finish simplifying, we rewrite the product and simplify $\sqrt{49u}$:

$$\sqrt{196 t^2 u} = 2t\sqrt{49u} \qquad \text{Rewriting the last step above}$$
$$= 2t\sqrt{49}\sqrt{u} \qquad \text{Identifying a perfect-square factor of } 49u$$
$$= 2t \cdot 7\sqrt{u}$$
$$= 14t\sqrt{u}.$$

> TRY EXERCISE 29

In Chapter 9, we will need to evaluate expressions similar to the next example as we solve certain equations.

EXAMPLE 3   Evaluate $\sqrt{b^2 - 4ac}$ for $a = 3, b = 6$, and $c = -5$.

SOLUTION   We have

$$\sqrt{b^2 - 4ac} = \sqrt{6^2 - 4 \cdot 3(-5)} \quad \text{Substituting}$$
$$= \sqrt{36 - 12(-5)}$$
$$= \sqrt{36 + 60}$$
$$= \sqrt{96}$$
$$= \sqrt{2 \cdot 2 \cdot 2 \cdot 2 \cdot 2 \cdot 3} \quad \text{The prime factorization always allows us to identify the largest perfect-square factor.}$$
$$= \sqrt{16} \cdot \sqrt{6}$$
$$= 4\sqrt{6}.$$

TRY EXERCISE 47

Simplifying Square Roots of Powers

To take the square root of an even power such as x^{10}, note that $x^{10} = (x^5)^2$. Then

$$\sqrt{x^{10}} = \sqrt{(x^5)^2} = x^5. \quad \text{Remember that we assume } x \geq 0.$$

The exponent of the square root is half the exponent of the radicand. That is,

$$\sqrt{x^{10}} = x^5. \longleftarrow \tfrac{1}{2}(10) = 5$$

EXAMPLE 4   Simplify: **(a)** $\sqrt{x^6}$; **(b)** $\sqrt{p^{12}}$; **(c)** $\sqrt{t^{22}}$.

SOLUTION

a) $\sqrt{x^6} = \sqrt{(x^3)^2} = x^3$ Half of 6 is 3.

b) $\sqrt{p^{12}} = \sqrt{(p^6)^2} = p^6$ Half of 12 is 6.

c) $\sqrt{t^{22}} = \sqrt{(t^{11})^2} = t^{11}$ Half of 22 is 11.

TRY EXERCISE 53

If a radicand is an odd power, we can simplify by factoring. For square roots of powers, after we have simplified, the radicand never contains an exponent greater than 1.

EXAMPLE 5   Simplify: **(a)** $\sqrt{x^9}$; **(b)** $\sqrt{32p^{19}}$.

SOLUTION

a) $\sqrt{x^9} = \sqrt{x^8 \cdot x}$ x^8 is the largest perfect-square factor of x^9.
$$= \sqrt{x^8}\sqrt{x}$$
$$= x^4\sqrt{x}$$

> *CAUTION!*   The square root of x^9 is *not* x^3.

b) $\sqrt{32p^{19}} = \sqrt{16p^{18}2p}$ 16 is the largest perfect-square factor of 32; p^{18} is the largest perfect-square factor of p^{19}.
$$= \sqrt{16}\sqrt{p^{18}}\sqrt{2p}$$
$$= 4p^9\sqrt{2p} \quad \text{Simplifying. We assume } p \text{ is positive. Since } 2p \text{ has no perfect-square factor, we are done.}$$

TRY EXERCISE 57

Multiplying and Simplifying

Sometimes we can simplify after multiplying. To do so, we again try to identify any perfect-square factors of the radicand.

EXAMPLE 6   Multiply and, if possible, simplify. Remember that all variables are assumed to represent nonnegative numbers.

a) $\sqrt{2}\ \sqrt{14}$

b) $\sqrt{5t}\ \sqrt{6t}$

c) $\sqrt{n^2}\ \sqrt{n^3}$

d) $\sqrt{2x^8}\ \sqrt{9x^3}$

SOLUTION

a) $\sqrt{2}\sqrt{14} = \sqrt{2 \cdot 14}$ Multiplying

$\qquad\qquad = \sqrt{2 \cdot 2 \cdot 7}$ Writing the prime factorization

$\qquad\qquad = \sqrt{2^2}\sqrt{7}$ Note that $2 \cdot 2$, or 4, is a perfect-square factor.

$\qquad\qquad = 2\sqrt{7}$ Simplifying

b) $\sqrt{5t}\sqrt{6t} = \sqrt{30t^2}$ Multiplying

$\qquad\qquad = \sqrt{t^2}\sqrt{30}$ ⎫

$\qquad\qquad = t\sqrt{30}$ ⎬ Simplifying

c) $\sqrt{n^2}\sqrt{n^3} = \sqrt{n^5}$ Multiplying

$\qquad\qquad = \sqrt{n^4 \cdot n}$ n^4 is a perfect square.

$\qquad\qquad = \sqrt{n^4}\sqrt{n}$ Factoring

$\qquad\qquad = n^2\sqrt{n}$ Simplifying

Note that both $\sqrt{n^2}$ and $\sqrt{n^3}$ can be simplified. If we had simplified before multiplying, the result would be the same.

d) Before multiplying, note that we can simplify both $\sqrt{2x^8}$ and $\sqrt{9x^3}$.

$\sqrt{2x^8}\sqrt{9x^3} = \sqrt{2 \cdot x^8}\sqrt{9 \cdot x^2 \cdot x}$ $x^8, 9$, and x^2 are perfect squares.

$\qquad\qquad = \sqrt{2}\sqrt{x^8}\sqrt{9}\sqrt{x^2}\sqrt{x}$

$\qquad\qquad = \sqrt{2} \cdot x^4 \cdot 3 \cdot x \cdot \sqrt{x}$ Simplifying

$\qquad\qquad = 3 \cdot x^4 \cdot x \cdot \sqrt{2} \cdot \sqrt{x}$ Using a commutative law

$\qquad\qquad = 3x^5\sqrt{2x}$ Multiplying

We could also multiply first and then simplify. The result would be the same.

TRY EXERCISE 67

8.2   EXERCISE SET

🡒 *Concept Reinforcement*   *In each of Exercises 1–10, match the expression with the equivalent expression from the column on the right. Assume that all variables represent nonnegative numbers.*

1. ___ $\sqrt{3} \cdot \sqrt{7}$
2. ___ $\sqrt{9} \cdot \sqrt{7}$
3. ___ $\sqrt{49 \cdot 3}$
4. ___ $\sqrt{49 \cdot 9}$
5. ___ $\sqrt{a^2}$
6. ___ $\sqrt{a^9}$
7. ___ $\sqrt{a^6}$
8. ___ $\sqrt{a^4}$
9. ___ $\sqrt{ab} \cdot \sqrt{bc}$
10. ___ $\sqrt{ac} \cdot \sqrt{bc}$

a) $7 \cdot 3$
b) $\sqrt{(a^3)^2}$
c) $\sqrt{b^2ac}$
d) $a^4 \cdot \sqrt{a}$
e) $\sqrt{c^2ab}$
f) $7 \cdot \sqrt{3}$
g) a^2
h) $\sqrt{21}$
i) a
j) $3\sqrt{7}$

Multiply.

11. $\sqrt{2}\sqrt{5}$
12. $\sqrt{3}\sqrt{11}$
13. $\sqrt{4}\sqrt{3}$
14. $\sqrt{2}\sqrt{9}$
15. $\sqrt{\frac{3}{5}}\sqrt{\frac{7}{8}}$
16. $\sqrt{\frac{3}{8}}\sqrt{\frac{1}{5}}$
17. $\sqrt{10}\sqrt{10}$
18. $\sqrt{15}\sqrt{15}$
19. $\sqrt{25}\sqrt{3}$
20. $\sqrt{36}\sqrt{2}$
21. $\sqrt{2}\sqrt{x}$
22. $\sqrt{3}\sqrt{a}$
23. $\sqrt{7}\sqrt{2a}$
24. $\sqrt{5}\sqrt{7t}$
25. $\sqrt{3x}\sqrt{7y}$
26. $\sqrt{3m}\sqrt{5n}$
27. $\sqrt{3a}\sqrt{2bc}$
28. $\sqrt{3x}\sqrt{yz}$

Simplify by factoring. Assume that all variables represent nonnegative numbers.

29. $\sqrt{12}$
30. $\sqrt{28}$
31. $\sqrt{75}$
32. $\sqrt{45}$
33. $\sqrt{500}$
34. $\sqrt{300}$
35. $\sqrt{16t}$
36. $\sqrt{64a}$
37. $\sqrt{20z}$
38. $\sqrt{40m}$
39. $\sqrt{100y^2}$
40. $\sqrt{9x^2}$
41. $\sqrt{13x^2}$
42. $\sqrt{29t^2}$
43. $\sqrt{27b^2}$
44. $\sqrt{125a^2}$
45. $\sqrt{144x^2y}$
46. $\sqrt{256u^2v}$

In Exercises 47–52, evaluate $\sqrt{b^2 - 4ac}$ for the values of a, b, and c given.

47. $a = 2$, $b = 4$, $c = -1$
48. $a = 3$, $b = 2$, $c = -4$
49. $a = 1$, $b = 4$, $c = 4$
50. $a = 1$, $b = -3$, $c = -10$
51. $a = 3$, $b = -6$, $c = -4$
52. $a = 1$, $b = -8$, $c = -3$

Simplify. Assume that all variables represent nonnegative numbers.

53. $\sqrt{a^{18}}$
54. $\sqrt{t^{20}}$
55. $\sqrt{x^{16}}$
56. $\sqrt{p^{14}}$
57. $\sqrt{r^5}$
58. $\sqrt{a^7}$
59. $\sqrt{t^{15}}$
60. $\sqrt{x^{25}}$
61. $\sqrt{40a^3}$
62. $\sqrt{250y^3}$
63. $\sqrt{45x^5}$
64. $\sqrt{20b^7}$
65. $\sqrt{200p^{25}}$
66. $\sqrt{99m^{33}}$

Multiply and, if possible, simplify.

67. $\sqrt{2}\sqrt{10}$
68. $\sqrt{3}\sqrt{6}$
69. $\sqrt{3} \cdot \sqrt{27}$
70. $\sqrt{2} \cdot \sqrt{8}$
71. $\sqrt{3x}\sqrt{12y}$
72. $\sqrt{2c}\sqrt{50d}$
73. $\sqrt{17}\sqrt{17x}$
74. $\sqrt{11}\sqrt{11x}$
75. $\sqrt{10b}\sqrt{50b}$
76. $\sqrt{6a}\sqrt{18a}$

Aha! 77. $\sqrt{12t}\sqrt{12t}$
78. $\sqrt{10a}\sqrt{10a}$
79. $\sqrt{ab}\sqrt{ac}$
80. $\sqrt{xy}\sqrt{xz}$
81. $\sqrt{x^7}\sqrt{x^{10}}$
82. $\sqrt{y^3}\sqrt{y^9}$
83. $\sqrt{7m^5}\sqrt{14m}$
84. $\sqrt{15m^7}\sqrt{5m}$
85. $\sqrt{x^2y^3}\sqrt{xy^4}$
86. $\sqrt{x^3y^2}\sqrt{xy}$
87. $\sqrt{6ab}\sqrt{12a^2b^5}$
88. $\sqrt{5xy^2}\sqrt{10x^2y^3}$

🖩 89. *Water flow.*   The required water flow *f* from a fire hose, in number of gallons per minute, is given by

$$f = 400\sqrt{p},$$

where *p* is the population, in thousands, of a community. Estimate the required flow for a community with a population of 20,000.
Source: inetdocs.loudoun.gov

90. *Falling object.* The number of seconds t that it takes for an object to fall d meters when thrown down at a velocity of 9.5 meters per second (m/sec) can be estimated by

$$t = 0.45\sqrt{d + 4.6} - 1.$$

A rock is thrown at 9.5 m/sec from the overlook in Great Bluffs State Park 190 m above the Mississippi River in Minnesota. After how many seconds will the rock hit the water?

Speed of a skidding car. The formula

$$r = 2\sqrt{5L}$$

can be used to approximate the speed r, in miles per hour, of a car that has left a skid mark L feet long.

91. What was the speed of a car that left skid marks of 20 ft? of 150 ft?

92. What was the speed of a car that left skid marks of 30 ft? of 70 ft?

93. How would you convince someone that the following equation is not true?

$$\sqrt{x^2 - 49} = \sqrt{x^2} - \sqrt{49} = x - 7$$
$$\uparrow$$
This is not true.

94. Explain why the rules for manipulating exponents are important when simplifying radical expressions.

Skill Review

To prepare for Section 8.3, review the quotient rule for exponents (Section 4.1).

Simplify. [4.1]

95. $\dfrac{x^5 y^6}{x^2 y}$

96. $\dfrac{a^7 b^2}{a^5 b}$

97. $\dfrac{7a}{8b} \cdot \dfrac{3a}{2b}$

98. $\dfrac{5x}{6y} \cdot \dfrac{7y}{2y}$

99. $\dfrac{2r^3}{7t} \cdot \dfrac{rt}{rt}$

100. $\dfrac{5x^7}{11y} \cdot \dfrac{xy}{xy}$

Synthesis

101. Explain why $\sqrt{16x^4} = 4x^2$, but $\sqrt{4x^{16}} \neq 2x^4$.

102. Simplify $\sqrt{49}$, $\sqrt{490}$, $\sqrt{4900}$, $\sqrt{49{,}000}$, and $\sqrt{490{,}000}$; then describe the pattern you see.

Simplify.

103. $\sqrt{0.01}$

104. $\sqrt{0.49}$

105. $\sqrt{0.0625}$

106. $\sqrt{0.000001}$

Use the proper symbol (> , < , or =) between each pair of values to make a true sentence. Do not use a calculator.

107. $4\sqrt{14}$ ▨ 15

108. $3\sqrt{11}$ ▨ $7\sqrt{2}$

109. $\sqrt{450}$ ▨ $15\sqrt{2}$

110. 16 ▨ $\sqrt{15}\,\sqrt{17}$

111. 8 ▨ $\sqrt{15} + \sqrt{17}$

112. $5\sqrt{7}$ ▨ $4\sqrt{11}$

Multiply and then simplify by factoring.

113. $\sqrt{54(x + 1)}\sqrt{6y(x + 1)^2}$

114. $\sqrt{18(x - 2)}\sqrt{20(x - 2)^3}$

115. $\sqrt{x^9}\sqrt{2x}\sqrt{10x^5}$

116. $\sqrt{2^{109}}\sqrt{x^{306}}\sqrt{x^{11}}$

Fill in the blank.

117. $\sqrt{21x^9} \cdot$ _____ $= 7x^{14}\sqrt{6x^7}$

118. $\sqrt{35x^7} \cdot$ _____ $= 5x^5\sqrt{14x^3}$

Simplify.

119. $\sqrt{x^{16n}}$

120. $\sqrt{0.04x^{4n}}$

121. Simplify $\sqrt{y^n}$, when n is an even natural number.

122. Simplify $\sqrt{y^n}$, when n is an odd whole number greater than or equal to 3.

Dividing Radical Expressions   ■   Rationalizing Denominators

In this section, we divide radical expressions and simplify quotients containing radicals. Again, in this section we assume that no radicands were formed by squaring negative quantities.

Dividing Radical Expressions

To see how to divide with radical notation, consider the following:

$$\frac{\sqrt{100}}{\sqrt{4}} = \frac{10}{2} = 5 \quad \text{since } \sqrt{100} = 10 \text{ and } \sqrt{4} = 2;$$

$$\sqrt{\frac{100}{4}} = \sqrt{25} = 5 \quad \text{since } \frac{100}{4} = 25 \text{ and } \sqrt{25} = 5.$$

Both results are the same, so we see that $\dfrac{\sqrt{100}}{\sqrt{4}} = \sqrt{\dfrac{100}{4}}$.

> ### The Quotient Rule for Square Roots
> For any real numbers $\sqrt{A}$ and $\sqrt{B}$ with $B \neq 0$,
>
> $$\frac{\sqrt{A}}{\sqrt{B}} = \sqrt{\frac{A}{B}}.$$
>
> (To divide two square roots, divide the radicands and take the square root.)

EXAMPLE 1   Divide and simplify: **(a)** $\dfrac{\sqrt{27}}{\sqrt{3}}$; **(b)** $\dfrac{\sqrt{8a^7}}{\sqrt{2a}}$.

SOLUTION

a) $\dfrac{\sqrt{27}}{\sqrt{3}} = \sqrt{\dfrac{27}{3}}$ We can now simplify the radicand: $\dfrac{27}{3} = 9$.

     $= \sqrt{9} = 3$

b) $\dfrac{\sqrt{8a^7}}{\sqrt{2a}} = \sqrt{\dfrac{8a^7}{2a}}$ Now $\dfrac{8a^7}{2a}$ can be simplified.

     $= \sqrt{4a^6} = 2a^3$ We assume $a > 0$.   **TRY EXERCISE** ▶ 5

The quotient rule for square roots can also be read from right to left:

$$\sqrt{\frac{A}{B}} = \frac{\sqrt{A}}{\sqrt{B}}.$$

EXAMPLE 2   Simplify by taking square roots in the numerator and the denominator separately.

a) $\sqrt{\dfrac{36}{25}}$ **b)** $\sqrt{\dfrac{1}{16}}$ **c)** $\sqrt{\dfrac{49}{t^2}}$

SOLUTION

a) $\sqrt{\dfrac{36}{25}} = \dfrac{\sqrt{36}}{\sqrt{25}} = \dfrac{6}{5}$ Taking the square root of the numerator and the square root of the denominator. This is sometimes done mentally, in one step.

b) $\sqrt{\dfrac{1}{16}} = \dfrac{\sqrt{1}}{\sqrt{16}} = \dfrac{1}{4}$ Taking the square root of the numerator and the square root of the denominator

c) $\sqrt{\dfrac{49}{t^2}} = \dfrac{\sqrt{49}}{\sqrt{t^2}} = \dfrac{7}{t}$ We assume $t > 0$.   **TRY EXERCISE** ▸ 23

Sometimes a rational expression can be simplified to one that has a perfect-square numerator and/or a perfect-square denominator.

EXAMPLE **3**   Simplify: **(a)** $\sqrt{\dfrac{14}{50}}$; **(b)** $\sqrt{\dfrac{48x^3}{3x^7}}$.

SOLUTION

a) $\sqrt{\dfrac{14}{50}} = \sqrt{\dfrac{7 \cdot 2}{25 \cdot 2}}$

$\phantom{\sqrt{\dfrac{14}{50}}} = \sqrt{\dfrac{7 \cdot 2}{25 \cdot 2}}$ Removing a factor equal to 1: $\dfrac{2}{2} = 1$

$\phantom{\sqrt{\dfrac{14}{50}}} = \dfrac{\sqrt{7}}{\sqrt{25}} = \dfrac{\sqrt{7}}{5}$

b) $\sqrt{\dfrac{48x^3}{3x^7}} = \sqrt{\dfrac{16 \cdot 3x^3}{x^4 \cdot 3x^3}}$ Removing a factor equal to 1: $\dfrac{3x^3}{3x^3} = 1$

$\phantom{\sqrt{\dfrac{48x^3}{3x^7}}} = \dfrac{\sqrt{16}}{\sqrt{x^4}} = \dfrac{4}{x^2}$ We assume $x \neq 0$.   **TRY EXERCISE** ▸ 29

Rationalizing Denominators

The expressions

$$\dfrac{1}{\sqrt{2}} \quad \text{and} \quad \dfrac{\sqrt{2}}{2}$$

are equivalent, but the second expression does not have an irrational number in the denominator. We can **rationalize the denominator** of a radical expression by multiplying by a carefully chosen form of 1. Before calculators, this was done to make long division involving decimal approximations easier to perform.

EXAMPLE **4**   Rationalize each denominator: **(a)** $\dfrac{1}{\sqrt{2}}$; **(b)** $\sqrt{\dfrac{5}{a}}$.

SOLUTION

a) We multiply by 1, using the fact that $\sqrt{2} \cdot \sqrt{2} = 2$ to choose the form of 1:

$$\dfrac{1}{\sqrt{2}} = \dfrac{1}{\sqrt{2}} \cdot \dfrac{\sqrt{2}}{\sqrt{2}} \qquad \text{Multiplying by 1, using the denominator, } \sqrt{2}, \text{ to write 1}$$

$$\phantom{\dfrac{1}{\sqrt{2}}} = \dfrac{\sqrt{2}}{2}. \qquad \sqrt{2} \cdot \sqrt{2} = 2. \text{ The denominator is now rational.}$$

b) $\sqrt{\dfrac{5}{a}} = \dfrac{\sqrt{5}}{\sqrt{a}}$ The square root of a quotient is the quotient of the square roots.

$= \dfrac{\sqrt{5}}{\sqrt{a}} \cdot \dfrac{\sqrt{a}}{\sqrt{a}}$ Multiplying by 1, using the denominator, $\sqrt{a}$, to write 1

$= \dfrac{\sqrt{5a}}{a}$ $\sqrt{a} \cdot \sqrt{a} = a.$ We assume $a > 0.$ ▸ TRY EXERCISE ▸ 35

It is usually easiest to rationalize a denominator after the expression has been simplified.

EXAMPLE   5   Rationalize each denominator: **(a)** $\dfrac{\sqrt{2}}{\sqrt{45}}$; **(b)** $\sqrt{\dfrac{7a^2}{12a}}$.

STUDENT NOTES

Be careful when writing radical expressions. There is a *big* difference between the expressions

$$\dfrac{\sqrt{10}}{15} \quad \text{and} \quad \sqrt{\dfrac{10}{15}}.$$

SOLUTION

a) $\dfrac{\sqrt{2}}{\sqrt{45}} = \dfrac{\sqrt{2}}{\sqrt{9}\sqrt{5}}$ Simplifying the denominator. Note that 9 is a perfect square.

$= \dfrac{\sqrt{2}}{3\sqrt{5}}$

$= \dfrac{\sqrt{2}}{3\sqrt{5}} \cdot \dfrac{\sqrt{5}}{\sqrt{5}}$ Multiplying by 1, using $\sqrt{5}$ to write 1

$= \dfrac{\sqrt{10}}{3 \cdot 5} = \dfrac{\sqrt{10}}{15}$

b) $\sqrt{\dfrac{7a^2}{12a}} = \sqrt{\dfrac{7a}{12}}$ Simplifying the radicand

$= \dfrac{\sqrt{7a}}{\sqrt{12}}$ The square root of a quotient is the quotient of the square roots.

$= \dfrac{\sqrt{7a}}{\sqrt{4}\sqrt{3}}$
$= \dfrac{\sqrt{7a}}{2\sqrt{3}}$ Simplifying the denominator. Note that 4 is a perfect square.

$= \dfrac{\sqrt{7a}}{2\sqrt{3}} \cdot \dfrac{\sqrt{3}}{\sqrt{3}}$ Multiplying by 1

$= \dfrac{\sqrt{21a}}{2 \cdot 3}$

$= \dfrac{\sqrt{21a}}{6}$ ▸ TRY EXERCISE ▸ 43

CAUTION!   Our solutions in Example 5 cannot be simplified any further. A common mistake is to remove a factor of 1 that does not exist. For example, $\dfrac{\sqrt{10}}{15}$ *cannot* be simplified to $\dfrac{\sqrt{2}}{3}$ because $\sqrt{10}$ and 15 do not share a common factor.

8.3 EXERCISE SET

For Extra Help
MyMathLab   Math XL PRACTICE   WATCH   DOWNLOAD

↪ **Concept Reinforcement** *Classify each statement as either true or false.*

1. To divide one square root by another, we can divide one radicand by the other and then take the square root.

2. The square root of a quotient can be found by dividing the square root of the numerator by the square root of the denominator.

3. To rationalize the denominator of a fraction, we square both the numerator and the denominator.

4. It is usually easiest to rationalize a denominator before the expression has been simplified.

Simplify. Assume that all variables represent positive numbers.

5. $\dfrac{\sqrt{500}}{\sqrt{5}}$

6. $\dfrac{\sqrt{50}}{\sqrt{2}}$

7. $\dfrac{\sqrt{40}}{\sqrt{10}}$

8. $\dfrac{\sqrt{72}}{\sqrt{2}}$

9. $\dfrac{\sqrt{55}}{\sqrt{5}}$

10. $\dfrac{\sqrt{18}}{\sqrt{3}}$

11. $\dfrac{\sqrt{5}}{\sqrt{20}}$

12. $\dfrac{\sqrt{2}}{\sqrt{18}}$

13. $\dfrac{\sqrt{18}}{\sqrt{32}}$

14. $\dfrac{\sqrt{12}}{\sqrt{75}}$

15. $\dfrac{\sqrt{8x}}{\sqrt{2x}}$

16. $\dfrac{\sqrt{18b}}{\sqrt{2b}}$

17. $\dfrac{\sqrt{63y^3}}{\sqrt{7y}}$

18. $\dfrac{\sqrt{48x^3}}{\sqrt{3x}}$

19. $\dfrac{\sqrt{500a^{10}}}{\sqrt{5a^2}}$

20. $\dfrac{\sqrt{27x^5}}{\sqrt{3x}}$

21. $\dfrac{\sqrt{21a^9}}{\sqrt{7a^3}}$

22. $\dfrac{\sqrt{35t^{13}}}{\sqrt{5t^5}}$

23. $\sqrt{\dfrac{4}{25}}$

24. $\sqrt{\dfrac{9}{49}}$

25. $\sqrt{\dfrac{49}{16}}$

26. $\sqrt{\dfrac{100}{49}}$

27. $-\sqrt{\dfrac{25}{81}}$

28. $-\sqrt{\dfrac{25}{64}}$

29. $\sqrt{\dfrac{2a^5}{50a}}$

30. $\sqrt{\dfrac{7a^5}{28a}}$

31. $\sqrt{\dfrac{6x^7}{32x}}$

32. $\sqrt{\dfrac{4x^3}{50x}}$

33. $\sqrt{\dfrac{21t^9}{28t^3}}$

34. $\sqrt{\dfrac{10t^9}{18t^5}}$

Form an equivalent expression by rationalizing each denominator.

35. $\dfrac{1}{\sqrt{3}}$

36. $\dfrac{2}{\sqrt{3}}$

37. $\dfrac{5}{\sqrt{7}}$

38. $\dfrac{3}{\sqrt{11}}$

39. $\dfrac{\sqrt{16}}{\sqrt{27}}$

40. $\dfrac{\sqrt{25}}{\sqrt{8}}$

41. $\dfrac{\sqrt{6}}{\sqrt{5}}$

42. $\dfrac{\sqrt{5}}{\sqrt{7}}$

43. $\dfrac{\sqrt{3}}{\sqrt{50}}$

44. $\dfrac{\sqrt{5}}{\sqrt{18}}$

45. $\dfrac{\sqrt{2a}}{\sqrt{45}}$

46. $\dfrac{\sqrt{3a}}{\sqrt{32}}$

47. $\sqrt{\dfrac{12}{5}}$

48. $\sqrt{\dfrac{8}{3}}$

49. $\sqrt{\dfrac{7}{z}}$

50. $\sqrt{\dfrac{13}{p}}$

51. $\sqrt{\dfrac{a}{200}}$

52. $\sqrt{\dfrac{t}{32}}$

53. $\sqrt{\dfrac{x}{90}}$

54. $\sqrt{\dfrac{y}{40}}$

Aha! 55. $\sqrt{\dfrac{3a}{25}}$

56. $\sqrt{\dfrac{5t}{16}}$

57. $\sqrt{\dfrac{5x^3}{12x}}$

58. $\sqrt{\dfrac{7t^3}{32t}}$

🖩 *Depreciation of a vehicle.* *The resale value of certain cars and trucks that originally sold for $18,000 can be estimated using the formula*

$$V = \dfrac{18{,}500}{\sqrt{t + 1.0565}},$$

where V is the value of the vehicle, in dollars, when it is t years old.

59. Estimate the vehicle's value when it is 3 yr old.

60. Estimate the vehicle's value when it is 9 yr old.

61. Estimate the vehicle's value when it is 1 yr old.

62. Estimate the vehicle's value when it is 4 yr old.

Period of a swinging pendulum. *The period T of a pendulum is the time it takes to move from one side to the other and back. A formula for the period is*

$$T = 2\pi\sqrt{\dfrac{L}{32}},$$

where T is in seconds and L is in feet. Use 3.14 for π.

63. Find the periods of pendulums of lengths 32 ft and 50 ft.

64. Find the periods of pendulums of lengths 8 ft and 2 ft.

65. The pendulum of a mantle clock is $2/\pi^2$ ft long. How long does it take to swing from one side to the other and back?

66. The pendulum of a grandfather clock is $98/\pi^2$ ft long. How long does it take to swing from one side to the other and back?

67. Ingrid is swinging at the bottom of a 72-ft long bungee cord. How long does it take her to complete one swing back and forth? (*Hint*: $\sqrt{2.25} = 1.5$.)

68. Don is swinging back and forth on an 18-ft long rope swing near the Green River Reservoir. How long does it take him to complete one swing back and forth?

69. Why is it important to know how to multiply radical expressions before learning how to divide them?

70. Describe a method that could be used to rationalize the *numerator* of a radical expression.

Skill Review

To prepare for Section 8.4, review combining like terms and multiplying polynomials (Sections 4.2, 4.4, and 4.5).

Simplify. [4.2]

71. $9x + 6x$

72. $10x + x$

73. $9y - z + 8y$

74. $3t - 4s - 7t$

Multiply. [4.4], [4.5]

75. $9x(2x - 7)$

76. $4n(3n^2 + 6n + 1)$

77. $(2x - 3)(2x + 3)$

78. $(5t + 7)(5t - 7)$

Synthesis

79. When calculating approximations by hand using long division, what makes it easier to compute $\sqrt{2}/2$ than $1/\sqrt{2}$?

80. Is it always best to rewrite an expression of the form $\sqrt{a}/\sqrt{b}$ as $\sqrt{a/b}$ before simplifying? Why or why not?

Rationalize each denominator and, if possible, simplify to form an equivalent expression.

81. $\sqrt{\dfrac{7}{1000}}$

82. $\sqrt{\dfrac{3}{800}}$

83. $\sqrt{\dfrac{3x^2}{8x^7y^3}}$

84. $\sqrt{\dfrac{3x^2y}{a^2x^5}}$

85. $\sqrt{\dfrac{1}{5zw^2}}$

86. $\sqrt{\dfrac{2a}{5b^3c^9}}$

Aha! **87.** $\dfrac{2}{\sqrt{\sqrt{5}}}$

88. $\dfrac{3}{\sqrt{\sqrt{7}}}$

Simplify. Assume that $0 < x \le y$ and $0 < z \le 1$.

89. $\sqrt{\dfrac{1}{x^2} - \dfrac{2}{xy} + \dfrac{1}{y^2}}$

90. $\sqrt{2 - \dfrac{4}{z^2} + \dfrac{2}{z^4}}$

91. Solve: $\sqrt{\dfrac{2x - 3}{8}} = \dfrac{5}{2}$.

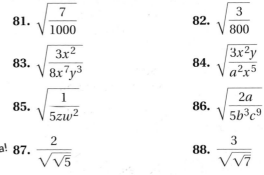

8.4 Radical Expressions with Several Terms

Adding and Subtracting Radical Expressions ● More with Multiplication ●
More with Rationalizing Denominators

We now consider addition and subtraction of radical expressions as well as some new types of multiplication and simplification.

Adding and Subtracting Radical Expressions

The sum of a rational number and an irrational number, like $5 + \sqrt{2}$, *cannot* be simplified. However, the sum of **like radicals**—that is, radical expressions that have the same radical factor—*can* be simplified.

EXAMPLE   1   Add or subtract, as indicated: **(a)** $3\sqrt{5} + 4\sqrt{5}$; **(b)** $\sqrt{x} - 7\sqrt{x}$.

SOLUTION

a) Recall that to simplify an expression like $3x + 4x$, we use the distributive law, as follows:

$$3x + 4x = (3 + 4)x = 7x.$$ The middle step is usually performed mentally.

In this example, x is replaced with $\sqrt{5}$:

$$3\sqrt{5} + 4\sqrt{5} = (3 + 4)\sqrt{5}$$ Using the distributive law to factor out $\sqrt{5}$
$$= 7\sqrt{5}.$$ $3\sqrt{5}$ and $4\sqrt{5}$ are like radicals.

b) $\sqrt{x} - 7\sqrt{x} = 1 \cdot \sqrt{x} - 7\sqrt{x}$ $\sqrt{x} = 1 \cdot \sqrt{x}$
$$= (1 - 7)\sqrt{x}$$ Using the distributive law. Try to do this mentally.
$$= -6\sqrt{x}$$

TRY EXERCISE 5

An expression may contain like radicals after individual radicals are themselves simplified.

EXAMPLE   2   Simplify: **(a)** $4\sqrt{2} - \sqrt{18}$; **(b)** $\sqrt{5} + \sqrt{20} + \sqrt{7}$.

SOLUTION

a) $4\sqrt{2} - \sqrt{18} = 4\sqrt{2} - \sqrt{9 \cdot 2}$
$$= 4\sqrt{2} - \sqrt{9}\sqrt{2}$$ Simplifying $\sqrt{18}$
$$= 4\sqrt{2} - 3\sqrt{2}$$ We now have like radicals.
$$= \sqrt{2}$$ Using the distributive law mentally:
$$4\sqrt{2} - 3\sqrt{2} = (4 - 3)\sqrt{2} = 1\sqrt{2} = \sqrt{2}$$

b) $\sqrt{5} + \sqrt{20} + \sqrt{7} = \sqrt{5} + \sqrt{4}\sqrt{5} + \sqrt{7}$ Simplifying $\sqrt{20}$
$$= \sqrt{5} + 2\sqrt{5} + \sqrt{7}$$ We now have 2 like radicals.
$$= 3\sqrt{5} + \sqrt{7}$$ Adding like radicals; $3\sqrt{5} + \sqrt{7}$ cannot be simplified.

TRY EXERCISE 23

> **CAUTION!**   It is *not true* that the sum of two square roots is the square root of the sum: $\sqrt{A} + \sqrt{B} \neq \sqrt{A + B}$. For example, $\sqrt{9} + \sqrt{16} \neq \sqrt{9 + 16}$ since $3 + 4 \neq 5$.

More with Multiplication

Radical expressions with more than one term are multiplied in much the same way that polynomials with more than one term are multiplied.

EXAMPLE **3**

Multiply.

a) $\sqrt{2}(\sqrt{3} + \sqrt{10})$ **b)** $(4 + \sqrt{7})(2 + \sqrt{7})$

c) $(2 - \sqrt{5})(2 + \sqrt{5})$ **d)** $(2 + \sqrt{3})(5 - 4\sqrt{3})$

SOLUTION

a) $\sqrt{2}(\sqrt{3} + \sqrt{10}) = \sqrt{2}\sqrt{3} + \sqrt{2}\sqrt{10}$ Using the distributive law

$\qquad\qquad\qquad = \sqrt{6} + \sqrt{20}$ Using the product rule for square roots

$\qquad\qquad\qquad = \sqrt{6} + 2\sqrt{5}$ Simplifying

b) $(4 + \sqrt{7})(2 + \sqrt{7}) = 4 \cdot 2 + 4 \cdot \sqrt{7} + \sqrt{7} \cdot 2 + \sqrt{7} \cdot \sqrt{7}$ Using FOIL

$\qquad\qquad\qquad = 8 + 4\sqrt{7} + 2\sqrt{7} + 7$

$\qquad\qquad\qquad = 15 + 6\sqrt{7}$ Combining like terms

c) Note that $(2 - \sqrt{5})(2 + \sqrt{5})$ is of the form $(A - B)(A + B)$:

$(2 - \sqrt{5})(2 + \sqrt{5}) = 2 \cdot 2 + 2 \cdot \sqrt{5} - \sqrt{5} \cdot 2 - \sqrt{5} \cdot \sqrt{5}$

$\qquad\qquad\qquad = 4 + 2\sqrt{5} - 2\sqrt{5} - 5$ The middle terms are opposites.

$\qquad\qquad\qquad = 4 - 5$ We can use $(A - B)(A + B) = A^2 - B^2$.

$\qquad\qquad\qquad = -1.$

d) $(2 + \sqrt{3})(5 - 4\sqrt{3}) = 2 \cdot 5 - 2 \cdot 4\sqrt{3} + \sqrt{3} \cdot 5 - \sqrt{3} \cdot 4\sqrt{3}$

$\qquad\qquad\qquad = 10 - 8\sqrt{3} + 5\sqrt{3} - 4 \cdot 3$ $2 \cdot 5 = 10; 2 \cdot 4 = 8;$ and $\sqrt{3} \cdot \sqrt{3} = 3$

$\qquad\qquad\qquad = 10 - 3\sqrt{3} - 12$ Adding like radicals

$\qquad\qquad\qquad = -2 - 3\sqrt{3}$ **TRY EXERCISES** 37 and 45

More with Rationalizing Denominators

Note in Example 3(c) that the result has no radicals. This will happen whenever expressions like $\sqrt{a} + \sqrt{b}$ and $\sqrt{a} - \sqrt{b}$ are multiplied:

$$(\sqrt{a} + \sqrt{b})(\sqrt{a} - \sqrt{b}) = (\sqrt{a})^2 - (\sqrt{b})^2 = a - b.$$

Expressions such as $\sqrt{3} - \sqrt{5}$ and $\sqrt{3} + \sqrt{5}$ are said to be **conjugates** of each other. So too are expressions like $2 + \sqrt{7}$ and $2 - \sqrt{7}$. Once the conjugate of a denominator has been found, it can be used to rationalize the denominator.

EXAMPLE **4**

Rationalize each denominator and, if possible, simplify.

a) $\dfrac{3}{2 + \sqrt{5}}$ **b)** $\dfrac{2}{\sqrt{7} - \sqrt{3}}$

SOLUTION

a) We multiply by a form of 1, using the conjugate of $2 + \sqrt{5}$, which is $2 - \sqrt{5}$, as the numerator and the denominator:

$\dfrac{3}{2 + \sqrt{5}} = \dfrac{3}{2 + \sqrt{5}} \cdot \dfrac{2 - \sqrt{5}}{2 - \sqrt{5}}$ Multiplying by 1

$\qquad = \dfrac{3(2 - \sqrt{5})}{(2 + \sqrt{5})(2 - \sqrt{5})}.$

TECHNOLOGY CONNECTION

Parentheses are needed to specify the radicand whenever the radicand has more than one term. For example, on most graphing calculators, we would enter the expression $\sqrt{7-2} + 8$ as $\sqrt{\ (7-2)} + 8$. This will give us a different result from $\sqrt{\ }7 - 2 + 8$ or $\sqrt{\ }(7-2+8)$.

Use a graphing calculator to approximate each of the following. Remember to use parentheses to enclose numerators and denominators with more than one term.

1. $\dfrac{-3 + \sqrt{8}}{2}$

2. $\dfrac{\sqrt{2} + \sqrt{5}}{3}$

3. $\sqrt{\sqrt{7} + 5}$

4. Use a graphing calculator to check Examples 4(a) and 4(b) by comparing the values of the original and simplified expressions.

Then

$$\frac{3(2 - \sqrt{5})}{(2 + \sqrt{5})(2 - \sqrt{5})} = \frac{3(2 - \sqrt{5})}{2^2 - (\sqrt{5})^2} \qquad \text{Using } (A + B)(A - B) = A^2 - B^2$$

$$= \frac{3(2 - \sqrt{5})}{-1} \qquad \begin{array}{l}\text{Simplifying the denominator.}\\ \text{See Example 3(c): } 4 - 5 = -1.\end{array}$$

$$= \frac{6 - 3\sqrt{5}}{-1}$$

$$= -6 + 3\sqrt{5}. \qquad \begin{array}{l}\text{Dividing } both \text{ terms in the numerator}\\ \text{by } -1\end{array}$$

b) $\dfrac{2}{\sqrt{7} - \sqrt{3}} = \dfrac{2}{\sqrt{7} - \sqrt{3}} \cdot \dfrac{\sqrt{7} + \sqrt{3}}{\sqrt{7} + \sqrt{3}} \qquad \begin{array}{l}\text{Multiplying by 1, using } \sqrt{7} + \sqrt{3},\\ \text{the conjugate of } \sqrt{7} - \sqrt{3}\end{array}$

$$= \frac{2(\sqrt{7} + \sqrt{3})}{(\sqrt{7} - \sqrt{3})(\sqrt{7} + \sqrt{3})}$$

$$= \frac{2(\sqrt{7} + \sqrt{3})}{(\sqrt{7})^2 - (\sqrt{3})^2} \qquad \text{Using } (A - B)(A + B) = A^2 - B^2$$

$$= \frac{2(\sqrt{7} + \sqrt{3})}{7 - 3} \qquad \text{The denominator is free of radicals.}$$

$$= \frac{2(\sqrt{7} + \sqrt{3})}{4} \qquad \begin{array}{l}\text{Since 2 is a common factor of both}\\ \text{the numerator and the denominator,}\\ \text{we simplify.}\end{array}$$

$$\left.\begin{array}{l}= \dfrac{2(\sqrt{7} + \sqrt{3})}{2 \cdot 2}\\[2mm] = \dfrac{\sqrt{7} + \sqrt{3}}{2}\end{array}\right\} \qquad \begin{array}{l}\text{Factoring and removing a factor}\\ \text{equal to 1: } \dfrac{2}{2} = 1\end{array}$$

 **TRY EXERCISE** 57

8.4 EXERCISE SET

For Extra Help   MyMathLab   Math XL PRACTICE   WATCH   DOWNLOAD

↪ **Concept Reinforcement** *In each of Exercises 1–4, match the item with the most appropriate choice from the column on the right.*

1. ____ The conjugate of $\sqrt{7} + \sqrt{5}$

2. ____ An example of like radicals

3. ____ The product of $\sqrt{5} - \sqrt{7}$ and its conjugate

4. ____ The conjugate of $\sqrt{7} - \sqrt{5}$

a) $\sqrt{7} + \sqrt{5}$

b) $\sqrt{7} - \sqrt{5}$

c) $3\sqrt{7}$ and $-5\sqrt{7}$

d) -2

Add or subtract. Simplify by combining like radical terms, if possible.

5. $3\sqrt{10} + 8\sqrt{10}$

6. $5\sqrt{7} + \sqrt{7}$

7. $4\sqrt{2} - \sqrt{2}$

8. $8\sqrt{3} - 5\sqrt{3}$

9. $4\sqrt{t} + 9\sqrt{t}$

10. $6\sqrt{t} + 4\sqrt{t}$

11. $7\sqrt{x} - 8\sqrt{x}$

12. $10\sqrt{a} - 15\sqrt{a}$

13. $5\sqrt{2a} + 3\sqrt{2a}$

14. $5\sqrt{6x} + 2\sqrt{6x}$

15. $9\sqrt{10y} - \sqrt{10y}$

16. $12\sqrt{14y} - \sqrt{14y}$

17. $6\sqrt{7} + 2\sqrt{7} + 4\sqrt{7}$

18. $2\sqrt{5} + 7\sqrt{5} + 5\sqrt{5}$

19. $5\sqrt{2} - 9\sqrt{2} + 8\sqrt{2}$

20. $3\sqrt{6} - 7\sqrt{6} + 2\sqrt{6}$

21. $5\sqrt{3} + \sqrt{8}$

22. $2\sqrt{5} + \sqrt{45}$

23. $\sqrt{x} - \sqrt{16x}$

24. $\sqrt{100a} - \sqrt{a}$

25. $2\sqrt{3} - 4\sqrt{75}$

26. $7\sqrt{50} - 3\sqrt{2}$

27. $6\sqrt{18} + 5\sqrt{8}$

28. $3\sqrt{12} + 2\sqrt{300}$

29. $\sqrt{72} + \sqrt{98}$

30. $\sqrt{45} + \sqrt{80}$

Aha! 31. $9\sqrt{8} + \sqrt{72} - 9\sqrt{8}$

32. $4\sqrt{12} + \sqrt{27} - \sqrt{12}$

33. $5\sqrt{18} - 2\sqrt{32} - \sqrt{50}$

34. $7\sqrt{12} - 2\sqrt{27} + \sqrt{75}$

35. $\sqrt{16a} - 4\sqrt{a} + \sqrt{25a}$

36. $\sqrt{9x} + \sqrt{49x} - 9\sqrt{x}$

Multiply.

37. $\sqrt{3}(\sqrt{2} + \sqrt{7})$ **38.** $\sqrt{2}(\sqrt{7} + \sqrt{5})$

39. $\sqrt{5}(\sqrt{6} - \sqrt{10})$ **40.** $\sqrt{6}(\sqrt{15} - \sqrt{11})$

41. $(3 + \sqrt{2})(4 + \sqrt{2})$ **42.** $(5 + \sqrt{11})(3 + \sqrt{11})$

43. $(\sqrt{7} - 2)(\sqrt{7} - 5)$ **44.** $(\sqrt{10} + 4)(\sqrt{10}-7)$

45. $(\sqrt{6} + 5)(\sqrt{6} - 5)$ **46.** $(2 + \sqrt{3})(2 - \sqrt{3})$

47. $(\sqrt{7} - \sqrt{3})(\sqrt{7} + \sqrt{3})$

48. $(\sqrt{2} + \sqrt{5})(\sqrt{2} - \sqrt{5})$

49. $(2 + 3\sqrt{2})(3 - \sqrt{2})$ **50.** $(8 - \sqrt{7})(3 + 2\sqrt{7})$

51. $(7 + \sqrt{3})^2$ **52.** $(2 - \sqrt{5})^2$

53. $(1 - 2\sqrt{3})^2$ **54.** $(6 + 3\sqrt{5})^2$

55. $(\sqrt{x} - \sqrt{10})^2$ **56.** $(\sqrt{a} - \sqrt{6})^2$

Rationalize each denominator and, if possible, simplify to form an equivalent expression.

57. $\dfrac{2}{5 + \sqrt{2}}$ **58.** $\dfrac{5}{3 - \sqrt{7}}$

59. $\dfrac{2}{1 - \sqrt{3}}$ **60.** $\dfrac{4}{2 + \sqrt{5}}$

61. $\dfrac{2}{\sqrt{7} + 5}$ **62.** $\dfrac{6}{\sqrt{10} + 3}$

63. $\dfrac{\sqrt{10}}{\sqrt{10} + 4}$ **64.** $\dfrac{\sqrt{15}}{\sqrt{15} - 5}$

65. $\dfrac{\sqrt{7}}{\sqrt{7} - \sqrt{3}}$ **66.** $\dfrac{\sqrt{11}}{\sqrt{11} + \sqrt{7}}$

67. $\dfrac{\sqrt{3}}{\sqrt{5} - \sqrt{3}}$ **68.** $\dfrac{\sqrt{6}}{\sqrt{7} + \sqrt{6}}$

69. $\dfrac{2}{\sqrt{7} + \sqrt{2}}$ **70.** $\dfrac{6}{\sqrt{5} - \sqrt{3}}$

71. $\dfrac{\sqrt{6} - \sqrt{x}}{\sqrt{6} + \sqrt{x}}$ **72.** $\dfrac{\sqrt{10} - \sqrt{x}}{\sqrt{10} + \sqrt{x}}$

73. Why does the product of a pair of conjugates contain no radicals?

74. Describe a method that could be used to rationalize a numerator that is the sum of two radical expressions.

Skill Review

To prepare for Section 8.5, review solving linear equations and quadratic equations and the five-step problem-solving strategy (Sections 2.2, 2.5, and 5.6).

Solve.

75. $3x + 5 + 2(x - 3) = 4 - 6x$ [2.2]

76. $4(x - 3) - 2 = 5(2x - 1)$ [2.2]

77. $x^2 - 5x = 6$ [5.6]

78. $x^2 + 10 = 7x$ [5.6]

79. *Cell phones.* Recently AT&T offered two Push-To-Talk walkie-talkie plans. The pay-per-use plan cost 15¢ per minute, and the unlimited calling plan cost $9.99 per month. For how many minutes of use in one month are the costs the same? [2.5]
Source: www.mobiledia.com

80. *Leisure spending.* Even after adjusting all figures to 2005 dollars, the average American spent $63 more on movie, theater, opera, and sporting event admissions in 2005 than in 1984. The amount spent in 2005 was $3 less than twice the amount spent in 1984. How much, in 2005 dollars, was spent each year? [2.5], [7.2]
Source: U.S. Census Bureau

Synthesis

81. Monica believes that since $(\sqrt{x})^2 = x$, the square of a sum of radical expressions is not a radical expression. Is she correct? Why or why not?

82. Why must you know how to add and subtract radical expressions before you can rationalize denominators with two terms?

Add or subtract and, if possible, simplify.

83. $\sqrt{\dfrac{25}{x}} + \dfrac{\sqrt{x}}{2x} - \dfrac{5}{\sqrt{2}}$

84. $5\sqrt{\dfrac{1}{2}} + \dfrac{7}{2}\sqrt{18} - 4\sqrt{98}$

85. $\sqrt{8x^6y^3} - x\sqrt{2y^7} - \dfrac{x}{3}\sqrt{18x^2y^9}$

86. $a\sqrt{a^{17}b^9} - b\sqrt{a^{13}b^{11}} + a\sqrt{a^9b^{15}}$

87. $7x\sqrt{12xy^2} - 9y\sqrt{27x^3} + 5\sqrt{300x^3y^2}$

88. For which pairs of nonnegative numbers a and b does $\sqrt{a} + \sqrt{b} = \sqrt{a + b}$?

89. Three students were asked to simplify $\sqrt{10} + \sqrt{50}$. Their answers were $\sqrt{10}(1 + \sqrt{5})$, $\sqrt{10} + 5\sqrt{2}$, and $\sqrt{2}(5 + \sqrt{5})$. Which answer(s), if any, is correct?

8.5   Radical Equations

Solving Radical Equations ■ **Problem Solving and Applications**

An equation in which a variable appears in a radicand is called a **radical equation**. The following are examples:

$$\sqrt{2x} - 4 = 7, \quad 2\sqrt{x+2} = \sqrt{x+10}, \quad \text{and} \quad 3 + \sqrt{27 - 3x} = x.$$

To solve a radical equation, we use a new equation-solving principle: the principle of squaring. Use of this principle does *not* always produce equivalent equations, so checking solutions will be more important than ever.

Solving Radical Equations

An equation with a square root can be rewritten without the radical by using *the principle of squaring*.

The Principle of Squaring
If $a = b$, then $a^2 = b^2$.

The principle of squaring does *not* say that if $a^2 = b^2$, then $a = b$. Indeed, if a is replaced with -5 and b with 5, then $a^2 = b^2$ is true (since $(-5)^2 = 5^2$), but the equation $a = b$ is false (since $-5 \neq 5$). Although the principle of squaring *can* lead us to a solution of a radical equation, it *may* lead us to numbers that are not solutions.

EXAMPLE 1   Solve: **(a)** $\sqrt{x} + 3 = 7$; **(b)** $\sqrt{2x} + 5 = 0$.

SOLUTION

a) Our plan is to isolate the radical term on one side of the equation and then use the principle of squaring:

$$\sqrt{x} + 3 = 7$$
$$\sqrt{x} = 4 \qquad \text{Subtracting 3 to get the radical alone on one side}$$
$$(\sqrt{x})^2 = 4^2 \qquad \text{Squaring both sides (using the principle of squaring)}$$
$$x = 16$$

Check:
$$\begin{array}{c|c} \sqrt{x} + 3 = 7 \\ \hline \sqrt{16} + 3 & 7 \\ 4 + 3 & \\ 7 \overset{?}{=} 7 & \text{TRUE} \end{array}$$

The solution is 16.

b) $\sqrt{2x} + 5 = 0$

Isolate the radical.

$$\sqrt{2x} = -5 \qquad \text{Subtracting 5 to isolate the radical}$$

Square both sides.

$$(\sqrt{2x})^2 = (-5)^2 \qquad \text{Squaring both sides (using the principle of squaring)}$$

Solve the equation.

$$\left.\begin{aligned} 2x &= 25 \\ x &= \tfrac{25}{2} \end{aligned}\right\} \qquad \text{Solving for } x$$

Check the possible solution.

Check:

$$\begin{array}{c|c} \sqrt{2x} + 5 = 0 \\ \hline \sqrt{2\left(\frac{25}{2}\right)} + 5 & 0 \\ \sqrt{25} + 5 & \\ 10 \overset{?}{=} 0 & \text{FALSE} \end{array}$$

There is no solution.

TRY EXERCISE 7

> *CAUTION!*   When the principle of squaring is used to solve an equation, all possible solutions *must* be checked in the original equation!

EXAMPLE 2   Solve: $3\sqrt{x} = \sqrt{x + 32}$.

SOLUTION   We have

$$3\sqrt{x} = \sqrt{x + 32}$$

$$(3\sqrt{x})^2 = (\sqrt{x + 32})^2 \qquad \text{Squaring both sides (using the principle of squaring)}$$

$$3^2(\sqrt{x})^2 = x + 32 \qquad \text{Squaring the product on the left; simplifying on the right}$$

$$9x = x + 32 \qquad \text{Simplifying on the left}$$

$$\left.\begin{aligned} 8x &= 32 \\ x &= 4. \end{aligned}\right\} \qquad \text{Solving for } x$$

Check:

$$\begin{array}{c|c} 3\sqrt{x} = \sqrt{x + 32} \\ \hline 3\sqrt{4} & \sqrt{4 + 32} \\ 3 \cdot 2 & \sqrt{36} \\ 6 \overset{?}{=} 6 & \text{TRUE} \end{array}$$

The number 4 checks. The solution is 4.

TRY EXERCISE 15

We have been using the following strategy.

To Solve a Radical Equation

1. Isolate a radical term.
2. Use the principle of squaring (square both sides).
3. Solve the new equation.
4. Check all possible solutions in the original equation.

In some cases, we apply the principle of zero products (see Section 5.6) after squaring.

EXAMPLE 3   Solve: **(a)** $x - 5 = \sqrt{x + 7}$; **(b)** $3 + \sqrt{27 - 3x} = x$.

SOLUTION

a)

$$x - 5 = \sqrt{x + 7}$$

$$(x - 5)^2 = (\sqrt{x + 7})^2 \qquad \text{Using the principle of squaring}$$

$$x^2 - 10x + 25 = x + 7 \qquad \text{Squaring a binomial on the left side}$$

$$x^2 - 11x + 18 = 0 \qquad \text{Adding } -x - 7 \text{ to both sides}$$

$$(x - 9)(x - 2) = 0 \qquad \text{Factoring}$$

$$x - 9 = 0 \quad or \quad x - 2 = 0 \qquad \text{Using the principle of zero products}$$

$$x = 9 \quad or \qquad x = 2$$

Check: For 9:

$$\frac{x - 5 = \sqrt{x + 7}}{9 - 5 \mid \sqrt{9 + 7}}$$
$$4 \overset{?}{=} 4 \qquad \text{TRUE}$$

For 2:

$$\frac{x - 5 = \sqrt{x + 7}}{2 - 5 \mid \sqrt{2 + 7}}$$
$$-3 \overset{?}{=} 3 \qquad \text{FALSE}$$

The number 9 checks, but 2 does not. Thus the solution is 9.

b) $3 + \sqrt{27 - 3x} = x$

$$\sqrt{27 - 3x} = x - 3 \qquad \text{Subtracting 3 to isolate the radical}$$

$$(\sqrt{27 - 3x})^2 = (x - 3)^2 \qquad \text{Using the principle of squaring}$$

$$27 - 3x = x^2 - 6x + 9$$

$$0 = x^2 - 3x - 18 \qquad \text{Adding } 3x - 27 \text{ to both sides}$$

$$0 = (x - 6)(x + 3) \qquad \text{Factoring}$$

$$x - 6 = 0 \quad or \quad x + 3 = 0 \qquad \text{Using the principle of zero products}$$

$$x = 6 \quad or \qquad x = -3$$

Check: For 6:

$$\frac{3 + \sqrt{27 - 3x} = x}{3 + \sqrt{27 - 3 \cdot 6} \mid 6}$$
$$3 + \sqrt{9}$$
$$3 + 3 \mid$$
$$6 \overset{?}{=} 6 \quad \text{TRUE}$$

For −3:

$$\frac{3 + \sqrt{27 - 3x} = x}{3 + \sqrt{27 - 3 \cdot (-3)} \mid -3}$$
$$3 + \sqrt{27 + 9}$$
$$3 + \sqrt{36}$$
$$3 + 6 \mid$$
$$9 \overset{?}{=} -3 \quad \text{FALSE}$$

The number 6 checks, but −3 does not. The solution is 6.

> **TRY EXERCISE** 21

TECHNOLOGY CONNECTION

To visualize the solution of Example 3(a), we let $y_1 = x - 5$ (the left side of $x - 5 = \sqrt{x + 7}$) and $y_2 = \sqrt{x + 7}$ (the right side of $x - 5 = \sqrt{x + 7}$). Using the INTERSECT option of CALC, we see that $(9, 4)$ is on both curves. This indicates that when $x = 9$, y_1 and y_2 are both 4. No intersection occurs at $x = 2$, which confirms our check.

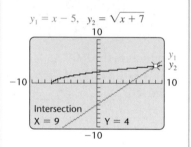

$y_1 = x - 5, \quad y_2 = \sqrt{x + 7}$

1. Use a graphing calculator to visualize the solution of Example 2.
2. Use a graphing calculator to visualize the solution of Example 3(b).

Problem Solving and Applications

Many applications translate to radical equations. For example, at a temperature of t degrees Fahrenheit, sound travels s feet per second, where

$$s = 21.9\sqrt{5t + 2457}.$$

EXAMPLE 4   *Musical performances.*   The band U2 often performs outdoors for crowds in excess of 20,000 people. An officer stationed at a U2 concert used a radar gun to determine that the sound of the band was traveling at a rate of 1170 ft/sec. What was the air temperature at the concert?

STUDENT NOTES

Be careful when checking answers. You don't want to discard a correct answer because of a careless mistake.

SOLUTION

1. **Familiarize.** If we did not already know the formula relating speed of sound to air temperature (see above), we could consult a reference book, a person in the field, or perhaps the Internet. There is no need to memorize this formula.

2. **Translate.** We substitute 1170 for s in the formula $s = 21.9\sqrt{5t + 2457}$:

$$1170 = 21.9\sqrt{5t + 2457}.$$

3. **Carry out.** We solve the equation for t:

$$1170 = 21.9\sqrt{5t + 2457}$$

$$\frac{1170}{21.9} = \sqrt{5t + 2457} \qquad \text{Dividing both sides by 21.9}$$

$$\left(\frac{1170}{21.9}\right)^2 = (\sqrt{5t + 2457})^2 \qquad \text{Using the principle of squaring}$$

$$2854.2 \approx 5t + 2457 \qquad \text{Simplifying}$$

$$397.2 \approx 5t \qquad \text{Subtracting 2457 from both sides}$$

$$79.4 \approx t. \qquad \text{Dividing both sides by 5}$$

4. **Check.** A temperature of 79.4°F seems reasonable. A complete check is left to the student.

5. **State.** The temperature at the concert was about 79.4°F.

TRY EXERCISE 39

8.5   EXERCISE SET

For Extra Help   *MyMathLab*   Math XL PRACTICE   WATCH   DOWNLOAD

 **Concept Reinforcement**   *Classify each statement as either true or false.*

1. If $a = b$ is a true statement, then $a^2 = b^2$ must be true.

2. If $a^2 = b^2$ is a true statement, then $a = b$ must be true.

3. Every radical equation has at least one solution.

4. To use the principle of squaring, we square both sides of the equation.

Solve.

5. $\sqrt{x} = 6$

6. $\sqrt{x} = 11$

7. $\sqrt{x} - 3 = 9$

8. $\sqrt{x} - 7 = 2$

9. $\sqrt{3x + 1} = 8$

10. $\sqrt{5x - 1} = 7$

11. $2 + \sqrt{3 - y} = 9$

12. $3 + \sqrt{1 - a} = 5$

13. $10 - 2\sqrt{3n} = 0$

14. $4 - 2\sqrt{5n} = 0$

15. $\sqrt{8t + 3} = \sqrt{6t + 7}$

16. $\sqrt{7t - 9} = \sqrt{t + 3}$

Aha! 17. $5\sqrt{y} = -2$

18. $3\sqrt{y} + 1 = 0$

19. $\sqrt{6 - 4t} = \sqrt{2 - 5t}$

20. $\sqrt{11 - 3t} = \sqrt{1 - 5t}$

21. $\sqrt{3x + 1} = x - 3$

22. $x - 7 = \sqrt{x - 5}$

23. $a - 9 = \sqrt{a - 3}$

24. $\sqrt{t + 18} = t - 2$

25. $x + 1 = 6\sqrt{x - 7}$

26. $x - 5 = \sqrt{15 - 3x}$

27. $\sqrt{5x + 21} = x + 3$

28. $\sqrt{22 - x} = x - 2$

29. $t + 4 = 4\sqrt{t + 1}$

30. $1 + 2\sqrt{y - 1} = y$

31. $\sqrt{x^2 + 6} - x + 3 = 0$

32. $\sqrt{x^2 + 7} - x + 2 = 0$

33. $\sqrt{(4x + 5)(x + 4)} = 2x + 5$

34. $\sqrt{(p + 6)(p + 1)} - 2 = p + 1$

35. $\sqrt{8 - 3x} = \sqrt{13 + x}$

36. $\sqrt{3 - 7x} = \sqrt{5 - 2x}$

37. $x = 1 + \sqrt{1 - x}$

38. $x - 1 = \sqrt{(x + 1)(x - 2)}$

Speed of a skidding car.   *The formula* $r = 2\sqrt{5L}$ *can be used to approximate the speed r, in miles per hour, of a car that has left a skid mark of length L, in feet.*

39. How far will a car skid at 40 mph? at 60 mph?

40. How far will a car skid at 48 mph? at 80 mph?

Temperature and the speed of sound.   *Solve, using the formula* $s = 21.9\sqrt{5t + 2457}$ *from Example 4.*

41. During blasting for avalanche control in Utah's Wasatch Mountains, sound traveled at a rate of 1113 ft/sec. What was the temperature at the time?

42. At a recent concert by the Dave Matthews Band, sound traveled at a rate of 1176 ft/sec. What was the temperature at the time?

Sighting to the horizon.   *At a height of h meters, one can see V kilometers to the horizon, where* $V = 3.5\sqrt{h}$.

43. A scout can see 56 km to the horizon from atop a firetower. What is the altitude of the scout's eyes?

44. Alejandro can see 420 km to the horizon from an airplane window. How high is the airplane?

Speed of surface waves.   *The speed v, in meters/second, of a wave on the surface of the ocean can be approximated by the formula* $v = 3.1\sqrt{d}$, *where d is the depth of the water, in meters.*
Source: myweb.dal.ca

45. A wave is traveling 62 m/sec. What is the water depth?

46. A wave is traveling 9.3 m/sec. What is the water depth?

Period of a swinging pendulum.   *The formula* $T = 2\pi\sqrt{L/32}$ *can be used to find the period T, in seconds, of a pendulum of length L, in feet.*

47. A playground swing has a period of 4.4 sec. Find the length of the swing's chain. Use 3.14 for π.

48. The pendulum in Cheri's regulator clock has a period of 2.0 sec. Find the length of the pendulum. Use 3.14 for π.

49. Do you believe that the principle of squaring can be extended to powers other than 2? That is, if $a = b$, does it follow that $a^n = b^n$ for any integer n? Why or why not?

50. Explain in your own words why possible solutions of radical equations must be checked.

Skill Review

Review graphing.

Graph.

51. $y = 2x - 1$ [3.6]

52. $2x - 3y = 12$ [3.3]

53. $x = 3$ [3.3]

54. $x = 1 - y$ [3.2]

55. $y < \frac{1}{3}x - 2$ [7.5]

56. $x + y \geq 2,$
$x - 2y \leq 2$ [7.6]

Synthesis

57. Mike believes that a radical equation can never have a negative number as a solution. Is he correct? Why or why not?

58. Explain what would have happened in Example 1(a) if we had not isolated the radical before squaring. Could we still have solved the equation? Why or why not?

59. Find a number such that the opposite of three times its square root is -33.

60. Find a number such that 1 less than the square root of twice the number is 7.

Sometimes the principle of squaring must be used more than once in order to solve an equation. Solve Exercises 61–68 by using the principle of squaring as often as necessary. (Hint: Isolate a radical before each use of the principle.)

61. $1 + \sqrt{x} = \sqrt{x + 9}$

62. $5 - \sqrt{x} = \sqrt{x - 5}$

63. $\sqrt{t + 4} = 1 - \sqrt{3t + 1}$

64. $\sqrt{y + 8} - \sqrt{y} = 2$

65. $\sqrt{y + 1} - \sqrt{y - 2} = \sqrt{2y - 5}$

66. $3 + \sqrt{19 - x} = 5 + \sqrt{4 - x}$

67. $2\sqrt{x - 1} - \sqrt{x - 9} = \sqrt{3x - 5}$

68. $x + (2 - x)\sqrt{x} = 0$

69. *Changing elevations.*   A mountain climber pauses to rest and view the horizon. Using the formula $V = 3.5\sqrt{h}$, the climber computes the distance to the horizon and then climbs another 100 m. At this higher elevation, the horizon is 20 km farther than before. At what height was the climber when the first computation was made? (*Hint*: Use a system of equations.)

70. Solve $A = \sqrt{1 + \sqrt{a/b}}$ for b.

Graph. Use at least three ordered pairs.

71. $y = \sqrt{x}$ **72.** $y = \sqrt{x - 4}$

73. $y = \sqrt{x + 2}$ **74.** $y = \sqrt{x} + 1$

75. *Safe driving.*   The formula $r = 2\sqrt{5L}$ can be used to find the speed, in miles per hour, that a car was

traveling when it leaves skid marks L feet long. Police often recommend that drivers allow a space in front of their vehicle of one car length for each 10 miles per hour of speed. Thus a driver traveling 65 mph should leave 6.5 car lengths between his or her vehicle and the car in front of it. Many of today's small cars are approximately 15 ft long. For what speed would the "one car length per 10 mph of speed" rule require a distance equal to that of the skid marks that would be made at that speed?

76. Graph $y = x - 7$ and $y = \sqrt{x - 5}$ using the same set of axes. Determine where the graphs intersect in order to estimate a solution of $x - 7 = \sqrt{x - 5}$.

77. Graph $y = 1 + \sqrt{x}$ and $y = \sqrt{x + 9}$ using the same set of axes. Determine where the graphs intersect in order to estimate a solution of $1 + \sqrt{x} = \sqrt{x + 9}$.

Use a graphing calculator to solve Exercises 78 and 79. Round answers to the nearest hundredth.

78. $\sqrt{x + 3} = 2x - 1$

79. $-\sqrt{x + 3} = 2x - 1$

CONNECTING the CONCEPTS

We attempt to *simplify expressions* and *solve equations*. The principle of squaring applies to *equations*, not expressions.

We can *write equivalent radical expressions* by

- finding the square root of an expression that is a perfect square: $\sqrt{36x^2} = 6x$. (We assume $x \geq 0$.)
- factoring and simplifying: $\sqrt{48a^7} = \sqrt{16 \cdot 3 \cdot a^6 \cdot a} = \sqrt{16} \cdot \sqrt{a^6} \cdot \sqrt{3a} = 4a^3\sqrt{3a}$. (We assume $a \geq 0$.)
- dividing and simplifying: $\dfrac{\sqrt{50t^5}}{\sqrt{2t^{11}}} = \sqrt{\dfrac{50t^5}{2t^{11}}} = \sqrt{\dfrac{25}{t^6}} = \dfrac{5}{t^3}$.   (We assume $t > 0$.)
- combining like terms: $\sqrt{75x} + \sqrt{12x} = 5\sqrt{3x} + 2\sqrt{3x} = 7\sqrt{3x}$.
- multiplying and simplifying: $(\sqrt{3} + 2)(\sqrt{6} - 5) = \sqrt{18} - 5\sqrt{3} + 2\sqrt{6} - 10$
  $$= 3\sqrt{2} - 5\sqrt{3} + 2\sqrt{6} - 10.$$

We can solve *radical equations* by using the principle of squaring:

$$\sqrt{x + 2} + 1 = 7$$

$$\sqrt{x + 2} = 6 \qquad \text{Isolating the radical}$$

$$(\sqrt{x + 2})^2 = 6^2 \qquad \text{Squaring both sides}$$

$$\left.\begin{array}{l} x + 2 = 36 \\ x = 34. \end{array}\right\} \text{Solving for } x$$

The solution is 34.

Check:

$$\begin{array}{c|c} \sqrt{x + 2} + 1 = 7 \\ \hline \sqrt{34 + 2} + 1 & 7 \\ \sqrt{36} + 1 & \\ 6 + 1 & \\ 7 \overset{?}{=} 7 & \text{TRUE} \end{array}$$

MIXED REVIEW

When simplifying, assume that all radicands represent nonnegative numbers. Thus no absolute-value signs are needed in an answer.

1. Simplify: $\sqrt{100t^2}$.

2. Simplify: $\sqrt{17x^2}$.

3. Solve: $\sqrt{x} + 1 = 7$.

4. Simplify: $\dfrac{\sqrt{20}}{\sqrt{45}}$.

5. Add and simplify: $\sqrt{15t} + 4\sqrt{15t}$.

6. Solve: $\sqrt{2x - 1} = 5$.

7. Multiply and simplify: $\sqrt{6}(\sqrt{10} - \sqrt{33})$.

8. Solve: $\sqrt{5x} = -7$.

9. Subtract and simplify: $2\sqrt{3} - 5\sqrt{12}$.

10. Multiply and simplify: $(\sqrt{5} + 3)(\sqrt{5} - 3)$.

11. Solve: $\sqrt{4n - 5} = \sqrt{n + 1}$.

12. Simplify: $\sqrt{\dfrac{3a^{10}}{12a^4}}$.

13. Solve: $4 = 3\sqrt{2x}$.

14. Solve: $x + 5 = \sqrt{x + 11}$.

15. Multiply and simplify: $(\sqrt{5} + 1)(\sqrt{10} + 3)$.

16. Add and simplify: $\sqrt{500} + \sqrt{125}$.

17. Solve: $\sqrt{5x + 6} = x + 2$.

18. Simplify: $\sqrt{28a^{19}}$.

19. Solve: $x - 1 = 2\sqrt{x - 2}$.

20. Multiply and simplify: $\sqrt{12x^2y^5} \cdot \sqrt{75x^6y^7}$.

8.6 Applications Using Right Triangles

Right Triangles • Problem Solving • The Distance Formula

Radicals frequently occur in problem-solving situations in which the Pythagorean theorem is used. In Section 5.7, when we first used the Pythagorean theorem, we had not yet studied square roots. We now know that if $x^2 = n$, then x is a square root of n.

Right Triangles

For convenience, we restate the Pythagorean theorem.

The Pythagorean Theorem*

In any right triangle, if a and b are the lengths of the legs and c is the length of the hypotenuse, then

$$a^2 + b^2 = c^2.$$

*Recall that the converse of the Pythagorean theorem also holds. That is, if a, b, and c are the lengths of the sides of a triangle and $a^2 + b^2 = c^2$, then the triangle is a right triangle.

EXAMPLE 1   Find the length of the hypotenuse of the triangle shown. Give an exact answer in radical notation, as well as a decimal approximation to the nearest thousandth.

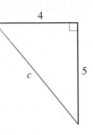

SOLUTION   We have

$$a^2 + b^2 = c^2$$
$$4^2 + 5^2 = c^2 \qquad \text{Substituting the lengths of the legs}$$
$$16 + 25 = c^2$$
$$41 = c^2.$$

We now use the fact that if $x^2 = n$, then $x = \sqrt{n}$ or $x = -\sqrt{n}$. In this case, since c is a length, it follows that c is the positive square root of 41:

$$c = \sqrt{41} \qquad \text{This is an exact answer.}$$
$$c \approx 6.403. \qquad \text{Using a calculator for an approximation}$$

TRY EXERCISE 5

EXAMPLE 2   Find the length of the indicated leg in each triangle. In each case, give an exact answer in radical notation, as well as a decimal approximation to the nearest thousandth.

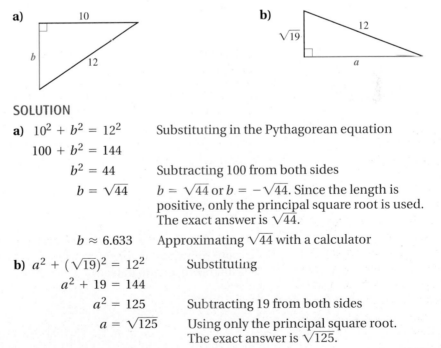

SOLUTION

a) $10^2 + b^2 = 12^2 \qquad$ Substituting in the Pythagorean equation

$\qquad 100 + b^2 = 144$

$\qquad\qquad b^2 = 44 \qquad$ Subtracting 100 from both sides

$\qquad\qquad b = \sqrt{44} \qquad b = \sqrt{44}$ or $b = -\sqrt{44}$. Since the length is positive, only the principal square root is used. The exact answer is $\sqrt{44}$.

$\qquad\qquad b \approx 6.633 \qquad$ Approximating $\sqrt{44}$ with a calculator

b) $a^2 + (\sqrt{19})^2 = 12^2 \qquad$ Substituting

$\qquad a^2 + 19 = 144$

$\qquad\qquad a^2 = 125 \qquad$ Subtracting 19 from both sides

$\qquad\qquad a = \sqrt{125} \qquad$ Using only the principal square root. The exact answer is $\sqrt{125}$.

$\qquad\qquad a \approx 11.180 \qquad$ Using a calculator

TRY EXERCISE 11

Problem Solving

The five-step process and the Pythagorean theorem can be used for problem solving.

EXAMPLE 3

Reach of a ladder.   A ladder is leaning against a house. The bottom of the ladder is 7 ft from the house, and the top of the ladder reaches 25 ft high. How long is the ladder? Give an exact answer in radical notation and a decimal approximation to the nearest tenth of a foot.

SOLUTION

1. **Familiarize.** First we make a drawing. In it, there is a right triangle. We label the unknown length l.

2. **Translate.** We use the Pythagorean theorem, substituting 7 for a, 25 for b, and l for c:

$$7^2 + 25^2 = l^2.$$

3. **Carry out.** We solve the equation:

$$7^2 + 25^2 = l^2$$
$$49 + 625 = l^2$$
$$674 = l^2$$
$$\sqrt{674} = l \qquad \text{Using only the principal square root. This answer is exact.}$$
$$26.0 \approx l. \qquad \text{Approximating with a calculator}$$

4. **Check.** We check by substituting 7, 25, and $\sqrt{674}$:

$$
\begin{array}{c|c}
a^2 + b^2 & = c^2 \\
\hline
7^2 + 25^2 & (\sqrt{674})^2 \\
49 + 625 & 674 \\
674 \overset{?}{=} 674 & \text{TRUE}
\end{array}
$$

5. **State.** The ladder is about 26.0 ft long.

TRY EXERCISE 23

EXAMPLE 4

Emergency communications.   During disaster relief operations, the Red Cross uses a radio antenna in the shape of an "inverted vee." The length of the antenna depends on the frequency being used. For a common frequency, the length of the antenna from the top of the pole to the ground should be 42 ft. If an antenna wire is fixed at the top of a 16-ft pole, how far from the base of the pole should the wire be attached to the ground?

Source: http://www.w0ipl.net

SOLUTION

1. **Familiarize.** We first make a drawing and label the known distances. The antenna and the pole form two sides of a right triangle. We let d represent the unknown length, in feet.

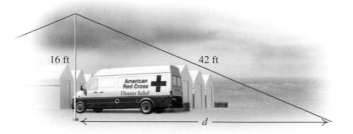

2. **Translate.** We use the Pythagorean theorem, substituting 16 for a, d for b, and 42 for c:

$$16^2 + d^2 = 42^2.$$

3. **Carry out.** We solve the equation:

$$16^2 + d^2 = 42^2$$
$$256 + d^2 = 1764$$
$$d^2 = 1508$$
$$d = \sqrt{1508} \qquad \text{Using only the principal square root. The exact answer is } \sqrt{1508} \text{ ft.}$$
$$d \approx 38.833. \qquad \text{Approximating using a calculator}$$

4. **Check.** To check, we substitute into the Pythagorean equation:

$$a^2 + b^2 = c^2$$

$16^2 + (\sqrt{1508})^2$	42^2
$256 + 1508$	1764
$1764 \overset{?}{=} 1764$	TRUE

5. **State.** For a 42-ft antenna, the end of the wire should be attached to the ground about 38.8 ft from the base of the pole.   `TRY EXERCISE` 25

The Distance Formula

We can use the Pythagorean theorem to find the distance between two points on a plane.

To find the distance between two points on a number line, we subtract. Depending on the order in which we subtract, the difference may be positive or negative. However, if we take the absolute value of the difference, we always obtain a positive value for the distance.

$$|4 - (-3)| = |7| = 7;$$
$$|-3 - 4| = |-7| = 7$$

To find the distance between two points on a plane, we use a right triangle and the idea of distance on a number line. Consider the points $(1, 3)$ and $(4, 5)$ and let the distance between the points be d.

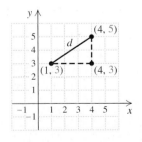

The length of the horizontal leg of the triangle is found by subtracting the x-coordinates: $|4 - 1| = 3$. The length of the vertical leg of the triangle is found by subtracting the y-coordinates: $|5 - 3| = 2$. Then, by the Pythagorean theorem, we have

$$d^2 = 2^2 + 3^2 \qquad c^2 = a^2 + b^2$$
$$d^2 = 4 + 9 \qquad \text{Squaring}$$
$$d^2 = 13$$
$$d = \sqrt{13}. \qquad d = -\sqrt{13} \text{ or } d = \sqrt{13}: \text{ Using the principal}$$
$$\text{square root since distance is never negative}$$

If we follow the same reasoning for two general points on the plane (x_1, y_1) and (x_2, y_2), we can develop a formula for the distance between any two points.

The Distance Formula

The distance d between any two points (x_1, y_1) and (x_2, y_2) is given by

$$d = \sqrt{(x_2 - x_1)^2 + (y_2 - y_1)^2}.$$

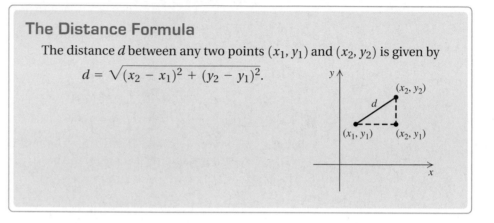

EXAMPLE 5

Find the distance between $(2, 3)$ and $(8, -1)$. Find an exact answer and an approximation to three decimal places.

SOLUTION   Since we will square the differences in the formula, it does not matter which point we regard as (x_1, y_1) and which we regard as (x_2, y_2). We substitute into the distance formula:

$$d = \sqrt{(x_2 - x_1)^2 + (y_2 - y_1)^2}$$
$$d = \sqrt{(8 - 2)^2 + (-1 - 3)^2} \qquad \text{Substituting}$$
$$d = \sqrt{6^2 + (-4)^2}$$
$$d = \sqrt{36 + 16}$$
$$d = \sqrt{52} \qquad \text{This is exact.}$$
$$d \approx 7.211. \qquad \text{Using a calculator for an approximation}$$

TRY EXERCISE 37

1. *Coin mixture.* A collection of nickels and quarters is worth $9.35. There are 59 coins in all. How many of each coin are there?

2. *Diagonal of a square.* Find the length of a diagonal of a square whose sides are 8 ft long.

3. *Shoveling time.* It takes Ian 55 min to shovel 4 in. of snow from his driveway. It takes Eric 75 min to do the same job. How long would it take if they worked together?

4. *Angles of a triangle.* The second angle of a triangle is three times as large as the first. The third is 17° less than the sum of the other angles. Find the measures of the angles.

5. *Perimeter.* The perimeter of a rectangle is 568 ft. The length is 26 ft greater than the width. Find the length and the width.

Translating for Success

Translate each word problem to an equation or a system of equations and select the most appropriate translation from equations A–O.

A. $5x + 25y = 9.35,$
$x + y = 59$

B. $4^2 + x^2 = 8^2$

C. $x(x + 26) = 568$

D. $8 = x \cdot 24$

E. $\dfrac{75}{x} = \dfrac{105}{x + 5}$

F. $\dfrac{75}{x} = \dfrac{55}{x + 5}$

G. $2x + 2(x + 26) = 568$

H. $x + 3x + (x + 3x - 17) = 180$

I. $x + 3x + (3x - 17) = 180$

J. $0.05x + 0.25y = 9.35,$
$x + y = 59$

K. $8^2 + 8^2 = x^2$

L. $x^2 + (x + 26)^2 = 568$

M. $\dfrac{1}{4} + \dfrac{1}{x} = \dfrac{1}{55}$

N. $\dfrac{1}{55} + \dfrac{1}{75} = \dfrac{1}{x}$

O. $x + 5\% \cdot x = 8568$

Answers on page A-33

An additional, animated version of this activity appears in MyMathLab. To use MyMathLab, you need a course ID and a student access code. Contact your instructor for more information.

6. *Bicycle travel.* One biker travels 75 km in the same time that a biker traveling 5 km/h faster travels 105 km. Find the speed of each biker.

7. *Money borrowed.* Emma borrows some money at 5% simple interest. After 1 yr, $8568 pays off her loan. How much did she originally borrow?

8. *TV time.* The average amount of time per day that TV sets in the United States are turned on is 8 hr. What percent of the time are our TV sets on?
Source: Nielsen Media Research

9. *Obstacle course.* As part of an obstacle course, an 8-ft plank is leaning against a tree. The bottom of the plank is 4 ft from the tree. How high is the top of the plank?

10. *Lengths of a rectangle.* The area of a rectangle is 568 ft². The length is 26 ft greater than the width. Find the length and the width.

8.6   EXERCISE SET

🐦 *Concept Reinforcement*   *Complete each of the following sentences.*

1. In any right triangle, the longest side is opposite the 90° angle and is called the _____.

2. In any right triangle, the two shortest sides, which form the 90° angle, are called the _____.

3. The _____ theorem states that if a and b are the lengths of the legs of a right triangle and c is the length of the third side, then $a^2 + b^2 = c^2$.

4. The formula $d = \sqrt{(x_2 - x_1)^2 + (y_2 - y_1)^2}$ is used to find the _____ between two points.

Find the length of the third side of each triangle. If an answer is not a whole number, use radical notation to give the exact answer and decimal notation for an approximation to the nearest thousandth.

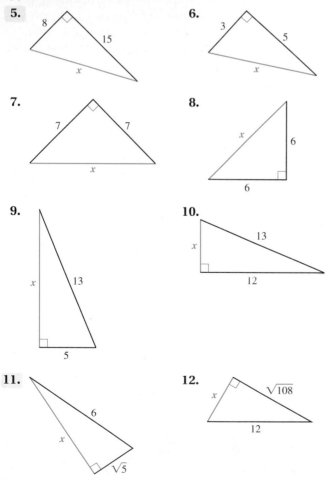

5.

8, 15, x

6.

3, 5, x

7.

7, 7, x

8.

x, 6, 6

9.

x, 13, 5

10.

x, 13, 12

11.

6, x, $\sqrt{5}$

12.

x, $\sqrt{108}$, 12

In a right triangle, find the length of the side not given. If an answer is not a whole number, use radical notation to give the exact answer and decimal notation for an approximation to the nearest thousandth. Regard a and b as the lengths of the legs and c as the length of the hypotenuse.

13. $a = 12$, $b = 5$

14. $a = 24$, $b = 10$

15. $a = 9$, $c = 15$

16. $a = 18$, $c = 30$

17. $b = 1$, $c = \sqrt{10}$

18. $b = 1$, $c = \sqrt{2}$

19. $a = 1$, $c = \sqrt{3}$

20. $a = \sqrt{2}$, $b = \sqrt{6}$

21. $c = 10$, $b = 5\sqrt{3}$

22. $a = 5$, $b = 5$

Solve. Don't forget to make drawings. If an answer is not a rational number, use radical notation to give the exact answer and decimal notation for an approximation to the nearest thousandth.

23. *Decorating.*   Roberto and Ava want to run a string of blinking lights from the top of a 15-ft sign on their restaurant to the edge of a surrounding garden 8 ft away. How long does the string of lights need to be?

15 ft

8 ft

24. *Decorating.*   To decorate their boat, Keisha and Bill plan to run a line with flags attached from the top of a 40-ft tall mast to the boat's bow 12 ft from the base of the mast. How long must the line be to span that distance?

40 ft

12 ft

25. *Rollerblading.*   Doug and Landon are building a rollerblade jump with a base that is 30 in. long and a ramp that is 33 in. long. How high will the back of the jump be?

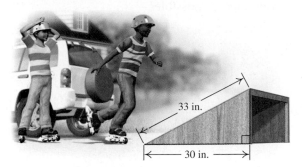

33 in.

30 in.

26. *Masonry.*   Find the length of a diagonal of a square tile that has sides 4 cm long.

27. *Plumbing.*   A new water pipe is being prepared so that it will run diagonally under a kitchen floor. If the kitchen is 8 ft wide and 12 ft long, how long should the pipe be?

28. *Guy wires.*   How long must a guy wire be to reach from the top of a 13-m telephone pole to a point on the ground 9 m from the foot of the pole?

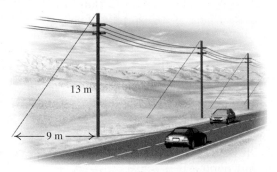

13 m

9 m

29. *Baseball.*   A baseball diamond is a square 90 ft on each side. How far is it from first base to third base?

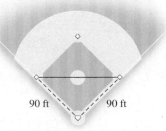

90 ft   90 ft

30. *Softball.*   A softball diamond is a square 65 ft on each side. How far is it from home plate to second base?

31. *Lacrosse.*   A regulation lacrosse field is 60 yd wide and 110 yd long. Find the length of a diagonal of such a field.

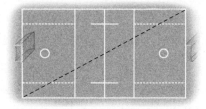

32. *Wiring.*   JR Electric is installing a security system with a wire that will run diagonally over the "drop" ceiling of a 10-ft by 16-ft room. How long will that section of wire need to be?

33. *Reptiles.*   A snake's cage should be large enough for the snake to stretch out to its full length diagonally. Kristin has a 30-cm by 60-cm rectangular cage. What is the length of the longest snake she should keep in the cage?

34. *Soccer fields.*   The largest regulation soccer field is 100 yd wide and 130 yd long. Find the length of a diagonal of such a field.

35. *Home maintenance.*   The hose on a power washer is 32 ft long. If the compressor to which the hose is attached is located at the corner of a 24-ft wide house, how high can the hose reach up the far corner without moving the compressor?

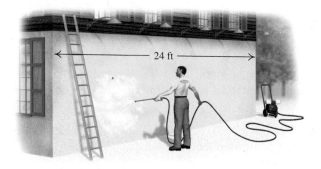

24 ft

36. *Surveying.*   A surveyor had poles located at points *P*, *Q*, and *R*. The distances that she was able to measure are marked in the figure. What is the approximate length of the lake?

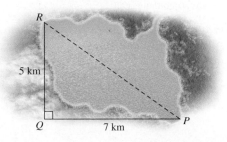

R

5 km

Q   7 km   *P*

Find the distance between each pair of points. If an answer is not a whole number, use radical notation to give the exact answer and decimal notation for an approximation to the nearest thousandth.

37. $(2, 3)$ and $(6, 10)$ **38.** $(1, 0)$ and $(7, 3)$

39. $(0, 3)$ and $(4, 0)$ **40.** $(-2, -8)$ and $(6, 7)$

41. $(-3, 2)$ and $(-1, 5)$ **42.** $(2, -7)$ and $(1, -4)$

43. $(-2, 4)$ and $(-8, -4)$ **44.** $(0, -5)$ and $(-12, 0)$

45. In an *isosceles triangle*, two sides have the same length. Can a right triangle be isosceles? Why or why not?

46. In an *equilateral triangle*, all sides have the same length. Can a right triangle be equilateral? Why or why not?

Skill Review

To prepare for Section 8.7, review powers of numbers (Section 4.1).

Simplify. [4.1]

47. $(-2)^5$ **48.** 2^5

49. $\left(\frac{2}{3}\right)^3$ **50.** $\left(\frac{1}{4}\right)^2$

51. 3^4 **52.** $(-3)^4$

53. 2^4 **54.** 3^3

55. $(-4)^3$ **56.** 5^4

Synthesis

57. Should a homeowner use a 28-ft ladder to repair clapboard that is 27 ft above ground level? Why or why not?

58. Can the length of a triangle's hypotenuse ever equal the combined lengths of the two legs? Why or why not?

59. The ancient Egyptians used a rope containing 12 evenly spaced knots as a construction tool. Explain how they could use this rope to construct a triangle with sides of lengths 3, 4, and 5, and thus a right angle.
Source: www.geom.uius.edu

60. *Holiday decorations.* From the peak of a 25-ft tall spruce tree in her front yard, Julia plans to run 10 strands of holiday lights to points in her lawn 15 ft from the tree's center. The lights cost $5.99 for 25 ft, $6.99 for 35 ft, and $11.99 for 68 ft. What should Julia purchase to make the decorations as economical as possible?

61. *Archaeology.* The base of the Khafre pyramid in Egypt is a square with sides measuring 704 ft. The length of each sloping side of the pyramid is 588 ft. How tall is the pyramid?

62. *Cordless telephones.* Brittany's AT&T E250 cordless telephone has a range of 1000 ft. Her apartment is a corner unit, located as shown in the figure below. Will Brittany be able to use the phone at the community pool?

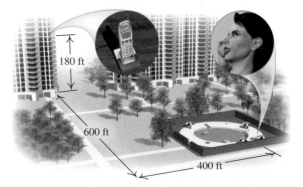

63. Aviation. A pilot is instructed to descend from 32,000 ft to 21,000 ft over a horizontal distance of 5 mi. What distance will the plane travel during this descent?

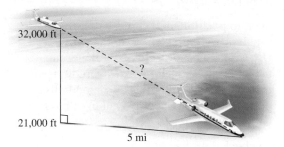

32,000 ft

?

21,000 ft

5 mi

64. The diagonal of a square has a length of $8\sqrt{2}$ ft. Find the length of a side of the square.

65. Find the length of a side of a square that has an area of 7 m².

66. A right triangle has sides with lengths that are consecutive even integers. Find the lengths of the sides.

67. Figure *ABCD* is a square. Find the length of a diagonal, $\overline{AC}$.

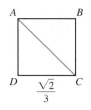

A   B

D $\dfrac{\sqrt{2}}{3}$ C

68. Find the length of the diagonal of a cube with sides of length *s*.

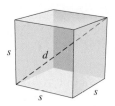

s   d

s   s

69. The area of square *PQRS* is 100 ft², and *A, B, C,* and *D* are midpoints of the sides on which they lie. Find the area of square *ABCD*.

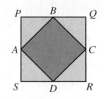

P   B   Q

A   C

S   D   R

70. Express the height *h* of an equilateral triangle in terms of the length of a side *a*.

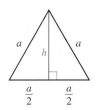

a   h   a

$\dfrac{a}{2}$ $\dfrac{a}{2}$

71. Racquetball. A racquetball court is 20 ft by 20 ft by 40 ft. What is the longest straight-line distance that can be measured in this racquetball court?

72. Distance driven. Two cars leave a service station at the same time. One car travels east at a speed of 50 mph, and the other travels south at a speed of 60 mph. After one half hour, how far apart are they?

73. Solve for *x*.

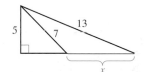

5   7   13

x

74. Ranching. If 2 mi of fencing encloses a square plot of land with an area of 160 acres, how large a square, in acres, will 4 mi of fencing enclose?

COLLABORATIVE

CORNER

Pythagorean Triples

Focus: Pythagorean theorem

Time: 15 minutes

Group size: 2–4

Materials: Tape measure and chalk; string and scissors

We mentioned in the footnote on p. 512 that the converse of the Pythagorean theorem is also true: If *a*, *b*, and *c* are the lengths of the sides of a triangle and $a^2 + b^2 = c^2$, then the triangle is a right triangle. Each such set of three numbers is called a *Pythagorean triple*. Since $3^2 + 4^2 = 5^2$ and $5^2 + 12^2 = 13^2$, then (3, 4, 5) and (5, 12, 13) are both Pythagorean triples. Such numbers provide carpenters, masons, archaeologists, and others with a handy way of locating a line that forms a 90° angle with another line.

ACTIVITY

1. Suppose that the group is in the process of building a deck. They determine that they need to position an 8-ft piece of lumber so that it forms a precise 90° angle with a wall. Use a tape measure, chalk, Pythagorean triples, and—if desired—string and scissors to construct a right angle at a specified point at the base of some wall in or near your classroom.

2. To check that the angle formed in part (1) is truly 90°, repeat the procedure using a different Pythagorean triple.

8.7   Higher Roots and Rational Exponents

Higher Roots • Products and Quotients Involving Higher Roots • Rational Exponents • Calculators

In this section, we study *higher* roots, such as cube roots or fourth roots, and exponents that are not integers.

Higher Roots

Recall that *c* is a square root of *a* if $c^2 = a$. A similar definition exists for *cube roots*.

> **Cube Root**
> The number *c* is the *cube root* of *a* if $c^3 = a$.

We have used $\sqrt{a}$ to represent the square root of *a*. Similarly, we use $\sqrt[3]{a}$ to represent the cube root of *a*. In the radical $\sqrt[3]{a}$, the number 3 is called the **index**, and *a* is called the **radicand**. The index of a square root $\sqrt{a}$ is 2, although it is generally not written.

EXAMPLE 1 Find the cube root of each number: **(a)** 8; **(b)** -125.

SOLUTION

a) The cube root of 8 is the number whose cube is 8. Since $2^3 = 2 \cdot 2 \cdot 2 = 8$, the cube root of 8 is 2: $\sqrt[3]{8} = 2$.

b) The cube root of -125 is the number whose cube is -125. Since $(-5)^3 = (-5)(-5)(-5) = -125$, the cube root of -125 is -5: $\sqrt[3]{-125} = -5$.

The symbol $\sqrt[n]{a}$ represents the principal nth root of a. For example, $\sqrt[5]{32}$ represents the fifth root of 32. Here the index is 5 and the radicand is 32.

---

nth Root

For n a natural number greater than 1:

- The number c is an nth *root* of a if $c^n = a$.
- If n is odd, there is only one nth root and $\sqrt[n]{a}$ represents that root.
- If n is even and a is positive, there are two nth roots, and $\sqrt[n]{a}$ represents the nonnegative nth root.
- If n is even and a is negative, $\sqrt[n]{a}$ is not a real number.

EXAMPLE 2 Find each root: **(a)** $\sqrt[4]{16}$; **(b)** $\sqrt[5]{-32}$; **(c)** $\sqrt[4]{-16}$; **(d)** $-\sqrt[3]{64}$.

SOLUTION

a) $\sqrt[4]{16} = 2$ Since $2^4 = 2 \cdot 2 \cdot 2 \cdot 2 = 16$

b) $\sqrt[5]{-32} = -2$ Since $(-2)^5 = (-2)(-2)(-2)(-2)(-2) = -32$

c) $\sqrt[4]{-16}$ is not a real number, because it is an even root of a negative number.

d) $-\sqrt[3]{64} = -(\sqrt[3]{64})$ This is the opposite of $\sqrt[3]{64}$.

   $\quad\quad\quad = -4$ $4^3 = 4 \cdot 4 \cdot 4 = 64$

TRY EXERCISE 5

Some roots occur so frequently that you may want to memorize them.

Square Roots		Cube Roots	Fourth Roots	Fifth Roots
$\sqrt{1} = 1$	$\sqrt{4} = 2$	$\sqrt[3]{1} = 1$	$\sqrt[4]{1} = 1$	$\sqrt[5]{1} = 1$
$\sqrt{9} = 3$	$\sqrt{16} = 4$	$\sqrt[3]{8} = 2$	$\sqrt[4]{16} = 2$	$\sqrt[5]{32} = 2$
$\sqrt{25} = 5$	$\sqrt{36} = 6$	$\sqrt[3]{27} = 3$	$\sqrt[4]{81} = 3$	$\sqrt[5]{243} = 3$
$\sqrt{49} = 7$	$\sqrt{64} = 8$	$\sqrt[3]{64} = 4$	$\sqrt[4]{256} = 4$	
$\sqrt{81} = 9$	$\sqrt{100} = 10$	$\sqrt[3]{125} = 5$	$\sqrt[4]{625} = 5$	
$\sqrt{121} = 11$	$\sqrt{144} = 12$	$\sqrt[3]{216} = 6$		

STUDY SKILLS

Abbrevs. cn help u go fst

If you take notes and have trouble keeping up with your instructor, use abbreviations to speed up your work. Consider standard abbreviations like "Ex" for "Example," "$\approx$" for "is approximately equal to," "$\therefore$" for "therefore," and "$\Rightarrow$" for "implies." Feel free to create your own abbreviations as well.

Products and Quotients Involving Higher Roots

The rules for working with products and quotients of square roots can be extended to products and quotients of nth roots. Prime factorizations can again be a useful aid when simplifying.

> ### The Product and Quotient Rules
> For any real numbers $\sqrt[n]{A}$ and $\sqrt[n]{B}$,
> $$\sqrt[n]{AB} = \sqrt[n]{A}\,\sqrt[n]{B} \quad \text{and} \quad \sqrt[n]{\frac{A}{B}} = \frac{\sqrt[n]{A}}{\sqrt[n]{B}} \quad (B \neq 0).$$

EXAMPLE **3**

Simplify: **(a)** $\sqrt[3]{40}$; **(b)** $\sqrt[3]{\frac{125}{27}}$; **(c)** $\sqrt[4]{1250}$; **(d)** $\sqrt[5]{\frac{2}{243}}$.

SOLUTION

a) $\sqrt[3]{40} = \sqrt[3]{8 \cdot 5}$ Note that $40 = 2 \cdot 2 \cdot 2 \cdot 5 = 8 \cdot 5$ and 8 is a perfect cube.

$= \sqrt[3]{8} \cdot \sqrt[3]{5}$

$= 2\sqrt[3]{5}$

b) $\sqrt[3]{\frac{125}{27}} = \frac{\sqrt[3]{125}}{\sqrt[3]{27}}$ $125 = 5 \cdot 5 \cdot 5$ and $27 = 3 \cdot 3 \cdot 3$, so 125 and 27 are perfect cubes. See the chart on p. 523.

$= \frac{5}{3}$

c) $\sqrt[4]{1250} = \sqrt[4]{625 \cdot 2}$ Note that $1250 = 2 \cdot 5 \cdot 5 \cdot 5 \cdot 5 = 2 \cdot 625$ and 625 is a perfect fourth power.

$= \sqrt[4]{625} \cdot \sqrt[4]{2}$

$= 5\sqrt[4]{2}$

d) $\sqrt[5]{\frac{2}{243}} = \frac{\sqrt[5]{2}}{\sqrt[5]{243}}$ $243 = 3 \cdot 3 \cdot 3 \cdot 3 \cdot 3$, so 243 is a perfect fifth power. See the chart on p. 523.

$= \frac{\sqrt[5]{2}}{3}$

> **TRY EXERCISE** 29

STUDENT NOTES

Note that

$$2^3\sqrt{5} \quad \text{and} \quad 2\sqrt[3]{5}$$

represent different numbers. Be very careful to place an exponent and an index correctly.

Rational Exponents

Recall that rational numbers can be written as ratios of integers, such as $\frac{1}{3}$, $\frac{5}{2}$, and $\frac{-3}{4}$. These numbers can also be used as exponents. Expressions like $8^{1/3}$, $4^{5/2}$, and $81^{-3/4}$ are defined in such a way that the laws of exponents still hold. For example, if the product rule, $a^m \cdot a^n = a^{m+n}$, is to hold, then

$$a^{1/2} \cdot a^{1/2} = a^{1/2+1/2}$$
$$= a^1 = a.$$

This says that $a^{1/2}$ times itself is a, which means that $a^{1/2}$ is a square root of a. This idea is generalized as follows.

> ## The Exponent $1/n$
>
> $a^{1/n}$ means $\sqrt[n]{a}$.
>
> $a^{1/2}$ is written $\sqrt{a}$. The index 2 is understood.
>
> If a is negative, then $a^{1/n}$ is a real number only when n is odd.

Thus, $a^{1/2}$ means $\sqrt{a}$, $a^{1/3}$ means $\sqrt[3]{a}$, and so on.

EXAMPLE **4**   Simplify: **(a)** $8^{1/3}$; **(b)** $100^{1/2}$; **(c)** $81^{1/4}$; **(d)** $(-243)^{1/5}$.

SOLUTION

a) $8^{1/3} = \sqrt[3]{8} = 2$

b) $100^{1/2} = \sqrt{100} = 10$

c) $81^{1/4} = \sqrt[4]{81} = 3$

d) $(-243)^{1/5} = \sqrt[5]{-243} = -3$

> **TRY EXERCISE** 41

In order to continue multiplying exponents when raising a power to a power, we must have $a^{2/3} = (a^{1/3})^2$ and $a^{2/3} = (a^2)^{1/3}$. This suggests both that $a^{2/3} = (\sqrt[3]{a})^2$ and that $a^{2/3} = \sqrt[3]{a^2}$.

> ## Positive Rational Exponents
>
> For any natural numbers m and n ($n \neq 1$) and any real number a for which $\sqrt[n]{a}$ exists,
>
> $$a^{m/n} \text{ means } (\sqrt[n]{a})^m, \text{ or equivalently, } a^{m/n} \text{ means } \sqrt[n]{a^m}.$$

In most cases, it is easiest to simplify using $(\sqrt[n]{a})^m$, because smaller numbers are involved.

EXAMPLE **5**   Simplify: **(a)** $27^{2/3}$; **(b)** $8^{5/3}$; **(c)** $81^{3/4}$.

SOLUTION

a) $27^{2/3} = (27^{1/3})^2 = (\sqrt[3]{27})^2 = 3^2 = 9$

b) $8^{5/3} = (8^{1/3})^5 = (\sqrt[3]{8})^5 = 2^5 = 32$

c) $81^{3/4} = (81^{1/4})^3 = (\sqrt[4]{81})^3 = 3^3 = 27$

> **TRY EXERCISE** 47

STUDENT NOTES

If you are unsure of the rules for when to add or multiply exponents, this would be a good time to review them in Section 4.1.

Negative rational exponents are defined in much the same way that negative integer exponents are.

> ## Negative Rational Exponents
>
> For any rational number m/n and any nonzero real number a for which $a^{m/n}$ exists,
>
> $$a^{-m/n} = \frac{1}{a^{m/n}}.$$

EXAMPLE 6   Simplify: **(a)** $16^{-1/2}$; **(b)** $27^{-1/3}$; **(c)** $32^{-2/5}$; **(d)** $64^{-3/2}$.

SOLUTION

a) $16^{-1/2} = \dfrac{1}{16^{1/2}} = \dfrac{1}{\sqrt{16}} = \dfrac{1}{4}$

b) $27^{-1/3} = \dfrac{1}{27^{1/3}} = \dfrac{1}{\sqrt[3]{27}} = \dfrac{1}{3}$

c) $32^{-2/5} = \dfrac{1}{32^{2/5}} = \dfrac{1}{(32^{1/5})^2} = \dfrac{1}{(\sqrt[5]{32})^2} = \dfrac{1}{2^2} = \dfrac{1}{4}$

d) $64^{-3/2} = \dfrac{1}{64^{3/2}} = \dfrac{1}{(\sqrt{64})^3} = \dfrac{1}{8^3} = \dfrac{1}{512}$

▶ TRY EXERCISE 57

> *CAUTION!* A negative exponent does not indicate that the expression in which it appears is negative.

Calculators

A calculator with a key for finding powers can be used to approximate numbers like $\sqrt[5]{8}$. Generally such keys are labeled $\boxed{x^y}$, $\boxed{a^x}$, or ⬤.

We can approximate $\sqrt[5]{8}$ by entering the radicand, 8, pressing the power key, entering the exponent, 0.2 or (1/5), and pressing $\boxed{=}$ or $\boxed{\text{ENTER}}$ to get $\sqrt[5]{8} \approx 1.515716567$. Note that the parentheses in (1/5) are necessary. Consult an owner's manual or your instructor if your calculator works differently.

8.7   EXERCISE SET

For Extra Help
MyMathLab Math XL PRACTICE WATCH DOWNLOAD

🔖 *Concept Reinforcement* *Classify each statement as either true or false.*

1. $4^{1/2}$ is the same as $\sqrt{4}$, or 2.

2. $a^{m/n}$ is the same as $\sqrt[n]{a^m}$.

3. $a^{m/n}$ is the same as $(\sqrt[n]{a})^m$.

4. $5^{-1/2}$ is a negative number.

Simplify. If an expression does not represent a real number, state this.

5. $\sqrt[3]{-8}$ 6. $\sqrt[3]{-64}$ 7. $\sqrt[3]{-1000}$

8. $\sqrt[3]{-27}$ 9. $\sqrt[3]{125}$ 10. $\sqrt[3]{8}$

11. $-\sqrt[3]{216}$ 12. $-\sqrt[3]{-343}$ 13. $\sqrt[4]{625}$

14. $\sqrt[4]{81}$ 15. $\sqrt[5]{0}$ 16. $\sqrt[5]{1}$

17. $\sqrt[5]{-1}$ 18. $\sqrt[5]{-243}$ 19. $\sqrt[4]{-81}$

20. $\sqrt[4]{-1}$ Aha! 21. $\sqrt[4]{10,000}$ 22. $\sqrt[5]{100,000}$

Aha! 23. $\sqrt[3]{6^3}$ 24. $\sqrt[4]{2^4}$ 25. $\sqrt[8]{1}$

26. $\sqrt[6]{64}$ 27. $\sqrt[7]{a^7}$ 28. $\sqrt[5]{t^5}$

29. $\sqrt[3]{54}$ 30. $\sqrt[3]{24}$ 31. $\sqrt[4]{48}$

32. $\sqrt[5]{160}$ 33. $\sqrt[3]{\dfrac{64}{125}}$ 34. $\sqrt[3]{\dfrac{125}{27}}$

35. $\sqrt[5]{\dfrac{32}{243}}$ 36. $\sqrt[4]{\dfrac{625}{256}}$ 37. $\sqrt[3]{\dfrac{7}{8}}$

38. $\sqrt[5]{\dfrac{15}{32}}$ 39. $\sqrt[4]{\dfrac{14}{81}}$ 40. $\sqrt[3]{\dfrac{10}{27}}$

Simplify.

41. $49^{1/2}$ 42. $36^{1/2}$ 43. $1000^{1/3}$

44. $27^{1/3}$ 45. $16^{1/4}$ 46. $32^{1/5}$

47. $16^{3/4}$ 48. $8^{4/3}$ 49. $16^{3/2}$

50. $4^{3/2}$ 51. $64^{2/3}$ 52. $32^{2/5}$

53. $1000^{4/3}$ 54. $16^{5/4}$ 55. $100^{5/2}$

56. $36^{3/2}$

57. $9^{-1/2}$

58. $32^{-1/5}$

59. $256^{-1/4}$

60. $81^{-1/2}$

61. $16^{-3/4}$

62. $81^{-3/4}$

63. $81^{-5/4}$

64. $32^{-2/5}$

65. $125^{-2/3}$

66. $8^{-4/3}$

67. Expressions of the form $a^{m/n}$ can be rewritten as $(\sqrt[n]{a})^m$ or $\sqrt[n]{a^m}$. Which radical expression would you use when simplifying $25^{3/2}$ and why?

68. Explain in your own words why $\sqrt[n]{a}$ is negative when n is odd and a is negative.

Skill Review

To prepare for Chapter 9, review solving quadratic equations (Section 5.6).

Solve. [5.6]

69. $x^2 + 7x + 12 = 0$

70. $x^2 - 6x - 7 = 0$

71. $16t^2 - 9 = 0$

72. $4t^2 - 25 = 0$

73. $3x^2 - x - 10 = 0$

74. $5x^2 - 19x + 12 = 0$

Synthesis

75. If $a > b$, does it follow that $a^{1/n} > b^{1/n}$? Why or why not?

76. Under what condition(s) will $a^{-3/5}$ be negative?

Using a calculator, approximate each of the following to three decimal places.

77. $8^{4/5}$

78. $20^{1/4}$

79. $48^{5/8}$

80. $12^{5/2}$

Simplify.

81. $a^{1/4}a^{3/2}$

82. $(x^{2/3})^{7/3}$

83. $m^{-2/3}m^{1/4}m^{3/2}$

84. $\dfrac{p^{5/6}}{p^{2/3}}$

Graph.

85. $y = \sqrt[3]{x}$

86. $y = \sqrt[4]{x}$

87. *Herpetology.* The daily number of calories c needed by a reptile of weight w pounds can be approximated by $c = 10w^{3/4}$. Find the daily calorie requirement of a green iguana weighing 16 lb.
Source: www.anapsid.org

88. Use a graphing calculator to draw the graphs of $y_1 = x^{2/3}$, $y_2 = x^1$, $y_3 = x^{5/4}$, and $y_4 = x^{3/2}$. Use the window $[-1, 17, -1, 32]$ and the SIMULTANEOUS mode. Then determine which curve corresponds to each equation.

Study Summary

KEY TERMS AND CONCEPTS	EXAMPLES

SECTION 8.1: INTRODUCTION TO SQUARE ROOTS AND RADICAL EXPRESSIONS

The following is an example of a **radical expression**:

$$\sqrt{7} \quad \begin{array}{l}\longleftarrow \text{ Radical sign} \\ \longleftarrow \text{ Radicand}\end{array}$$

If $c^2 = a$, then c is a **square root** of a. Every positive number has two square roots. The notation $\sqrt{a}$ indicates the **principal**, or nonnegative, square root of a. If a is not a perfect square, then $\sqrt{a}$ is irrational.

The square roots of 25 are -5 and 5.

$$\sqrt{25} = 5$$
$\sqrt{25}$ is rational.
$\sqrt{10}$ is irrational.

For any real number A,
$$\sqrt{A^2} = |A|.$$
If we assume that $A \geq 0$, then
$$\sqrt{A^2} = A.$$

Assume that x is any real number.
$$\sqrt{(3x)^2} = |3x| = 3|x|$$
Assume that x is a nonnegative number.
$$\sqrt{(3x)^2} = 3x$$

SECTION 8.2: MULTIPLYING AND SIMPLIFYING RADICAL EXPRESSIONS

The Product Rule for Square Roots
For any real numbers $\sqrt{A}$ and $\sqrt{B}$,
$$\sqrt{A} \cdot \sqrt{B} = \sqrt{A \cdot B}.$$

$$\sqrt{2x} \cdot \sqrt{3y} = \sqrt{6xy}$$

We can use the product rule in reverse to factor and simplify radical expressions:
$$\sqrt{A \cdot B} = \sqrt{A} \cdot \sqrt{B}.$$

$$\sqrt{75} = \sqrt{25 \cdot 3} = \sqrt{25} \cdot \sqrt{3} = 5\sqrt{3}$$

Every even power is a perfect square.

$$\sqrt{x^{12}} = x^6;$$
$$\sqrt{p^{21}} = \sqrt{p^{20} \cdot p} = \sqrt{p^{20}} \cdot \sqrt{p} = p^{10}\sqrt{p}$$

After multiplying, we should simplify if possible.

$$\sqrt{6x} \cdot \sqrt{15xy} = \sqrt{6 \cdot 15 \cdot x^2 \cdot y} = \sqrt{2 \cdot 3 \cdot 3 \cdot 5 \cdot x^2 \cdot y}$$
$$= \sqrt{3^2} \cdot \sqrt{x^2} \cdot \sqrt{2 \cdot 5 \cdot y} = 3x\sqrt{10y} \quad \text{(We assume } x \geq 0.\text{)}$$

SECTION 8.3: QUOTIENTS INVOLVING SQUARE ROOTS

The Quotient Rule for Square Roots
For any real numbers $\sqrt{A}$ and $\sqrt{B}$ with $B \neq 0$,

$$\frac{\sqrt{A}}{\sqrt{B}} = \sqrt{\frac{A}{B}}.$$

$$\frac{\sqrt{18a^9}}{\sqrt{2a^3}} = \sqrt{\frac{18a^9}{2a^3}} = \sqrt{9a^6} = 3a^3 \quad \text{(We assume } a > 0.\text{)}$$

$$\sqrt{\frac{81x^2}{100}} = \frac{\sqrt{81x^2}}{\sqrt{100}} = \frac{9x}{10} \quad \text{(We assume } x \geq 0.\text{)}$$

We can **rationalize a denominator** by multiplying by 1.

$$\frac{\sqrt{3}}{\sqrt{5}} = \frac{\sqrt{3}}{\sqrt{5}} \cdot \frac{\sqrt{5}}{\sqrt{5}} = \frac{\sqrt{15}}{5}$$

SECTION 8.4: RADICAL EXPRESSIONS WITH SEVERAL TERMS

Like radicals are expressions such as $2\sqrt{3}$ and $5\sqrt{3}$. These expressions have the same radical factor and can be combined.

$$\sqrt{12} + 5\sqrt{3} = \sqrt{4 \cdot 3} + 5\sqrt{3}$$
$$= 2\sqrt{3} + 5\sqrt{3} = 7\sqrt{3}$$

Radical expressions are multiplied in much the same way that polynomials are multiplied.

$$(1 + 5\sqrt{2})(4 - \sqrt{6}) = 1 \cdot 4 - 1\sqrt{6} + 4 \cdot 5\sqrt{2} - 5\sqrt{2} \cdot \sqrt{6}$$
$$= 4 - \sqrt{6} + 20\sqrt{2} - 5\sqrt{2 \cdot 2 \cdot 3}$$
$$= 4 - \sqrt{6} + 20\sqrt{2} - 10\sqrt{3}$$

To rationalize a denominator containing two terms, we use the **conjugate** of the denominator to write a form of 1.

$$\frac{2}{1 - \sqrt{3}} = \frac{2}{1 - \sqrt{3}} \cdot \frac{1 + \sqrt{3}}{1 + \sqrt{3}} = \frac{2 + 2\sqrt{3}}{1 - 3}$$
$$= \frac{2(1 + \sqrt{3})}{-2} = -1 - \sqrt{3}$$

SECTION 8.5: RADICAL EQUATIONS

The Principle of Squaring

If $a = b$, then $a^2 = b^2$.

Solutions found using the principle of squaring must be checked in the original equation.

Solve: $\sqrt{2x - 3} = 5$.

$$(\sqrt{2x - 3})^2 = 5^2$$
$$2x - 3 = 25$$
$$2x = 28$$
$$x = 14$$

Check:

$$\begin{array}{c|c} \sqrt{2x - 3} = 5 \\ \hline \sqrt{2 \cdot 14 - 3} & 5 \\ \sqrt{25} & \\ 5 \overset{?}{=} 5 & \text{TRUE} \end{array}$$

The solution is 14.

SECTION 8.6: APPLICATIONS USING RIGHT TRIANGLES

The Pythagorean Theorem

In any right triangle, the sum of the squares of the lengths of the **legs** is the square of the **hypotenuse**.

$$a^2 + b^2 = c^2$$

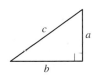

Find the length of the hypotenuse. Give an exact answer in radical notation, as well as an approximation to the nearest thousandth using a calculator.

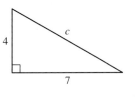

$$a^2 + b^2 = c^2$$
$$4^2 + 7^2 = c^2$$
$$16 + 49 = c^2$$
$$65 = c^2$$
$$\sqrt{65} = c \qquad \text{Exact answer}$$
$$8.062 \approx c \qquad \text{Approximation}$$

The Distance Formula

The distance d between any two points (x_1, y_1) and (x_2, y_2) is given by

$$d = \sqrt{(x_2 - x_1)^2 + (y_2 - y_1)^2}.$$

Find the distance d between $(-2, 5)$ *and* $(-3, -1)$.

$$d = \sqrt{(-3 - (-2))^2 + (-1 - 5)^2}$$
$$d = \sqrt{(-1)^2 + (-6)^2}$$
$$d = \sqrt{37} \qquad \text{Exact answer}$$
$$d \approx 6.083 \qquad \text{Approximation}$$

SECTION 8.7: HIGHER ROOTS AND RATIONAL EXPONENTS

The number c is the **cube root** of a if $c^3 = a$.

The number c is the ***n*th root** of a if $c^n = a$.

The *n*th root of a is written $\sqrt[n]{a}$, and n is called the **index**.

$$\sqrt[3]{125} = 5$$
$$\sqrt[3]{-125} = -5$$
$$\sqrt[4]{16} = 2$$

$\sqrt[4]{-16}$ is not a real number.

$a^{1/n}$ means $\sqrt[n]{a}$.

$a^{m/n}$ means $(\sqrt[n]{a})^m$ or, equivalently, $\sqrt[n]{a^m}$.

$$a^{-m/n} = \frac{1}{a^{m/n}}$$

$$64^{1/2} = \sqrt{64} = 8$$
$$125^{2/3} = (\sqrt[3]{125})^2 = 5^2 = 25$$

$$8^{-1/3} = \frac{1}{8^{1/3}} = \frac{1}{2}$$

Review Exercises: Chapter 8

Classify each statement as either true or false.

1. The sum of two square roots is the square root of the sum of the radicands. [8.4]

2. The product of two square roots is the square root of the product of the radicands. [8.2]

3. The quotient of two cube roots is the cube root of the quotient of the radicands. [8.7]

4. The principal square root of a number is never negative. [8.1]

5. The cube root of a number is never negative. [8.7]

6. The conjugate of $5 - \sqrt{3}$ is $1/(5 - \sqrt{3})$. [8.4]

7. For any triangle, if the lengths of two sides are known, the Pythagorean theorem can be used to determine the length of the third side. [8.6]

8. The rules for adding or multiplying integer exponents also apply to rational exponents. [8.7]

9. The distance formula is an application of the Pythagorean theorem. [8.6]

10. The principle of squaring does not always lead to a solution. [8.5]

Find the square roots of each number. [8.1]

11. 16

12. 64

13. 400

14. 225

Simplify. [8.1]

15. $\sqrt{144}$

16. $\sqrt{81}$

17. $-\sqrt{4}$

18. $-\sqrt{169}$

Identify each radicand. [8.1]

19. $3x\sqrt{5x^3y}$

20. $a\sqrt{\dfrac{a}{b}}$

Determine whether each square root is rational or irrational. [8.1]

21. $-\sqrt{36}$

22. $-\sqrt{45}$

23. $\sqrt{99}$

24. $\sqrt{25}$

Use a calculator to approximate each square root. Round to three decimal places. [8.1]

25. $\sqrt{5}$

26. $\sqrt{90}$

Simplify. Assume that x can be any real number. [8.1]

27. $\sqrt{(5x)^2}$

28. $\sqrt{(x + 2)^2}$

Assume for Exercises 29–61 that all variables represent nonnegative numbers.

Simplify. [8.1]

29. $\sqrt{p^2}$

30. $\sqrt{(3x)^2}$

31. $\sqrt{49n^2}$

32. $\sqrt{(ac)^2}$

Simplify by factoring. [8.2]

33. $\sqrt{48}$

34. $\sqrt{300t^2}$

35. $\sqrt{32p}$

36. $\sqrt{x^{16}}$

37. $\sqrt{12a^{13}}$

38. $\sqrt{36m^{15}}$

Multiply and, if possible, simplify. [8.2]

39. $\sqrt{5}\,\sqrt{11}$

40. $\sqrt{6}\,\sqrt{10}$

41. $\sqrt{3s}\,\sqrt{7t}$

42. $\sqrt{3a}\,\sqrt{8a}$

43. $\sqrt{5x}\,\sqrt{10xy^2}$

44. $\sqrt{20a^3b}\,\sqrt{5a^2b^2}$

Simplify.

45. $\dfrac{\sqrt{35}}{\sqrt{45}}$ [8.3]

46. $\dfrac{\sqrt{30y^9}}{\sqrt{54y}}$ [8.3]

47. $\sqrt{\dfrac{49}{64}}$ [8.3]

48. $\sqrt{\dfrac{20}{45}}$ [8.3]

49. $\sqrt{\dfrac{64t}{t^7}}$ [8.3]

50. $10\sqrt{5} + 3\sqrt{5}$ [8.4]

51. $\sqrt{80} - \sqrt{45}$ [8.4]

52. $2\sqrt{x} - \sqrt{25x}$ [8.4]

53. $(2 + \sqrt{3})^2$ [8.4]

54. $(1 + \sqrt{7})(1 - \sqrt{7})$ [8.4]

55. $(1 + 2\sqrt{7})(3 - \sqrt{7})$ [8.4]

Write an equivalent expression by rationalizing each denominator.

56. $\sqrt{\dfrac{1}{5}}$ [8.3]

57. $\dfrac{\sqrt{5}}{\sqrt{8}}$ [8.3]

58. $\sqrt{\dfrac{7}{y}}$ [8.3]

59. $\dfrac{2}{\sqrt{3}}$ [8.3]

60. $\dfrac{4}{2 + \sqrt{3}}$ [8.4]

61. $\dfrac{1 + \sqrt{5}}{2 - \sqrt{5}}$ [8.4]

Solve. [8.5]

62. $\sqrt{x - 5} = 7$

63. $\sqrt{5x + 3} = \sqrt{2x - 1}$

64. $\sqrt{x + 5} = x - 1$

65. $1 + x = \sqrt{1 + 5x}$

66. *Length of a rope swing.* The formula $T = 2\pi\sqrt{L/32}$ can be used to find the period T, in seconds, of a pendulum of length L, in feet. A rope swing over a river has a period of 6.6 sec. Find the length of the rope. Use 3.14 for π. [8.5]

In a right triangle, find the length of the side not given. If the answer is not a whole number, use radical notation to give an exact answer and decimal notation to give an approximation to the nearest thousandth. Keep in mind that a and b are the lengths of the legs and c is the length of the hypotenuse. [8.6]

67. $a = 15$, $c = 25$

68. $a = 1$, $b = \sqrt{2}$

69. Four telephone poles form a square that is 48 ft on each side, with the intersection of Blakely Rd and Rt 7 in the center. A traffic light is to be hung from a wire running diagonally from two of the poles. How long should the wire be? [8.6]

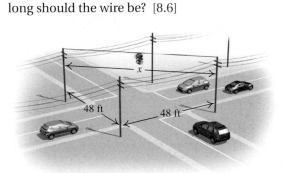

70. Find the distance between the points $(-1, 5)$ and $(-4, 9)$. [8.6]

Simplify. If an expression does not represent a real number, state this. [8.7]

71. $\sqrt[5]{-32}$

72. $\sqrt[4]{-625}$

73. $\sqrt[3]{\dfrac{8}{27}}$

74. $\sqrt[3]{1000}$

Simplify. [8.7]

75. $81^{1/2}$

76. $100^{-1/2}$

77. $25^{3/2}$

78. $32^{-4/5}$

Synthesis

79. When are absolute-value signs necessary for simplifying radical expressions? [8.1]

80. Why should you simplify each term in a radical expression before attempting to combine like radical terms? [8.4]

81. Simplify: $\sqrt{\sqrt{\sqrt{256}}}$. [8.1]

82. Solve: $\sqrt{x^2} = -10$. [8.1]

83. Use square roots to factor $x^2 - 5$. [8.4]

84. Solve $A = \sqrt{a^2 + b^2}$ for b. [8.5]

Test: Chapter 8

Step-by-step test solutions are found on the video CD in the front of this book.

1. Find the square roots of 49.

Simplify.

2. $\sqrt{16}$

3. $-\sqrt{25}$

4. Identify the radicand in $3t\sqrt{t^2 + 1}$.

Determine whether each square root is rational or irrational.

5. $\sqrt{44}$

6. $\sqrt{49}$

Approximate using a calculator. Round to three decimal places.

7. $\sqrt{2}$

8. $\sqrt{50}$

Simplify. Assume that $a, y \geq 0$.

9. $\sqrt{a^2}$

10. $\sqrt{64y^2}$

Simplify by factoring. For Exercises 11–29, assume that all variables represent nonnegative numbers.

11. $\sqrt{60}$

12. $\sqrt{27x^6}$

13. $\sqrt{36t^{11}}$

Perform the indicated operation and, if possible, simplify.

14. $\sqrt{5}\sqrt{6}$

15. $\sqrt{5}\sqrt{15}$

16. $\sqrt{7x}\sqrt{2y}$

17. $\sqrt{2t}\sqrt{8t}$

18. $\sqrt{3ab}\sqrt{6ab^3}$

19. $\dfrac{\sqrt{28}}{\sqrt{63}}$

20. $\dfrac{\sqrt{35x}}{\sqrt{80xy^2}}$

21. $\sqrt{\dfrac{27}{12}}$

22. $\dfrac{\sqrt{144}}{a^2}$

23. $3\sqrt{18} - 5\sqrt{18}$

24. $\sqrt{27} + 2\sqrt{12}$

25. $(4 - \sqrt{5})^2$

26. $(4 - \sqrt{5})(4 + \sqrt{5})$

Rationalize each denominator to form an equivalent expression.

27. $\sqrt{\dfrac{2}{5}}$

28. $\dfrac{2x}{\sqrt{y}}$

29. $\dfrac{10}{4 - \sqrt{5}}$

30. Find the distance between the points $(1, -3)$ and $(4, -1)$.

31. *Guy wires.* One wire steadying a radio tower stretches from a point 100 ft high on the tower to a point on the ground 25 ft from the base of the tower. How long is the wire?

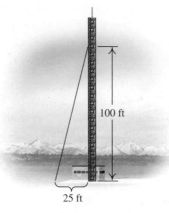

100 ft

25 ft

Solve.

32. $\sqrt{2x} - 3 = 7$

33. $\sqrt{6x + 13} = x + 3$

34. Valerie calculates that she can see 247.49 km to the horizon from an airplane window. How high is the airplane? Use the formula $V = 3.5\sqrt{h}$, where h is the altitude, in meters, and V is the distance to the horizon, in kilometers.

Simplify. If an expression does not represent a real number, state this.

35. $\sqrt[4]{16}$

36. $-\sqrt[6]{1}$

37. $\sqrt[3]{-64}$

38. $\sqrt[4]{-81}$

39. $9^{1/2}$

40. $27^{-1/3}$

41. $100^{3/2}$

42. $16^{-5/4}$

Synthesis

43. Solve: $\sqrt{1 - x} + 1 = \sqrt{6 - x}$.

44. Simplify: $\sqrt{y^{16n}}$.

Cumulative Review: Chapters 1–8

Simplify.

1. $-\frac{1}{2} - \frac{1}{3}$ [1.6]

2. $24 - 2^3 \div 2 \cdot 4 + 3$ [1.8]

3. $(-2x^5)^3$ [4.1]

4. $\left(\frac{a}{b}\right)^{-1}$ [4.8]

5. $\dfrac{6a^{-2}}{8a}$ [4.8]

6. $\dfrac{2x^2 + 3x - 2}{x^2 - 4}$ [6.1]

7. $\sqrt[3]{-27}$ [8.7]

8. $\sqrt{121}$ [8.1]

9. Determine the degree of the polynomial
$$-2x^4 + 3x^2 - x + 7. \quad [4.2]$$

Perform the indicated operation and, if possible, simplify.

10. $(2x^3 - x - 5) - (5x^3 + 3x^2 - x)$ [4.3]

11. $-5x^2(3x^3 - x^2 + 2x)$ [4.4]

12. $(8t + 3)(8t - 3)$ [4.5]

13. $(9x^2 + 1)^2$ [4.5]

14. $(x^3 - x^2 + 2) \div (x - 1)$ [4.7]

15. $\dfrac{x^2 - 1}{x - 2} \cdot \dfrac{6x - 12}{x + 1}$ [6.2]

16. $\dfrac{3}{t - 2} + \dfrac{5}{t}$ [6.4]

17. $\dfrac{2x^2 - x}{x^2 + 2x + 1} \div \dfrac{4x^3 + 4x^2}{2x^2 + x - 2}$ [6.2]

18. $\sqrt{10} \cdot \sqrt{15}$ [8.2]

19. $\sqrt{\dfrac{4x^2}{81}}$ (Assume $x \geq 0$.) [8.3]

20. $(3 - \sqrt{7})(3 + \sqrt{7})$ [8.4]

Factor completely.

21. $40x^2 - 90$ [5.4]

22. $t^2 - 11t + 10$ [5.2]

23. $6z^3 - 8z^2 + 10z$ [5.1]

24. $6x^2 - 13x + 2$ [5.3]

25. $5a^2b^2 + 10ab + 5$ [5.4]

26. $16 - t^4$ [5.4]

Graph.

27. $y = 3x - 2$ [3.6]

28. $y = \frac{1}{3}x$ [3.2]

29. $3x - 2y = 12$ [3.3]

30. $y = -3$ [3.3]

31. $x - 2y < 2$ [7.5]

32. $x + y \leq 1$,
$x - y \geq 2$ [7.6]

33. In which quadrant is the point $\left(-\frac{1}{2}, \frac{1}{3}\right)$ located? [3.1]

34. Find the slope of the line containing the points $(2, 5)$ and $(-1, 6)$. [3.5]

35. Find the slope of the line given by $y = 5x - 3$. [3.6]

36. Write the slope–intercept equation of the line with slope $-\frac{1}{2}$ and y-intercept $(0, -1)$. [3.6]

Solve.

37. $3(x + 5) = 5(x - 1)$ [2.2]

38. $t^2 - t = 6$ [5.6]

39. $5 - 2n \geq 1 - (n - 1)$ [2.6]

40. $\sqrt{2x - 5} = 4$ [8.5]

41. $x + y = 5$,
$x - y = -3$ [7.3]

42. $2x - 5 = y$,
$x - y = 7$ [7.2]

43. $6x^2 = 24$ [5.6]

44. $\dfrac{1}{2} - \dfrac{1}{2t} = \dfrac{3}{t}$ [6.6]

45. $\sqrt{x + 2} - 2 = x$ [8.5]

46. $6y^2 + 1 = 5y$ [5.6]

47. $\dfrac{1}{x + 1} - \dfrac{9}{4x} = \dfrac{1}{5}$ [6.6]

48. $3x - 2(5x - 1) = 5 - 4x$ [2.2]

Solve each formula for the given letter.

49. $a = 2bc$, for b [2.3]

50. $z = \dfrac{x + y}{2}$, for x [2.3]

Solve.

51. *Music downloads.* It took Ty 48 min to download two movies. The second movie took three times as long to download as the first movie. How long did it take to download each movie? [2.5]

52. *Amusement parks.* In 2005, about 176 million people visited North America's 50 most popular amusement parks. This was a 4.2% increase over the number of visitors in 2004. How many people visited these amusement parks in 2004? [2.4]
Source: *Deseret News*, 12/27/2005

53. *Metallurgy.* Alan stocks two alloys that are different purities of gold. The first is three-fourths pure gold and the second is five-twelfths pure gold. How many ounces of each should be combined in order to obtain a 60-oz mixture that is two-thirds pure gold? [7.4]

54. *Construction.* Cindi is designing rafters for a house. The rise of the rafter will be 4 ft and the run 12 ft. How long is the rafter length? [8.6]

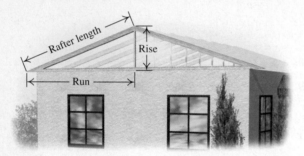

55. *Construction.* What is the slope of the roof described in Exercise 54? [3.5]

56. *Construction.* By checking work records, Cindi finds that Brady can roof her house in 20 hr. Devin can roof Cindi's house in 16 hr. How long would it take if they worked together? [6.7]

57. *Driving distances.* A car leaves Valley Heights traveling west at a speed of 56 km/h. Another car leaves Valley Heights one hour later, traveling the same route at 84 km/h. How far from Valley Heights will the second car overtake the first? [7.3]

58. *Rectangle dimensions.* The length of a rectangle is 1 cm less than three times the width. The area of the rectangle is 24 cm^2. Find the dimensions of the rectangle. [5.7]

Synthesis

59. Solve: $2|n| - 5 = 11$. [2.2]

60. Solve: $(x^2 - 9)(x^2 + x - 20) = 0$. [5.6]

61. Find all x-values for which $\dfrac{x + 1}{\dfrac{x}{x - 3}}{x + 4}$ is undefined. [6.5]

62. Simplify: $d^{1/3}\, d^{5/6}$. [8.7]

Quadratic Equations

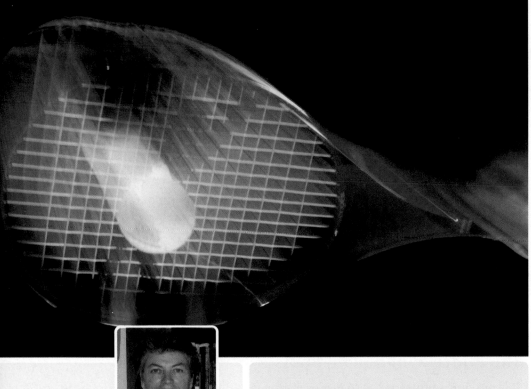

MICHELE M. HOWARD
MASTER RACQUET
TECHNICIAN, USTPA
Kyle, Texas

As a master racquet technician, I am often asked to customize tennis rackets. This may involve knowing just how many grams of weight to add to different areas of the racket to get the right proportion of weight distribution. The "sweet spot" can change depending on how much weight is added to a racket and where. Using math concepts is an integral part of my profession.

AN APPLICATION

The "sweet spot," or center of percussion, of a tennis racquet is located q centimeters from the hand, where $q = \dfrac{I}{mr}$ and I is the swing weight of the racquet, in kg · cm², m is the mass of the racquet, in kilograms, and r is the distance from the hand to the center of mass, in centimeters. Solve for m.

Source: www.racquetresearch.com

This problem appears as Example 1 in Section 9.4.

535

uadratic equations first appeared in Section 5.6. At that time, we used the principle of zero products because all of the equations could be solved by factoring. In this chapter, we develop methods for solving *any* quadratic equation. These methods are then used in applications and in graphing.

9.1   Solving Quadratic Equations: The Principle of Square Roots

The Principle of Square Roots   •   Solving Quadratic Equations of the Type $(x + k)^2 = p$

The following are examples of quadratic equations:

$$x^2 - 7x + 9 = 0, \qquad 5t^2 - 4t = 8, \qquad 6y^2 = -9y, \qquad m^2 = 49.$$

We saw in Chapter 5 that one way to solve an equation like $m^2 = 49$ is to subtract 49 from both sides, factor, and then use the principle of zero products:

$$m^2 - 49 = 0$$
$$(m + 7)(m - 7) = 0$$
$$m + 7 = 0 \quad or \quad m - 7 = 0$$
$$m = -7 \quad or \qquad m = 7.$$

Another equation-solving principle, the *principle of square roots*, allows us to solve equations like $m^2 = 49$ without factoring.

The Principle of Square Roots

It is possible to solve $m^2 = 49$ just by noting that m must be a square root of 49, namely, -7 or 7. This approach was used in Section 8.6, although there we used only the positive square root since we were dealing with length and distance.

> ### The Principle of Square Roots
> For any nonnegative real number p,
> $$x^2 = p \quad \text{is equivalent to} \quad x = \sqrt{p} \text{ or } x = -\sqrt{p}.$$

The notation $x = \pm\sqrt{p}$ is often used to indicate $x = \sqrt{p}$ or $x = -\sqrt{p}$.

EXAMPLE   1   Solve: $x^2 = 16$.

SOLUTION   We use the principle of square roots:

$$x^2 = 16$$
$$x = \sqrt{16} \quad or \quad x = -\sqrt{16} \qquad \text{Using the principle of square roots}$$
$$x = 4 \qquad or \quad x = -4. \qquad \text{Simplifying}$$

We check mentally that $4^2 = 16$ and $(-4)^2 = 16$. The solutions are 4 and -4.

▶ TRY EXERCISE   5

Unlike the principle of zero products, the principle of square roots can be used to solve quadratic equations that have irrational solutions.

EXAMPLE **2**

Solve: **(a)** $x^2 = 17$; **(b)** $5t^2 = 15$; **(c)** $-3x^2 + 7 = 0$.

SOLUTION

a)
$$x^2 = 17$$
$$x = \sqrt{17} \quad or \quad x = -\sqrt{17} \qquad \text{Using the principle of square roots}$$

Check: For $\sqrt{17}$:

$$\frac{x^2 = 17}{(\sqrt{17})^2 \mid 17}$$
$$17 \stackrel{?}{=} 17 \quad \text{TRUE}$$

For $-\sqrt{17}$:

$$\frac{x^2 = 17}{(-\sqrt{17})^2 \mid 17}$$
$$17 \stackrel{?}{=} 17 \quad \text{TRUE}$$

The solutions are $\sqrt{17}$ and $-\sqrt{17}$.

b)
$$5t^2 = 15$$
$$t^2 = 3 \qquad \text{Dividing both sides by 5 to isolate } t^2$$
$$t = \sqrt{3} \quad or \quad t = -\sqrt{3} \qquad \text{Using the principle of square roots}$$

We leave the check to the student. The solutions are $\sqrt{3}$ and $-\sqrt{3}$.

c)
$$-3x^2 + 7 = 0$$
$$7 = 3x^2 \qquad \text{Adding } 3x^2 \text{ to both sides}$$
$$\frac{7}{3} = x^2 \qquad \text{Dividing both sides by 3 to isolate } x^2$$
$$x = \sqrt{\frac{7}{3}} \quad or \quad x = -\sqrt{\frac{7}{3}} \qquad \text{Using the principle of square roots}$$

(These answers can also be written $\dfrac{\sqrt{21}}{3}$ and $-\dfrac{\sqrt{21}}{3}$ by rationalizing denominators.)

Check: For $\sqrt{\frac{7}{3}}$:

$$\frac{-3x^2 + 7 = 0}{-3\left(\sqrt{\frac{7}{3}}\right)^2 + 7 \mid 0}$$
$$-3 \cdot \frac{7}{3} + 7$$
$$-7 + 7$$
$$0 \stackrel{?}{=} 0 \quad \text{TRUE}$$

For $-\sqrt{\frac{7}{3}}$:

$$\frac{-3x^2 + 7 = 0}{-3\left(-\sqrt{\frac{7}{3}}\right)^2 + 7 \mid 0}$$
$$-3 \cdot \frac{7}{3} + 7$$
$$-7 + 7$$
$$0 \stackrel{?}{=} 0 \quad \text{TRUE}$$

The solutions of $-3x^2 + 7 = 0$ are $\sqrt{\frac{7}{3}}$ and $-\sqrt{\frac{7}{3}}$.

TRY EXERCISE 13

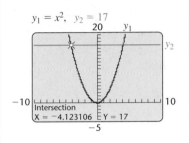
Solving Quadratic Equations of the Type $(x + k)^2 = p$

Equations like $(x - 5)^2 = 9$ or $(t + 2)^2 = 7$ are of the form $(x + k)^2 = p$. The principle of square roots can be used to solve such equations.

EXAMPLE **3**

Solve: **(a)** $(x - 5)^2 = 9$; **(b)** $(t + 2)^2 = 7$.

SOLUTION

a)
$$(x - 5)^2 = 9$$
$$x - 5 = 3 \quad or \quad x - 5 = -3 \qquad \begin{array}{l}\text{Using the principle of square roots:}\\ \text{If } A^2 = 9, \text{ then } A = 3 \text{ or } A = -3.\end{array}$$
$$x = 8 \quad or \qquad x = 2 \qquad \text{Adding 5 to both sides}$$

The solutions are 8 and 2. We leave the check to the student.

b) $(t + 2)^2 = 7$

$t + 2 = \sqrt{7}$ *or* $t + 2 = -\sqrt{7}$ Using the principle of square roots

$t = -2 + \sqrt{7}$ *or* $t = -2 - \sqrt{7}$ Adding -2 to both sides

Check: For $-2 + \sqrt{7}$:

$$\frac{(t + 2)^2 = 7}{(-2 + \sqrt{7} + 2)^2 \,\Big|\, 7}$$
$$(\sqrt{7})^2 \,\Big|$$
$$7 \overset{?}{=} 7 \quad \text{TRUE}$$

For $-2 - \sqrt{7}$:

$$\frac{(t + 2)^2 = 7}{(-2 - \sqrt{7} + 2)^2 \,\Big|\, 7}$$
$$(-\sqrt{7})^2 \,\Big|$$
$$7 \overset{?}{=} 7 \quad \text{TRUE}$$

The solutions are $-2 + \sqrt{7}$ and $-2 - \sqrt{7}$, or simply $-2 \pm \sqrt{7}$ (read "-2 plus or minus $\sqrt{7}$").

TRY EXERCISE 21

In Example 3, the left sides of the equations are squares of binomials. Sometimes factoring can be used to express an equation in that form.

EXAMPLE **4** Solve by factoring and using the principle of square roots.

a) $x^2 + 8x + 16 = 49$ **b)** $x^2 - 6x + 9 = 10$

SOLUTION

a) $x^2 + 8x + 16 = 49$ The left side is a perfect-square trinomial.

$(x + 4)^2 = 49$ Factoring

$x + 4 = 7$ *or* $x + 4 = -7$ Using the principle of square roots

$x = 3$ *or* $x = -11$

The solutions are 3 and -11. We leave the check to the student.

b) $x^2 - 6x + 9 = 10$ The left side is a perfect-square trinomial.

$(x - 3)^2 = 10$ Factoring

$x - 3 = \sqrt{10}$ *or* $x - 3 = -\sqrt{10}$ Using the principle of square roots

$x = 3 + \sqrt{10}$ *or* $x = 3 - \sqrt{10}$

The solutions are $3 + \sqrt{10}$ and $3 - \sqrt{10}$, or simply $3 \pm \sqrt{10}$. We leave the check to the student.

TRY EXERCISE 39

9.1 EXERCISE SET

For Extra Help
MyMathLab   Math XL PRACTICE   WATCH   DOWNLOAD

Concept Reinforcement *Classify each statement as either true or false.*

1. To solve $3x^2 - 2 = 5$ using the principle of square roots, we must first isolate x^2.

2. The notation $x = \pm\sqrt{5}$ means $x = -\sqrt{5}$ *or* $x = \sqrt{5}$.

3. The principle of square roots can be used only to find rational solutions.

4. The principle of square roots cannot be used when factoring is possible.

Solve. Use the principle of square roots.

5. $t^2 = 81$

6. $n^2 = 64$

7. $x^2 = 1$

8. $x^2 = 100$

9. $a^2 = 11$

10. $t^2 = 15$

11. $10x^2 = 40$

12. $5x^2 = 45$

13. $3t^2 = 6$

14. $7t^2 = 21$

15. $4 - 9x^2 = 0$

16. $25 - 4a^2 = 0$

Aha! **17.** $12y^2 + 1 = 1$

18. $4y^2 - 3 = 9$

19. $15x^2 - 25 = 0$

20. $4x^2 - 14 = 0$

21. $(x - 1)^2 = 49$

22. $(x - 2)^2 = 25$

23. $(t + 6)^2 = 4$

24. $(t + 9)^2 = 100$

25. $(m + 3)^2 = 6$

26. $(m - 4)^2 = 21$

Aha! **27.** $(a - 7)^2 = 0$

28. $(a + 12)^2 = 81$

29. $(5 - x)^2 - 14$

30. $(7 - x)^2 = 12$

31. $(t + 1)^2 = 1$

32. $(x - 5)^2 = 25$

33. $\left(y - \frac{3}{4}\right)^2 = \frac{17}{16}$

34. $\left(x + \frac{3}{2}\right)^2 = \frac{13}{4}$

35. $x^2 - 10x + 25 = 100$

36. $x^2 - 6x + 9 = 64$

37. $p^2 + 8p + 16 = 1$

38. $y^2 + 14y + 49 = 4$

39. $t^2 - 16t + 64 = 7$

40. $m^2 - 2m + 1 = 5$

41. $x^2 + 12x + 36 = 18$

42. $x^2 + 4x + 4 = 12$

43. Under what conditions is it easier to use the principle of square roots rather than the principle of zero products to solve a quadratic equation?

44. Under what conditions is it easier to use the principle of zero products rather than the principle of square roots to solve a quadratic equation?

Skill Review

To prepare for Section 9.2, review perfect-square trinomials (Sections 4.5 and 5.4).

Multiply. [4.5]

45. $(x - 3)^2$

46. $(x + a)^2$

47. $\left(x + \frac{1}{2}\right)^2$

48. $\left(x - \frac{3}{2}\right)^2$

Factor. [5.4]

49. $x^2 + 2x + 1$

50. $x^2 + 10x + 25$

51. $x^2 - 4x + 4$

52. $x^2 - 20x + 100$

Synthesis

53. Under what conditions does a quadratic equation have only one solution?

54. Is it possible for a quadratic equation with rational coefficients to have $5 + \sqrt{2}$ as a solution, but not $5 - \sqrt{2}$? Why or why not?

Factor the left side of each equation. Then solve.

55. $x^2 - 5x + \frac{25}{4} = \frac{13}{4}$

56. $x^2 + \frac{7}{3}x + \frac{49}{36} = \frac{7}{36}$

57. $t^2 + 3t + \frac{9}{4} = \frac{49}{4}$

58. $m^2 - \frac{3}{2}m + \frac{9}{16} = \frac{17}{16}$

59. $x^2 + 2.5x + 1.5625 = 9.61$

60. $a^2 - 3.8a + 3.61 = 27.04$

Use the graph of

$$y = (x + 3)^2$$

to solve each equation.

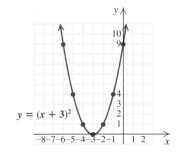

61. $(x + 3)^2 = 1$

62. $(x + 3)^2 = 4$

63. $(x + 3)^2 = 9$

64. $(x + 3)^2 = 0$

65. *Gravitational force.* Newton's law of gravitation states that the gravitational force f between objects of mass M and m, at a distance d from each other, is given by

$$f = \frac{kMm}{d^2},$$

where k is a constant. Solve for d.

CORNER

How Big is Half a Pizza?

Focus: Principle of square roots; formulas

Time: 15–25 minutes

Group size: 3

Materials: Calculators are optional.

Frankie, Johnnie, & Luigi, Too! has the best pizza in the Palo Alto area, according to a poll of Stanford University students. Pizzas there have 12-in., 14-in., and 16-in. diameters.

ACTIVITY

1. Let each member of the group play the role of either Frankie, Johnnie, or Luigi. As part of a promotion, Frankie is offering a "mini" pizza pie that has half the area of a 12-in. pie, Johnnie is offering a "personal" pie that has half the area of a 14-in. pie, and Luigi is offering a "junior" pie that has half the area of a 16-in. pie. Each group member, according to the selected role, should calculate the area of both the original pie and the new offering.

2. Each group member, according to the chosen role, should calculate the radius and then the diameter of the "new" pie. Summarize the group's findings in a table similar to the one below.

	Diameter of Original Pie	Area of Original Pie	Area of New Pie	Radius of New Pie	Diameter of New Pie
Frankie					
Johnnie					
Luigi					

3. After checking each other's work, group members should develop a formula that can be used to determine the diameter of a circle that has half the area of a given circle. That is, if a circle has diameter d, what is the diameter of a circle with half the area?

4. If the diameter of one pizza is twice the diameter of another pizza, should the pizza cost twice as much? Why or why not?

9.2   # Solving Quadratic Equations: Completing the Square

Completing the Square   ■   Solving by Completing the Square

In Section 9.1, we solved equations like $(x - 5)^2 = 7$ using the principle of square roots. Equations like $x^2 + 8x + 16 = 12$ were similarly solved because the left side of the equation is a perfect-square trinomial. We now learn to solve equations like $x^2 - 8x = 2$, in which the left side is not (yet!) a perfect-square trinomial. The new procedure involves *completing the square* and enables us to solve *any* quadratic equation.

Completing the Square

Recall that

$$(x + 3)^2 = (x + 3)(x + 3)$$
$$= x^2 + 3x + 3x + 9$$
$$= x^2 + 6x + 9 \qquad \text{This is a perfect-square trinomial.}$$

and, in general,

$$(x + a)^2 = x^2 + 2ax + a^2. \qquad \text{This is also a perfect-square trinomial.}$$

In $x^2 + 6x + 9$, note that 9 is the square of half of the coefficient of x: $\frac{1}{2} \cdot 6 = 3$, and $3^2 = 9$. Similarly, in $x^2 + 2ax + a^2$, note that a^2 is also the square of half of the coefficient of x: $\frac{1}{2} \cdot 2a = a$ and $a \cdot a = a^2$.

Consider the quadratic equation

$$x^2 + 10x = 4.$$

We would like to add a number to both sides that will make the left side a perfect-square trinomial. Such a number is described above as the square of half of the coefficient of x: $\frac{1}{2} \cdot 10 = 5$, and $5^2 = 25$. Thus we add 25 to both sides:

$$x^2 + 10x = 4 \qquad \textit{Think:} \text{ Half of 10 is 5; } 5^2 = 25.$$
$$x^2 + 10x + 25 = 4 + 25 \qquad \text{Adding 25 to both sides}$$
$$(x + 5)^2 = 29. \qquad \text{Factoring the perfect-square trinomial}$$

By adding 25 to $x^2 + 10x$, we have *completed the square*. The resulting equation contains the square of a binomial on one side. Solutions can then be found using the principle of square roots, as in Section 9.1:

$$(x + 5)^2 = 29$$
$$x + 5 = \sqrt{29} \qquad or \quad x + 5 = -\sqrt{29} \qquad \text{Using the principle of square roots}$$
$$x = -5 + \sqrt{29} \quad or \qquad x = -5 - \sqrt{29}.$$

The solutions are $-5 \pm \sqrt{29}$.

Completing the Square

To *complete the square* for an expression like $x^2 + bx$, add half of the coefficient of x, squared. That is, add $(b/2)^2$.

A visual interpretation of completing the square is sometimes helpful.

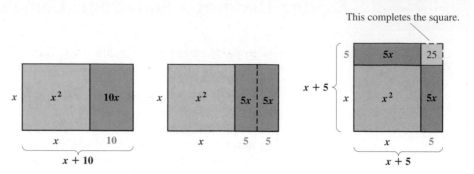

This completes the square.

In each figure above, the sum of the pink area and the purple area is $x^2 + 10x$. However, by splitting the purple area in half, we can "complete" a square by adding the blue area. The blue area is $5 \cdot 5$, or 25 square units.

CAUTION!   Completing the square does *not* produce an equivalent expression. It will be used with the addition principle to produce equivalent equations.

EXAMPLE 1

For each expression, find the number that will complete the square. Check by multiplying.

a) $x^2 - 12x$ **b)** $x^2 + 5x$

STUDENT NOTES

Keep in mind that each time we complete the square, as in Example 1, we are forming a new expression that is *not equivalent* to the original expression.

SOLUTION

a) To complete the square for $x^2 - 12x$, note that the coefficient of x is -12. Half of -12 is -6 and $(-6)^2$ is 36. Thus, $x^2 - 12x$ becomes a perfect-square trinomial when 36 is added:

$$x^2 - 12x + 36 \quad \text{is the square of} \quad x - 6.$$

The number 36 completes the square.

Check: $(x - 6)^2 = (x - 6)(x - 6) = x^2 - 6x - 6x + 36 = x^2 - 12x + 36.$

b) To complete the square for $x^2 + 5x$, we take half of the coefficient of x and square it:

$$\left(\tfrac{5}{2}\right)^2 = \tfrac{25}{4}. \qquad \text{Half of 5 is } \tfrac{5}{2}; \left(\tfrac{5}{2}\right)^2 = \tfrac{5}{2} \cdot \tfrac{5}{2} = \tfrac{25}{4}.$$

Thus, $x^2 + 5x + \tfrac{25}{4}$ is the square of $x + \tfrac{5}{2}$. The number $\tfrac{25}{4}$ completes the square.

Check: $\left(x + \tfrac{5}{2}\right)^2 = \left(x + \tfrac{5}{2}\right)\left(x + \tfrac{5}{2}\right) = x^2 + \tfrac{5}{2}x + \tfrac{5}{2}x + \tfrac{25}{4} = x^2 + 5x + \tfrac{25}{4}.$

TRY EXERCISE 7

CAUTION!   In Example 1, we are neither solving an equation nor writing an equivalent expression. Instead, we are learning how to find the number that completes the square. In Examples 2 and 3, we use the number that completes the square, along with the addition principle, to solve equations.

Solving by Completing the Square

The concept of completing the square can now be used to solve equations like $x^2 + 10x = 4$ much as we did on p. 541.

EXAMPLE 2   Solve by completing the square.

a) $x^2 + 6x = -8$ **b)** $x^2 - 10x + 14 = 0$

SOLUTION

a) To solve $x^2 + 6x = -8$, we take half of 6 and square it, to get 9. Then we add 9 to both sides of the equation. This makes the left side the square of a binomial:

$$x^2 + 6x + 9 = -8 + 9$$ Adding 9 to both sides to complete the square on the left side

$$(x + 3)^2 = 1$$ Factoring

$$x + 3 = 1 \quad or \quad x + 3 = -1$$ Using the principle of square roots. If $A^2 = 1$, then $A = 1$ or $A = -1$.

$$x = -2 \quad or \quad x = -4.$$

The solutions are -2 and -4. The check is left to the student.

b) We have

$$x^2 - 10x + 14 = 0$$

$$x^2 - 10x = -14$$ Subtracting 14 from both sides in preparation for completing the square

$$x^2 - 10x + 25 = -14 + 25$$ Adding 25 to both sides to complete the square: $(-10/2)^2 = 25$

$$(x - 5)^2 = 11$$ Factoring

$$x - 5 = \sqrt{11} \quad or \quad x - 5 = -\sqrt{11}$$ Using the principle of square roots

$$x = 5 + \sqrt{11} \quad or \quad x = 5 - \sqrt{11}.$$

The solutions are $5 + \sqrt{11}$ and $5 - \sqrt{11}$, or simply $5 \pm \sqrt{11}$. The check is left to the student.

TRY EXERCISE 21

To complete the square, the coefficient of x^2 must be 1. When the x^2 coefficient is not 1, we can multiply or divide on both sides to find an equivalent equation with an x^2-coefficient of 1.

EXAMPLE 3   Solve by completing the square.

a) $3x^2 + 24x = 3$ **b)** $2x^2 - 5x + 1 = 0$

SOLUTION

a)

$$3x^2 + 24x = 3$$

$$\left. \begin{array}{l} \frac{1}{3}(3x^2 + 24x) = \frac{1}{3} \cdot 3 \\ x^2 + 8x = 1 \end{array} \right\}$$ We multiply by $\frac{1}{3}$ (or divide by 3) on both sides to ensure an x^2-coefficient of 1.

$$x^2 + 8x + 16 = 1 + 16$$ Adding 16 to both sides to complete the square: $\left(\frac{8}{2}\right)^2 = 16$

$$(x + 4)^2 = 17$$ Factoring

$$x + 4 = \sqrt{17} \quad or \quad x + 4 = -\sqrt{17}$$ Using the principle of square roots

$$x = -4 + \sqrt{17} \quad or \quad x = -4 - \sqrt{17}$$

The solutions are $-4 \pm \sqrt{17}$. The check is left to the student.

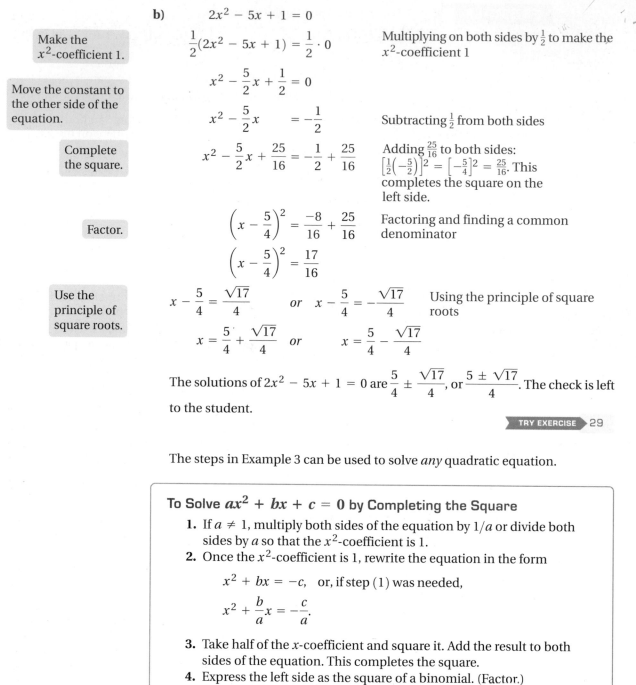

b)

$$2x^2 - 5x + 1 = 0$$

Make the x^2-coefficient 1.

$$\frac{1}{2}(2x^2 - 5x + 1) = \frac{1}{2} \cdot 0$$ Multiplying on both sides by $\frac{1}{2}$ to make the x^2-coefficient 1

$$x^2 - \frac{5}{2}x + \frac{1}{2} = 0$$

Move the constant to the other side of the equation.

$$x^2 - \frac{5}{2}x = -\frac{1}{2}$$ Subtracting $\frac{1}{2}$ from both sides

Complete the square.

$$x^2 - \frac{5}{2}x + \frac{25}{16} = -\frac{1}{2} + \frac{25}{16}$$ Adding $\frac{25}{16}$ to both sides: $\left[\frac{1}{2}\left(-\frac{5}{2}\right)\right]^2 = \left[-\frac{5}{4}\right]^2 = \frac{25}{16}$. This completes the square on the left side.

Factor.

$$\left(x - \frac{5}{4}\right)^2 = \frac{-8}{16} + \frac{25}{16}$$ Factoring and finding a common denominator

$$\left(x - \frac{5}{4}\right)^2 = \frac{17}{16}$$

Use the principle of square roots.

$$x - \frac{5}{4} = \frac{\sqrt{17}}{4} \quad or \quad x - \frac{5}{4} = -\frac{\sqrt{17}}{4}$$ Using the principle of square roots

$$x = \frac{5}{4} + \frac{\sqrt{17}}{4} \quad or \quad x = \frac{5}{4} - \frac{\sqrt{17}}{4}$$

The solutions of $2x^2 - 5x + 1 = 0$ are $\frac{5}{4} \pm \frac{\sqrt{17}}{4}$, or $\frac{5 \pm \sqrt{17}}{4}$. The check is left to the student.

TRY EXERCISE 29

The steps in Example 3 can be used to solve *any* quadratic equation.

To Solve $ax^2 + bx + c = 0$ by Completing the Square

1. If $a \neq 1$, multiply both sides of the equation by $1/a$ or divide both sides by a so that the x^2-coefficient is 1.
2. Once the x^2-coefficient is 1, rewrite the equation in the form

$$x^2 + bx = -c, \quad \text{or, if step (1) was needed,}$$

$$x^2 + \frac{b}{a}x = -\frac{c}{a}.$$

3. Take half of the x-coefficient and square it. Add the result to both sides of the equation. This completes the square.
4. Express the left side as the square of a binomial. (Factor.)
5. Use the principle of square roots and complete the solution.

9.2 EXERCISE SET

↪ Concept Reinforcement *In each of Exercises 1–6, match the equation with the equation from the column on the right to which it is equivalent.*

1. ___ $x^2 + 6x = 2$ **a)** $(x + 3)^2 = 10$

2. ___ $x^2 - 6x + 9 = 10$ **b)** $x^2 - 6x + 9 = 11$

3. ___ $x^2 + 6x + 9 = 10$ **c)** $x^2 + 6x + 9 = 11$

4. ___ $x^2 - 6x = 2$ **d)** $(x + 4)^2 = 18$

5. ___ $x^2 + 8x = 2$ **e)** $(x - 3)^2 = 10$

6. ___ $x^2 - 8x = 2$ **f)** $(x - 4)^2 = 18$

Determine the number that will complete the square. Check by multiplying.

7. $x^2 + 8x$ **8.** $x^2 + 4x$ **9.** $x^2 - 2x$

10. $x^2 - 20x$ **11.** $x^2 - 3x$ **12.** $x^2 - 9x$

13. $t^2 + t$ **14.** $a^2 - a$ **15.** $x^2 + \frac{3}{2}x$

16. $x^2 + \frac{2}{3}x$ **17.** $m^2 - \frac{8}{3}m$ **18.** $y^2 + \frac{3}{4}y$

Solve by completing the square.

19. $x^2 + 7x + 10 = 0$ **20.** $x^2 - 3x - 10 = 0$

21. $x^2 - 24x + 21 = 0$ **22.** $x^2 - 10x + 20 = 0$

23. $t^2 + 10t + 12 = 0$ **24.** $t^2 - 16t + 50 = 0$

25. $2x^2 - 8x = 14$ **26.** $3x^2 + 6x = 18$

27. $x^2 + 3x - 3 = 0$ **28.** $x^2 - x - 3 = 0$

29. $2t^2 + t - 1 = 0$ **30.** $3t^2 + t - 2 = 0$

31. $x^2 - \frac{3}{2}x - 2 = 0$ **32.** $x^2 + \frac{5}{2}x - 2 = 0$

33. $2x^2 - 5x - 10 = 0$ **34.** $2x^2 - 7x - 10 = 0$

35. $3t^2 + 4t - 5 = 0$ **36.** $3t^2 - 8t + 1 = 0$

37. $2x^2 = 5 + 9x$ **38.** $3x^2 = 22x - 7$

39. $6x^2 + 11x = 10$ **40.** $4x^2 + 12x = 7$

41. How does completing the square allow us to solve equations that we could not otherwise have solved?

42. Explain how the addition principle, the multiplication principle, and the square-root principle were used in this section.

Skill Review

To prepare for Section 9.3, review evaluating algebraic expressions and simplifying radical expressions (Sections 1.8 and 8.4).

Evaluate. [1.8]

43. $-b$, for $b = 7$ **44.** $-b$, for $b = -7$

45. $b^2 - 4ac$, for $a = 2$, $b = 6$, and $c = 3$

46. $b^2 - 4ac$, for $a = 3$, $b = -1$, and $c = -1$

Simplify. [8.4]

47. $\dfrac{2(1 - \sqrt{5})}{2}$ **48.** $\dfrac{4(3 + \sqrt{7})}{2}$

49. $\dfrac{24 - 3\sqrt{5}}{9}$ **50.** $\dfrac{35 - 7\sqrt{6}}{14}$

Synthesis

51. Sal states that "since solving a quadratic equation by completing the square relies on the principle of square roots, the solutions are always opposites of each other." Is Sal correct? Why or why not?

52. When completing the square, what determines if the number being added is a whole number or a fraction?

Find b such that each trinomial is a square.

53. $x^2 + bx + 25$ **54.** $x^2 + bx + 81$

55. $x^2 + bx + 45$ **56.** $x^2 + bx + 50$

57. $4x^2 + bx + 16$ **58.** $x^2 - bx + 48$

Solve each of the following by letting y_1 represent the left side of each equation, letting y_2 represent the right side, and graphing y_1 and y_2 on the same set of axes. INTERSECT *can then be used to determine the x-coordinate at any point of intersection. Find solutions accurate to two decimal places.*

59. $(x + 4)^2 = 13$ **60.** $(x + 6)^2 = 2$

61. $x^2 - 24x + 21 = 0$ **62.** $x^2 + 3x - 3 = 0$
   (see Exercise 21)   (see Exercise 27)

63. $6x^2 + 11x = 10$ **64.** $2x^2 = 5 + 9x$
   (see Exercise 39)   (see Exercise 37)

Aha! 65. What is the best way to solve $x^2 + 8x = 0$? Why?

CORNER

Using Areas to Complete the Square

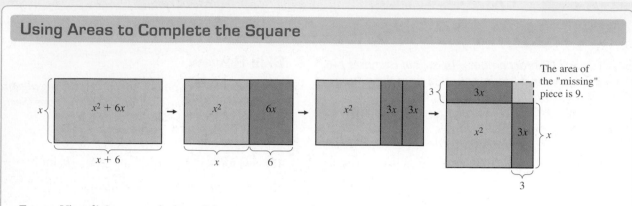

The area of the "missing" piece is 9.

Focus: Visualizing completion of the square

Time: 15–25 minutes

Group size: 2

Materials: Rulers and graph paper may be helpful.

To draw a representation of completing the square, we use areas and the fact that the area of any rectangle is given by multiplying the length and the width. For example, the above sequence of figures can be drawn to explain why 9 completes the square for $x^2 + 6x$.

ACTIVITY

1. Draw a sequence of four figures, similar to those shown above, to complete the square for $x^2 + 8x$. Group members should take turns, so that each person draws and labels two of the figures.

2. Repeat part (1) to complete the square for $x^2 + 14x$. The person who drew the first drawing in part (1) should take the second turn this time.

3. When we add the area of the missing piece, we increase the original area. For this reason, to complete the square we use the addition principle, adding the "missing" number to both sides to form an equivalent *equation*. Use the work in parts (1) and (2) to solve the equations $x^2 + 8x = 9$ and $x^2 + 14x = 15$.

4. Each equation in part (3) has two solutions. Can both be represented geometrically? Why or why not?

9.3   The Quadratic Formula and Applications

The Quadratic Formula ▪ Problem Solving

When mathematicians use a procedure repeatedly, they often try to find a formula for the procedure. The *quadratic formula* condenses into one calculation the many steps used to solve a quadratic equation by completing the square.

STUDY SKILLS

Crunch Time for the Final

It is always best to study for a final exam over several days or more. If you have only one or two days of study time, however, begin by studying the formulas, problems, properties, and procedures in each chapter's Study Summary. Then do the exercises in the Cumulative Reviews. Make sure to attend a review session if one is offered.

The Quadratic Formula

Consider a quadratic equation in *standard form*, $ax^2 + bx + c = 0$, with $a > 0$. Our plan is to solve this equation for x by completing the square. As the steps are performed, compare them with those in Example 3(b) on p. 544:

$$ax^2 + bx + c = 0$$

$$\frac{1}{a}(ax^2 + bx + c) = \frac{1}{a} \cdot 0 \qquad \text{Multiplying by } \frac{1}{a} \text{ to make the } x^2\text{-coefficient 1}$$

$$x^2 + \frac{b}{a}x + \frac{c}{a} = 0$$

$$x^2 + \frac{b}{a}x \qquad = -\frac{c}{a} \qquad \text{Adding } -\frac{c}{a} \text{ to both sides}$$

$$x^2 + \frac{b}{a}x + \frac{b^2}{4a^2} = -\frac{c}{a} + \frac{b^2}{4a^2} \qquad \text{Adding } \frac{b^2}{4a^2} \text{ to both sides:}$$

$$\left[\frac{1}{2} \cdot \frac{b}{a}\right]^2 = \left[\frac{b}{2a}\right]^2 = \frac{b^2}{4a^2}$$

$$\left(x + \frac{b}{2a}\right)^2 = -\frac{4ac}{4a^2} + \frac{b^2}{4a^2} \qquad \text{Factoring and finding a common denominator}$$

$$\left(x + \frac{b}{2a}\right)^2 = \frac{b^2 - 4ac}{4a^2}$$

$$x + \frac{b}{2a} = \sqrt{\frac{b^2 - 4ac}{4a^2}} \quad \text{or} \quad x + \frac{b}{2a} = -\sqrt{\frac{b^2 - 4ac}{4a^2}} \qquad \text{Using the principle of square roots}$$

$$x + \frac{b}{2a} = \frac{\sqrt{b^2 - 4ac}}{2a} \quad \text{or} \quad x + \frac{b}{2a} = -\frac{\sqrt{b^2 - 4ac}}{2a} \qquad \sqrt{4a^2} = 2a \text{ because } a > 0.$$

$$x = -\frac{b}{2a} + \frac{\sqrt{b^2 - 4ac}}{2a} \quad \text{or} \quad x = -\frac{b}{2a} - \frac{\sqrt{b^2 - 4ac}}{2a} \qquad \text{Adding } -b/(2a) \text{ to both sides}$$

$$x = -\frac{b}{2a} \pm \frac{\sqrt{b^2 - 4ac}}{2a}, \quad \text{or} \quad x = \frac{-b \pm \sqrt{b^2 - 4ac}}{2a}.$$

This last equation is the result we sought. A similar proof would show that this formula also holds when $a < 0$. Unless $b^2 - 4ac$ is 0, the formula represents two solutions.

> ## The Quadratic Formula
>
> The solutions of $ax^2 + bx + c = 0$, $a \neq 0$, are given by
>
> $$x = \frac{-b \pm \sqrt{b^2 - 4ac}}{2a}.$$

The quadratic formula is so useful that it is worth memorizing.

 **1**

Solve using the quadratic formula.

a) $4x^2 + 5x - 6 = 0$ **b)** $t^2 = 4t + 7$ **c)** $x^2 + x = -1$

STUDENT NOTES

After identifying which numbers to use as a, b, and c, be careful to substitute exactly for the *letters* in the quadratic formula. Using parentheses, as shown in Example 1, can help in this regard. Also, reading "$-b$" as "the opposite of b" will serve as a reminder of how to substitute correctly.

SOLUTION

a) We identify a, b, and c and substitute into the quadratic formula:

$$4x^2 + 5x - 6 = 0;$$
$$\qquad\uparrow\qquad\uparrow\qquad\uparrow$$
$$\qquad a\qquad b\qquad c$$

$$x = \frac{-b \pm \sqrt{b^2 - 4ac}}{2a}$$

$$x = \frac{-(5) \pm \sqrt{(5)^2 - 4(4)(-6)}}{2(4)} \qquad \text{Substituting for } a, b, \text{ and } c$$

> Be sure to write the fraction bar all the way across.

$$x = \frac{-5 \pm \sqrt{25 - (-96)}}{8}$$

$$x = \frac{-5 \pm \sqrt{121}}{8}$$

$$x = \frac{-5 \pm 11}{8}$$

$$x = \frac{-5 + 11}{8} \quad or \quad x = \frac{-5 - 11}{8}$$

$$x = \frac{6}{8} \quad or \quad x = \frac{-16}{8}$$

$$x = \frac{3}{4} \quad or \quad x = -2.$$

The solutions are $\frac{3}{4}$ and -2.

b) We rewrite $t^2 = 4t + 7$ in standard form, identify a, b, and c, and solve using the quadratic formula:

$$1t^2 - 4t - 7 = 0; \qquad \text{Subtracting } 4t + 7 \text{ from both sides}$$
$$\uparrow\qquad\uparrow\qquad\uparrow$$
$$a\qquad b\qquad c$$

$$t = \frac{-(-4) \pm \sqrt{(-4)^2 - 4(1)(-7)}}{2 \cdot 1} \qquad \text{Substituting into the quadratic formula}$$

$$t = \frac{4 \pm \sqrt{16 + 28}}{2} = \frac{4 \pm \sqrt{44}}{2}$$

$$t = \frac{4}{2} \pm \frac{\sqrt{44}}{2}$$

$$\left.\begin{array}{l} t = 2 \pm \dfrac{\sqrt{4}\sqrt{11}}{2} \\[2mm] t = 2 \pm \dfrac{2\sqrt{11}}{2} \end{array}\right\} \quad \text{Simplifying } \sqrt{44}$$

$$t = 2 \pm \sqrt{11}. \qquad \text{Removing a factor equal to 1: } \frac{2}{2} = 1$$

The solutions are $2 + \sqrt{11}$ and $2 - \sqrt{11}$, or $2 \pm \sqrt{11}$.

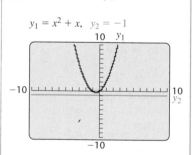

c) We rewrite $x^2 + x = -1$ in standard form and use the quadratic formula:

$$1x^2 + 1x + 1 = 0; \qquad \text{Adding 1 to both sides}$$

$$\overset{\uparrow}{a} \qquad \overset{\uparrow}{b} \qquad \overset{\uparrow}{c}$$

$$x = \frac{-1 \pm \sqrt{1^2 - 4 \cdot 1 \cdot 1}}{2 \cdot 1} \qquad \text{Substituting into the quadratic formula}$$

$$x = \frac{-1 \pm \sqrt{1 - 4}}{2}$$

$$x = \frac{-1 \pm \sqrt{-3}}{2}.$$

Since the radicand, -3, is negative, there are no real-number solutions. In Section 9.5, we will study a number system in which solutions of this equation can be found. For now we simply state, "No real-number solution exists."

> **TRY EXERCISE** 15

Problem Solving

EXAMPLE **2**   *Diagonals in a polygon.*   The number of diagonals d in a polygon that has n sides is given by the formula

$$d = \frac{n^2 - 3n}{2}.$$

If a polygon has 27 diagonals, how many sides does it have?

SOLUTION

1. **Familiarize.** A sketch can help us to become familiar with the problem. We draw a hexagon (6 sides) and count the diagonals. As the formula predicts, for $n = 6$, there are

$$\frac{6^2 - 3 \cdot 6}{2} = \frac{36 - 18}{2}$$

$$= \frac{18}{2} = 9 \text{ diagonals.}$$

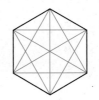

Clearly, the polygon in question must have more than 6 sides. We might suspect that tripling the number of diagonals requires tripling the number of sides. Evaluating the above formula for $n = 18$, you can confirm that this is *not* the case. Rather than continue guessing, we proceed to a translation.

2. **Translate.** Since the number of diagonals is 27, we substitute 27 for d:

$$27 = \frac{n^2 - 3n}{2}.$$

This gives us a translation.

3. **Carry out.** We solve the equation for n, first reversing the equation for convenience:

$$\frac{n^2 - 3n}{2} = 27$$

$$n^2 - 3n = 54 \qquad \text{Multiplying both sides by 2 to clear fractions}$$

$$n^2 - 3n - 54 = 0 \qquad \text{Subtracting 54 from both sides}$$

$$(n - 9)(n + 6) = 0 \qquad \text{Factoring. There is no need for the quadratic formula here.}$$

$$n - 9 = 0 \quad or \quad n + 6 = 0$$

$$n = 9 \quad or \qquad n = -6.$$

4. **Check.** Since the number of sides cannot be negative, -6 cannot be a solution. We leave it to the student to show by substitution that 9 checks.

5. **State.** The polygon has 9 sides (it is a nonagon). **TRY EXERCISE** 45

EXAMPLE **3**

Free-falling objects. The highest residence in the United States is 1110 ft high on the 92nd floor of the John Hancock Center in Chicago, Illinois. How many seconds will it take a golf ball to fall from that height? Round to the nearest hundredth.

SOLUTION

1. **Familiarize.** If we did not know anything about this problem, we might consider looking up a formula in a mathematics or physics book. A formula that fits this situation is

$$s = 16t^2, \qquad \text{It is useful to remember this formula.}$$

where s is the distance, in feet, traveled by a body falling freely from rest in t seconds. This formula is actually an approximation because it does not account for air resistance. In this problem, we know the distance s to be 1110 ft. We want to determine the time t that it takes the golf ball to reach the ground.

2. **Translate.** The distance is 1110 ft and we need to solve for t. We substitute 1110 for s in the formula above to get the following translation:

$$1110 = 16t^2.$$

3. **Carry out.** Because there is no t-term, we can use the principle of square roots to solve:

$$1110 = 16t^2$$

$$\frac{1110}{16} = t^2 \qquad \text{Solving for } t^2$$

$$\sqrt{\frac{1110}{16}} = t \quad or \quad -\sqrt{\frac{1110}{16}} = t \qquad \text{Using the principle of square roots}$$

$$\frac{\sqrt{1110}}{4} = t \quad or \quad \frac{-\sqrt{1110}}{4} = t$$

$$8.33 \approx t \quad or \qquad -8.33 \approx t. \qquad \text{Using a calculator and rounding to the nearest hundredth}$$

4. **Check.** The number -8.33 cannot be a solution because time cannot be negative in this situation. We substitute 8.33 in the original equation:

$$s = 16(8.33)^2 = 16(69.3889) = 1110.2224.$$

This is close. Remember that we approximated a solution.

5. **State.** It takes about 8.33 sec for the golf ball to fall to the ground from the 92nd floor of the John Hancock Center.

TRY EXERCISE 47

EXAMPLE 4

Right triangles. The hypotenuse of a right triangle is 6 m long. One leg is 1 m longer than the other. Find the lengths of the legs. Round to the nearest hundredth.

SOLUTION

1. **Familiarize.** We first make a drawing and label it. We let $s =$ the length, in meters, of one leg. Then $s + 1 =$ the length, in meters, of the other leg.

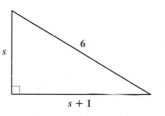

STUDENT NOTES

The a, b, and c in the Pythagorean equation represent sides of the triangle. These are not the same as the a, b, and c used in the quadratic formula.

Because we have a right triangle, the Pythagorean equation $a^2 + b^2 = c^2$ holds. Note that if $s = 3$, then $s + 1 = 4$ and $3^2 + 4^2 = 25 \neq 6^2$. Thus we see that $s \neq 3$. Another guess, $s = 4$, is too big since $4^2 + (4 + 1)^2 = 41 \neq 6^2$. Although we have not guessed the solution, we expect s to be between 3 and 4.

2. **Translate.** To translate, we use the Pythagorean theorem:

$$s^2 + (s + 1)^2 = 6^2.$$

3. **Carry out.** We solve the equation:

$$s^2 + (s + 1)^2 = 6^2$$
$$s^2 + s^2 + 2s + 1 = 36$$
$$2s^2 + 2s - 35 = 0$$

$a \quad b \quad c$

We cannot factor this so we use the quadratic formula: $s = \dfrac{-b \pm \sqrt{b^2 - 4ac}}{2a}$.

$$s = \frac{-2 \pm \sqrt{2^2 - 4 \cdot 2(-35)}}{2 \cdot 2}$$

$$s = \frac{-2 \pm \sqrt{4 + 280}}{4} = \frac{-2 \pm \sqrt{284}}{4}$$

$$s \approx 3.71 \quad or \quad s \approx -4.71. \qquad \text{Using a calculator and rounding to the nearest hundredth}$$

4. **Check.** Length cannot be negative, so -4.71 does not check. Note that if the shorter leg is 3.71 m, the other leg is 4.71 m. Then

$$(3.71)^2 + (4.71)^2 = 13.7641 + 22.1841 = 35.9482$$

and since $35.9482 \approx 6^2$, our answer checks. Also, note that the value of s, 3.71, is between 3 and 4, as predicted in step (1).

5. **State.** One leg is about 3.71 m long; the other is about 4.71 m long.

TRY EXERCISE 51

9.3 EXERCISE SET

Concept Reinforcement *Classify each statement as either true or false.*

1. A quadratic equation $ax^2 + bx + c = 0$, $a > 0$, is said to be in standard form.

2. The quadratic formula is derived by completing the square.

3. The quadratic formula always yields two different solutions of a quadratic equation.

4. If $b^2 - 4ac$ is negative, there is no real-number solution of $ax^2 + bx + c = 0$.

Solve. If no real-number solutions exist, state this.

5. $x^2 - 8x = 20$
6. $x^2 + 3x = 40$
7. $t^2 = 2t - 1$
8. $t^2 = 10t - 25$
9. $3y^2 + 7y + 4 = 0$
10. $3y^2 + 2y - 8 = 0$
11. $4x^2 - 12x = 7$
12. $4x^2 + 4x = 15$
13. $p^2 = 25$
14. $r^2 = 1$
15. $x^2 + 4x - 7 = 0$
16. $x^2 + 2x - 2 = 0$
17. $y^2 - 10y + 19 = 0$
18. $y^2 + 6y - 2 = 0$

Aha! 19. $x^2 - 10x + 25 = 3$
20. $x^2 + 2x + 1 = 3$
21. $3t^2 + 8t + 2 = 0$
22. $3t^2 - 4t - 2 = 0$
23. $2x^2 - 5x = 1$
24. $2x^2 + 2x = 3$
25. $4y^2 + 2y - 3 = 0$
26. $4y^2 - 4y - 3 = 0$
27. $2m^2 - m + 3 = 0$
28. $3p^2 + 2p + 5 = 0$
29. $3x^2 - 5x = 4$
30. $2x^2 + 3x = 1$
31. $2y^2 - 6y = 10$
32. $5m^2 = 3 + 11m$
33. $6t^2 + 26t = 20$
34. $7x^2 + 2 = 6x$
35. $5t^2 - 7t = -4$
36. $15t^2 + 10t = 0$
37. $5y^2 = 60$
38. $4y^2 = 200$

Solve using the quadratic formula. Use a calculator to approximate the solutions to the nearest thousandth.

39. $x^2 + 3x - 2 = 0$
40. $x^2 - 6x - 1 = 0$

41. $y^2 - 5y - 1 = 0$
42. $y^2 + 7y + 3 = 0$
43. $4x^2 + 4x = 1$
44. $4x^2 = 4x + 1$

Solve. If an irrational answer occurs, round to the nearest hundredth.

45. A polygon has 35 diagonals. How many sides does it have? (Use $d = (n^2 - 3n)/2$.)

46. A polygon has 20 diagonals. How many sides does it have? (Use $d = (n^2 - 3n)/2$.)

47. *Free-fall time.* At 1482 ft, the Petronas Towers in Malaysia is the world's tallest twin towers. How long would it take a marble to fall from the top? (Use $s = 16t^2$.)

48. *Free-fall time.* The roof of the Sears Tower in Chicago is 1454 ft high. How long would it take a golf ball to fall from the top? (Use $s = 16t^2$.)

49. *Skateboarding.* On April 6, 2006, skateboarder Danny Way set a free-fall record by dropping 28 ft onto a ramp. For how long was Way airborne? (Use $s = 16t^2$.)
    Source: skateboard.about.com

BUILD RAMPS NOT BOMBS

50. *Free-fall record.* The women's free-fall world record is held by Corrie Jansen, who leaped 182 ft off a cliff as a movie stunt. For how long was she airborne? (Use $s = 16t^2$.)
    Source: www.answers.com

51. *Right triangles.*   The hypotenuse of a right triangle is 25 ft long. One leg is 17 ft longer than the other. Find the lengths of the legs.

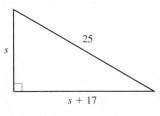

52. *Right triangles.*   The hypotenuse of a right triangle is 26 yd long. One leg is 14 yd longer than the other. Find the lengths of the legs.

53. *Area of a rectangle.*   The length of a rectangle is 4 cm greater than the width. The area is 60 cm². Find the length and the width.

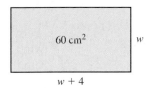

54. *Area of a rectangle.*   The length of a rectangle is 3 m greater than the width. The area is 70 m². Find the length and the width.

55. *Plumbing.*   A pipe runs diagonally under a rectangular yard that is 6 m longer than it is wide. If the pipe is 30 m long, determine the dimensions of the yard.

56. *Guy wires.*   A 26-ft long guy wire is anchored 10 ft from the base of a telephone pole. How far up the pole does the wire reach?

57. *Right triangles.*   The area of a right triangle is 26 cm². One leg is 5 cm longer than the other. Find the lengths of the legs.

58. *Right triangles.*   The area of a right triangle is 31 m². One leg is 2.4 m longer than the other. Find the lengths of the legs.

59. *Area of a rectangle.*   The length of a rectangle is 5 ft greater than the width. The area is 25 ft². Find the length and the width.

60. *Area of a rectangle.*   The length of a rectangle is 3 in. greater than the width. The area is 30 in². Find the length and the width.

61. *Area of a rectangle.*   The length of a rectangle is twice the width. The area is 16 m². Find the length and the width.

62. *Area of a rectangle.*   The length of a rectangle is twice the width. The area is 20 cm². Find the length and the width.

Investments.   *The formula $A = P(1 + r)^t$ is used to find the value A to which P dollars grows when invested for t years at an annual interest rate r. In Exercises 63–66, find the interest rate for the given information.*

63. $2560 grows to $3610 in 2 years

64. $2000 grows to $2880 in 2 years

65. $6000 grows to $6615 in 2 years

66. $3125 grows to $3645 in 2 years

67. *Environmental science.*   In 1995, oil leaking from the burning tanker *Sea Prince* formed a circular slick with an area of 100 mi². What was the diameter of the slick?
Source: www.foe.org

68. *Gardening.*   Laurie has enough mulch to cover 250 ft² of garden space. How wide is the largest circular flower garden that Laurie can cover with mulch?

69. Under what condition(s) is the quadratic formula *not* the easiest way to solve a quadratic equation?

70. Roy claims to be able to solve any quadratic equation by completing the square. He also claims to be incapable of understanding why the quadratic formula works. Does this strike you as odd? Why or why not?

Skill Review

To prepare for Section 9.4, review solving formulas (Section 2.3).

Solve. [2.3]

71. $a = \frac{1}{3} b$, for b

72. $n = \frac{y}{4}$, for y

73. $t = \frac{c + d}{2}$, for d

74. $a = \frac{4n}{p}$, for p

75. $y = Ax + B$, for x

76. $y = Ax + Bx$, for x

Synthesis

77. Anna can tell from the value of $b^2 - 4ac$ whether or not the solutions of a quadratic equation are rational. How can she do this?

78. Where does the $\pm$ symbol in the quadratic formula come from?

Solve.

79. $5x = -x(x + 6)$

80. $x(3x + 7) = 3x$

81. $3 - x(x - 3) = 4$

82. $x(5x - 7) = 1$

83. $(y + 4)(y + 3) = 15$

84. $x^2 + (x + 2)^2 = 7$

85. $\dfrac{x^2}{x + 3} = \dfrac{11}{x + 3}$

86. $\dfrac{x^2}{x + 5} - \dfrac{3}{x + 5} = 0$

87. $\dfrac{1}{x} + \dfrac{1}{x + 1} = \dfrac{1}{3}$

88. $\dfrac{1}{x} + \dfrac{1}{x + 6} = \dfrac{1}{5}$

89. Find *r* in this figure. Round to the nearest hundredth.

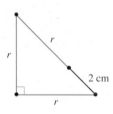

90. *Area of a square.*   Find the area of a square for which the diagonal is 3 units longer than the length of the sides.

91. *Golden rectangle.*   The so-called *golden rectangle* is said to be extremely pleasing visually and was used often by ancient Greek and Roman architects. The length of a golden rectangle is approximately 1.6 times the width. Find the dimensions of a golden rectangle if its area is 9000 m².

92. *Flagpoles.*   A 20-ft flagpole is struck by lightning and, while not completely broken, falls over and touches the ground 10 ft from the bottom of the pole. How high up did the pole break?

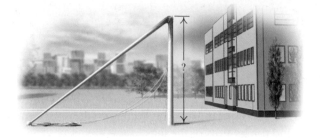

93. *Investments.* $4000 is invested at interest rate *r* for 2 yr. After 1 yr, an additional $2000 is invested, again at interest rate *r*. What is the interest rate if $6510 is in the account at the end of 2 yr?

94. *Investments.*   Dar needs $5000 in 2 yr to pay for dental work. How much does he need to invest now if he can get an interest rate of 5.75% compounded annually?

95. *Enlarged strike zone.*   In baseball, a batter's strike zone is a rectangular area about 15 in. wide and 40 in. high. Many batters subconsciously enlarge this area by 40% when fearful that if they don't swing, the umpire will call the pitch a strike. Assuming that the strike zone is enlarged by an invisible band of uniform width around the actual zone, find the dimensions of the enlarged strike zone.

96. Use a graph to approximate to the nearest thousandth the solutions of Exercises 39–44. Compare your answers with those found using a calculator.

CONNECTING the CONCEPTS

We have now studied four different ways of solving quadratic equations. Each of these methods has certain advantages and disadvantages, as outlined in the chart below. Note that although the quadratic formula can be used to solve *any* quadratic equation, the other methods are sometimes faster and easier to use.

Method	Advantages	Disadvantages
The quadratic formula	Can be used to solve *any* quadratic equation.	Can be slower than factoring or the principle of square roots.
Completing the square	Works well on equations of the form $x^2 + bx = -c$, where b is even.	Can be complicated when b is odd or a fraction.
The principle of square roots	Fastest way to solve equations of the form $ax^2 = p$, or $(x + k)^2 = p$. Can be used to solve *any* quadratic equation.	Can be slow when completing the square is required.
Factoring	Can be very fast.	Can be used only on certain equations. Many equations are difficult or impossible to solve by factoring.

MIXED REVIEW

Solve. Examine each problem carefully, and try to solve using the easiest method. If no real-number solution exists, state this.

1. $(x + 2)(x - 1) = 0$

2. $x^2 = 100$

3. $x^2 + 3x - 5 = 0$

4. $x^2 + 10x = 7$

5. $2x^2 - 5x - 3 = 0$

6. $x^2 + 2x + 1 = 0$

7. $x^2 + x + 1 = 0$

8. $(x - 2)(x - 3) = 12$

9. $x^2 = 5x$

10. $5x^2 = x + 2$

11. $x^2 + x = 6$

12. $(x - 2)^2 = 5$

13. $x^2 + 8x + 1 = 0$

14. $3x^2 + x - 2 = 0$

15. $121x^2 - 1 = 0$

16. $2x^2 - x - 5 = 0$

17. $(x + 1)^2 - 3 = 0$

18. $2x^2 - x + 2 = 0$

19. $(x + 2)(x - 1) = 3$

20. $9x^2 = x$

9.4 Formulas

Solving Formulas

Formulas arise frequently in such areas as the natural and social sciences, business, engineering, and health care. We often need to solve a formula for a variable.

Solving Formulas

Formulas can be linear, rational, radical, or quadratic equations, as well as other types of equations that are beyond the scope of this text. To solve formulas, we use the same steps that we use to solve equations. Probably the greatest difference is that whereas the solution of an equation is a number, the solution of a formula is generally a variable expression.

EXAMPLE 1

Sports engineering. The "sweet spot," or center of percussion, of a tennis racquet is located q centimeters from the hand, where $q = \dfrac{I}{mr}$ and I is the swing weight of the racquet, in kg · cm², m is the mass of the racquet, in kilograms, and r is the distance from the hand to the center of mass, in centimeters. Solve for m.

Source: www.racquetresearch.com

SOLUTION   We have

$$q = \frac{I}{mr}$$

$$mr \cdot q = mr \cdot \frac{I}{mr} \qquad \text{Multiplying both sides by } mr$$

$$mrq = I$$

$$m = \frac{I}{rq}. \qquad \text{Dividing both sides by } rq$$

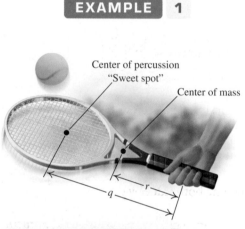

Center of percussion
"Sweet spot"

Center of mass

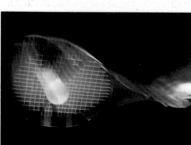

> TRY EXERCISE 7

EXAMPLE 2

A work formula. The formula $\dfrac{t}{a} + \dfrac{t}{b} = 1$ was used in Section 6.7. Solve this formula for t.

SOLUTION   We have

$$\frac{t}{a} + \frac{t}{b} = 1$$

$$ab\left(\frac{t}{a} + \frac{t}{b}\right) = ab \cdot 1 \qquad \text{Multiplying by the LCD, } ab, \text{ to clear fractions}$$

$$\left.\begin{array}{c} \dfrac{abt}{a} + \dfrac{abt}{b} = ab \\[2mm] bt + at = ab. \end{array}\right\} \quad \begin{array}{l} \text{Multiplying to remove parentheses and} \\ \text{removing factors equal to 1: } \dfrac{a}{a} = 1 \text{ and } \dfrac{b}{b} = 1 \end{array}$$

If the last equation were $3t + 2t = 6$, we would simply combine like terms ($3t + 2t$ is $5t$) and then divide by the coefficient of t (which would be 5). Since $bt + at$ *cannot* be combined, we factor instead:

$$(b + a)t = ab \qquad \text{Factoring out } t, \text{ the letter for which we are solving}$$

$$t = \frac{ab}{b + a}. \qquad \text{Dividing both sides by } b + a$$

TRY EXERCISE ▸ 13

The answer to Example 2 can be used when the times required to do a job independently (a and b) are known and t represents the time required to complete the task working together.

EXAMPLE 3

Temperature of a fluid. The temperature T of a fluid being heated by a steady flame or burner can be determined by the formula $T = a + b\sqrt{ct}$, where a is the initial temperature of the fluid, b and c are constants determined by the intensity of the heat and the type of fluid used, and t is time. Solve for t.

SOLUTION   We have

$$T = a + b\sqrt{ct}$$

$$T - a = b\sqrt{ct} \qquad \text{Subtracting } a \text{ from both sides}$$

$$\frac{T - a}{b} = \sqrt{ct} \qquad \text{Dividing both sides by } b$$

$$\left(\frac{T - a}{b}\right)^2 = ct \qquad \begin{array}{l}\text{Using the principle of powers:} \\ \text{squaring both sides}\end{array}$$

$$\frac{T^2 - 2Ta + a^2}{b^2} = ct \qquad \text{Squaring } T - a \text{ and } b$$

$$\frac{T^2 - 2Ta + a^2}{b^2 c} = t. \qquad \text{Dividing both sides by } c$$

This equation can be used to determine the heating time required for an experiment.

TRY EXERCISE ▸ 11

In some cases, the quadratic formula is needed.

EXAMPLE 4

A physics formula. The distance s of an object that is steadily accelerating from some established point can be determined using the formula $s = s_0 + v_0 t + \frac{1}{2}at^2$, where s_0 is the object's original distance from the point, v_0 is the object's original velocity, a is the object's acceleration, and t is time. Solve for t.

SOLUTION   Note first that t appears in two terms, raised to the first power and then the second power. Thus the equation is "quadratic in t," and the quadratic formula is needed:

$$\tfrac{1}{2}at^2 + v_0 t + s_0 = s \qquad \text{Rewriting the equation from left to right}$$

$$\underset{a}{\underbrace{\tfrac{1}{2}at^2}} + \underset{b}{\underbrace{v_0 t}} + \underset{c}{\underbrace{s_0 - s}} = 0; \qquad \begin{array}{l}\text{Subtracting } s \text{ from both sides and} \\ \text{identifying the coefficients needed for} \\ \text{the quadratic formula}\end{array}$$

$$t = \frac{-v_0 \pm \sqrt{(v_0)^2 - 4\left(\tfrac{1}{2}a\right)(s_0 - s)}}{2 \cdot \tfrac{1}{2}a} \qquad \begin{array}{l}\text{Substituting into the quadratic} \\ \text{formula}\end{array}$$

$$t = \frac{-v_0 \pm \sqrt{(v_0)^2 - 2a(s_0 - s)}}{a}$$

$$t = \frac{-v_0 \pm \sqrt{(v_0)^2 - 2as_0 + 2as}}{a}.$$

Simplifying

TRY EXERCISE 33

9.4   EXERCISE SET

For Extra Help
MyMathLab   Math XL   PRACTICE   WATCH   DOWNLOAD

🔖 **Concept Reinforcement** *Match each formula
with the process that should be used to solve for t.*

1. ___ $a = \sqrt{t}$

2. ___ $a_2 t^2 + a_1 t + a_0 = 0$

3. ___ $v = \dfrac{t}{a}$

4. ___ $a = t^2$

a) Use the principle of
   square roots.

b) Square both sides.

c) Multiply both sides
   by a.

d) Use the quadratic
   formula.

Solve each formula for the specified variable.

5. $A = \frac{1}{2} bh$, for h
   (The area of a triangle)

6. $S = 2\pi rh$, for h
   (A formula for surface area)

7. $Q = \dfrac{100m}{c}$, for c
   (A formula for intelligence quotient)

8. $y = \dfrac{k}{x}$, for x
   (A formula for inverse variation)

9. $A = P(1 + rt)$, for t
   (An investment formula)

10. $A = \pi r(s + r)$, for s
    (The surface area of a cone)

11. $d = c\sqrt{h}$, for h
    (A formula for distance to the horizon)

12. $n = c + \sqrt{t}$, for t
    (A formula for population)

13. $\dfrac{1}{R} = \dfrac{1}{r_1} + \dfrac{1}{r_2}$, for R
    (An electricity formula)

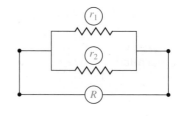

14. $\dfrac{1}{p} + \dfrac{1}{q} = \dfrac{1}{f}$, for f
    (An optics formula)

Aha! 15. $ax^2 + bx + c = 0$, for x

16. $(x - d)^2 = k$, for x

17. $Ax + By = C$, for y
    (An equation for a line)

18. $T = mg + mf$, for m

19. $S = 2\pi r(r + h)$, for h
    (The surface area of a
    right circular cylinder)

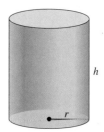

20. $xy - z = wy$, for y

21. $\dfrac{M - g}{t} = r + s$, for t

22. $\dfrac{m}{n} = p - q$, for n

23. $ab = ac + d$, for a

24. $\dfrac{x}{y} = \dfrac{z}{x}$, for x

25. $s = \frac{1}{2}gt^2$, for t
(A physics formula for distance)

26. $E = \frac{1}{2}mv^2$, for v
(A formula for kinetic energy)

27. $\dfrac{s}{h} = \dfrac{h}{t}$, for h
(A geometry formula)

28. $v = 8s(t - s)$, for t

29. $d - (r - w)t$, for w

30. $x - y = a(x + y)$, for y

31. $\sqrt{2x - y} + 1 = 6$, for x

32. $\sqrt{n - 2m} - 4 = 3$, for m

33. $mt^2 + nt - p = 0$, for t
(*Hint*: Use the quadratic formula.)

34. $rs^2 - ts + p = 0$, for s

35. $m + t = \dfrac{n}{m}$, for m

36. $x - y = \dfrac{z}{y}$, for y

37. $n = p - 3\sqrt{t + c}$, for t

38. $M - m\sqrt{ct} = r$, for t

39. $\sqrt{2t + s} = \sqrt{s - t}$, for t

40. $\sqrt{m - n} = \sqrt{3t}$, for n

41. Is it easier to solve
$$\dfrac{1}{25} + \dfrac{1}{23} = \dfrac{1}{x} \text{ for } x,$$
or to solve
$$\dfrac{1}{p} + \dfrac{1}{q} = \dfrac{1}{f} \text{ for } f?$$
Explain why.

42. Explain why someone might want to solve
$A = \frac{1}{2}bh$ for h.
(See Exercise 5.)

Skill Review

To prepare for Section 9.5, review multiplying and simplifying radical expressions (Section 8.2).

Simplify. [8.2]

43. $\sqrt{48}$

44. $\sqrt{250}$

45. $\sqrt{6} \cdot \sqrt{15}$

46. $\sqrt{10} \cdot \sqrt{35}$

47. $\sqrt{20} \cdot \sqrt{30}$

48. $\sqrt{75} \cdot \sqrt{12}$

Synthesis

49. As a step in solving a formula for a certain variable, a student takes the reciprocal of both sides of the equation. Is a mistake being made? Why or why not?

50. Describe a situation in which the result of Example 2,
$$t = \dfrac{ab}{b + a},$$
would be especially useful.

51. *Health care.* Young's rule for determining the size of a particular child's medicine dosage c is
$$c = \dfrac{a}{a + 12} \cdot d,$$
where a is the child's age and d is the typical adult dosage. If a child receives 8 mg of antihistamine when the typical adult receives 24 mg, how old is the child?
Source: Olsen, June Looby, Leon J. Ablon, and Anthony Patrick Giangrasso, *Medical Dosage Calculations*, 6th ed., p. A-31

Solve.

52. $Sr = \dfrac{rl - a}{r - l}$, for r

53. $fm = \dfrac{gm - t}{m}$, for m

54. $\dfrac{n_1}{p_1} + \dfrac{n_2}{p_2} = \dfrac{n_2 - n_1}{R}$, for n_2

55. *Economics.* The formula
$$V = \dfrac{k}{\sqrt{at + b}} + c$$
can be used to estimate the value of certain items that depreciate over time. Solve for t.

56. *Marine biology.* The formula
$$N = \dfrac{(b + d)f_1 - v}{(b - v)f_2}$$
is used when monitoring the water in fisheries. Solve for v.

57. *Meteorology.* The formula
$$C = \frac{5}{9}(F - 32)$$
is used to convert the Fahrenheit temperature F to the Celsius temperature C. At what temperature are the Fahrenheit and Celsius readings the same?

9.5   Complex Numbers as Solutions of Quadratic Equations

The Complex-Number System   ▪   Solutions of Equations

The Complex-Number System

Because negative numbers do not have square roots that are real numbers, mathematicians have devised a larger set of numbers known as *complex numbers*. In the complex-number system, the number i is used to represent the square root of -1.

> #### The Number i
> i is the unique number for which $i^2 = -1$ and $i = \sqrt{-1}$.

The square root of any negative number can be written using i. If $p \geq 0$, then

$$\sqrt{-p} = \sqrt{-1}\sqrt{p} = i\sqrt{p}, \text{ or } \sqrt{p}i.$$

EXAMPLE 1   Express in terms of i.

a) $\sqrt{-3}$
b) $\sqrt{-25}$
c) $-\sqrt{-10}$
d) $\sqrt{-24}$

SOLUTION

a) $\sqrt{-3} = \sqrt{-1 \cdot 3} = \sqrt{-1} \cdot \sqrt{3} = i\sqrt{3}, \text{ or } \sqrt{3}i$ ⟵ *i is not under the radical.*

b) $\sqrt{-25} = \sqrt{-1 \cdot 25} = \sqrt{-1} \cdot \sqrt{25} = i \cdot 5 = 5i$

c) $-\sqrt{-10} = -\sqrt{-1 \cdot 10} = -\sqrt{-1} \cdot \sqrt{10} = -i\sqrt{10}, \text{ or } -\sqrt{10}i$

d) $\sqrt{-24} = \sqrt{4(-1)6} = \sqrt{4}\sqrt{-1}\sqrt{6} = 2i\sqrt{6}, \text{ or } 2\sqrt{6}i$ **TRY EXERCISE** ❭ 7

> #### Imaginary Numbers
> An *imaginary number* is a number that can be written in the form $a + bi$, where a and b are real numbers and $b \neq 0$.

The following are examples of imaginary numbers:

$$3 + 8i, \quad \sqrt{7} - 2i, \quad 4 + \sqrt{6}i, \quad \text{and} \quad 4i \text{ (here } a = 0\text{)}.$$

Numbers like $4i$ are often called *pure imaginary* because they are of the form $0 + bi$. Imaginary numbers can be thought of as "nonreal." Together, the imaginary numbers and the real numbers form the set of **complex numbers**.*

> #### Complex Numbers
> A *complex number* is any number that can be written as $a + bi$, where a and b are real numbers. (Note that a and b both can be 0.)

───────────────

*The names "imaginary" and "complex" should not lead you to believe that these numbers are unimportant or complicated. Imaginary and complex numbers have important applications in engineering and the physical sciences.

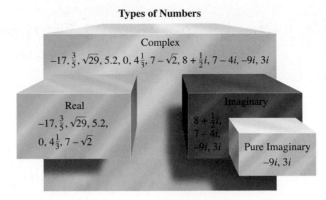

Types of Numbers

STUDENT NOTES

Do not confuse the a and the b in the expression $a + bi$ with the a and the b used in the quadratic formula.

It may help to remember that every real number is a complex number, but not every complex number is real. Numbers like 2 and $\sqrt{3}$, where $b = 0$, are complex and real. Numbers like $7 + 3i$ and $-5i$, where $b \neq 0$, are complex and not real.

Solutions of Equations

As we saw in Example 1(c) of Section 9.3, not all quadratic equations have real-number solutions. All quadratic equations *do* have complex-number solutions. These solutions are usually written in the form $a + bi$ unless a or b is zero.

EXAMPLE 2

Solve: **(a)** $x^2 = -100$; **(b)** $x^2 + 3x + 4 = 0$; **(c)** $x^2 + 2 = 2x$.

SOLUTION

a) We use the principle of square roots:

$$x^2 = -100$$
$$x = \sqrt{-100} \quad or \quad x = -\sqrt{-100} \qquad \text{Using the principle of square roots}$$
$$x = \sqrt{-1}\sqrt{100} \quad or \quad x = -\sqrt{-1}\sqrt{100}$$
$$x = 10i \qquad\qquad or \quad x = -10i. \qquad \sqrt{-1} = i \text{ and } \sqrt{100} = 10$$

The solutions are $10i$ and $-10i$.

b) We use the quadratic formula:

$$1x^2 + 3x + 4 = 0$$
$$x = \frac{-3 \pm \sqrt{3^2 - 4 \cdot 1 \cdot 4}}{2 \cdot 1} \qquad x = \frac{-b \pm \sqrt{b^2 - 4ac}}{2a}$$
$$x = \frac{-3 \pm \sqrt{-7}}{2} \qquad\qquad \text{Simplifying}$$
$$\left.\begin{array}{l} x = \dfrac{-3 \pm \sqrt{-1}\sqrt{7}}{2} \\[2mm] x = \dfrac{-3 \pm i\sqrt{7}}{2} \end{array}\right\} \qquad \text{Rewriting } \sqrt{-7} \text{ as } i\sqrt{7}$$
$$x = -\frac{3}{2} \pm \frac{\sqrt{7}}{2} i. \qquad\qquad \text{Writing in the form } a + bi$$

The solutions are $-\dfrac{3}{2} + \dfrac{\sqrt{7}}{2} i$ and $-\dfrac{3}{2} - \dfrac{\sqrt{7}}{2} i$.

c) We have

$$x^2 + 2 = 2x$$

$$1x^2 - 2x + 2 = 0 \qquad \text{Rewriting in standard form}$$

$$x = \frac{-(-2) \pm \sqrt{(-2)^2 - 4 \cdot 1 \cdot 2}}{2 \cdot 1} \qquad x = \frac{-b \pm \sqrt{b^2 - 4ac}}{2a}$$

$$x = \frac{2 \pm \sqrt{-4}}{2} \qquad \text{Simplifying}$$

$$x = \frac{2 \pm \sqrt{-1}\sqrt{4}}{2}$$

$$x = \frac{2 \pm 2i}{2} \qquad \sqrt{-1} = i \text{ and } \sqrt{4} = 2$$

$$x = \frac{2}{2} \pm \frac{2}{2}i \qquad \text{Rewriting in the form } a + bi$$

$$x = 1 \pm i. \qquad \text{Simplifying}$$

The solutions are $1 + i$ and $1 - i$.

TRY EXERCISE ▶ 25

9.5   EXERCISE SET

🦢 **Concept Reinforcement**   *Classify each statement as either true or false.*

1. Every complex number is a real number.

2. Many complex numbers are real numbers.

3. Many complex numbers are imaginary numbers.

4. Every complex number is an imaginary number.

Express in terms of i.

5. $\sqrt{-1}$

6. $\sqrt{-4}$

7. $\sqrt{-49}$

8. $\sqrt{-100}$

9. $\sqrt{-5}$

10. $\sqrt{-15}$

11. $\sqrt{-45}$

12. $\sqrt{-18}$

13. $-\sqrt{-50}$

14. $-\sqrt{-12}$

15. $4 + \sqrt{-49}$

16. $7 + \sqrt{-4}$

17. $3 - \sqrt{-9}$

18. $-8 - \sqrt{-36}$

19. $-2 + \sqrt{-75}$

20. $5 - \sqrt{-20}$

Solve.

Aha! 21. $x^2 + 25 = 0$

22. $x^2 + 16 = 0$

23. $x^2 = -28$

24. $x^2 = -48$

25. $t^2 + 4t + 5 = 0$

26. $t^2 - 4t + 6 = 0$

27. $(x - 4)^2 = -9$

28. $(x + 3)^2 = -4$

29. $x^2 + 5 = 2x$

30. $x^2 + 3 = -2x$

31. $t^2 + 7 - 4t = 0$

32. $t^2 + 8 + 4t = 0$

33. $5y^2 + 4y + 1 = 0$

34. $4y^2 + 3y + 2 = 0$

35. $1 + 2m + 3m^2 = 0$

36. $4p^2 + 3 = 6p$

📝 37. Is it possible for a quadratic equation to have one imaginary-number solution and one real-number solution? Why or why not?

📝 38. Under what condition(s) will an equation of the form $x^2 = c$ have imaginary-number solutions?

Skill Review

To prepare for Section 9.6, review graphing (Sections 3.2, 3.3, and 3.6).

Graph. [3.2], [3.3], [3.6]

39. $y = \frac{3}{5}x$

40. $2x - 3y = 10$

41. $y = -4x$

42. $x = 2$

43. $y = \frac{1}{2}x - 3$

44. $y = -1$

Synthesis

45. When using the quadratic formula to solve an equation, if $b^2 < 4ac$, are the solutions imaginary? Why or why not?

46. Can imaginary-number solutions of a quadratic equation be found using the method of completing the square? Why or why not?

Solve.

47. $(x + 1)^2 + (x + 3)^2 = 0$

48. $(p + 5)^2 + (p + 1)^2 = 0$

49. $\dfrac{2x - 1}{5} - \dfrac{2}{x} = \dfrac{x}{2}$

50. $\dfrac{1}{a - 1} - \dfrac{2}{a - 1} = 3a$

51. Use a graphing calculator to confirm that there are no real-number solutions of Examples 2(a), 2(b), and 2(c).

CONNECTING the CONCEPTS

We have now completed our study of the various types of equations that appear in elementary algebra. Below are examples of each type, along with their solutions.

Linear Equations

$$5x + 3 = 2x + 9$$

$\quad 3x = 6$ Adding $-2x - 3$ to both sides

$\quad\ \ x = 2$ Dividing both sides by 3

The solution is 2.

Radical Equations

$$\sqrt{2x + 1} - 5 = 2$$

$\quad \sqrt{2x + 1} = 7$ Adding 5 to both sides

$\quad 2x + 1 = 49$ Squaring both sides

$\quad\quad\ \ 2x = 48$ Subtracting 1 from both sides

$\quad\quad\quad x = 24$ Dividing both sides by 2

The solution is 24.

Rational Equations

$$\frac{5}{2x} + \frac{4}{3x} = 2 \qquad \text{Note that } x \neq 0.$$

$$6x\left(\frac{5}{2x} + \frac{4}{3x}\right) = 6x \cdot 2 \qquad \begin{array}{l}\text{Multiplying}\\ \text{both sides by}\\ \text{the LCD, } 6x\end{array}$$

$$\frac{6x \cdot 5}{2x} + \frac{6x \cdot 4}{3x} = 12x \qquad \begin{array}{l}\text{Using the}\\ \text{distributive law}\end{array}$$

$$15 + 8 = 12x \qquad \text{Simplifying}$$

$$23 = 12x$$

$$\frac{23}{12} = x$$

The solution is $\frac{23}{12}$.

Quadratic Equations

$$2x^2 + 3x = 1$$

$$2x^2 + 3x - 1 = 0 \qquad \text{Subtracting 1 from both sides}$$

$$x = \frac{-3 \pm \sqrt{3^2 - 4(2)(-1)}}{2 \cdot 2} \qquad \begin{array}{l}\text{Using the}\\ \text{quadratic}\\ \text{formula}\end{array}$$

$$x = \frac{-3 \pm \sqrt{17}}{4}$$

The solutions are $\dfrac{-3}{4} + \dfrac{\sqrt{17}}{4}$ and $\dfrac{-3}{4} - \dfrac{\sqrt{17}}{4}$.

You should always check solutions in the original equation in case an error was made in the solution process. For rational and radical equations, a check is necessary because the equation-solving principles used may not yield equivalent equations. The checks for the equations above are left to the student.

(continued)

MIXED REVIEW

Solve.

1. $3x + 8 = 7x + 4$

2. $x^2 = 5x - 6$

3. $3 + \sqrt{x} = 8$

4. $\dfrac{5}{x} + \dfrac{3}{4} = 2$

5. $11 = 4\sqrt{3x + 1} - 5$

6. $2t^2 - 7t + 3 = 0$

7. $n^2 + 2n - 2 = 0$

8. $n^2 + 2n + 2 = 0$

9. $\dfrac{1}{4t} + \dfrac{t}{6} = 2$

10. $4x - 7 = 2(5x - 3)$

11. $2\sqrt{5t + 3} = 3\sqrt{2t - 1}$

12. $\dfrac{3}{10x} - \dfrac{4}{5x} = 6$

13. $\sqrt{x + 3} = x + 1$

14. $x + 1 = \dfrac{5}{x - 3}$

15. $(x + 3)^2 = 7$

16. $t^2 + 100 = 0$

17. $t^2 - 100 = 0$

18. $2x^2 + 5x = 1$

19. $5t - 3 = 2 - 4(1 - t)$

20. $x^2 + 6x + 2 = 0$

9.6 Graphs of Quadratic Equations

Graphing Equations of the Form $y = ax^2$ ▪ Graphing Equations of the Form $y = ax^2 + bx + c$

In this section, we will graph quadratic equations like

$$y = \tfrac{1}{2}x^2, \qquad y = x^2 + 2x - 3, \quad \text{and} \quad y = -5x^2 + 4.$$

Such equations, of the form $y = ax^2 + bx + c$ with $a \neq 0$, have graphs that are cupped either upward or downward. These graphs are symmetric with respect to an **axis of symmetry**, as shown below. When folded along its axis, the graph has two halves that match exactly.

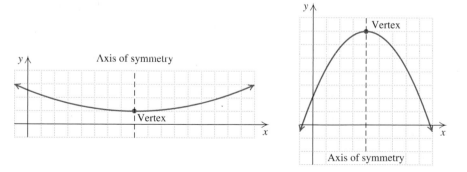

The point at which the graph of a quadratic equation crosses its axis of symmetry is called the **vertex** (plural, vertices). The y-coordinate of the vertex is the graph's largest value of y (if the curve opens downward) or smallest value of y (if the curve opens upward). Graphs of quadratic equations of the form $y = ax^2 + bx + c$, with $a \neq 0$, are called **parabolas**.

Graphing Equations of the Form $y = ax^2$

The simplest parabolas to sketch are given by equations of the form $y = ax^2$.

EXAMPLE 1

Graph: $y = x^2$.

SOLUTION   We choose numbers for x and find the corresponding values for y.

If $x = -2$, then $y = (-2)^2 = 4$.   We get the pair $(-2, 4)$.
If $x = -1$, then $y = (-1)^2 = 1$.   We get the pair $(-1, 1)$.
If $x = 0$,   then $y = 0^2 = 0$.   We get the pair $(0, 0)$.
If $x = 1$,   then $y = 1^2 = 1$.   We get the pair $(1, 1)$.
If $x = 2$,   then $y = 2^2 = 4$.   We get the pair $(2, 4)$.

The following table lists these solutions of $y = x^2$. After several ordered pairs are found, we plot them and connect them with a smooth curve.

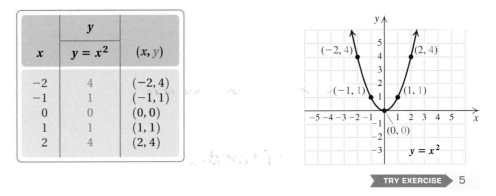

x	$y = x^2$	(x, y)
-2	4	$(-2, 4)$
-1	1	$(-1, 1)$
0	0	$(0, 0)$
1	1	$(1, 1)$
2	4	$(2, 4)$

TRY EXERCISE ▸ 5

In Example 1, the vertex is $(0, 0)$ and the axis of symmetry is the y-axis. This will be the case for any parabola having an equation of the form $y = ax^2$.

EXAMPLE 2

Graph: $y = -\frac{1}{2}x^2$.

SOLUTION   We select numbers for x, find the corresponding y-values, plot the resulting ordered pairs, and connect them with a smooth curve.

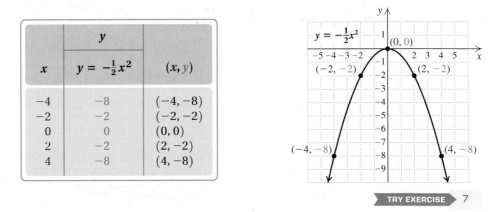

x	$y = -\frac{1}{2}x^2$	(x, y)
-4	-8	$(-4, -8)$
-2	-2	$(-2, -2)$
0	0	$(0, 0)$
2	-2	$(2, -2)$
4	-8	$(4, -8)$

TRY EXERCISE ▸ 7

Graphing Equations of the Form $y = ax^2 + bx + c$

Recall from our work with lines that it is often very useful to know the x- and y-intercepts of a line. Intercepts are also useful when graphing parabolas. To find the intercepts of a parabola, we use the same approach we used with lines.

> ### The Intercepts of a Parabola
>
> To find the y-intercept of the graph of $y = ax^2 + bx + c$, replace x with 0 and solve for y. The result will be c.
>
> To find the x-intercept(s) of the graph of $y = ax^2 + bx + c$, if any exist, replace y with 0 and solve for x. To do this, factor or use the quadratic formula. If no real solution exists, there are no x-intercepts.

EXAMPLE 3   Find all y- and x-intercepts of the graph of $y = 2x^2 - x - 28$.

SOLUTION   To find the y-intercept, we replace x with 0 and solve for y:

$$y = 2 \cdot 0^2 - 0 - 28 \qquad \text{At a } y\text{-intercept, } x = 0.$$
$$y = 0 - 0 - 28 = -28.$$

When x is 0, we have $y = -28$. Thus the y-intercept is $(0, -28)$.

To find the x-intercept(s), we replace y with 0 and solve for x:

$$0 = 2x^2 - x - 28. \qquad \text{At an } x\text{-intercept, } y = 0.$$

The quadratic formula could be used, but factoring is faster:

$$0 = (2x + 7)(x - 4) \qquad \text{Factoring}$$
$$2x + 7 = 0 \quad or \quad x - 4 = 0$$
$$2x = -7 \quad or \qquad x = 4$$
$$x = -\tfrac{7}{2} \quad or \qquad x = 4.$$

STUDENT NOTES

The graphs of some parabolas do not cross the horizontal axis. For example, if the parabola in Example 3 had the same vertex, but opened downward instead of upward, no x-intercepts would exist.

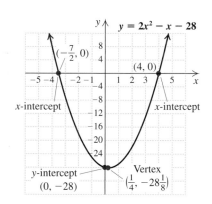

The x-intercepts are $(4, 0)$ and $\left(-\tfrac{7}{2}, 0\right)$. The y-intercept is $(0, -28)$.

Although we were not asked to graph the equation in Example 3, we did so to show that the x-coordinate of the vertex, $\tfrac{1}{4}$, is exactly midway between the x-intercepts. The quadratic formula, $x = \dfrac{-b \pm \sqrt{b^2 - 4ac}}{2a}$, can be used to locate x-intercepts when an equation of the form $y = ax^2 + bx + c$ is graphed. If one x-intercept is determined by $\dfrac{-b - \sqrt{b^2 - 4ac}}{2a}$ and the other by $\dfrac{-b + \sqrt{b^2 - 4ac}}{2a}$, then the average of these two values can be used to find the x-coordinate of the vertex. In Exercise 46, you are asked to show that this x-value is $\dfrac{-b}{2a}$. A more complicated approach can be used to show that the x-value of the vertex is $\dfrac{-b}{2a}$ even when no x-intercepts exist.

> ### The Vertex of a Parabola
>
> For a parabola given by the quadratic equation $y = ax^2 + bx + c$:
>
> **1.** The x-coordinate of the vertex is $-\dfrac{b}{2a}$.
>
> **2.** The y-coordinate of the vertex is found by substituting $-\dfrac{b}{2a}$ for x and solving for y.

EXAMPLE 4   Graph: $y = x^2 + 2x - 3$.

SOLUTION   Our plan is to plot the vertex and some points on either side of the vertex. We will then draw a parabola passing through these points.

To locate the vertex, we use $-b/(2a)$ to find its x-coordinate:

$$x\text{-coordinate of the vertex} = -\frac{b}{2a} = -\frac{2}{2 \cdot (1)}$$
$$= -1.$$

We substitute -1 for x to find the y-coordinate of the vertex:

$$y\text{-coordinate of the vertex} = (-1)^2 + 2(-1) - 3 = 1 - 2 - 3$$
$$= -4.$$

The vertex is $(-1, -4)$. The axis of symmetry is $x = -1$.

Next, we choose some x-values on both sides of the vertex and graph the parabola.

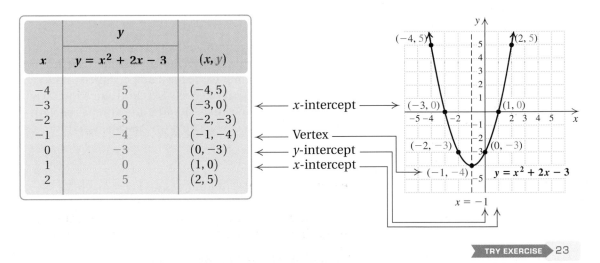

	y	
x	$y = x^2 + 2x - 3$	(x, y)
-4	5	$(-4, 5)$
-3	0	$(-3, 0)$
-2	-3	$(-2, -3)$
-1	-4	$(-1, -4)$
0	-3	$(0, -3)$
1	0	$(1, 0)$
2	5	$(2, 5)$

⟵ x-intercept ⟶
⟵ Vertex
⟵ y-intercept
⟵ x-intercept

TRY EXERCISE 23

One tip for graphing quadratic equations involves the coefficient of x^2. The sign of a in $y = ax^2 + bx + c$ tells us whether the graph opens upward or downward. When a is positive, as in Examples 1, 3, and 4, the graph opens upward; when a is negative, as in Example 2, the graph opens downward.

EXAMPLE   5   Graph: $y = -2x^2 + 4x + 1$.

SOLUTION   Since the coefficient of x^2 is negative, we know that the graph opens downward. To locate the vertex, we first find its x-coordinate:

$$x\text{-coordinate of the vertex} = -\frac{b}{2a} = -\frac{4}{2 \cdot (-2)}$$
$$= 1.$$

We substitute 1 for x to find the y-coordinate of the vertex:

$$y\text{-coordinate of the vertex} = -2 \cdot 1^2 + 4 \cdot 1 + 1 = -2 + 4 + 1$$
$$= 3.$$

The vertex is $(1, 3)$. The axis of symmetry is $x = 1$.

We choose some x-values on both sides of the vertex, calculate their corresponding y-values, and graph the parabola.

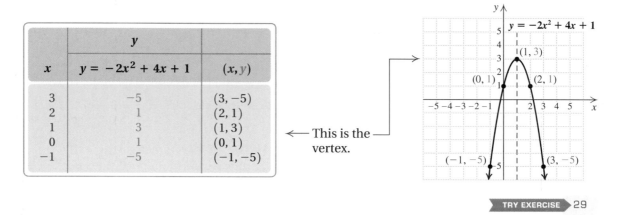

x	$y = -2x^2 + 4x + 1$	(x, y)
3	-5	$(3, -5)$
2	1	$(2, 1)$
1	3	$(1, 3)$
0	1	$(0, 1)$
-1	-5	$(-1, -5)$

← This is the vertex.

TRY EXERCISE 29

A second tip for graphing quadratic equations can cut our calculation time in half. In Examples 1–5, note that any x-value to the left of the vertex is paired with the same y-value as an x-value the same distance to the right of the vertex. Thus, since the vertex for Example 5 is $(1, 3)$ and since the x-values -1 and 3 are both 2 units from 1, we know that -1 and 3 are both paired with the same y-value. This symmetry provides a useful check and allows us to plot *two* points after calculating just *one*.

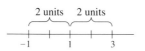

Guidelines for Graphing Quadratic Equations

1. Graphs of quadratic equations, $y = ax^2 + bx + c$, are parabolas. They are cupped upward if $a > 0$ and downward if $a < 0$.
2. Use the formula $x = -b/(2a)$ to find the x-coordinate of the vertex. After calculating the y-coordinate, plot the vertex and some points on either side of it.
3. After a point has been graphed, a second point with the same y-coordinate can be found on the opposite side of the axis of symmetry.
4. Graph the y-intercept and, if requested, any x-intercepts.

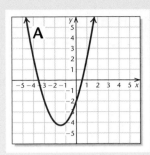

A

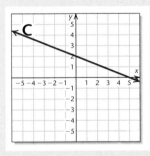

B

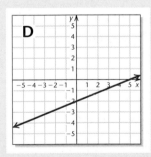

C

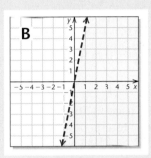

D

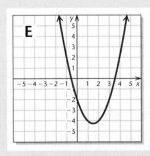

E

Visualizing for Success

Match each equation or inequality with its graph.

1. $y = -4 + 4x - x^2$

2. $y = 5 - x^2$

3. $5x + 2y = -10$

4. $5x + 2y \leq 10$

5. $y < 5x$

6. $y = x^2 - 3x - 2$

7. $2x - 5y = 10$

8. $5x - 2y = 10$

9. $2x + 5y = 10$

10. $y = x^2 + 3x - 2$

Answers on page A-36

An additional, animated version of this activity appears in MyMathLab. To use MyMathLab, you need a course ID and a student access code. Contact your instructor for more information.

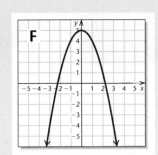

F

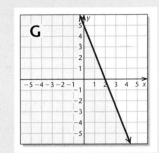

G

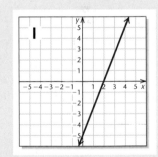

H

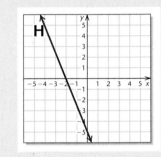

I

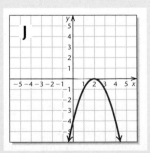

J

9.6   EXERCISE SET

For Extra Help
MyMathLab   Math XL PRACTICE   WATCH   DOWNLOAD

ↄ **Concept Reinforcement** *Classify each statement as either true or false.*

1. The graph of every quadratic equation has two x-intercepts.

2. The sign of the constant a in $y = ax^2 + bx + c$ indicates whether the graph of the equation opens upward or downward.

3. The constant b in $y = ax^2 + bx + c$ gives the x-coordinate of the x-intercepts.

4. The constant c in $y = ax^2 + bx + c$ gives the y-coordinate of the y-intercept.

Graph each quadratic equation, labeling the vertex and the y-intercept.

5. $y = 2x^2$

6. $y = 3x^2$

7. $y = -2x^2$

8. $y = -1 \cdot x^2$

9. $y = -\frac{1}{3}x^2$

10. $y = \frac{1}{4}x^2$

11. $y = x^2 - 2$

12. $y = x^2 + 1$

13. $y = x^2 - 2x + 1$

14. $y = x^2 + 4x + 4$

15. $y = x^2 + 3x - 10$

16. $y = x^2 - 2x - 3$

17. $y = -2x^2 + 12x - 13$

18. $y = -3x^2 + 12x - 11$

19. $y = -\frac{1}{2}x^2 + 5$

20. $y = \frac{1}{2}x^2 - 7$

21. $y = x^2 - 3x$

22. $y = -x^2 + 2x$

Graph each equation, labeling the vertex, the y-intercept, and any x-intercepts. If an x-intercept is irrational, round to three decimal places.

23. $y = x^2 + 2x - 8$

24. $y = x^2 + x - 6$

25. $y = 2x^2 - 6x$

26. $y = 2x^2 - 7x$

27. $y = -x^2 - x + 12$

28. $y = -x^2 - 3x + 10$

29. $y = 3x^2 - 6x + 1$

30. $y = 3x^2 + 12x + 11$

31. $y = x^2 + 2x + 3$

32. $y = -x^2 - 2x - 3$

33. $y = 3 - 4x - 2x^2$

34. $y = 1 - 4x - 2x^2$

35. Why is it helpful to know the coordinates of the vertex when graphing a parabola?

36. Suppose that both x-intercepts of a parabola are known. What is the easiest way to find the coordinates of the vertex?

Skill Review

To prepare for Section 9.7, review evaluating polynomials (Section 4.2).

Evaluate. [4.2]

37. $3x^2 - 4x$, for $x = -1$

38. $5a^3 + 3a$, for $a = -1$

39. $6 - t^3$, for $t = -2$

40. $3t^4 + t^2$, for $t = 10$

41. $(a - 9)^2$, for $a = 8$

42. $(a - 9)^2$, for $a = -3$

Synthesis

43. Describe a method that could be used to find an equation for a parabola that has x-intercepts $(p, 0)$ and $(q, 0)$.

44. What effect does the size of $|a|$ have on the graph of $y = ax^2 + bx + c$?

45. *Height of a golf ball.* The height H, in feet, of a golf ball with an initial velocity of 96 ft/sec is given by the equation

$$H = -16t^2 + 96t,$$

where t is the number of seconds from launch. Use the graph of this equation, shown below, or any equation-solving technique to answer the following.

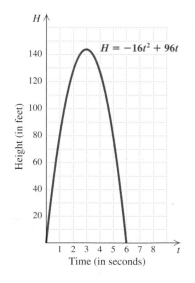

a) How many seconds after launch is the golf ball 128 ft above ground?

b) When does the golf ball reach its maximum height?

Aha! c) How many seconds after launch does the golf ball return to the ground?

46. Show that the average of
$$\frac{-b - \sqrt{b^2 - 4ac}}{2a} \quad \text{and} \quad \frac{-b + \sqrt{b^2 - 4ac}}{2a}$$
is $-\dfrac{b}{2a}$.

47. *Stopping distance.*   In how many feet can a car stop if it is traveling at a speed of r miles per hour? One estimate, developed in Britain, is as follows. The distance d, in feet, is given by

$$d = \underbrace{\text{Thinking distance}}_{\text{(in feet)}} + \underbrace{\text{Stopping distance}}_{\text{(in feet)}}$$

$$d = \quad r \quad + \quad 0.05r^2.$$

   a) How many feet would it take to stop a car traveling 25 mph? 40 mph? 55 mph? 65 mph? 75 mph? 100 mph?

   b) Graph the equation, assuming $r \geq 0$.

48. On one set of axes, graph $y = x^2$, $y = (x - 3)^2$, and $y = (x + 1)^2$. Describe the effect that h has on the graph of $y = (x - h)^2$.

49. On one set of axes, graph $y = x^2$, $y = x^2 - 5$, and $y = x^2 + 2$. Describe the effect that k has on the graph of $y = x^2 + k$.

50. *Seller's supply.*   As the price of a product increases, the seller is willing to sell, or *supply*, more of the product. Suppose that the supply for a certain product is given by
$$S = p^2 + p + 10,$$
where p is the price in dollars and S is the number supplied, in thousands, at that price. Graph the equation for values of p such that $0 \leq p \leq 6$.

51. *Consumer's demand.*   As the price of a product increases, consumers purchase, or *demand*, less of the product. Suppose that the demand for a certain product is given by
$$D = (p - 6)^2,$$
where p is the price in dollars and D is the number demanded, in thousands, at that price. Graph the equation for values of p such that $0 \leq p \leq 6$.

52. *Equilibrium point.*   The price p at which the consumer and the seller agree determines the *equilibrium point*. Find p such that
$$D = S$$
for the demand and supply curves in Exercises 50 and 51. How many units of the product will be sold at that price?

53. Use a graphing calculator to draw the graph of $y = x^2 - 5$ and then, using the graph, estimate $\sqrt{5}$ to four decimal places.

9.7   Functions

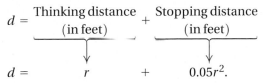

Identifying Functions • Functions Written as Formulas • Function Notation •
Graphs of Functions • Recognizing Graphs of Functions

Functions are enormously important in modern mathematics and science. The more mathematics and science you study, the more you will use functions.

Identifying Functions

Functions appear regularly in many contexts, although they are usually not referred to as such. To motivate understanding of functions, consider the following table relating heart rate to average life span.

	Large Whale	Elephant	Horse	Human	Small Dog	Cat	Monkey	Chicken	Hamster
Heart Rate (in number of beats per minute)	20	30	44	60	100	150	190	275	450
Average Life Span (in years)	80	70	40	70	10	15	15	15	3

Note that to each heart rate there corresponds *exactly one* average life span. A correspondence of this type is called a **function**.

> ## Function
>
> A *function* is a correspondence (or rule) that assigns to each member of some set (called the *domain*) exactly one member of another set (called the *range*). The domain and the range may be identical sets.

The members of the domain are sometimes called **inputs**, and the members of the range **outputs**. For the information above, we have the following.

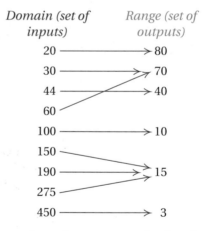

Domain (set of inputs)   *Range (set of outputs)*

The function above can be written as a set of ordered pairs. Note that each input has exactly *one* output, even though some outputs are used more than once:

$$\{(20, 80), \quad (30, 70), \quad (44, 40), \quad (60, 70), \quad (100, 10), \quad (150, 15),$$
$$(190, 15), \quad (275, 15), \quad (450, 3)\}.$$

Correspondences are often named using letters.

EXAMPLE 1   Determine whether or not each of the following correspondences is a function.

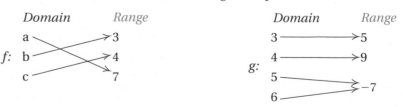

	Domain	*Range*		*Domain*	*Range*
h:	New York	Mets	*p:*	Cubs	Chicago
		Yankees		White Sox	
	Florida	Marlins		Orioles	Baltimore
	Texas	Rangers		Padres	San Diego

SOLUTION   Correspondence *f* is a function because each member of the domain is matched to just one member of the range.

Correspondence *g* is also a function because each member of the domain is matched to just one member of the range.

Correspondence *h* is *not* a function because one member of the domain, New York, is matched to more than one member of the range.

Correspondence *p* is a function because each member of the domain is paired with just one member of the range.

> **TRY EXERCISE** ▶ 5

Functions Written as Formulas

Many functions are described by formulas. Equations like $y = x + 3$ and $y = 4x^2$ are examples of such formulas. Outputs are found by substituting members of the domain for x.

EXAMPLE 2

Thunderstorm distance.   During a thunderstorm, it is possible to calculate how far away lightning is by using the formula

$$M = \tfrac{1}{5}t.$$

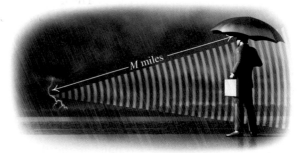

Here M is the distance, in miles, that a storm is from an observer when the sound of thunder arrives t seconds after the lightning has been sighted.

Complete the following table for this function.

t (in seconds)	0	1	2	3	4	5	6	10
M (in miles)	0	$\tfrac{1}{5}$						

SOLUTION   To complete the table, we substitute values of t and compute M.

For $t = 2$, $M = \tfrac{1}{5} \cdot 2 = \tfrac{2}{5}.$ For $t = 3$, $M = \tfrac{1}{5} \cdot 3 = \tfrac{3}{5}.$

For $t = 4$, $M = \tfrac{1}{5} \cdot 4 = \tfrac{4}{5}.$ For $t = 5$, $M = \tfrac{1}{5} \cdot 5 = 1.$

For $t = 6$, $M = \tfrac{1}{5} \cdot 6 = \tfrac{6}{5},$ or $1\tfrac{1}{5}.$ For $t = 10$, $M = \tfrac{1}{5} \cdot 10 = 2.$

t (in seconds)	0	1	2	3	4	5	6	10
M (in miles)	0	$\tfrac{1}{5}$	$\tfrac{2}{5}$	$\tfrac{3}{5}$	$\tfrac{4}{5}$	1	$1\tfrac{1}{5}$	2

Function Notation

In Example 2, it was somewhat time-consuming to repeatedly write "For $t = $ ■, $M = \frac{1}{5} \cdot$ ■." *Function notation* clearly and concisely presents inputs and outputs together. If we name the above function M, the notation $M(t)$, read "M of t," denotes the output that is paired with the input t by the function M. Thus, for Example 2,

$$M(2) = \tfrac{1}{5} \cdot 2 = \tfrac{2}{5}, \qquad M(3) = \tfrac{1}{5} \cdot 3 = \tfrac{3}{5}, \quad \text{and, in general,} \quad M(t) = \tfrac{1}{5} \cdot t.$$

The notation $M(4) = \frac{4}{5}$ means "when 4 is the input, $\frac{4}{5}$ is the output."

> **CAUTION!** $M(4)$ *does not* mean M times 4 and should not be read or pronounced that way.

Equations for nonvertical lines can be written in function notation. For example, $f(x) = x + 2$, read "f of x equals x plus 2," can be used instead of $y = x + 2$ when we are discussing functions, although both equations describe the same correspondence.

The variable x in $f(x) = x + 2$ is called the **independent variable**. To find a particular function value $f(a)$, we replace every occurrence of x with a.

EXAMPLE 3

For the function given by $f(x) = x + 2$, find each of the following.

a) $f(8)$ **b)** $f(-3)$ **c)** $f(0)$

SOLUTION

a) $f(8) = 8 + 2$, or 10 This function adds 2 to each input.

b) $f(-3) = -3 + 2$, or -1 $f(-3)$ is read "f of -3."

c) $f(0) = 0 + 2$, or 2 $f(0)$ *does not mean* $f \cdot 0$.   **TRY EXERCISE** 17

It is sometimes helpful to think of a function as a machine that gives an output for each input that enters the machine. For example, the function $g(t) = 2t^2 + t$ pairs an input with the sum of the input and twice its square. The following diagram is one way in which the function given by $g(t) = 2t^2 + t$ can be illustrated.

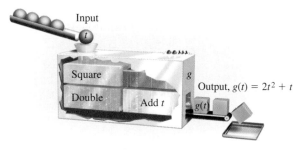

Input

Square

Double

Add t

Output, $g(t) = 2t^2 + t$

EXAMPLE   4

For the function $g(t) = 2t^2 + t$, find each of the following.

a) $g(3)$ **b)** $g(0)$ **c)** $g(-2)$

SOLUTION

a) $g(3) = 2 \cdot 3^2 + 3$ Using 3 for each occurrence of t

$\quad\quad\quad = 2 \cdot 9 + 3$

$\quad\quad\quad = 21$

b) $g(0) = 2 \cdot 0^2 + 0$ Using 0 for each occurrence of t

$\quad\quad\quad = 0$

c) $g(-2) = 2(-2)^2 + (-2)$ Using -2 for each occurrence of t

$\quad\quad\quad = 2 \cdot 4 - 2$

$\quad\quad\quad = 6$

> **TRY EXERCISE** ▸ 19

TECHNOLOGY CONNECTION

There are several ways in which to check Example 4. We can enter $y_1 = 2x^2 + x$ and use TRACE, and then (on most calculators) enter the desired input. We can also use CALC and VALUE, or the TABLE feature. Yet another way to check is to use the VARS or Y-VARS key to access the function y_1. Parentheses can then be used to complete the writing of $y_1(3)$. When **ENTER** is pressed, we see the output 21. Use this approach to find $y_1(0)$ and $y_1(-2)$.

Outputs are also called **function values**. In Example 4, $g(-2) = 6$. We can say that the "function value at -2 is 6," or "when x is -2, the value of the function is 6." Most often we simply say "g of -2 is 6."

Graphs of Functions

To graph a function, we usually calculate ordered pairs of the form (x, y) or $(x, f(x))$, plot them, and connect the points. The symbols y and $f(x)$ are often used interchangeably when working with functions and their graphs.

EXAMPLE   5

Graph: $f(x) = x + 2$.

SOLUTION   A list of some function values is shown in the following table. We plot the points and connect them. The graph is a straight line.

x	$f(x)$
-4	-2
-3	-1
-2	0
-1	1
0	2
1	3
2	4
3	5
4	6

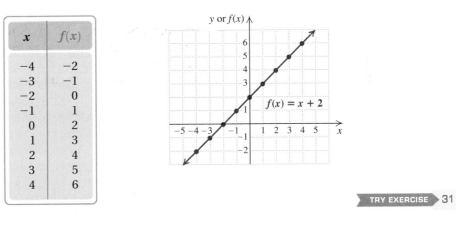

> **TRY EXERCISE** ▸ 31

EXAMPLE   6

Graph: $g(x) = 4 - x^2$.

SOLUTION   Recall from Section 9.6 that the graph is a parabola. We calculate some function values and draw the curve.

$$g(0) = 4 - 0^2 = 4 - 0 = 4,$$
$$g(-1) = 4 - (-1)^2 = 4 - 1 = 3,$$
$$g(2) = 4 - (2)^2 = 4 - 4 = 0,$$
$$g(-3) = 4 - (-3)^2 = 4 - 9 = -5$$

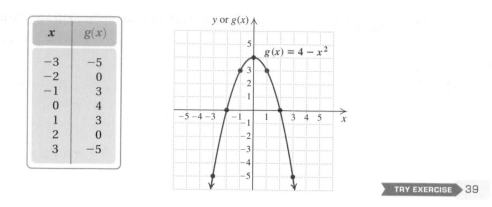

x	g(x)
−3	−5
−2	0
−1	3
0	4
1	3
2	0
3	−5

TRY EXERCISE 39

EXAMPLE 7 Graph: $h(x) = |x|$.

SOLUTION A list of some function values is shown in the following table. We plot the points and connect them. The graph is V-shaped and symmetric, rising on either side of the vertical axis.

x	h(x)
−3	3
−2	2
−1	1
0	0
1	1
2	2
3	3

TRY EXERCISE 37

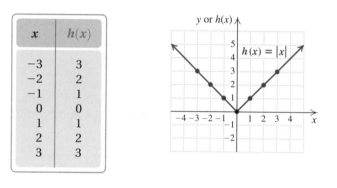

Recognizing Graphs of Functions

Consider the function f described by $f(x) = x^2 - 5$. Its graph is shown at left. It is also the graph of the equation $y = x^2 - 5$.

To find a function value, like $f(3)$, from a graph, we locate the input on the horizontal axis, move vertically to the graph of the function, and then horizontally to find the output on the vertical axis, where members of the range are found.

Recall that when one member of the domain is paired with two or more different members of the range, the correspondence is not a function. Thus, when a graph contains two or more points with the same first coordinate, it cannot represent a function. Points sharing a common first coordinate are vertically above and below each other, as shown in the following figure.

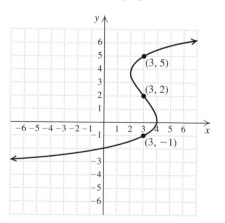

Since 3 is paired with more than one member of the range, the graph does not represent a function.

This observation leads to the *vertical-line test*.

> ### The Vertical-Line Test
> A graph represents a function if it is impossible to draw a vertical line that intersects the graph more than once.

EXAMPLE 8 Determine whether each of the following is the graph of a function.

a)

b)

c)

d)

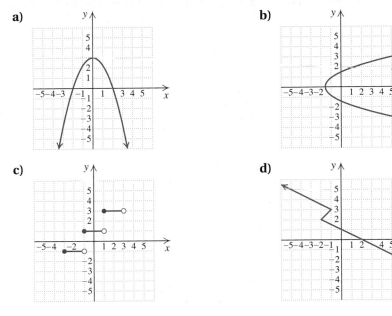

STUDENT NOTES

To apply the vertical-line test, you can lay a ruler, pencil, or ID card parallel to the *y*-axis at the left side of the graph. As you move the straightedge from left to right, the vertical lines formed cannot cross the graph more than once.

SOLUTION

a) The graph *is* that of a function because it is not possible to draw a vertical line that crosses the graph more than once. This can be confirmed with a ruler or straightedge.

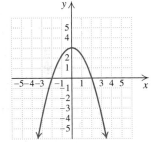

b) The graph is *not* that of a function because it does not pass the vertical-line test. The line $x = 1$ is one of many vertical lines that cross the graph at more than one point.

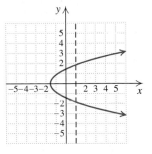

STUDY SKILLS

Ask to See Your Final Exam

Once the course is over, many students neglect to find out how they fared on the final exam. Please don't overlook this valuable opportunity to extend your learning. It is important for you to find out what mistakes you may have made and to be certain no grading errors have occurred.

c) The graph *is* that of a function because it is not possible to draw a vertical line that crosses the graph more than once. Note that the open dots indicate the absence of a point.

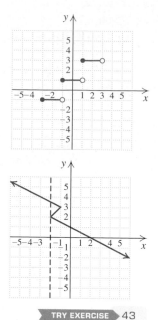

d) The graph *is not* that of a function because it does not pass the vertical-line test. For example, as shown, the line $x = -2$ crosses the graph at more than one point.

TRY EXERCISE ▶ 43

9.7   EXERCISE SET

For Extra Help
MyMathLab   Math XL   PRACTICE   WATCH   DOWNLOAD

✎ **Concept Reinforcement**   *Classify each statement as either true or false.*

1. When we are discussing functions, the notation $f(3)$ does not mean $f \cdot 3$.

2. If $f(x) = x^2$, then $f(-5) = 25$.

3. In order to pass the vertical-line test, a function must score over 70%.

4. If a graph includes both $(5, 9)$ and $(5, 7)$, it cannot represent a function.

Determine whether each correspondence is a function.

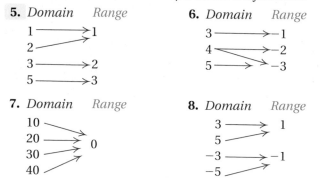

5. Domain   Range
1 ⟶ 1
2
3 ⟶ 2
5 ⟶ 3

6. Domain   Range
3 ⟶ 1
4 ⟶ 2
5 ⟶ 3

7. Domain   Range
10
20 ⟶ 0
30
40

8. Domain   Range
3 ⟶ 1
5
-3 ⟶ -1
-5

9. Domain   *Range*
Texas ⟶ Austin
⟶ Houston
⟶ Dallas
Ohio ⟶ Cleveland
⟶ Toledo
⟶ Cincinnati

10. Domain   *Range*
Austin
Houston ⟶ Texas
Dallas
Cleveland
Toledo ⟶ Ohio
Cincinnati

11. *Domain*
(Where college spending money goes, nationally*)
Food ⟶ 78%
Transportation ⟶ 7%
Books
Clothes ⟶ 3%
Cigarettes ⟶ 1%
Social activities ⟶ 10%
Personal items

Range
(Percentage of spending money)

---

*Due to rounding, the total exceeds 100% (*Source: USA Today*).

12.

Domain	Range
(Brand of single-serving pizza)	(Number of calories)

Old Chicago Pizza-lite ──────→324
Smart Ones Cheese ╲
Banquet Zap Cheese ──────→310
Lean Cuisine Cheese ╱
Pizza Hut Supreme Personal Pan ──→647
Celeste Suprema Pizza-For-One ──→678

Find the indicated outputs.

13. $f(4), f(7),$ and $f(-2)$

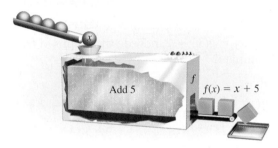

14. $g(1), g(6),$ and $g(13)$

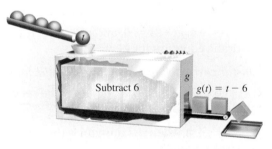

15. $h(p) = -3p$; find $h(-7), h(5),$ and $h(10)$.

16. $f(p) = 10p$; find $f(6), f\left(-\frac{1}{2}\right),$ and $f(20)$.

17. $g(s) = 3s - 2$; find $g(1), g(-5),$ and $g(2.5)$.

18. $h(t) = 12$; find $h(6), h(-25),$ and $h(14.6)$.

19. $F(x) = 2x^2 + x$; find $F(-1), F(0),$ and $F(2)$.

20. $P(x) = x^3 + x$; find $P(0), P(-1),$ and $P(5)$.

21. $f(t) = (t - 3)^2$; find $f(-2), f(8),$ and $f\left(\frac{1}{2}\right)$.

22. $g(t) = |t - 3|$; find $g(-2), g(8),$ and $g\left(\frac{1}{2}\right)$.

23. $h(x) = |2x| - x$; find $h(0), h(-4),$ and $h(4)$.

24. $f(x) = x^4 - x$; find $f(-1), f(2),$ and $f(0)$.

25. *Life span.* The function given by $l(x) = \dfrac{1700}{x}$ can be used to approximate the life span of an animal with a pulse rate of x beats per minute.

   a) Find the approximate life span of a horse with a pulse rate of 50 beats per minute.

   b) Find the approximate life span of a seal with a pulse rate of 85 beats per minute.

26. *Temperature as a function of depth.* The function given by $T(d) = 10d + 20$ gives the temperature, in degrees Celsius, inside the earth as a function of the depth d, in kilometers. Find the temperature at 5 km, 20 km, and 1000 km.

27. *Predicting heights.* An anthropologist can estimate the height of a male or a female, given the lengths of certain bones. A *humerus* is the bone from the elbow to the shoulder. The height, in centimeters, of a female with a humerus of x centimeters is given by

$$F(x) = 2.75x + 71.48.$$

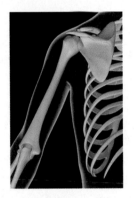

If a humerus is known to be from a female, how tall was the female if the bone is **(a)** 32 cm long? **(b)** 30 cm long?

28. When a humerus (see Exercise 27) is from a male, the function given by $M(x) = 2.89x + 70.64$ is used to find the male's height, in centimeters. If a humerus is known to be from a male, how tall was the male if the bone is **(a)** 30 cm long? **(b)** 35 cm long?

29. *Temperature conversions.* The function given by $C(F) = \frac{5}{9}(F - 32)$ determines the Celsius temperature that corresponds to F degrees Fahrenheit. Find the Celsius temperature that corresponds to 62°F, 77°F, and 23°F.

30. *Pressure at sea depth.* The function given by $P(d) = 1 + (d/33)$ gives the pressure, in *atmospheres* (atm), at a depth of d feet, in the sea. Note that $P(0) = 1$ atm, $P(33) = 2$ atm, and so on. Find the pressure at 20 ft, 30 ft, and 100 ft.

Graph each function.

31. $f(x) = 2x - 3$

32. $g(x) = 2x + 5$

33. $g(x) = -x + 4$

34. $f(x) = -\frac{1}{2}x + 2$

35. $f(x) = \frac{1}{2}x + 1$

36. $f(x) = -\frac{3}{4}x - 2$

37. $g(x) = 2|x|$

38. $h(x) = -|x|$

39. $g(x) = x^2$

40. $f(x) = x^2 - 1$

41. $f(x) = x^2 - x - 2$

42. $g(x) = x^2 + 6x + 5$

Determine whether each graph is that of a function.

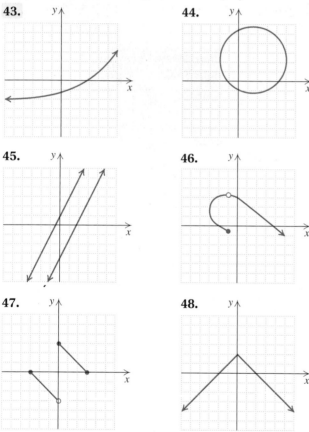

43.

44.

45.

46.

47.

48.

49. Is it possible for a function to have more numbers in the range than in the domain? Why or why not?

50. Is it possible for a function to have more numbers in the domain than in the range? Why or why not?

Skill Review

Review graphing.

Graph on a plane.

51. $y = \frac{4}{5}x$ [3.2]

52. $y = -\frac{1}{2}x + 3$ [3.6]

53. $x - 2y < 4$ [7.5]

54. $y = -3$ [3.3]

55. $y = x^2 - 6x + 1$ [9.6]

56. $3x - y = 6$ [3.3]

Synthesis

57. Explain in your own words how the vertical-line test works.

58. If $f(x) = g(x) + 2$, how do the graphs of f and g compare?

Graph.

59. $g(x) = x^3$

60. $f(x) = 2 + \sqrt{x}$

61. $f(x) = |x| + x$

62. $g(x) = |x| - x$

63. Sketch a graph that is not that of a function.

64. If $f(-1) = -7$ and $f(3) = 8$, find a linear equation for $f(x)$.

65. If $g(0) = -4, g(-2) = 0$, and $g(2) = 0$, find a quadratic equation for $g(x)$.

Find the range of each function for the given domain.

66. $f(x) = 5 - 3x$, when the domain is the set of whole numbers less than 4

67. $g(t) = t^2 - t$, when the domain is the set of integers between -4 and 2

68. $f(m) = m^3 + 1$, when the domain is the set of integers between -3 and 3

69. $h(x) = |x| - x$, when the domain is the set of integers between -2 and 20

70. Use a graphing calculator to check your answers to Exercises 59–62 and 64–69.

Study Summary

KEY TERMS AND CONCEPTS	EXAMPLES

SECTION 9.1: SOLVING QUADRATIC EQUATIONS: THE PRINCIPLE OF SQUARE ROOTS

A **quadratic equation in standard form** is written $ax^2 + bx + c = 0$, with a, b, and c constant and $a \neq 0$.

Some quadratic equations can be solved by factoring and using the principle of zero products.

$$x^2 - 3x - 10 = 0$$
$$(x + 2)(x - 5) = 0 \qquad \text{Factoring}$$
$$x + 2 = 0 \quad \text{or} \quad x - 5 = 0 \qquad \text{Using the principle of zero products}$$
$$x = -2 \quad \text{or} \qquad x = 5$$

Some quadratic equations can be solved using the principle of square roots.

The Principle of Square Roots

For any nonnegative real number p,
$x^2 = p$ is equivalent to
$x = \sqrt{p} \quad \text{or} \quad -\sqrt{p}.$

$$x^2 - 8x + 16 = 25 \qquad \text{The left side is a perfect-square trinomial.}$$
$$(x - 4)^2 = 25$$
$$x - 4 = -5 \quad \text{or} \quad x - 4 = 5 \qquad \text{Using the principle of square roots}$$
$$x = -1 \quad \text{or} \qquad x = 9$$

SECTION 9.2: SOLVING QUADRATIC EQUATIONS: COMPLETING THE SQUARE

Any quadratic equation can be solved by completing the square. To complete the square for an equation like $x^2 + bx = k$, add half of the coefficient of x, squared, or $(b/2)^2$, to both sides.

$$x^2 + 6x = 1$$
$$x^2 + 6x + \left(\frac{6}{2}\right)^2 = 1 + \left(\frac{6}{2}\right)^2 \qquad \text{Completing the square on the left side}$$
$$x^2 + 6x + 9 = 1 + 9$$
$$(x + 3)^2 = 10 \qquad \text{Factoring}$$
$$x + 3 = \pm\sqrt{10} \qquad \text{Using the principle of square roots}$$
$$x = -3 \pm \sqrt{10}$$

SECTION 9.3: THE QUADRATIC FORMULA AND APPLICATIONS

Any quadratic equation can be solved using the quadratic formula.

The Quadratic Formula

The solutions of a quadratic equation $ax^2 + bx + c = 0$ are given by
$$x = \frac{-b \pm \sqrt{b^2 - 4ac}}{2a}.$$

$$3x^2 - 2x - 5 = 0$$
$$x = \frac{-(-2) \pm \sqrt{(-2)^2 - 4 \cdot 3(-5)}}{2 \cdot 3} \qquad a = 3, b = -2, c = -5$$
$$x = \frac{2 \pm \sqrt{4 + 60}}{6}$$
$$x = \frac{2 \pm \sqrt{64}}{6}$$
$$x = \frac{2 \pm 8}{6}$$
$$x = \frac{10}{6} = \frac{5}{3} \quad \text{or} \quad x = \frac{-6}{6} = -1$$

SECTION 9.4: FORMULAS

To solve formulas, we use the same steps that we use to solve equations.

Solve $\sqrt{3m - n} - 1 = 9$ for m.

$$\sqrt{3m - n} - 1 = 9$$
$$\sqrt{3m - n} = 10 \qquad \text{Isolating the radical expression}$$
$$(\sqrt{3m - n})^2 = 10^2 \qquad \text{Using the principle of squaring}$$
$$3m - n = 100$$
$$3m = 100 + n$$
$$m = \frac{100 + n}{3} \qquad \text{We have solved for } m.$$

SECTION 9.5: COMPLEX NUMBERS AS SOLUTIONS OF QUADRATIC EQUATIONS

The square root of a negative number can be written using i.

$$i^2 = -1 \quad \text{and} \quad i = \sqrt{-1}$$
$$\sqrt{-p} = i\sqrt{p}$$

$$\sqrt{-25} = \sqrt{-1}\sqrt{25} = i \cdot 5 = 5i;$$
$$-\sqrt{-3} = -\sqrt{-1}\sqrt{3} = -i\sqrt{3} = -\sqrt{3}i$$

Solutions of quadratic equations may be imaginary numbers.

$$x^2 - x + 1 = 0$$
$$x = \frac{-(-1) \pm \sqrt{(-1)^2 - 4(1)(1)}}{2(1)} \qquad x = \frac{-b \pm \sqrt{b^2 - 4ac}}{2a}$$
$$x = \frac{1 \pm \sqrt{1 - 4}}{2}$$
$$x = \frac{1 \pm \sqrt{-3}}{2}$$
$$x = \frac{1 \pm i\sqrt{3}}{2}$$
$$x = \frac{1}{2} + \frac{\sqrt{3}}{2}i \quad or \quad x = \frac{1}{2} - \frac{\sqrt{3}}{2}i$$

SECTION 9.6: GRAPHS OF QUADRATIC EQUATIONS

The graph of a quadratic equation $y = ax^2 + bx + c$ is a **parabola**. The graph opens upward for $a > 0$ and downward for $a < 0$. The **vertex** and the **axis of symmetry** occur at $x = -\frac{b}{2a}$.

$$y = x^2 - 2x - 8$$

x-coordinate of vertex: $-\dfrac{b}{2a} = -\dfrac{-2}{-2(1)} = 1$

y-coordinate of vertex: $y = (1)^2 - 2(1) - 8 = -9$

x	y $y = x^2 - 2x - 8$
-2	0
-1	-5
0	-8
1	-9
2	-8
3	-5
4	0

SECTION 9.7: FUNCTIONS

A **function** is a correspondence that assigns to each member of the **domain** exactly one member of the **range**. For a function f, the notation $f(a)$ represents the **output** that is paired with the **input** a.

If $f(x) = x^2 - x + 3$, then

$$f(-1) = (-1)^2 - (-1) + 3$$
$$= 1 + 1 + 3$$
$$= 5.$$

The Vertical-Line Test

A graph represents a function if it is impossible to draw a vertical line that intersects the graph more than once.

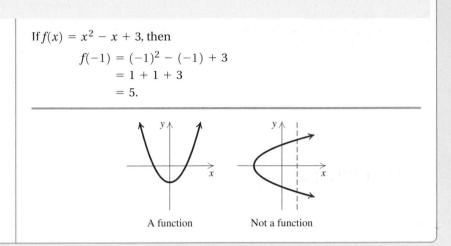

A function   Not a function

Review Exercises: Chapter 9

▶ *Concept Reinforcement* In each of Exercises 1–8, match the item with the appropriate item from the column on the right.

1. ____ An equation most easily solved by factoring [9.3]

2. ____ An equation most easily solved using the principle of square roots [9.1]

3. ____ An equation most easily solved by completing the square [9.2]

4. ____ An equation most easily solved by the quadratic formula [9.3]

5. ____ A parabola that opens upward [9.6]

6. ____ A parabola that opens downward [9.6]

7. ____ An expression for "the input 2 is paired with the output 5" [9.7]

8. ____ An expression for "the input 5 is paired with the output 2" [9.7]

a) $f(2) = 5$

b) $y = 3x^2 + 4x - 7$

c) $f(5) = 2$

d) $y = -3x^2 + 4x - 7$

e) $3x^2 - 4x - 5 = 0$

f) $(x - 7)^2 = 64$

g) $x^2 + 8x = 3$

h) $x^2 - 7x + 6 = 0$

Solve by completing the square. [9.2]

9. $x^2 - 10x = 1$

10. $2x^2 + 3x - 2 = 0$

Solve.

11. $5x^2 = 30$ [9.1]

12. $5x^2 - 8x + 3 = 0$ [9.2]

13. $x^2 - 2x - 10 = 0$ [9.3]

14. $(x + 3)^2 + 4 = 0$ [9.5]

15. $3y^2 + 5y = 2$ [9.2]

16. $11t^2 = t$ [9.3]

17. $x^2 + 1 = 0$ [9.5]

18. $2y^2 - 6y = 20$ [9.3]

19. $(p + 10)^2 = 12$ [9.1]   **20.** $x^2 + 6x = 9$ [9.3]

21. $x^2 + x + 1 = 0$ [9.5]   **22.** $1 + 4x^2 = 8x$ [9.3]

23. $t^2 + 9 = 6t$ [9.3]   **24.** $40 = 5y^2$ [9.1]

25. $3m = 4 + 5m^2$ [9.5]   **26.** $6x^2 + 11x = 35$ [9.2]

Solve. [9.4]

27. $p = ct - bt$, for t **28.** $\sqrt{2a} + 3 = 8$, for a

29. $m + n = \dfrac{p}{n}$, for n

30. $\dfrac{1}{r} + \dfrac{1}{s} = \dfrac{1}{t}$, for t

Approximate the solutions to the nearest thousandth. [9.3]

31. $x^2 + 3 = 5x$ **32.** $4y^2 + 8y + 1 = 0$

33. *Right triangles.*   The hypotenuse of a right triangle is 7 m long. One leg is 3 m longer than the other. Find the lengths of the legs. Round to the nearest thousandth. [9.3]

34. *Investments.* $1200 is invested at interest rate r, compounded annually. In 2 yr, it grows to $1297.92. What is the interest rate? [9.3]

35. *Area of a rectangle.*   The length of a rectangle is 3 m greater than the width. The area is 108 m². Find the length and the width. [9.3]

36. *Free-fall time.*   The height of the Lake Point Towers in Chicago is 645 ft. How long would it take an object to fall from the top? (Use $s = 16t^2$.) [9.3]

Express in terms of i. [9.5]

37. $\sqrt{-9}$ **38.** $-\sqrt{-125}$

Graph. Label the vertex, the y-intercept, and any x-intercepts. [9.6]

39. $y = 2 - x^2$ **40.** $y = x^2 + 4x - 1$

41. $y = 3x^2 - 10x$ **42.** $y = x^2 - 2x + 1$

43. If $f(t) = 3t - 2$, find $f(4), f(-2)$, and $f(1.5)$. [9.7]

44. If $g(x) = |x + 2|$, find $g(-3), g(-1)$, and $g(0)$. [9.7]

45. *Calories.*   Each day a moderately active person needs about 15 calories per pound of body weight. The function given by $C(p) = 15p$ approximates the number of calories that are needed to maintain body weight p, in pounds. How many calories are needed to maintain a body weight of 130 lb? [9.7]

Graph. [9.7]

46. $g(x) = x - 6$ **47.** $f(x) = x^2 + 2$

48. $h(x) = 3|x|$

Determine whether each graph is that of a function. [9.7]

49. 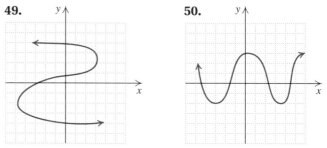   **50.**

Synthesis

51. Brody is shown the graphs of a line and a parabola, and told that one is the graph of $y = 3x + 2$ and the other is the graph of $y = 3x^2 + 2$. Explain how he can match each graph with its equation without calculating any points. [9.6]

52. How can $b^2 - 4ac$ be used to determine how many x-intercepts the graph of $y = ax^2 + bx + c$ has? [9.6]

53. A graph consisting of three points is not the graph of a function. Where must the points be in relation to each other? [9.7]

54. Two consecutive integers have squares that differ by 63. Find the integers. [9.3]

55. Find b such that the trinomial $x^2 + bx + 49$ is the square of a binomial. [9.2]

56. Solve: $x - 4\sqrt{x} - 5 = 0$. [9.3]

57. A square with sides of length s has the same area as a circle with radius 5 in. Find s. [9.3]

Solve.

1. $8x^2 = 80$

2. $x^2 = -9$

3. $2t^2 - 5t = 0$

4. $(t - 1)^2 = 8$

5. $50 = p^2 - 5p$

6. $3m^2 + 13m = 10$

7. $y^2 = y + 5$

8. $x^2 - 4x = -4$

9. $x^2 - 4x = -5$

10. $m^2 - 4m = 2$

11. $10 = 4x + x^2$

12. $3x^2 - 7x + 1 = 0$

13. Solve by completing the square:
$$x^2 - 4x - 10 = 0.$$

Solve.

14. $3 = n + 2\sqrt{p + 5}$, for p

15. $1 + t = \dfrac{a}{b}$, for b

16. Approximate the solutions to the nearest thousandth:
$$x^2 - 3x - 8 = 0.$$

17. *Area of a rectangle.*   The width of a rectangle is 4 m less than the length. The area is 16.25 m². Find the length and the width.

18. *Diagonals of a polygon.*   A polygon has 44 diagonals. How many sides does it have? (*Hint*: $d = (n^2 - 3n)/2$.)

Express in terms of i.

19. $\sqrt{-200}$

20. $-\sqrt{-100}$

Graph. Label the vertex, the y-intercept, and any x-intercepts.

21. $y = -x^2 + x - 5$

22. $y = x^2 + 2x - 15$

23. If $f(x) = \frac{1}{2}x + 1$, find $f(0), f(1),$ and $f(2)$.

24. If $g(t) = -2t^2 + 5t$, find $g(-1), g(0),$ and $g(3)$.

25. *World records.*   The world record for the 10,000-m run has been decreasing steadily since 1940. The function given by $R(t) = 30.18 - 0.06t$ estimates the record, in minutes, t years after 1940. Predict what the record will be in 2010.

Graph.

26. $h(x) = x - 4$

27. $g(x) = x^2 - 4$

Determine whether each graph is that of a function.

28.

29.

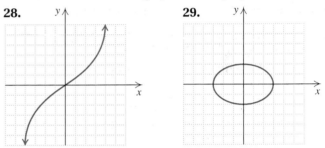

Synthesis

30. Find the area of a square whose diagonal is 5 ft longer than a side.

31. Solve this system for x. Use the substitution method.
$$x - y = 2,$$
$$xy = 4$$

Cumulative Review/Final Exam: Chapters 1–9

1. Evaluate $(1 - x)^2 + 3x$ for $x = -2$. [1.8]

2. Find the absolute value: $|-13|$. [1.4]

3. Find the LCM of 15 and 24. [6.3]

4. Remove parentheses and simplify:
   $2n - 5 - (5n - 6)$. [1.8]

Compute and simplify.

5. $-6 + 12 + (-4) + 7$ [1.5]

6. $2.8 - (-12.2)$ [1.6]

7. $-\frac{3}{8} \div \frac{5}{2}$ [1.7]

8. $13 \cdot 6 \div 3 \cdot 2 \div 13$ [1.8]

Simplify. Write the result using scientific notation.

9. $(2.1 \times 10^7)(1.3 \times 10^{-12})$ [4.8]

10. $\dfrac{5.2 \times 10^{-1}}{2.6 \times 10^{-15}}$ [4.8]

Simplify.

11. $x^{-6} \cdot x^2$ [4.8]

12. $\dfrac{y^3}{y^{-4}}$ [4.8]

13. $(3m^4)^2$ [4.1]

14. Combine like terms and arrange in descending order:
   $2x - 3 + 5x^3 - 2x^3 + 7x^3 + x$. [4.2]

Perform the indicated operation and simplify.

15. $(4x^3 + 3x^2 - 5) + (3x^3 - 5x^2 + 4x - 12)$ [4.3]

16. $(6x^2 - 4x + 1) - (-6x^2 - 4x + 7)$ [4.3]

17. $-2y^2(4y^2 - 3y + 1)$ [4.4]

18. $(2t - 3)(3t^2 - 4t + 2)$ [4.4]

19. $\left(t - \frac{1}{4}\right)\left(t + \frac{1}{4}\right)$ [4.5]

20. $(3m - 2)^2$ [4.5]

21. $(2x + 7)(3x - 8)$ [4.5]

22. $(12x^2y + 10xy^2 - 7) - (3x^2y^2 - 2xy^2 - 5)$ [4.6]

23. $(5p^2 + 2q)^2$ [4.6]

24. $\dfrac{4}{2x - 6} \cdot \dfrac{x - 3}{x + 3}$ [6.2]

25. $\dfrac{3a^4}{a^2 - 1} \div \dfrac{2a^3}{a^2 - 2a + 1}$ [6.2]

26. $\dfrac{3}{3x - 1} + \dfrac{4}{5x}$ [6.4]

27. $\dfrac{2}{x^2 - 16} - \dfrac{x - 3}{x^2 - 9x + 20}$ [6.4]

28. $(x^3 + 7x^2 - 2x + 3) \div (x - 2)$ [4.7]

Factor.

29. $18x^2 - 8$ [5.4]

30. $y^3 - 6y^2 - 5y + 30$ [5.1]

31. $m^2 + 10m + 25$ [5.4]

32. $10x^3 - 11x^2 - 6x$ [5.3]

33. $25x^3 - 20x^2 + 5x$ [5.1]

34. $t^2 - t - 6$ [5.2]

35. $49p^2 + 50p + 1$ [5.3]

36. $100u^2v^2 - 1$ [5.4]

37. $4x^2 - 20xy + 25y^2$ [5.4]

38. $3y^4 - 39y^3 + 120y^2$ [5.2]

Simplify.

39. $\dfrac{\dfrac{3}{x} + \dfrac{1}{2x}}{\dfrac{1}{3x} - \dfrac{3}{4x}}$ [6.5]

40. $\sqrt{64}$ [8.1]

41. $-\sqrt[3]{-125}$ [8.7]

42. $\sqrt{49t^2}$ [8.1]
   (Assume $t \geq 0$.)

43. $16^{-3/4}$ [8.7]

44. $\sqrt{250x^6y^7}$ [8.2]
   (Assume $x, y \geq 0$.)

45. $\sqrt{\dfrac{16}{9}}$ [8.3]

46. $(1 + 3\sqrt{2})(1 - 3\sqrt{2})$ [8.4]

47. $5\sqrt{12} + 2\sqrt{48}$ [8.4]

48. Multiply: $\sqrt{a + b}\sqrt{a - b}$. [8.2]

49. Multiply and simplify: $\sqrt{32ab}\sqrt{6a^4b^2}$ $(a, b \geq 0)$.
   [8.2]

50. Divide and simplify: $\dfrac{\sqrt{72}}{\sqrt{45}}$. [8.3]

Solve. If an equation has no solution, state this.

51. $-5x = 30$ [2.1]

52. $-5x > 30$ [2.6]

53. $3(y - 1) - 2(y + 2) = 0$ [2.2]

54. $x^2 - 8x + 15 = 0$ [5.6]

55. $y - x = 1,$
$\quad y = 3 - x$ [7.2]

56. $\dfrac{x}{x + 1} = \dfrac{3}{2x} + 1$ [6.6]

57. $4x - 3y = 3,$
$\quad 3x - 2y = 4$ [7.3]

58. $x^2 - x - 6 = 0$ [5.6]

59. $x^2 - 2x = 2$ [9.3]

60. $3 - x = \sqrt{x^2 - 3}$ [8.5]

61. $4 - 9x \le 18 + 5x$ [2.6]

62. $-\frac{7}{8}x + 7 = \frac{3}{8}x - 3$ [2.2]

63. $0.6x - 1.8 = 1.2x$ [2.2]

64. $x + y = 15,$
$\quad x - y = 15$ [7.3]

65. $x^2 + 2x + 5 = 0$ [9.5]

66. $3y^2 = 30$ [9.1]

67. $(x - 3)^2 = 6$ [9.1]

68. $\dfrac{t}{t + 2} = \dfrac{1}{t}$ [6.6]

69. $12x^2 + x = 20$ [5.6]

70. $\dfrac{2x}{x + 3} + \dfrac{6}{x} + 7 = \dfrac{18}{x^2 + 3x}$ [6.6]

71. $\sqrt{x + 9} = \sqrt{2x - 3}$ [8.5]

Solve each formula for the given letter.

72. $ay = by - ax$, for a [9.4]

73. $\dfrac{1}{t} = \dfrac{1}{m} - \dfrac{1}{n}$, for m [9.4]

74. Solve $9x^2 - 12x - 2 = 0$ by completing the square. [9.2]

75. Approximate the solutions of $4x^2 = 4x + 1$ to the nearest thousandth. [9.3]

Graph on a plane.

76. $y = \frac{1}{3}x - 5$ [3.6]

77. $2x + 3y = -6$ [3.3]

78. $y = -2$ [3.3]

79. $4x - 3y > 12$ [7.5]

80. $y = x^2 + 2x + 1$ [9.6]

81. $x \ge -3$ [7.5]

82. Graph $y = x^2 + 2x - 5$. Label the vertex, the y-intercept, and the x-intercepts. [9.6]

83. Graph the following system of inequalities: [7.6]
$\quad x + y \le 6,$
$\quad x + y \ge 2,$
$\quad\quad x \le 3,$
$\quad\quad x \ge 1.$

84. Find the slope and the y-intercept:
$\quad -6x + 3y = -24.$ [3.6]

85. Find the slope of the line containing the points $(-5, -6)$ and $(-4, 9)$. [3.5]

86. Find a point–slope equation for the line containing $(1, -3)$ and having slope $m = -\frac{1}{2}$. [3.7]

87. Determine whether the graphs of the following equations are parallel:
$\quad y - x = 4,$
$\quad 3y = 5 + 3x.$ [3.6]

88. For the function f described by
$\quad f(x) = 2x^2 + 7x - 4,$
find $f(0), f(-4),$ and $f\left(\frac{1}{2}\right)$. [9.7]

89. Simplify: $-\sqrt{-1}$. [9.5]

Solve.

90. *Carbon dioxide emissions.* If practices do not change, worldwide carbon dioxide emissions from burning fossil fuels are expected to increase from 28 billion tons in 2005 to 42 billion tons in 2030. Find the rate at which emissions will increase. [3.4]
Source: U.S. Energy Information Administration

91. *Carbon dioxide emissions.* Between 2005 and 2006, energy-related carbon dioxide emissions in the United States fell to 5877 million tons, a drop of 1.3%. How many tons of carbon dioxide were emitted in the United States in 2005? [2.4]
Source: U.S. Energy Information Administration

92. *Alternative-fueled vehicles.* In 2005, approximately 891,000 alternative-fueled vehicles were made available in the United States. Of these, there were five times as many E-85 flexible fuel vehicles as there were all other types of vehicles. How many E-85 flexible-fuel vehicles were made available in 2005? [2.5]
Source: U.S. Energy Information Administration

93. *Candy blends.* The Best Chocolate in Town mixes dark-chocolate cheesecake truffles worth $28.50 per pound with milk-chocolate hazelnut truffles worth $24.50 per pound to make 20 lb of a mixture worth $26.90 per pound. How many pounds of each kind of chocolate do they use? [7.4]

94. *Sighting to the horizon.* At a height of h meters, one can see V kilometers to the horizon, where $V = 3.5\sqrt{h}$. Shari can see 35 km to the horizon from the top of a cell-phone tower. What is the height of the tower? [8.5]

95. *Hemoglobin.* A normal 10-cc specimen of human blood contains 1.2 g of hemoglobin. How much hemoglobin would 16 cc of the same blood contain? [6.7]

96. The length of a rectangle is 4 ft longer than twice the width. The area of the rectangle is 30 ft^2. Find the dimensions of the rectangle. [5.7]

97. The length of a rectangle is 7 m more than the width. The length of a diagonal is 13 m. Find the dimensions of the rectangle. [9.3]

98. The hypotenuse of a right triangle is 40 cm long and the length of a leg is 10 cm. Find the length of the other leg. Give an exact answer and an approximation to three decimal places. [8.6]

Synthesis

99. Find b such that the trinomial $x^2 - bx + 225$ is the square of a binomial. [9.2]

100. Find x. [8.6]

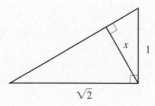

Determine whether each pair of expressions is equivalent.

101. $x^2 - 9$, $(x - 3)(x + 3)$ [4.5]

102. $\dfrac{n^2 + 2n}{2n + 3}$, $\dfrac{n^2}{3}$ [6.1]

103. $(x + 5)^2$, $x^2 + 25$ [4.5]

104. $\sqrt{(t - 3)^2}$, $t - 3$ [8.1]

105. $\sqrt{9x^4}$, $3x^2$ [8.1]

Appendixes

A   Factoring Sums or Differences of Cubes

Factoring a Sum of Two Cubes • Factoring a Difference of Two Cubes

It is possible to factor both a sum and a difference of two cubes. To see how, consider the following:

$$(A + B)(A^2 - AB + B^2) = A(A^2 - AB + B^2) + B(A^2 - AB + B^2)$$
$$= A^3 - A^2B + AB^2 + A^2B - AB^2 + B^3$$
$$= A^3 + B^3$$

and

$$(A - B)(A^2 + AB + B^2) = A(A^2 + AB + B^2) - B(A^2 + AB + B^2)$$
$$= A^3 + A^2B + AB^2 - A^2B - AB^2 - B^3$$
$$= A^3 - B^3.$$

These equations show how we can factor a sum or a difference of two cubes.

> ### To Factor a Sum or a Difference of Cubes
> $$A^3 + B^3 = (A + B)(A^2 - AB + B^2),$$
> $$A^3 - B^3 = (A - B)(A^2 + AB + B^2)$$

Remembering this list of cubes may prove helpful when factoring.

N	0.2	0.1	0	1	2	3	4	5	6
N^3	0.008	0.001	0	1	8	27	64	125	216

EXAMPLE   **1**   Factor: $x^3 - 8$.

SOLUTION   We have

$$x^3 - 8 = x^3 - 2^3 = (x - 2)(x^2 + x \cdot 2 + 2^2).$$
$$A^3 - B^3 = (A - B)(A^2 + A \ B + B^2)$$

This tells us that $x^3 - 8 = (x - 2)(x^2 + 2x + 4)$. Note that we cannot factor $x^2 + 2x + 4$. (It is not a perfect-square trinomial nor can it be factored by trial and error or grouping.) The check is left to the student.   **TRY EXERCISE**   5

EXAMPLE   2   Factor: $x^3 + 125$.

SOLUTION   We have

$$x^3 + 125 = x^3 + 5^3 = (x + 5)\,(x^2 - x \cdot 5 + 5^2).$$
$$\uparrow \quad \uparrow \quad \uparrow \quad \uparrow \quad \uparrow \quad \uparrow \; \uparrow \quad \uparrow$$
$$A^3 + B^3 = (A + B)(A^2 - A\ B + B^2)$$

Thus, $x^3 + 125 = (x + 5)(x^2 - 5x + 25)$. The check is left to the student.

> TRY EXERCISE   1

EXAMPLE   3   Factor: $250 + 128t^3$.

SOLUTION   We first look for a common factor:

$$250 + 128t^3 = 2[125 + 64t^3]$$
$$= 2[5^3 + (4t)^3] \qquad \text{This is of the form } A^3 + B^3, \text{ where}$$
$$ A = 5 \text{ and } B = 4t.$$
$$= 2[(5 + 4t)(25 - 20t + 16t^2)].$$

We check using the distributive law:

$$2(5 + 4t)(25 - 20t + 16t^2)$$
$$= 2[5(25 - 20t + 16t^2) + 4t(25 - 20t + 16t^2)]$$
$$= 2[125 - 100t + 80t^2 + 100t - 80t^2 + 64t^3]$$
$$= 2[125 + 64t^3]$$
$$= 250 + 128t^3.$$

We have a check. The factorization is

$$2(5 + 4t)(25 - 20t + 16t^2).$$

> TRY EXERCISE   21

EXAMPLE   4   Factor: $y^3 - 0.001$.

SOLUTION   Since $0.001 = (0.1)^3$, we have a difference of cubes:

$$y^3 - 0.001 = (y - 0.1)(y^2 + 0.1y + 0.01).$$

The check is left to the student.

> TRY EXERCISE   29

Remember the following about factoring sums or differences of squares and cubes:

Difference of cubes: $A^3 - B^3 = (A - B)(A^2 + AB + B^2),$

Sum of cubes: $A^3 + B^3 = (A + B)(A^2 - AB + B^2),$

Difference of squares: $A^2 - B^2 = (A + B)(A - B).$

There is no formula for factoring a sum of squares.

A   EXERCISE SET

Factor completely.

1. $t^3 + 8$

2. $p^3 + 27$

3. $x^3 + 1$

4. $w^3 - 1$

5. $z^3 - 125$

6. $a^3 + 64$

7. $8a^3 - 1$

8. $27x^3 - 1$

9. $y^3 - 27$

10. $p^3 - 64$

11. $64 + 125x^3$

12. $8 + 27b^3$

13. $125p^3 + 1$

14. $64w^3 + 1$

15. $27m^3 - 64$

16. $8t^3 - 27$

17. $p^3 - q^3$

18. $a^3 + b^3$

19. $x^3 + \frac{1}{8}$

20. $y^3 - \frac{1}{27}$

21. $2y^3 - 128$

22. $3z^3 - 375$

23. $24a^3 + 3$

24. $54x^3 + 2$

25. $rs^3 - 125r$

26. $a^2b^3 + 64a^2$

27. $5x^3 + 40z^3$

28. $2y^3 - 54z^3$

29. $x^3 + 0.008$

30. $y^3 - 0.125$

Synthesis

31. Dino incorrectly believes that
$$a^3 - b^3 = (a - b)(a^2 + b^2).$$
How could you convince him that he is wrong?

32. If $x^3 + c$ is prime, what can you conclude about c? Why?

Factor. Assume that variables in exponents represent natural numbers.

33. $125c^6 + 8d^6$

34. $64x^6 + 8t^6$

35. $3x^{3a} - 24y^{3b}$

36. $\frac{8}{27}x^3 - \frac{1}{64}y^3$

37. $\frac{1}{24}x^3y^3 + \frac{1}{3}z^3$

38. $\frac{1}{16}x^{3a} + \frac{1}{2}y^{6a}z^{9b}$

B   Mean, Median, and Mode

Mean • Median • Mode

One way to analyze data is to look for a single representative number, called a **center point** or **measure of central tendency**. Those most often used are the **mean** (or **average**), the **median**, and the **mode**.

Mean

Let's first consider the *mean,* or *average.*

> #### Mean, or Average
> The *mean,* or *average,* of a set of numbers is the sum of the numbers divided by the number of addends.

EXAMPLE **1**

Consider the following data on revenue, in billions of dollars, for Starbucks Corporation in five recent years:

$$\$2.2, \quad \$2.6, \quad \$3.3, \quad \$4.1, \quad \$5.3.$$

What is the mean of the numbers?

SOLUTION   First, we add the numbers:

$$2.2 + 2.6 + 3.3 + 4.1 + 5.3 = 17.5.$$

Then we divide by the number of addends, 5:

$$\frac{(2.2 + 2.6 + 3.3 + 4.1 + 5.3)}{5} = \frac{17.5}{5} = 3.5.$$

The mean, or average, revenue of Starbucks for those five years is $3.5 billion.

Note that $3.5 + 3.5 + 3.5 + 3.5 + 3.5 = 17.5$. If we use this center point, 3.5, repeatedly as the addend, we get the same sum that we do when adding individual data numbers.

Median

The *median* is useful when we wish to de-emphasize extreme scores. For example, suppose five workers in a technology company manufactured the following number of computers during one day's work:

Sarah: 88
Matt: 92
Pat: 66
Jen: 94
Mark: 91

Let's first list the scores in order from smallest to largest:

66   88   91   92   94.

↑
Middle number

The middle number—in this case, 91—is the **median**.

> ## Median
> Once a set of data has been arranged from smallest to largest, the *median* of the set of data is the middle number if there is an odd number of data numbers. If there is an even number of data numbers, then there are two middle numbers and the median is the *average* of the two middle numbers.

EXAMPLE **2**

Find the median of the following set of household incomes:

$$\$76,000, \quad \$58,000, \quad \$87,000, \quad \$32,500, \quad \$64,800, \quad \$62,500.$$

SOLUTION   We first rearrange the numbers in order from smallest to largest.

$$\$32,500, \quad \$58,000, \quad \$62,500, \quad \$64,800, \quad \$76,000, \quad \$87,000$$

↑
Median

There is an even number of numbers. We look for the middle two, which are $62,500 and $64,800. The median is the average of $62,500 and $64,800:

$$\frac{\$62,500 + \$64,800}{2} = \$63,650.$$

Mode

The last center point we consider is called the *mode*. A number that occurs most often in a set of data is sometimes considered a representative number or center point.

> ### Mode
> The *mode* of a set of data is the number or numbers that occur most often. If each number occurs the same number of times, then there is *no* mode.

EXAMPLE 3   Find the mode of the following data:

23, 24, 27, 18, 19, 27.

SOLUTION   The number that occurs most often is 27. Thus the mode is 27.

It is easier to find the mode of a set of data if the data are ordered.

EXAMPLE 4   Find the mode of the following data:

83, 84, 84, 84, 85, 86, 87, 87, 87, 88, 89, 90.

SOLUTION   There are two numbers that occur most often, 84 and 87. Thus the modes are 84 and 87.

EXAMPLE 5   Find the mode of the following data:

115, 117, 211, 213, 219.

SOLUTION   Each number occurs the same number of times. The set of data has *no* mode.

B   EXERCISE SET

For each set of numbers, find the mean (average), the median, and any modes that exist.

1. 13, 21, 18, 13, 20

2. 5, 2, 8, 10, 7, 1, 9

3. 3, 8, 20, 3, 20, 10

4. 19, 19, 8, 16, 8, 7

5. 4.7, 2.3, 4.6, 4.9, 3.8

6. 13.4, 13.4, 12.6, 42.9

7. 234, 228, 234, 228, 234, 278

8. $29.95, $28.79, $30.95, $29.95

9. *Hurricanes.* The following bar graph shows the number of hurricanes that struck the United States by month from 1851 to 2006. What is the average number for the 8 months given? the median? the mode?

Atlantic Storms and Hurricanes

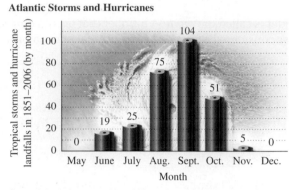

Source: Atlantic Oceanographic and Meteorological Laboratory

10. *iPod prices.* A price comparison showed the following online prices for an Apple iPod Nano:

$199, $197.97, $249.99, $179, $197.97.

What was the average price? the median price? the mode?

11. *NBA tall men.* The following lists the heights, in inches, of the tallest men in the NBA in a recent year. Find the mean, the median, and the mode.

Zydrunas Ilgauskas	87
Yao Ming	90
Dikembe Mutombo	86
Kosta Perovic	86

Source: National Basketball Association

12. *Coffee consumption.* The following lists the annual coffee consumption, in number of cups per person, for various countries. Find the mean, the median, and the mode.

Germany	1113
United States	610
Switzerland	1215
France	798
Italy	750

Source: Beverage Marketing Corporation

13. *PBA scores.* Kelly Kulick rolled scores of 254, 202, 184, 269, 151, 223, 258, 222, and 202 in a recent tour trial for the Professional Bowlers Association. What was her average? her median? her mode?

Source: Professional Bowlers Association

14. *Salmon prices.* The following prices per pound of Atlantic salmon were found at six fish markets:

$8.99, $8.49, $8.99, $9.99, $9.49, $7.99.

What was the average price per pound? the median price? the mode?

Synthesis

15. *Hank Aaron.* Hank Aaron averaged $34\frac{7}{22}$ home runs per year over a 22-yr career. After 21 yr, Aaron had averaged $35\frac{10}{21}$ home runs per year. How many home runs did Aaron hit in his final year?

16. *Length of pregnancy.* Marta was pregnant 270 days, 259 days, and 272 days for her first three pregnancies. In order for Marta's average length of pregnancy to equal the worldwide average of 266 days, how long must her fourth pregnancy last?

Source: David Crystal (ed.), *The Cambridge Factfinder.* Cambridge CB2 1RP: Cambridge University Press, 1993, p. 84.

17. The ordered set of data 18, 21, 24, a, 36, 37, b has a median of 30 and an average of 32. Find a and b.

18. *Male height.* Jason's brothers are 174 cm, 180 cm, 179 cm, and 172 cm tall. The average male is 176.5 cm tall. How tall is Jason if he and his brothers have an average height of 176.5 cm?

C   Sets

Naming Sets   ▪   Membership   ▪   Subsets   ▪   Intersections   ▪   Unions

A **set** is a collection of objects. In mathematics the objects, or **elements**, of a set are generally numbers. This section provides an introduction to sets and how to combine them.

Naming Sets

To name the set of whole numbers less than 6, we can use *roster notation*, as follows:

$$\{0, 1, 2, 3, 4, 5\}.$$

The set of real numbers x for which x is less than 6 cannot be named by listing all its members because there is an infinite number of them. We name such a set using *set-builder notation*, as follows:

$$\{x \mid x < 6\}.$$

This is read

> "The set of all x such that x is less than 6."

See Section 2.6 for more on this notation.

Membership

The symbol $\in$ means *is a member of* or *belongs to*, or *is an element of*. Thus,

$$x \in A$$

means

> x is a member of A,   or x belongs to A,   or x is an element of A.

EXAMPLE **1**   Classify each of the following as true or false.

a) $1 \in \{1, 2, 3\}$

b) $1 \in \{2, 3\}$

c) $4 \in \{x \mid x \text{ is an even whole number}\}$

d) $5 \in \{x \mid x \text{ is an even whole number}\}$

SOLUTION

a) Since 1 is listed as a member of the set, $1 \in \{1, 2, 3\}$ is true.

b) Since 1 is *not* a member of $\{2, 3\}$, the statement $1 \in \{2, 3\}$ is false.

c) Since 4 is an even whole number, $4 \in \{x \mid x \text{ is an even whole number}\}$ is true.

d) Since 5 is *not* even, $5 \in \{x \mid x \text{ is an even whole number}\}$ is false.

TRY EXERCISE ▶ 7

Set membership can be illustrated with a diagram, as shown below.

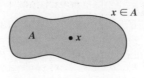

Subsets

If every element of A is also an element of B, then A is a *subset* of B. This is denoted $A \subseteq B$.

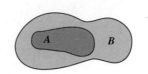

The set of whole numbers is a subset of the set of integers. The set of rational numbers is a subset of the set of real numbers.

EXAMPLE 2   Classify each of the following as true or false.

a) $\{1, 2\} \subseteq \{1, 2, 3, 4\}$

b) $\{p, q, r, w\} \subseteq \{a, p, r, z\}$

c) $\{x | x < 6\} \subseteq \{x | x \leq 11\}$

SOLUTION

a) Since every element of $\{1, 2\}$ is in the set $\{1, 2, 3, 4\}$, it follows that $\{1, 2\} \subseteq \{1, 2, 3, 4\}$ is true.

b) Since $q \in \{p, q, r, w\}$, but $q \notin \{a, p, r, z\}$, it follows that $\{p, q, r, w\} \subseteq \{a, p, r, z\}$ is false.

c) Since every number that is less than 6 is also less than 11, the statement $\{x | x < 6\} \subseteq \{x | x \leq 11\}$ is true.

TRY EXERCISE 15

Intersections

The *intersection* of sets A and B, denoted $A \cap B$, is the set of members common to both sets.

EXAMPLE 3   Find each intersection.

a) $\{0, 1, 3, 5, 25\} \cap \{2, 3, 4, 5, 6, 7, 9\}$

b) $\{a, p, q, w\} \cap \{p, q, t\}$

SOLUTION

a) $\{0, 1, 3, 5, 25\} \cap \{2, 3, 4, 5, 6, 7, 9\} = \{3, 5\}$

b) $\{a, p, q, w\} \cap \{p, q, t\} = \{p, q\}$

TRY EXERCISE 19

Set intersection can be illustrated with a diagram, as shown below.

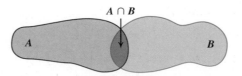

The set without members is known as the *empty set*, and is written $\varnothing$ and sometimes { }. Each of the following is a description of the empty set:

The set of all 12-ft–tall people;

$\{2, 3\} \cap \{5, 6, 7\}$;

$\{x \mid x$ is an even natural number$\} \cap \{x \mid x$ is an odd natural number$\}$.

Unions

Two sets A and B can be combined to form a set that contains the members of both A and B. The new set is called the *union* of A and B, denoted $A \cup B$.

EXAMPLE 4   Find each union.

a) $\{0, 5, 7, 13, 27\} \cup \{0, 2, 3, 4, 5\}$

b) $\{a, c, e, g\} \cup \{b, d, f\}$

SOLUTION

a) $\{0, 5, 7, 13, 27\} \cup \{0, 2, 3, 4, 5\} = \{0, 2, 3, 4, 5, 7, 13, 27\}$

Note that the 0 and the 5 are *not* listed twice in the solution.

b) $\{a, c, e, g\} \cup \{b, d, f\} = \{a, b, c, d, e, f, g\}$ **TRY EXERCISE** 25

Set union can be illustrated with a diagram, as shown below.

$A \cup B$ is shaded.

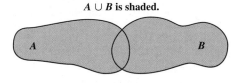

C   EXERCISE SET

Name each set using the roster method.

1. The set of whole numbers 8 through 11

2. The set of whole numbers 83 through 89

3. The set of odd numbers between 40 and 50

4. The set of multiples of 5 between 10 and 40

5. $\{x \mid$ the square of x is 9$\}$

6. $\{x \mid x$ is the cube of $\frac{1}{2}\}$

Classify each statement as either true or false.

7. $5 \in \{x \mid x$ is an odd number$\}$

8. $8 \in \{x \mid x$ is an odd number$\}$

9. Skiing $\in$ The set of all sports

10. Pharmacist $\in$ The set of all professions requiring a college degree

11. $3 \in \{-4, -3, 0, 1\}$

12. $0 \in \{-4, -3, 0, 1\}$

13. $\frac{2}{3} \in \{x \mid x$ is a rational number$\}$

14. $\frac{2}{3} \in \{x \mid x$ is a real number$\}$

15. $\{-1, 0, 1\} \subseteq \{-3, -2, -1, 1\,2, 3\}$

16. The set of vowels $\subseteq$ The set of consonants

17. The set of integers $\subseteq$ The set of rational numbers

18. $\{2, 4, 6\} \subseteq \{1, 2, 3, 4, 5, 6, 7\}$

Find each intersection.

19. $\{a, b, c, d, e\} \cap \{c, d, e, f, g\}$

20. $\{a, e, i, o, u\} \cap \{q, u, i, c, k\}$

21. $\{1, 2, 3, 4, 6, 12\} \cap \{1, 2, 3, 6, 9, 18\}$

22. $\{1, 2, 3, 4, 6, 12\} \cap \{1, 5, 7, 35\}$

23. $\{2, 4, 6, 8\} \cap \{1, 3, 5, 7\}$

24. $\{a, e, i, o, u\} \cap \{m, n, f, g, h\}$

Find each union.

25. $\{a, e, i, o, u\} \cup \{q, u, i, c, k\}$

26. $\{a, b, c, d, e\} \cup \{c, d, e, f, g\}$

27. $\{1, 2, 3, 4, 6, 12\} \cup \{1, 2, 3, 6, 9, 18\}$

28. $\{1, 2, 3, 4, 6, 12\} \cup \{1, 5, 7, 35\}$

29. $\{2, 4, 6, 8\} \cup \{1, 3, 5, 7\}$

30. $\{a, e, i, o, u\} \cup \{m, n, f, g, h\}$

31. What advantage(s) does set-builder notation have over roster notation?

32. What advantage(s) does roster notation have over set-builder notation?

Synthesis

33. Find the union of the set of integers and the set of whole numbers.

34. Find the intersection of the set of odd integers and the set of even integers.

35. Find the union of the set of rational numbers and the set of irrational numbers.

36. Find the intersection of the set of even integers and the set of positive rational numbers.

37. Find the intersection of the set of rational numbers and the set of irrational numbers.

38. Find the union of the set of negative integers, the set of positive integers, and the set containing 0.

39. For a set A, find each of the following.
   a) $A \cup \varnothing$
   b) $A \cup A$
   c) $A \cap A$
   d) $A \cap \varnothing$

Classify each statement as either true or false.

40. The empty set can be written $\varnothing$, $\{\ \}$, or $\{0\}$.

41. For any set A, $\varnothing \subseteq A$.

42. For any set A, $A \subseteq A$.

43. For any sets A and B, $A \cap B \subseteq A$.

44. A set is *closed* under an operation if, when the operation is performed on its members, the result is in the set. For example, the set of real numbers is closed under the operation of addition since the sum of any two real numbers is a real number.
   a) Is the set of even numbers closed under addition?
   b) Is the set of odd numbers closed under addition?
   c) Is the set $\{0, 1\}$ closed under addition?
   d) Is the set $\{0, 1\}$ closed under multiplication?
   e) Is the set of real numbers closed under multiplication?
   f) Is the set of integers closed under division?

45. Experiment with sets of various types and determine whether the following distributive law for sets is true:
$$A \cap (B \cup C) = (A \cap B) \cup (A \cap C).$$

Tables

TABLE 1 Fraction and Decimal Equivalents

Fraction Notation	$\frac{1}{10}$	$\frac{1}{8}$	$\frac{1}{6}$	$\frac{1}{5}$	$\frac{1}{4}$	$\frac{3}{10}$	$\frac{1}{3}$	$\frac{3}{8}$	$\frac{2}{5}$	$\frac{1}{2}$
Decimal Notation	0.1	0.125	$0.16\overline{6}$	0.2	0.25	0.3	$0.333\overline{3}$	0.375	0.4	0.5
Percent Notation	10%	12.5%, or $12\frac{1}{2}\%$	$16.6\overline{6}\%$, or $16\frac{2}{3}\%$	20%	25%	30%	$33.3\overline{3}\%$, or $33\frac{1}{3}\%$	37.5%, or $37\frac{1}{2}\%$	40%	50%
Fraction Notation	$\frac{3}{5}$	$\frac{5}{8}$	$\frac{2}{3}$	$\frac{7}{10}$	$\frac{3}{4}$	$\frac{4}{5}$	$\frac{5}{6}$	$\frac{7}{8}$	$\frac{9}{10}$	$\frac{1}{1}$
Decimal Notation	0.6	0.625	$0.666\overline{6}$	0.7	0.75	0.8	$0.83\overline{3}$	0.875	0.9	1
Percent Notation	60%	62.5%, or $62\frac{1}{2}\%$	$66.6\overline{6}\%$, or $66\frac{2}{3}\%$	70%	75%	80%	$83.3\overline{3}\%$, or $83\frac{1}{3}\%$	87.5%, or $87\frac{1}{2}\%$	90%	100%

TABLE 2   Squares and Square Roots with Approximations to Three Decimal Places

N	$\sqrt{N}$	N^2	N	$\sqrt{N}$	N^2	N	$\sqrt{N}$	N^2	N	$\sqrt{N}$	N^2
1	1	1	26	5.099	676	51	7.141	2601	76	8.718	5776
2	1.414	4	27	5.196	729	52	7.211	2704	77	8.775	5929
3	1.732	9	28	5.292	784	53	7.280	2809	78	8.832	6084
4	2	16	29	5.385	841	54	7.348	2916	79	8.888	6241
5	2.236	25	30	5.477	900	55	7.416	3025	80	8.944	6400
6	2.449	36	31	5.568	961	56	7.483	3136	81	9	6561
7	2.646	49	32	5.657	1024	57	7.550	3249	82	9.055	6724
8	2.828	64	33	5.745	1089	58	7.616	3364	83	9.110	6889
9	3	81	34	5.831	1156	59	7.681	3481	84	9.165	7056
10	3.162	100	35	5.916	1225	60	7.746	3600	85	9.220	7225
11	3.317	121	36	6	1296	61	7.810	3721	86	9.274	7396
12	3.464	144	37	6.083	1369	62	7.874	3844	87	9.327	7569
13	3.606	169	38	6.164	1444	63	7.937	3969	88	9.381	7744
14	3.742	196	39	6.245	1521	64	8	4096	89	9.434	7921
15	3.873	225	40	6.325	1600	65	8.062	4225	90	9.487	8100
16	4	256	41	6.403	1681	66	8.124	4356	91	9.539	8281
17	4.123	289	42	6.481	1764	67	8.185	4489	92	9.592	8464
18	4.243	324	43	6.557	1849	68	8.246	4624	93	9.644	8649
19	4.359	361	44	6.633	1936	69	8.307	4761	94	9.695	8836
20	4.472	400	45	6.708	2025	70	8.367	4900	95	9.747	9025
21	4.583	441	46	6.782	2116	71	8.426	5041	96	9.798	9216
22	4.690	484	47	6.856	2209	72	8.485	5184	97	9.849	9409
23	4.796	529	48	6.928	2304	73	8.544	5329	98	9.899	9604
24	4.899	576	49	7	2401	74	8.602	5476	99	9.950	9801
25	5	625	50	7.071	2500	75	8.660	5625	100	10	10,000

Answers

The complete step-by-step solutions for the exercises listed below can be found in the *Student's Solutions Manual,* ISBN 0-321-56733-1/978-0-321-56733-8, which can be purchased online or at your bookstore.

CHAPTER 1

Technology Connection, p. 7

1. 3438   **2.** 47,531

Translating for Success, p. 9

1. H   **2.** E   **3.** K   **4.** B   **5.** O   **6.** L   **7.** M   **8.** C
9. D   **10.** F

Exercise Set 1.1, pp. 10–12

1. Expression   **2.** Equation   **3.** Equation
4. Expression   **5.** Equation   **6.** Equation
7. Expression   **8.** Equation   **9.** Equation
10. Expression   **11.** Expression   **12.** Expression
13. 45   **15.** 8   **17.** 5   **19.** 4   **21.** 5   **23.** 3
25. 24 ft^2 **27.** 15 cm^2 **29.** 0.345   **31.** Let r represent Ron's age; $r + 5$, or $5 + r$ **33.** $6b$, or $b \cdot 6$ **35.** $c - 9$
37. $6 + q$, or $q + 6$ **39.** Let m represent Mai's speed;
$8m$, or $m \cdot 8$ **41.** $y - x$ **43.** $x : w$, or $\dfrac{x}{w}$ **45.** Let l
represent the length of the box and h represent the height;
$l + h$, or $h + l$ **47.** $9 \cdot 2m$, or $2m \cdot 9$ **49.** Let y
represent "some number"; $\dfrac{1}{4}y - 13$, or $\dfrac{y}{4} - 13$ **51.** Let a
and b represent the two numbers; $5(a - b)$ **53.** Let w
represent the number of women attending; 64% of w,
or $0.64w$ **55.** Yes   **57.** No   **59.** Yes   **61.** Yes
63. Let x represent the unknown number; $73 + x = 201$
65. Let x represent the unknown number; $42x = 2352$
67. Let s represent the number of unoccupied squares;
$s + 19 = 64$ **69.** Let w represent the amount of solid
waste generated, in millions of tons; 32% of $w = 79$, or
$0.32w = 79$ **71.** (f)   **73.** (d)   **75.** (g)   **77.** (e)
79. ✍   **81.** ✍   **83.** $450   **85.** 2   **87.** 6   **89.** $w + 4$
91. $l + w + l + w$, or $2l + 2w$ **93.** $t + 8$ **95.** ✍

Exercise Set 1.2, pp. 18–20

1. Commutative   **2.** Associative   **3.** Associative
4. Commutative   **5.** Distributive   **6.** Associative
7. Associative   **8.** Commutative   **9.** Commutative
10. Distributive   **11.** $t + 11$ **13.** $8x + 4$
15. $3y + 9x$ **17.** $5(1 + a)$ **19.** $x \cdot 7$ **21.** ts

23. $5 + ba$ **25.** $(a + 1)5$ **27.** $x + (8 + y)$
29. $(u + v) + 7$ **31.** $ab + (c + d)$ **33.** $8(xy)$
35. $(2a)b$ **37.** $(3 \cdot 2)(a + b)$
39. $(s + t) + 6$; $(t + 6) + s$ **41.** $17(ab)$; $b(17a)$
43. $(1 + x) + 2 = (x + 1) + 2$ Commutative law
$ = x + (1 + 2)$ Associative law
$ = x + 3$ Simplifying
45. $(m \cdot 3)7 = m(3 \cdot 7)$ Associative law
$ = m \cdot 21$ Simplifying
$ = 21m$ Commutative law
47. $2x + 30$ **49.** $4 + 4a$ **51.** $24 + 8y$
53. $90x + 60$ **55.** $5r + 10 + 15t$ **57.** $2a + 2b$
59. $5x + 5y + 10$ **61.** $x, xyz, 1$ **63.** $2a, \dfrac{a}{3b}, 5b$
65. x, y **67.** $4x, 4y$ **69.** $2(a + b)$ **71.** $7(1 + y)$
73. $4(8x + 1)$ **75.** $5(x + 2 + 3y)$ **77.** $7(a + 5b)$
79. $11(4x + y + 2z)$ **81.** $5, n$ **83.** $3, (x + y)$
85. $7, a, b$ **87.** $(a - b), (x - y)$ **89.** ✍   **91.** Let k
represent Kara's salary; $\dfrac{1}{2}k$, or $\dfrac{k}{2}$ **92.** $2(m + 3)$, or
$2(3 + m)$ **93.** ✍   **95.** Yes; distributive law
97. No; for example, let $m = 1$. Then $7 \div 3 \cdot 1 = \frac{7}{3}$ and
$1 \cdot 3 \div 7 = \frac{3}{7}$.   **99.** No; for example, let $x = 1$ and
$y = 2$. Then $30 \cdot 2 + 1 \cdot 15 = 60 + 15 = 75$ and
$5[2(1 + 3 \cdot 2)] = 5[2(7)] = 5 \cdot 14 = 70$.   **101.** ✍

Exercise Set 1.3, pp. 27–29

1. (b)   **2.** (c)   **3.** (d)   **4.** (a)   **5.** Composite
7. Prime   **9.** Composite   **11.** Prime   **13.** Neither
15. $1 \cdot 50$; $2 \cdot 25$; $5 \cdot 10$; $1, 2, 5, 10, 25, 50$
17. $1 \cdot 42$; $2 \cdot 21$; $3 \cdot 14$; $6 \cdot 7$; $1, 2, 3, 6, 7, 14, 21, 42$
19. $3 \cdot 13$ **21.** $2 \cdot 3 \cdot 5$ **23.** $3 \cdot 3 \cdot 3$ **25.** $2 \cdot 3 \cdot 5 \cdot 5$
27. $2 \cdot 2 \cdot 2 \cdot 5$ **29.** Prime   **31.** $2 \cdot 3 \cdot 5 \cdot 7$ **33.** $5 \cdot 23$
35. $\frac{3}{5}$ **37.** $\frac{2}{7}$ **39.** $\frac{1}{4}$ **41.** 4   **43.** $\frac{1}{4}$ **45.** 6
47. $\frac{21}{25}$ **49.** $\frac{60}{41}$ **51.** $\frac{15}{7}$ **53.** $\frac{3}{10}$ **55.** 6   **57.** $\frac{1}{2}$
59. $\frac{7}{6}$ **61.** $\dfrac{3b}{7a}$ **63.** $\dfrac{10}{n}$ **65.** $\frac{5}{6}$ **67.** 1   **69.** $\frac{5}{18}$
71. 0   **73.** $\frac{35}{18}$ **75.** $\frac{10}{3}$ **77.** 27   **79.** 1   **81.** $\frac{6}{35}$
83. 18   **85.** ✍   **87.** $5(3 + x)$; answers may vary
88. $7 + (b + a)$, or $(a + b) + 7$ **89.** ✍
91. Row 1: 7, 2, 36, 14, 8, 8; row 2: 9, 18, 2, 10, 12, 21
93. $\frac{2}{5}$ **95.** $\dfrac{5q}{t}$ **97.** $\frac{6}{25}$ **99.** $\dfrac{5ap}{2cm}$ **101.** $\dfrac{23r}{18t}$

103. $\frac{28}{45}\,\text{m}^2$ **105.** $14\frac{2}{9}\,\text{m}$ **107.** $27\frac{3}{5}\,\text{cm}$

Technology Connection, p. 33

1. 2.236067977   **2.** 2.645751311   **3.** 3.605551275
4. 5.196152423   **5.** 6.164414003   **6.** 7.071067812

Exercise Set 1.4, pp. 35–37

1. Repeating   **2.** Terminating   **3.** Integer
4. Whole number   **5.** Rational number
6. Irrational number   **7.** Natural number
8. Absolute value   **9.** $-10{,}500,\ 27{,}482$ **11.** $136,\ -4$
13. $-554,\ 499.19$ **15.** $650,\ -180$ **17.** $8,\ -5$
19.

$$\frac{10}{3}$$

21.

$$-4.3$$

23.

25. 0.875   **27.** -0.75
29. $-1.1\overline{6}$ **31.** $0.\overline{6}$ **33.** -0.5 **35.** $0.1\overline{3}$
37.

$$\sqrt{5}$$

39. $-\sqrt{22}$

41. $>$ **43.** $<$ **45.** $<$ **47.** $>$ **49.** $<$ **51.** $<$
53. $x < -2$ **55.** $y \ge 10$ **57.** True   **59.** False
61. True   **63.** 58   **65.** 12.2   **67.** $\sqrt{2}$ **69.** $\frac{9}{7}$ **71.** 0
73. 8   **75.** $-83,\ -4.7,\ 0,\ \frac{5}{9},\ 2.\overline{16},\ 62$ **77.** $-83,\ 0,\ 62$
79. $-83,\ -4.7,\ 0,\ \frac{5}{9},\ 2.\overline{16},\ \pi,\ \sqrt{17},\ 62$ **81.** 🖵   **83.** 42
84. $ba + 5$, or $5 + ab$ **85.** 🖵   **87.** 🖵
89. $-23,\ -17,\ 0,\ 4$ **91.** $-\frac{4}{3},\frac{4}{9},\frac{4}{8},\frac{4}{6},\frac{4}{5},\frac{4}{3},\frac{4}{2}$ **93.** $<$ **95.** $=$
97. $-19,\ 19$ **99.** $-4,\ -3,\ 3,\ 4$ **101.** $\frac{3}{3}$ **103.** $\frac{70}{9}$
105. $x \le 0$ **107.** $|t| \ge 20$ **109.** 🖵

Exercise Set 1.5, pp. 41–43

1. (f)   **2.** (d)   **3.** (e)   **4.** (a)   **5.** (b)   **6.** (c)
7. -3 **9.** 4   **11.** -7 **13.** -8 **15.** -35 **17.** -8
19. 0   **21.** -41 **23.** 0   **25.** 9   **27.** -2 **29.** 11
31. -43 **33.** 0   **35.** 18   **37.** -45 **39.** 0   **41.** 16
43. -0.8 **45.** -9.1 **47.** $\frac{3}{5}$ **49.** $\frac{-6}{7}$ **51.** $-\frac{1}{15}$
53. $\frac{2}{9}$ **55.** -3 **57.** 0   **59.** The price rose 29¢.
61. Her new balance was $95.   **63.** The total gain was 20 yd.
65. The lake rose $\frac{3}{10}$ ft.   **67.** Logan owes $85.   **69.** $17a$
71. $9x$ **73.** $25t$ **75.** $-2m$ **77.** $-10y$ **79.** $1 - 2x$
81. $12x + 17$ **83.** $7r + 8t + 16$ **85.** $18n + 16$
87. 🖵   **89.** $21z + 14y + 7$ **90.** $\frac{28}{3}$ **91.** 🖵
93. $451.70   **95.** $-5y$ **97.** $-7m$ **99.** $-7t,\ -23$
101. 1 under par

Exercise Set 1.6, pp. 48–51

1. (d)   **2.** (g)   **3.** (f)   **4.** (h)   **5.** (a)   **6.** (c)
7. (b)   **8.** (e)   **9.** Six minus ten
11. Two minus negative twelve   **13.** Nine minus the
opposite of t **15.** The opposite of x minus y
17. Negative three minus the opposite of n **19.** -51
21. $\frac{11}{3}$ **23.** 3.14   **25.** 45   **27.** $\frac{14}{3}$ **29.** -0.101

31. 37   **33.** $-\frac{2}{5}$ **35.** 1   **37.** -15 **39.** -3 **41.** -6
43. -3 **45.** -7 **47.** -6 **49.** 0   **51.** -5
53. -10 **55.** -11 **57.** 0   **59.** 0   **61.** 8   **63.** -11
65. 16   **67.** -19 **69.** 1   **71.** 17   **73.** 3   **75.** -3
77. -21 **79.** 10   **81.** -8 **83.** -60 **85.** -23
87. -7.3 **89.** 1.1   **91.** -5.5 **93.** -0.928 **95.** $-\frac{7}{11}$
97. $-\frac{4}{5}$ **99.** $\frac{5}{17}$ **101.** $3.8 - (-5.2);\ 9$
103. $114 - (-79);\ 193$ **105.** -40 **107.** 43   **109.** 32
111. -62 **113.** -139 **115.** 0   **117.** $-3y,\ -8x$
119. $9,\ -5t,\ -3st$ **121.** $-3x$ **123.** $-5a + 4$
125. $-n - 9$ **127.** $-3x - 6$ **129.** $-8t - 7$
131. $-12x + 3y + 9$ **133.** $8x + 66$ **135.** 214°F
137. 30,347 ft   **139.** 116 m   **141.** 🖵   **143.** 432 ft²
144. $2 \cdot 2 \cdot 2 \cdot 2 \cdot 2 \cdot 3 \cdot 3 \cdot 3$ **145.** 🖵
147. 11:00 P.M., August 14   **149.** False. For example,
let $m = -3$ and $n = -5$. Then $-3 > -5$, but
$-3 + (-5) = -8 \not> 0$.   **151.** True. For example, for
$m = 4$ and $n = -4,\ 4 = -(-4)$ and $4 + (-4) = 0$; for
$m = -3$ and $n = 3,\ -3 = -3$ and $-3 + 3 = 0$.
153. (−) 9 − (−) 7 ENTER

Exercise Set 1.7, pp. 56–58

1. 1   **2.** 0   **3.** 0   **4.** 1   **5.** 0   **6.** 1   **7.** 1   **8.** 0
9. 1   **10.** 0   **11.** -40 **13.** -56 **15.** -40 **17.** 72
19. -42 **21.** 45   **23.** 190   **25.** -132 **27.** 1200
29. -126 **31.** 11.5   **33.** 0   **35.** $-\frac{2}{7}$ **37.** $\frac{1}{12}$
39. -11.13 **41.** $-\frac{5}{12}$ **43.** 252   **45.** 0   **47.** $\frac{1}{28}$
49. 150   **51.** 0   **53.** -720 **55.** $-30{,}240$ **57.** -9
59. -4 **61.** -7 **63.** 4   **65.** -9 **67.** 5.1   **69.** $\frac{100}{11}$
71. -8 **73.** Undefined   **75.** -4 **77.** 0   **79.** 0
81. $-\frac{8}{3};\frac{8}{-3}$ **83.** $-\frac{29}{35};\frac{-29}{35}$ **85.** $\frac{-7}{3};\frac{7}{-3}$ **87.** $-\frac{x}{2};\frac{x}{-2}$
89. $-\frac{5}{4}$ **91.** $-\frac{10}{51}$ **93.** $-\frac{1}{10}$ **95.** $\frac{1}{4.3}$, or $\frac{10}{43}$ **97.** $-\frac{4}{9}$
99. Does not exist   **101.** $\frac{21}{20}$ **103.** -1 **105.** 1
107. $\frac{3}{11}$ **109.** $-\frac{7}{4}$ **111.** 1   **113.** $\frac{1}{10}$ **115.** $-\frac{7}{6}$
117. Undefined   **119.** $-\frac{14}{15}$ **121.** 🖵   **123.** $\frac{22}{39}$
124. $12x - y - 9$ **125.** 🖵   **127.** $\dfrac{1}{a + b}$
129. $-(a + b)$ **131.** $x = -x$ **133.** For 2 and 3,
the reciprocal of the sum is $1/(2 + 3)$, or $1/5$. But
$1/5 \ne 1/2 + 1/3$.   **135.** 5°F   **137.** Positive
139. Positive   **141.** Positive   **143.** Distributive law;
law of opposites; multiplicative property of zero

Connecting the Concepts, p. 59

1. -10 **2.** 16   **3.** 4   **4.** -6 **5.** -120 **6.** -7
7. -23 **8.** -3 **9.** -1 **10.** -3 **11.** -0.8
12. -3.77 **13.** -7 **14.** -4.1 **15.** -12 **16.** $\frac{5}{3}$
17. 100   **18.** 77   **19.** 180   **20.** -52

Exercise Set 1.8, pp. 66–68

1. (a) Division; (b) subtraction; (c) addition;
(d) multiplication; (e) subtraction; (f) multiplication

2. (a) Multiplication; **(b)** subtraction; **(c)** addition; **(d)** subtraction; **(e)** division; **(f)** multiplication
3. x^6 **5.** $(-5)^3$ **7.** $(3t)^5$ **9.** $2n^4$ **11.** 16   **13.** 9
15. -9 **17.** 64   **19.** 625   **21.** 7   **23.** -32
25. $81t^4$ **27.** $-343x^3$ **29.** 26   **31.** 51   **33.** -6
35. 1   **37.** 298   **39.** 11   **41.** -36 **43.** 1291
45. 152   **47.** 36   **49.** 1   **51.** -44 **53.** 41   **55.** -10
57. -5 **59.** -19 **61.** -3 **63.** -75 **65.** 9   **67.** 30
69. 6   **71.** -17 **73.** $-9x - 1$ **75.** $7n - 8$
77. $-4a + 3b - 7c$ **79.** $-3x^2 - 5x + 1$ **81.** $2x - 7$
83. $-9x + 6$ **85.** $21t - r$ **87.** $9y - 25z$ **89.** $x^2 + 6$
91. $-t^3 + 4t$ **93.** $37a^2 - 23ab + 35b^2$
95. $-22t^3 - t^2 + 9t$ **97.** $2x - 25$ **99.** 📝   **101.** Let n represent the number; $2n - 9$ **102.** Let m and n represent the two numbers; $\frac{1}{2}(m + n)$ **103.** 📝
105. $-6r - 5t + 21$ **107.** $-2x - f$ **109.** 📝
111. True   **113.** False   **115.** 0   **117.** 17
119. 39,000   **121.** $44x^3$

Review Exercises: Chapter 1, pp. 73–74

1. True   **2.** True   **3.** False   **4.** True   **5.** False
6. False   **7.** True   **8.** False   **9.** False   **10.** True
11. 24   **12.** 4   **13.** -16 **14.** -15 **15.** $y - 7$
16. $xz + 10$, or $10 + xz$ **17.** Let b represent Brandt's speed and w represent the wind speed; $15(b - w)$
18. No   **19.** Let d represent the number of digital prints, in billions, made in 2006; $14.1 = d + 3.2$ **20.** $t \cdot 3 + 5$
21. $2x + (y + z)$ **22.** $(4x)y, 4(yx), (4y)x$; answers may vary   **23.** $18x + 30y$ **24.** $40x + 24y + 16$
25. $3(7x + 5y)$ **26.** $11(2a + 9b + 1)$ **27.** $2 \cdot 2 \cdot 2 \cdot 7$
28. $\frac{5}{12}$ **29.** $\frac{9}{4}$ **30.** $\frac{19}{24}$ **31.** $\frac{3}{16}$ **32.** $\frac{3}{5}$ **33.** $\frac{27}{25}$
34. $-3600, 1350$ **35.**

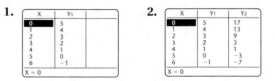

36. $x > -3$ **37.** True   **38.** False   **39.** $-0.\overline{4}$ **40.** 1
41. -12 **42.** -10 **43.** $-\frac{7}{12}$ **44.** 0   **45.** -5 **46.** 8
47. $-\frac{7}{5}$ **48.** -7.9 **49.** 63   **50.** -9.18 **51.** $-\frac{2}{7}$
52. -140 **53.** -7 **54.** -3 **55.** $\frac{9}{4}$ **56.** 48
57. 168   **58.** $\frac{21}{8}$ **59.** 18   **60.** 53   **61.** $\frac{103}{17}$
62. $7a - b$ **63.** $-4x + 5y$ **64.** 7   **65.** $-\frac{1}{7}$
66. $(2x)^4$ **67.** $-125x^3$ **68.** $-3a + 9$ **69.** $11b - 27$
70. $3x^4 + 10x$ **71.** $17n^2 + m^2 + 20mn$ **72.** $5x + 28$
73. 📝 The value of a constant never varies. A variable can represent a variety of numbers.   **74.** 📝 A term is one of the parts of an expression that is separated from the other parts by plus signs. A factor is part of a product.   **75.** 📝 The distributive law is used in factoring algebraic expressions, multiplying algebraic expressions, combining like terms, finding the opposite of a sum, and subtracting algebraic expressions.   **76.** 📝 A negative number raised to an even power is positive; a negative number raised to an odd power is negative.   **77.** 25,281   **78. (a)** $\frac{3}{11}$; **(b)** $\frac{10}{11}$
79. $-\frac{5}{8}$ **80.** -2.1 **81.** (i)   **82.** (j)   **83.** (a)
84. (h)   **85.** (k)   **86.** (b)   **87.** (c)   **88.** (e)
89. (d)   **90.** (f)   **91.** (g)

Test: Chapter 1, p. 75

1. [1.1] 4   **2.** [1.1] Let x and y represent the numbers; $xy - 9$ **3.** [1.1] 240 ft^2 **4.** [1.2] $q + 3p$
5. [1.2] $(x \cdot 4) \cdot y$ **6.** [1.1] No   **7.** [1.1] Let p represent the maximum production capability; $p - 4250 = 45,950$
8. [1.2] $35 + 7x$ **9.** [1.7] $-5y + 10$
10. [1.2] $11(1 + 4x)$ **11.** [1.2] $7(x + 1 + 7y)$
12. [1.3] $2 \cdot 2 \cdot 3 \cdot 5 \cdot 5$ **13.** [1.3] $\frac{3}{7}$ **14.** [1.4] $<$
15. [1.4] $>$ **16.** [1.4] $\frac{9}{4}$ **17.** [1.4] 3.8   **18.** [1.6] $\frac{2}{3}$
19. [1.7] $-\frac{7}{4}$ **20.** [1.6] 10   **21.** [1.4] $-5 \geq x$
22. [1.6] 7.8   **23.** [1.5] -8 **24.** [1.6] -2.5 **25.** [1.6] $-\frac{7}{8}$
26. [1.7] -48 **27.** [1.7] $\frac{2}{9}$ **28.** [1.7] -6 **29.** [1.7] $\frac{3}{4}$
30. [1.7] -9.728 **31.** [1.8] -173 **32.** [1.6] 15
33. [1.8] -64 **34.** [1.8] 448   **35.** [1.6] $21a + 22y$
36. [1.8] $16x^4$ **37.** [1.8] $x + 7$ **38.** [1.8] $9a - 12b - 7$
39. [1.8] $-y - 16$ **40.** [1.1] 5
41. [1.8] $9 - (3 - 4) + 5 = 15$ **42.** [1.8] 15
43. [1.8] $4a$ **44.** [1.8] False

CHAPTER 2

Exercise Set 2.1, pp. 83–85

1. (c)   **2.** (b)   **3.** (f)   **4.** (a)   **5.** (d)   **6.** (e)   **7.** (d)
8. (b)   **9.** (c)   **10.** (a)   **11.** 11   **13.** -25 **15.** -31
17. 41   **19.** 19   **21.** -6 **23.** $\frac{7}{3}$ **25.** $-\frac{1}{10}$ **27.** $\frac{41}{24}$
29. $-\frac{1}{20}$ **31.** 9.1   **33.** -5 **35.** 7   **37.** 12   **39.** -38
41. 8   **43.** -7 **45.** 8   **47.** 88   **49.** 20   **51.** -54
53. $-\frac{5}{9}$ **55.** 1   **57.** $\frac{9}{2}$ **59.** -7.6 **61.** -2.5 **63.** -15
65. -5 **67.** $-\frac{7}{6}$ **69.** -128 **71.** $-\frac{1}{2}$ **73.** -15 **75.** 9
77. 310.756   **79.** 📝   **81.** -6 **82.** 2   **83.** 1
84. -16 **85.** 📝   **87.** Identity   **89.** 0
91. Contradiction   **93.** Contradiction   **95.** 11.6   **97.** 2
99. $-23, 23$ **101.** 9000   **103.** 250

Technology Connection, p. 88

1.

X	Y₁
0	5
1	4
2	3
3	2
4	1
5	0
6	−1

X = 0

2.

X	Y₁	Y₂
0	5	17
1	4	13
2	3	9
3	2	5
4	1	1
5	0	−3
6	−1	−7

X = 0

3. 4; not reliable because, depending on the choice of ΔTbl, it is easy to scroll past a solution without realizing it.

Exercise Set 2.2, pp. 90–91

1. (c)   **2.** (e)   **3.** (a)   **4.** (f)   **5.** (b)   **6.** (d)   **7.** 8
9. 7   **11.** 5   **13.** $\frac{10}{3}$ **15.** -7 **17.** -5 **19.** -4
21. 19   **23.** -2.8 **25.** 3   **27.** 15   **29.** -6 **31.** $-\frac{25}{2}$
33. 4   **35.** -3 **37.** -6 **39.** 2   **41.** 0   **43.** 6
45. $-\frac{1}{2}$ **47.** 0   **49.** 10   **51.** 4   **53.** 0   **55.** $\frac{5}{2}$
57. -8 **59.** $\frac{1}{6}$ **61.** 2   **63.** $\frac{16}{3}$ **65.** $\frac{2}{5}$ **67.** 1
69. -4 **71.** $1.\overline{6}$ **73.** $-\frac{60}{37}$ **75.** 11   **77.** 8   **79.** $\frac{16}{15}$

81. $-\frac{1}{31}$ **83.** 2   **85.** ✍   **87.** -7 **88.** 15
89. -15 **90.** -28 **91.** ✍   **93.** $\frac{1136}{909}$, or $1.\overline{2497}$
95. Contradiction   **97.** Identity   **99.** $\frac{2}{3}$ **101.** 0
103. 0   **105.** -2

Technology Connection, p. 93

1. 800

Exercise Set 2.3, pp. 96–99

1. 309.6 m   **3.** 1423 students   **5.** 8.4734   **7.** 255 mg
9. $b = \dfrac{A}{h}$ **11.** $r = \dfrac{d}{t}$ **13.** $P = \dfrac{I}{rt}$ **15.** $m = 65 - H$
17. $l = \dfrac{P - 2w}{2}$, or $l = \dfrac{P}{2} - w$ **19.** $\pi = \dfrac{A}{r^2}$
21. $h = \dfrac{2A}{b}$ **23.** $c^2 = \dfrac{E}{m}$ **25.** $d = 2Q - c$
27. $b = 3A - a - c$ **29.** $r = wf$ **31.** $C = \frac{5}{9}(F - 32)$
33. $y = 2x - 1$ **35.** $y = -\frac{2}{5}x + 2$ **37.** $y = \frac{4}{3}x - 2$
39. $y = -\frac{9}{8}x + \frac{1}{2}$ **41.** $y = \frac{3}{5}x - \frac{8}{5}$
43. $x = \dfrac{z - 13}{2} - y$, or $x = \dfrac{z - 13 - 2y}{2}$
45. $l = 4(t - 27) + w$ **47.** $t = \dfrac{A}{a + b}$ **49.** $h = \dfrac{2A}{a + b}$
51. $L = W - \dfrac{N(R - r)}{400}$, or $L = \dfrac{400W - NR + Nr}{400}$
53. ✍   **55.** -10 **56.** -196 **57.** 0   **58.** -32
59. -13 **60.** 65   **61.** ✍   **63.** 40 yr   **65.** 27 in³
67. $a = \dfrac{w}{c} \cdot d$ **69.** $c = \dfrac{d}{a - b}$ **71.** $a = \dfrac{c}{3 + b + d}$
73. $K = 9.632w + 19.685h - 10.54a + 102.3$

Exercise Set 2.4, pp. 103–107

1. (d)   **2.** (c)   **3.** (e)   **4.** (b)   **5.** (c)   **6.** (d)   **7.** (f)
8. (a)   **9.** (b)   **10.** (e)   **11.** 0.49   **13.** 0.01
15. 0.041   **17.** 0.2   **19.** 0.0625   **21.** 0.002   **23.** 1.75
25. 38%   **27.** 3.9%   **29.** 45%   **31.** 70%   **33.** 0.09%
35. 106%   **37.** 180%   **39.** 60%   **41.** 32%   **43.** 25%
45. 26%   **47.** $46\frac{2}{3}$%, or $\frac{140}{3}$ **49.** 2.5   **51.** 10,000
53. 125%   **55.** 0.8   **57.** 50%   **59.** $33.\overline{3}$%, or $33\frac{1}{3}$%
61. 2.85 million Americans   **63.** 23.37 million Americans
65. 75 credits   **67.** 595 at bats   **69.** (a) 16%; (b) $29
71. $33.\overline{3}$%, or $33\frac{1}{3}$%; $66.\overline{6}$%, or $66\frac{2}{3}$%   **73.** $168
75. 285 women   **77.** $19.20 an hour   **79.** The actual
cost was 43.7% more than the estimate.   **81.** $45
83. $148.50   **85.** About 31.5 lb   **87.** About 2.45 billion
pieces of mail   **89.** About 165 calories   **91.** ✍
93. Let l represent the length and w represent the width;
$2l + 2w$ **94.** $0.05 \cdot 180$ **95.** Let p represent the number
of points Tino scored; $p - 5$ **96.** $15 + 1.5x$ **97.** $10\left(\frac{1}{2}a\right)$
98. Let n represent the number; $3n + 10$ **99.** Let l
represent the length and w represent the width; $w = l - 2$
100. Let x represent the first number and y represent the
second number; $x = 4y$ **101.** ✍   **103.** 18,500 people
105. About 6 ft 7 in.   **107.** About 27%   **109.** ✍

Exercise Set 2.5, pp. 115–120

1. 11   **3.** $\frac{11}{2}$ **5.** $150   **7.** $130   **9.** About 78.4 mi
11. 160 mi   **13.** 1204 and 1205   **15.** 285 and 287
17. 32, 33, 34   **19.** Man: 103 yr; woman: 101 yr
21. Non-spam: 25 billion messages; spam: 100 billion
messages   **23.** 140 and 141   **25.** Width: 100 ft; length:
160 ft; area: 16,000 ft²   **27.** Width: 21 m; length: 25 m
29. $1\frac{3}{4}$ in. by $3\frac{1}{2}$ in.   **31.** 30°, 90°, 60°   **33.** 70°
35. Bottom: 144 ft; middle: 72 ft; top: 24 ft   **37.** 8.75 mi,
or $8\frac{3}{4}$ mi   **39.** $128\frac{1}{3}$ mi   **41.** 65°, 25°   **43.** 140°, 40°
45. Length: 27.9 cm; width: 21.6 cm   **47.** $6600
49. 830 points   **51.** $125,000   **53.** 160 chirps per
minute   **55.** ✍   **57.** <   **58.** >   **59.** >   **60.** <
61. $-4 \le x$ **62.** $5 > x$ **63.** $y < 5$ **64.** $t \ge -10$
65. ✍   **67.** $37   **69.** 20   **71.** Half-dollars: 5;
quarters: 10; dimes: 20; nickels: 60   **73.** $95.99
75. 5 DVDs   **77.** 6 mi   **79.** ✍   **81.** Width: 23.31 cm;
length: 27.56 cm

Exercise Set 2.6, pp. 126–128

1. $\ge$ **2.** $\le$ **3.** $<$ **4.** $>$ **5.** Equivalent
6. Equivalent   **7.** Equivalent   **8.** Not equivalent
9. (a) Yes; (b) no; (c) no   **11.** (a) Yes; (b) no; (c) yes
13. (a) Yes; (b) yes; (c) yes   **15.** (a) No; (b) yes; (c) no
17. $y < 2$ **19.** $x \ge -1$
21. $0 \le t$ **23.** $-5 \le x < 2$
25. $-4 < x < 0$ **27.** $\{x \mid x > -4\}$
29. $\{x \mid x \le 2\}$ **31.** $\{x \mid x < -1\}$ **33.** $\{x \mid x \ge 0\}$
35. $\{y \mid y > 3\}$,
37. $\{x \mid x \le -21\}$,
39. $\{n \mid n < 17\}$,
41. $\{x \mid x \le -9\}$,
43. $\{y \mid y \le \frac{1}{2}\}$,
45. $\{t \mid t > \frac{5}{8}\}$,
47. $\{x \mid x < 0\}$,
49. $\{t \mid t < 23\}$,
51. $\{y \mid y \ge 22\}$,
53. $\{x \mid x < 7\}$,
55. $\{x \mid x > -\frac{13}{7}\}$,
57. $\{t \mid t < -3\}$,
59. $\{n \mid n \ge -1.5\}$,
61. $\{y \mid y \ge -\frac{1}{10}\}$,

63. $\left\{x \mid x < -\frac{4}{5}\right\}$,

65. $\{x \mid x < 6\}$ **67.** $\{t \mid t \le 7\}$
69. $\{y \mid y \le -1\}$ **71.** $\{x \mid x > -4\}$ **73.** $\left\{y \mid y < -\frac{10}{3}\right\}$
75. $\{x \mid x > -10\}$ **77.** $\{y \mid y < 0\}$ **79.** $\left\{y \mid y \ge \frac{3}{2}\right\}$
81. $\{x \mid x > -4\}$ **83.** $\{x \mid x > -4\}$ **85.** $\{t \mid t > 1\}$
87. $\{x \mid x \le -9\}$ **89.** $\{t \mid t < 14\}$ **91.** $\{y \mid y \le -4\}$
93. $\left\{t \mid t < -\frac{5}{3}\right\}$ **95.** $\{r \mid r > -3\}$ **97.** $\{x \mid x \le 7\}$
99. $\left\{x \mid x > -\frac{5}{32}\right\}$ **101.** ✍   **103.** $17x - 6$
104. $2m - 16n$ **105.** $7x - 8y - 46$ **106.** $-47t + 1$
107. $35a - 20b - 17$ **108.** $-21x + 32$ **109.** ✍
111. $\{x \mid x \text{ is a real number}\}$ **113.** $\left\{x \mid x \le \frac{5}{6}\right\}$
115. $\{x \mid x \le -4a\}$ **117.** $\left\{x \mid x > \dfrac{y - b}{a}\right\}$
119. $\{x \mid x \text{ is a real number}\}$

Connecting the Concepts, pp. 128–129

1. 21   **2.** $\{x \mid x \le 21\}$ **3.** -6 **4.** $\{x \mid x > -6\}$
5. $\{x \mid x < 6\}$ **6.** 3   **7.** $-\frac{1}{3}$ **8.** $\{y \mid y < 3\}$
9. $\{t \mid t \le -16\}$ **10.** $\frac{11}{2}$ **11.** $\{a \mid a < -1\}$
12. $\{x \mid x < 10.75\}$ **13.** $\{x \mid x \ge -11\}$ **14.** 105
15. -4.24 **16.** 15   **17.** $\{y \mid y \ge 3\}$ **18.** $\frac{14}{3}$
19. $\left\{x \mid x > \frac{22}{5}\right\}$ **20.** $\{a \mid a \ge 0\}$

Translating for Success, p. 132

1. F   **2.** I   **3.** C   **4.** E   **5.** D   **6.** J   **7.** O   **8.** M
9. B   **10.** L

Exercise Set 2.7, pp. 133–136

1. $b \le a$ **2.** $b < a$ **3.** $a \le b$ **4.** $a < b$ **5.** $b \le a$
6. $a \le b$ **7.** $b < a$ **8.** $a < b$ **9.** Let n represent the
number; $n < 10$ **11.** Let t represent the temperature;
$t \le -3$ **13.** Let d represent the number of years of
driving experience; $d \ge 5$ **15.** Let a represent the age of
the altar; $a > 1200$ **17.** Let h represent Tania's hourly
wage; $12 < h < 15$ **19.** Let w represent the wind speed;
$w > 50$ **21.** Let c represent the cost of a room; $c \le 120$
23. More than 2.375 hr   **25.** At least 2.25   **27.** Scores
greater than or equal to 97   **29.** 8 credits or more
31. At least 3 plate appearances   **33.** Lengths greater than
6 cm   **35.** Depths less than 437.5 ft   **37.** Blue-book value
is greater than or equal to $10,625   **39.** Lengths less than
55 in.   **41.** Temperatures greater than 37°C
43. No more than 3 ft tall   **45.** A serving contains at least
16 g of fat.   **47.** Dates after September 16   **49.** No more
than 134 text messages   **51.** Years after 2012
53. Mileages less than or equal to 225   **55.** ✍   **57.** -14
58. $-\frac{2}{3}$ **59.** -60 **60.** -11.1 **61.** 0   **62.** 5
63. -2 **64.** -1 **65.** ✍   **67.** Temperatures between
$-15°C$ and $-9\frac{4}{9}°C$ **69.** Lengths less than or equal to 8 cm
71. They contain at least 7.5 g of fat per serving.
73. At least $42   **75.** ✍

Review Exercises: Chapter 2, pp. 140–141

1. True   **2.** False   **3.** True   **4.** True   **5.** True
6. False   **7.** True   **8.** True   **9.** -25 **10.** 7
11. -65 **12.** 3   **13.** -20 **14.** 1.11   **15.** $\frac{1}{2}$ **16.** $-\frac{3}{2}$
17. -8 **18.** -4 **19.** $-\frac{1}{3}$ **20.** 4   **21.** 3   **22.** 4
23. 16   **24.** 1   **25.** $-\frac{7}{5}$ **26.** 0   **27.** 4   **28.** $d = \dfrac{C}{\pi}$
29. $B = \dfrac{3V}{h}$ **30.** $y = \frac{5}{2}x - 5$ **31.** $x = \dfrac{b}{t - a}$
32. 0.012   **33.** 44%   **34.** 70%   **35.** 140   **36.** No
37. Yes   **38.** No   **39.**

40.

41.

42. $\left\{t \mid t \ge -\frac{1}{2}\right\}$
43. $\{x \mid x \ge 7\}$ **44.** $\{y \mid y > 3\}$ **45.** $\{y \mid y \le -4\}$
46. $\{x \mid x < -11\}$ **47.** $\{y \mid y > -7\}$ **48.** $\{x \mid x > -6\}$
49. $\left\{x \mid x > -\frac{9}{11}\right\}$ **50.** $\{t \mid t \le -12\}$ **51.** $\{x \mid x \le -8\}$
52. About 48%   **53.** 8 ft, 10 ft   **54.** Japanese students:
41,000; Chinese students: 62,000   **55.** 57, 59
56. Width: 11 cm; length: 17 cm   **57.** $160   **58.** About 109
million subscribers   **59.** 35°, 85°, 60°   **60.** No more than
55 g of fat   **61.** 14 or fewer copies   **62.** ✍ Multiplying
both sides of an equation by *any* nonzero number results in
an equivalent equation. When multiplying on both sides of
an inequality, the sign of the number being multiplied by
must be considered. If the number is positive, the direction
of the inequality symbol remains unchanged; if the number
is negative, the direction of the inequality symbol must be
reversed to produce an equivalent inequality.   **63.** ✍
The solutions of an equation can usually each be checked.
The solutions of an inequality are normally too numerous
to check. Checking a few numbers from the solution set
found cannot guarantee that the answer is correct, although
if any number does not check, the answer found is
incorrect.   **64.** About 1 hr 36 min   **65.** Nile: 4160 mi;
Amazon: 4225 mi   **66.** $16   **67.** $-23, 23$ **68.** $-20, 20$
69. $a = \dfrac{y - 3}{2 - b}$ **70.** $F = \dfrac{0.3(12w)}{9}$, or $F = 0.4w$

Test: Chapter 2, p. 142

1. [2.1] 9   **2.** [2.1] 15   **3.** [2.1] -3 **4.** [2.1] 49
5. [2.1] -12 **6.** [2.2] 2   **7.** [2.1] -8 **8.** [2.2] $-\frac{23}{67}$
9. [2.2] 7   **10.** [2.2] $-\frac{5}{3}$ **11.** [2.2] $\frac{23}{3}$
12. [2.6] $\{x \mid x > -5\}$ **13.** [2.6] $\{x \mid x > -13\}$
14. [2.6] $\{y \mid y \le -13\}$ **15.** [2.6] $\left\{y \mid y \le -\frac{15}{2}\right\}$
16. [2.6] $\{n \mid n < -5\}$ **17.** [2.6] $\{x \mid x < -7\}$
18. [2.6] $\{t \mid t \ge -1\}$ **19.** [2.6] $\{x \mid x \le -1\}$
20. [2.3] $r = \dfrac{A}{2\pi h}$ **21.** [2.3] $l = 2w - P$ **22.** [2.4] 2.3
23. [2.4] 0.3%   **24.** [2.4] 14.8   **25.** [2.4] 44%
26. [2.6]

27. [2.6]

28. [2.5] Width: 7 cm; length: 11 cm   **29.** [2.5] 525 mi from Springer Mountain and 1575 mi from Mt. Katahdin   **30.** [2.5] 81 mm, 83 mm, 85 mm   **31.** [2.4] $65   **32.** [2.7] More than 36 one-way trips per month   **33.** [2.3] $d = \dfrac{a}{3}$ **34.** [1.4], [2.2] $-15, 15$ **35.** [2.7] Let $h =$ the number of hours of sun each day; $4 \le h \le 6$ **36.** [2.5] 60 tickets

Cumulative Review: Chapters 1–2, pp. 143–144

1. -12 **2.** $\frac{3}{4}$ **3.** -4.2 **4.** 10   **5.** 134   **6.** 149   **7.** $2x + 1$ **8.** $-10t + 12$ **9.** $-21n + 36$ **10.** $-\frac{5}{2}$ (number line: $-5\,-4\,-3\,-2\,-1\ 0\ 1\ 2\ 3\ 4\ 5$)   **11.** 27   **12.** $6(2x + 3y + 5z)$ **13.** -6 **14.** 16   **15.** 9   **16.** $\frac{13}{18}$ **17.** 1   **18.** $-\frac{7}{2}$ **19.** $z = \dfrac{x}{4y}$ **20.** $y = \frac{4}{9}x - \frac{1}{9}$ **21.** $n = \dfrac{p}{a + r}$ **22.** 1.83   **23.** 37.5%   **24.** $-\frac{5}{2}$, $t > -\frac{5}{2}$ (number line: $-5\,-4\,-3\,-2\,-1\ 0\ 1\ 2\ 3\ 4\ 5$)   **25.** $\{t \mid t \le -2\}$ **26.** $\{t \mid t < 3\}$ **27.** $\{x \mid x < 30\}$ **28.** $\{n \mid n \ge 2\}$ **29.** 48 million   **30.** 3.2 hr   **31.** $14\frac{1}{3}$ m   **32.** 9 ft, 15 ft   **33.** No more than $52   **34.** About 194%   **35.** 105°   **36.** For widths greater than 27 cm   **37.** $4t$ **38.** $-5, 5$ **39.** $1025

CHAPTER 3

Exercise Set 3.1, pp. 152–155

1. (a)   **2.** (c)   **3.** (b)   **4.** (d)   **5.** 2 drinks   **7.** The person weighs more than 140 lb.   **9.** About $5156   **11.** $231,856,000   **13.** About 29.2 million tons   **15.** About 3.2 million tons   **17.** About $12 billion   **19.** 2001

21.

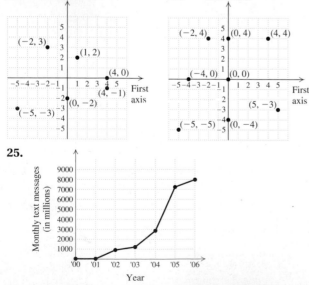

25.

27. $A(-4, 5); B(-3, -3); C(0, 4); D(3, 4); E(3, -4)$
29. $A(4, 1); B(0, -5); C(-4, 0); D(-3, -2); E(3, 0)$

31.

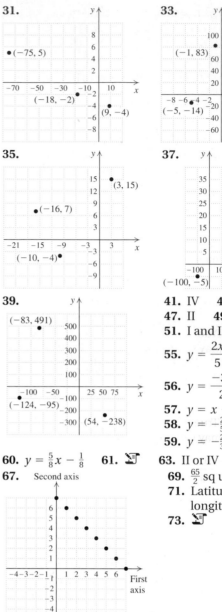

33.

35.

37.

39.

41. IV   **43.** III   **45.** I   **47.** II   **49.** I and IV   **51.** I and III   **53.** ✍   **55.** $y = \dfrac{2x}{5}$, or $y = \frac{2}{5}x$ **56.** $y = \dfrac{-3x}{2}$, or $y = -\frac{3}{2}x$ **57.** $y = x - 8$ **58.** $y = -\frac{2}{5}x + 2$ **59.** $y = -\frac{2}{3}x + \frac{5}{3}$ **60.** $y = \frac{5}{8}x - \frac{1}{8}$ **61.** ✍   **63.** II or IV   **65.** $(-1, -5)$ **67.**   **69.** $\frac{65}{2}$ sq units   **71.** Latitude 27° North; longitude 81° West   **73.** ✍

Technology Connection, p. 162

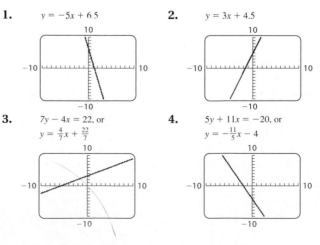

1. $y = -5x + 6.5$ **2.** $y = 3x + 4.5$

3. $7y - 4x = 22$, or $y = \frac{4}{7}x + \frac{22}{7}$ **4.** $5y + 11x = -20$, or $y = -\frac{11}{5}x - 4$

5. $2y - x^2 = 0$, or
$y = 0.5x^2$

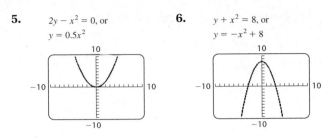

6. $y + x^2 = 8$, or
$y = -x^2 + 8$

Exercise Set 3.2, pp. 163–165

1. False   **2.** True   **3.** True   **4.** True   **5.** True
6. False   **7.** Yes   **9.** No   **11.** No

13.

$$\begin{array}{c|c} y = x + 3 \\ \hline 2 & -1 + 3 \\ 2 \overset{?}{=} 2 & \text{True} \end{array}$$

$$\begin{array}{c|c} y = x + 3 \\ \hline 7 & 4 + 3 \\ 7 \overset{?}{=} 7 & \text{True} \end{array}$$

$(2, 5)$; answers may vary

15.

$$\begin{array}{c|c} y = \frac{1}{2}x + 3 \\ \hline 5 & \frac{1}{2} \cdot 4 + 3 \\ & 2 + 3 \\ 5 \overset{?}{=} 5 & \text{True} \end{array}$$

$$\begin{array}{c|c} y = \frac{1}{2}x + 3 \\ \hline 2 & \frac{1}{2}(-2) + 3 \\ & -1 + 3 \\ 2 \overset{?}{=} 2 & \text{True} \end{array}$$

$(0, 3)$; answers may vary

17.

$$\begin{array}{c|c} y + 3x = 7 \\ \hline 1 + 3 \cdot 2 & 7 \\ 1 + 6 & \\ 7 \overset{?}{=} 7 & \text{True} \end{array}$$

$$\begin{array}{c|c} y + 3x = 7 \\ \hline -5 + 3 \cdot 4 & 7 \\ -5 + 12 & \\ 7 \overset{?}{=} 7 & \text{True} \end{array}$$

$(1, 4)$; answers may vary

19.

$$\begin{array}{c|c} 4x - 2y = 10 \\ \hline 4 \cdot 0 - 2(-5) & 10 \\ 0 + 10 & \\ 10 \overset{?}{=} 10 & \text{True} \end{array}$$

$$\begin{array}{c|c} 4x - 2y = 10 \\ \hline 4 \cdot 4 - 2 \cdot 3 & 10 \\ 16 - 6 & \\ 10 \overset{?}{=} 10 & \text{True} \end{array}$$

$(2, -1)$; answers may vary

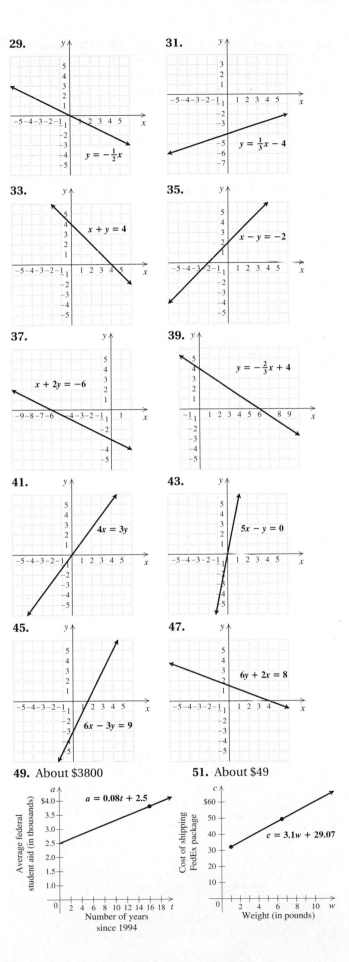

49. About \$3800

51. About \$49

53. About $96

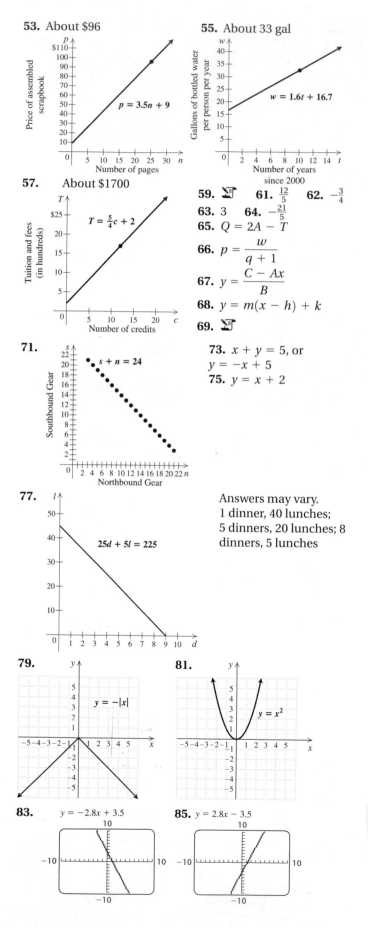

55. About 33 gal

57. About $1700

71.

73. $x + y = 5$, or $y = -x + 5$
75. $y = x + 2$

77.

Answers may vary.
1 dinner, 40 lunches;
5 dinners, 20 lunches; 8
dinners, 5 lunches

59. ✍ **61.** $\frac{12}{5}$ **62.** $-\frac{3}{4}$
63. 3 **64.** $-\frac{21}{5}$
65. $Q = 2A - T$
66. $p = \dfrac{w}{q + 1}$
67. $y = \dfrac{C - Ax}{B}$
68. $y = m(x - h) + k$
69. ✍

79. $y = -|x|$

81. $y = x^2$

83. $y = -2.8x + 3.5$

85. $y = 2.8x - 3.5$

87. $y = x^2 + 4x + 1$

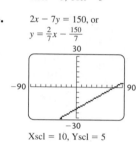

89. $56.62; 16.2 gal

Technology Connection, p. 169

1. $y = -0.72x - 15$

Xscl = 5, Yscl = 5

2. $y - 2.13x = 27$, or $y = 2.13x + 27$

Xscl = 5, Yscl = 5

3. $5x + 6y = 84$, or $y = -\frac{5}{6}x + 14$

Xscl = 5, Yscl = 5

4. $2x - 7y = 150$, or $y = \frac{2}{7}x - \frac{150}{7}$

Xscl = 10, Yscl = 5

5. $19x - 17y = 200$, or $y = \frac{19}{17}x - \frac{200}{17}$

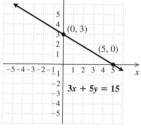

6. $6x + 5y = 159$, or $y = -\frac{6}{5}x + \frac{159}{5}$

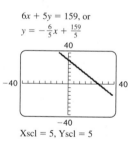

Xscl = 5, Yscl = 5

Exercise Set 3.3, pp. 171–173

1. (f) **2.** (e) **3.** (d) **4.** (c) **5.** (b) **6.** (a)
7. (a) $(0, 5)$; (b) $(2, 0)$ **9.** (a) $(0, -4)$; (b) $(3, 0)$
11. (a) $(0, -2)$; (b) $(-3, 0), (3, 0)$ **13.** (a) $(0, 0)$;
(b) $(-2, 0), (0, 0), (5, 0)$ **15.** (a) $(0, 3)$; (b) $(5, 0)$
17. (a) $(0, -18)$; (b) $(4, 0)$ **19.** (a) $(0, 16)$; (b) $(-20, 0)$
21. (a) None; (b) $(12, 0)$ **23.** (a) $(0, -9)$; (b) none
25.

$3x + 5y = 15$

27.

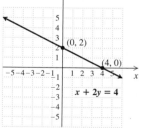

$x + 2y = 4$

29.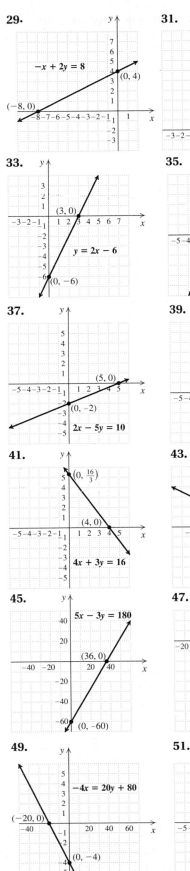

31.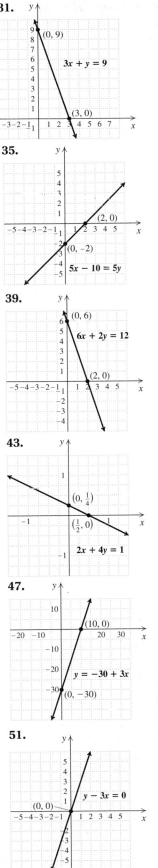

33.

35.

37.

39.

41.

43.

45.

47.

49.

51.

53.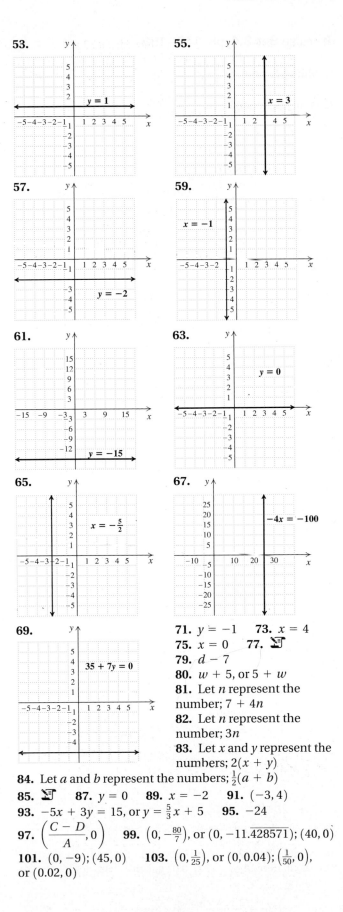

55.

57.

59.

61.

63.

65.

67.

69.

71. $y = -1$ **73.** $x = 4$

75. $x = 0$ **77.** ✍

79. $d - 7$

80. $w + 5$, or $5 + w$

81. Let n represent the number; $7 + 4n$

82. Let n represent the number; $3n$

83. Let x and y represent the numbers; $2(x + y)$

84. Let a and b represent the numbers; $\frac{1}{2}(a + b)$

85. ✍   **87.** $y = 0$ **89.** $x = -2$ **91.** $(-3, 4)$

93. $-5x + 3y = 15$, or $y = \frac{5}{3}x + 5$ **95.** -24

97. $\left(\dfrac{C - D}{A}, 0\right)$ **99.** $\left(0, -\frac{80}{7}\right)$, or $(0, -11.\overline{428571})$; $(40, 0)$

101. $(0, -9)$; $(45, 0)$ **103.** $\left(0, \frac{1}{25}\right)$, or $(0, 0.04)$; $\left(\frac{1}{50}, 0\right)$, or $(0.02, 0)$

Exercise Set 3.4, pp. 177–182

1. Miles per hour, or $\dfrac{\text{miles}}{\text{hour}}$

2. Hours per chapter, or $\dfrac{\text{hours}}{\text{chapter}}$

3. Dollars per mile, or $\dfrac{\text{dollars}}{\text{mile}}$

4. Petunias per foot, or $\dfrac{\text{petunias}}{\text{foot}}$

5. Minutes per errand, or $\dfrac{\text{minutes}}{\text{errand}}$

6. Cups of flour per cake, or $\dfrac{\text{cups of flour}}{\text{cake}}$

7. (a) 30 mpg; **(b)** \$39.33/day; **(c)** 130 mi/day; **(d)** 30¢/mi
9. (a) 7 mph; **(b)** \$7.50/hr; **(c)** \$1.07/mi
11. (a) \$22/hr; **(b)** 20.6 pages/hr; **(c)** \$1.07/page
13. \$568.4 billion/yr **15. (a)** 14.5 floors/min;
(b) 4.14 sec/floor **17. (a)** 23.42 ft/min; **(b)** 0.04 min/ft
19. **21.**

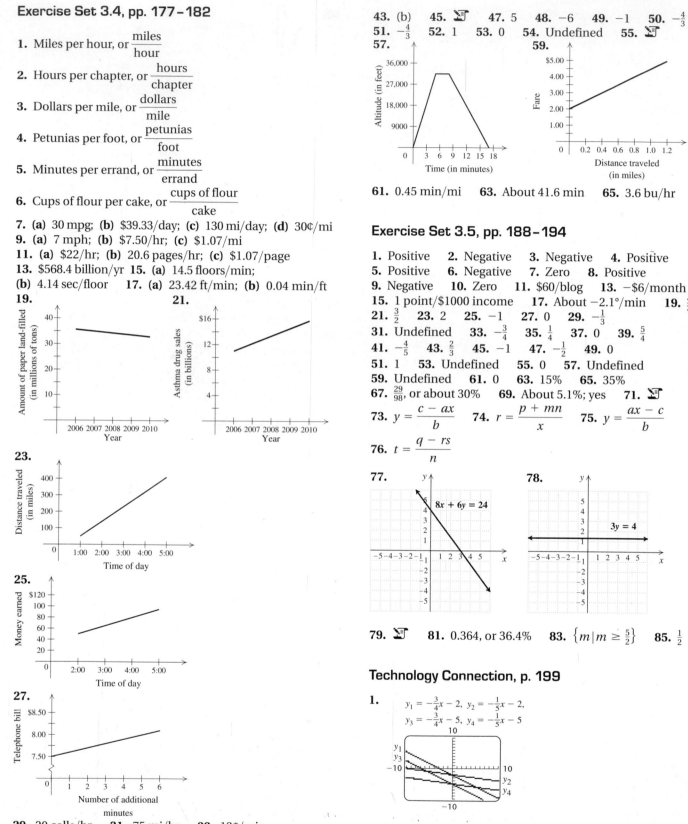

29. 20 calls/hr **31.** 75 mi/hr **33.** 12¢/min
35. −2000 people/yr **37.** 0.02 gal/mi **39. (e)** **41. (d)**

43. (b) **45.** ✑ **47.** 5 **48.** −6 **49.** −1 **50.** $-\frac{4}{3}$
51. $-\frac{4}{3}$ **52.** 1 **53.** 0 **54.** Undefined **55.** ✑
57. **59.**

61. 0.45 min/mi **63.** About 41.6 min **65.** 3.6 bu/hr

Exercise Set 3.5, pp. 188–194

1. Positive **2.** Negative **3.** Negative **4.** Positive
5. Positive **6.** Negative **7.** Zero **8.** Positive
9. Negative **10.** Zero **11.** \$60/blog **13.** −\$6/month
15. 1 point/\$1000 income **17.** About −2.1°/min **19.** $\frac{4}{3}$
21. $\frac{3}{2}$ **23.** 2 **25.** −1 **27.** 0 **29.** $-\frac{1}{3}$
31. Undefined **33.** $-\frac{3}{4}$ **35.** $\frac{1}{4}$ **37.** 0 **39.** $\frac{5}{4}$
41. $-\frac{4}{5}$ **43.** $\frac{2}{3}$ **45.** −1 **47.** $-\frac{1}{2}$ **49.** 0
51. 1 **53.** Undefined **55.** 0 **57.** Undefined
59. Undefined **61.** 0 **63.** 15% **65.** 35%
67. $\frac{29}{98}$, or about 30% **69.** About 5.1%; yes **71.** ✑
73. $y = \dfrac{c - ax}{b}$ **74.** $r = \dfrac{p + mn}{x}$ **75.** $y = \dfrac{ax - c}{b}$
76. $t = \dfrac{q - rs}{n}$

77. **78.**

79. ✑ **81.** 0.364, or 36.4% **83.** $\left\{m \mid m \ge \frac{5}{2}\right\}$ **85.** $\frac{1}{2}$

Technology Connection, p. 199

1.

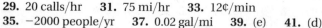

Exercise Set 3.6, pp. 199–201

1. (f) **2.** (b) **3.** (d) **4.** (c) **5.** (e) **6.** (a)

7. **9.**

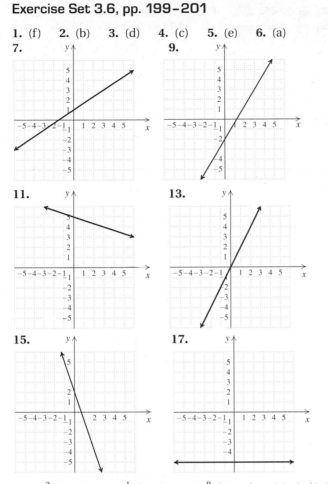

11. **13.**

15. **17.**

19. $-\frac{2}{7}$; $(0, 5)$ **21.** $\frac{1}{3}$; $(0, 7)$ **23.** $\frac{9}{5}$; $(0, -4)$ **25.** 3; $(0, 7)$
27. -2; $(0, 4)$ **29.** 0; $(0, 3)$ **31.** $\frac{2}{5}$; $\left(0, \frac{8}{5}\right)$ **33.** $\frac{9}{8}$; $(0, 0)$
35. $y = 5x + 7$ **37.** $y = \frac{7}{8}x - 1$ **39.** $y = -\frac{5}{3}x - 8$
41. $y = \frac{1}{3}$ **43.** $y = \frac{3}{2}x + 17$, where y is the number of gallons per person and x is the number of years since 2000
45. $y = \frac{2}{5}x + 15$, where y is the number of jobs, in millions, and x is the number of years since 2000

47. **49.**

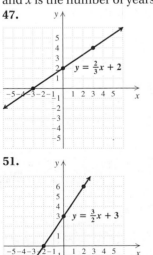

$y = \frac{2}{3}x + 2$ $y = -\frac{2}{3}x + 3$

51. **53.**

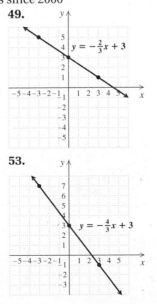

$y = \frac{3}{2}x + 3$ $y = -\frac{4}{3}x + 3$

55. **57.**

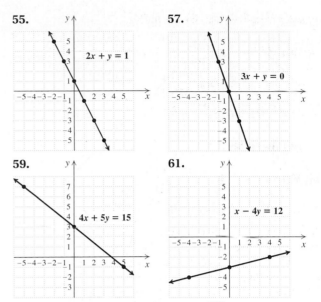

$2x + y = 1$ $3x + y = 0$

59. **61.**

$4x + 5y = 15$ $x - 4y = 12$

63. Yes   **65.** No   **67.** Yes   **69.** ✍
71. $y = m(x - h) + k$ **72.** $y = -2(x + 4) + 9$
73. -7 **74.** 13   **75.** -9 **76.** -11 **77.** ✍
79. When $x = 0, y = b$, so $(0, b)$ is on the line. When $x = 1, y = m + b$, so $(1, m + b)$ is on the line. Then

$$\text{slope} = \frac{(m + b) - b}{1 - 0} = m.$$

81. $y = \frac{1}{3}x + 3$ **83.** $y = -\frac{4}{5}x + 4$ **85.** $y = \frac{1}{2}x$
87. Yes   **89.** Yes   **91.** No   **93.** $y = -\frac{5}{3}x + 3$
95. $y = \frac{5}{2}x + 1$

Connecting the Concepts, p. 202

1. (a) Yes;
(b)

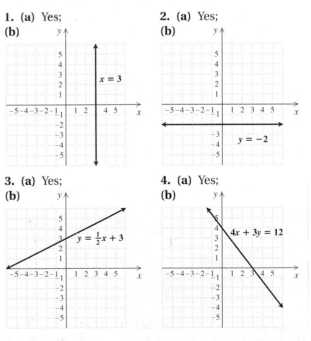

$x = 3$

2. (a) Yes;
(b)

$y = -2$

3. (a) Yes;
(b)

$y = \frac{1}{2}x + 3$

4. (a) Yes;
(b)

$4x + 3y = 12$

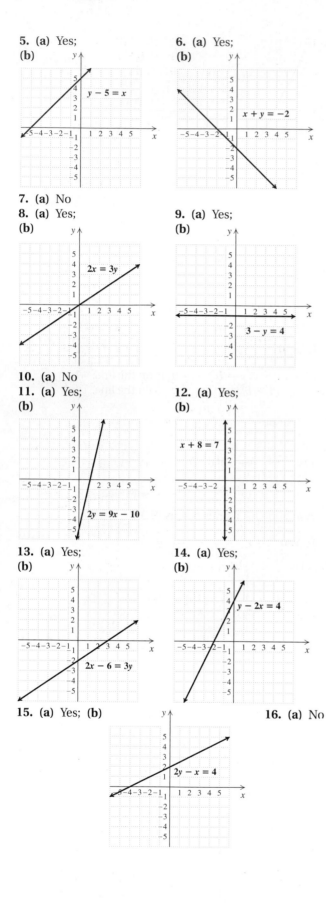

5. (a) Yes;
(b)
$y - 5 = x$

6. (a) Yes;
(b)
$x + y = -2$

7. (a) No
8. (a) Yes;
(b)
$2x = 3y$

9. (a) Yes;
(b)
$3 - y = 4$

10. (a) No
11. (a) Yes;
(b)
$2y = 9x - 10$

12. (a) Yes;
(b)
$x + 8 = 7$

13. (a) Yes;
(b)
$2x - 6 = 3y$

14. (a) Yes;
(b)
$y - 2x = 4$

15. (a) Yes; **(b)**
$2y - x = 4$

16. (a) No

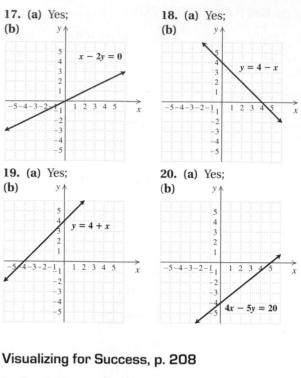

17. (a) Yes;
(b)
$x - 2y = 0$

18. (a) Yes;
(b)
$y = 4 - x$

19. (a) Yes;
(b)
$y = 4 + x$

20. (a) Yes;
(b)
$4x - 5y = 20$

Visualizing for Success, p. 208

1. C **2.** G **3.** F **4.** B **5.** D **6.** A **7.** I
8. H **9.** J **10.** E

Exercise Set 3.7, pp. 209–212

1. (g) **2.** (b) **3.** (d) **4.** (h) **5.** (e) **6.** (a)
7. (f) **8.** (c) **9.** (c) **10.** (b) **11.** (d) **12.** (a)
13. $y - 6 = 3(x - 1)$ **15.** $y - 8 = \frac{3}{5}(x - 2)$
17. $y - 1 = -4(x - 3)$ **19.** $y - (-4) = \frac{3}{2}(x - 5)$
21. $y - 6 = -\frac{5}{4}(x - (-2))$
23. $y - (-1) = -2(x - (-4))$
25. $y - 8 = 1(x - (-2))$ **27.** $y = 4x - 7$
29. $y = \frac{7}{4}x - 9$ **31.** $y = -2x + 1$ **33.** $y = -4x - 9$
35. $y = \frac{2}{3}x + \frac{8}{3}$ **37.** $y = -\frac{5}{6}x + 4$
39.

41.

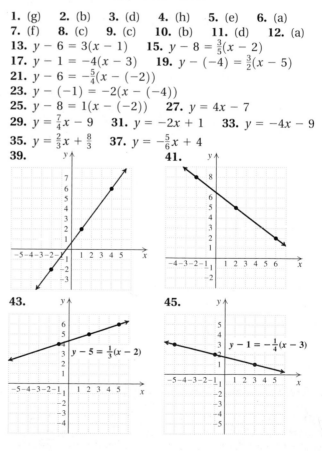

43.
$y - 5 = \frac{1}{3}(x - 2)$

45.
$y - 1 = -\frac{1}{4}(x - 3)$

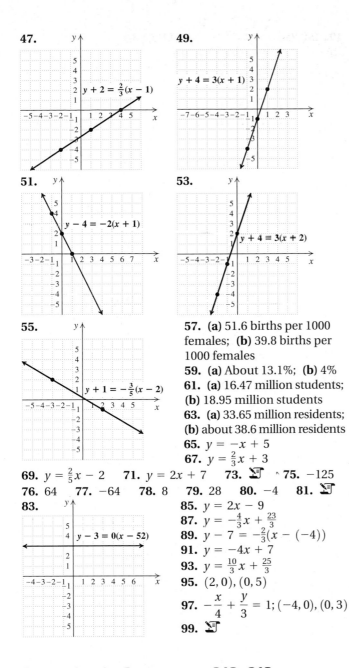

47.

$y + 2 = \frac{2}{3}(x - 1)$

49.

$y + 4 = 3(x + 1)$

51.

$y - 4 = -2(x + 1)$

53.

$y + 4 = 3(x + 2)$

55.

$y + 1 = -\frac{3}{5}(x - 2)$

57. **(a)** 51.6 births per 1000 females; **(b)** 39.8 births per 1000 females
59. **(a)** About 13.1%; **(b)** 4%
61. **(a)** 16.47 million students; **(b)** 18.95 million students
63. **(a)** 33.65 million residents; **(b)** about 38.6 million residents
65. $y = -x + 5$
67. $y = \frac{2}{3}x + 3$
69. $y = \frac{2}{5}x - 2$ **71.** $y = 2x + 7$ **73.** 🖘   ⁀**75.** -125
76. 64   **77.** -64 **78.** 8   **79.** 28   **80.** -4 **81.** 🖘
83.

$y - 3 = 0(x - 52)$

85. $y = 2x - 9$
87. $y = -\frac{4}{3}x + \frac{23}{3}$
89. $y - 7 = -\frac{2}{3}(x - (-4))$
91. $y = -4x + 7$
93. $y = \frac{10}{3}x + \frac{25}{3}$
95. $(2, 0), (0, 5)$
97. $-\dfrac{x}{4} + \dfrac{y}{3} = 1; (-4, 0), (0, 3)$
99. 🖘

Connecting the Concepts, pp. 212–213

1. Slope–intercept form   **2.** Standard form
3. None of these   **4.** Standard form
5. Point–slope form   **6.** None of these
7. $2x - 5y = 10$ **8.** $x - y = 2$
9. $-2x + y = 7$, or $2x - y = -7$ **10.** $\frac{1}{2}x + y = 3$
11. $3x - y = -23$, or $-3x + y = 23$ **12.** $x + 0y = 18$
13. $y = \frac{2}{7}x - \frac{8}{7}$ **14.** $y = -x - 8$ **15.** $y = 8x - 3$
16. $y = -\frac{3}{5}x + 3$ **17.** $y = -\frac{8}{9}x + \frac{5}{9}$ **18.** $y = -x$
19. $y = -\frac{1}{2}$ **20.** $y = \frac{1}{2}x$

Review Exercises: Chapter 3, pp. 216–217

1. True   **2.** True   **3.** False   **4.** False   **5.** True
6. True   **7.** True   **8.** False   **9.** True   **10.** True
11. About 1.3 billion searches   **12.** About 1137 searches

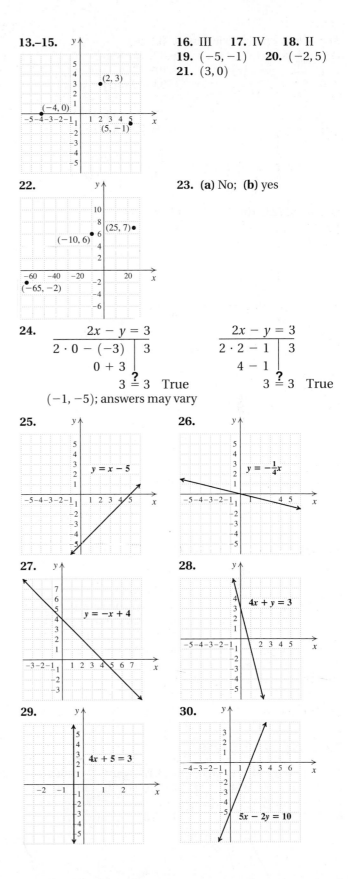

13.–15.

16. III   **17.** IV   **18.** II
19. $(-5, -1)$ **20.** $(-2, 5)$
21. $(3, 0)$

22.

23. **(a)** No; **(b)** yes

24.

$$\begin{array}{c|c} 2x - y = 3 & 3 \\ \hline 2 \cdot 0 - (-3) & \\ 0 + 3 & \\ 3 \overset{?}{=} 3 \quad \text{True} \end{array} \qquad \begin{array}{c|c} 2x - y = 3 & 3 \\ \hline 2 \cdot 2 - 1 & \\ 4 - 1 & \\ 3 \overset{?}{=} 3 \quad \text{True} \end{array}$$

$(-1, -5)$; answers may vary

25.

$y = x - 5$

26.

$y = -\frac{1}{4}x$

27.

$y = -x + 4$

28.

$4x + y = 3$

29.

$4x + 5 = 3$

30.

$5x - 2y = 10$

31. About 7 million viewers

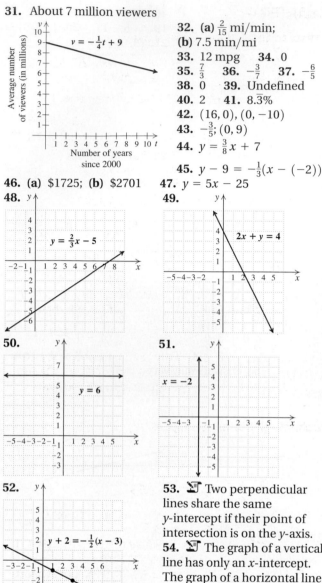

32. (a) $\frac{2}{15}$ mi/min;
(b) 7.5 min/mi
33. 12 mpg   **34.** 0
35. $\frac{7}{3}$ **36.** $-\frac{3}{7}$ **37.** $-\frac{6}{5}$
38. 0   **39.** Undefined
40. 2   **41.** 8.$\overline{3}$%
42. $(16, 0), (0, -10)$
43. $-\frac{3}{5}; (0, 9)$
44. $y = \frac{3}{8}x + 7$
45. $y - 9 = -\frac{1}{3}(x - (-2))$
46. (a) \$1725; **(b)** \$2701   **47.** $y = 5x - 25$
48.

49.

50.

51.

52.

53. ✍ Two perpendicular lines share the same y-intercept if their point of intersection is on the y-axis.
54. ✍ The graph of a vertical line has only an x-intercept. The graph of a horizontal line has only a y-intercept. The graph of a nonvertical, non-horizontal line will have only one intercept if it passes through the origin: $(0, 0)$ is both the x-intercept and the y-intercept.   **55.** -1
56. 19   **57.** Area: 45 sq units; perimeter: 28 units
58. $(0, 4), (1, 3), (-1, 3)$; answers may vary

Test: Chapter 3, p. 218

1. [3.1] About 95 students   **2.** [3.1] About 137 students
3. [3.1] III   **4.** [3.1] II   **5.** [3.1] $(3, 4)$ **6.** [3.1] $(0, -4)$
7. [3.1] $(-5, 2)$
8. [3.2]

9. [3.3]

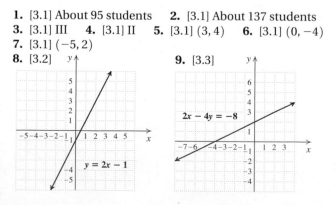

10. [3.3]

11. [3.2]

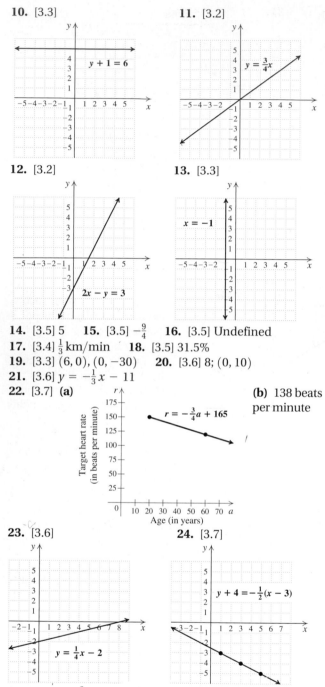

12. [3.2]

13. [3.3]

14. [3.5] 5   **15.** [3.5] $-\frac{9}{4}$ **16.** [3.5] Undefined
17. [3.4] $\frac{1}{3}$ km/min   **18.** [3.5] 31.5%
19. [3.3] $(6, 0), (0, -30)$ **20.** [3.6] 8; $(0, 10)$
21. [3.6] $y = -\frac{1}{3}x - 11$
22. [3.7] **(a)**   **(b)** 138 beats per minute

23. [3.6]   **24.** [3.7]

25. [3.6] $y = \frac{2}{5}x + 9$ **26.** [3.1] Area: 25 sq units;
perimeter: 20 units   **27.** [3.2], [3.7] $(0, 12), (-3, 15), (5, 7)$

Cumulative Review: Chapters 1–3, pp. 219–220

1. 7   **2.** $12a - 6b + 18$ **3.** $4(2x - y + 1)$ **4.** $2 \cdot 3^3$
5. -0.15 **6.** 37   **7.** $\frac{1}{10}$ **8.** -10 **9.** 0.367   **10.** $\frac{11}{60}$
11. 2.6   **12.** 7.28   **13.** $-\frac{5}{12}$ **14.** -3 **15.** 27
16. $-2y - 7$ **17.** $5x + 11$ **18.** -2.6 **19.** -27
20. 16   **21.** -6 **22.** 2   **23.** $\frac{7}{9}$ **24.** -17 **25.** 2
26. $\{x | x < 16\}$ **27.** $\{x | x \le -\frac{11}{8}\}$ **28.** $h = \dfrac{A - \pi r^2}{2\pi r}$

29. IV   **30.**

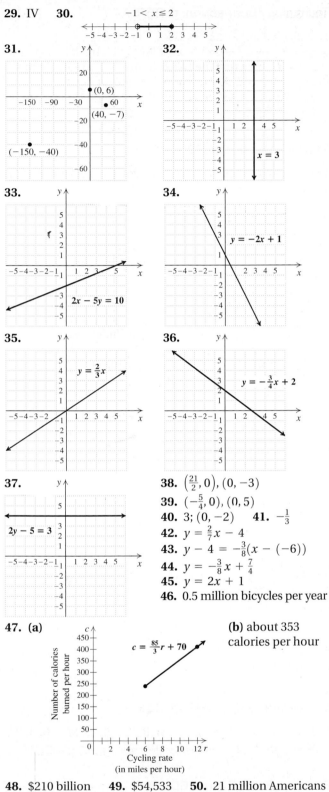

$-1 < x \le 2$

31.

32. $x = 3$

33. $2x - 5y = 10$

34. $y = -2x + 1$

35. $y = \frac{2}{3}x$

36. $y = -\frac{3}{4}x + 2$

37. $2y - 5 = 3$

38. $\left(\frac{21}{2}, 0\right), (0, -3)$

39. $\left(-\frac{5}{4}, 0\right), (0, 5)$

40. 3; $(0, -2)$ **41.** $-\frac{1}{3}$

42. $y = \frac{2}{7}x - 4$

43. $y - 4 = -\frac{3}{8}(x - (-6))$

44. $y = -\frac{3}{8}x + \frac{7}{4}$

45. $y = 2x + 1$

46. 0.5 million bicycles per year

47. (a)

Number of calories burned per hour

$c = \frac{85}{3}r + 70$

Cycling rate (in miles per hour)

(b) about 353 calories per hour

48. $210 billion   **49.** $54,533   **50.** 21 million Americans
51. $120   **52.** 50 m, 53 m, 40 m   **53.** 4 hr
54. $25,000   **55.** $-4, 4$ **56.** 2   **57.** -5 **58.** 3
59. No solution   **60.** $Q = \dfrac{2 - pm}{p}$

61. $y = -\frac{7}{3}x + 7; y = -\frac{7}{3}x - 7; y = \frac{7}{3}x - 7; y = \frac{7}{3}x + 7$

CHAPTER 4

Exercise Set 4.1, pp. 228–229

1. (e)   **2.** (f)   **3.** (b)   **4.** (h)   **5.** (g)   **6.** (a)
7. (c)   **8.** (d)   **9.** Base: $2x$; exponent: 5

11. Base: x; exponent: 3   **13.** Base: $\dfrac{4}{y}$; exponent: 7

15. d^{13} **17.** a^7 **19.** 6^{15} **21.** $(3y)^{12}$ **23.** $(8n)^{10}$
25. a^5b^9 **27.** $(x + 3)^{13}$ **29.** r^{10} **31.** m^4n^9 **33.** 7^3
35. t^7 **37.** $5a$ **39.** 1   **41.** $(r + s)^8$ **43.** $\frac{4}{5}d^7$
45. $4a^7b^6$ **47.** $x^{12}y^7$ **49.** 1   **51.** 5   **53.** 2   **55.** -4
57. x^{33} **59.** 5^{32} **61.** t^{80} **63.** $100x^2$ **65.** $-8a^3$
67. $25n^{14}$ **69.** $a^{14}b^7$ **71.** $r^{17}t^{11}$ **73.** $24x^{19}$

75. $\dfrac{x^3}{125}$ **77.** $\dfrac{49}{36n^2}$ **79.** $\dfrac{a^{18}}{b^{48}}$ **81.** $\dfrac{x^8y^4}{z^{12}}$ **83.** $\dfrac{a^{12}}{16b^{20}}$

85. $-\dfrac{125x^{21}y^3}{8z^{12}}$ **87.** 1   **89.** 🖫   **91.** $8x$

92. $-3a - 6b$ **93.** $-2x - 7$ **94.** $-4t - r - 5$
95. 1004   **96.** 9   **97.** 🖫   **99.** 🖫   **101.** Let $x = 1$;
then $3x^2 = 3$, but $(3x)^2 = 9$.   **103.** Let $t = -1$; then
$\dfrac{t^6}{t^2} = 1$, but $t^3 = -1$.   **105.** y^{6x} **107.** x^t **109.** 13

111. <   **113.** <   **115.** >   **117.** 4,000,000; 4,194,304;
194,304   **119.** 2,000,000,000; 2,147,483,648; 147,483,648
121. 1,536,000 bytes, or approximately 1,500,000 bytes

Technology Connection, p. 234

1. 9.8

Exercise Set 4.2, pp. 234–238

1. (b)   **2.** (f)   **3.** (h)   **4.** (d)   **5.** (g)   **6.** (e)
7. (a)   **8.** (c)   **9.** $8x^3, -11x^2, 6x, 1$
11. $-t^6, -3t^3, 9t, -4$ **13.** Coefficients: 8, 2; degrees: 4, 1
15. Coefficients: 9, -3, 4; degrees: 2, 1, 0
17. Coefficients: 6, 9, 1; degrees: 5, 1, 3   **19.** Coefficients:
1, -1, 4, -3; degrees: 4, 3, 1, 0   **21. (a)** 1, 3, 4; **(b)** $8t^4$, 8;
(c) 4   **23. (a)** 2, 0, 4; **(b)** $2a^4$, 2; **(c)** 4
25. (a) 0, 2, 1, 5; **(b)** $-x^5, -1$; **(c)** 5
27.

Term	Coefficient	Degree of the Term	Degree of the Polynomial
$8x^5$	8	5	
$-\frac{1}{2}x^4$	$-\frac{1}{2}$	4	
$-4x^3$	-4	3	5
$7x^2$	7	2	
6	6	0	

29. Trinomial **31.** Polynomial with no special name
33. Binomial **35.** Monomial **37.** $11n^2 + n$ **39.** $4a^4$
41. $7x^3 + x^2 - 6x$ **43.** $11b^3 + b^2 - b$ **45.** $-x^4 - x^3$
47. $\frac{1}{15}x^4 + 10$ **49.** $-1.1a^2 + 5.3a - 7.5$ **51.** $-3; 21$
53. $16; 34$ **55.** $-38; 148$ **57.** $159; 165$ **59.** $-39; 21$
61. $1.93 billion **63.** 1112 ft **65.** 62.8 cm
67. 153.86 m^2 **69.** About 75 donations
71. About 9 words **73.** About 6 **75.** About 16;
about 19 **77.** ✑ **79.** $x - 3$ **80.** $-2x - 6$
81. $6a - 3$ **82.** $-t - 1$ **83.** $-t^4 + 17t$
84. $0.4a^2 - a + 11$ **85.** ✑
87. $2x^5 + 4x^4 + 6x^3 + 8$; answers may vary **89.** $2510
91. $3x^6$ **93.** 3 and 8 **95.** 85.0
97.

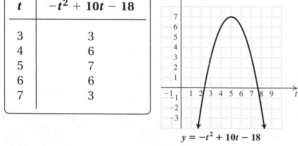

t	$-t^2 + 10t - 18$
3	3
4	6
5	7
6	6
7	3

$y = -t^2 + 10t - 18$

Technology Connection, p. 242

1. In each case, let $y_1 = $ the expression before the addition or subtraction has been performed, $y_2 = $ the simplified sum or difference, and $y_3 = y_2 - y_1$; and note that the graph of y_3 coincides with the x-axis. That is, $y_3 = 0$.

Exercise Set 4.3, pp. 243–245

1. x^2 **2.** -6 **3.** $-$ **4.** $+$ **5.** $4x + 9$ **7.** $-6t + 8$
9. $-3x^2 + 6x - 2$ **11.** $9t^2 + t + 3$
13. $8m^3 - 3m - 7$ **15.** $7 + 13a - a^2 + 7a^3$
17. $2x^6 + 9x^4 + 2x^3 - 4x^2 + 5x$
19. $x^4 + \frac{1}{4}x^3 - \frac{3}{4}x^2 - \frac{5}{6}x + 3$
21. $4.2t^3 + 3.5t^2 - 6.4t - 1.8$ **23.** $-4x^3 + 4x^2 + 6x$
25. $1.3x^4 + 0.35x^3 + 9.53x^2 + 2x + 0.96$
27. $-(-3t^3 + 4t^2 - 7); 3t^3 - 4t^2 + 7$
29. $-(x^4 - 8x^3 + 6x); -x^4 + 8x^3 - 6x$
31. $-9x + 10$ **33.** $-3a^4 + 5a^2 - 1.2$
35. $4x^4 - 6x^2 - \frac{3}{4}x + 8$ **37.** $-2x - 7$
39. $-t^2 - 12t + 13$ **41.** $8a^3 + 8a^2 + a - 10$
43. $4.6x^3 + 9.2x^2 - 3.8x - 23$ **45.** 0
47. $1 + a + 12a^2 - 3a^3$ **49.** $\frac{9}{8}x^3 - \frac{1}{2}x$
51. $0.05t^3 - 0.07t^2 + 0.01t + 1$ **53.** $2x^2 + 6$
55. $-3x^4 - 8x^3 - 7x^2$ **57. (a)** $5x^2 + 4x$; **(b)** 145; 273
59. $16y + 26$ **61.** $(r + 11)(r + 9); 9r + 99 + 11r + r^2$
63. $(x + 3)^2; x^2 + 3x + 9 + 3x$ **65.** $m^2 - 40$
67. $\pi r^2 - 49$ **69.** $(x^2 - 12)$ ft^2 **71.** $(z^2 - 36\pi)$ ft^2
73. $\left(144 - \frac{d^2}{4}\pi\right)$ m^2 **75.** ✑ **77.** $2x^2 - 2x + 6$
78. $-15x^2 + 10x + 35$ **79.** x^8 **80.** y^7 **81.** $2n^3$
82. $-6n^{12}$ **83.** ✑ **85.** $9t^2 - 20t + 11$
87. $-6x + 14$ **89.** $250.591x^3 + 2.812x$ **91.** $20w + 42$
93. $2x^2 + 20x$ **95.** $8x + 24$ **97.** ✑

Technology Connection, p. 250

1. Let $y_1 = (5x^4 - 2x^2 + 3x)(x^2 + 2x)$ and $y_2 = 5x^6 + 10x^5 - 2x^4 - x^3 + 6x^2$. With the table set in AUTO mode, note that the values in the Y_1- and Y_2-columns match, regardless of how far we scroll up or down.
2. Use TRACE, a table, or a boldly drawn graph to confirm that y_3 is always 0.

Exercise Set 4.4, pp. 251–253

1. (c) **2.** (d) **3.** (d) **4.** (a) **5.** (c) **6.** (b)
7. $21x^5$ **9.** $-x^7$ **11.** x^8 **13.** $36t^4$ **15.** $-0.12x^9$
17. $-\frac{1}{20}x^{12}$ **19.** $5n^3$ **21.** $-44x^{10}$ **23.** $72y^{10}$
25. $20x^2 + 5x$ **27.** $3a^2 - 27a$ **29.** $x^5 + x^2$
31. $-6n^3 + 24n^2 - 3n$ **33.** $-15t^3 - 30t^2$
35. $4a^9 - 8a^7 - \frac{5}{12}a^4$ **37.** $x^2 + 7x + 12$
39. $t^2 + 4t - 21$ **41.** $a^2 - 1.3a + 0.42$ **43.** $x^2 - 9$
45. $28 - 15x + 2x^2$ **47.** $t^2 + \frac{17}{6}t + 2$
49. $\frac{3}{16}a^2 + \frac{5}{4}a - 2$

51.

x^2	$5x$

x ↕ ... x — 5 ... $x + 5$

53.

x	2
x^2	$2x$

$x + 1$; $x + 2$

55.

$3x$	15
x^2	$5x$

$x + 3$; $x + 5$

57. $x^3 + 2x + 3$

59. $2a^3 - a^2 - 11a + 10$
61. $3y^6 - 21y^4 + y^3 + 2y^2 - 7y - 14$
63. $33x^2 + 25x + 2$ **65.** $x^4 + 4x^3 - 3x^2 + 16x - 3$
67. $10t^4 + 3t^3 - 20t^2 + \frac{9}{2}t - 2$
69. $x^4 + 8x^3 + 12x^2 + 9x + 4$ **71.** ✑ **73.** 0 **74.** 0
75. 8 **76.** 7 **77.** 32 **78.** 50 **79.** ✑
81. $75y^2 - 45y$ **83.** 5 **85.** $V = (4x^3 - 48x^2 + 144x)$ in^3;
$S = (-4x^2 + 144)$ in^2 **87.** $(x^3 - 5x^2 + 8x - 4)$ cm^3
89. 16 ft by 8 ft **91.** 0 **93.** $x^3 + x^2 - 22x - 40$ **95.** 0

Visualizing for Success, p. 259

1. E, F **2.** B, O **3.** S, K **4.** R, G **5.** D, M **6.** J, P
7. C, L **8.** N, Q **9.** A, H **10.** I, T

Exercise Set 4.5, pp. 260–262

1. True **2.** False **3.** False **4.** True
5. $x^3 + 3x^2 + 2x + 6$ **7.** $t^5 + 7t^4 - 2t - 14$
9. $y^2 - y - 6$ **11.** $9x^2 + 21x + 10$
13. $5x^2 + 17x - 12$ **15.** $15 - 13t + 2t^2$
17. $x^4 - 4x^2 - 21$ **19.** $p^2 - \frac{1}{16}$ **21.** $x^2 - 0.6x + 0.09$
23. $-3n^2 - 19n + 14$ **25.** $x^2 + 20x + 100$
27. $1 - 3t + 5t^2 - 15t^3$ **29.** $x^5 + 3x^3 - x^2 - 3$

31. $3x^6 - 2x^4 - 6x^2 + 4$ **33.** $4t^6 + 20t^3 + 25$
35. $8x^5 + 16x^3 + 5x^2 + 10$ **37.** $100x^4 - 9$
39. $x^2 - 64$ **41.** $4x^2 - 1$ **43.** $25m^4 - 16$
45. $81a^6 - 1$ **47.** $x^8 - 0.01$ **49.** $t^2 - \frac{9}{16}$
51. $x^2 + 6x + 9$ **53.** $49x^6 - 14x^3 + 1$
55. $a^2 - \frac{4}{5}a + \frac{4}{25}$ **57.** $t^8 + 6t^4 + 9$
59. $4 - 12x^4 + 9x^8$ **61.** $25 + 60t^2 + 36t^4$
63. $49x^2 - 4.2x + 0.09$ **65.** $14n^5 - 7n^3$
67. $a^3 - a^2 - 10a + 12$ **69.** $49 - 42x^4 + 9x^8$
71. $5x^3 + 30x^2 - 10x$ **73.** $q^{10} - 1$
75. $15t^5 - 3t^4 + 3t^3$ **77.** $36x^8 - 36x^5 + 9x^2$
79. $18a^4 + 0.8a^3 + 4.5a + 0.2$ **81.** $\frac{1}{25} - 36x^8$
83. $a^3 + 1$ **85.** $x^2 + 6x + 9$ **87.** $t^2 + 7t + 12$
89. $a^2 + 10a + 25$ **91.** $x^2 + 10x + 21$
93. $a^2 + 8a + 7$ **95.** $25t^2 + 20t + 4$
97.

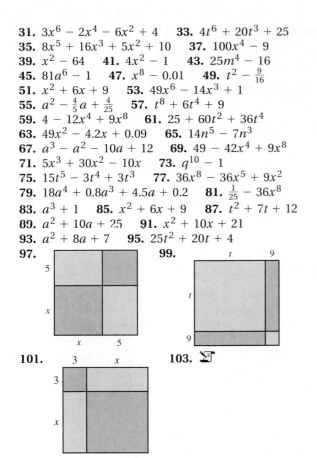

99.

101.   **103.** 🔄

105. Washing machine: 9 kWh/mo; refrigerator: 189 kWh/mo; freezer: 99 kWh/mo   **106.** About
$5.8 billion   **107.** $y = \dfrac{8}{5x}$ **108.** $a = \dfrac{c}{3b}$
109. $x = \dfrac{by + c}{a}$ **110.** $y = \dfrac{ax - c}{b}$ **111.** 🔄
113. $16x^4 - 81$ **115.** $81t^4 - 72t^2 + 16$
117. $t^{24} - 4t^{18} + 6t^{12} - 4t^6 + 1$ **119.** 396   **121.** -7
123. $17F + 7(F - 17)$, $F^2 - (F - 17)(F - 7)$; other equivalent expressions are possible.   **125.** $(y + 1)(y - 1)$, $y(y + 1) - y - 1$; other equivalent expressions are possible.
127. $y^2 - 4y + 4$ **129.** 〰

Connecting the Concepts, p. 263

1. Addition; $3x^2 + 3x + 3$ **2.** Subtraction; $6x + 13$
3. Multiplication; $48x^5 - 42x^3$ **4.** Multiplication;
$6x^2 + x - 2$ **5.** Subtraction; $9x^3 - 5x^2 - 7x + 13$
6. Multiplication; $2x^3 + 3x^2 - 5x - 3$ **7.** Multiplication;
$81x^2 - 1$ **8.** Addition; $9x^4 + 2x^3 - 5x$
9. $-6x^2 + 2x - 12$ **10.** $9x^2 + 45x + 56$
11. $40x^9 - 48x^8 + 16x^5$ **12.** $t^9 + 5t^7$
13. $4m^2 - 4m + 1$ **14.** $x^3 - 1$ **15.** $5x^3 + 3$
16. $c^2 - 9$ **17.** $16y^6 + 56y^3 + 49$
18. $3a^4 - 13a^3 - 13a^2 - 4$ **19.** $16t^4 - 25$
20. $a^8 - 5a^4 - 24$

Technology Connection, p. 267

1. 36.22   **2.** 22,312

Exercise Set 4.6, pp. 268–271

1. (a)   **2.** (b)   **3.** (b)   **4.** (a)   **5.** (c)   **6.** (c)
7. (a)   **8.** (a)   **9.** -13 **11.** -68 **13.** 3.51 L
15. 1889 calories   **17.** 73.005 in²   **19.** 66.4 m
21. Coefficients: 3, -5, 2, -11; degrees: 3, 2, 2, 0; 3
23. Coefficients: 7, -1, 1, 9; degrees: 0, 3, 3, 3; 3
25. $2r - 6s$ **27.** $5xy^2 - 2x^2y + x + 3x^2$
29. $9u^2v - 11uv^2 + 11u^2$ **31.** $6a^2c - 7ab^2 + a^2b$
33. $11x^2 - 10xy - y^2$ **35.** $-6a^4 - 8ab + 7ab^2$
37. $-6r^2 - 5rt - t^2$ **39.** $3x^3 - x^2y + xy^2 - 3y^3$
41. $-2y^4x^3 - 3y^3x$ **43.** $-8x + 8y$
45. $12c^2 + 5cd - 2d^2$ **47.** $x^2y^2 + 4xy - 5$
49. $4a^2 - b^2$ **51.** $20r^2t^2 - 23rt + 6$
53. $m^6n^2 + 2m^3n - 48$ **55.** $30x^2 - 28xy + 6y^2$
57. $0.01 - p^2q^2$ **59.** $x^2 + 2xh + h^2$
61. $16a^2 - 40ab + 25b^2$ **63.** $a^2b^2 - c^2d^4$
65. $x^3y^2 + x^2y^3 + 2x^2y^2 + 2xy^3 + 3xy + 3y^2$
67. $a^2 + 2ab + b^2 - c^2$ **69.** $a^2 - b^2 - 2bc - c^2$
71. $x^2 + 2xy + y^2$ **73.** $\frac{1}{2}a^2b^2 - 2$
75. $a^2 + c^2 + ab + 2ac + ad + bc + bd + cd$
77. $m^2 - n^2$
79. We draw a rectangle with   **81.**
dimensions $r + s$ by $u + v$.

83. 🔄   **85.** $x^2 - 8x - 4$
86. $2x^3 - x^2 - x + 4$
87. $-2x + 5$ **88.** $5x^2 + x$ **89.** $13x^2 + 1$
90. $-x - 3$ **91.** 🔄   **93.** $4xy - 4y^2$
95. $2\pi ab - \pi b^2$ **97.** $x^3 + 2y^3 + x^2y + xy^2$
99. $2x^2 - 2\pi r^2 + 4xh + 2\pi rh$ **101.** 🔄   **103.** 40
105. $P + 2Pr + Pr^2$ **107.** $15,638.03

Exercise Set 4.7, pp. 276–277

1. $8x^6 - 5x^3$ **3.** $1 - 2u + u^6$ **5.** $6t^2 - 8t + 2$
7. $7x^3 - 6x + \frac{3}{2}$ **9.** $-4t^2 - 2t + 1$ **11.** $4x - 5 + \dfrac{1}{2x}$
13. $x + 2x^3y + 3$ **15.** $-3rs - r + 2s$ **17.** $x - 6$
19. $t - 5 + \dfrac{-45}{t - 5}$ **21.** $2x - 1 + \dfrac{1}{x + 6}$
23. $t^2 - 3t + 9$ **25.** $a + 5 + \dfrac{4}{a - 5}$
27. $x - 3 - \dfrac{3}{5x - 1}$ **29.** $3a + 1 + \dfrac{3}{2a + 5}$
31. $t^2 - 3t + 1$ **33.** $x^2 + 1$ **35.** $t^2 - 1 + \dfrac{3t - 1}{t^2 + 5}$
37. $3x^2 - 3 + \dfrac{x - 1}{2x^2 + 1}$ **39.** 🔄   **41.** y^{13} **42.** y^{40}
43. y^3 **44.** $8p^6q^{15}$ **45.** a^5b^5 **46.** 1   **47.** $\dfrac{9}{64n^6}$
48. $-\dfrac{8x^{15}y^6}{27z^9}$ **49.** 🔄   **51.** $5x^{6k} - 16x^{3k} + 14$

53. $3t^{2h} + 2t^h - 5$ **55.** $a + 3 + \dfrac{5}{5a^2 - 7a - 2}$

57. $2x^2 + x - 3$ **59.** 3   **61.** -1

Technology Connection, p. 283

1. 1.71×10^{17} **2.** $5.\overline{370} \times 10^{-15}$ **3.** 3.68×10^{16}

Exercise Set 4.8, pp. 284–286

1. (c)   **2.** (d)   **3.** (a)   **4.** (b)   **5.** $\dfrac{1}{2^3} = \dfrac{1}{8}$

7. $\dfrac{1}{(-2)^6} = \dfrac{1}{64}$ **9.** $\dfrac{1}{t^9}$ **11.** $\dfrac{x}{y^2}$ **13.** $\dfrac{t}{r^5}$ **15.** a^8

17. $\dfrac{1}{7}$ **19.** $\left(\dfrac{5}{3}\right)^2 = \dfrac{25}{9}$ **21.** $\left(\dfrac{2}{x}\right)^5 = \dfrac{32}{x^5}$ **23.** $\left(\dfrac{t}{s}\right)^7 = \dfrac{t^7}{s^7}$

25. 9^{-2} **27.** y^{-3} **29.** 5^{-1} **31.** t^{-1} **33.** 2^3, or 8

35. $\dfrac{1}{x^{12}}$ **37.** $\dfrac{1}{t^2}$ **39.** $\dfrac{1}{n^{15}}$ **41.** t^{18} **43.** $\dfrac{1}{t^{12}}$

45. $\dfrac{1}{m^7 n^7}$ **47.** $\dfrac{9}{x^8}$ **49.** $\dfrac{25t^6}{r^8}$ **51.** t^{14} **53.** $\dfrac{1}{y^4}$

55. $5y^3$ **57.** $2x^5$ **59.** $\dfrac{-3b^9}{2a^7}$ **61.** 1   **63.** $\dfrac{8y^7 z}{x^3}$

65. $3s^2 t^4 u^4$ **67.** $\dfrac{1}{x^{12} y^{15}}$ **69.** $\dfrac{m^{10} n^6}{9}$ **71.** $\dfrac{b^5 c^4}{a^8}$

73. $\dfrac{9}{a^8}$ **75.** $\dfrac{n^{12}}{m^3}$ **77.** $\dfrac{27b^{12}}{8a^6}$ **79.** 1   **81.** 4920

83. 0.00892   **85.** 904,000,000   **87.** 0.000003497

89. 42,090,000   **91.** 3.6×10^7 **93.** 5.83×10^{-3}

95. 7.8×10^{10} **97.** 5.27×10^{-7} **99.** 1.032×10^{-6}

101. 1.094×10^{15} **103.** 6×10^{13} **105.** 2.47×10^8

107. 3.915×10^{-16} **109.** 2.5×10^{13} **111.** 5.0×10^{-6}

113. 3×10^{-21} **115.** 🖬

117.

118.

$y = -\dfrac{2}{3}x + 4$

$3x - 4y = 12$

119.

$3y - 2 = 7$

120.

$8x = 4y$

121. $-\dfrac{3}{10}$ **122.** Slope: 4; y-intercept: $\left(0, \dfrac{7}{2}\right)$

123. $y = -5x - 10$ **124.** $y = \dfrac{5}{4}x - \dfrac{9}{2}$ **125.** 🖬

127. 8×10^5 **129.** 2^{-12} **131.** 5   **133.** 5^6

135. $\dfrac{1}{3} + \dfrac{1}{4} = \dfrac{7}{12}$ **137.** 9   **139.** $6.304347826 \times 10^{25}$

141. $1.19140625 \times 10^{-15}$ **143.** 3×10^8 mi

145. $\$2.31 \times 10^8$ **147.** 2.277×10^{10} min

Connecting the Concepts, p. 287

1. x^{14} **2.** $\dfrac{1}{x^{14}}$ **3.** $\dfrac{1}{x^{14}}$ **4.** x^{14} **5.** x^{40} **6.** x^{40}

7. c^8 **8.** $\dfrac{1}{c^8}$ **9.** $16x^{12} y^4$ **10.** $\dfrac{1}{16x^{12} y^4}$ **11.** 1

12. a^5 **13.** $\dfrac{a^{15}}{b^{20}}$ **14.** $\dfrac{b^{20}}{a^{15}}$ **15.** $\dfrac{5x^3}{2y^4}$ **16.** $\dfrac{6a^2}{7b^5}$

17. $\dfrac{7t^6}{xp^5}$ **18.** $\dfrac{16a^2}{9b^6}$ **19.** $18p^4 q^{14}$ **20.** $\dfrac{9x^3}{2y^5}$

Review Exercises: Chapter 4, pp. 291–292

1. True   **2.** True   **3.** True   **4.** False   **5.** False

6. False   **7.** True   **8.** True   **9.** n^{12} **10.** $(7x)^{10}$

11. t^6 **12.** 4^3, or 64   **13.** 1   **14.** $-9c^4 d^2$

15. $-8x^3 y^6$ **16.** $18x^5$ **17.** $a^7 b^6$ **18.** $\dfrac{4t^{10}}{9s^8}$

19. $8x^2, -x, \dfrac{2}{3}$ **20.** $-4y^5, 7y^2, -3y, -2$ **21.** $9, -1, 7$

22. $7, -\dfrac{5}{6}, -4, 10$ **23.** (a) $2, 0, 5$; (b) $15t^5, 15$; (c) 5

24. (a) $5, 0, 2, 1$; (b) $-2x^5, -2$; (c) 5   **25.** Trinomial

26. Polynomial with no special name   **27.** Monomial

28. $-x^2 + 7x$ **29.** $-\dfrac{1}{4}x^3 + 4x^2 + 7$ **30.** $t - 1$

31. $14a^5 - 2a^2 - a - \dfrac{2}{3}$ **32.** -24 **33.** 16

34. $x^5 + 8x^4 + 6x^3 - 2x - 9$ **35.** $6a^5 - a^3 - 12a^2$

36. $-3y^2 + 8y + 3$ **37.** $x^5 - 3x^3 - 2x^2 + 8$

38. $\dfrac{3}{4}x^4 + \dfrac{1}{4}x^3 - \dfrac{1}{3}x^2 - \dfrac{7}{4}x + \dfrac{3}{8}$

39. $-x^5 + x^4 - 5x^3 - 2x^2 + 2x$ **40.** (a) $4w + 6$;

(b) $w^2 + 3w$ **41.** $-30x^5$ **42.** $49x^2 + 14x + 1$

43. $a^2 - 3a - 28$ **44.** $d^2 - 64$

45. $12x^3 - 23x^2 + 13x - 2$ **46.** $x^2 - 16x + 64$

47. $15t^5 - 6t^4 + 12t^3$ **48.** $4a^2 - 81$

49. $x^2 - 1.3x + 0.4$

50. $x^7 + x^5 - 3x^4 + 3x^3 - 2x^2 + 5x - 3$

51. $16y^6 - 40y^3 + 25$ **52.** $2t^4 - 11t^2 - 21$

53. $a^2 + \dfrac{1}{6}a - \dfrac{1}{3}$ **54.** $-49 + 4n^2$ **55.** 49

56. Coefficients: $1, -7, 9, -8$; degrees: $6, 2, 2, 0$; 6

57. Coefficients: $1, -1, 1$; degrees: $13, 22, 15$; 22

58. $-4u + 4v - 7$ **59.** $6m^3 + 4m^2 n - mn^2$

60. $2a^2 - 16ab$ **61.** $11x^3 y^2 - 8x^2 y - 6x^2 - 6x + 6$

62. $2x^2 - xy - 15y^2$ **63.** $25a^2 b^2 - 10abcd^2 + c^2 d^4$

64. $\dfrac{1}{2}x^2 - \dfrac{1}{2}y^2$ **65.** $y^4 - \dfrac{1}{3}y + 4$

66. $3x^2 - 7x + 4 + \dfrac{1}{2x + 3}$ **67.** $t^3 + 2t - 3$ **68.** $\dfrac{1}{8^6}$

69. a^{-9} **70.** $\dfrac{1}{4^2}$, or $\dfrac{1}{16}$ **71.** $\dfrac{2b^9}{a^{13}}$ **72.** $\dfrac{1}{w^{15}}$

73. $\dfrac{x^6}{4y^2}$ **74.** $\dfrac{y^3}{8x^3}$ **75.** 470,000,000   **76.** 1.09×10^{-5}

77. 2.09×10^4 **78.** 5.12×10^{-5} **79.** 🖬 In the expression $5x^3$, the exponent refers only to the x. In the expression $(5x)^3$, the entire expression $5x$ is the base.

80. 🖬 It is possible to determine two possibilities for the binomial that was squared by using the equation $(A - B)^2 = A^2 - 2AB + B^2$ in reverse. Since, in

$x^2 - 6x + 9$, $A^2 = x^2$ and $B^2 = 9$, or 3^2, the binomial that was squared was $A - B$, or $x - 3$. If the polynomial is written $9 - 6x + x^2$, then $A^2 = 9$ and $B^2 = x^2$, so the binomial that was squared was $3 - x$. We cannot determine without further information whether the binomial squared was $x - 3$ or $3 - x$.   **81. (a)** 9; **(b)** 28   **82.** $64x^{16}$
83. $8x^4 + 4x^3 + 5x - 2$ **84.** $-16x^6 + x^2 - 10x + 25$
85. $\frac{94}{13}$ **86.** 2.28×10^{11} platelets

Test: Chapter 4, p. 293

1. [4.1] x^{13} **2.** [4.1] 3   **3.** [4.1] 1   **4.** [4.1] t^{45}
5. [4.1] $-27y^6$ **6.** [4.1] $-40x^{19}y^4$ **7.** [4.1] $\frac{6}{5}a^5b^3$
8. [4.1] $\dfrac{16p^2}{25q^6}$ **9.** [4.2] Binomial   **10.** [4.2] $3, -1, \frac{1}{9}$
11. [4.2] Degrees of terms: 3, 1, 5, 0; leading term: $7t^5$; leading coefficient: 7; degree of polynomial: 5
12. [4.2] -7 **13.** [4.2] $5a^2 - 6$ **14.** [4.2] $\frac{7}{4}y^2 - 4y$
15. [4.2] $4x^3 + 4x^2 + 3$
16. [4.3] $4x^5 + x^4 + 5x^3 - 8x^2 + 2x - 7$
17. [4.3] $5x^4 + 5x^2 + x + 5$
18. [4.3] $-2a^4 + 3a^3 - a - 7$
19. [4.3] $-t^4 + 2.5t^3 - 0.6t^2 - 9$
20. [4.4] $-6x^4 + 6x^3 + 10x^2$ **21.** [4.5] $x^2 - \frac{2}{3}x + \frac{1}{9}$
22. [4.5] $25t^2 - 49$ **23.** [4.5] $6b^2 + 7b - 5$
24. [4.5] $x^{14} - 4x^8 + 4x^6 - 16$
25. [4.5] $48 + 34y - 5y^2$ **26.** [4.4] $6x^3 - 7x^2 - 11x - 3$
27. [4.5] $64a^6 + 48a^3 + 9$ **28.** [4.6] 24
29. [4.6] $-4x^3y - x^2y^2 + xy^3 - y^3 + 19$
30. [4.6] $8a^2b^2 + 6ab + 6ab^2 + ab^3 - 4b^3$
31. [4.6] $9x^{10} - y^2$ **32.** [4.7] $4x^2 + 3x - 5$
33. [4.7] $2x^2 - 4x - 2 + \dfrac{17}{3x + 2}$ **34.** [4.8] $\dfrac{1}{y^7}$
35. [4.8] 5^{-6} **36.** [4.8] $\dfrac{1}{t^9}$ **37.** [4.8] $\dfrac{3y^5}{x^5}$
38. [4.8] $\dfrac{b^4}{16a^{12}}$ **39.** [4.8] $\dfrac{c^3}{a^3b^3}$ **40.** [4.8] 3.06×10^9
41. [4.8] 0.00000005   **42.** [4.8] 1.75×10^{17}
43. [4.8] 1.296×10^{22} **44.** [4.4], [4.5]
$V = l(l - 2)(l - 1) = l^3 - 3l^2 + 2l$ **45.** [2.2], [4.5] $\frac{100}{21}$
46. [4.8] $\frac{1}{2} - \frac{1}{4} = \frac{1}{4}$ **47.** [4.8] About 1.4×10^7 hr

Cumulative Review: Chapters 1–4, pp. 294–295

1. 6   **2.** -8 **3.** $-\frac{7}{45}$ **4.** 6   **5.** $y + 10$ **6.** t^{12}
7. $-4x^5y^3$ **8.** $50a^4b^7$ **9.** $2(5a - 3b + 6)$ **10.** $\frac{11}{16}$
11. 6   **12.** $\frac{1}{4}t^3 - 5t^2 - 0.35$ **13.** II
14.

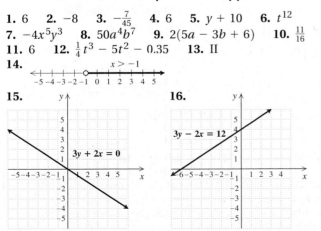

15.   **16.**

17.   **18.**

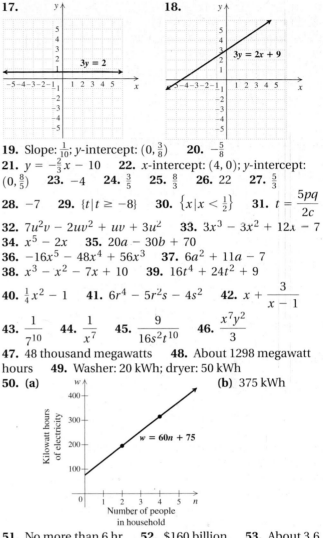

19. Slope: $\frac{1}{10}$; y-intercept: $(0, \frac{3}{8})$ **20.** $-\frac{5}{8}$
21. $y = -\frac{2}{3}x - 10$ **22.** x-intercept: $(4, 0)$; y-intercept:
$(0, \frac{8}{5})$ **23.** -4 **24.** $\frac{3}{5}$ **25.** $\frac{8}{3}$ **26.** 22   **27.** $\frac{5}{3}$
28. -7 **29.** $\{t | t \geq -8\}$ **30.** $\{x | x < \frac{1}{2}\}$ **31.** $t = \dfrac{5pq}{2c}$
32. $7u^2v - 2uv^2 + uv + 3u^2$ **33.** $3x^3 - 3x^2 + 12x - 7$
34. $x^5 - 2x$ **35.** $20a - 30b + 70$
36. $-16x^5 - 48x^4 + 56x^3$ **37.** $6a^2 + 11a - 7$
38. $x^3 - x^2 - 7x + 10$ **39.** $16t^4 + 24t^2 + 9$
40. $\frac{1}{4}x^2 - 1$ **41.** $6r^4 - 5r^2s - 4s^2$ **42.** $x + \dfrac{3}{x - 1}$
43. $\dfrac{1}{7^{10}}$ **44.** $\dfrac{1}{x^7}$ **45.** $\dfrac{9}{16s^2t^{10}}$ **46.** $\dfrac{x^7y^2}{3}$
47. 48 thousand megawatts   **48.** About 1298 megawatt hours   **49.** Washer: 20 kWh; dryer: 50 kWh
50. (a)   **(b)** 375 kWh

51. No more than 6 hr   **52.** $160 billion   **53.** About 3.6
54. $\frac{12}{7}$ billion per year   **55.** 0   **56.** No solution
57. $\frac{22}{9}$ **58.** $y = \frac{1}{2}x + 4$ **59.** $\frac{1}{7} + 1 = \frac{8}{7}$ **60.** $15x^{12}$

CHAPTER 5

Technology Connection, p. 304

1. Correct   **2.** Correct   **3.** Not correct
4. Not correct   **5.** Not correct   **6.** Correct
7. Not correct   **8.** Correct

Exercise Set 5.1, pp. 304–305

1. (h)   **2.** (f)   **3.** (b)   **4.** (e)   **5.** (c)   **6.** (g)
7. (d)   **8.** (a)   **9.** Answers may vary. $(14x)(x^2)$, $(7x^2)(2x), (-2)(-7x^3)$ **11.** Answers may vary. $(-15)(a^4)$, $(-5a)(3a^3), (-3a^2)(5a^2)$ **13.** Answers may vary. $(5t^2)(5t^3), (25t)(t^4), (-5t)(-5t^4)$
15. $8(x + 3)$ **17.** $6(x - 5)$ **19.** $2(x^2 + x - 4)$
21. $t(3t + 1)$ **23.** $-5y(y + 2)$ **25.** $x^2(x + 6)$
27. $8a^2(2a^2 - 3)$ **29.** $-t^2(6t^4 - 9t^2 + 4)$

31. $6x^2(x^6 + 2x^4 - 4x^2 + 5)$
33. $x^2y^2(x^3y^3 + x^2y + xy - 1)$
35. $-5a^2b^2(7ab^2 - 2b + 3a)$ **37.** $(n - 6)(n + 3)$
39. $(x + 3)(x^2 - 7)$ **41.** $(2y - 9)(y^2 + 1)$
43. $(x + 2)(x^2 + 5)$ **45.** $(a + 3)(5a^2 + 2)$
47. $(3n - 2)(3n^2 + 1)$ **49.** $(t - 5)(4t^2 + 3)$
51. $(7x + 5)(x^2 - 3)$ **53.** $(6a + 7)(a^2 + 1)$
55. $(x + 6)(2x^2 - 5)$ **57.** Not factorable by grouping
59. $(y + 8)(y^2 - 2)$ **61.** $(x - 4)(2x^2 - 9)$ **63.** ✍
65. $x^2 + 9x + 14$ **66.** $x^2 - 9x + 14$ **67.** $x^2 - 5x - 14$
68. $x^2 + 5x - 14$ **69.** $a^2 - 4a + 3$ **70.** $t^2 + 8t + 15$
71. $t^2 + 5t - 50$ **72.** $a^2 - 2a - 24$ **73.** ✍
75. $(2x^3 + 3)(2x^2 + 3)$ **77.** $2x(x + 1)(x^2 - 2)$
79. $(x - 1)(5x^4 + x^2 + 3)$ **81.** Answers may vary.
$8x^4y^3 - 24x^2y^4 + 16x^3y^4$

Exercise Set 5.2, pp. 311–312

1. Positive; positive   **2.** Negative; negative
3. Negative; positive   **4.** Positive; positive
5. Positive   **6.** Negative   **7.** $(x + 4)(x + 4)$
9. $(x + 1)(x + 10)$ **11.** $(x + 3)(x + 7)$
13. $(t - 2)(t - 7)$ **15.** $(b - 4)(b - 1)$
17. $(a - 3)(a - 4)$ **19.** $(d - 2)(d - 5)$
21. $(x - 5)(x + 3)$ **23.** $(x + 5)(x - 3)$
25. $2(x + 2)(x - 9)$ **27.** $-x(x + 2)(x - 8)$
29. $(y - 5)(y + 9)$ **31.** $(x - 6)(x + 12)$
33. $-5(b - 3)(b + 10)$ **35.** $x^3(x - 2)(x + 1)$
37. Prime   **39.** $(t + 4)(t + 8)$ **41.** $(x + 9)(x + 11)$
43. $3x(x - 25)(x + 4)$ **45.** $-2(x - 24)(x + 3)$
47. $(y - 12)(y - 8)$ **49.** $-a^4(a - 6)(a + 15)$
51. $\left(t + \frac{1}{3}\right)^2$ **53.** Prime   **55.** $(p - 5q)(p - 2q)$
57. Prime   **59.** $(s - 6t)(s + 2t)$ **61.** $6a^8(a - 2)(a + 7)$
63. ✍   **65.** $6x^2 + 17x + 12$ **66.** $6x^2 + x - 12$
67. $6x^2 - x - 12$ **68.** $6x^2 - 17x + 12$
69. $5x^2 - 36x + 7$ **70.** $3x^2 + 13x - 30$
71. ✍   **73.** $-5, 5, -23, 23, -49, 49$
75. $(y + 0.2)(y - 0.4)$ **77.** $-\frac{1}{3}a(a - 3)(a + 2)$
79. $(x^m + 4)(x^m + 7)$ **81.** $(a + 1)(x + 2)(x + 1)$
83. $(x + 3)^3$, or $(x^3 + 9x^2 + 27x + 27)$ cubic meters
85. $x^2\left(\frac{3}{4}\pi + 2\right)$, or $\frac{1}{4}x^2(3\pi + 8)$ **87.** $x^2\left(9 - \frac{1}{2}\pi\right)$
89. $(x + 4)(x + 5)$

Exercise Set 5.3, pp. 321–322

1. (c)   **2.** (a)   **3.** (d)   **4.** (b)   **5.** $(2x - 1)(x + 4)$
7. $(3x + 1)(x - 6)$ **9.** $(2t + 1)(2t + 5)$
11. $(5a - 3)(3a - 1)$ **13.** $(3x + 4)(2x + 3)$
15. $2(3x + 1)(x - 2)$ **17.** $t(7t + 1)(t + 2)$
19. $(4x - 5)(3x - 2)$ **21.** $-1(7x + 4)(5x + 2)$, or
$-(7x + 4)(5x + 2)$ **23.** Prime   **25.** $(5x + 4)^2$
27. $(20y - 1)(y + 3)$ **29.** $(7x + 5)(2x + 9)$
31. $-1(x - 3)(2x + 5)$, or $-(x - 3)(2x + 5)$
33. $-3(2x + 1)(x + 5)$ **35.** $2(a + 1)(5a - 9)$
37. $4(3x - 1)(x + 6)$ **39.** $(3x + 1)(x + 1)$
41. $(x + 3)(x - 2)$ **43.** $(4t - 3)(2t - 7)$

45. $(3x + 2)(2x + 5)$ **47.** $(y + 4)(2y - 1)$
49. $(3a - 4)(2a - 1)$ **51.** $(16t + 7)(t + 1)$
53. $-1(3x + 1)(3x + 5)$, or $-(3x + 1)(3x + 5)$
55. $10(x^2 + 3x - 7)$ **57.** $3x(3x - 1)(2x + 3)$
59. $(x + 1)(25x + 64)$ **61.** $3x(7x + 1)(8x + 1)$
63. $-t^2(2t - 3)(7t + 1)$ **65.** $2(2y + 9)(8y - 3)$
67. $(2a - b)(a - 2b)$ **69.** $2(s + t)(4s + 7t)$
71. $3(3x - 4y)^2$ **73.** $-2(3a - 2b)(4a - 3b)$
75. $x^2(2x + 3)(7x - 1)$ **77.** $a^6(3a + 4)(3a + 2)$
79. ✍   **81.** $x^2 - 4x + 4$ **82.** $x^2 + 4x + 4$
83. $x^2 - 4$ **84.** $25t^2 - 30t + 9$ **85.** $16a^2 + 8a + 1$
86. $4n^2 - 49$ **87.** $9c^2 - 60c + 100$
88. $1 - 10a + 25a^2$ **89.** $64n^2 - 9$ **90.** $81 - y^2$
91. ✍   **93.** $(3xy + 2)(6xy - 5)$ **95.** Prime
97. $(4t^5 - 1)^2$ **99.** $-1(5x^m - 2)(3x^m - 4)$, or
$-(5x^m - 2)(3x^m - 4)$ **101.** $(3a^{3n} + 1)(a^{3n} - 1)$
103. $[7(t - 3)^n - 2][(t - 3)^n + 1]$

Connecting the Concepts, pp. 322–323

1. $6x^2(x^3 - 3)$ **2.** $(x + 2)(x + 8)$ **3.** $(x + 7)(2x - 1)$
4. $(x + 3)(x^2 + 2)$ **5.** $5(x - 2)(x + 10)$ **6.** Prime
7. $7y(x - 4)(x + 1)$ **8.** $3a^2(5a^2 - 9b^2 + 7b)$
9. $(b - 7)^2$ **10.** $(3x - 1)(4x + 1)$
11. $(c + 1)(c + 2)(c - 2)$ **12.** $2(x - 5)(x + 20)$
13. Prime   **14.** $15(d^2 - 2d + 5)$
15. $(3p + 2q)(5p + 2q)$ **16.** $-2t(t + 2)(t + 3)$
17. $(x + 11)(x - 7)$ **18.** $10(c + 1)^2$
19. $-1(2x - 5)(x + 1)$ **20.** $2n(m - 5)(m^2 - 3)$

Exercise Set 5.4, pp. 328–329

1. Prime polynomial   **2.** Difference of squares
3. Difference of squares   **4.** None of these
5. Perfect-square trinomial   **6.** Perfect-square trinomial
7. None of these   **8.** Prime polynomial
9. Difference of squares   **10.** Perfect-square trinomial
11. Yes   **13.** No   **15.** No   **17.** Yes   **19.** $(x + 8)^2$
21. $(x - 5)^2$ **23.** $5(p + 2)^2$ **25.** $(1 - t)^2$, or $(t - 1)^2$
27. $2(3x + 1)^2$ **29.** $(7 - 4y)^2$, or $(4y - 7)^2$
31. $-x^3(x - 9)^2$ **33.** $2n(n + 10)^2$ **35.** $5(2x + 5)^2$
37. $(7 - 3x)^2$, or $(3x - 7)^2$ **39.** $(4x + 3)^2$
41. $2(1 + 5x)^2$, or $2(5x + 1)^2$ **43.** $(3p + 2q)^2$
45. Prime   **47.** $-1(8m + n)^2$, or $-(8m + n)^2$
49. $-2(4s - 5t)^2$ **51.** Yes   **53.** No   **55.** Yes
57. $(x + 5)(x - 5)$ **59.** $(p + 3)(p - 3)$
61. $(7 + t)(-7 + t)$, or $(t + 7)(t - 7)$
63. $6(a + 2)(a - 2)$ **65.** $(7x - 1)^2$
67. $2(10 + t)(10 - t)$ **69.** $-5(4a + 3)(4a - 3)$
71. $5(t + 4)(t - 4)$ **73.** $2(2x + 9)(2x - 9)$
75. $x(6 + 7x)(6 - 7x)$ **77.** Prime
79. $(t^2 + 1)(t + 1)(t - 1)$ **81.** $-3x(x - 4)^2$
83. $3t(5t + 3)(5t - 3)$ **85.** $a^6(a - 1)^2$
87. $10(a + b)(a - b)$ **89.** $(4x^2 + y^2)(2x + y)(2x - y)$
91. $2(3t + 2s)(3t - 2s)$ **93.** ✍   **95.** $-\frac{1}{5}$ **96.** $\frac{1}{4}$

97.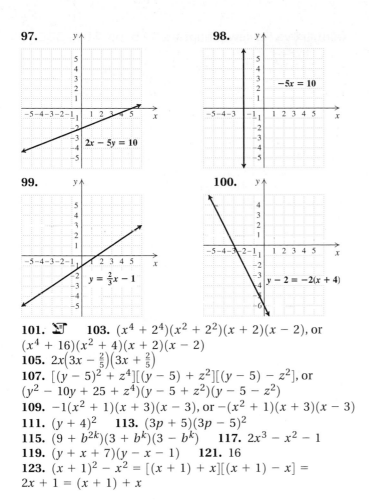

98.

99.

100.

101. ⟰   **103.** $(x^4 + 2^4)(x^2 + 2^2)(x + 2)(x - 2)$, or
$(x^4 + 16)(x^2 + 4)(x + 2)(x - 2)$
105. $2x\left(3x - \frac{2}{5}\right)\left(3x + \frac{2}{5}\right)$
107. $[(y - 5)^2 + z^4][(y - 5) + z^2][(y - 5) - z^2]$, or
$(y^2 - 10y + 25 + z^4)(y - 5 + z^2)(y - 5 - z^2)$
109. $-1(x^2 + 1)(x + 3)(x - 3)$, or $-(x^2 + 1)(x + 3)(x - 3)$
111. $(y + 4)^2$ **113.** $(3p + 5)(3p - 5)^2$
115. $(9 + b^{2k})(3 + b^k)(3 - b^k)$ **117.** $2x^3 - x^2 - 1$
119. $(y + x + 7)(y - x - 1)$ **121.** 16
123. $(x + 1)^2 - x^2 = [(x + 1) + x][(x + 1) - x] = 2x + 1 = (x + 1) + x$

Exercise Set 5.5, pp. 333–334

1. Common factor   **2.** Perfect-square trinomial
3. Grouping   **4.** Multiplying   **5.** $5(a + 5)(a - 5)$
7. $(y - 7)^2$ **9.** $(3t + 7)(t + 3)$ **11.** $x(x + 9)^2$
13. $(x - 5)^2(x + 5)$ **15.** $3t(3t + 1)(3t - 1)$
17. $3x(3x - 5)(x + 3)$ **19.** Prime   **21.** $6(y - 5)(y + 8)$
23. $-2a^4(a - 2)^2$ **25.** $5x(x^2 + 4)(x + 2)(x - 2)$
27. $(t^2 + 3)(t^2 - 3)$ **29.** $-x^4(x^2 - 2x + 7)$
31. $(p + q)(p - q)$ **33.** $a(x^2 + y^2)$ **35.** $2\pi r(h + r)$
37. $(a + b)(5a + 3b)$ **39.** $(x + 1)(x + y)$
41. $(a - 2)(a - y)$ **43.** $(3x - 2y)(x + 5y)$
45. $8mn(m^2 - 4mn + 3)$ **47.** $(a - 2b)^2$ **49.** $(4x + 3y)^2$
51. Prime   **53.** $(a^2b^2 + 4)(ab + 2)(ab - 2)$
55. $4c(4d - c)(5d - c)$ **57.** $(3b - a)(b + 6a)$
59. $-1(xy + 2)(xy + 6)$, or $-(xy + 2)(xy + 6)$
61. $5(pq + 6)(pq - 1)$ **63.** $b^4(ab - 4)(ab + 8)$
65. $x^4(x + 2y)(x - y)$ **67.** $\left(6a - \frac{5}{4}\right)^2$ **69.** $\left(\frac{1}{9}x - \frac{4}{3}\right)^2$
71. $(1 + 4x^6y^6)(1 + 2x^3y^3)(1 - 2x^3y^3)$ **73.** $(2ab + 3)^2$
75. $z(z + 6)(z^2 - 6)$ **77.** ⟰   **79.** $\frac{9}{8}$ **80.** $-\frac{5}{3}$
81. $-\frac{7}{2}$ **82.** $\frac{1}{4}$ **83.** 3   **84.** 11   **85.** -1 **86.** 3
87. ⟰   **89.** $-x(x^2 + 9)(x^2 - 2)$
91. $-1(x^2 + 2)(x + 3)(x - 3)$, or
$-(x^2 + 2)(x + 3)(x - 3)$ **93.** $(y + 1)(y - 7)(y + 3)$
95. $(y + 4 + x)^2$ **97.** $(a + 3)^2(2a + b + 4)(a - b + 5)$
99. $(7x^2 + 1 + 5x^3)(7x^2 + 1 - 5x^3)$

Technology Connection, p. 340

1. $-4.65, 0.65$ **2.** $-0.37, 5.37$ **3.** $-8.98, -4.56$
4. No solution   **5.** $0, 2.76$

Exercise Set 5.6, pp. 340–343

1. (c)   **2.** (a)   **3.** (d)   **4.** (b)   **5.** $-9, -2$ **7.** $-1, 8$
9. $-6, \frac{3}{2}$ **11.** $\frac{1}{7}, \frac{3}{10}$ **13.** $0, 7$ **15.** $\frac{1}{21}, \frac{18}{11}$ **17.** $-\frac{8}{3}, 0$
19. $50, 70$ **21.** $1, 6$ **23.** $-7, 3$ **25.** $-9, -2$ **27.** $0, 10$
29. $-6, 0$ **31.** $-6, 6$ **33.** $-\frac{7}{2}, \frac{7}{2}$ **35.** -5 **37.** 8
39. $0, 2$ **41.** $-\frac{5}{4}, 3$ **43.** 3   **45.** $0, \frac{4}{3}$ **47.** $-\frac{7}{6}, \frac{7}{6}$
49. $-4, -\frac{2}{3}$ **51.** $-3, 1$ **53.** $-\frac{5}{2}, \frac{4}{3}$ **55.** $-1, 4$
57. $-3, 2$ **59.** $(-2, 0), (3, 0)$ **61.** $(-4, 0), (2, 0)$
63. $(-3, 0), \left(\frac{3}{2}, 0\right)$ **65.** ⟰
67. Let m and n represent the numbers; $(m + n)^2$
68. Let m and n represent the numbers; $m^2 + n^2$
69. Let x represent the first integer; then $x + 1$ represents
the second integer; $x(x + 1)$ **70.** Mother's Day:
\$13.8 billion; Father's Day: \$9 billion   **71.** $140°, 35°, 5°$
72. Length: 64 in.; width: 32 in.   **73.** ⟰   **75.** $-7, -\frac{8}{3}, \frac{11}{2}$
77. (a) $x^2 - x - 20 = 0$;   (b) $x^2 - 6x - 7 = 0$;
(c) $4x^2 - 13x + 3 = 0$;   (d) $6x^2 - 5x + 1 = 0$;
(e) $12x^2 - 17x + 6 = 0$;   (f) $x^3 - 4x^2 + x + 6 = 0$
79. $-5, 4$ **81.** $-\frac{3}{5}, \frac{3}{5}$ **83.** $-4, 2$
85. (a) $2x^2 + 20x - 4 = 0$;   (b) $x^2 - 3x - 18 = 0$;
(c) $(x + 1)(5x - 5) = 0$;   (d) $(2x + 8)(2x - 5) = 0$;
(e) $4x^2 + 8x + 36 = 0$;   (f) $9x^2 - 12x + 24 = 0$
87. ⟰   **89.** $2.33, 6.77$ **91.** $-9.15, -4.59$ **93.** $-3.76, 0$

Connecting the Concepts, p. 343

1. Expression   **2.** Equation   **3.** Equation
4. Expression   **5.** Expression   **6.** Equation
7. $2x^3 + x^2 - 8x$ **8.** $-2x^2 - 11$ **9.** $-10, 10$
10. $6a^2 - 19a + 10$ **11.** $(n - 1)(n - 9)$ **12.** $2, 8$
13. $-\frac{5}{2}$ **14.** $7x^3 + 2x - 7$ **15.** $(4x + 9)(4x - 9)$
16. $-3, 8$ **17.** $-4a^2 - a - 11$
18. $2x^2(9x^2 - 12x + 10)$ **19.** $-1, -\frac{2}{3}$
20. $8x^5 - 20x^4 + 12x^2$

Translating for Success, p. 350

1. O   **2.** M   **3.** K   **4.** I   **5.** G   **6.** E   **7.** C
8. A   **9.** H   **10.** B

Exercise Set 5.7, pp. 351–355

1. $-2, 3$ **3.** 6 m, 8 m, 10 m   **5.** 11, 12
7. -14 and -12; 12 and 14   **9.** Length: 30 ft; width: 6 ft
11. Length: 6 cm; width: 4 cm   **13.** Base: 12 in.; height: 9 in.
15. Foot: 7 ft; height: 12 ft   **17.** 1 min, 3 min   **19.** In 2008
21. 16 teams   **23.** 66 handshakes   **25.** 12 players
27. 9 ft   **29.** 32 ft   **31.** 300 ft by 400 ft by 500 ft
33. Dining room: 12 ft by 12 ft; kitchen: 12 ft by 10 ft
35. 20 ft   **37.** 1 sec, 2 sec   **39.** ⟰   **41.** $-\frac{12}{35}$ **42.** $-\frac{21}{20}$
43. -1 **44.** $-\frac{7}{4}$ **45.** $\frac{1}{4}$ **46.** $\frac{4}{5}$ **47.** $\frac{53}{168}$ **48.** $\frac{19}{18}$
49. ⟰   **51.** \$180   **53.** 39 cm   **55.** 4 in., 6 in.
57. 35 ft   **59.** 2 hr, 4.2 hr   **61.** 3 hr

Review Exercises: Chapter 5, pp. 359–360

1. False   **2.** True   **3.** True   **4.** False   **5.** False
6. True   **7.** True   **8.** False   **9.** Answers may vary.
$(4x)(5x^2), (-2x^2)(-10x), (x^3)(20)$ **10.** Answers may
vary. $(-3x^2)(6x^3), (2x^4)(-9x), (-18x)(x^4)$
11. $6x^3(2x - 3)$ **12.** $4a(2a - 3)$
13. $(10t + 1)(10t - 1)$ **14.** $(x + 4)(x - 3)$
15. $(x + 7)^2$ **16.** $3x(2x + 1)^2$ **17.** $(2x + 3)(3x^2 + 1)$
18. $(6a - 5)(a + 1)$ **19.** $(5t - 3)^2$
20. $2(24t^2 - 14t + 3)$ **21.** $(9a^2 + 1)(3a + 1)(3a - 1)$
22. $3x(3x - 5)(x + 3)$ **23.** $3(x + 3)(x - 3)$
24. $(x + 4)(x^3 - 2)$ **25.** $(ab^2 + 8)(ab^2 - 8)$
26. $-4x^4(2x^2 - 8x + 1)$ **27.** $3(2x - 5)^2$
28. Prime   **29.** $-t(t + 6)(t - 7)$ **30.** $(2x + 5)(2x - 5)$
31. $(n + 6)(n - 10)$ **32.** $5(z^2 - 6z + 2)$
33. $(4t + 5)(t + 2)$ **34.** $(2t + 1)(t - 4)$
35. $7x(x + 1)(x + 4)$ **36.** $5x(x + 2)(x + 5)$
37. $5(2x - 1)^2$ **38.** $-6x(x + 5)(x - 5)$
39. $(5 - x)(3 - x)$ **40.** Prime   **41.** $(xy + 8)(xy - 2)$
42. $3(2a + 7b)^2$ **43.** $(m + 5)(m + t)$
44. $32(x^2 + 2y^2z^2)(x^2 - 2y^2z^2)$ **45.** $(2m + n)(3m + n)$
46. $(3r + 5s)(2r - 3s)$ **47.** $-11, 9$ **48.** $-7, 5$
49. $-\frac{3}{4}, \frac{3}{4}$ **50.** $\frac{2}{3}, 1$ **51.** $-\frac{5}{2}, 6$ **52.** $-2, 3$ **53.** $0, \frac{3}{5}$
54. 1 **55.** $-3, 4$ **56.** 10 teams   **57.** $(-1, 0), \left(\frac{5}{2}, 0\right)$
58. Height: 14 ft; base: 14 ft   **59.** 10 holes
60. ✍ Answers may vary. Because Celia did not first factor
out the largest common factor, 4, her factorization will not be
"complete" until she removes a common factor of 2 from each
binomial. The answer should be $4(x - 5)(x + 5)$. Awarding 3
to 7 points would seem reasonable.   **61.** ✍ The equations
solved in this chapter have an x^2-term (are quadratic),
whereas those solved previously have no x^2-term (are
linear). The principle of zero products is used to solve
quadratic equations and is not used to solve linear
equations.   **62.** 2.5 cm   **63.** $0, 2$ **64.** Length: 12 cm;
width: 6 cm   **65.** 100 cm², 225 cm²   **66.** $-3, 2, \frac{5}{2}$
67. No real solution

Test: Chapter 5, pp. 360–361

1. [5.1] Answers may vary. $(3x^2)(4x^2), (-2x)(-6x^3),$
$(12x^3)(x)$ **2.** [5.2] $(x - 4)(x - 9)$ **3.** [5.4] $(x - 5)^2$
4. [5.1] $2y^2(2y^2 - 4y + 3)$ **5.** [5.1] $(x + 1)(x^2 + 2)$
6. [5.1] $t^5(t^2 - 3)$ **7.** [5.2] $a(a + 4)(a - 1)$
8. [5.3] $2(5x - 6)(x + 4)$ **9.** [5.4] $(2t + 5)(2t - 5)$
10. [5.2] $(x + 2)(x - 3)$ **11.** [5.3] $-3m(2m + 1)(m + 1)$
12. [5.4] $3(w + 5)(w - 5)$ **13.** [5.4] $5(3r + 2)^2$
14. [5.4] $3(x^2 + 4)(x + 2)(x - 2)$ **15.** [5.4] $(7t + 6)^2$
16. [5.1] $(x + 2)(x^3 - 3)$ **17.** [5.2] Prime
18. [5.3] $(2x + 3)(2x - 5)$ **19.** [5.3] $3t(2t + 5)(t - 1)$
20. [5.3] $3(m - 5n)(m + 2n)$ **21.** [5.6] $1, 5$
22. [5.6] $-\frac{3}{2}, 5$ **23.** [5.6] $0, \frac{2}{5}$ **24.** [5.6] $-\frac{1}{5}, \frac{1}{5}$
25. [5.6] $-4, 5$ **26.** [5.6] $(-1, 0), \left(\frac{8}{3}, 0\right)$
27. [5.7] Length: 10 m; width: 4 m   **28.** [5.7] 10 people
29. [5.7] 5 ft   **30.** [5.7] 15 cm by 30 cm
31. [5.2] $(a - 4)(a + 8)$ **32.** [5.6] $-\frac{8}{3}, 0, \frac{2}{5}$

Cumulative Review: Chapters 1–5, pp. 361–362

1. $\frac{1}{2}$ **2.** $\frac{9}{32}$ **3.** $\frac{9}{8}$ **4.** 29   **5.** $\frac{1}{9x^4y^6}$ **6.** t^{10}
7. $3x^4 + 5x^3 + x - 10$ **8.** $-3a^2b - ab^2 + 2b^3$
9. $\frac{t^8}{4s^2}$ **10.** $-\frac{8x^6y^3}{27z^{12}}$ **11.** 8   **12.** -8 **13.** $-x^4$
14. $2x^3 - 5x^2 + \frac{1}{2}x - 1$ **15.** $-4t^{11} + 8t^9 + 20t^8$
16. $9x^2 - 30x + 25$ **17.** $100x^{10} - y^2$
18. $x^3 - 2x^2 + 1$ **19.** $(c + 1)(c - 1)$
20. $5(x + y)(1 + 2x)$ **21.** $(x - 2)(x - 12)$
22. $(2r - t)^2$ **23.** $(2x - 5)(3x - 2)$ **24.** $10(y^2 + 4)$
25. $y(x - 1)(x - 2)$ **26.** $(3x - 2y)(4x + y)$ **27.** $\frac{1}{12}$
28. 1   **29.** $-\frac{9}{4}$ **30.** $\{x | x \geq 1\}$ **31.** $-3, 1$ **32.** $-4, 3$
33. $-2, 2$ **34.** $0, 4$ **35.** $c = \frac{a}{b + d}$ **36.** 0
37. $-2; (0, 5)$ **38.** $y = 5x - \frac{1}{3}$ **39.** $y = 5x + \frac{5}{3}$
40.

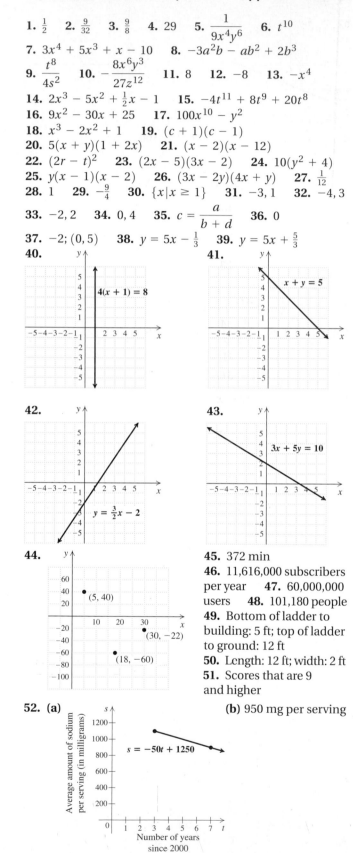

41.

42.

43.

44.

45. 372 min
46. 11,616,000 subscribers
per year   **47.** 60,000,000
users   **48.** 101,180 people
49. Bottom of ladder to
building: 5 ft; top of ladder
to ground: 12 ft
50. Length: 12 ft; width: 2 ft
51. Scores that are 9
and higher
52. (a)   **(b)** 950 mg per serving

53. $b = \dfrac{2}{a+1}$ **54.** $y = -8$

55. (a) $9y^2 + 12y + 4 - x^2$; (b) $(3y + 2 + x)(3y + 2 - x)$

56. $-1, 0, \frac{1}{3}$

CHAPTER 6

Technology Connection, p. 367

1. Correct   **2.** Correct   **3.** Not correct   **4.** Not correct

Exercise Set 6.1, pp. 369-370

1. (e)   **2.** (a)   **3.** (d)   **4.** (b)   **5.** (c)   **6.** (f)

7. 0   **9.** -5 **11.** 5   **13.** $-4, 7$ **15.** $-6, \frac{1}{2}$ **17.** $\dfrac{5a}{4b^2}$

19. $\dfrac{t+2}{t-3}$ **21.** $\frac{7}{8}$ **23.** $\dfrac{a-3}{a+1}$ **25.** $-\dfrac{2x^3}{3}$ **27.** $\dfrac{y-3}{4y}$

29. $\dfrac{3(2a+1)}{7(a+1)}$ **31.** $\dfrac{t-4}{t-5}$ **33.** $\dfrac{a+4}{2(a-4)}$ **35.** $\dfrac{x-4}{x+4}$

37. $t-1$ **39.** $\dfrac{y^2+4}{y+2}$ **41.** $\frac{1}{2}$ **43.** $\dfrac{y}{2y+1}$ **45.** $\dfrac{2x-3}{5x+2}$

47. -1 **49.** -7 **51.** $-\frac{1}{4}$ **53.** $-\frac{3}{2}$ **55.** -1 **57.** 📑

59. $-\frac{4}{21}$ **60.** $-\frac{5}{3}$ **61.** $-\frac{15}{4}$ **62.** $-\frac{21}{16}$ **63.** $\frac{13}{63}$ **64.** $\frac{5}{48}$

65. 📑   **67.** $-(2y+x)$ **69.** $\dfrac{x^3+4}{(x^3+2)(x^2+2)}$

71. $\dfrac{(t-1)(t-9)^2}{(t+1)(t^2+9)}$ **73.** $\dfrac{(x-y)^3}{(x+y)^2(x-5y)}$ **75.** 📑

Technology Connection, pp. 373-374

1. Let $y_1 = ((x^2 + 3x + 2)/(x^2 + 4))/(5x^2 + 10x)$ and $y_2 = (x + 1)/((x^2 + 4)(5x))$. With the tables set in AUTO mode, note that the values in the Y_1- and Y_2-columns match except for $x = -2$.   **2.** ERROR messages occur when division by 0 is attempted. Since the simplified expression has no factor of $x + 5$ or $x + 1$ in a denominator, no ERROR message occurs in Y2 for $x = -5$ or -1.

Exercise Set 6.2, pp. 374-376

1. $\dfrac{3x(x+2)}{8(5x-1)}$ **3.** $\dfrac{(a-4)(a+2)}{(a+6)^2}$ **5.** $\dfrac{(2x+3)(x+1)}{4(x-5)}$

7. $\dfrac{(n-4)(n+4)}{(n^2+4)(n^2-4)}$ **9.** $\dfrac{(y+6)(y-3)}{(1+y)(y+3)}$ **11.** $\dfrac{6t}{5}$

13. $\dfrac{4}{c^2 d}$ **15.** $\dfrac{(x-5)(x+2)}{x-2}$ **17.** $\dfrac{(n-5)(n-1)(n-6)}{(n+6)(n^2+36)}$

19. $\dfrac{7(a+3)}{a(a+4)}$ **21.** $\dfrac{3v}{v-2}$ **23.** $\dfrac{t-5}{t+5}$ **25.** $\dfrac{9(y-5)}{10(y-1)}$

27. 1   **29.** $\dfrac{(t-2)(t+5)}{(t-5)(t-5)}$ **31.** $(2x-1)^2$ **33.** $\dfrac{9}{2x}$

35. $\dfrac{1}{a^4+3a}$ **37.** $\dfrac{x^2-4x+7}{x^2+2x-5}$ **39.** $\frac{20}{27}$ **41.** $\dfrac{x^2}{20}$

43. $\dfrac{a^3}{b^3}$ **45.** $\dfrac{4(t-3)}{3(t+1)}$ **47.** $4(y-2)$ **49.** $-\dfrac{a}{b}$

51. $\dfrac{(n+3)(n+3)}{n-2}$ **53.** $\frac{15}{16}$ **55.** $\dfrac{-x^2-4}{x+2}$

57. $\dfrac{a-5}{3(a-1)}$ **59.** $\dfrac{(2x-1)(2x+1)}{x-5}$

61. $\dfrac{(w-7)(w-8)}{(2w-7)(3w+1)}$ **63.** $\dfrac{1}{(c-5)(5c-3)}$

65. $\dfrac{(x-4y)(x-y)}{(x+y)^3}$ **67.** 📑   **69.** $\frac{19}{12}$ **70.** $\frac{41}{24}$ **71.** $\frac{1}{18}$

72. $-\frac{1}{6}$ **73.** x^2+3 **74.** $-2x^2-4x+1$ **75.** 📑

77. $\dfrac{3}{7x}$ **79.** 1   **81.** $\dfrac{1}{(x+y)^3(3x+y)}$ **83.** $\dfrac{a^2-2b}{a^2+3b}$

85. $\dfrac{(z+4)^3}{3(z-4)^2}$ **87.** $\dfrac{x(x^2+1)}{3(x+y-1)}$ **89.** $\dfrac{a-3b}{c}$ **91.** 📈

Exercise Set 6.3, pp. 383-385

1. Numerators; denominator   **2.** Term   **3.** Least common denominator; LCD   **4.** Factorizations; denominators   **5.** $\dfrac{8}{t}$ **7.** $\dfrac{3x+5}{12}$ **9.** $\dfrac{9}{a+3}$

11. $\dfrac{8}{4x-7}$ **13.** $\dfrac{2y+7}{2y}$ **15.** 6   **17.** $\dfrac{4(x-1)}{x+3}$

19. $a+5$ **21.** $y-7$ **23.** 0   **25.** $\dfrac{1}{x+2}$ **27.** $\dfrac{t-4}{t+3}$

29. $\dfrac{y+2}{y-4}$ **31.** $-\dfrac{5}{x-4}$, or $\dfrac{5}{4-x}$ **33.** $-\dfrac{1}{x-1}$, or $\dfrac{1}{1-x}$

35. 180   **37.** 72   **39.** 60   **41.** $18t^5$ **43.** $30a^4b^8$

45. $6(y-3)$ **47.** $(x-5)(x+3)(x-3)$

49. $t(t-4)(t+2)^2$ **51.** $120x^2y^3z^2$

53. $(a+1)(a-1)^2$ **55.** $(2n-1)(n+1)(n+2)$

57. $12x^3(x-5)(x-3)(x-1)$ **59.** $\dfrac{15}{18t^4}, \dfrac{st^2}{18t^4}$

61. $\dfrac{21y}{9x^4y^3}, \dfrac{4x^3}{9x^4y^3}$ **63.** $\dfrac{2x(x+3)}{(x-2)(x+2)(x+3)}$, $\dfrac{4x(x-2)}{(x-2)(x+2)(x+3)}$ **65.** 📑   **67.** $\frac{-5}{8}, \frac{5}{-8}$ **68.** $\frac{-4}{11}, -\frac{4}{11}$

69. $-x+y$, or $y-x$ **70.** $-3+a$, or $a-3$

71. $-2x+7$, or $7-2x$ **72.** $-a+b$, or $b-a$ **73.** 📑

75. $\dfrac{18x+5}{x-1}$ **77.** $\dfrac{x}{3x+1}$ **79.** 30 strands

81. 60 strands   **83.** $(2x+5)(2x-5)(3x+4)^4$

85. 30 sec   **87.** 7:55 A.M.   **89.** 📑

Exercise Set 6.4, pp. 390-392

1. LCD   **2.** Missing; denominator   **3.** Numerators; LCD

4. Simplify   **5.** $\dfrac{3+5x}{x^2}$ **7.** $-\dfrac{5}{24r}$ **9.** $\dfrac{3u^2+4v}{u^3v^2}$

11. $\dfrac{-2(xy+9)}{3x^2y^3}$ **13.** $\dfrac{7x+1}{24}$ **15.** $\dfrac{-x-4}{6}$

17. $\dfrac{a^2+13a-5}{15a^2}$ **19.** $\dfrac{7z-12}{12z}$ **21.** $\dfrac{(3c-d)(c+d)}{c^2d^2}$

23. $\dfrac{4x^2-13xt+9t^2}{3x^2t^2}$ **25.** $\dfrac{6x}{(x+2)(x-2)}$

27. $\dfrac{(t-3)(t+1)}{(t-1)(t+3)}$ **29.** $\dfrac{11x+2}{3x(x+1)}$ **31.** $\dfrac{-5t+3}{2t(t-1)}$

33. $\dfrac{a^2}{(a-3)(a+3)}$ **35.** $\dfrac{16}{3(z+4)}$ **37.** $\dfrac{5q-3}{(q-1)^2}$

39. $\dfrac{3x-14}{(x-2)(x+2)}$ **41.** $\dfrac{9a}{4(a-5)}$ **43.** 0

45. $\dfrac{10}{(a-3)(a+2)}$ **47.** $\dfrac{x-5}{(x+5)(x+3)}$

49. $\dfrac{3z^2+19z-20}{(z-2)^2(z+3)}$ **51.** $\dfrac{-7}{x^2+25x+24}$ **53.** $\dfrac{3x-1}{2}$

55. $y+3$ **57.** 0   **59.** $\dfrac{p^2+7p+1}{(p-5)(p+5)}$

61. $\dfrac{(x+1)(x+3)}{(x-4)(x+4)}$ **63.** $\dfrac{-a-2}{(a+1)(a-1)}$, or $\dfrac{a+2}{(1+a)(1-a)}$ **65.** $\dfrac{2(5x+3y)}{(x-y)(x+y)}$ **67.** $\dfrac{2x-3}{2-x}$

69. 3   **71.** 0   **73.** ✍   **75.** $-\frac{3}{22}$ **76.** $\frac{7}{9}$ **77.** $\frac{9}{10}$

78. $\frac{16}{27}$ **79.** $\frac{2}{3}$ **80.** $\dfrac{(x-3)(x+2)}{(x-2)(x+3)}$ **81.** ✍

83. Perimeter: $\dfrac{2(5x-7)}{(x-5)(x+4)}$; area: $\dfrac{6}{(x-5)(x+4)}$

85. $\dfrac{x}{3x+1}$ **87.** $\dfrac{x^4+4x^3-5x^2-126x-441}{(x+2)^2(x+7)^2}$

89. $\dfrac{5(a^2+2ab-b^2)}{(a-b)(3a+b)(3a-b)}$ **91.** $\dfrac{a}{a-b}+\dfrac{3b}{b-a}$; answers may vary.   **93.** ✍, 〜

Connecting the Concepts, pp. 392–393

1. Addition; $\dfrac{3x+10}{5x^2}$ **2.** Multiplication; $\dfrac{6}{5x^3}$

3. Division; $\dfrac{3x}{10}$ **4.** Subtraction; $\dfrac{3x-10}{5x^2}$

5. Multiplication; $\dfrac{x-3}{15(x-2)}$ **6.** Multiplication; $\dfrac{6}{(x+3)(x+4)}$

7. Division; $\frac{1}{3}$ **8.** Subtraction; $\dfrac{x^2-2x-2}{(x-1)(x+2)}$

9. Addition; $\dfrac{5x+17}{(x+3)(x+4)}$

10. Addition; -5 **11.** Subtraction; $\dfrac{5}{x-4}$ **12.** Division; $\dfrac{(2x+3)(x+3)}{(x+1)^2}$

13. Subtraction; $\frac{1}{6}$ **14.** Multiplication; $\dfrac{x(x+4)}{(x-1)^2}$

15. Addition; $\dfrac{x+7}{(x-5)(x+1)}$ **16.** Division; $\dfrac{9(u-1)}{16}$

17. Addition; $\dfrac{7t+8}{30}$ **18.** Multiplication; $(t+5)^2$

19. Division; $\dfrac{a-1}{(a+2)(a-2)^2}$

20. Subtraction; $\dfrac{-3x^2+9x-14}{2x}$

Technology Connection, p. 397

1. $(1-1/x)/(1-1/x^2)$ **2.** Parentheses are needed to group separate terms into factors. When a fraction bar is replaced with a division sign, we need parentheses to preserve the groupings that had been created by the fraction bar. This holds for denominators and numerators alike.

Exercise Set 6.5, pp. 397–399

1. (d)   **2.** (a)   **3.** (b)   **4.** (c)   **5.** $\frac{5}{11}$ **7.** 10

9. $\dfrac{5x^2}{4(x^2+4)}$ **11.** $\dfrac{-10t}{5t-2}$ **13.** $\dfrac{2(2a-5)}{a-7}$

15. $\dfrac{x^2-18}{2(x+3)}$ **17.** $-\frac{1}{5}$ **19.** $\dfrac{1+t^2}{t(1-t)}$ **21.** $\dfrac{x}{x-y}$

23. $\dfrac{c(4c+7)}{3(2c^2-1)}$ **25.** $\dfrac{15(4-a^3)}{14a^2(9+2a)}$ **27.** 1

29. $\dfrac{3a^2+4b^3}{b^3(5-3a^2)}$ **31.** $\dfrac{x^2(2x^2-3)}{4x^4+9}$ **33.** $\dfrac{(t-3)(t+3)}{t^2+4}$

35. $\dfrac{a(3ab^3+4)}{b^2(3+a)}$ **37.** $\dfrac{t^2+5t+3}{(t+1)^2}$ **39.** $\dfrac{x^2-2x-1}{x^2-5x-4}$

41. ✍   **43.** -4 **44.** -4 **45.** $\frac{19}{3}$ **46.** $-\frac{14}{27}$ **47.** 3, 4

48. $-15, 2$ **49.** ✍   **51.** 6, 7, 8   **53.** $-3, -\frac{4}{5}, 3$

55. $\dfrac{A}{B}\div\dfrac{C}{D}=\dfrac{\frac{A}{B}}{\frac{C}{D}}=\dfrac{\frac{A}{B}}{\frac{C}{D}}\cdot\dfrac{BD}{BD}=\dfrac{AD}{BC}=\dfrac{A}{B}\cdot\dfrac{D}{C}$

57. $\dfrac{x^2+5x+15}{-x^2+10}$ **59.** 0   **61.** $\dfrac{2z(5z-2)}{(z+2)(13z-6)}$

63. ✍

Exercise Set 6.6, pp. 405–406

1. False   **2.** True   **3.** True   **4.** True   **5.** $-\frac{2}{5}$ **7.** $\frac{6}{7}$

9. $\frac{24}{5}$ **11.** $-4, -1$ **13.** $-6, 6$ **15.** 12   **17.** $\frac{14}{3}$

19. -10 **21.** $-4, -3$ **23.** $\frac{23}{2}$ **25.** $\frac{5}{2}$ **27.** -1

29. No solution   **31.** -10 **33.** $-\frac{7}{3}$ **35.** $-2, \frac{7}{3}$

37. $-3, 13$ **39.** 2   **41.** -8 **43.** No solution

45. $-6, 6$ **47.** ✍   **49.** 137, 139   **50.** 14 yd

51. Base: 9 cm; height: 12 cm   **52.** $-8, -6; 6, 8$

53. 0.06 cm per day   **54.** 0.28 in. per day   **55.** ✍

57. -2 **59.** 3   **61.** 4   **63.** 4   **65.** -2 **67.** 〜

Connecting the Concepts, pp. 406–407

1. Expression; $\dfrac{x-2}{x+1}$ **2.** Expression; $\dfrac{19n-2}{5n(2n-1)}$

3. Equation; 8   **4.** Expression; $\dfrac{(z+1)^2(z-1)}{z}$

5. Equation; $-\frac{1}{2}$ **6.** Expression; $\dfrac{8(t+1)}{(t-1)(2t-1)}$

7. Expression; $\dfrac{2a}{a-1}$ **8.** Equation; $-5, 3$

9. Expression; $\dfrac{3x}{10}$ **10.** Equation; $-10, 10$

Translating for Success, p. 415

1. K   **2.** E   **3.** C   **4.** N   **5.** D   **6.** O   **7.** F   **8.** H
9. B   **10.** A

Exercise Set 6.7, pp. 416–420

1. $\frac{40}{9}$ hr, or $4\frac{4}{9}$ hr   **3.** $\frac{300}{7}$ hr, or $42\frac{6}{7}$ hr   **5.** 6 hr   **7.** $\frac{120}{11}$ hr, or $10\frac{10}{11}$ hr   **9.** $\frac{42}{13}$ min, or $3\frac{3}{13}$ min
11.

	Distance (in km)	Speed (in km/h)	Time (in hours)
B & M	330	$r - 14$	$\dfrac{330}{r - 14}$
AMTRAK	400	r	$\dfrac{400}{r}$

AMTRAK: 80 km/h; B & M: 66 km/h   **13.** Rita: 50 mph; Sean: 65 mph   **15.** Tau: 5 km/h; Baruti: 9 km/h
17. 3 hr   **19.** 10.5   **21.** $\frac{8}{3}$ **23.** $3\frac{3}{4}$ in.   **25.** 20 ft
27. 15 ft   **29.** 3 cm   **31.** 12.6   **33.** 1440 messages
35. About 231,000 people   **37.** 18 gal   **39.** $26\frac{2}{3}$ cm
41. 7   **43.** 52 bulbs   **45.** $7\frac{1}{2}$ oz   **47.** 184 moose
49. 90 whales   **51.** (a) 1.92 T; (b) 28.8 lb
53. ✍

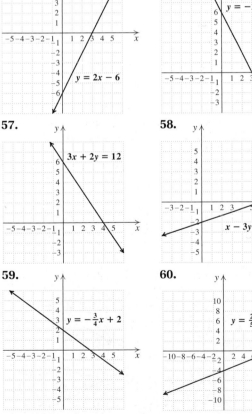

55.   **56.**   **57.**   **58.**   **59.**   **60.**
61. ✍

63. Equation 1: $\frac{1}{a} \cdot t + \frac{1}{b} \cdot t = 1$;

Equation 2: $\left(\frac{1}{a} + \frac{1}{b}\right)t = 1$;

Equation 3: $\frac{t}{a} + \frac{t}{b} = 1$;

Equation 4: $\frac{1}{a} + \frac{1}{b} = \frac{1}{t}$

$\dfrac{1}{a} \cdot t + \dfrac{1}{b} \cdot t = 1$	Equation 1
$t\left(\dfrac{1}{a} + \dfrac{1}{b}\right) = 1$	Factoring out t; equation 1 = equation 2
$t \cdot \dfrac{1}{a} + t \cdot \dfrac{1}{b} = 1$	Using the distributive law
$\dfrac{t}{a} + \dfrac{t}{b} = 1$	Multiplying; equation 2 = equation 3
$\dfrac{1}{t} \cdot \left(\dfrac{t}{a} + \dfrac{t}{b}\right) = \dfrac{1}{t} \cdot 1$	Multiplying both sides by $\dfrac{1}{t}$
$\dfrac{1}{t} \cdot \dfrac{t}{a} + \dfrac{1}{t} \cdot \dfrac{t}{b} = \dfrac{1}{t} \cdot 1$	Using the distributive law
$\dfrac{1}{a} + \dfrac{1}{b} = \dfrac{1}{t}$	Multiplying; equation 3 = equation 4

65. Michelle: 6 hr; Sal: 3 hr; Kristen: 4 hr   **67.** 12 hr
69. About 9.7 sec   **71.** 2 mph   **73.** $66\frac{2}{3}$ ft   **75.** $27\frac{3}{11}$ min
77. ✍

Review Exercises: Chapter 6, pp. 425–426

1. False   **2.** True   **3.** False   **4.** True   **5.** True
6. False   **7.** False   **8.** False   **9.** 0   **10.** -5
11. $-6, 6$ **12.** $-6, 5$ **13.** -2 **14.** $\dfrac{x - 3}{x + 5}$ **15.** $\dfrac{7x + 3}{x - 3}$
16. $\dfrac{3(y - 3)}{2(y + 3)}$ **17.** $-5(x + 2y)$ **18.** $\dfrac{a - 6}{5}$
19. $\dfrac{3(y - 2)^2}{4(2y - 1)(y - 1)}$ **20.** $-32t$ **21.** $\dfrac{2x(x - 1)}{x + 1}$
22. $\dfrac{(x^2 + 1)(2x + 1)}{(x - 2)(x + 1)}$ **23.** $\dfrac{(t + 4)^2}{t + 1}$ **24.** $60a^5b^8$
25. $x^4(x - 1)(x + 1)$ **26.** $(y - 2)(y + 2)(y + 1)$
27. $\dfrac{15 - 3x}{x + 3}$ **28.** $\dfrac{4}{x - 4}$ **29.** $\dfrac{x + 5}{2x}$
30. $\dfrac{2a^2 - 21ab + 15b^2}{5a^2b^2}$ **31.** $y - 4$ **32.** $\dfrac{t(t - 2)}{(t - 1)(t + 1)}$
33. $d + 2$ **34.** $\dfrac{-x^2 + x + 26}{(x + 1)(x - 5)(x + 5)}$ **35.** $\dfrac{2(x - 2)}{x + 2}$
36. $\dfrac{3(7t + 2)}{4t(3t + 2)}$ **37.** $\dfrac{z}{1 - z}$ **38.** $\dfrac{10x}{3x^2 + 16}$ **39.** $c - d$
40. 4   **41.** $\frac{7}{2}$ **42.** $-6, -1$ **43.** $5\frac{1}{7}$ hr
44. Car: 105 km/h; train: 90 km/h   **45.** 55 seals
46. 6   **47.** 72 radios   **48.** ✍ The LCM of denominators is used to clear fractions when simplifying a complex rational expression using the method of multiplying by the

LCD, and when solving rational equations. **49.** 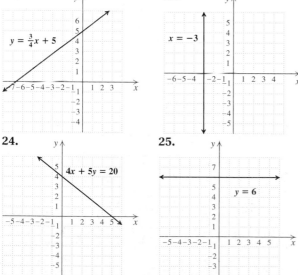 Although multiplying the denominators of the expressions being added results in a common denominator, it is often not the *least* common denominator. Using a common denominator other than the LCD makes the expressions more complicated, requires additional simplifying after the addition has been performed, and leaves more room for error.

50. $\dfrac{5(a+3)^2}{a}$ **51.** $\dfrac{10a}{(a-b)(b-c)}$ **52.** 0 **53.** 44%

Test: Chapter 6, pp. 426–427

1. [6.1] 0 **2.** [6.1] -8 **3.** [6.1] $-1, 1$ **4.** [6.1] $1, 2$

5. [6.1] $\dfrac{3x+7}{x+3}$ **6.** [6.2] $\dfrac{2t(t+3)}{3(t-1)}$

7. [6.2] $\dfrac{(5y+1)(y+1)}{3y(y+2)}$ **8.** [6.2] $\dfrac{(2a+1)(4a^2+1)}{4a^2(2a-1)}$

9. [6.2] $(x+3)(x-3)$ **10.** [6.3] $(y-3)(y+3)(y+7)$

11. [6.3] $\dfrac{-3x+9}{x^3}$ **12.** [6.3] $\dfrac{-2t+8}{t^2+1}$ **13.** [6.4] 1

14. [6.4] $\dfrac{3x-5}{x-3}$ **15.** [6.4] $\dfrac{11t-8}{t(t-2)}$

16. [6.4] $\dfrac{y^2+3}{(y-1)(y+3)^2}$ **17.** [6.4] $\dfrac{x^2+2x-7}{(x+1)(x-1)^2}$

18. [6.5] $\dfrac{3y+1}{y}$ **19.** [6.5] $x-8$ **20.** [6.6] $\frac{8}{3}$

21. [6.6] $-3, 5$ **22.** [6.7] 12 min **23.** [6.7] $1\frac{1}{4}$ mi
24. [6.7] Ryan: 65 km/h; Alicia: 45 km/h **25.** [6.7] Rema: 4 hr; Pe'rez: 10 hr **26.** [6.5] a **27.** [6.7] -1

Cumulative Review: Chapters 1–6, pp. 427–428

1. $a+cb$ **2.** -25 **3.** 25 **4.** $-3x+33$ **5.** 10
6. $-7, 7$ **7.** $-\frac{5}{3}$ **8.** $\frac{8}{5}$ **9.** 4 **10.** $-10, -1$ **11.** -8
12. $1, 4$ **13.** $\left\{ y \mid y \le -\frac{2}{3} \right\}$ **14.** -17 **15.** $-4, \frac{1}{2}$
16. $\{ x \mid x > 43 \}$ **17.** 5 **18.** $-\frac{7}{2}, 5$ **19.** -13

20. $b=3a-c+9$ **21.** $y=\dfrac{4z-3x}{6}$

26. -2 **27.** Slope: $\frac{1}{2}$; y-intercept: $\left(0, -\frac{1}{4}\right)$

28. $y=-\frac{5}{8}x-4$ **29.** $\dfrac{1}{x^2}$ **30.** y^{-7}, or $\dfrac{1}{y^7}$

31. $-4a^4b^{14}$ **32.** $-y^3-2y^2-2y+7$

33. $-15a+10b-5c$ **34.** $2x^5+x^3-6x^2-x+3$
35. $36x^2-60xy+25y^2$ **36.** $3n^2-13n-10$
37. $4x^6-1$ **38.** $2x(3-x-12x^3)$
39. $(4x+9)(4x-9)$ **40.** $(t-4)(t-6)$
41. $(4x+3)(2x+1)$ **42.** $2(3x-2)(x-4)$
43. $(5t+4)^2$ **44.** $(xy-5)(xy+4)$

45. $(x+2)(x^3-3)$ **46.** $\dfrac{4}{t+4}$ **47.** 1

48. $\dfrac{a^2+7ab+b^2}{(a+b)(a-b)}$ **49.** $\dfrac{2x+5}{4-x}$ **50.** $\dfrac{x}{x-2}$

51. $\dfrac{y(3y^2+2)}{y^3-3}$ **52.** $6x^2-5x+2+\dfrac{4}{x}+\dfrac{1}{x^2}$

53. $15x^3-57x^2+177x-529+\dfrac{1605}{x+3}$ **54.** 15 books

55. About 340.8 billion admissions **56.** $33{,}339{,}507$ and $33{,}339{,}508$ **57.** 12 ft **58.** $\frac{75}{4}$ min, or $18\frac{3}{4}$ min
59. 90 antelope **60.** 150 calories **61.** $-144, 144$
62. $16y^6-y^4+6y^2-9$ **63.** $-7, 4, 12$ **64.** 18

CHAPTER 7

Technology Connection, p. 433

1. 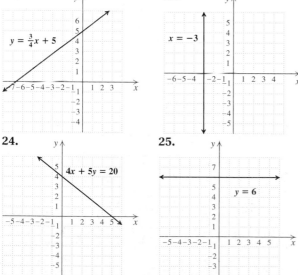 **2.** $y_1-y_2=y_3=-2$

Exercise Set 7.1, pp. 434–435

1. Both **2.** Intersect **3.** Consistent **4.** Dependent
5. Yes **7.** No **9.** Yes **11.** $(3, 1)$ **13.** $(1, 3)$
15. $(-3, 5)$ **17.** No solution **19.** $(4, -1)$ **21.** $(5, -2)$
23. $(-2, 3)$ **25.** Infinite number of solutions
27. $(-6, -2)$ **29.** Infinite number of solutions
31. $(2, -6)$ **33.** $(-8, -7)$ **35.** $(3, -4)$ **37.** $(2, 6)$
39. 2; **41.** 5;

43. 2; **45.** -4;

22.

23.

24.

25.

47. 📈   **49.** $\frac{13}{5}$ **50.** 13   **51.** $\frac{12}{11}$ **52.** $\frac{7}{10}$
53. $y = \frac{3}{4}x - \frac{1}{2}$ **54.** $x = -\frac{3}{2}y - \frac{1}{2}$ **55.** 📈
57. Exercises 18, 25, 26, 29   **59.** Exercises 17 and 30
61. $2x + y = 2$. Answers may vary.   **63.** $A = 2; B = 2$
65. **(a)** Copy card: $y = 20$; per page: $y = 0.06x$
(b)

(c) more than 333 copies
67. **(a)** Full-time: $n = 8t + 632$; part-time: $n = 13t + 543$;
(b) about 2021   **69.** 2.24

Technology Connection, p. 438

1. 📈   **2.** Both equations change: The first becomes $y = 6 - \dfrac{x}{-2}$ and the second becomes $y = 4 - \dfrac{3x}{2}$.

Exercise Set 7.2, pp. 440–443

1. False   **2.** True   **3.** True   **4.** Ture   **5.** $(4, 5)$
7. $(2, 1)$ **9.** $(1, 4)$ **11.** $(-20, 5)$ **13.** $(-3, 2)$
15. No solution   **17.** $(-1, -4)$ **19.** $\left(\frac{17}{3}, \frac{2}{3}\right)$
21. Infinite number of solutions   **23.** Infinite number of solutions   **25.** $\left(-\frac{13}{22}, \frac{7}{11}\right)$ **27.** $(0, -6)$ **29.** $(10, 5)$
31. No solution   **33.** $(-3, -4)$ **35.** No solution
37. 28, 35   **39.** 19, 32   **41.** 8, 10   **43.** 55°, 125°
45. 28°, 62°   **47.** Length: 172 ft; width: 54 ft
49. Length: 365 mi; width: 275 mi   **51.** Length: 90 yd;
width: 50 yd   **53.** Height: 20 ft; width: 5 ft   **55.** 📈
57. $2x + y$ **58.** $-11x$ **59.** $-11y$ **60.** $-6y - 5$
61. $26x$ **62.** $23x$ **63.** 📈   **65.** $(7, -1)$
67. $(4.38, 4.33)$ **69.** 34 yr   **71.** $(30, 50, 100)$ **73.** 📈

Exercise Set 7.3, pp. 448–450

1. False   **2.** True   **3.** False   **4.** False   **5.** $(8, 5)$
7. $(5, 1)$ **9.** $(2, -3)$ **11.** $(-1, 3)$ **13.** $\left(-1, \frac{1}{5}\right)$
15. Infinite number of solutions   **17.** $(-2, -1)$
19. $(4, 5)$ **21.** $(3, 10)$ **23.** No solution   **25.** $\left(\frac{1}{2}, -\frac{1}{2}\right)$
27. $(-3, -1)$ **29.** No solution   **31.** $\left(\frac{1}{2}, 30\right)$
33. $(-2, 2)$ **35.** $(2, -1)$ **37.** $\left(\frac{231}{202}, \frac{117}{202}\right)$ **39.** 40 mi
41. 32°, 58°   **43.** 40 min   **45.** 37°, 143°   **47.** Riesling:
340 acres; Chardonnay: 480 acres   **49.** White: 92 loaves;
whole-wheat: 83 loaves   **51.** 📈   **53.** 0.122   **54.** 0.005
55. 40%   **56.** 3.06   **57.** Let $n =$ the number of liters:
$0.12n$ **58.** Let $x =$ the number of pounds; $0.105x$

59. 📈   **61.** $(-1, 1)$ **63.** $(0, 4)$ **65.** $(0, 3)$
67. $x = \dfrac{-b - c}{a}; y = 1$ **69.** Rabbits: 12; pheasants: 23
71. Man: 45 yr; his daughter: 10 yr

Connecting the Concepts, p. 451

1. $(2, 2)$ **2.** $(4, 1)$ **3.** $\left(\frac{5}{3}, \frac{8}{3}\right)$ **4.** $\left(2, \frac{2}{3}\right)$ **5.** $(-2, -3)$
6. $\left(-\frac{1}{2}, \frac{13}{4}\right)$ **7.** $\left(\frac{11}{7}, \frac{13}{14}\right)$ **8.** No solution   **9.** $(-1, 5)$
10. $(-11, -17)$ **11.** No solution   **12.** $\left(\frac{5}{4}, \frac{1}{8}\right)$
13. Infinite number of solutions   **14.** $\left(\frac{5}{6}, 2\right)$ **15.** $\left(2, \frac{4}{11}\right)$
16. $(7, 3)$ **17.** $(0, 0)$ **18.** $(-9, -5)$ **19.** $(10, 20)$
20. $(12, 8)$

Exercise Set 7.4, pp. 457–460

1. Two-pointers: 3; three-pointers: 7   **3.** Two-pointers: 33;
three-pointers: 6   **5.** 3-credit courses: 19; 4-credit
courses: 8   **7.** 5-cent bottles or cans: 336; 10-cent bottles
or cans: 94   **9.** 2220 motorcycles   **11.** 127 adult admissions   **13.** Private lessons: 7 students; group lessons:
5 students   **15.** 9 Xbox 360 games; 2 PS3 games
17. Brazilian: 200 kg; Turkish: 100 kg   **19.** Cashews: 6 kg;
pecans: 4 kg
21.

Type of Solution	50%-Acid	80%-Acid	68%-Acid Mix
Amount of Solution	x	y	200
Percent Acid	50%	80%	68%
Amount of Acid in Solution	$0.5x$	$0.8y$	136

80 mL of 50%; 120 mL of 80%   **23.** 128 L of 80%;
72 L of 30%   **25.** 87-octane: 4 gal; 93-octane: 8 gal
27. 1300-word pages: 7; 1850-word pages: 5   **29.** Foul
shots: 28; two-pointers: 36   **31.** Kinney's: $33\frac{1}{3}$ fl oz;
Coppertone: $16\frac{2}{3}$ fl oz   **33.** Breadshop Supernatural:
20 lb; New England Natural Bakers Muesli: 20 lb   **35.** 📈
37. $\{x | x < 3\}$
38. $\{x | x \geq -2\}$
39. $\{x | x < -5\}$
40. $\{x | x \geq 2\}$
41. $\{x | x \leq 3\}$
42. $\{x | x > -6\}$
43. 📈   **45.** $19\frac{2}{7}$ lb   **47.** 1.8 L   **49.** 2.5 gal   **51.** $13
53. 10 workers at $20/hr; 5 workers at $25/hr   **55.** 54
57. Tweedledum: 120 lb; Tweedledee: 121 lb

Technology Connection, p. 464

1. $y_1 > \frac{2}{3}x - 2$

2. $y_1 \le x + 5$

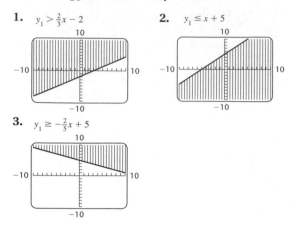

3. $y_1 \ge -\frac{2}{5}x + 5$

Exercise Set 7.5, pp. 464–465

1. True **2.** False **3.** True **4.** False **5.** No **7.** Yes

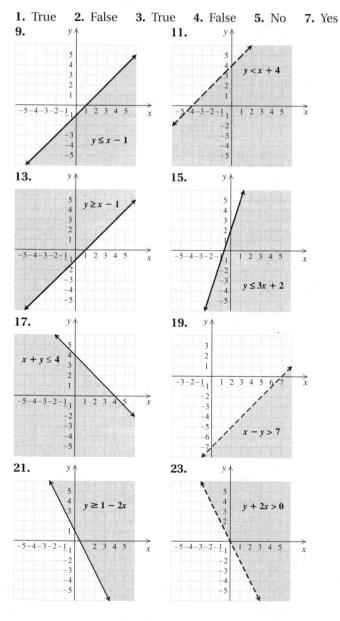

9. $y \le x - 1$

11. $y < x + 4$

13. $y \ge x - 1$

15. $y \le 3x + 2$

17. $x + y \le 4$

19. $x - y > 7$

21. $y \ge 1 - 2x$

23. $y + 2x > 0$

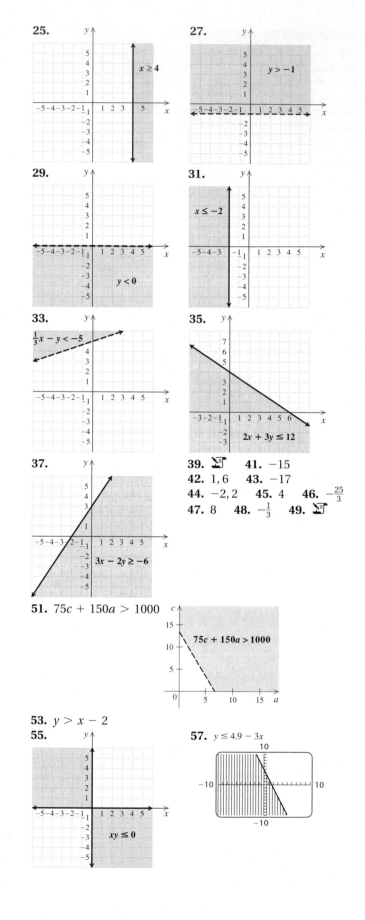

25. $x \ge 4$

27. $y > -1$

29. $y < 0$

31. $x \le -2$

33. $\frac{1}{3}x - y < -5$

35. $2x + 3y \le 12$

37. $3x - 2y \ge -6$

39. ✍ **41.** -15

42. $1, 6$ **43.** -17

44. $-2, 2$ **45.** 4 **46.** $-\frac{25}{3}$

47. 8 **48.** $-\frac{1}{3}$ **49.** ✍

51. $75c + 150a > 1000$

$75c + 150a > 1000$

53. $y > x - 2$

55. $xy \le 0$

57. $y \le 4.9 - 3x$

Visualizing for Success, p. 468

1. G   **2.** B   **3.** F   **4.** H   **5.** C   **6.** D   **7.** A
8. J   **9.** E   **10.** I

Exercise Set 7.6, p. 469

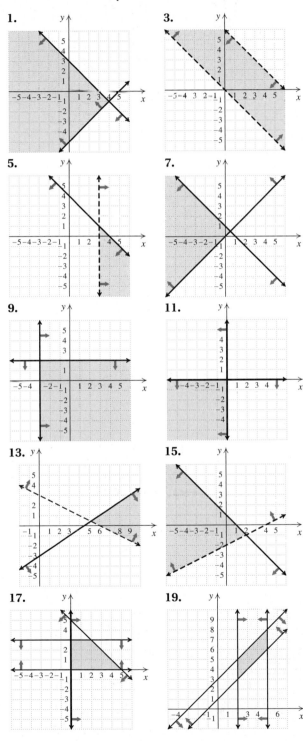

1.

3.

5.

7.

9.

11.

13.

15.

17.

19.

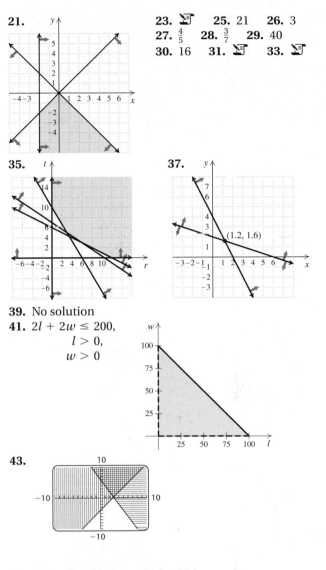

21.

23. 📊   **25.** 21   **26.** 3
27. $\frac{4}{5}$ **28.** $\frac{3}{7}$ **29.** 40
30. 16   **31.** 📊   **33.** 📊

35.

37.

(1.2, 1.6)

39. No solution
41. $2l + 2w \le 200,$
  $l > 0,$
  $w > 0$

43.

Exercise Set 7.7, pp. 474–476

1. Inverse variation   **2.** Direct variation   **3.** Direct variation   **4.** Inverse variation   **5.** Inverse variation
6. Direct variation   **7.** $y = 5x$ **9.** $y = 7x$
11. $y = \frac{3}{5}x$ **13.** $y = \frac{2}{3}x$ **15.** $y = \frac{120}{x}$ **17.** $y = \frac{1}{x}$
19. $y = \frac{20}{x}$ **21.** $y = \frac{210}{x}$ **23.** \$207   **25.** 27.75 kWh
27. 99.86% gold   **29.** 3.2 hr   **31.** 1.6 ft   **33.** 20 people
35. 📊   **37.** 📊   **39.** 100   **40.** 100   **41.** $25x^2$
42. $49t^2$ **43.** a^2b^4 **44.** s^6t^8 **45.** $x^2 + 2xy + y^2$
46. $4a^2 - 4ab + b^2$ **47.** 📊   **49.** $P = kv^3$
51. $S = kv^6$ **53.** $N = k \cdot \frac{T}{P}$ **55.** $P = 8S$ **57.** $A = \pi r^2$

Review Exercises: Chapter 7, pp. 478–479

1. Ordered   **2.** Infinite   **3.** Parallel   **4.** Parentheses
5. Coefficients   **6.** Below   **7.** Intersection
8. Decreases   **9.** No   **10.** Yes   **11.** $(1, 3)$

12. $(6, -2)$ **13.** No solution **14.** Infinite number of solutions **15.** $(-5, 9)$ **16.** $\left(\frac{10}{3}, \frac{4}{3}\right)$ **17.** No solution **18.** $(-3, 9)$ **19.** $\left(4, -\frac{1}{2}\right)$ **20.** Infinite number of solutions **21.** $(10, 15)$ **22.** $(5, -3)$ **23.** Infinite number of solutions **24.** $\left(-\frac{1}{2}, \frac{1}{3}\right)$ **25.** $(-1, -6)$ **26.** $(3, 2)$ **27.** $\left(\frac{18}{7}, \frac{4}{7}\right)$ **28.** No solution **29.** Length: 25 ft; width: 8 ft **30.** Two-pointers: 8; three-pointers: 3 **31.** Turkey subs: 35; veggie subs: 15 **32.** Café Rich: $66\frac{2}{3}$ g; Café Light: $133\frac{1}{3}$ g

33.

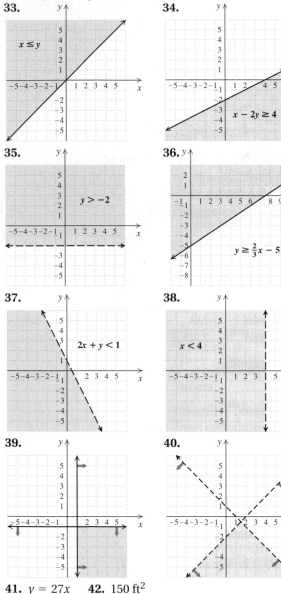

34.

35.

36.

37.

38.

39.

40.

41. $y = 27x$ **42.** 150 ft^2
43. ✍ A solution of a system of two equations is an ordered pair that makes both equations true. The graph of an equation represents all ordered pairs that make that equation true. So in order for an ordered pair to make *both* equations true, it must be on both graphs. **44.** ✍ The solution sets of linear inequalities are regions, not lines. Thus the solution sets can intersect even if the boundary lines do not. **45.** $C = 1, D = 3$ **46.** $(2, 1, -2)$ **47.** $(2, 0)$ **48.** 24 **49.** $1800

Test: Chapter 7, p. 480

1. [7.1] Yes **2.** [7.1] $(1, 3)$ **3.** [7.1] No solution
4. [7.2] $(-3, 5)$ **5.** [7.2] $\left(\frac{1}{2}, 3\right)$ **6.** [7.2] Infinite number of solutions **7.** [7.3] $(3, 2)$ **8.** [7.3] $(12, -6)$
9. [7.3] $(0, 1)$ **10.** [7.3] $\left(-\frac{1}{5}, \frac{14}{5}\right)$ **11.** [7.2] 23°, 67°
12. [7.4] 40%: 20 L; 25%: 40 L **13.** [7.4] $1\frac{7}{8}$ mi
14. [7.5] **15.** [7.5]

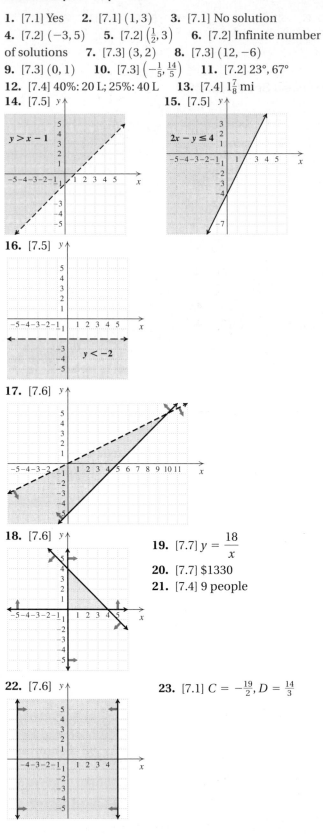

16. [7.5]

17. [7.6]

18. [7.6]
19. [7.7] $y = \dfrac{18}{x}$
20. [7.7] $1330
21. [7.4] 9 people

22. [7.6] **23.** [7.1] $C = -\frac{19}{2}, D = \frac{14}{3}$

Cumulative Review: Chapters 1–7, pp. 481–482

1. 24   **2.** $\frac{1}{10}$ **3.** -10 **4.** $13x^2 - y$ **5.** 7   **6.** $\frac{2}{3}$
7. -24 **8.** 2   **9.** $\frac{9}{7}$ **10.** $-3, -2$ **11.** 2, 3   **12.** $\frac{1}{9}$
13. $\left\{y | y \geq -\frac{2}{3}\right\}$ **14.** $-10, 10$ **15.** $(1, 2)$ **16.** -11
17. $-\frac{1}{2}, \frac{2}{3}$ **18.** $(7, 3)$ **19.** $\left(\frac{9}{11}, \frac{17}{22}\right)$ **20.** $-\frac{5}{4}$
21. $p = \frac{3t}{q}$ **22.** $s = 2A - r$ **23.** $3a^3 + 8a^2 - 14$
24. 3   **25.** $\dfrac{m(m + 2n)}{(m - n)(2m + n)}$ **26.** $-3a^3 + 8a^2 - 12a$
27. $\dfrac{-6p}{p + q}$ **28.** $\dfrac{2x + 4}{x - 2}$ **29.** $25a^4 - b^2$
30. $-15x^5 + 18x^4 + 6x^3 - 3x^2$ **31.** $4n^2 + 20n + 25$
32. $8t^5 + 5t^3 + 32t^2 + 20$ **33.** $\dfrac{(x - 1)(x + 2)}{(x - 2)^2}$
34. $2x + 1$ **35.** $\dfrac{x}{2(x + 1)(x + 2)}$ **36.** $-5(t - 2s + 3)$
37. $(x + 1)(x^2 + 2)$ **38.** $(m^2 + 1)(m + 1)(m - 1)$
39. $2x(x + 4)(x + 5)$ **40.** $(x^2 - 3)^2$
41. $2(2x^2 - x - 5)$ **42.** $(2x - 5)(5x - 2)$
43. $a^2b^2(3 + 4b)(3 - 4b)$
44.   **45.**

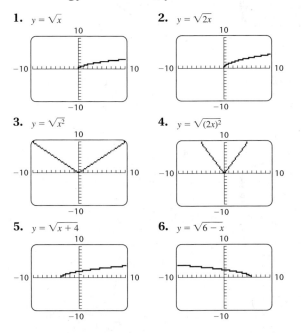

46.   **47.**

48.   **49.**

50. x-intercept: $(6, 0)$; y-intercept: $(0, -4)$
51. $y = -2x + \frac{4}{7}$ **52.** y^{-4}, or $\dfrac{1}{y^4}$ **53.** $9a^9$ **54.** $\dfrac{x^{10}}{2}$
55. ab^3 **56.** \$1260   **57.** At least 35 patients
58. Dog sled: 12 mph; snowmobile: 52 mph
59. 11 and 13   **60.** 100 min   **61.** 32.5°, 65°, 82.5°

62. 42 hr   **63.** $x = \dfrac{t}{p - q}$ **64.** $y = 2x + 7$
65. $\dfrac{3t^2(2t^2 - t + 1)}{4(t - 1)}$
66.   **67.** Length: 6 in.; width: $4\frac{1}{4}$ in

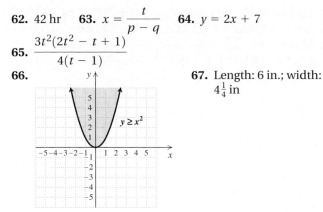

CHAPTER 8

Technology Connection, p. 486

1. $\sqrt{8} \approx 2.828427125$; $\sqrt{11} \approx 3.31662479$;
$\sqrt{48} \approx 6.92820323$

Technology Connection, p. 487

1. $y = \sqrt{x}$ **2.** $y = \sqrt{2x}$

3. $y = \sqrt{x^2}$ **4.** $y = \sqrt{(2x)^2}$

5. $y = \sqrt{x + 4}$ **6.** $y = \sqrt{6 - x}$

Exercise Set 8.1, pp. 488–489

1. (c)   **2.** (e)   **3.** (a)   **4.** (f)   **5.** (b)   **6.** (d)
7. True   **9.** False   **11.** $-10, 10$ **13.** $-6, 6$ **15.** $-1, 1$
17. $-12, 12$ **19.** 10   **21.** -1 **23.** 0   **25.** -11
27. 30   **29.** -12 **31.** $10a$ **33.** $t^3 - 2$ **35.** $\dfrac{7}{x + y}$
37. Rational   **39.** Irrational   **41.** Irrational
43. Rational   **45.** Rational   **47.** 2.828   **49.** 3.873
51. 9.110   **53.** $|x|$ **55.** $|10x|$, or $10|x|$ **57.** $|x + 7|$
59. $|5 - 2x|$ **61.** x **63.** $5y$ **65.** $4t$ **67.** $n + 7$

69. (a) 15; (b) 14   **71.** 0.899 sec   **73.** 📊   **75.** t^{20}
76. p^{10} **77.** $50a^{13}$ **78.** $45x^7$ **79.** $300m^9$
80. $50y^{15}$ **81.** 📊   **83.** $\frac{1}{10}$ **85.** 7   **87.** 5
89. $-6, -5$ **91.** $-4, 4$ **93.** $-3, 3$ **95.** $\{x \mid x < 10\}$
97. $\frac{2x^4}{y^3}$ **99.** $\frac{20}{m^8}$ **101.** (a) 1.7; (b) 2.2; (c) 2.6
103. 1097.1 ft/sec
105. $y_1 = \sqrt{x-2};\ y_2 = \sqrt{x+7};$ **107.** $\{a \mid a < 1\}$
$y_3 = 5 + \sqrt{x};\ y_4 = -4 + \sqrt{x}$

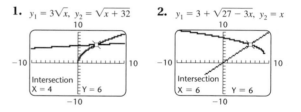

Exercise Set 8.2, pp. 495–496

1. (h)   **2.** (j)   **3.** (f)   **4.** (a)   **5.** (i)   **6.** (d)   **7.** (b)
8. (g)   **9.** (c)   **10.** (e)   **11.** $\sqrt{10}$ **13.** $\sqrt{12}$, or $2\sqrt{3}$
15. $\sqrt{\frac{21}{40}}$, or $\frac{1}{2}\sqrt{\frac{21}{10}}$ **17.** 10   **19.** $\sqrt{75}$, or $5\sqrt{3}$
21. $\sqrt{2x}$ **23.** $\sqrt{14a}$ **25.** $\sqrt{21xy}$ **27.** $\sqrt{6abc}$
29. $2\sqrt{3}$ **31.** $5\sqrt{3}$ **33.** $10\sqrt{5}$ **35.** $4\sqrt{t}$ **37.** $2\sqrt{5z}$
39. $10y$ **41.** $x\sqrt{13}$ **43.** $3b\sqrt{3}$ **45.** $12x\sqrt{y}$
47. $2\sqrt{6}$ **49.** 0   **51.** $2\sqrt{21}$ **53.** a^9 **55.** x^8
57. $r^2\sqrt{r}$ **59.** $t^7\sqrt{t}$ **61.** $2a\sqrt{10a}$ **63.** $3x^2\sqrt{5x}$
65. $10p^{12}\sqrt{2p}$ **67.** $2\sqrt{5}$ **69.** 9   **71.** $6\sqrt{xy}$
73. $17\sqrt{x}$ **75.** $10b\sqrt{5}$ **77.** $12t$ **79.** $a\sqrt{bc}$
81. $x^8\sqrt{x}$ **83.** $7m^3\sqrt{2}$ **85.** $xy^3\sqrt{xy}$ **87.** $6ab^3\sqrt{2a}$
89. 1789 gal/min   **91.** 20 mph; 54.8 mph   **93.** 📊
95. x^3y^5 **96.** a^2b **97.** $\frac{21a^2}{16b^2}$ **98.** $\frac{35x}{12y}$ **99.** $\frac{2r^3}{7t}$
100. $\frac{5x^7}{11y}$ **101.** 📊   **103.** 0.1   **105.** 0.25   **107.** $<$
109. $=$ **111.** $>$ **113.** $18(x+1)\sqrt{y(x+1)}$
115. $2x^7\sqrt{5x}$ **117.** $\sqrt{14x^{26}}$ **119.** x^{8n} **121.** $y^{n/2}$

Exercise Set 8.3, pp. 500–501

1. True   **2.** True   **3.** False   **4.** False   **5.** 10   **7.** 2
9. $\sqrt{11}$ **11.** $\frac{1}{2}$ **13.** $\frac{3}{4}$ **15.** 2   **17.** $3y$ **19.** $10a^4$
21. $a^3\sqrt{3}$ **23.** $\frac{2}{5}$ **25.** $\frac{7}{4}$ **27.** $-\frac{5}{9}$ **29.** $\frac{a^2}{5}$
31. $\frac{x^3\sqrt{3}}{4}$ **33.** $\frac{t^3\sqrt{3}}{2}$ **35.** $\frac{\sqrt{3}}{3}$ **37.** $\frac{5\sqrt{7}}{7}$ **39.** $\frac{4\sqrt{3}}{9}$
41. $\frac{\sqrt{30}}{5}$ **43.** $\frac{\sqrt{6}}{10}$ **45.** $\frac{\sqrt{10a}}{15}$ **47.** $\frac{2\sqrt{15}}{5}$ **49.** $\frac{\sqrt{7z}}{z}$
51. $\frac{\sqrt{2a}}{20}$ **53.** $\frac{\sqrt{10x}}{30}$ **55.** $\frac{\sqrt{3a}}{5}$ **57.** $\frac{x\sqrt{15}}{6}$
59. \$9185   **61.** \$12,901   **63.** 6.28 sec; 7.85 sec
65. 0.5 sec   **67.** 9.42 sec   **69.** 📊   **71.** $15x$ **72.** $11x$
73. $17y - z$ **74.** $-4t - 4s$ **75.** $18x^2 - 63x$
76. $12n^3 + 24n^2 + 4n$ **77.** $4x^2 - 9$ **78.** $25t^2 - 49$
79. 📊   **81.** $\frac{\sqrt{70}}{100}$ **83.** $\frac{\sqrt{6xy}}{4x^3y^2}$ **85.** $\frac{\sqrt{5z}}{5wz}$
87. $\frac{2\sqrt{5\sqrt{5}}}{5}$ **89.** $\frac{1}{x} - \frac{1}{y}$ **91.** $\frac{53}{2}$

1. -0.086 **2.** 1.217   **3.** 2.765

Exercise Set 8.4, pp. 504–505

1. (b)   **2.** (c)   **3.** (d)   **4.** (a)   **5.** $11\sqrt{10}$ **7.** $3\sqrt{2}$
9. $13\sqrt{t}$ **11.** $-\sqrt{x}$ **13.** $8\sqrt{2a}$ **15.** $8\sqrt{10y}$
17. $12\sqrt{7}$ **19.** $4\sqrt{2}$ **21.** $5\sqrt{3} + 2\sqrt{2}$ **23.** $-3\sqrt{x}$
25. $-18\sqrt{3}$ **27.** $28\sqrt{2}$ **29.** $13\sqrt{2}$ **31.** $6\sqrt{2}$
33. $2\sqrt{2}$ **35.** $5\sqrt{a}$ **37.** $\sqrt{6} + \sqrt{21}$ **39.** $\sqrt{30} - 5\sqrt{2}$
41. $14 + 7\sqrt{2}$ **43.** $17 - 7\sqrt{7}$ **45.** -19 **47.** 4
49. $7\sqrt{2}$ **51.** $52 + 14\sqrt{3}$ **53.** $13 - 4\sqrt{3}$
55. $x - 2\sqrt{10x} + 10$ **57.** $\frac{10 - 2\sqrt{2}}{23}$ **59.** $-1 - \sqrt{3}$
61. $\frac{5 - \sqrt{7}}{9}$ **63.** $\frac{-5 + 2\sqrt{10}}{3}$ **65.** $\frac{7 + \sqrt{21}}{4}$
67. $\frac{\sqrt{15} + 3}{2}$ **69.** $\frac{2\sqrt{7} - 2\sqrt{2}}{5}$ **71.** $\frac{6 - 2\sqrt{6x} + x}{6 - x}$
73. 📊   **75.** $\frac{5}{11}$ **76.** $-\frac{3}{2}$ **77.** $-1, 6$ **78.** 2, 5
79. 66.6 min   **80.** 1984: \$66; 2005: \$129   **81.** 📊
83. $\frac{11\sqrt{x} - 5x\sqrt{2}}{2x}$ **85.** $xy\sqrt{2y}(2x^2 - y^2 - xy^3)$
87. $37xy\sqrt{3x}$ **89.** All three are correct.

1. $y_1 = 3\sqrt{x},\ y_2 = \sqrt{x + 32}$ **2.** $y_1 = 3 + \sqrt{27 - 3x},\ y_2 = x$

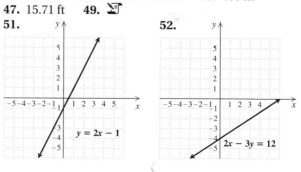

Exercise Set 8.5, pp. 509–511

1. True   **2.** False   **3.** False   **4.** True   **5.** 36   **7.** 144
9. 21   **11.** -46 **13.** $\frac{25}{3}$ **15.** 2   **17.** No solution
19. -4 **21.** 8   **23.** 12   **25.** 11, 23   **27.** 3   **29.** 0, 8
31. No solution   **33.** 5   **35.** $-\frac{5}{4}$ **37.** 1   **39.** 80 ft;
180 ft   **41.** 25.2°F   **43.** 256 m   **45.** 400 m
47. 15.71 ft   **49.** 📊
51.   **52.**

53.

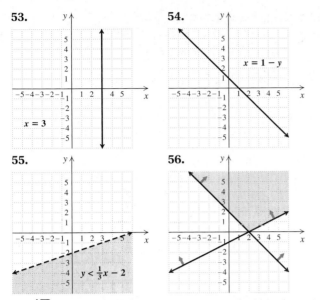

54.

55.

56.

57. ✏️   **59.** 121   **61.** 16   **63.** No solution   **65.** 3
67. 10   **69.** $34\frac{569}{784}$ m

71.   **73.**

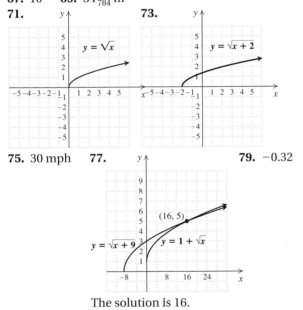

75. 30 mph   **77.**   **79.** −0.32

The solution is 16.

Connecting the Concepts, pp. 511–512

1. $10t$ **2.** $x\sqrt{17}$ **3.** 36   **4.** $\frac{2}{3}$ **5.** $5\sqrt{15t}$
6. 13   **7.** $2\sqrt{15} - 3\sqrt{22}$ **8.** No solution   **9.** $-8\sqrt{3}$
10. −4   **11.** 2   **12.** $\dfrac{a^3}{2}$ **13.** $\frac{8}{9}$ **14.** −2
15. $5\sqrt{2} + 3\sqrt{5} + \sqrt{10} + 3$ **16.** $15\sqrt{5}$ **17.** −1, 2
18. $2a^9\sqrt{7a}$ **19.** 3   **20.** $30x^4y^6$

Translating for Success, p. 517

1. J   **2.** K   **3.** N   **4.** H   **5.** G   **6.** E   **7.** O
8. D   **9.** B   **10.** C

Exercise Set 8.6, pp. 518–521

1. Hypotenuse   **2.** Legs   **3.** Pythagorean
4. Distance   **5.** 17   **7.** $\sqrt{98}$, or $7\sqrt{2}$; approximately
9.899   **9.** 12   **11.** $\sqrt{31} \approx 5.568$ **13.** 13   **15.** 12
17. 3   **19.** $\sqrt{2} \approx 1.414$ **21.** 5   **23.** 17 ft
25. $\sqrt{189}$ in. ≈ 13.748 in.   **27.** $\sqrt{208}$ ft ≈ 14.422 ft
29. $\sqrt{16,200}$ ft ≈ 127.279 ft
31. $\sqrt{15,700}$ yd ≈ 125.300 yd
33. $\sqrt{4500}$ cm ≈ 67.082 cm   **35.** $\sqrt{448}$ ft ≈ 21.166 ft
37. $\sqrt{65} \approx 8.062$ **39.** 5   **41.** $\sqrt{13} \approx 3.606$ **43.** 10
45. ✏️   **47.** −32   **48.** 32   **49.** $\frac{8}{27}$ **50.** $\frac{1}{16}$ **51.** 81
52. 81   **53.** 16   **54.** 27   **55.** −64   **56.** 625
57. ✏️   **59.** ✏️   **61.** $\sqrt{221,840}$ ft ≈ 471 ft
63. 28,600 ft   **65.** $\sqrt{7}$ m   **67.** $\frac{2}{3}$ **69.** 50 ft²
71. $20\sqrt{6}$ ft ≈ 48.990 ft   **73.** $12 - 2\sqrt{6} \approx 7.101$

Exercise Set 8.7, pp. 526–527

1. True   **2.** True   **3.** True   **4.** False   **5.** −2
7. −10   **9.** 5   **11.** −6   **13.** 5   **15.** 0   **17.** −1
19. Not a real number   **21.** 10   **23.** 6   **25.** 1   **27.** a
29. $3\sqrt[3]{2}$ **31.** $2\sqrt[4]{3}$ **33.** $\frac{4}{5}$ **35.** $\frac{2}{3}$ **37.** $\dfrac{\sqrt[3]{7}}{2}$
39. $\dfrac{\sqrt[4]{14}}{3}$ **41.** 7   **43.** 10   **45.** 2   **47.** 8   **49.** 64
51. 16   **53.** 10,000   **55.** 100,000   **57.** $\frac{1}{3}$ **59.** $\frac{1}{4}$
61. $\frac{1}{8}$ **63.** $\frac{1}{243}$ **65.** $\frac{1}{25}$ **67.** ✏️   **69.** −4, −3
70. −1, 7   **71.** $-\frac{3}{4}, 3$ **72.** $-\frac{5}{2}, \frac{5}{2}$ **73.** $-\frac{5}{3}, 2$ **74.** $\frac{4}{5}, 3$
75. ✏️   **77.** 5.278   **79.** 11.240   **81.** $a^{7/4}$
83. $m^{13/12}$

85.

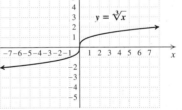

87. 80 calories

Review Exercises: Chapter 8, pp. 530–531

1. False   **2.** True   **3.** True   **4.** True   **5.** False
6. False   **7.** False   **8.** True   **9.** True   **10.** True
11. 4, −4   **12.** 8, −8   **13.** 20, −20   **14.** 15, −15
15. 12   **16.** 9   **17.** −2   **18.** −13   **19.** $5x^3y$
20. $\dfrac{a}{b}$ **21.** Rational   **22.** Irrational   **23.** Irrational
24. Rational   **25.** 2.236   **26.** 9.487   **27.** $|5x|$, or $5|x|$
28. $|x + 2|$ **29.** p **30.** $3x$ **31.** $7n$ **32.** ac
33. $4\sqrt{3}$ **34.** $10t\sqrt{3}$ **35.** $4\sqrt{2p}$ **36.** x^8
37. $2a^6\sqrt{3a}$ **38.** $6m^7\sqrt{m}$ **39.** $\sqrt{55}$ **40.** $2\sqrt{15}$
41. $\sqrt{21st}$ **42.** $2a\sqrt{6}$ **43.** $5xy\sqrt{2}$ **44.** $10a^2b\sqrt{ab}$
45. $\dfrac{\sqrt{7}}{3}$ **46.** $\dfrac{y^4\sqrt{5}}{3}$ **47.** $\frac{7}{8}$ **48.** $\frac{2}{3}$ **49.** $\dfrac{8}{t^3}$

50. $13\sqrt{5}$ **51.** $\sqrt{5}$ **52.** $-3\sqrt{x}$ **53.** $7 + 4\sqrt{3}$
54. -6 **55.** $-11 + 5\sqrt{7}$ **56.** $\dfrac{\sqrt{5}}{5}$ **57.** $\dfrac{\sqrt{10}}{4}$
58. $\dfrac{\sqrt{7y}}{y}$ **59.** $\dfrac{2\sqrt{3}}{3}$ **60.** $8 - 4\sqrt{3}$ **61.** $-7 - 3\sqrt{5}$
62. 54   **63.** No solution   **64.** 4   **65.** 0, 3
66. 35.344 ft   **67.** 20   **68.** $\sqrt{3} \approx 1.732$
69. $48\sqrt{2}$ ft ≈ 67.882 ft   **70.** 5   **71.** -2 **72.** Not a
real number   **73.** $\frac{2}{3}$ **74.** 10   **75.** 9   **76.** $\frac{1}{10}$ **77.** 125
78. $\frac{1}{16}$ **79.** ✍ Absolute-value signs may be necessary
when simplifying a radical expression with an even index.
For n an even number, if it is possible that A is negative,
then $\sqrt[n]{A^n} = |A|$.   **80.** ✍ Some radical terms that are like
terms may not appear to be so until they are in simplified
form.   **81.** 2   **82.** No solution
83. $(x + \sqrt{5})(x - \sqrt{5})$
84. $b = \sqrt{A^2 - a^2}$ or $b = -\sqrt{A^2 - a^2}$

Test: Chapter 8, p. 532

1. [8.1] 7, -7 **2.** [8.1] 4   **3.** [8.1] -5 **4.** [8.1] $t^2 + 1$
5. [8.1] Irrational   **6.** [8.1] Rational   **7.** [8.1] 1.414
8. [8.1] 7.071   **9.** [8.1] a **10.** [8.1] $8y$ **11.** [8.2] $2\sqrt{15}$
12. [8.2] $3x^3\sqrt{3}$ **13.** [8.2] $6t^5\sqrt{t}$ **14.** [8.2] $\sqrt{30}$
15. [8.2] $5\sqrt{3}$ **16.** [8.2] $\sqrt{14xy}$ **17.** [8.2] $4t$
18. [8.2] $3ab^2\sqrt{2}$ **19.** [8.3] $\frac{2}{3}$ **20.** [8.3] $\dfrac{\sqrt{7}}{4y}$
21. [8.3] $\frac{3}{2}$ **22.** [8.3] $\dfrac{12}{a}$ **23.** [8.4] $-6\sqrt{2}$
24. [8.4] $7\sqrt{3}$ **25.** [8.4] $21 - 8\sqrt{5}$ **26.** [8.4] 11
27. [8.3] $\dfrac{\sqrt{10}}{5}$ **28.** [8.3] $\dfrac{2x\sqrt{y}}{y}$ **29.** [8.4] $\dfrac{40 + 10\sqrt{5}}{11}$
30. [8.6] $\sqrt{13} \approx 3.606$ **31.** [8.6] $\sqrt{10,625}$ ft ≈ 103.078 ft
32. [8.5] 50   **33.** [8.5] $-2, 2$ **34.** [8.5] About 5000 m
35. [8.7] 2   **36.** [8.7] -1 **37.** [8.7] -4 **38.** [8.7] Not a
real number   **39.** [8.7] 3   **40.** [8.7] $\frac{1}{3}$ **41.** [8.7] 1000
42. [8.7] $\frac{1}{32}$ **43.** [8.5] -3 **44.** [8.2] y^{8n}

Cumulative Review: Chapters 1–8, pp. 533–534

1. $-\frac{5}{6}$ **2.** 11   **3.** $-8x^{15}$ **4.** $\dfrac{b}{a}$ **5.** $\dfrac{3}{4a^3}$ **6.** $\dfrac{2x - 1}{x - 2}$
7. -3 **8.** 11   **9.** 4   **10.** $-3x^3 - 3x^2 - 5$
11. $-15x^5 + 5x^4 - 10x^3$ **12.** $64t^2 - 9$
13. $81x^4 + 18x^2 + 1$ **14.** $x^2 + \dfrac{2}{x - 1}$ **15.** $6(x - 1)$
16. $\dfrac{2(4t - 5)}{t(t - 2)}$ **17.** $\dfrac{(2x - 1)(2x^2 + x - 2)}{4x(x + 1)^3}$ **18.** $5\sqrt{6}$
19. $\dfrac{2x}{9}$ **20.** 2   **21.** $10(2x + 3)(2x - 3)$
22. $(t - 1)(t - 10)$ **23.** $2z(3z^2 - 4z + 5)$
24. $(x - 2)(6x - 1)$ **25.** $5(ab + 1)^2$

26. $(4 + t^2)(2 + t)(2 - t)$

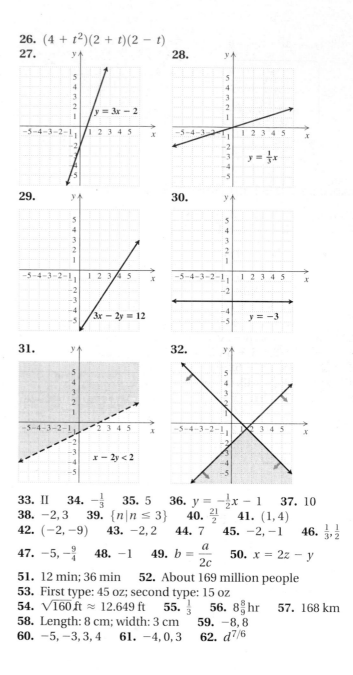

27. $y = 3x - 2$ **28.** $y = \frac{1}{3}x$
29. $3x - 2y = 12$ **30.** $y = -3$
31. $x - 2y < 2$ **32.**

33. II   **34.** $-\frac{1}{3}$ **35.** 5   **36.** $y = -\frac{1}{2}x - 1$ **37.** 10
38. $-2, 3$ **39.** $\{n \mid n \le 3\}$ **40.** $\frac{21}{2}$ **41.** $(1, 4)$
42. $(-2, -9)$ **43.** $-2, 2$ **44.** 7   **45.** $-2, -1$ **46.** $\frac{1}{3}, \frac{1}{2}$
47. $-5, -\frac{9}{4}$ **48.** -1 **49.** $b = \dfrac{a}{2c}$ **50.** $x = 2z - y$
51. 12 min; 36 min   **52.** About 169 million people
53. First type: 45 oz; second type: 15 oz
54. $\sqrt{160}$ ft ≈ 12.649 ft   **55.** $\frac{1}{3}$ **56.** $8\frac{8}{9}$ hr   **57.** 168 km
58. Length: 8 cm; width: 3 cm   **59.** $-8, 8$
60. $-5, -3, 3, 4$ **61.** $-4, 0, 3$ **62.** $d^{7/6}$

CHAPTER 9

Technology Connection, p. 537

1.

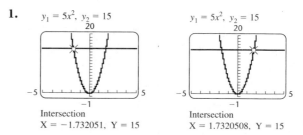

$y_1 = 5x^2, \ y_2 = 15$
Intersection
X = -1.732051, Y = 15

$y_1 = 5x^2, \ y_2 = 15$
Intersection
X = 1.7320508, Y = 15

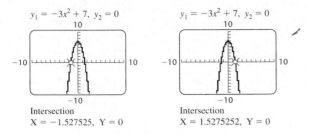

Intersection
X = −1.527525, Y = 0

Intersection
X = 1.5275252, Y = 0

Exercise Set 9.1, pp. 538–539

1. True **2.** True **3.** False **4.** False **5.** $-9, 9$
7. $-1, 1$ **9.** $-\sqrt{11}, \sqrt{11}$ **11.** $-2, 2$ **13.** $-\sqrt{2}, \sqrt{2}$
15. $-\frac{2}{3}, \frac{2}{3}$ **17.** 0 **19.** $-\frac{\sqrt{15}}{3}, \frac{\sqrt{15}}{3}$ **21.** $-6, 8$
23. $-8, -4$ **25.** $-3 - \sqrt{6}, -3 + \sqrt{6}$, or $-3 \pm \sqrt{6}$
27. 7 **29.** $5 - \sqrt{14}, 5 + \sqrt{14}$, or $5 \pm \sqrt{14}$ **31.** $-2, 0$
33. $\frac{3}{4} - \frac{\sqrt{17}}{4}, \frac{3}{4} + \frac{\sqrt{17}}{4}$, or $\frac{3}{4} \pm \frac{\sqrt{17}}{4}$ **35.** $-5, 15$
37. $-5, -3$ **39.** $8 - \sqrt{7}, 8 + \sqrt{7}$, or $8 \pm \sqrt{7}$
41. $-6 - 3\sqrt{2}, -6 + 3\sqrt{2}$, or $-6 \pm 3\sqrt{2}$
43. 📜 **45.** $x^2 - 6x + 9$ **46.** $x^2 + 2a + a^2$
47. $x^2 + x + \frac{1}{4}$ **48.** $x^2 - 3x + \frac{9}{4}$ **49.** $(x + 1)^2$
50. $(x + 5)^2$ **51.** $(x - 2)^2$ **52.** $(x - 10)^2$
53. 📜 **55.** $\frac{5}{2} - \frac{\sqrt{13}}{2}, \frac{5}{2} + \frac{\sqrt{13}}{2}$, or $\frac{5}{2} \pm \frac{\sqrt{13}}{2}$
57. $-5, 2$ **59.** $-4.35, 1.85$ **61.** $-4, -2$ **63.** $-6, 0$
65. $d = \dfrac{\sqrt{kMmf}}{f}$

Exercise Set 9.2, p. 545

1. (c) **2.** (e) **3.** (a) **4.** (b) **5.** (d) **6.** (f)
7. $16; (x + 4)^2 = x^2 + 8x + 16$
9. $1; (x - 1)^2 = x^2 - 2x + 1$
11. $\frac{9}{4}; \left(x - \frac{3}{2}\right)^2 = x^2 - 3x + \frac{9}{4}$
13. $\frac{1}{4}; \left(t + \frac{1}{2}\right)^2 = t^2 + t + \frac{1}{4}$
15. $\frac{9}{16}; \left(x + \frac{3}{4}\right)^2 = x^2 + \frac{3}{2}x + \frac{9}{16}$
17. $\frac{16}{9}; \left(m - \frac{4}{3}\right)^2 = m^2 - \frac{8}{3}m + \frac{16}{9}$
19. $-5, -2$ **21.** $12 \pm \sqrt{123}$ **23.** $-5 \pm \sqrt{13}$
25. $2 \pm \sqrt{11}$ **27.** $-\frac{3}{2} \pm \frac{\sqrt{21}}{2}$, or $\frac{-3 \pm \sqrt{21}}{2}$ **29.** $-1, \frac{1}{2}$
31. $\frac{3}{4} \pm \frac{\sqrt{41}}{4}$, or $\frac{3 \pm \sqrt{41}}{4}$ **33.** $\frac{5}{4} \pm \frac{\sqrt{105}}{4}$, or $\frac{5 \pm \sqrt{105}}{4}$
35. $-\frac{2}{3} \pm \frac{\sqrt{19}}{3}$, or $\frac{-2 \pm \sqrt{19}}{3}$ **37.** $-\frac{1}{2}, 5$ **39.** $-\frac{5}{2}, \frac{2}{3}$
41. 📜 **43.** -7 **44.** 7 **45.** 12 **46.** 13
47. $1 - \sqrt{5}$ **48.** $2(3 + \sqrt{7})$ **49.** $\dfrac{8 - \sqrt{5}}{3}$
50. $\dfrac{5 - \sqrt{6}}{2}$ **51.** 📜 **53.** $-10, 10$ **55.** $-6\sqrt{5}, 6\sqrt{5}$
57. $-16, 16$ **59.** $-7.61, -0.39$ **61.** $0.91, 23.09$
63. $-2.5, 0.67$ **65.** 📜

Technology Connection, p. 548

1. An ERROR message appears. **2.** The graph has no
x-intercepts, so there is no value of x for which
$x^2 + x + 1 = 0$, or, equivalently, for which $x^2 + x = -1$.

Exercise Set 9.3, pp. 552–554

1. True **2.** True **3.** False **4.** True **5.** $-2, 10$
7. 1 **9.** $-\frac{4}{3}, -1$ **11.** $-\frac{1}{2}, \frac{7}{2}$ **13.** $-5, 5$
15. $-2 \pm \sqrt{11}$ **17.** $5 \pm \sqrt{6}$ **19.** $5 \pm \sqrt{3}$
21. $-\frac{4}{3} \pm \frac{\sqrt{10}}{3}$, or $\frac{-4 \pm \sqrt{10}}{3}$
23. $\frac{5}{4} \pm \frac{\sqrt{33}}{4}$, or $\frac{5 \pm \sqrt{33}}{4}$
25. $-\frac{1}{4} \pm \frac{\sqrt{13}}{4}$, or $\frac{-1 \pm \sqrt{13}}{4}$
27. No real-number solution **29.** $\frac{5}{6} \pm \frac{\sqrt{73}}{6}$, or $\frac{5 \pm \sqrt{73}}{6}$
31. $\frac{3}{2} \pm \frac{\sqrt{29}}{2}$, or $\frac{3 \pm \sqrt{29}}{2}$ **33.** $-5, \frac{2}{3}$ **35.** No real-
number solution **37.** $\pm 2\sqrt{3}$ **39.** $-3.562, 0.562$
41. $-0.193, 5.193$ **43.** $-1.207, 0.207$ **45.** 10 sides
47. 9.62 sec **49.** 1.32 sec **51.** 7 ft, 24 ft **53.** Length:
10 cm; width: 6 cm **55.** Length: 24 m; width: 18 m
57. 5.13 cm, 10.13 cm **59.** Length: 8.09 ft; width: 3.09 ft
61. Length: 5.66 m; width: 2.83 m **63.** 18.75% **65.** 5%
67. 11.28 mi **69.** 📜 **71.** $b = 3a$ **72.** $y = 4n$
73. $d = 2t - c$ **74.** $p = \dfrac{4n}{a}$ **75.** $x = \dfrac{y - B}{A}$
76. $x = \dfrac{y}{A + B}$ **77.** 📜 **79.** $-11, 0$ **81.** $\frac{3}{2} \pm \frac{\sqrt{5}}{2}$, or
$\dfrac{3 \pm \sqrt{5}}{2}$ **83.** $-\frac{7}{2} \pm \frac{\sqrt{61}}{2}$, or $\dfrac{-7 \pm \sqrt{61}}{2}$ **85.** $\pm\sqrt{11}$
87. $\frac{5}{2} \pm \frac{\sqrt{37}}{2}$, or $\dfrac{5 \pm \sqrt{37}}{2}$ **89.** 4.83 cm
91. Length: 120 m; width: 75 m **93.** 5%
95. Length: 44.06 in.; width: 19.06 in.

Connecting the Concepts, p. 555

1. $-2, 1$ **2.** $-10, 10$ **3.** $-\frac{3}{2} \pm \frac{\sqrt{29}}{2}$, or $\dfrac{-3 \pm \sqrt{29}}{2}$
4. $-5 \pm 4\sqrt{2}$ **5.** $-\frac{1}{2}, 3$ **6.** -1
7. No real-number solution **8.** $-1, 6$ **9.** $0, 5$
10. $\frac{1}{10} \pm \frac{\sqrt{41}}{10}$, or $\dfrac{1 \pm \sqrt{41}}{10}$ **11.** $-3, 2$ **12.** $2 \pm \sqrt{5}$
13. $-4 \pm \sqrt{15}$ **14.** $-1, \frac{2}{3}$ **15.** $-\frac{1}{11}, \frac{1}{11}$
16. $\frac{1}{4} \pm \frac{\sqrt{41}}{4}$, or $\dfrac{1 \pm \sqrt{41}}{4}$ **17.** $-1 \pm \sqrt{3}$
18. No real-number solution
19. $-\frac{1}{2} \pm \frac{\sqrt{21}}{2}$, or $\dfrac{-1 \pm \sqrt{21}}{2}$ **20.** $0, \frac{1}{9}$

Exercise Set 9.4, pp. 558–559

1. (b)   **2.** (d)   **3.** (c)   **4.** (a)   **5.** $h = \dfrac{2A}{b}$

7. $c = \dfrac{100m}{Q}$ **9.** $t = \dfrac{A}{Pr} - \dfrac{1}{r}$, or $t = \dfrac{A - P}{Pr}$

11. $h = \dfrac{d^2}{c^2}$ **13.** $R = \dfrac{r_1 r_2}{r_1 + r_2}$

15. $x = \dfrac{-b \pm \sqrt{b^2 - 4ac}}{2a}$ **17.** $y = \dfrac{C - Ax}{B}$

19. $h = \dfrac{S - 2\pi r^2}{2\pi r}$ **21.** $t = \dfrac{M - g}{r + s}$ **23.** $a = \dfrac{d}{b - c}$

25. $t = \pm\sqrt{\dfrac{2s}{g}}$ **27.** $h = \pm\sqrt{st}$

29. $w = r - \dfrac{d}{t}$, or $w = \dfrac{rt - d}{t}$ **31.** $x = \dfrac{25 + y}{2}$

33. $t = \dfrac{-n \pm \sqrt{n^2 + 4mp}}{2m}$

35. $m = \dfrac{-t \pm \sqrt{t^2 + 4n}}{2}$ **37.** $t = \dfrac{p^2 - 2pn + n^2 - 9c}{9}$

39. $t = 0$ **41.** 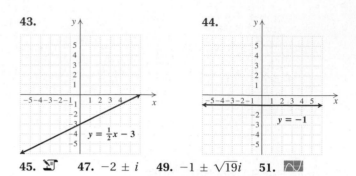   **43.** $4\sqrt{3}$ **44.** $5\sqrt{10}$ **45.** $3\sqrt{10}$
46. $5\sqrt{14}$ **47.** $10\sqrt{6}$ **48.** 30   **49.**   **51.** 6 yr

53. $m = \dfrac{g \pm \sqrt{g^2 - 4ft}}{2f}$ **55.** $t = -\dfrac{b}{a} + \dfrac{k^2}{a(V - c)^2}$

57. $-40°$

Exercise Set 9.5, pp. 562–563

1. False   **2.** True   **3.** True   **4.** False   **5.** i
7. $7i$ **9.** $\sqrt{5}i$, or $i\sqrt{5}$ **11.** $3\sqrt{5}i$, or $3i\sqrt{5}$
13. $-5\sqrt{2}i$, or $-5i\sqrt{2}$ **15.** $4 + 7i$ **17.** $3 - 3i$
19. $-2 + 5\sqrt{3}i$, or $-2 + 5i\sqrt{3}$ **21.** $\pm 5i$
23. $\pm 2\sqrt{7}i$, or $\pm 2i\sqrt{7}$ **25.** $-2 \pm i$ **27.** $4 \pm 3i$
29. $1 \pm 2i$ **31.** $2 \pm \sqrt{3}i$, or $2 \pm i\sqrt{3}$ **33.** $-\dfrac{2}{5} \pm \dfrac{1}{5}i$
35. $-\dfrac{1}{3} \pm \dfrac{\sqrt{2}}{3}i$ **37.**

39. 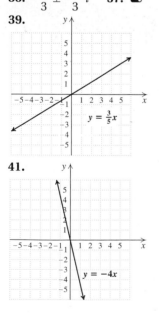   **40.**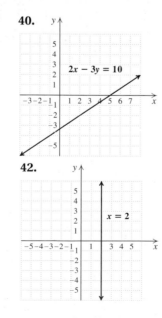

41.   **42.**

Connecting the Concepts, pp. 563–564

1. 1   **2.** 2, 3   **3.** 25   **4.** 4   **5.** 5   **6.** $\dfrac{1}{2}$, 3
7. $-1 \pm \sqrt{3}$ **8.** $-1 \pm i$ **9.** $6 \pm \dfrac{\sqrt{138}}{2}$, or $\dfrac{12 \pm \sqrt{138}}{2}$
10. $-\dfrac{1}{6}$ **11.** No solution   **12.** $-\dfrac{1}{12}$ **13.** 1   **14.** $-2, 4$
15. $-3 \pm \sqrt{7}$ **16.** $\pm 10i$ **17.** $-10, 10$
18. $-\dfrac{5}{4} \pm \dfrac{\sqrt{33}}{4}$, or $\dfrac{-5 \pm \sqrt{33}}{4}$ **19.** 1   **20.** $-3 \pm \sqrt{7}$

Visualizing for Success, p. 569

1. J   **2.** F   **3.** H   **4.** G   **5.** B   **6.** E   **7.** D   **8.** I
9. C   **10.** A

Exercise Set 9.6, pp. 570–571

1. False   **2.** True   **3.** False   **4.** True
5.   **7.**
9.   **11.**
13.   **15.**

17.

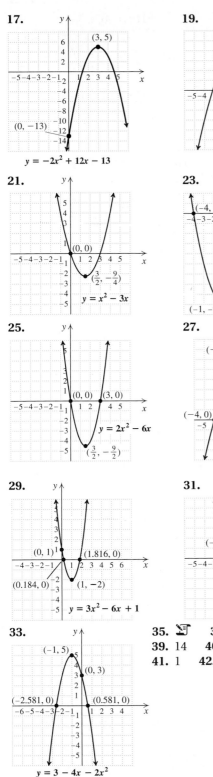

$y = -2x^2 + 12x - 13$

(3, 5)

(0, −13)

19.

(0, 5)

$y = -\frac{1}{2}x^2 + 5$

21.

(0, 0)

$\left(\frac{3}{2}, -\frac{9}{4}\right)$

$y = x^2 - 3x$

23.

(−4, 0)   (2, 0)

(0, −8)

$y = x^2 + 2x - 8$

(−1, −9)

25.

(0, 0)   (3, 0)

$y = 2x^2 - 6x$

$\left(\frac{3}{2}, -\frac{9}{2}\right)$

27.

$\left(-\frac{1}{2}, \frac{49}{4}\right)$ (0, 12)

(−4, 0)   (3, 0)

$y = -x^2 - x + 12$

29.

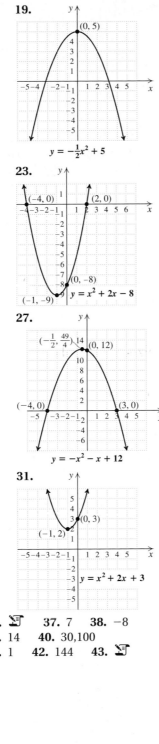

(0, 1)   (1.816, 0)

(0.184, 0)

(1, −2)

$y = 3x^2 - 6x + 1$

31.

(0, 3)

(−1, 2)

$y = x^2 + 2x + 3$

33.

(−1, 5)

(0, 3)

(−2.581, 0)   (0.581, 0)

$y = 3 - 4x - 2x^2$

35. 🖵   **37.** 7   **38.** −8

39. 14   **40.** 30,100

41. 1   **42.** 144   **43.** 🖵

45. (a) 2 sec after launch and 4 sec after launch; **(b)** 3 sec after launch; **(c)** 6 sec after launch   **47. (a)** 56.25 ft, 120 ft, 206.25 ft, 276.25 ft, 356.25 ft, 600 ft;

(b)

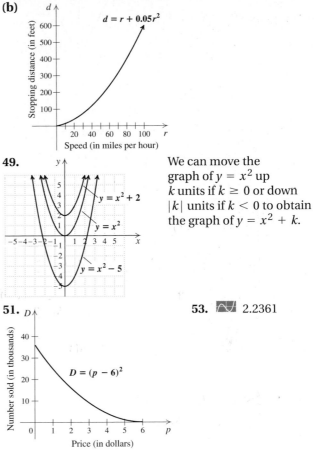

$d = r + 0.05r^2$

Stopping distance (in feet)

Speed (in miles per hour)

49.

$y = x^2 + 2$

$y = x^2$

$y = x^2 - 5$

We can move the graph of $y = x^2$ up k units if $k \geq 0$ or down $|k|$ units if $k < 0$ to obtain the graph of $y = x^2 + k$.

51.

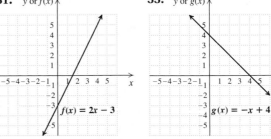

Number sold (in thousands)

$D = (p - 6)^2$

Price (in dollars)

53. 〰️ 2.2361

Exercise Set 9.7, pp. 578–580

1. True   **2.** True   **3.** False   **4.** True   **5.** Yes   **7.** Yes
9. No   **11.** Yes   **13.** 9; 12; 3   **15.** 21; −15; −30
17. 1; −17; 5.5   **19.** 1; 0; 10   **21.** 25; 25; $\frac{25}{4}$ **23.** 0; 12; 4
25. (a) 34 yr; **(b)** 20 yr   **27. (a)** 159.48 cm; **(b)** 153.98 cm
29. 16.$\overline{6}$°C; 25°C; −5°C

31.

y or $f(x)$

$f(x) = 2x - 3$

33.

y or $g(x)$

$g(x) = -x + 4$

35.

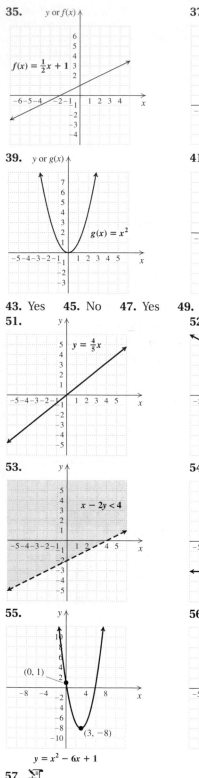

$f(x) = \frac{1}{2}x + 1$

37.   y or $g(x)$

$g(x) = 2|x|$

39.   y or $g(x)$

$g(x) = x^2$

41.   y or $f(x)$

$f(x) = x^2 - x - 2$

43. Yes   **45.** No   **47.** Yes   **49.**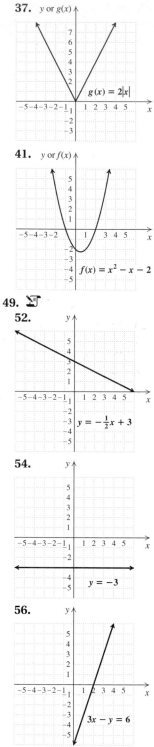

51. $y = \frac{4}{5}x$

52. $y = -\frac{1}{2}x + 3$

53. $x - 2y < 4$

54. $y = -3$

55.

$(0, 1)$

$(3, -8)$

$y = x^2 - 6x + 1$

56.

$3x - y = 6$

57.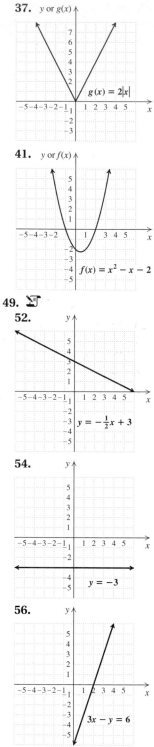

59.   y or $g(x)$

$g(x) = x^3$

61.   y or $f(x)$

$f(x) = |x| + x$

63. Answers may vary.   **65.** $g(x) = x^2 - 4$
67. $\{0, 2, 6, 12\}$ **69.** $\{0, 2\}$

Review Exercises: Chapter 9, pp. 583–584

1. (h)   **2.** (f)   **3.** (g)   **4.** (e)   **5.** (b)   **6.** (d)   **7.** (a)
8. (c)   **9.** $5 \pm \sqrt{26}$ **10.** $-2, \frac{1}{2}$ **11.** $\pm\sqrt{6}$ **12.** $\frac{3}{5}, 1$
13. $1 \pm \sqrt{11}$ **14.** $-3 \pm 2i$ **15.** $-2, \frac{1}{3}$ **16.** $0, \frac{1}{11}$
17. $\pm i$ **18.** $-2, 5$ **19.** $-10 \pm 2\sqrt{3}$ **20.** $-3 \pm 3\sqrt{2}$
21. $-\frac{1}{2} \pm \frac{\sqrt{3}}{2}i$ **22.** $1 \pm \frac{\sqrt{3}}{2}$, or $\frac{2 \pm \sqrt{3}}{2}$ **23.** 3

24. $\pm 2\sqrt{2}$ **25.** $\frac{3}{10} \pm \frac{\sqrt{71}}{10}i$ **26.** $-\frac{7}{2}, \frac{5}{3}$

27. $t = \frac{p}{c - b}$ **28.** $a = \frac{25}{2}$ **29.** $n = \frac{-m \pm \sqrt{m^2 + 4p}}{2}$

30. $t = \frac{rs}{r + s}$ **31.** $0.697, 4.303$ **32.** $-1.866, -0.134$

33. 3.217 m, 6.217 m   **34.** 4%   **35.** Length: 12 m; width: 9 m
36. 6.349 sec   **37.** $3i$ **38.** $-5\sqrt{5}i$, or $-5i\sqrt{5}$

39.

$y = 2 - x^2$
$(0, 2)$
$(-1.414, 0)$
$(1.414, 0)$

40.

$(-4.236, 0)$
$(0.236, 0)$
$(0, -1)$
$y = x^2 + 4x - 1$
$(-2, -5)$

41.

$(0, 0)$
$(\frac{10}{3}, 0)$
$(\frac{5}{3}, -\frac{25}{3})$
$y = 3x^2 - 10x$

42.

$(0, 1)$
$(1, 0)$
$y = x^2 - 2x + 1$

43. $10; -8; 2.5$ **44.** $1; 1; 2$ **45.** 1950 calories

46.

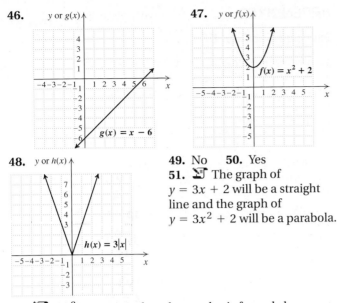

$g(x) = x - 6$

47. $f(x) = x^2 + 2$

48. $h(x) = 3|x|$

49. No   **50.** Yes
51. 📝 The graph of $y = 3x + 2$ will be a straight line and the graph of $y = 3x^2 + 2$ will be a parabola.

52. 📝 If $b^2 - 4ac$ is 0, then the quadratic formula becomes $x = -b/(2a)$; thus there is only one x-intercept. If $b^2 - 4ac$ is negative, then there are no real-number solutions and thus no x-intercepts. If $b^2 - 4ac$ is positive, then

$$x = \frac{-b + \sqrt{b^2 - 4ac}}{2a} \text{ or } x = \frac{-b - \sqrt{b^2 - 4ac}}{2a} \text{ so there}$$

must be two x-intercepts.   **53.** 📝 At least two of the points must be above and below each other on the graph.
54. -32 and -31; 31 and 32   **55.** ± 14 **56.** 25
57. $5\sqrt{\pi}$ in.

Test: Chapter 9, p. 585

1. [9.1] $\pm\sqrt{10}$ **2.** [9.5] $\pm 3i$ **3.** [9.3] $0, \frac{5}{2}$
4. [9.1] $1 \pm 2\sqrt{2}$ **5.** [9.3] $-5, 10$ **6.** [9.3] $-5, \frac{2}{3}$
7. [9.3] $\frac{1}{2} \pm \frac{\sqrt{21}}{2}$, or $\frac{1 \pm \sqrt{21}}{2}$ **8.** [9.1] 2
9. [9.5] $2 \pm i$ **10.** [9.3] $2 \pm \sqrt{6}$ **11.** [9.3] $-2 \pm \sqrt{14}$
12. [9.3] $\frac{7}{6} \pm \frac{\sqrt{37}}{6}$, or $\frac{7 \pm \sqrt{37}}{6}$ **13.** [9.2] $2 \pm \sqrt{14}$
14. [9.4] $p = \frac{n^2 - 6n - 11}{4}$ **15.** [9.4] $b = \frac{a}{1 + t}$
16. [9.3] $-1.702, 4.702$ **17.** [9.3] 6.5 m, 2.5 m
18. [9.3] 11   **19.** [9.5] $10\sqrt{2}i$, or $10i\sqrt{2}$ **20.** [9.5] $-10i$
21. [9.6]   **22.** [9.6]

$y = -x^2 + x - 5$ $(\frac{1}{2}, -\frac{19}{4})$ $(0, -5)$

$y = x^2 + 2x - 15$ $(-5, 0)$ $(3, 0)$ $(0, -15)$ $(-1, -16)$

23. [9.7] $1, \frac{3}{2}, 2$ **24.** [9.7] $-7, 0, -3$ **25.** [9.7] 25.98 min

26. [9.7]

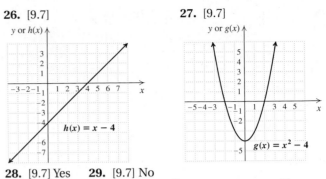

$h(x) = x - 4$

27. [9.7]

$g(x) = x^2 - 4$

28. [9.7] Yes   **29.** [9.7] No
30. [9.3] $75 + 50\sqrt{2} \approx 145.7 \text{ ft}^2$ **31.** [9.3] $1 \pm \sqrt{5}$

Cumulative Review/Final Exam: Chapters 1–9, pp. 586–588

1. 3   **2.** 13   **3.** 120   **4.** $-3n + 1$ **5.** 9   **6.** 15
7. $-\frac{3}{20}$ **8.** 4   **9.** 2.73×10^{-5} **10.** 2.0×10^{14}
11. x^{-4}, or $\frac{1}{x^4}$ **12.** y^7 **13.** $9m^8$ **14.** $10x^3 + 3x - 3$
15. $7x^3 - 2x^2 + 4x - 17$ **16.** $12x^2 - 6$
17. $-8y^4 + 6y^3 - 2y^2$ **18.** $6t^3 - 17t^2 + 16t - 6$
19. $t^2 - \frac{1}{16}$ **20.** $9m^2 - 12m + 4$ **21.** $6x^2 + 5x - 56$
22. $12x^2y - 3x^2y^2 + 12xy^2 - 2$
23. $25p^4 + 20p^2q + 4q^2$ **24.** $\frac{2}{x + 3}$ **25.** $\frac{3a(a - 1)}{2(a + 1)}$
26. $\frac{27x - 4}{5x(3x - 1)}$ **27.** $\frac{-x^2 + x + 2}{(x + 4)(x - 4)(x - 5)}$
28. $x^2 + 9x + 16 + \frac{35}{x - 2}$ **29.** $2(3x + 2)(3x - 2)$
30. $(y - 6)(y^2 - 5)$ **31.** $(m + 5)^2$
32. $x(2x - 3)(5x + 2)$ **33.** $5x(5x^2 - 4x + 1)$
34. $(t - 3)(t + 2)$ **35.** $(49p + 1)(p + 1)$
36. $(10uv + 1)(10uv - 1)$ **37.** $(2x - 5y)^2$
38. $3y^2(y - 8)(y - 5)$ **39.** $-\frac{42}{5}$ **40.** 8   **41.** 5
42. $7t$ **43.** $\frac{1}{8}$ **44.** $5x^3y^3\sqrt{10y}$ **45.** $\frac{4}{3}$ **46.** -17
47. $18\sqrt{3}$ **48.** $\sqrt{a^2 - b^2}$ **49.** $8a^2b\sqrt{3ab}$
50. $\frac{2\sqrt{10}}{5}$ **51.** -6 **52.** $\{x \mid x < -6\}$ **53.** 7   **54.** $3, 5$
55. $(1, 2)$ **56.** $-\frac{3}{5}$ **57.** $(6, 7)$ **58.** $3, -2$
59. $1 \pm \sqrt{3}$ **60.** 2   **61.** $\{x \mid x \geq -1\}$ **62.** 8   **63.** -3
64. $(15, 0)$ **65.** $-1 \pm 2i$ **66.** $\pm\sqrt{10}$ **67.** $3 \pm \sqrt{6}$
68. $-1, 2$ **69.** $\frac{5}{4}, -\frac{4}{3}$ **70.** No solution   **71.** 12
72. $a = \frac{by}{x + y}$ **73.** $m = \frac{tn}{t + n}$ **74.** $\frac{2}{3} \pm \frac{\sqrt{6}}{3}$,
or $\frac{2 \pm \sqrt{6}}{3}$ **75.** $-0.207, 1.207$
76.

$y = \frac{1}{3}x - 5$

77.

$2x + 3y = -6$

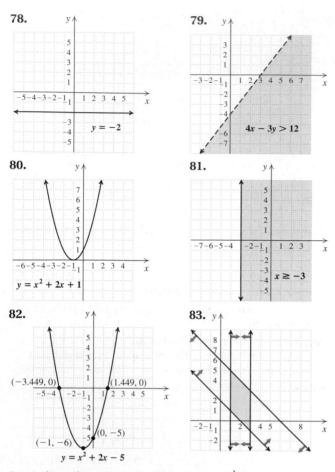

78. $y = -2$

79. $4x - 3y > 12$

80. $y = x^2 + 2x + 1$

81. $x \geq -3$

82. $y = x^2 + 2x - 5$ $(-3.449, 0)$ $(1.449, 0)$ $(0, -5)$ $(-1, -6)$

83.

84. 2; $(0, -8)$ **85.** 15 **86.** $y + 3 = -\frac{1}{2}(x - 1)$
87. Parallel **88.** -4; 0; 0 **89.** $-i$ **90.** $\frac{14}{25}$ billion tons per year **91.** Approximately 5954.4 million tons
92. 742,500 vehicles **93.** Dark-chocolate cheesecake truffles: 12 lb; milk-chocolate hazelnut truffles: 8 lb
94. 100 m **95.** 1.92 g **96.** Length: 10 ft; width: 3 ft
97. Length: 12 m; width: 5 m

98. $\sqrt{1500}$ cm ≈ 38.730 cm **99.** 30, -30 **100.** $\dfrac{\sqrt{6}}{3}$

101. Yes **102.** No **103.** No **104.** No **105.** Yes

APPENDIXES

Exercise Set A, p. 591

1. $(t + 2)(t^2 - 2t + 4)$ **3.** $(x + 1)(x^2 - x + 1)$
5. $(z - 5)(z^2 + 5z + 25)$ **7.** $(2a - 1)(4a^2 + 2a + 1)$
9. $(y - 3)(y^2 + 3y + 9)$ **11.** $(4 + 5x)(16 - 20x + 25x^2)$
13. $(5p + 1)(25p^2 - 5p + 1)$
15. $(3m - 4)(9m^2 + 12m + 16)$
17. $(p - q)(p^2 + pq + q^2)$ **19.** $\left(x + \frac{1}{2}\right)\left(x^2 - \frac{1}{2}x + \frac{1}{4}\right)$
21. $2(y - 4)(y^2 + 4y + 16)$
23. $3(2a + 1)(4a^2 - 2a + 1)$ **25.**
$r(s - 5)(s^2 + 5s + 25)$ **27.** $5(x + 2z)(x^2 - 2xz + 4z^2)$
29. $(x + 0.2)(x^2 - 0.2x + 0.04)$ **31.** ✑
33. $(5c^2 + 2d^2)(25c^4 - 10c^2d^2 + 4d^4)$
35. $3(x^a - 2y^b)(x^{2a} + 2x^a y^b + 4y^{2b})$
37. $\frac{1}{3}\left(\frac{1}{2}xy + z\right)\left(\frac{1}{4}x^2y^2 - \frac{1}{2}xyz + z^2\right)$

Exercise Set B, pp. 593–594

1. Mean: 17; median: 18; mode: 13 **3.** Mean: $10.\overline{6}$;
median: 9; mode: 3, 20 **5.** Mean: 4.06; median: 4.6;
mode: none **7.** Mean: $239.\overline{3}$; median: 234; mode: 234
9. Average: 34.875; median: 22; mode: 0 **11.** Mean: 87.25;
median: 86.5; mode: 86 **13.** Average: $218.\overline{3}$; median: 222;
mode: 202 **15.** 10 home runs **17.** $a = 30, b = 58$

Exercise Set C, pp. 597–598

1. $\{8, 9, 10, 11\}$ **3.** $\{41, 43, 45, 47, 49\}$ **5.** $\{-3, 3\}$
7. True **9.** True **11.** False **13.** True **15.** False
17. True **19.** $\{c, d, e\}$ **21.** $\{1, 2, 3, 6\}$ **23.** $\varnothing$
25. $\{a, e, i, o, u, q, c, k\}$ **27.** $\{1, 2, 3, 4, 6, 9, 12, 18\}$
29. $\{1, 2, 3, 4, 5, 6, 7, 8\}$ **31.** ✑ **33.** The set of integers
35. The set of real numbers **37.** $\varnothing$
39. **(a)** A; **(b)** A; **(c)** A; **(d)** $\varnothing$ **41.** True **43.** True
45. True

Glossary

A

absolute value [1.4]   A number's distance from 0 on the number line.

additive inverses [1.6]   Two numbers whose sum is 0; opposites.

algebraic expression [1.1]   Variables and/or numerals, often with operation signs $(+, -, \cdot, \div, (\ )^n, \text{ or } \sqrt[n]{(\ )})$ and grouping symbols.

ascending order [4.3]   A polynomial written with the terms arranged according to the degree of one variable, from least to greatest.

associative law of addition [1.2]   The statement that when three numbers are added, regrouping the addends gives the same sum.

associative law of multiplication [1.2]   The statement that when three numbers are multiplied, regrouping the factors gives the same product.

average [2.7]   Most commonly, the mean of a set of numbers.

axes (singular, axis) [3.1]   Two perpendicular number lines used to identify points in a plane.

axis of symmetry [9.6]   A line that can be drawn through a graph such that the part of the graph on one side of the line is an exact reflection of the part on the other side.

B

bar graph [3.1]   A graphic display of data using bars proportional in length to the numbers represented.

base [1.8]   In exponential notation, the number or expression being raised to a power. In expressions of the form a^n, a is the base.

binomial [4.2]   A polynomial composed of two terms.

C

circle graph [3.1]   A graphic display of data using sectors of a circle to represent percents.

circumference [2.3], [4.2]   The distance around a circle.

coefficient [2.1]   The numerical multiplier of a variable or variables.

commutative law of addition [1.2]   The statement that when two numbers are added, changing the order in which the numbers are added does not affect the sum.

commutative law of multiplication [1.2]   The statement that when two numbers are multiplied, changing the order in which the numbers are multiplied does not affect the product.

complementary angles [2.5], [7.2]   A pair of angles, the sum of whose measures is 90°.

completing the square [9.2]   To complete the square for an expression like $x^2 + bx$, add half of the coefficient of x, squared.

complex number [9.5]   Any number that can be written as $a + bi$, where a and b are real numbers.

complex rational expression [6.5]   A rational expression that has one or more rational expressions within its numerator and/or denominator.

composite number [1.3]   A natural number that has more than two different natural numbers as factors.

conjugates [8.4]   Expressions of the form $a\sqrt{b} + c\sqrt{d}$ and $a\sqrt{b} - c\sqrt{d}$ are called conjugates. Their product does not contain a radical term.

consistent system of equations [7.1]   A system of equations that has at least one solution.

constant [1.1]   A known number.

constant of proportionality [7.7]   In an equation of the form $y = kx$ or $y = \dfrac{k}{x}$ $(k > 0)$, with x and y variables, k is the constant of proportionality.

contradiction [2.1]   An equation that has no solution and is never true.

coordinates [3.1]   The numbers in an ordered pair.

cube root [8.7]   The number c is the cube root of a if $c^3 = a$.

D

degree of a polynomial [4.2]   The degree of the term of highest degree of a polynomial.

degree of a term [4.2]   The number of variable factors in a term.

denominator [1.3]   The number below the fraction bar in a fraction.

dependent equations [7.1]   For two equations, when one equation can be obtained by multiplying both sides of the other equation by a constant.

descending order [4.2]   A polynomial written with the terms arranged according to the degree of one variable, from greatest to least.

difference of squares [4.5], [5.4]   An expression that can be written in the form $a^2 - b^2$.

direct variation [7.7]   A situation that translates to an equation of the form $y = kx$, with k a constant. The equation $y = kx$ is an equation of direct variation.

distributive law [1.2] The statement that multiplying a factor by the sum of two numbers gives the same result as multiplying the factor by each of the two numbers and then adding.

domain [9.7] The set of all first coordinates of the ordered pairs in a function.

E

equation [1.1] A number sentence with the verb =.

equivalent equations [2.1] Equations that have the same solutions.

equivalent expressions [1.2] Expressions that have the same value for all allowable replacements.

equivalent inequalities [2.6] Inequalities that have the same solution set.

evaluate [1.1] To substitute a value for each occurrence of a variable in an expression and calculate the result.

exponent [1.8] In expressions of the form a^n, the number n is an exponent. For n a natural number, a^n represents n factors of a.

exponential notation [1.8] A representation of a number using a base raised to an exponent.

extrapolation [3.7] The process of estimating a value that goes beyond the given data.

F

factor [1.2] *Verb*: To write an equivalent expression that is a product. *Noun*: A number being multiplied.

FOIL [4.5] To multiply two binomials by multiplying the **F**irst terms, the **O**uter terms, the **I**nner terms, and the **L**ast terms, and then adding the results.

formula [2.3] An equation that uses numbers and/or letters to represent a relationship between two or more quantities.

fraction notation [1.3] A quotient represented with a bar under (or slash following) the dividend and over (or before) the divisor.

function [9.7] A correspondence that assigns to each member of a set called the domain exactly one member of a set called the range.

function values [9.7] Outputs from equations in function notation.

G

grade [3.5] The grade of a road is the measure of the road's steepness expressed as a percent.

graph [1.4], [2.6], [3.1] A graph can be a picture or a diagram of the data in a table. A graph can also be a set of points that form a line, a curve, or a plane that represent all the solutions of an equation or an inequality.

H

higher roots [8.7] Cube roots, fourth roots, and, in general, roots of the form $\sqrt[n]{\ }$, with n a natural number and $n > 2$.

hypotenuse [5.7] In a right triangle, the side opposite the right angle.

I

i [9.5] The square root of -1. That is, $i = \sqrt{-1}$ and $i^2 = -1$.

identity [2.1] An equation that is always true.

identity property of 0 [1.5] The statement that the sum of a number and 0 is always the original number.

identity property of 1 [1.3] The statement that the product of a number and 1 is always the original number.

imaginary number [9.5] A number that can be written in the form $a + bi$, where a and b are real numbers and $b \neq 0$.

inconsistent system of equations [7.1] A system of equations for which there is no solution.

independent equations [7.1] Equations that are not dependent.

index [8.7] In $\sqrt[n]{a}$, the number n is called the index.

inequality [1.4] A mathematical sentence using $>$, $<$, $\geq$, $\leq$, or $\neq$.

input [9.7] A member of the domain of a function.

integers [1.4] The whole numbers and their opposites.

interpolation [3.7] The process of estimating a value between given values.

inverse variation [7.7] A situation that translates to an equation of the form $y = k/x$, with k a constant. The equation $y = k/x$ is an equation of inverse variation.

irrational number [1.4] A number that cannot be named as a ratio of two integers. In decimal form, they neither terminate nor repeat.

L

largest common factor [5.1] The common factor of the terms of a polynomial with the largest possible coefficient and the largest possible exponent(s) of the variable factors.

leading coefficient [4.2] The coefficient of the term of highest degree in a polynomial.

leading term [4.2] A polynomial's term of highest degree.

least common denominator [6.3] The least common multiple of the denominators.

least common multiple [6.3] The multiple of all expressions under consideration that has the smallest positive coefficient and the least possible degree.

legs [5.7] In a right triangle, the sides that form the right angle.

like radicals [8.4] Radical expressions that have the same common radical factor.

like terms [1.5], [1.8] Terms that have exactly the same variable factors.

line graph [3.1] A graph in which quantities are represented as points connected by straight-line segments.

linear equation [3.2]   Any equation that can be written in the form $y = mx + b$ or $Ax + By = C$, where x and y are variables, and m, b, A, B, and C are constants.

linear inequality [7.5]   An inequality whose related equation is a linear equation.

M

mean [2.7]   The sum of a set of numbers divided by the number of addends.

monomial [4.2]   A constant, a variable, or a product of a constant and one or more variables.

motion problem [6.7]   A problem that deals with distance, speed, and time.

multiplicative inverses [1.3]   Two numbers whose product is 1; reciprocals.

multiplicative property of zero [1.7]   The product of 0 and any real number is 0.

N

natural numbers [1.3]   The counting numbers: 1, 2, 3, 4, 5, ...

nonlinear equation (in two variables) [3.2]   An equation that when graphed is not a line.

numerator [1.3]   The number above the fraction bar in a fraction.

O

opposites [1.6]   Two expressions whose sum is 0; additive inverses.

opposites, law of [1.6]   The sum of two opposites is 0.

origin [3.1]   The point on a coordinate plane where the two axes intersect.

output [9.7]   A member of the range of a function.

P

parabola [9.6]   The graph of any equation of the form $y = ax^2 + bx + c$, with a, b, and c constants and $a \neq 0$.

percent [2.4]   A ratio or fraction with a denominator of 100. $n\%$ is read "n percent" and means n per hundred.

perfect square [8.1]   Any number that is the square of a rational number.

perfect-square trinomial [4.5], [5.4]   A trinomial that is the square of a binomial.

point–slope equation [3.7]   An equation of the form $y - y_1 = m(x - x_1)$, where x and y are variables and m is the slope of the line and (x_1, y_1) is a point on the line.

polynomial [4.2]   A monomial or a sum of monomials.

prime factorization [1.3]   The factorization of a whole number into a product of its prime factors.

prime number [1.3]   A natural number that has exactly two different factors: the number itself and 1.

principal square root [8.1]   The nonnegative square root of a number.

principle of zero products [5.6]   The statement that an equation $AB = 0$ is true if and only if $A = 0$ or $B = 0$.

proportion [6.7]   An equation stating that two ratios are equal.

Pythagorean theorem [5.7]   In any right triangle, if a and b are the lengths of the legs and c is the length of the hypotenuse, then $a^2 + b^2 = c^2$.

Q

quadrants [3.1]   The four regions into which the axes divide a plane.

quadratic equation [5.6]   An equation equivalent to one of the form $ax^2 + bx + c = 0$, where a, b, and c are constants and $a \neq 0$.

quadratic formula [9.3]   For $a \neq 0$, the solutions of $ax^2 + bx + c = 0$ are given by $x = \dfrac{-b \pm \sqrt{b^2 - 4ac}}{2a}$.

R

radical equation [8.5]   An equation in which a variable appears in a radicand.

radical expression [8.1]   An algebraic expression that contains at least one radical sign.

radical sign [8.1]   A symbol, $\sqrt{\ }$, which indicates the principal square root or $\sqrt[n]{\ }$, where $n > 2$, which denotes a higher root.

radicand [8.1]   The expression under the radical sign.

range [9.7]   The set of all the second coordinates of the ordered pairs in a function.

rate [3.4]   A ratio that indicates how two quantities change with respect to each other.

ratio [6.7]   The ratio of a to b is a/b, also written $\dfrac{a}{b}$ or $a{:}b$.

rational equation [6.6]   An equation containing one or more rational expressions.

rational expression [6.1]   A quotient of two polynomials.

rational numbers [1.4]   Numbers that can be written in the form $\dfrac{a}{b}$, where a and b are integers and $b \neq 0$. In decimal notation, a rational number repeats or terminates.

rationalizing the denominator [8.3]   A procedure for finding an equivalent expression without a radical expression in the denominator.

real number [1.4]   Any number that is either rational or irrational.

reciprocals [1.3]   Any two numbers whose product is 1; multiplicative inverses.

repeating decimal [1.4]   A decimal in which a block of digits repeats indefinitely.

right triangle [5.7]   A triangle that includes a 90° angle.

S

scientific notation [4.8]   A number written in the form $N \times 10^m$, where m is an integer, $1 \leq N < 10$, and N is expressed in decimal notation.

set [1.4]   A collection of objects.

set-builder notation [2.6]   A set named by describing basic characteristics of the elements in the set.

similar triangles [6.7]   Triangles in which corresponding angles have the same measure and corresponding sides are proportional.

slope [3.5]   The ratio of the rise to the run for any two points on a line.

slope–intercept equation [3.6]   An equation of the form $y = mx + b$, where x and y are variables, m is the slope of the line, and $(0, b)$ is the y-intercept.

solution [1.1]   A replacement or substitution that makes an equation or inequality true.

solution set [2.6]   The set of all solutions of an equation or inequality.

solve [2.1]   To find all solutions of an equation, inequality, or problem.

square root [8.1]   The number c is a square root of a if $c^2 = a$.

substitute [1.1]   To replace a variable with a number or an expression that represents a number.

supplementary angles [2.5], [7.2]   A pair of angles, the sum of whose measures is 180°.

system of equations [7.1]   A set of two or more equations that are to be solved simultaneously.

system of linear inequalities [7.6]   A set of two or more linear inequalities that are to be solved simultaneously.

T

term [1.2]   A number, a variable, a product, or a quotient of numbers and/or variables.

terminating decimal [1.4]   A decimal number that can be written using a finite number of decimal places.

trinomial [4.2]   A polynomial that is composed of three terms.

U

undefined [1.7]   An expression that has no meaning attached to it.

V

variable [1.1]   A letter that represents an unknown number.

vertex (plural, vertices) [9.6]   The point at which a parabola crosses its axis of symmetry.

vertical-line test [9.7]   The statement that a graph represents a function if it is impossible to draw a vertical line that intersects the graph more than once.

W

whole numbers [1.3]   The set of numbers: 0, 1, 2, 3, 4, ...

X

x-intercept [3.3]   A point at which a graph crosses the x-axis.

Y

y-intercept [3.3]   A point at which a graph crosses the y-axis.

Index

Index of Applications

Astronomy

Composition of the sun, 104
Jupiter's atmosphere, 104
Lunar weight, 475
Meteorology, 559
Orbit time, 10
Rocket sections, 117
Weight on Mars, 419
Weight on the moon, 419

Automotive

Alternative fuels, 363, 412–414, 588
Automobile maintenance, 460
Automobile prices, 102–103
Car cleaning, 420
Car depreciation, 182
Driving under the influence, 152
Fuel efficiency, 161–162, 165
Gas consumption, 174
Gas mileage, 180, 181, 217, 418
Gasoline custom blending pump, 461
Insurance-covered repairs, 134
Octane ratings, 459, 460
Safe driving, 511
Speed of a skidding car, 496, 510
Stopping distance, 571

Biology

Animal population, 419, 426, 428
Cricket chirps and temperature, 119
Ecology, 475
Herpetology, 483, 527
Horticulture, 458
Life span of an animal, 579
Marine biology, 559
Number of humpback whales, 419
Plant species, 104
Predicting heights, 579
Reptiles, 519
Threatened species, 197
Veterinary science, 419
Weight of a fish, 98
Wildlife population, 414
Wing aspect ratio, 418
Zoology, 10

Business

Advertising, 261
Bicycle sales, 220
Billboards, 441
Blogging, 189
Budget overruns, 106, 119
Call center, 179
Copiers, 118, 163, 385, 416, 427
Defective radios, 426
Digitizing books, 183–185, 195
Direct mail, 106

Discount stores, 102
Employee theft, 140
Event promotion, 92–93
Framing, 450
Hairdresser, 179
Hotel management, 286
Library mistakenly charged sales tax, 481
Manufacturing, 345, 592
Markup, 106
Number of cell-phone subscribers, 362
Packaging, 29
Printing, 459
Production, 182, 458, 459
Profits and losses, 42
Ranching, 521
Retail losses due to crime, 295
Retail sales, 189
Returnable bottles, 457
Revenue, cost, and profit, 237
Sales meeting attendance, 350
Sales tax, 106, 220, 350
Search engine ads, 415
Selling at an auction, 113–114
Selling a guitar, 144
Selling a home, 119
Starbuck's revenue, 592
Store credit, 107
Storekeeper requesting change, 119
Total profit, 245

Chemistry

Acid mixtures, 458, 460, 477
Gas volume, 474
Gold temperatures, 135
Karat rating, 475
Metallurgy, 533
Mixing solutions, 458
Temperature of a fluid, 557
Zinc and copper in pennies, 418

Construction

Architecture, 412, 417, 469
Blueprints, 141, 412, 417
Concrete walk, 416
Corner on a building's foundation, 361
Cutting a beam, ribbon, or wire, 140, 143, 220, 350
Depths of a gutter, 355
Diagonal braces in a lookout tower, 353
Dimensions of a room, 353
Fencing, 132
Folding sheet metal, 355
Grade of a stairway, 193, 194
Hancock Building dimensions, 117
Heavy machinery, 416
Home remodeling, 116, 220
Length of a diagonal brace, 359

Length of a wire, 531
Masonry, 416, 519
Painting, 416
Plumbing, 519, 553
Reach of a ladder, 353, 358, 362, 514
Roofing, 114–115, 193, 354, 420, 534
Sanding oak floors, 425
Slope of land, 193
Two by four, 117, 441
Wiring, 420, 519
World's tallest buildings, 7

Consumer

Appliances, 385
Bills, 42, 142, 179
Bottled water consumption, 164, 200
Broadband cable and DSL
  subscribers, 297
Catering costs, 130
Cell-phone costs, 189, 505
Coffee consumption, 594
Conserving energy and money, 436
Cordless telephones, 520
Cost, 12, 135, 163, 180, 200, 220, 435,
  447–448, 475, 479
Deducting sales tax, 106
Discount, 108, 119, 136
eBay purchases, 120
Energy use, 104, 261, 295, 474
Frequent buyer bonus, 136
Furnace repairs, 133
Holiday spending, 342
Home maintenance, 420, 519
Home video spending, 153
iPod prices, 594
Leisure spending, 505
Ordering books, 428
Parking fees, 136
Phone rates, 449
Prepaid calling card, 482
Prices, 116, 141, 164, 189, 460, 594
Purchasing stamps, 453–454
Sales, 108
Taxi fares, 118, 120, 181, 480
Tipping, 105, 106
Toll charges, 136
U.S. wireless-phone subscribers, 141
Vehicle rental, 118, 174–175, 177, 178, 449

Economics

Consumer's demand, 571
Crude oil imports, 105
Currency exchange, 376
Depreciation, 180, 182, 271, 500, 559
Equilibrium point, 571
Federal funds rate, 96
Gasoline prices, 42
National debt, 178

Geometric Formulas

Plane Geometry

Rectangle
Area: $A = lw$
Perimeter: $P = 2l + 2w$

Square
Area: $A = s^2$
Perimeter: $P = 4s$

Triangle
Area: $A = \frac{1}{2}bh$

Triangle
Sum of Angle Measures:
$A + B + C = 180°$

Right Triangle
Pythagorean Theorem
(Equation):
$a^2 + b^2 = c^2$

Parallelogram
Area: $A = bh$

Trapezoid
Area: $A = \frac{1}{2}h(b_1 + b_2)$

Circle
Area: $A = \pi r^2$
Circumference:
$C = \pi d = 2\pi r$
$\left(\frac{22}{7}\right.$ and 3.14 are different
approximations for π $\left.\right)$

Solid Geometry

Rectangular Solid
Volume: $V = lwh$

Cube
Volume: $V = s^3$

Right Circular Cylinder
Volume: $V = \pi r^2 h$
Total Surface Area:
$S = 2\pi rh + 2\pi r^2$

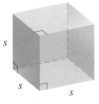

Right Circular Cone
Volume: $V = \frac{1}{3}\pi r^2 h$
Total Surface Area:
$S = \pi r^2 + \pi rs$
Slant Height:
$s = \sqrt{r^2 + h^2}$

Sphere
Volume: $V = \frac{4}{3}\pi r^3$
Surface Area: $S = 4\pi r^2$

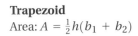

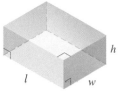

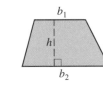

Selected Keys of the Scientific Calculator

This secondary function takes the square root of number displayed.

Squares number displayed.

Activates secondary functions printed above certain keys. Also denoted INV or 2nd.

Used when entering numbers in scientific notation. Also denoted EXP.

Finds reciprocal of number displayed.

Used to raise any base to a power. Also denoted y^x, a^x, or ∧.

Stores number displayed in memory. Also denoted MIN or M.

Recalls number stored in memory. Also denoted MR.

Clears last number displayed but not preceding operations.

Used when entering decimal notation.

This secondary function raises 10 to any power entered.

Clears all preceding numbers and operations. Also used to turn calculator on.

Used as an approximation for pi.

Used to perform indicated operation.

Used to control order in which certain operations are performed.

Used to change sign of number displayed.

3.141592654

AC
ON

√ 10^x

SHIFT x^2 log   π   ÷

1/x   EE   (   )   ×

x^y 7   8   9   −

STO   4   5   6   +

RCL   1   2   3

CE/C   0   •   +/−   =